Laminar-Turbulent Transition

International Union of Theoretical and Applied Mechanics

Laminar-Turbulent Transition

Symposium, Novosibirsk, USSR
July 9–13, 1984

Editor
V. V. Kozlov

With 501 Figures

Springer-Verlag
Berlin Heidelberg New York Tokyo

Dr. V. V. Kozlov
Institute of Theoretical and Applied Mechanics
Institutskaya, 4/1
Novosibirsk-90, USSR

ISBN 3-540-15250-4 Springer-Verlag Berlin Heidelberg New York Tokyo
ISBN 0-387-15250-4 Springer-Verlag New York Heidelberg Berlin Tokyo

Library of Congress Cataloging in Publication Data.

Main entry under title:
Laminar-turbulent transition.
At head of title: International Union of Theoretical and Applied Mechanics.
"Second IUTAM Symposium on Laminar-Turbulent Transition in Novosibirsk"-Introd.
1. Laminar flow-Congresses.
2. Turbulence-Congresses.
3. Boundary layer-Congresses.
I. Kozlov, V. V. (Viktor Vladimirovich).
II. International Union of Theoretical and Applied Mechanics.
III. IUTAM Symposium on Laminar-Turbulent Transition (2nd : 1984 : Novosibirsk, R.S.F.S.R.).
TA 357.L35 1985 532'.516 85-4762

Printed in Germany

Offsetprinting: Color-Druck, G. Baucke, Berlin
Bookbinding: Mikolai, Berlin
2061/3020 5 4 3 2 1 0

Dedicated to the Memory
of Professor N. N. Yanenko

Introduction

At the suggestion of the USSR National Committee of Theoretical and Applied Mechanics the International Union of Theoretical and Applied Mechanics made a decision to organize the Second IUTAM Symposium on Laminar-Turbulent Transition in Novosibirsk.

Professor Yanenko was entrusted to be the chairman of the scientific committee. It was proposed that about 100 scientists would participate in the Symposium and that 60 lectures and invited lectures would be delivered; as for the theme of the Symposium, it was devoted to experimental and theoretical studies of laminar-turbulent transition in fluids and gases, i.e. the physical problems of transition and mathematical modeling in shear flows. It was planned to lay a special emphasis on fundamental mechanisms of the transition process. The scientific committee approved of the scope of the Symposium having considered the changes in the directions of the studies which took place since the First Symposium. For example, there was singled out the problem of transformation of external disturbances into eigenoscillations of shear flows, i.e. the receptivity problem which was only slightly touched upon at the First Symposium. A new section appeared devoted to the interrelation of the flow separation phenomenon and its stability.

It was planned to focus the attention on the secondary instability at transition to turbulence at the non-linear stage of the flow breakdown as well as on the loss of stability and the disturbances development in 3-D flows.

How all these projects were realized the reader can find out in the present proceedings.

Contributed papers were selected by the scientific committee from numerous abstracts. To select the papers was rather complicated because the ones presented were highly qualified. It became clear that the traditional planning of the Symposium won't permit to hear all the lectures, therefore it was decided to divide them into oral and those presented in the poster session. The introduction of the poster session enabled to increase the number of the participants, and apart from the known specialists a great number of young scientists engaged in the given themes took part in the Symposium. Nevertheless, unfortunately, a series of papers was not accepted not on account of the contents but because the finally accepted ones were more relevant to the scope of the Symposium.

Due to the able guidance of the session chairmen the discussions at the Symposium were animated and went on during poster sessions for which special time was allotted.

To our deep regret, the chairman of the Symposium, professor Yanenko, deceased on the 16th of January 1984 in the prime of his creative activities. To organize the second scientific meeting of the specialists professor Yanenko made a great contribution, and that's why the proposal to dedicate the Symposium and its ptoceedings to his memory was endorsed unanimously.

The Symposium was concluded by a final discussion which summed up the 5 day work and showed that only mutual international cooperation of the representatives of all the schools and all the leading countries can facilitate the solution of complicated problems in the given branch of science.

Dr. V.Kozlov

Novosibirsk

Scientific Committee

N.N.Yanenko , USSR (Chairman)

R.Eppler, FRG
V.V.Kozlov, USSR
M.Landahl, Sweden
V.Ya.Levchenko, USSR
J.Mathieu, France
E.Reshotko, USA
W.S.Saric, USA
J.T.Stuart, England
I.Tani, Japan

Local Organizing Committee

V.G.Dulov (Chairman)
V.N.Zhigulev
V.E.Nakoryakov
V.V.Kozlov (Scientific Secretary)
V.V.Demidenko
V.Ya.Levchenko
G.K.Mikhailov
V.A.Proskuryakov
M.I.Rabinovich
O.S.Ryzhov

Symposium Sponsors

International Union of Theoretical and Applied Mechanics (IUTAM)

National Committee of Theoretical and Applied Mechanics (NCTAM)

USSR Academy of Sciences
Siberian Division
Institute of Theoretical and
Applied Mechanics

Participants

Afendikov, A.L.	USSR	Institute of Applied Mathematics, USSR Academy of Sciences, A-47 Moscow
Aihara, Y.	JAP	Department of Aeronautics, Faculty of Engineering, University of Tokyo, Tokyo
Arnal, D.	F	ONERA/Cert-Derat, 2 Avenue Eduard Belin
Asai, M.	JAP	College of Engineering, University of Osaka Prefecture, Sakai, 591 Osaka
Atta, van, C.W.	USA	Department of Applied Mechanics and Engineering Sciences, University of California, San Diego, La Jolla, California 92093
Babenko, V.V.	USSR	Institute of Hydromechanics, Ukrainian Academy of Sciences, ul. Zheliabova 8/4, 252057 Kiev
Belyaev, Yu.N.	USSR	Institute of Mechanics, Moscow State University, 117415 Moscow
Bozduganov, I.	BULG	Institute of Mechanics and Biomechanics, Bulgarian Academy of Sciences, ul. Boncheva, bl.8, 1113 Sofia
Bune, A.V.	USSR	Institute for Problems in Mechanics, USSR Academy of Sciences, Prospect Vernadskogo, 117526 Moscow
Burnat, M.	POL	Institute of Mathematics, Warsaw P.K. i Nauki 9p., 00901 Warszawa
Bushnell, D.M.	USA	NASA Langley Research Center, Hampton, Virginia, 23665
Chernykh, A.I.	USSR	Institute of Automatics and Electrometry, USSR Academy of Sciences, 630090 Novosibirsk
Christov, C.I.	BULG	Institute of Mechanics and Biomechanics, Bulgarian Academy of Sciences, ul. Boncheva, bl.8 1113 Sofia
Coustols, E.	F	ONERA/Cert-Derat, 31055, Toulouse 2 Avenue Eduard Belin
Dallmann, U.Ch.	FRG	Institut für Theoretische Strömungsmechanik 1034 Göttingen, Bonsenst
Dovgal, A.V.	USSR	Institute of Theoretical and Applied Mechanics, USSR Academy of Sciences, 630090 Novosibirsk

Dulov, V.G.	USSR	Institute of Theoretical and Applied Mechanics, USSR Academy of Sciences, 630090 Novosibirsk
Dvorak, R.	CSSR	Ustav Termomechaniky CSAN, 16000 Praha 6, Puskinovo nám. 9
Eppler, R.	FRG	University of Stuttgart, Institut A für Mechanik, Pfaffenwaldring, 9 7000 Stuttgart 80
Ezersky, A.B.	USSR	Institute of Applied Physics, USSR Academy of Sciences, 603163 Gorky
Fedorov, A.V.	USSR	Moscow Physical Technical Institute, 417000 Moscow
Gaponov, S.A.	USSR	Institute of Theoretical and Applied Mechanics, USSR Academy of Sciences, 630090 Novosibirsk
Gaster, M.	UK	Tettington, Middlesek TW11 OLW
Gershuni, G.L.	USSR	Perm State University, 614039 Perm
Gertsenstein, S.Y.	USSR	Institute of Mechanics, Moscow State University, 117415 Moscow
Gilev, V.M.	USSR	Institute of Theoretical and Applied Mechanics, USSR Academy of Sciences, 630090 Novosibirsk
Glaznev, V.N.	USSR	Institute of Theoretical and Applied Mechanics, USSR Academy of Sciences, 630090 Novosibirsk
Goldfeld, M.A.	USSR	Institute of Theoretical and Applied Mechanics, USSR Academy of Sciences, 630090 Novosibirsk
Goldshtik, M.A.	USSR	Institute of Thermophysics, USSR Academy of Sciences, 630090 Novosibirsk
Grek, H.R.	USSR	Institute of Theoretical and Applied Mechanics, USSR Academy of Sciences, 630090 Novosibirsk
Hackeschmidt, M.	GDR	"Friedrick List" University of Transport and Communications, PSF 103, 8072 Dresden
Herbert, T.	USA	Department of Engineering Science and Mechanics, Virginia Polytechnic Institute and State University, Blacksburg, Virginia 24061
Isataev, S.I.	USSR	S.M.Kirov Kazakh State University, Alma-Ata

Ito, A.	JAP	Department of Mechanical Engineering, Faculty of Engineering, Meiji University, Kanagawa 214
Johansson, A.V.	S	The Royal Institute of Technology, 10044 Stockholm 70
Kachanov, Yu.S.	USSR	Institute of Theoretical and Applied Mechanics, USSR Academy of Sciences, 630090 Novosibirsk
Kharitonov, A.M.	USSR	Institute of Theoretical and Applied Mechanics, USSR Academy of Sciences, 630090 Novosibirsk
Kleiser, L.	FRG	**DFVLR**, Institut für Theoretische Strömungsmechanik, D-3400 Göttingen
Kobayashi, R.	JAP	Institute of High Speed Mechanics, Tohoku University, Sendai
Kohama, Y.	JAP	Institute of High Speed Mechanics, Tohoku University, Sendai
Kornilov, V.I.	USSR	Institute of Theoretical and Applied Mechanics, USSR Academy of Sciences, 630090 Novosibirsk
Kosinov, A.D.	USSR	Institute of Theoretical and Applied Mechanics, USSR Academy of Sciences, 630090 Novosibirsk
Kosorygin, V.A.	USSR	Institute of Theoretical and Applied Mechanics, USSR Academy of Sciences, 630090 Novosibirsk
Kovalenko, V.M.	USSR	Institute of Theoretical and Applied Mechanics, USSR Academy of Sciences, 630090 Novosibirsk
Kozlov, L.F.	USSR	Institute of Hydromechanics, Ukrainian Academy of Sciences, Kiev
Kozlov, V.V.	USSR	Institute of Theoretical and Applied Mechanics, USSR Academy of Sciences, 630090 Novosibirsk
Krause, E.	FRG	Aerodynamisches Institut der RWTH Aachen, Wüllnerstr. zw. 5 und 7, 5100 Aachen
Landahl, M.	S	The Royal Institute of Technology, S-10044, Stockholm 70
Lebiga, V.A.	USSR	Institute of Theoretical and Applied Mechanics, USSR Academy of Sciences, 630090 Novosibirsk
Leehey, P.	USA	Department of Ocean Engineering, Massachusetts Institute of Technology, Cambridge, Massachusetts 02139

Lessen, M.	USA	Department of Mechanical Engineering, University of Rochester, River Campus, Rochester, New York 14627
Levchenko, V.Ya.	USSR	Institute of Theoretical and Applied Mechanics, USSR Academy of Sciences, 630090 Novosibirsk
Levina, G.V.	USSR	Institute of Continuous Media Mechanics, USSR Academy of Sciences, the Ural Research Centre, Perm
Lindberg, P.A.	S	The Royal Institute of Technology, S-10044, Stockholm 70
Lukashchuk, S.N.	USSR	Institute of Automatics and Electrometry, USSR Academy of Sciences, 630090 Novosibirsk
Lvov, V.S.	USSR	Institute of Automatics and Electrometry, USSR Academy of Sciences, 630090 Novosibirsk
Lysenko, V.I.	USSR	Institute of Theoretical and Applied Mechanics, USSR Academy of Sciences, 630090 Novosibirsk
Mack, L.	USA	Jet Propulsion Laboratory, California Institute of Technology, Pasadena, CA 91103
Maslennikova, I.I.	USSR	Institute of Theoretical and Applied Mechanics, USSR Academy of Sciences, 630090 Novosibirsk
Maslov, A.A.	USSR	Institute of Theoretical and Applied Mechanics, USSR Academy of Sciences, 630090 Novosibirsk
Masuda, S.	JAP	Department of Mechanical Engineering, Faculty of Engineering, Keio University, 223 Keio
Matsui, T.	JAP	Faculty of Science and Technology, Meijo University, Yagoto, Tempaku, Nagoya 468
McCarthy, J.	USA	Head, Naval Hydromechanics Division, Bethesda, MD 20084
Mikhailov, K.	USSR	National Committee for Theoretical and Applied Mechanics, Moscow B-296, 117296
Miksad, R.W.	USA	College of Engineering, University of Texas at Austin, Austin,Texas
Moiseev, S.S.	USSR	Space Research Institute, USSR Academy of Sciences, Moscow
Monkewitz, P.A.	USA	Department of Mechanics and Structures, University of California, Los Angeles, CA 90024

Narasimha, R.	IND	Department of Aeronautical Engineering, Indian Institute of Science, Bangalore, 560012
Nartov, V.P.	USSR	Institute of Theoretical and Applied Mechanics, USSR Academy of Sciences, 630090 Novosibirsk
Nehring, U.	FRG	Institut für Theoretische Strömungsmechanik, Bunsenstr. 10, 34 Göttingen
Nepomnyashchy, A.A.	USSR	Institute of Continuous Media Mechanics, Ural Research Centre, USSR Academy of Sciences, Perm
Nishioka, M.	JAP	College of Engineering, University of Osaka Prefecture 4-804 Mozu-Umemachi, Sakai, 591
Novopashin, S.A.	USSR	Institute of Thermophysics, USSR Academy of Sciences, 630090 Novosibirsk
Okude, M.	JAP	Department of Mechanical Engineering, Faculty of Science and Technology, Meijo University, Yagoto, Tempaku, Nagoya 468
Pavlyuchenko, A.M.	USSR	Institute of Theoretical and Applied Mechanics, USSR Academy of Sciences, 630090 Novosibirsk
Petrov, G.V.	USSR	Institute of Theoretical and Applied Mechanics, USSR Academy of Sciences, 630090 Novosibirsk
Pfenninger, W.	USA	NASA, VA 23601, Newport News, 225 James River Dr.
Polezhaev, V.I.	USSR	Institute for Problems in Mechanics, USSR Academy of Sciences, Prospect Vernadskogo 101, 117526 Moscow
Polyakov, N.F.	USSR	Institute of Theoretical and Applied Mechanics, USSR Academy of Sciences, 630090 Novosibirsk
Predtechensky, A.A.	USSR	Institute of Automatics and Electrometry, USSR Academy of Sciences, 630090 Novosibirsk
Pungin, V.G.	USSR	Space Research Institute, USSR Academy of Sciences, Moscow
Prohoda, J.	CSSR	Ustav Termomechaniky CSAN, 16000 Praha 6, Puskinovo nam. 9
Rabinovich, M.I.	USSR	Institute of Applied Physics, USSR Academy of Sciences, 603600 Gorky

Ramazanov, M.P.	USSR	Institute of Theoretical and Applied Mechanics, USSR Academy of Sciences, 630090 Novosibirsk
Rannacher, J.	GDR	Polytechnical University of Dresden, Dresden
Reed, H.L.	USA	Department of Mechanical Engineering, Stanford University, Stanford, California 94305
Reutov, V.P.	USSR	Institute of Applied Physics, USSR Academy of Sciences, 603600 Gorky
Rozhdestvensky, B.L.	USSR	Keldysh Institute of Applied Mathematics, USSR Academy of Sciences, Moscow
Ruban, A.I.	USSR	TsaGI, Moscow
Rudyak, V.Ya.	USSR	Institute of Theoretical and Applied Mechanics, USSR Academy of Sciences, 630090 Novosibirsk
Ryzhov, O.S.	USSR	Computing Center of the USSR Academy of Sciences, 117967 Moscow
Sadovsky, V.S.	USSR	TsAGI, Moscow
Saric, W.S.	USA	Department of Mechanical and Aerospace Engineering, Arizona State University, Tempe, Arizona 85287
Savulescu, S.N.	RUM	Bucharest, INCREST
Schrauf, G.	USA	Applied Mathematics 217-50 California Institute of Technology Pasadena, California 91125
Scobelev, B.Yu.	USSR	Institute of Theoretical and Applied Mechanics, USSR Academy of Sciences, 630090 Novosibirsk
Seifert, G.	GDR	Institut für Mechanik der AdW der DDR, 1199 Berlin
Shidlovsky, V.P.	USSR	Computing Center of the USSR Academy of Sciences, Moscow
Shilein, E.H.	USSR	Institute of Theoretical and Applied Mechanics, USSR Academy of Sciences, 630090 Novosibirsk
Shkadov, V.Ya.	USSR	Institute of Mechanics, Moscow State University, 117415 Moscow
Shtern, V,N,	USSR	Institute of Thermophysics, USSR Academy of Sciences, 630090 Novosibirsk
Sidorenko, N.V.	USSR	Moscow Physical Technical Institute, 417000 Moscow

Simakin, I.N.	USSR	TsaGI, Moscow
Simanovsky, I.B.	USSR	Institute of Mechanics of Continuous Medium, Ural Research Center, USSR Academy of Sciences, Perm
Slavchev, S.	BULG	Institute of Mechanics and Biomechanics, Bulgarian Academy of Sciences, ul.Boncheva, bl.8, 1113 Sofia
Sosedko, Yu.P.	USSR	TsaGI, Moscow
Strunz, M.	FRG	Technische Universität Stuttgart, Stuttgart, 7253 Renningen 2
Suleimanov, Sh.	USSR	Institute of Theoretical and Applied Mechanics, USSR Academy of Sciences, 630090 Novosibirsk
Sushchik, M.M.	USSR	Institute of Applied Physics, USSR Academy of Sciences, Gorky
Suyasov, N.V.	USSR	Space Research Institute, USSR Academy of Sciences, Moscow
Taganov, G.I.	USSR	TsaGI, Moscow
Terekhova, N.M.	USSR	Institute of Theoretical and Applied Mechanics, USSR Academy of Sciences, 630090 Novosibirsk
Terentiev, Ye.D.	USSR	Computing Center of the USSR Academy of Sciences, 117967 Moscow
Tetyanko, V.A.	USSR	Krylov Shipbuilding Research Institute, Leningrad
Tomita, Yu.	JAP	Department of Mechanical Engineering, Kyushu Institute of Technology, Kitakyushu 804
Tumin, A.M.	USSR	Moscow Physical Technical Institute, Moscow
Vancoillie, G.	BELG	Department of Machinery, State University of Gent, Gent
Vasilyev, O.F.	USSR	Institute of Hydrodynamics, USSR Academy of Sciences, 630090 Novosibirsk
Veretentsev, A.N.	USSR	Institute of Theoretical and Applied Mechanics, USSR Academy of Sciences, 630090 Novosibirsk
Vladimirov, V.A.	USSR	Lavrentyev Institute of Hydrodynamics, USSR Academy of Sciences, 630090 Novosibirsk
Wadih, M.	F	Institut de Mécanique des Fluides, Université d'Aix-Marseille, rue Honnorat, 13003 Marseille

Yakovlev, V.N.	USSR	Institute of Theoretical and Applied Mechanics, USSR Academy of Sciences, 630090 Novosibirsk
Yavorskaya, I.M.	USSR	Space Research Institute, USSR Academy of Sciences, Moscow
Yosifov, K.Y.	BULG	Bulgarian Institute of Ship's Hydrodynamics, 9003 Varna
Yurchenko, N.F.	USSR	Institute of Hydromechanics, Ukrainian Academy of Sciences, ul. Zheliabova 8/4, 252057 Kiev
Zanin, B.Yu.	USSR	Institute of Theoretical and Applied Mechanics, USSR Academy of Sciences, 630090 Novosibirsk
Zapryanov, Z.	BULG	Bulgarian Institute of Ship's Hydrodynamics, 9003 Varna
Zatoloka, V.V.	USSR	Institute of Theoretical and Applied Mechanics, USSR Academy of Sciences, 630090 Novosibirsk
Zelman, M.B.	USSR	Institute of Theoretical and Applied Mechanics, USSR Academy of Sciences, 630090 Novosibirsk
Zharilkasinov, A.	USSR	Institute of Theoretical and Applied Mechanics, USSR Academy of Sciences, 630090 Novosibirsk
Zheltovodov, A.A.	USSR	Institute of Theoretical and Applied Mechanics, USSR Academy of Sciences, 630090 Novosibirsk
Zheltukhin, N.A.	USSR	Institute of Theoretical and Applied Mechanics, USSR Academy of Sciences, 630090 Novosibirsk
Zhigulev, V.N.	USSR	Moscow Physical Technical Institute, 417000 Moscow
Zhuk, V.I.	USSR	Computing Center of the USSR Academy of Sciences, 117967 Moscow
Zhukhovitsky, E.M.	USSR	Perm Pedagogical Institute, Perm
Zierep, J.	FRG	Institut für Strömungslehre und Strömungsmaschinen, Universität (TH) Karlsruhe, Kaiserstr. 12 D-7500 Karlsruhe 1
Zlatov, P.Z.	BULG	Bulgarian Institute of Ship's Hydrodynamics, 9003 Varna
Zwick, W.	GDR	Institut für Mechanik der AdW der DDR, Rudower Chaussee 5, 1199 Berlin

Contributors

Afendikov, A.L.	USSR	Institute of Applied Mathematics, USSR Academy of Sciences, A-47 Moscow
Aihara, Y.	JAP	Department of Aeronautics, Faculty of Engineering, University of Tokyo, Tokyo
Alfredsson, P.H.	S	The Royal Institute of Technology, 10044 Stockholm 70
Ariga, I.	JAP	Department of Mathematical Engineering, Keio University, Keio
Arnal, D.	F	ONERA/Cert-Derat, 2 Avenue Eduard Belin, 31055 Toulouse
Asai, M.		College of Engineering, University of Osaka Prefecture, Sakai, 591 Osaka
Babenko, K.I.	USSR	Institute of Applied Mathematics, USSR Academy of Sciences, A-47 Moscow
Babenko, V.V.	USSR	Hydromechanics Institute, Ukrainian Academy of Sciences, ul. Zheliabova 8/4, 252057 Kiev
Beckwith, I.E.	USA	NASA Langley Research Center, Hampton, Virginia 23665
Belyaev, Yu.N.	USSR	Space Research Institute, USSR Academy of Sciences, Moscow
Bragin, O.A.	USSR	Institute of Theoretical and Applied Mechanics, USSR Academy of Sciences, 630090 Novosibirsk
Bühler, K.	FRG	Institut für Strömungslehre und Strömungsmaschinen, Universität (TH) Karlsruhe, Kaiserstrasse 12, D-7500 Karlsruhe 1
Bune, A.V.	USSR	Institute for Problems in Mechanics, USSR Academy of Sciences, Prospekt Vernadskogo 101, 117526, Moscow
Burnat, M.	POL	Institute of Mathematics, Warsaw University, P.K. i Nauki 9p., 00901 Warszawa
Bushnell, D.M.	USA	NASA Langley Research Center, Hampton, Virginia 23665
Cayzac, R.	F	Institut de Mécanique des Fluides, Université d'Aix-Marseille, 1 rue Honnorat, 13003 Marseille
Chen, F.J.	USA	NASA Langley Research Center, Hampton, Virginia 23665
Chernykh, A.I.	USSR	Institute of Automatics and Electrometry, USSR Academy of Sciences, 630090 Novosibirsk

Christov, C.L.	Bulg	Institute of Mechanics and Biomechanics, Bulgarian Academy of Sciences, ul. Boncheva, bl.8, 1113 Sofia
Clarion, C.	F	Institut de Mécanique des Fluides, Université d'Aix-Marseille, 1 rue Honnorat, 13003 Marseille
Coustols, E.	F	Onera/Cert-Derat, 2 Avenue Eduard Belin, 31055 Toulouse
Craik, A.D.D.	UK	Mathematical Institute, University of St.Andrews, St.Andrews KY16 9SS, Scotland
Demekhin, V.A.	USSR	Institute of Mechanics, Moscow State University, Moscow
Dovgal, A.V.	USSR	Institute of Theoretical and Applied Mechanics, USSR Academy of Sciences, 630090 Novosibirsk
Dovgiy, S.A.	USSR	Hydromechanics Institute, Ukrainian Academy of Sciences, ul. Zheliabova 8/4, 252057 Kiev
Erofeev, E.A.	USSR	Moscow Physical Institute, 417000 Moscow
Ezerskii, A.B.	USSR	Institute of Applied Physics, USSR Academy of Sciences, 603600 Gorky
Fahlgren, E.M.	S	The Royal Institute of Technology, 10044 Stockholm 70
Fauchas, J.	F	Institut de Mécanique des Fluides, Université d'Aix-Marseille, 1 rue Honnorat, 13003 Marseille
Fedorov, A.V.	USSR	Moscow Physical Technical Institute, Moscow
Filjand, L.V.	USSR	Institute of Mechanics, Moscow State University, Moscow
Gaponov, S.A.	USSR	Institute of Theoretical and Applied Mechanics, USSR Academy of Sciences, 630090 Novosibirsk
Gasley, C.	USA	The Rand Corporation Santa Monica, California 90406
Gaster, M.	UK	National Maritime Institute, Queens Road, Teddington, Middlesex TW11 OLW
Gedney, C.J.	USA	Mechanical and Ocean Engineering Massachusetts Institute of Technology, Cambridge, Massachusetts 02139

Gertsenstein, S.Y.	USSR	Institute of Mechanics, Moscow State University, Moscow
Gershuni, G.Z.	USSR	Perm University, 614039 Perm
Gilev, V.M.	USSR	Institute of Theoretical and Applied Mechanics, USSR Academy of Sciences, 630090 Novosibirsk
Glaznev, V.N.	USSR	Institute of Theoretical and Applied Mechanics, USSR Academy of Sciences, 630090 Novosibirsk
Gnitetsky, N.A.	USSR	Hydromechanics Institute, Ukrainian Academy of Sciences, ul. Zheliabova 8/4, 252057 Kiev
Goldfeld, M.A.	USSR	Institute of Theoretical and Applied Mechanics, USSR Academy of Sciences, 630090 Novosibirsk
Goldshtik, M.A.	USSR	Institute of Thermophysics, USSR Academy of Sciences, 630090 Novosibirsk
Grek, H.R.	USSR	Institute of Theoretical and Applied Mechanics, USSR Academy of Sciences, 630090 Novosibirsk
Griaznov, V.L.	USSR	Institute for Problems in Mechanics, USSR Academy of Sciences, Moscow
Hackeschmidt, M.	GDR	"Friedrich List" University of Transport and Communications, PSF 103, 8072 Dresden
Her, J.Y.	USA	Mechanical and Ocean Engineering Massachusetts Institute of Technology, Cambridge, Massachusetts, 02139
Herbert, T.	USA	Department of Engineering Science and Mechanics, Virginia Polytechnic Institute and State University, Blacksburg, Virginia 24061
Hilbrich, H.D.	GDR	"Friedrich List" University of Transport and Communications, PSF 103, 8072 Dresden
Ho, Ch.M.	USA	Department of Aerospace Engineering, University of Southern California, Los Angeles, CA 90039-1454
Hussain, A.K.M.F.	USA	Department of Mechanical Engineering, University of Houston, TX
Isataev, S.I.	USSR	S.M.Kirov Kazakh State University, Alma-Ata
Ito, A.	JAP	Department of Mechanical Engineering, Faculty of Engineering, Meiji University, Kanagava 214

Itoh, N.	JAP	National Aerospace Laboratory, Chfu, Tokyo
Ivanov, V.P.	USSR	Hydromechanics Institute, Ukrainian Academy of Sciences, ul. Zheliabova 8/4, 252057 Kiev
Johansson, A.V.	S	The Royal Institute of Technology, 10044 Stockholm 70
Jones, F.L.	USA	College of Engineering, University of Texas at Austin, Austin, TX
Juillen, J.C.	F	Onera/Cert-Derat, 2 Avenue Eduard Belin, 31055 Toulouse
Kachanov, Yu.S.	USSR	Institute of Theoretical and Applied Mechanics, USSR Academy of Sciences, 630090 Novosibirsk
Kleiser, L.	FRG	Institut für Theoretische Strömungsmechanik, D-3400, Göttingen
Kobayashi, R.	JAP	Institute of High Speed Mechanics, Tohoku University, Sendai
Koh, P.H.	USA	Department of Engineering, University of Rochester, River Campus, Rochester, New York 14627
Kohama, Y.	JAP	Institute of High Speed Mechanics, Tohoku University, Sendai
Korobov, V.I.	USSR	Hydromechanics Institute, Ukrainian Academy of Sciences, ul. Zheliabova 8/4, 252057 Kiev
Kosinov, A.D.	USSR	Institute of Theoretical and Applied Mechanics, USSR Academy of Sciences, 630090 Novosibirsk
Kosorygin, V.S.	USSR	Institute of Theoretical and Applied Mechanics, USSR Academy of Sciences, 630090 Novosibirsk
Kozlov, L.F.	USSR	Hydromechanics Institute, Ukrainian Academy of Sciences, ul. Zheliabova 8/4, 252057 Kiev
Kozlov, V.V.	USSR	Institute of Theoretical and Applied Mechanics, USSR Academy of Sciences, 630090 Novosibirsk
Krause, E.	FRG	Aerodynamisches Institut der RWTH Aachen, Wüllnerstr. zw.5 und 7 5100 Aachen
Landahl, M.T.	S	The Royal Institute of Technology, S-10044, Stockholm
Laurien, E.	FRG	DFVLR Institut für Theoretische Strömungsmechanik, D-3400 Göttingen

Leehey, P.	USA	Department of Ocean Engineering, Room 1-207, M.I.T., Cambridge, MA 02139
Lessen, M.	USA	Department of Engineering, the University of Rochester, New York 14627
Levchenko, V.Ya.	USSR	Institute of Theoretical and Applied Mechanics, USSR Academy of Sciences, 630090 Novosibirsk
Levina, G.V.	USSR	Institute of Continuous Media Mechanics of the Ural Research Center, USSR Academy of Sciences, Perm
Lifschitz, A.M.	USSR	Institute of Thermophysics, 630090 Novosibirsk
Lindberg, P.A.	S	The Royal Institute of Technology, 10044 Stockholm 70
Lvov, V.S.	USSR	Institute of Automation and Electrometry, USSR Academy of Sciences, 630090 Novosibirsk
Lukaschuk, S.N.	USSR	Institute of Automation and Electrometry, USSR Academy of Sciences, 630090 Novosibirsk
Lysenko, V.I.	USSR	Institute of Theoretical and Applied Mechanics, USSR Academy of Sciences, 630090 Novosibirsk
Mack, L.	USA	Jet Propulsion Laboratory, California Institute of Technology, Pasadena, CA 91109
Maksimova, E.M.	USSR	Institute of Theoretical and Applied Mechanics, USSR Academy of Sciences, 630090 Novosibirsk
Malik, M.R.	USA	High Technology Corporation P.O.Box 7262, Hampton, VA 23666
Maslennikova, I.I.	USSR	Institute of Theoretical and Applied Mechanics, USSR Academy of Sciences, 630090 Novosibirsk
Maslov, A.A.	USSR	Institute of Theoretical and Applied Mechanics, USSR Academy of Sciences, 630090 Novosibirsk
Masuda, S.	JAP	Department of Mechanical Engineering, Faculty of Engineering, Keio University, 223 Keio
Matsui, T.	JAP	Department of Mechanical Engineering, Faculty of Science and Technology, Meijo University, Yagoto, Tempacu, Nagoya 468
Mautner, T.S.	USA	Dept. of Applied Mechanics and Engineering Sciences, University of California, San Diego, La Jolla, CA 92093

Michel, R.	F	ONERA/Cert-Derat, 2 Avenue Eduard, Belin, 31055 Toulouse
Miksad, R.W.	USA	College of Engineering, University of Texas at Austin, Austin TX
Moiseev, S.S.	USSR	Space Research Institute, USSR Academy of Sciences, Moscow
Monakhov, A.A.	USSR	Space Research Institute, USSR Academy of Sciences, Moscow
Monkewitz, P.	USA	Department of Mechanics and Structures, University of California, Los Angeles, CA 90024
Narasimha, R.	IND	Department of Aerospace Engineering, Indian Institute of Science, Bangalore 560 012
Nartov, V.P.	USSR	Institute of Theoretical and Applied Mechanics, USSR Academy of Sciences, 630090 Novosibirsk
Nehring, U.	FRG	Deutsche Forschungs- und Versuchsanstalt für Luft- und Raumfahrt e.V. Institut für Theoretische Strömungsmechanik Bunsenstr. 10, 34 Göttingen
Nepomnyaschy, A.A.	USSR	Institute of Mechanics of Continuous Medium, USSR Academy of Sciences, Perm
Nikitin, N.V.	USSR	Institute of Mechanics, Moscow State University, Moscow
Nishioka, M.	JAP	College of Engineering, University of Osaka Prefecture, 4-804 Mozu-Umemachi, Sakai, 591
Novopashin, S.A.	USSR	Institute of Thermophysics, USSR Academy of Sciences, 630090 Novosibirsk
Okamae, K.	JAP	Department of Mechanical Engineering, Keio University, Keio
Okude, M.	JAP	Department of Mechanical Engineering, Faculty of Science and Technology, Meijo University, Yagoto, Tempaku, Nagoya 468
Pavlov, A.A.	USSR	Institute of Theoretical and Applied Mechanics, USSR Academy of Sciences, 630090 Novosibirsk
Pavlyuchenko, A.M.	USSR	Institute of Theoretical and Applied Mechanics, USSR Academy of Sciences, 630090 Novosibirsk
Petrov, G.V.	USSR	Institute of Theoretical and Applied Mechanics, USSR Academy of Sciences, 630090 Novosibirsk

Polezhaev, V.I.	USSR	Institute for Problems in Mechanics, USSR Academy of Sciences, Prospekt Vernadskogo, 101 117526 Moscow
Poll, D.I.A.	UK	Aerodynamics Division, College of Aeronautics, Cranfield, Institute of Technology, Cranfield, Bedford, MK430AL
Polyakov, N.Ph.	USSR	Institute of Theoretical and Applied Mechanics, USSR Academy of Sciences, 630090 Novosibirsk
Powers, E.J.	USA	College of Engineering, University of Texas at Austin, Austin TX
Pozdeev, A.A.	USSR	Institute of Continuous Media Mechanics of the Ural Research Center, USSR Academy of Sciences, Perm
Predtechensky, A.A.	USSR	Institute of Automation and Electrometry, USSR Academy of Sciences, 630090 Novosibirsk
Pungin, V.G.	USSR	Space Research Institute, USSR Academy of Sciences, Moscow
Rabinovich, M.I.	USSR	Institute of Applied Physics, USSR Academy of Sciences, 603600 Gorky
Ramazanov, M.P.	USSR	Institute of Theoretical and Applied Mechanics, USSR Academy of Sciences, 630090 Novosibirsk
Reed, H.L.	USA	Department of Mechanical Engineering, Stanford University, Stanford, CA 94305
Repic, Ye.U.	USSR	TsaGI, Moscow
Reutov, V.P.	USSR	Institute of Applied Physics, USSR Academy of Sciences, 603600 Gorky
Rössler, M.	GDR	"Friedrich List" University of Transport and Communications, 8072 Dresden
Rozhdestvensky, B.L.	USSR	Keldysh Institute of Applied Mathematics, USSR Academy of Sciences, Moscow
Ruban, A.I.	USSR	TsaGI, Moscow
Rudnitsky, A.Y.	USSR	Institute of Mechanics, Moscow State University, Moscow
Rudyak, V.Ya.	USSR	Institute of Theoretical and Applied Mechanics, USSR Academy of Sciences, 630090 Novosibirsk
Ryzhov, O.S.	USSR	Computing Center of the USSR Academy of Sciences, 117967 Moscow

Sadovsky, V.S.	USSR	TsAGI, Moscow
Saric, W.S.	USA	Department of Mechanical and Aerospace Engineering, Arizona State University, Tempe, Arizona 85287
Savulescu, S.N.	RUM	INCREST, Bucuresti
Scherbakov, S.A.	USSR	Space Research Institute, USSR Academy of Sciences, Moscow
Schultz-Grunov, F.	FRG	Institute of Applied Mechanics, Technical University of Aachen, Aachen
Schrauf, G.	USA	Applied Mathematics 217-50 California Institute of Technology Pasadena, California 91125
Shilein, E.H.	USSR	Institute of Theoretical and Applied Mechanics, USSR Academy of Sciences, 630090 Novosibirsk
Shkadov, V.Ya.	USSR	Institute of Mechanics, Moscow State University, Moscow
Shtern, V.N.	USSR	Institute of Thermophysics, USSR Academy of Sciences, 630090 Novosibirsk
Scobelev, B.Yu.	USSR	Institute of Theoretical and Applied Mechanics, USSR Academy of Sciences, 630090 Novosibirsk
Sidorenko, N.V.	USSR	Moscow Physical Technical Institute, 417000 Moscow
Simakin, I.N.	USSR	TsaGI, Moscow
Simanovsky, I.B.	USSR	Perm Pedagogical Institute, Perm
Sosedko, Yu.P.	USSR	TsaGI, Moscow
Srivastava, K.M.	FRG	Institute for Theoretical Fluid Mechanics, DFVLR-AVA, Bunsenstr. 10, 3400 Göttingen
Stettler, J.C.	Switz	Institute of Biological Research, ETH-University of Zuerich, Zuerich
Sukhorukov, A.N.	USSR	Institute of Mechanics, Moscow State University, Moscow
Sushchik, M.M.	USSR	Institute of Applied Physics, USSR Academy of Sciences, 603600 Gorky
Suyasov, N.V.	USSR	Space Research Institute, USSR Academy of Sciences, Moscow
Taganov, G.I.	USSR	TsAGI, Moscow
Tarasov, V.F.	USSR	Lavrentiev Institute of Hydrodynamics, USSR Academy of Sciences, 630090 Novosibirsk

Terekhova, N.M.	USSR	Institute of Theoretical and Applied Mechanics, USSR Academy of Sciences, 630090 Novosibirsk
Terentiev, E.D.	USSR	Computing Center of the USSR Academy of Sciences, 117967 Moscow
Tetyanko, B.A.	USSR	Krylov Shipbuilding Research Institute, Leningrad
Tomita, Y.	JAP	Department of Mechanical Engineering, Faculty of Engineering, Kyushu Institute of Technology, Kitakyushu-shi, 804
Tumin, A.M.	USSR	Moscow Physical Technical Institute, Moscow
Tyutin, A.A.	USSR	Institute of Theoretical and Applied Mechanics, USSR Academy of Sciences, 630090 Novosibirsk
Van Atta, C.W.	USA	Department of Applied Mechanics and Engineering Sciences, University of California, San Diego, La Jolla, CA 92093
Vancoillie, G.	BELG	Department of Machinery, State University at Gent, Gent
Veretentsev, A.N.	USSR	Institute of Theoretical and Applied Mechanics, USSR Academy of Sciences, 630090 Novosibirsk
Vladimirov, V.A.	USSR	Lavrentiev Institute of Hydrodynamics, USSR Academy of Sciences, 630090 Novosibirsk
Wadih, M.	F	Institut de Mécanique des Fluides, Université d'Aix-Marseille, 1 rue Honnorat, 13003 Marseille
Wazzan, A.R.	USA	University of California, Los Angeles, CA 90024
Yakovlev, V.N.	USSR	Institute of Theoretical and Applied Mechanics, USSR Academy of Sciences, 630090 Novosibirsk
Yanenko, N.N.	USSR	Institute of Theoretical and Applied Mechanics, USSR Academy of Sciences, 630090 Novosibirsk
Yavorskaya, I.M.	USSR	Space Research Institute, USSR Academy of Sciences, Moscow
Yeats, L.G.	USA	Department of Engineering Science and Mechanics, Virginia Polytechnic Institute and State University, Blacksburg, Virginia 24061
Yurchenko, N.F.	USSR	Hydromechanics Institute, Ukrainian Academy of Sciences, ul.Zheliabova, 8/4, 252057 Kiev

Zanin, B.Yu.	USSR	Institute of Theoretical and Applied Mechanics, USSR Academy of Sciences, 630090 Novosibirsk
Zatoloka, V.V.	USSR	Institute of Theoretical and Applied Mechanics, USSR Academy of Sciences, 630090 Novosibirsk
Zelman, M.B.	USSR	Institute of Theoretical and Applied Mechanics, USSR Academy of Sciences, 630090 Novosibirsk
Zierep, J.	FRG	Institut für Strömungslehre und Strömungsmaschinen, Universität Karlsruhe, Kaiserstr. 12, D-7500 Karlsruhe 1
Zhang, Yu.Q.	USA	Department of Aerospace Engineering, University of Southern California, Los Angeles, CA 90089-1454
Zharilkasinov, A.	USSR	Institute of Theoretical and Applied Mechanics, USSR Academy of Sciences, 630090 Novosibirsk
Zhdanova, E.M.	USSR	Institute of Thermophysics, USSR Academy of Sciences, 630090 Novosibirsk
Zheligovsky, V.A.	USSR	Institute of Mechanics, Moscow State University, Moscow
Zheltovodov, A.A.	USSR	Institute of Theoretical and Applied Mechanics, USSR Academy of Sciences, 630090 Novosibirsk
Zheltukhin, N.A.	USSR	Institute of Theoretical and Applied Mechanics, USSR Academy of Sciences, 630090 Novosibirsk
Zhigulev, V.N.	USSR	Moscow Physical Technical Institute, 417000 Moscow
Zhuk, V.I.	USSR	Computing Center of the USSR Academy of Sciences, 117967
Zhukhovitsky, E.M.	USSR	Perm Pedagogical Institute, Perm
Zwick, W.	GDR	Institut für Mechanik der AdW der DDR, Rudower Chaussee 5, 1199 Berlin

Contents

Session 1

Chairman: R. Eppler

Session 2

Chairman: V.N. Zhigulev

Poster Session

Session 3

Chairman: O.C. Ryzhov

Session 4

Chairman: M.A. Goldshtik

Session 5

Chairman: M. Landahl

Session 6

Chairman: M. Gaster

Poster Session

Session 9

Chairman: T. Herbert

Session 10

Chairman: O.F. Vasiliev

Session 11

Chairman: M. Nishioka

Session 12

Chairman: V.Ya. Levchenko

Session 13

Chairman: D.M. Bushnell

Poster Session

Session 14

Chairman: C.W. Van Atta

Session 15

Chairman: M. Mack

Session 16

Chairman: M.I. Rabinovich

Session 17

Chairman: B.L. Rozhdestvensky

Poster Session

---

*) Invited lecture

+) Paper not available at time of printing

Resonant Interactions in Shear Flows

A. D. D. CRAIK

Department of Applied Mathematics
St Andrews University
St Andrews, Fife KY16 9SS, Scotland, U.K.

Raetz [1,2] and Stuart [3] first showed theoretically that resonant triads of Tollmein-Schlichting waves can occur in boundary-layer flows. At about the same time, interest arose in resonance (triads and quartets) among surface waves in water (Phillips [4], McGoldrick [5]) and among surface and internal waves in stratified fluid (Ball [6], Thorpe [7], Davis & Acrivos [8]). Meanwhile, similar advances were taking place in other fields, notably nonlinear optics and plasma physics. Subsequently, Kelly [9,10] examined subharmonic and three-wave resonance among neutrally-stable inviscid modes in jet and shear-layer profiles; while Craik [11] demonstrated that resonant triads of surface gravity waves exist if a sufficiently strong mean shear flow is present.

Craik [12] and Usher & Craik [13,14] developed the weakly-nonlinear theory of resonant wave triads in (quasi-parallel) boundary layer flows $U(z)$. They suggested that particularly strong interactions may be expected among the members of symmetric triads comprising a two-dimensional wave mode of periodicity $\exp[i\alpha(x - c_r t)]$ and oblique modes with periodicities $\exp[i\frac{1}{2}\alpha(x - c_r' t) \pm i\beta y]$. Normally, there is just one (positive) spanwise wavenumber β for which there occurs exact resonance with $c_r = c_r'$.

The classic boundary-layer experiments of Klebanoff, Tidstrom & Sargent [15] showed spontaneous growth of oblique modes when a virtually two-dimensional wave was driven by an oscillating ribbon. But these oblique modes had frequency close to that of the driven mode (and so was interpreted as a spanwise -periodic 'warping' of the waves) : not one-half of the frequency of the two-dimensional mode as required by the Craik-Usher model. This experimental data may be modelled as a special case of four-, rather than three-, wave interaction (see, particularly, Orszag & Kells [16], Orszag & Patera [17]).

In contrast, Miksad's [18,19] experiments on unstable free shear layers showed evidence of generation of subharmonics with half the forced frequency. For the flat-plate boundary layer, the first indication of the appearance of subharmonic modes, much as predicted by the Craik-Usher model, was found by Kachanov, Kozlov & Levchenko [20]. Subsequent work, in independent experimental facilities at Novosibirsk and at Blacksburg, Va, confirmed this finding : the fullest accounts are given by Kachanov & Levchenko [21] and Saric & Thomas [22]. There is now no doubt that three-wave resonance, or near-resonance, provides an alternative pathway for the growth of three-dimensionality (and subsequent transition) which is distinct from, and complementary to, that explored by Klebanoff, Tidstrom & Sargent. Typically the former mechanism dominates when the forced wave has very small amplitude and the latter at larger forcing amplitudes.

With severe truncation at $O(\varepsilon^2)$ - which may be hard to justify formally - where ε is a measure of each wave amplitude A_j $(j = 1,2,3)$, the general equations of three-wave resonance have normalized form

$$\begin{aligned}(\partial/\partial t + \underline{v}_1.\nabla)A_1 &= \sigma_1 A_1 + \gamma_1 A_2^* A_3^* \\ (\partial/\partial t + \underline{v}_2.\nabla)A_2 &= \sigma_2 A_2 + \gamma_2 A_3^* A_1^* \\ (\partial/\partial t + \underline{v}_3.\nabla)A_3 &= \sigma_3 A_3 + \gamma_3 A_1^* A_2^*\end{aligned} \tag{1}$$

If real, the σ_j denote linear growth or damping rates; but complex values of σ_j may be employed to incorporate detuning from resonance. The $\underline{v}_j$ are respective group velocities in x-y propagation space and the γ_j the quadratic coupling coefficients.

General solutions are known in conservative cases where $\underline{v}_j$, γ_j are real and $\sigma_j = 0$. Though mostly obtained by inverse-scattering and Bäcklund transform techniques (Zakharov [23], Kaup, Reiman & Bers [24], Kaup [25]), broad classes of solutions take simple analytic form (Craik [26]). These have a rich structure which cannot be described in detail here; but attention must be drawn to the existence, under suitable circumstances, of 'explosive' solutions which become singular after a finite time. An inviscid fluid configuration for which such solutions exist is three-layer Kelvin-Helmholtz flow (Craik & Adam [27]).

In non-conservative viscous flows, $\sigma_j \neq 0$ and the $\underline{v}_j$ and γ_j may be complex

Then, only a few particular solutions of (1) are known (see, e.g. Craik [28]). 'Explosive' solutions remain a likely possibility. Purely temporal modulations of not-quite-resonant triads comprising one linearly-amplified and two linearly-damped modes were examined numerically by Wersinger, Finn & Ott [29]. For various parameter ranges, they discovered families of bifurcating periodic solutions and chaotic 'strange-attractor' behaviour. There is a need for further work on these, and related, non-conservative equations.

For boundary layers, the suggestion of Craik [12], based on asymptotic analysis, that the interaction coefficients γ_k for oblique modes may often be much larger than that for a two-dimensional mode was in part substantiated by computations of Hendriks, reported by Usher & Craik [14], and of Volodin & Zel'man [30]. The latter authors considered spatial, rather than temporal, wave evolution and attempted to incorporate the effects of boundary-layer growth. Further theoretical work of Nayfeh & Bozatli [31,32], confirmed that exact resonance does not occur between purely two-dimensional modes with frequencies ω_1 and $\omega_2 = \frac{1}{2}\omega_1$ and that the spatial mismatch $\Delta = k_1 - 2k_2$ is then normally sufficiently large to suppress parametric instability of the subharmonic. Also, Nayfeh & Bozatli [33] considered resonant configurations similar to the Craik-Usher type, but with the additional complexity that the two-dimensional second (or sub-) harmonic lies close to a linear eigenmode. Effectively, they envisage two near-resonant triads : a degenerate one comprising a two-dimensional mode $(\frac{1}{2}\alpha,0)$ and its harmonic $(\alpha,0)$ and a three-dimensional triad $(\alpha,0)$, $(\frac{1}{2}\alpha,\beta)$, $(\frac{1}{2}\alpha,-\beta)$.

Equation (1) omits cubic and higher-order nonlinearities, but Usher & Craik [14] showed that large cubic interaction coefficients may be expected both in resonant and non-resonant cases. Such higher-order nonlinearities were incorporated in a series of computational studies by Herbert [34,35,36] both for plane Poiseuille flow and for Blasius flow. (For lowest eigenmode disturbances in the former flow, the coefficients γ_j are all, exceptionally, equal to zero and so a higher-order theory - or other modes - becomes essential.) These concern the temporal eigenvalue spectrum of infinitesimal, symmetric, oblique-wave pairs in the presence of a finite-amplitude two-dimensional wave. Herbert's [34] plane Poiseuille flow results are somewhat inaccurate; but they, together with the improved calculations of [36], show the expected parametric growth of oblique-wave modes at subharmonic frequencies.

Herbert's [35] computational results for the Blasius boundary layer (neglecting non-parallelism) display features reproduced by the following simpler model. Near-resonant infinitesimal oblique modes in the presence of a two-dimensional wave of constant finite amplitude A_0 satisfy

$$\begin{aligned} dA_1/dt &= \sigma A_1 + \gamma A_0^* A_2^* + \lambda |A_0|^2 A_1 \\ dA_2/dt &= \sigma A_2 + \gamma A_0^* A_1^* + \lambda |A_0|^2 A_2 \end{aligned} \tag{2}$$

on neglecting higher-order terms in $|A_0|^4 A_k$ $(k = 1,2)$, etc. (Herbert also employed series truncation at low order). In the reference frame of the wave A_0, $\sigma = \sigma_r + i\sigma_i$ incorporates a detuning from resonance as an imaginary part σ_i. Setting $A_1 \propto \exp st$, $A_2 \propto \exp s^*t$ yields eigenvalues

$$s = \sigma_r + \lambda_r |A_0|^2 \pm \{-(\sigma_i + \lambda_i |A_0|^2)^2 + |\gamma A_0|^2\}^{\frac{1}{2}}. \tag{3}$$

These eigenvalues are complex conjugates provided $|A_0|$ is sufficiently small. But, as $|A_0|$ is increased, the two roots become real provided $|\gamma|^2 > 4\sigma_i\lambda_i$: modes A_1 and A_2 are then phase-locked to the fundamental. At still larger $|A_0|$, the roots s become complex conjugates again. The roots denote unstable waves if, for the given A_0, $\mathrm{Re}\{s\} > 0$.

Herbert [35] also investigated the interaction of Orr-Sommerfeld (Tollmein-Schlichting) eigenmodes and linearly-damped 'Squire modes' associated with the equation for normal-to-wall (z-) vorticity. This, too, may be modelled much as in (2), by a set of four coupled evolution equations for oblique modes A_1, A_2, A_3, A_4 with A_0 as a parameter. Herbert found that a real root originating as a Squire mode at small $|A_0|$ is sometimes first to become unstable as $|A_0|$ is increased. Saric & Thomas [22] accordingly invoked Squire modes to interpret some recent experimental observations. However, it must be remembered that modes can exchange physical identities when their dispersion curves pass close together : and that pure Squire modes have zero velocity component normal to the wall.

Saric & Thomas claim to distinguish two distinct subharmonic mechanisms ('C-type' and 'H-type') in addition to the 'K-type' mechanism studied by Klebanoff *et al.* It is certainly true that the preferred spanwise wave-number of the supposed 'H-type' mechanism is considerably larger than that selected by the ('C-type') resonance condition applied to the linear dispersion relation. But it seems to the present writer that this discrepancy may be attributable to the stronger nonlinear dispersive effects present at larger forcing amplitudes. The possible importance of Squire modes, and also of higher Orr-Sommerfeld eigenmodes, certainly deserves further study (c.f. Gustavsson & Hultgren [37]) : but the suggestion of qualitatively distinct 'C-type' and 'H-type' mechanisms in boundary layers does not yet seem firmly established.

References

1. Raetz, G.S.: 1959 Norair Rep. NOR-59-383. Hawthorne, Calif. A new theory of the cause of transition in fluid flows.

2. Raetz, G.S.: 1964 Norair Rep. NOR-64-111. Hawthorne, Calif. Current status of resonance theory of transition.

3. Stuart, J.T.: 1962 Adv. Aero. Sci. 3-4, 121-142. On three-dimensional nonlinear effects in the stability of parallel flows.

4. Phillips, O.M.: 1960 J. Fluid Mech. 9, 193-217. On the dynamics of unsteady gravity waves of finite amplitude, Part 1. The elementary interactions.

5. McGoldrick, L.F.: 1965 J. Fluid Mech. 21, 305-331. Resonant interactions among capillary-gravity waves.

6. Ball, F.K.: 1964 J. Fluid Mech. 20, 457-479. Energy transfer between external and internal gravity waves.

7. Thorpe, S.A.: 1966 J. Fluid Mech. 24, 737-751. On wave interactions in a stratified fluid.

8. Davis, R.E. & Acrivos, A.: 1967 J. Fluid Mech. 30, 723-736. The stability of oscillatory internal waves.

9. Kelly, R.E.: 1967 J. Fluid Mech. 27, 657-689. On the stability of an inviscid shear layer which is periodic in space and time.

10. Kelly, R.E.: 1968 J. Fluid Mech. 31, 789-799. On the resonant interaction of neutral disturbances in two inviscid shear flows.

11. Craik, A.D.D.: 1968 J. Fluid Mech. 34, 531-549. Resonant gravity-wave interactions in a shear flow.

12. Craik, A.D.D.: 1971 J. Fluid Mech. 50, 393-413. Nonlinear resonant instability in boundary layers.

13. Usher, J.R. & Craik, A.D.D.: 1974 J. Fluid Mech. 66, 209-221. Nonlinear wave interactions in shear flows. Part 1. A variational formulation.

14. Usher, J.R. & Craik, A.D.D.: 1975 J. Fluid Mech. 70, 437-461. Nonlinear wave interactions in shear flows. Part 2. Third-order theory.

15. Kelbanoff, P.S., Tidstrom, K.D. & Sargent, L.M.: 1962 J. Fluid Mech. 12, 1-34. The three-dimensional nature of boundary-layer instability.

16. Orszag, S.A. & Kells, L.C.: 1980 J. Fluid Mech. 96, 159-205. Transition to turbulence in plane Poiseuille and plane Couette flow.

17. Orszag, S.A. & Patera, A.T.: 1983 J. Fluid Mech. 128, 347-385. Secondary instability of wall-bounded shear flows.

18. Miksad, R.W.: 1972 J. Fluid Mech. 56, 695-719. Experiments on the nonlinear stages of free-shear-layer transition.

19. Miksad, R.W.: 1973 J. Fluid Mech. 59, 1-21. Experiments on nonlinear interactions in the transition of a free-shear-layer.

20. Kachanov, Yu.S., Kozlov, V.V. & Levchenko, V.Ya.: 1977 Fluid Dyn. 3, 383-390. [Mekh. Zhid. i Gaza 3, 49-53.] Nonlinear development of a wave in a boundary layer (Engl. trs.)

21. Kachanov, Yu.S. & Levchenko, V.Ya.: 1983 J. Fluid Mech. 138, 209-247. The resonant interaction of disturbances at laminar-turbulent transition in a boundary layer.

22. Saric, W.S. & Thomas, A.S.W.: 1984 Proc. IUTAM Symp. 'Turbulence and Chaotic Phenomena in Fluids', Kyoto 1983. Experiments on the subharmonic route to turbulence in boundary layers.

23. Zakharov, V.E.: 1976 Dokl. Akad. Nauk. S.S.S.R. 228, 1314-1316 [Soviet Phys. Dokl. 21, 322-323]. Exact solutions to the problem of the parametric interaction of three-dimensional wave packets.

24. Kaup, D.J., Reiman, A. & Bers, A.: 1979 Rev. Mod. Phys. 51, 275-310. [Errata, in Rev. Mod. Phys. 51, 915, are corrected in reprints.] Space-time evolution of nonlinear three-wave interactions. I Interactions in a homogeneous medium.

25. Kaup, D.J.: 1981 J. Math. Phys. 22, 1176-1181. The lump solutions and the Backlund transformation for the three-dimensional three-wave resonant interaction.

26. Craik, A.D.D.: 1978 Proc. Roy. Soc. Lond. A363, 257-269. Evolution in space and time of resonant wave triads. II A class of exact solutions.

27. Craik, A.D.D. & Adam, J.A.: 1979 J. Fluid Mech. 92, 15-33. 'Explosive' resonant wave interactions in a three-layer fluid flow.

28. Craik, A.D.D.: 1983 In Nonlinear Waves (ed. L. Debnath) pp 116-132. Two- and three-wave resonance.

29. Wersinger, J-M., Finn, J.M. & Ott, E.: 1980 Phys. Fluids 23, 1142-1154. Bifurcation and 'strange' behaviour in instability saturation by nonlinear three-wave mode coupling.

30. Volodin, A.G. & Zel'man, M.B.: 1979 Fluid Dyn. (Trs. of Russian Mekh. Zhid. i Gaza 5, 81.) Three-wave resonant interaction of disturbances in a boundary layer.

31. Nayfeh, A.H. & Bozatli, A.N.: 1979a Phys. Fluids 22, 805-813. Secondary instability in boundary-layer flows.

32. Nayfeh, A.H. & Bozatli, A.N.: 1980 Phys. Fluids 23, 448-458. Nonlinear interaction of two waves in boundary-layer flows.

33. Nayfeh, A.H. & Bozatli, A.N.: 1979b Proc. AIAA 12th Fluid & Plasma Dyn. Conf., Williamsburg, Va.: AIAA Paper 79-1456. Nonlinear wave interactions in boundary layers.

34. Herbert, T.: 1983a Phys. Fluids 26, 871-874. Secondary instability of plane channel flow to subharmonic three-dimensional disturbances.

35. Herbert, T.: 1983b AIAA 16th Fluid and Plasma Dyn. Conf., Danvers, Mass. AIAA-83-1759. Subharmonic three-dimensional disturbances in unstable plane shear flows.

36. Herbert, T.: 1984 Proc. IUTAM Symp. 'Turbulence and Chaotic Phenomena in Fluids', Kyoto, 1983. Modes of secondary instability in plane Poiseuille flow.

37. Gustavsson, L.H. & Hultgren, L.S. 1980 J. Fluid Mech 98, 149-159. A resonance mechanism in plane Couette flow.

Secondary Instability of Plane Shear Flows - Theory and Application

Th. HERBERT

Department of Engineering Science and Mechanics
Virginia Polytechnic Institute and State University
Blacksburg, Virginia 24061, U.S.A.

Summary

Secondary instability in plane shear flows originates from parametric instability of the (almost) periodic flow in presence of finite-amplitude traveling waves. The instability is governed by a Floquet system of linear disturbance equations that allows for various classes of normal modes associated with different types of resonance. The analysis of the secondary instability appears as a new level consistently grafted on the classical stability analysis. Results for subharmonic and fundamental (peak-valley splitting) modes in plane Poiseuille flow and in the Blasius boundary layer are consistent with experimental data and prove the theory's scope for understanding, classification, and prediction of the observed secondary phenomena.

Introduction

Considerable research efforts have been devoted to those phenomena that occur subsequent to the onset of primary instability with respect to two-dimensional waves in shear flows. Prototypes of these phenomena are the vortex pairing in mixing layers [1], and the various types of three-dimensional disturbances observed in boundary layers [2]-[4] and in plane Poiseuille flow [5],[6]. The lasting imprint of these phenomena on the transition process and on the structure of the turbulent flow challenges explanation of their origin and analysis of their characteristics. This analysis represents a major step toward more rational schemes for predicting transition.

Earlier work based on weakly nonlinear models remained largely inconclusive. These theories rest on evaluating the lowest coefficients in perturbation expansions in the time-dependent amplitudes of a set of primary disturbance modes that are intuitively chosen to describe the essential features of an observed phenomenon. It is usually left open, whether the perturbation series are used within their radii of convergence and whether the low order of approximation provides reliable results. In principle, these questions can be answered by using high-order approximations [7] combined with techniques for the analysis and improvement of series [8]. However, the relevance of the chosen modes and the completeness of a given model cannot be secured in a rigorous way.

A new and more successful approach has been developed from a pragmatic review of experimental and computational results. In this approach, it is recognized that at the onset of secondary instability the original basic flow is modulated by a wave of small but finite amplitude. Disregarding for the time being the weak variation of the amplitude, the flow can be considered as streamwise periodic in a frame moving with the phase velocity. It was suggested [9] that the periodic modulation of the flow may cause parametric excitation of secondary disturbances. This linear secondary instability is governed by a Floquet system of disturbance equations with periodic coefficients. The main problems then are to identify classes and forms of solutions from known properties of such systems and to solve numerically the resulting high-order systems of ordinary differential equations.

Related Work

Floquet theory has been used primarily for time-periodic flows. For wave-type flows, Kelly [10] found the vortex pairing in an inviscid mixing layer "due to a secondary instability associated with the nearly periodic flow" but solved the equations in a weakly nonlinear framework. The two- and three-dimensional instability of an array of vortices in a mixing layer was studied by Pierrehumbert & Widnall [11]. Nagata & Busse [12] analyzed the secondary and tertiary instability in an inclined fluid layer heated from above. In these studies, however, the concept of the stability analysis is difficult to discern from the semi-analytical method of solution. Maseev [13] investigated the occurrence of peak-valley splitting in the boundary layer, but the potential of his approach was not clearly revealed.

A rudimentary theory of secondary instability has been simultaneously developed by Orszag & Patera [14],[15] and Herbert [16],[17] for nonlinear equilibrium states in plane Poiseuille flow. Based primarily on computer simulations, Orszag & Patera [18] emphasized the universality of the secondary instability mechanism in shear flows. Light was shed on the vortical nature of the mechanism and on the catalytic role of the primary wave in the parametric excitation of three-dimensional effects. Whereas this work concerned only the fundamental (peak-valley splitting) mode of instability, a second class of subharmonic modes was found by Herbert [19]. A summary of theory and results for plane Poiseuille flow is given in [20].

Guided by the vortical nature and the onset of secondary instability at low amplitudes, a simpler approach was suggested [21] that can be considered a consistent higher level of the classical stability theory. Analysis of the subharmonic instability in the Blasius boundary layer [22] has proved the potential of this approach. In the following, an outline of the theory is given together with some typical results for the Blasius flow.

Analysis of Secondary Instability

We consider a two-dimensional basic flow $\underset{\sim}{v}_2$ subject to three-dimensional disturbances $\underset{\sim}{v}_3 = (u_3,v_3,w_3)$ such that

$$\underset{\sim}{v} = \underset{\sim}{v}_2(x',y,t) + \varepsilon\underset{\sim}{v}_3(x',y,z,t), \quad \underset{\sim}{v}_2 = \underset{\sim}{v}_0(y) + A\underset{\sim}{v}_1(x',y,t), \tag{1}$$

where ε is sufficiently small for linearization. The steady component $\underset{\sim}{v}_0 = (u_0(y),0,0)$ is a parallel flow in the x' direction. The wave-like component $\underset{\sim}{v}_1$ is considered periodic in x' with wavelength $\lambda_x = 2\pi/\alpha$ and traveling with the phase velocity c. In a wave-fixed frame, $\underset{\sim}{v}_1$ satisfies

$$\underset{\sim}{v}_1(x',y,t) = \underset{\sim}{v}_1(x,y) = \underset{\sim}{v}_1(x+\lambda_x,y), \quad x = x' - ct. \tag{2}$$

The normalization of $\underset{\sim}{v}_1$ is such that the amplitude A measures directly the maximum streamwise r.m.s. fluctuation of the basic flow. It is convenient to express $\underset{\sim}{v}_1 = (\partial\psi_1/\partial y, -\partial\psi_1/\partial x, 0)$ in terms of a streamfunction ψ_1. We then follow the usual procedure for deriving a first set of disturbance equations: substitute (1) into the Navier-Stokes equations, collect all terms of order $O(\varepsilon)$, eliminate the pressure by taking the curl, introduce vorticity components $\nabla \times \underset{\sim}{v} = (\xi, \eta, \zeta)$, and finally eliminate w_3 by exploiting continuity. Since $\partial\eta_3/\partial z$ and $\partial\xi_3/\partial z$ can be expressed in terms of u_3, v_3, one obtains two coupled differential equations for u_3 and v_3,

$$\left[\frac{1}{R}\nabla^2 - (u_0 - c)\frac{\partial}{\partial x} - \frac{\partial}{\partial t}\right]\frac{\partial\eta_3}{\partial z} + \zeta_0\frac{\partial^2 v_3}{\partial z^2} + A\left\{\left(-\frac{\partial\psi_1}{\partial y}\frac{\partial}{\partial x} + \frac{\partial\psi_1}{\partial x}\frac{\partial}{\partial y} - \frac{\partial^2\psi_1}{\partial x\partial y}\right)\frac{\partial\eta_3}{\partial z}\right.$$
$$\left. + \frac{\partial^2\psi_1}{\partial x^2}\left(\frac{\partial^2 u_3}{\partial x\partial y} + \frac{\partial^2 v_3}{\partial y^2}\right) - \frac{\partial^2\psi_1}{\partial y^2}\frac{\partial^2 v_3}{\partial z^2}\right\} = 0, \tag{3}$$

$$\left[\frac{1}{R}\nabla^2 - (u_0 - c)\frac{\partial}{\partial x} - \frac{\partial}{\partial t}\right]\nabla^2 v_3 - \frac{\partial\zeta_0}{\partial y}\frac{\partial v_3}{\partial x} + A\left\{\left(-\frac{\partial\psi_1}{\partial y}\frac{\partial}{\partial x} + \frac{\partial\psi_1}{\partial x}\frac{\partial}{\partial y}\right)\nabla^2 v_3\right.$$
$$+ \frac{\partial^2\psi_1}{\partial x^2}\left(\frac{\partial\zeta_3}{\partial y} + \frac{\partial\eta_3}{\partial z}\right) - \frac{\partial^2\psi_1}{\partial x\partial y}\left(\frac{\partial\zeta_3}{\partial x} + \frac{\partial\xi_3}{\partial z}\right) - \frac{\partial\zeta_1}{\partial x}\left(2\frac{\partial u_3}{\partial x} + \frac{\partial v_3}{\partial y}\right) - \frac{\partial\zeta_1}{\partial y}\frac{\partial v_3}{\partial x}$$
$$\left. - \left(u_3\frac{\partial}{\partial x} + v_3\frac{\partial}{\partial y}\right)\frac{\partial\zeta_1}{\partial x}\right\} = 0, \tag{4}$$

subject to homogeneous boundary conditions on u_3, v_3 and $\partial v_3/\partial y$. R is the Reynolds number. For A=0, normal modes in x, z and t can be introduced. In this case, eq. (3) reduces to Squire's equation for the η-component of vorticity whereas eq. (4) provides the Orr-Sommerfeld equation for oblique waves in the basic flow $\underset{\sim}{v}_0$.

For A≠0, x-periodic coefficients appear in eqs. (3),(4). Insight into classes of solutions and their streamwise structure can be obtained from the Floquet theory of ordinary differential equations [20]. Here, we consider only two classes that evolve from the temporal growth concept in the form

$$\underset{\sim}{v}_f = e^{\sigma t+i\beta z}\,[\underset{\sim}{v}_0^+(y) + 2\sum_{n=1}^{\infty}\{\underset{\sim}{v}_{2n}^+(y)\cos n\alpha x - \underset{\sim}{v}_{2n}^-(y)\sin n\alpha x\}] \quad (5)$$

$$\underset{\sim}{v}_s = 2e^{\sigma t+i\beta z}\sum_{n=1}^{\infty}\{\underset{\sim}{v}_{2n-1}^+(y)\cos\frac{2n-1}{2}\alpha x - \underset{\sim}{v}_{2n-1}^-(y)\sin\frac{2n-1}{2}\alpha x\} \quad (6)$$

where $\beta=2\pi/\lambda_z$ is the spanwise wavenumber and $\sigma = \sigma_r + i\sigma_i$ is in general complex. The *fundamental* modes $\underset{\sim}{v}_f$ in (5) have streamwise wavelength λ_x and originate from primary resonance. The *subharmonic* modes $\underset{\sim}{v}_s$ in (6) have wavelength $2\lambda_x$ and arise from principal parametric resonance in the periodic basic flow $\underset{\sim}{v}_2$. The formal characteristics of these modes are consistent with observations [20]. Fundamental modes are associated with peak-valley splitting. The aperiodic terms represent a mean-flow distortion (u_0^+) as well as a longitudinal vortex system (v_0^+, w_0^+). No such terms appear in the subharmonic modes. Subharmonic modes are invariant under the translation $(x,z) \rightarrow (x+\lambda_x, z+\lambda_z/2)$ as the staggered pattern of Λ-vortices [4] and contribute peaks in frequency spectra at f/2, 3f/2, ... but not at the fundamental frequency f=αc. The two classes are also consistent with the modes found in the mixing layer [11]. Two-dimensional vortex pairing is contained as special case β=0 of the subharmonic modes (6).

Use of the Shape Assumption

Substituting (5) or (6) and a Fourier representation of ψ_1 into eqs. (3), (4) provides coupled systems of ordinary differential equations for the coefficient functions. Since the basic flow is real, these systems can be written in real form. Consequently, the systems support either real eigenvalues σ or complex conjugate pairs. The exponential time-dependence in (5), (6) is a direct consequence of assuming A to be constant. This assumption is strictly satisfied for nonlinear equilibrium states as they exist in plane Poiseuille flow for points on the neutral surface. Analysis of the secondary instability of these motions [20] has shown that the results are rather insensitive to the levels of Fourier truncation for the basic flow and the disturbances. While nonlinear effects are important for establishing the amplitude of the equilibrium state, they are negligible for the disturbances. At the low amplitudes of concern, the streamwise periodic vorticity field that is key to the secondary instability can be well represented by using the s h a p e assumption, i.e. by taking ψ_1 in the form,

$$\psi_1 = 2\{\phi^+(y) \cos \alpha x - \phi^-(y) \sin \alpha x\}, \tag{7}$$

where $\phi(y) = \phi^+ + i\phi^-$ is the eigenfunction associated with the principal mode $c_0 = c_r + ic_i$ of the Orr-Sommerfeld equation for given R,α and $\beta=0$. The phase velocity c in (2) is replaced by c_r. In order to neglect the weak variation of A (on a viscous time scale), we introduce the q u a s i - s t e a d y assumption which requires $|\alpha c_i| << |c_r|$. These approximations blur some detail at the onset of secondary instability but are well justified for situations of practical interest where strong growth of three-dimensional modes (on a convective time scale occurs. With the shape assumption, A and σ appear linear in the disturbance equations. This permits consideration of either $\sigma = \sigma(R,\alpha,\beta,A)$ or $A = A(R,\alpha,\beta,\sigma)$ as the eigenvalue of the problem. The latter case is of particular interest if A and σ are real.

Numerical Solution

Retaining in (5) only the modes up to n=1 provides a system of six coupled equations of the Squire type or Orr-Sommerfeld type for the fundamental modes. This number reduces to four for subharmonic modes. Many methods suitable for the Orr-Sommerfeld problem can be adapted for solving the secondary eigenvalue problems. However, in absence of any guidance, we prefer applying a boundary-value method. With a spectral collocation method based on Chebyshev polynomials (and an algebraic mapping to a finite domain for boundary layers), the system of equations is converted into an algebraic system. The access to the complete eigenvalue spectrum is very helpful for identifying the principal mode and obtaining the initial guess for a local procedure. The spectrum also helps in tracing the intricate analytical connections between modes of secondary instability and modes of primary instability as $A \to 0$.

Results for the Blasius Boundary Layer

We use $\delta_r = (U_\infty/\nu x')^{1/2}$ as a reference length such that $R = R_{x'}^{1/2}$. Moreover, we use the frequency parameter $F = 10^6 \cdot \alpha c_r/R$, and $b = 10^3 \cdot \beta/R$ in order to specify three-dimensional disturbances of a fixed dimensional wavelength in z as they travel downstream. The key experiments provide details on different aspects at different frequencies:

[2] : F=59, hot-wire data on peak-valley splitting,

[3] : F=124, hot-wire data on subharmonic disturbances,

[4] : F=83, flow visualization of various modes.

The results shown here are selected in favor of a comparison with hot-wire data, waiving the chance for a more complete and comparative picture at fixed frequency.

First, we consider s u b h a r m o n i c modes at F=124 and branch II of the neutral curve, where A reaches a maximum. With R_{II}=606 (α=0.2033), A=0.01 and b=0.184, the eigenvalue spectrum shows three unstable modes: $\sigma_0 = 0.007458$, $\sigma_{1,2} = 0.000895 \pm i0.011715$. This constellation is characteristic for the range of larger amplitudes and also has been found for fundamental modes in the boundary layer and in Poiseuille flow. The principal (most amplified) mode is associated with a real eigenvalue σ_0, i.e., this mode is synchronous with the basic flow. This result is consistent with the experiments [3, Fig. 21]. In the parameter range of interest, the growth rate σ_0 is always larger than σ_{1r}. Therefore, we consider in the following only the principal mode.

The variation of σ_0 with A and b is shown in Fig. 1. At very small amplitudes, subharmonic instability occurs in a small band of wavenumbers centered at the wavenumber b*=0.184 for resonance of Craik's wave triad [23]. A narrowing band of instability persists as A→0 since the subharmonic wave at b* is an unstable primary mode. These features are held responsible for the selection of disturbances with b≈b* (and large spanwise wavelengths

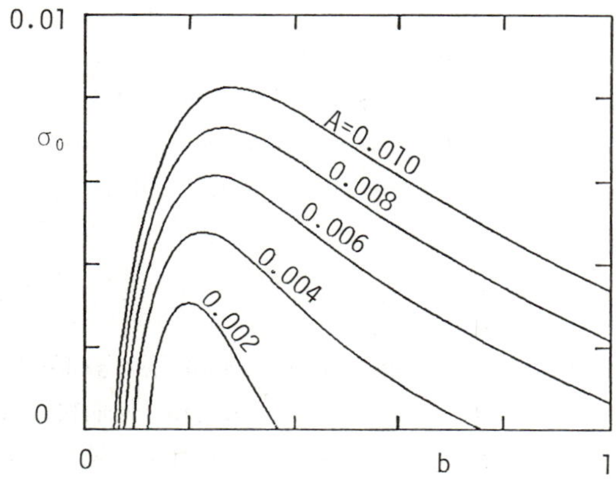

Fig. 1. Growth rate of the principal subharmonic mode as a function of the wavenumber b for various amplitudes A at F=124, R_{II} = 606.

$\lambda_z > \lambda_x$) at very low amplitude levels [4]. As A increases, the band of dangerous wavenumbers extends to larger values of b. The sharp cutoff at small b indicates the suppression of the two-dimensional pairing (β=0) in the neighborhood of a wall. The broad-band nature of the subharmonic instability at larger amplitudes originates from a near-resonant triad between the basic flow and Squire modes ("normal vorticity modes"). This mechanism is the sole cause for the principal mode of subharmonic instability in Poiseuille flow [19]. Here it appears in a complicated interplay with Craik's mechanism [21].

At A=0.01, the wavenumber for maximum growth is shifted from b* to b≈0.28. The experimental value b=0.33 [3] is well within the region of strong instability. However, the local observation can only be understood in context with the initial (background) amplitudes and the growth history for primary and secondary disturbances as they travel downstream. Given the initial amplitude A_0 at F=124 and branch I, R_I=381, the amplitude A(R) can be obtained by integrating the primary growth rate. The associated growth rate $\sigma_0(R)$ of the subharmonic mode then contains the effects of the downstream increase in R as well as the variation in A. Fig. 2 shows $\sigma_0(R)$ for different initial amplitudes and wavenumbers. In all cases, σ_0 increases strongly due to the simultaneous increase in A and R. Maximum growth occurs at $R > R_{II}$. As R further increases, σ decreases due to the decay of the two-dimensional wave. At the lower amplitude levels, A=0.002, the instability for b=0.19 is stronger and extends over a larger range of Reynolds numbers than for b=0.33. At the larger initial amplitude A_0=0.004, Fig. 2 shows no clear preference for either one of the two wavenumbers.

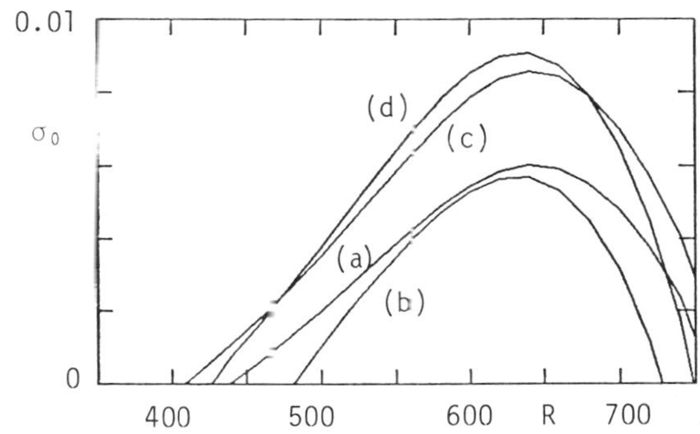

Fig. 2. Growth rate σ_0 as a function of the Reynolds number R for (a) A_0=0.002, b=0.19, (b) A_0=0.002, b=0.33, (c) A_0=0.004, b=0.19, (d) A_0=0.004, b=0.33.

For an evaluation of the total growth of the three-dimensional disturbances between onset of secondary instability at R_0 and some downstream position $R > R_0$, the temporal growth rates σ_0 can be converted into spatial growth rates γ_0 by using the crude transformation $\gamma_0=-\sigma_0/c_r$. The amplitude growth curve $\varepsilon(R)/\varepsilon_0$ for the subharmonic disturbance is then given by

$$\ell n \frac{\varepsilon(R)}{\varepsilon_0} = -2 \int_{R_0}^{R} \gamma_0 dR \tag{8}$$

where ε_0 is the initial amplitude at R_0. At small amplitudes of the two-dimensional wave, $A_0=0.002$, the wavenumber b=0.19 is clearly preferred. However, at $A_0=0.004$, the growth curves for the two wavenumbers b=0.19 and b=0.33 are very similar and reach maximum amplification factors $\varepsilon/\varepsilon_0$ of the order of 10^4. With the background disturbance levels in experiments, it is likely that nonlinear processes take over before $\varepsilon/\varepsilon_0$ reaches this maximum. The occurrence of b=0.33 in the experiments [3] can only be attributed to a larger initial amplitude ε_0 of this component in a nonuniform disturbance background.

The theoretical and experimental growth curves for the two-dimensional wave and the subharmonic disturbance with b=0.33 are shown in Fig. 3. The vibrating ribbon in the experiments was located at $R\approx400$. Since some distance from the ribbon is necessary for the subharmonic disturbance to fully develop, the initial amplitude ε_0 (at $R_0=414$) cannot be obtained from the experiment. Therefore, we have chosen ε_0 such that theoretical and experimental values of $\varepsilon(R)$ coincide at $R=R_{II}$. The subharmonic growth curves in Fig. 3 verify the theory's capability of predicting the growth rates in essential agreement with the observation.

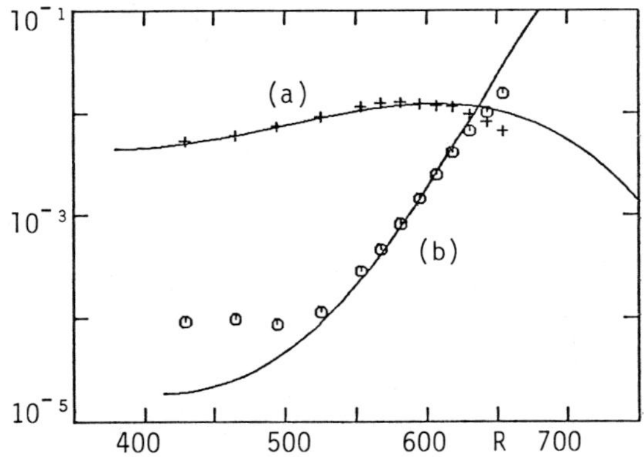

Fig. 3. Amplitude variation with R of (a) the two-dimensional wave with $A_0=0.004$, F=124, and (b) the subharmonic disturbance with $\varepsilon_0 = 1.85 \cdot 10^{-5}$, b=0.33. Comparison between theory (——) and experiment (+,o), Ref. [3], Fig. 23.

Beyond the eigenvalues σ, the analysis provides the associated eigensolution, and hence the disturbance velocity field (6) with n=1. Measurements have only been made of the streamwise r.m.s. velocity u' and phase at F=124, R=608, and b=0.33. For subharmonic disturbances, we obtain

$$u_s' = \varepsilon \; |\cos\beta z| \; \{8[(u_1^+)^2 + (u_1^-)^2]\}^{1/2}.$$

At fixed y, the spanwise variation is $u_3' \sim |\cos\beta z|$ with phase jumps by π at the positions where $\cos\beta z=0$. This variation is consistent with the observation [3, Fig. 20]. The y-distributions of the normalized u_s' velocity and phase are given in Fig. 4. together with the experimental data. Minor derivations, especially for the phase, may be due to the difficulty of measurements at the very low subharmonic amplitudes of less than 0.002. The u_s'- distribution with a sharp peak just above the critical layer at y_c is typical for a wide range of parameters. The maximum closely coincides with the location of the extrema of vorticity in the periodic basic flow.

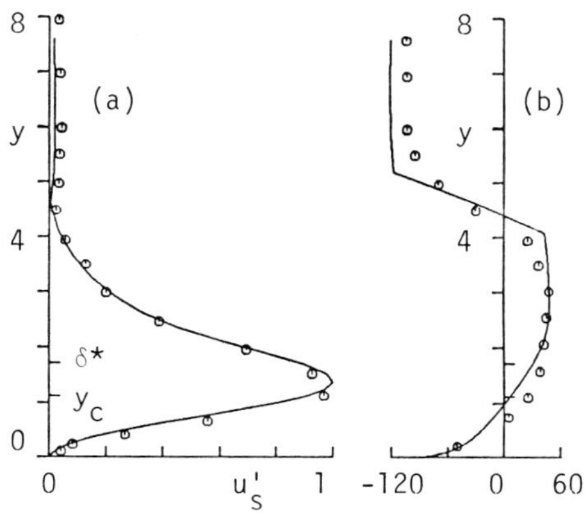

Fig. 4. Distribution of (a) the streamwise r.m.s. fluctuation u_s' and (b) the phase of the subharmonic disturbance at R=608, b=0.33. Comparison between theory (——) and experiment (o), Ref. [3], Fig. 22. The marks δ* and y_c indicate the displacement thickness and position of the critical layer, respectively.

The analysis of f u n d a m e n t a l (peak-valley splitting) modes in the Blasius boundary layer is considered preliminary since the nonparallelism of the flow is neglected. Nonparallelism may affect the aperiodic components $\underline{v}_0^+$ in (5) as pointed out in [9]. Detailed results have been obtained for F=59 as in the experiments of Klebanoff et al. [2] with a 145 c/s wave. We have not yet completed the true comparison with the experiments due to the necessary change, e.g. from maximum u' values to those at fixed dimensional distance from the flat plate. Nevertheless, the agreement is rather encouraging.

At amplitudes of order A=0.01, the principal mode of fundamental instability is associated with a real eigenvalue σ_0, indicating perfect synchronization between disturbances and basic flow as an optimum condition for energy transfer. The fundamental instability occurs in a broad band of spanwise wavenumbers. The growth rates are of the same order as those for subharmonic disturbances. In the absence of any resonance mechanism at A=0, however, the fundamental instability occurs only at sufficiently large amplitude of the two-dimensional wave, and therefore, peak-valley splitting is considered a threshold phenomenon as in Poiseuille flow. No relation has yet been found between the principal mode of fundamental instability and the weakly nonlinear Benney-Lin model [24]. The intricate interplay of fundamental and subharmonic disturbance growth will be discussed elsewhere.

The eigensolution for the principal mode shows the characteristic features of the observations regarding the u'-distributions at peak and valley as well as the associated longitudinal vortex system. Since no separate measurements of the three-dimensional component u_f' have been made, we have linearly superposed two- and three-dimensional disturbances with prescribed amplitudes A and ε, where ε measures the maximum of u_f'. Fig. 5a shows the u'-distributions at peak and valley for parameters that model the conditions near station B of the experiments [2]. With $\delta \simeq 5\delta_r$, our results resemble those shown in [2], Fig. 5b. The distribution of the spanwise component $\overline{w}$ of the mean velocity is given in Fig. 5b for comparison with the experimental data [2], Fig. 12. Some deviations may be due to the difficulty of measuring $\overline{w}$ at these low amplitudes. In fact, the shape of $\overline{w}$ matches closely the measurements at larger amplitudes (station C).

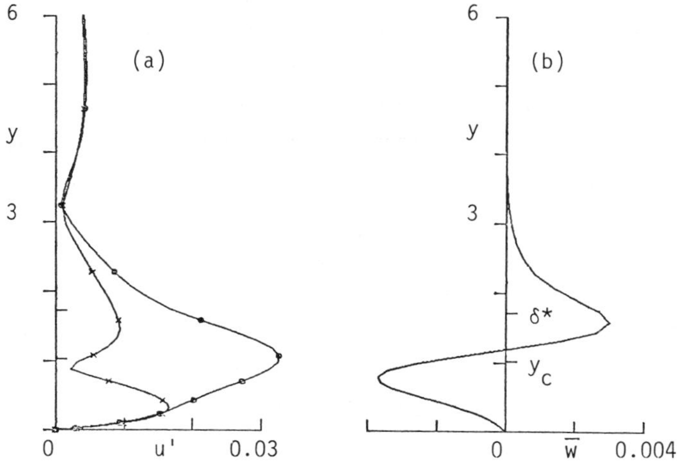

Fig. 5. Distribution of (a) the streamwise r.m.s. fluctuation u' at peak (o) and valley (+) and (b) the spanwise component $\overline{w}$ of the mean velocity between peak and valley in the presence of a fundamental disturbance at F=59, R_{II}=1065, b=0.24, A=0.018, ε=0.018. The parameters resemble the conditions for the 145 c/s wave at station B of the experiments [2].

The growth rate for the parameters of Fig. 5 is $\sigma_0=0.01$, i.e., the three-dimensional component would increase by a factor of 3.1 to a value of $\varepsilon=0.056$ between stations B and C. At this amplitude level, nonlinear effects are likely to occur and seem to enhance the growth of the two-dimensional component. However, the y-distributions [2, Figs. 5c, 12] and spanwise z-distributions [2, Fig. 11] of the velocity components at station C are very similar to the results of our linear theory, in agreement with the observation "that the wave motion still retains some semblance of linear behavior" [2, p.9].

Conclusions

The present theory explains the secondary instability with respect to three-dimensional disturbances as originating from parametrical instability of a basic flow that is periodic in space and time. The major shortcomings of weakly nonlinear methods are circumvented, and moreover, means are provided for identifying the relevant modes for constructing weakly nonlinear models. Results for subharmonic and peak-valley splitting modes of instability in plane Poiseuille flow and Blasius boundary layer indicate the theory's capability of quantitatively predicting growth rates and spatial disturbance field up to the stages immediately preceding breakdown and transition. The theory can be applied to a variety of other shear flows and is open to implementing the spatial growth concept, to modifications for weakly nonparallel flows and to accounting for nonlinear interactions of the three-dimensional disturbances with the periodic basic flow.

Acknowledgment

The various stages of this work was supported by the National Science Foundation under Grant MEA 81-20935, by the Office of Naval Research under Contracts N00014-83-K-0089 and N00014-84-K-0093, and by the Air Force Office of Scientific Research under Contract F49620-84-K-0002.

References

1. Ho, C. M. & Huerre, P.: Perturbed free shear layers. Ann. Rev. Fluid Mech. 16 (1984) 365-424.
2. Klebanoff, P. S., Tidstrom, K. D. & Sargent, L. M.: The three-dimensional nature of boundary-layer instability. J. Fluid Mech. 12 (1962) 1-34.
3. Kachanov, Yu. S. & Levchenko, V. Ya.: The resonant interaction of disturbances at laminar-turbulent transition in a boundary layer. J. Fluid Mech. 138 (1984) 209-247.

4. Saric, W. S. & Thomas, A. S. W.: Experiments on the subharmonic route to transition. Proc. IUTAM Symposium on Turbulence and Chaotic Phenomena in Fluids, Kyoto, Japan (1983).
5. Nishioka, M., Asai, M. & Iida, S.: An experimental investigation of secondary instability. In: Eppler, R. & Fasel, H. (eds.) Laminar-Turbulent Transition. Berlin: Springer-Verlag (1980).
6. Kozlov, V. V. & Ramazanov, M. P.: Formation of three-dimensional structures in Poiseuille flow by resonant interaction. Preprint No. 11-83, I.T.A.M., USSR Acad. Sci., Novosibirsk (1983).
7. Herbert, Th.: On perturbation methods in nonlinear stability theory. J. Fluid Mech. 126 (1983) 167-186.
8. Van Dyke, M.: Computer-extended series. Ann. Rev. Fluid Mech. 16 (1984) 287-309.
9. Herbert, Th. & Morkovin, M. V.: Dialogue on bridging some gaps in stability and transition research. In: Eppler, R. & Fasel, H. (eds.) Laminar-Turbulent Transition. Berlin: Springer-Verlag (1980).
10. Kelly, R. E.: On the stability of an inviscid shear layer which is periodic in space and time. J. Fluid Mech. 27 (1967) 657-689.
11. Pierrehumbert, R. T. & Widnall, S. E.: The two- and three-dimensional instabilities of a spatially periodic shear layer. J. Fluid Mech. 114 (1982) 59-82.
12. Nagata, M. & Busse, F. H.: Three-dimensional tertiary motions in a plane shear layer. J. Fluid Mech. 135 (1983) 1-28.
13. Maseev, L. M.: Occurrence of three-dimensional disturbances in boundary layers. Fluid Dyn. 3 (1968) 23-24.
14. Orszag, S. A. & Patera, A. T.: Subcritical transition to turbulence in plane channel flow. Phys. Rev. Lett. 45 (1980) 989-993.
15. Orszag, S. A. & Patera, A. T.: Subcritical transition to turbulence in plane shear flows. In: Meyer, R. E. (ed.) Transition and Turbulence. New York: Academic Press 1981.
16. Herbert, Th.: Stability of plane Poiseuille flow - theory and experiment. VPI Report VPI-E-81-35 (1981). See also: Fluid Dyn. Trans. 11 (1983) 77-126.
17. Herbert, Th.: A secondary instability mechanism in plane Poiseuille flow. Bull. Amer. Phys. Soc. 26 (1981) 1257.
18. Orszag, S. A. & Patera, A. T.: Secondary instability of wall-bounded shear flows. J. Fluid Mech. 128 (1983) 347-385.
19. Herbert, Th.: Secondary instability of plane channel flow to subharmonic three-dimensional disturbances. Phys. Fluids 26 (1983) 871-874.
20. Herbert, Th.: Modes of secondary instability in plane Poiseuille flow. Proc. IUTAM Symposium on Turbulence and Chaotic Phenomena in Fluids, Kyoto, Japan (1983).
21. Herbert, Th.: Subharmonic three-dimensional disturbances in unstable shear flows. AIAA Paper No. 83-1759 (1983).
22. Herbert, Th.: Analysis of the subharmonic route to transition in boundary layers. AIAA Paper No. 84-0009 (1984).
23. Craik, A. D. D.: Nonlinear resonant instability in boundary layers. J. Fluid Mech. 50 (1971) 393-413.
24. Benney, D. J. & Lin, C. C.: On the secondary motion induced by oscillations in a shear flow. Phys. Fluids 3 (1960) 656-657.

On Subharmonic-Type Laminar-Turbulent Transition in Boundary Layer

I.I. MASLENNIKOVA, M.B. ZELMAN

Institute of Theoretical and Applied Mechanics
Siberian Branch of the USSR Academy of Sciences
630090 Novosibirsk, USSR

Summary

Resonant Tollmien-Schlichting waves interactions are analyzed with purpose of modeling subharmonic-type transition in Blasius boundary layer.

The recent experiment [1] in boundary layer has shown the existence of subharmonic-type (C-type) of laminar-turbulent transition. Its essential features are the selection of the frequency half as much as fundamental, gradual filling of the low frequency spectrum, the formation of three-dimensional structures at low intensities. In [2] the selection of subharmonics was attributed to wave resonance in Craik's-type triads [3]. Models of three-dimensional resonant interaction in boundary layer were also investigated in [4-7]. The resonant features of the disturbances evolution were observed in experiments [8, 9]. The present investigation is concerned with further analysis of nonlinear interaction mechanism of Tollmien-Schlichting waves in Blasius boundary layer. These results are used for explanation of the initial stage of subharmonic route to transition. Connection with Klebanoff's type (K-type) of transition is discussed.

Here we consider the velocity field of flow in the form $u = [(U+\varepsilon u, \varepsilon v, \varepsilon w) + O(1/Re)]$ where U is Blasius profile, $\vec{V} = (u,v,w)$ is the disturbance velocity, $\varepsilon \ll 1$, Re is Reynolds number based on displacement thickness δ^*. Using averaging method (see [2]), in the range of weakly nonlinear theory validity the solution takes the form:

$$\bar{V} = \sum_j \hat{A}_j e^{\gamma_j t + i\theta_j} \bar{\varphi}_j^{(0)}(y, Re) + \sum_{n=1} \varepsilon^n \sum_r \bar{\varphi}_r^{(n)}(y, Re, \varepsilon x, \varepsilon z, \varepsilon t) e_r^{i\theta_r} \quad (1)$$

$$\left(\frac{\partial}{\partial t} + V_j \frac{\partial}{\partial x} + W_j \frac{\partial}{\partial z} - \gamma_j + \sigma_{1j} \frac{\partial^2}{\partial x^2} + \sigma_{2j} \frac{\partial^2}{\partial z^2} + \sigma_{3j} \frac{\partial^2}{\partial x \partial z}\right) A_j = \quad (2)$$

$$= \varepsilon \sum_{k\ell} \left\{ \left(S_{jk\ell} A_k + Q^{(1)}_{jk\ell} A_k \frac{\partial}{\partial x} + Q^{(2)}_{jk\ell} A_k \frac{\partial}{\partial z} \right) A_\ell e^{i\theta_{jk\ell}} + \sum_m \varepsilon D_{jk\ell m} A_\ell A_k A_m e^{i\theta_{jk\ell m}} \right\},$$

$$A_{j0} = A_j(t=0). \quad \theta_j = \int \alpha_j dx + \beta_j z - \omega_j t, \quad A_j(\varepsilon x, \varepsilon z, \varepsilon t) = \hat{A}_j e^{\gamma_j t}$$

are the phase and complex amplitude of wave with wave vector (α_j, β_j); $\theta_{jk\ell\ldots} = \theta_j - (\theta_k + \theta_\ell + \cdots)$; $\bar{\varphi}_j^{(0)}$, $\Omega(\alpha_j, \beta_j, Re) = \omega_j + i\gamma_j$ are determined by solving **Orr-Sommerfeld** (O-S) problem; $\alpha_r, \beta_r, \omega_r$ need not be necessary the natural oscillations; $\bar{\varphi}_r^{(n)}$ are derived from recurrence system of linear nonhomogeneous equations; coefficients in (2) are determined by $\bar{\varphi}_r^{(n)}$ and by the solution of adjoint O-S problem [2] . In accordance with vibrating ribbon experiments we shall analyse wave disturbances with $\partial\hat{A}_j/\partial z = \partial\hat{A}_j/\partial t = 0$. Omitting terms $O(\varepsilon^2)$, the three-wave resonance is the only interaction mechanism. The elementary case is realized by triad. For three waves $A_j(\omega_j, \alpha_j, \beta_j)$ $(j=1,2,3)$ (2) reduces to a system of three equations. The interaction in triad depends on wave vector (α_j, β_j) via S_{jke} and θ_{jke}.

We revealed that effective resonant interaction is possible only when $\beta_j \beta_k = -|\beta_j \beta_k|$. The extremely strong interaction exists in triads with $\omega_1/2 = \omega_2 = \omega_3$; $\alpha_1/2 \approx \alpha_2 = \alpha_3$; $\beta_1 = 0$, $\beta_2 = -\beta_3$ (in accordance with Craik's estimates [3]). Such triads behaviour was treated in [6] . Effective resonance takes place at $\max(|\gamma_1 A_1/A_2 A_3 S_{123}|, \gamma_1 A_1/S_{231}|) \leq 1$ and leads to spontaneous spanwise amplitude symmetrization. Figure 1a shows the growth of disturbance amplitudes $|A_j| = a_j$ for the case $a_2(x_0) > a_3(x_0)$. Intensities of oblique waves rapidly become equal and Craik's-type triad do realize. Initial phases variation (Φ_j = arg A_j) slows down the $a_{2,3}$ growth but does not destroy the symmetry (see Fig. 1a). The total feature of Craik's triad is the preferential growth of subharmonics. During parametric interaction

($a_1 > a_{2,3}$) we have $\ln a_{2,3} = \frac{V_1 \varepsilon}{V_2 \gamma_2} \exp \int \gamma_1 \frac{dx}{V_1}$ (see [2]) and when $a_1 \lesssim a_{2,3}$ the nonlinear interaction leads to the phase localization $(\Delta\Phi = [\Phi_1(x) - \Phi_2(x) - \Phi_3(x) + \int(\alpha_1 - 2\alpha_2)dx] \to 0$ and to explosive amplification of wave intensities. A broad band of possible three-dimensional disturbances grows as a result of interaction with ω_1 in the $x = x_0$ environment. Fig. 1b shows the behaviour of a_j with respect to various propagating angles $\xi(x) = \mathrm{arctg}(\beta_2/\alpha_2)$. For different ξ the synchronization condition holds true at different values of Re. It appeared that the case $\theta_{123} = 0$ corresponds to $\xi = 45°$ for various ω_1 , nevertheless the local maximum of growth rates shifts to large values of ξ : $45° \leq \xi \leq 52°$. This effect depends on S_{2-31} variation. X-evolution does not also destroy the strong interaction because of resonance spectral range $\gamma_{eff} \sim |\frac{1}{a_j}\frac{\partial a_j}{\partial x}|$ expansion. As a result the possibility exists for amplification of broad band of wave pairs in the subharmonic frequency environment in boundary layer. The resonance may occur also in asymmetric triads [10] . Figs. 2a,b show some pictures of such a_j evolution. The preferential growth of low frequency waves (a_2, a_3) keeps as before, but their amplification rates proved to be not so great as in Craik's triads. Nonlinear interaction mechanism may transfer energy from high to low frequencies. The triads evolution may qualitatively depend on initial conditions (see Fig. 2b) that is of interest in view of flow randomization. The isolated triad is only the elementary model of process. More realistic model needs to take into account the existence of coupled triads system (wave cluster). Let us consider the resonance between dominant two-dimensional fundamental $A_1(\omega_1, \alpha_1, 0)$ and a number of spanwise symmetric wave pairs $\{A_{2n}(\omega_{2n}, \alpha_{2n}, \beta_{2n}),\ A_{2n+1}(\omega_{2n}, \alpha_{2n}, -\beta_{2n}),\ n = 1, 2, \ldots\}$. Fig. 3 shows the disturbances amplification in 5 waves cluster ($n \leq 2$), when symmetric (A_1, A_2, A_3), (A_1, A_4, A_5) and asymmetric (A_1, A_2, A_5), (A_1, A_3, A_4) triads are in resonance. All wave intensities become equal and grow rapidly. Phases do localize

$$\{\Delta\Phi_{1,2} = [\Phi_1 - \Phi_{2,4} - \Phi_{3,5} + \int(\alpha_1 - 2\alpha_{2,4})dx] \to 0,$$

$$\Delta\Phi_{3,4} = [\Phi_1 - \Phi_{2,4} - \Phi_{3,5} \pm (\beta_4 - \beta_2)z + \int(\alpha_1 - \alpha_3 - \alpha_4)dx] \to 0\}.$$

As a result the system behaves as some isolated symmetric triad, but their ξ - and y-distributions differ from the ones investigated earlier. We suggest that three-dimensional structure observed in [8, 9] is the result of such cluster collectivization. Under experimental conditions when the only subharmonic is forced [9] , Craik's triad is sufficient for the description of the process. Fig. 4a shows the comparison of our calculations with experimental curves [9] .

Cluster collectivization effect is determined by crossinteraction inside asymmetric triads but not by the interaction of symmetric triads via A_1. When symmetric waves are not in resonance, the only crossinteraction forms the field structure. As we believe such conditions were realized in experiment [11] where two frequencies, ω_1 and ω_2 , were introduced in the flow by means of vibrating ribbon. It appeared that oscillations with $\omega_3 = \omega_1 - \omega_2$ and $\omega_4 = 2\omega_1 - \omega_2$ were anomalously increased in contrast with two-waves interaction theory. Taking into account that $\omega_1 = \omega_3 + \omega_4$, we attributed such behaviour to resonance in the system
$\{A_1(\omega_1,\alpha_1,0),\ A_{2,3}(\omega_3,\alpha_3,\pm\beta_3),\ A_{4,5}(\omega_4,\alpha_4,\pm\beta_4),\ \alpha_3\approx\alpha_1-\alpha_4,\ \beta_3\approx\beta_4\}$
(phase measurements have not been conducted). Oscillations with ω_2 are of importance only to initiate harmonic oscillations at $x = x_0$ since they rapidly decay in the investigated Re range. Results of our calculations are in good agreement with experimental data (see Fig. 4b). Resonant mechanism explains all inherent properties of the first stage of C-type transition. Three-dimensional quasisubharmonic oscillations packet is amplified by the dominant wave; cluster is formed; triad-like symmetric structure is formed and energy transfer to the low frequency spectrum range occurs. This energy transfer may lead to limitation of the intensities growth and flow stochasticity.

High-frequency waves generation is the effect of the next order of nonlinear theory. These effects lead to harmonic generation as well as to the mean flow deformation and additional resonant interactions. We calculated (2) for Craik's-type triad taking into account ε^2 -order terms in nearly parallel limit. Fig. 5

shows the y-distribution of $|\varphi_r^{(1)}|, |\frac{\partial}{\partial y}\varphi_r^{(1)}|$ ($\omega_r = 3/2\,\omega_1$, $\alpha_r = \alpha_1 + \alpha_2$, $\beta_r = \beta_2$) . Growth rates at initial stage of amplification and y-distribution of $\varphi_{3/2}^{(1)}$ function occur to be quite close to the ones measured in [8] . But preliminary calculations do not indicate stabilization of amplitudes growth contrary to experiments [8, 9] .

Nonlinear waves may produce new resonant connections arising from interaction between harmonics with background disturbances. In experiments fundamental wave feels self-interaction at initial stage of evolution. As an adequate model we used a system consisting of two-dimensional nonlinear wave $A_1(\omega_1, \alpha_1, 0)$, subharmonics $A_{2,3}(\omega_0 = \frac{\alpha_1}{2}, \alpha \approx \frac{\alpha_1}{2}, \mp\beta_0)$ and $A_{4,5}(\omega_1, \alpha \approx \alpha_1, \pm\beta_1)$ being in resonance with two-dimensional harmonic A_1. Calculations showed that direct interaction ($A_1 \leftrightarrow A_{2,3}$) plays dominant role (see Fig. 6). $A_{4,5}$ amplify only in the region of explosive growth of A_1 . This explosion may be a result of Craik's resonance or self-interaction, when $\varepsilon A_1 \gtrsim 10^{-2}$ and $A_{2,3} \ll A_1$. $A_{4,5}$ excitation is not effective in the range of validity of weakly nonlinear theory. It may be shown [6] that if we omit mean flow deformation and $A_1 \leftrightarrow A_{2,3}$ coupling considered system may be reduced to Craik's model [3, 4] of K-type transition [12] . Obtained results point that this model is hardly adequate to experimental observation [12] . In this connection K-regime seems to us to be the process in which introduced wave is strongly nonlinear already at the first stage of evolution. Direct connections in triads grow feeble when the secondary instability of modulation type becomes stronger.

The dominant role of Tollmien-Schlichting waves resonance appears to be the peculiarity of C-type transition in the boundary layer and explains specific features of its initial stage sufficiently well.

References

1. Kachanov, Yu.S.; Kozlov, V.V.; Levchenko, V.Ya.: Nonlinear development of a wave in a boundary layer. Izv. Akad. Nauk

SSSR, Mekh. Zhidk. i Gaza, 3 (1977) 49-58 (in Russian).

2. Volodin, A.G.; Zelman, M.B.: Three-wave resonance interaction of disturbances in a boundary layer. Izv. Akad. Nauk SSSR, Mekh. Zhidk. i Gaza, 5 (1978) 78-84 (in Russian).

3. Craik, A.D.D.: Non-linear resonant instability in boundary layers. J.Fluid Mech. 50 (1971) 393-413.

4. Nayfeh, A.H.; Bozatly, A.N.: Nonlinear wave interactions in boundary layers. AIAA Paper N79-1493 (1979).

5. Gertsenshtein, S.Ya.; Shtemler, Yu.M.: Nonlinear disturbances evolution and their stability in boundary layers. Dokl. Akad. Nauk SSSR 236 (1977) N°6 (in Russian).

6. Zelman, M.B.; Maslennikova, I.I.: On resonant interaction effects of wave disturbances in a boundary layer. Izv. Akad. Nauk SSSR, Mekh. Zhidk. i Gaza, 4 (1984) (in Russian).

7. Herbert, T.: Analysis of the subharmonic route to transition in boundary layers. AIAA-84-0009 (1984).

8. Kachanov, Yu.S.; Levchenko, V.Ya.: The resonant interaction of disturbances at laminar-turbulent transition in a boundary layer. J.Fluid Mech. 138 (1984) 209-247.

9. Kozlov, V.V.; Levchenko, V.Ya.; Saric, W.S. (USA): Three-dimensional structures forming in boundary-layer transition. Preprint N°10-83 (1983) Novosibirsk, ITPM SO AN SSSR (in Russian).

10. Zelman, M.B.; Maslennikova, I.I.: On resonant interaction of three-dimensional disturbances in a boundary layer. Prikl. Mekh. i Tekhn. Fiz. (1985) N°2 (in Russian).

11. Saric, W.S.; Reynolds, G.A.: Experiments on the stability of nonlinear waves in a boundary layer. Laminar-Turbulent Transition (1980) Springer Verlag, 125-131.

12. Klebanoff, P.S.; Tidstrom, K.D.; Sargent, L.M.: The three-dimensional nature of boundary-layer instability. J. Fluid Mech. 12 (1962) 1-41.

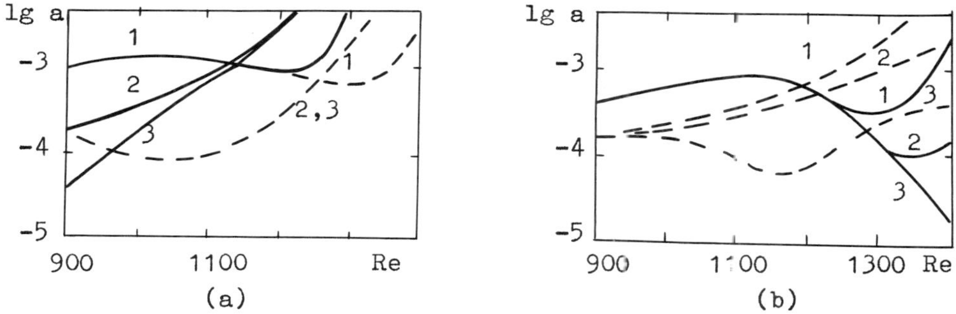

Figure 1. Amplification curves for symmetric triads. (a) 1-plane wave, 2,3- oblique waves; solid lines - $\Delta\Phi(x_o)=0$, dashed lines - $\Delta\Phi(x_o)\neq 0$; (b) 1- $=49°$, 2- $\xi=44°$, 3- $\xi=32°$ at Re=900; solid lines-plane wave, dashed lines - oblique waves. $F_1=115\cdot 10^{-6}$, $F_{2,3}=1/2\ F_1$.

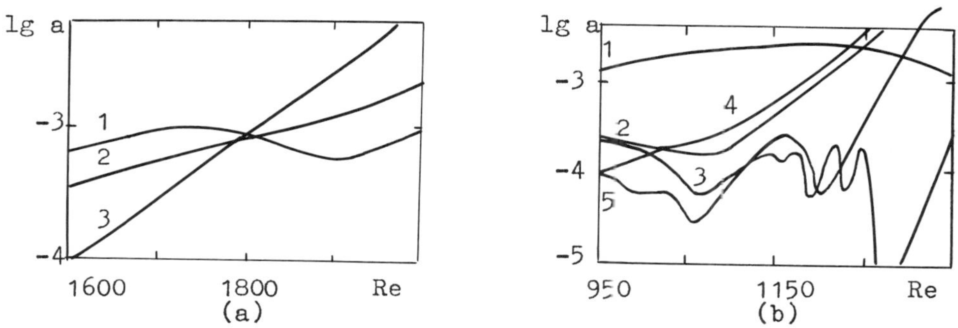

Figure 2. Amplification curves for asymmetric triads. (a) $F_1=60\cdot 10^{-6}$, $F_2=F_3=1/2\ F_1$, $\beta_1=\beta_2+\beta_3$, $\beta=0.15$, $\beta_2=0.2$ at Re=1650; (b) $F_{2,3}=38\cdot 10^{-6}$, $F_{4,5}=67\cdot 10^{-6}$, $F_1=F_{2,3}+F_{4,5}$, $\beta_2=-\beta_3=0.16$, $\beta_4=-\beta_5=0.25$, $\Delta\Phi=3/2\pi$ at Re=900.

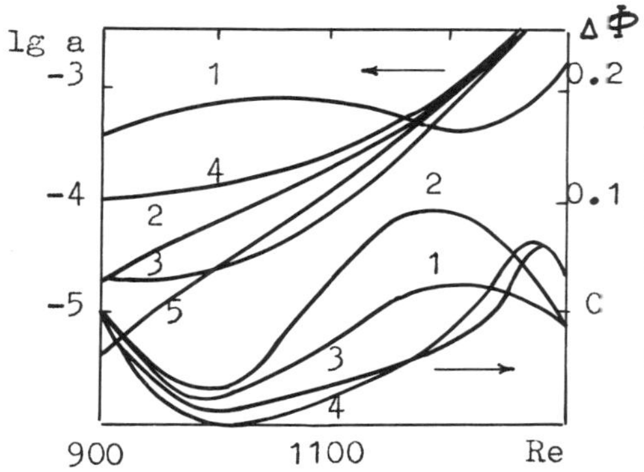

Figure 3. Amplitudes and phases in 5-waves cluster. 1-plane wave, 2, 3, 4, 5 - oblique waves. $F_1=115\cdot 10^{-6}$, $F_{2,3}=F_{4,5}=1/2\ F_1$; $\beta_2=-\beta_3=0.136$, $\beta_4=-\beta_5=0.154$ at Re=900.

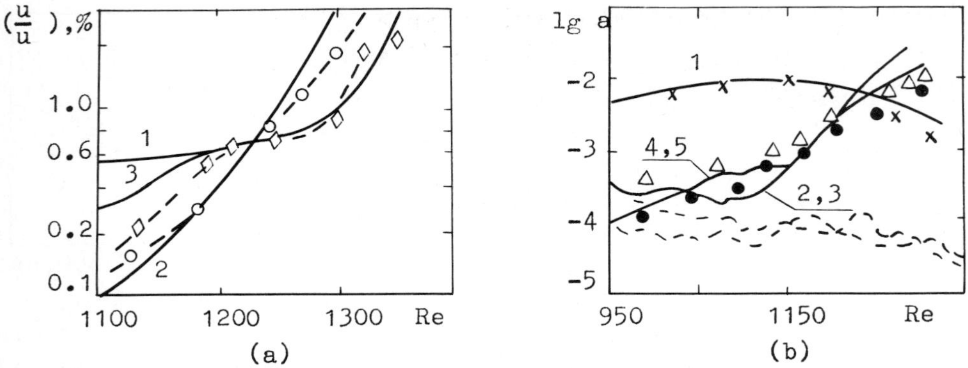

Figure 4. Comparison of theoretical and experimental curves. (a) (1, ◊)-plane wave, (2, o)-oblique waves, 3-plane wave with effect of nonparallelism. $F_1=106 \cdot 10^{-6}$, $\xi=61°$. (b) (1,x)-plane wave, (2,3, Δ) and (4,5, ●)-oblique waves, dashed lines-without crossinteraction. $F_1=F_{2,3}+F_{4,5}$, $F_{2,3}=38 \cdot 10^{-6}$, $F_{4,5}=67 \cdot 10^{-6}$, $\beta_2=0.16$, $\beta_4=0.25$ at Re=950.

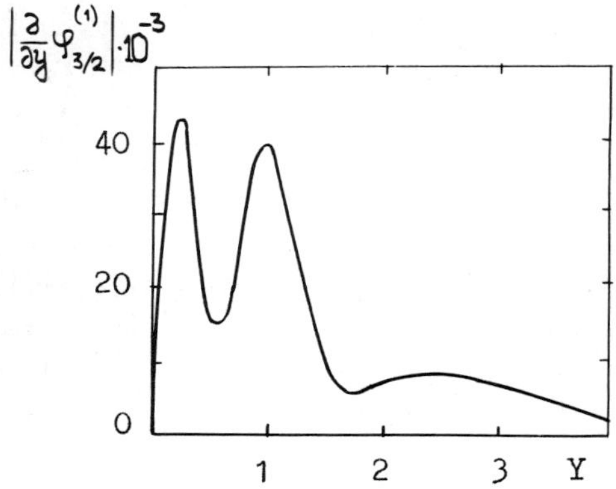

Figure 5. Y - distribution of $\left|\frac{\partial}{\partial y}\varphi^{(1)}_{3/2}\right|$. $F_1=115 \cdot 10^{-6}$, $\xi=43°$, Re=1300.

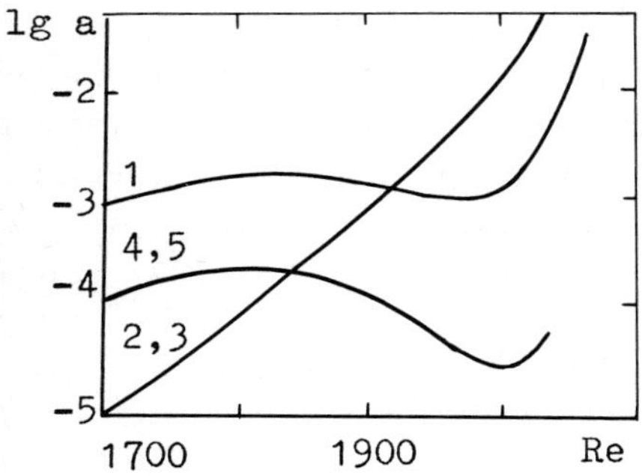

Figure 6. Amplification curves. 1-plane wave, 2-5-oblique waves. $F_1=59 \cdot 10^{-6}$, $F_{2,3}=1/2\ F_1$, $F_{4,5}=F_1$, $\beta_2=0.17$, $\beta_4=0.1$ at Re==1700.

Three-Dimensional Numerical Simulation of Laminar-Turbulent Transition and its Control by Periodic Disturbances

L. KLEISER, E. LAURIEN
DFVLR, Institute for Theoretical Fluid Mechanics
Bunsenstrasse 10, D-3400 Göttingen, FRG

Abstract

The laminar-turbulent transition process in plane Poiseuille flow is simulated by numerical integration of the three-dimensional time-dependent Navier-Stokes equations using a spectral method. The classical peak-valley splitting mode of secondary instability is considered. Detailed comparison with experiments has demonstrated very satisfactory agreement up to the 1-spike stage. Visualizations of the development of three-dimensional flow structures are presented including timelines of fluid markers. It is found that the Λ vortex structure appearing in flow visualizations does not consist of a vortex tube as the usual interpretation of the vortex loop concept suggests. A numerical study of transition control by superposition of purely two-dimensional waves shows that efficient control is possible at an early stage of the transition process but fails once significant three-dimensional disturbances are present.

1. Introduction

Transition to turbulence in shear flows proceeds through a sequence of increasingly complex stages initiated by flow instabilities. In low-disturbance channel and boundary layer flows the first step of the transition process consists of the amplification of two-dimensional (2D) Tollmien-Schlichting (TS) waves. While this primary instability has been well understood for a long time the subsequent three-dimensional (3D) secondary instability could be clarified only recently. Moreover, experimental and theoretical work has demonstrated the existence of several distinct secondary instability modes corresponding to different paths to turbulence (see [1-4] and references therein). The present work is concerned with the numerical simulation of the nonlinear, 3D stages of the transition process originating from the classical peak-valley splitting mode in plane Poiseuille flow. Experimental [5-7] and theoretical [3] evidence indicates close similarity of this process with observations for Blasius boundary layer flow. Transition in plane Poiseuille flow is reviewed from different viewpoints in [1,3,6,8]. In our previous work [9-12] we have made extensive comparisons of our simulation results with measurements of Nishioka et al. [5] and demonstrated good agreement up to the 1-spike stage. Similar agreement for boundary layer transition simulations with experiments has been obtained in [13]. The present contribution reports on a further investigation of the development of 3D flow structures obtained from velocity and vorticity fields and from flow visualizations by fluid markers. In addition results of a numerical study of transition control by periodic disturbance superposition are presented.

2. Mathematical model and numerical discretisation

Coordinates and integration domain of the simulation are shown in fig.1 (all variables are non-dimensionalized with channel half-width and undisturbed center-line velocity). For efficiency we use the model of a streamwise and spanwise periodic flow developing in time in a frame of reference $x_1^C = x_1 - Ct$ moving with the phase velocity of the TS wave. Accordingly the roles of x_1 and t are exchanged to compare experimental and numerical results. In analogy with vibrating ribbon experiments the evolution of disturbances $\underline{u}(\underline{x},t)$ (having non-zero mean $\bar{\underline{u}}$) from an initial disturbance $\underline{u}^0(\underline{x})$ superimposed on the basic flow $U(x_3) = 1-x_3^2$ is calculated by numerical integration of the 3D time-dependent Navier-Stokes equations. The spatial discretisation is based on a highly accurate spectral method with Fourier expansions in the horizontal coordinates,

$$\underline{u}(\underline{x},t) = \sum_{k_1}\sum_{k_2} \hat{\underline{u}}(k_1,k_2,x_3,t)\exp(ik_1\alpha_1 x_1 + ik_2\alpha_2 x_2) \qquad (1)$$

($\alpha_j = 2\pi/L_j$). In the normal direction Chebychev polynomial expansions are applied. More details are given in [9].

3. Three-dimensional simulation of transition

We discuss some results of a simulation with Reynolds number Re = 5000 and wavenumbers $\alpha_1 = 1.12$, $\alpha_2 = 2.1$ chosen to match the experimental conditions of [5]. The initial disturbance $\underline{u}^0 = A_{TS}\underline{u}_{TS} + \varepsilon\underline{u}_{3D}$ consists of a 2D Tollmien-Schlichting wave and a pair of oblique waves obtained as Orr-Sommerfeld eigensolutions normalized by $\max|u_{1,TS}| = \max|u_{1,3D}| = 1$. Fig.2 shows the development of the amplitudes $\hat{u}_{1max} = \max(x_3)|\hat{u}_1|$ of the Fourier modes in (1) excited initially, $(|k_1|,|k_2|) = (1,0)$ and $(1,1)$, for six runs with different values of A_{TS} and $\varepsilon = 0.001$. We see how after an initial transient the TS wave amplitude controls the growth or decay of the 3D disturbance. The growth rates agree well with the secondary instability theory [8]. The case with A_{TS} = 0.03 has been studied in great detail [9-12] and will be considered for the rest of this paper.

In fig.3 the development of the maximum (across channel) rms fluctuation $u_1'_{max}$ at different spanwise positions x_2 is shown indicating the well-known peak-valley splitting. For $t \geq 110$ two local maxima in $u_1'_{max}(x_2)$ appear in accordance with the experiment (see [9,10]). At $t \simeq 122$ the 1-spike stage occurs. The distribution of the rms fluctuation u_1' across the half-channel at $t = 110$ is given in fig.4 together with the experimental results at the $u_1'_{max}$(peak) = 6 % stage. Except for a slight deviation around $x_3 \simeq -0.6$ at the valley position the results agree well. A more thorough comparison of numerical and experimental results including rms fluctuation u_1', mean velocity $U+\bar{u}_1$ and instantaneous velocity distributions $U+u_1$ in several cross-sections is reported in [9-12]. Features like the formation of a peak-valley structure of u_1', mean longitudinal vortices and inflectional instantaneous velocity profiles connected with high-shear regions and low-velocity spike signals are reproduced. The degree of agreement obtained up to the 1-spike stage with available experimental information is remarkable if we consider that a pure initial value problem is solved for a streamwise periodic flow evolving in time from initial disturbances with purely antisymmetric $u_1^0(x_3)$. Beyond the 1-spike stage the accuracy of the simulation results deteriorates due to lack of sufficient numerical resolution (up to $N_1 \times N_2 \times N_3 = 32 \times 32 \times 41$ points have been used). Continuation of the simulation up to $t = 158$ showed that the solution becomes chaotic with mean properties similar to fully developed turbulent channel flow [12].

4. Visualization of three-dimensional flow structure evolution

Having confirmed the validity of our simulation results we investigated 3D flow structures developing in transition. Maps of velocity and vorticity fields and other derived quantities have been produced as well as flow visualizations by fluid markers. The time evolution is illustrated by a film. Here the formation of the vortical structure and of the high-shear layer will be considered. Fig.5 shows a set of timelines generated at a line parallel to x_2 located at $x_3 = -0.8$ slightly above the critical layer, $x_{3,crit} = -0.85$. Thirty lines of particles are released per period T_{TS}. The phase positions of the three frames at each stage differ by $T_{TS}/5$. At t = 80 the folding of the particle sheet [14,15] is seen where the particles on the high-speed side of the critical layer overtake those on the wall side. The wave front is slightly warped at this stage. At t = 122 the Λ vortex structure known from boundary layer transition visualizations (e.g. [15]) is visible. The time-lines are strongly distorted and extend now over a range $-0.65 \lesssim x_3 \lesssim -0.95$. We stress that these timelines have been obtained from the computed velocity field at a fixed computational time t_o which is strictly periodic in x_1-Ct. The downstream development of the structures appearing in fig.5 therefore must not be misinterpreted as indicating a development of the velocity field towards breakdown but is a pure integration effect [14].

The Λ structure is usually interpreted as a vortex loop consisting of a vortex tube, i.e. a bunch of vortex lines (see [16] for a discussion of the vortex loop concept). To check this idea let us look at a set of selected vortex lines, fig.6. In fig.6a the Λ structure (extended periodically to two wavelengths in x_1 and x_2) becomes visible from the strong warping of the vortex lines. However, as further evidenced by fig.6b and c, it does not consist of a vortex tube as a whole. Rather, the vortex lines point mainly in spanwise direction following the Λ vortex axis only locally when crossing the vortex region. The spanwise waviness outside the Λ vortex is connected with a mean longitudinal vortex system [12].

The evolution of surfaces of constant shear is shown in fig.7. The tips appearing around spanwise peak positions enclose the detached high-shear layer where later breakdown into hairpin eddies will occur. Quite similar structures have been found in the experiment [6].

5. Control of transition by periodic disturbance superposition

Recent experiments [17-19] have shown that transition to turbulence in a flat plate boundary layer can be delayed by superposition of 2D disturbances antiphased to oncoming instability waves. Linear secondary instability theory [1-3] provides a rational framework for understanding this phenomenon as long as 3D disturbances remain sufficiently small. The aim of our numerical study is not a naturalistic simulation of experiments rather than an investigation of the nonlinear 2D/3D disturbance interaction in order to obtain an indication of the theoretical potential and the limitations of this control scheme. For our purposes a crude modelling of the 2D disturbance input is sufficient. We have used alternatively a downstream-directed fluctuating mass force and periodic wall suction and blowing with zero average mass flux. Both types gave similar results. In our simulations control is applied over a certain time interval (typically of length $T_{TS}/2$) which corresponds to a damping action over a certain downstream distance experienced by a wave passing a fixed control station.

Fig.8 shows results of two cases with control. For comparison case I considered above is included. In cases II and III a body force was applied at the early stage $40 \le t \le 50$. In case III the 2D wave amplitude is damped below the threshold for secondary instability so that all disturbances will finally die out. In case II (with half force amplitude of case III) the 2D amplitude remains slightly above this threshold so that the 3D disturbance part continues to grow, though at a reduced rate. Thus transition is delayed in comparison with case I. Further studies have shown that delay (not to mention prevention) of transition by 2D control becomes increasingly difficult at higher Reynolds numbers and, in particular, at later stages of disturbance development [20]. At $t = 80$, where the 3D disturbances reach the order of the TS wave amplitude and significant higher harmonics have developed, transition could no longer be prevented, even not by artificially setting to zero all 2D disturbances. Even if control is successful and all fluctuations have decayed to a vanishing level the flow does not soon return to the undisturbed state (cf. [19]). As illustrated by fig.9, streaky mean longitudinal vortex structures generated during incipient transition remain in the flow field and are only very slowly damped.

We conclude that numerical simulations with careful data evaluation can contribute to an improved understanding of the transition process and offer a considerable potential for future investigations.

Our simulation results have been obtained using the computer code CHANSON developed by the first author at Kernforschungszentrum Karlsruhe. He expresses his gratitude to U. Schumann for the advice obtained during that period and to Th. Herbert for many stimulating discussions and for making available his Orr-Sommerfeld solver.

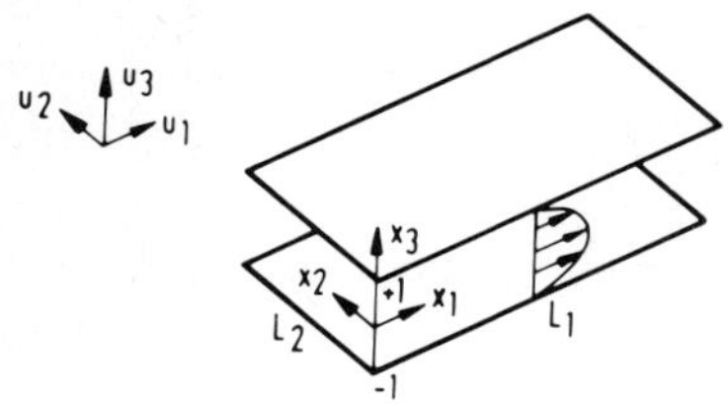

Fig. 1 Spatial integration domain and coordinates.

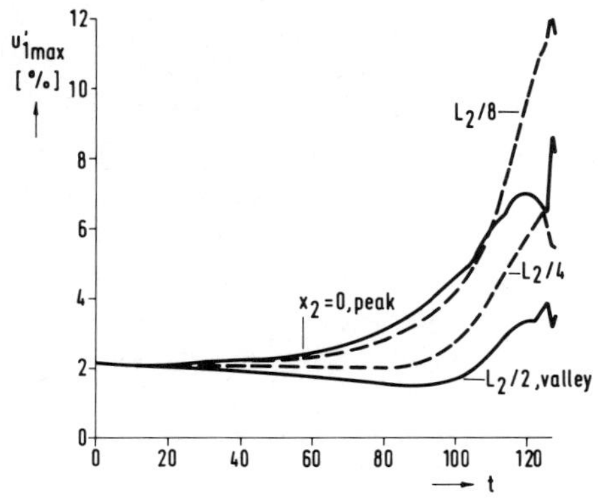

Fig. 3 Development of depth-maximum of rms fluctuation at different spanwise positions x_2.

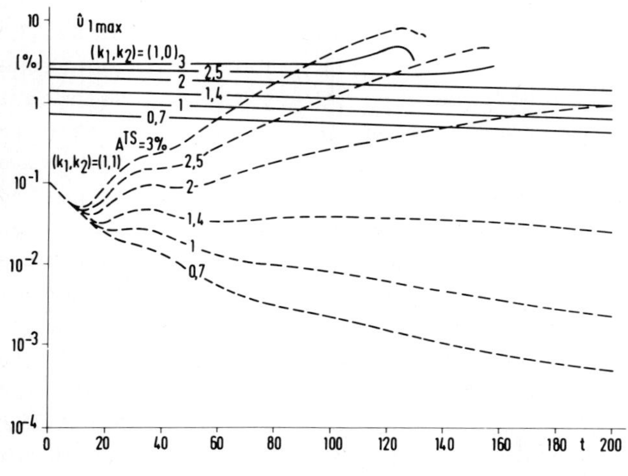

Fig. 2 Development of 2D (——) and 3D (---) Fourier amplitudes for different initial amplitudes A_{TS}.

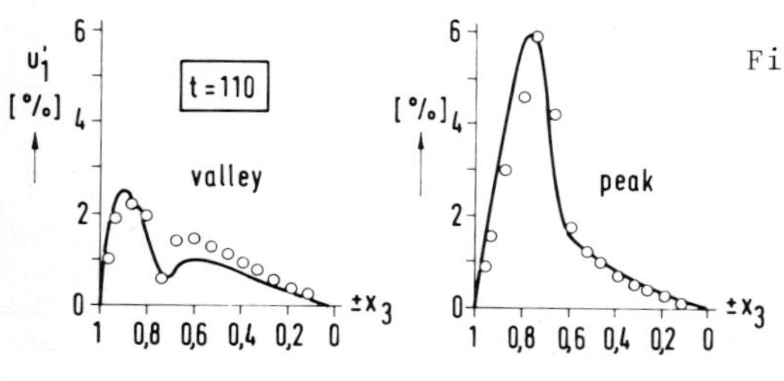

Fig. 4 Distribution of rms fluctuation at peak and valley position. —— simulation, ∘∘∘ experiment [5].

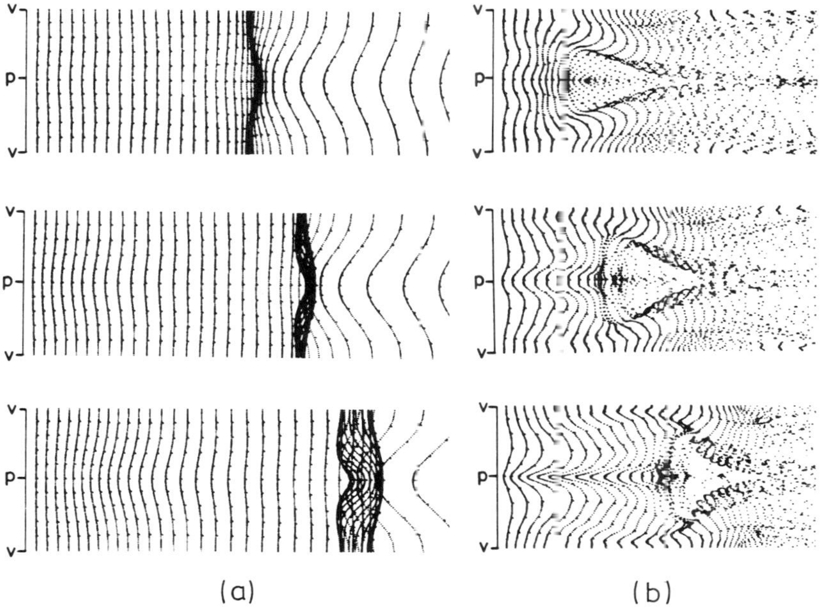

Fig. 5 Snapshots of time-lines at three phase positions
(a) t = 80, early 3D stage (b) t = 122, 1-spike stage.

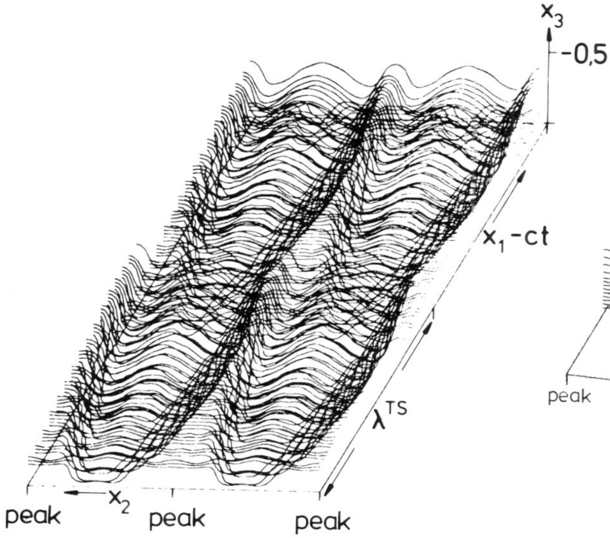

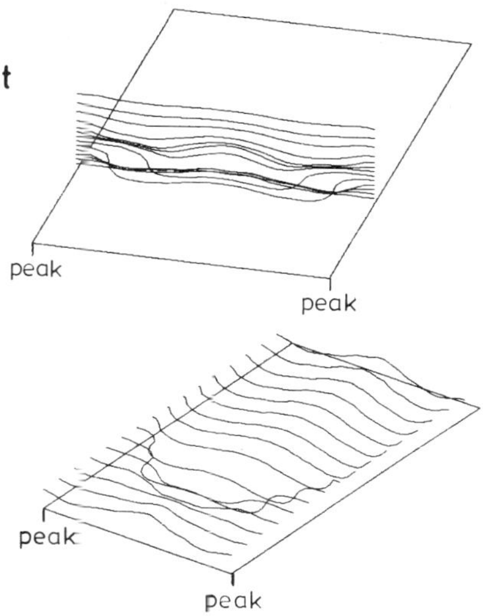

Fig. 6 Selected Vortex lines at t = 122,
1-spike stage.

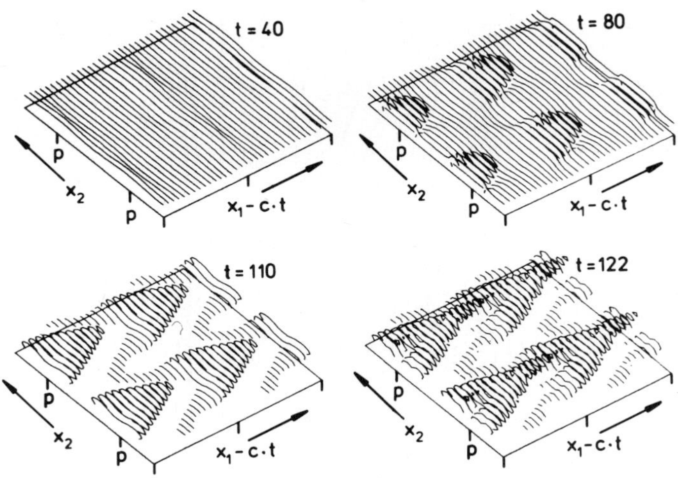

Fig. 7 Surfaces of constant shear $\partial/\partial x_3\,(U+u_1) = 1.65$. Near-wall lines suppressed at $t \geqq 110$ for visibility.

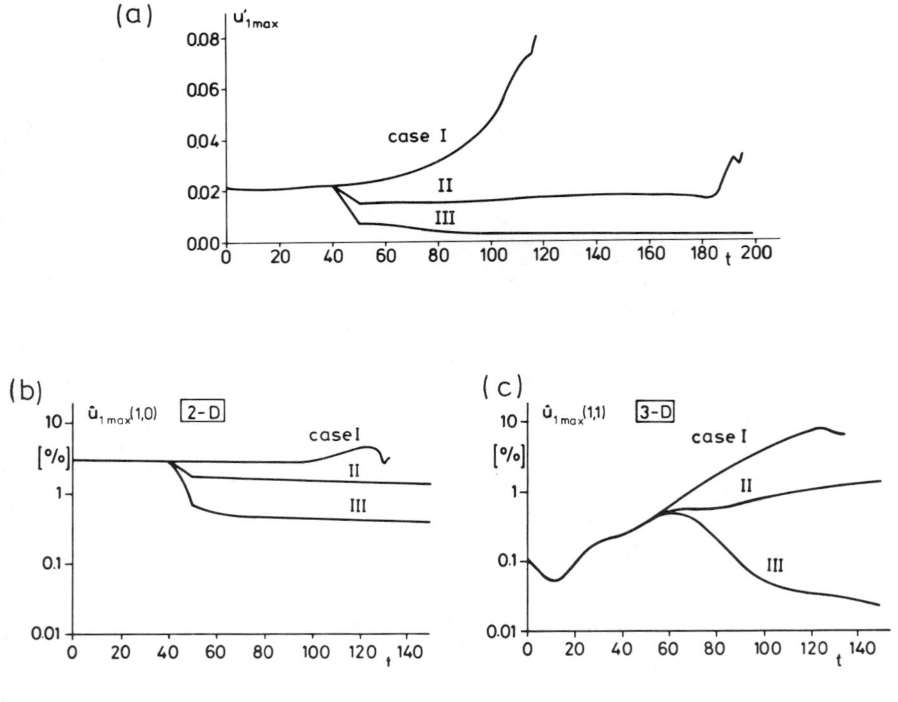

Fig. 8 Maximum rms fluctuation at peak (a) and 2D (b) and 3D (c) Fourier amplitudes for three cases: without control (I), superposition of anti-phased 2D wave of moderate amplitude (II) and of large amplitude (III).

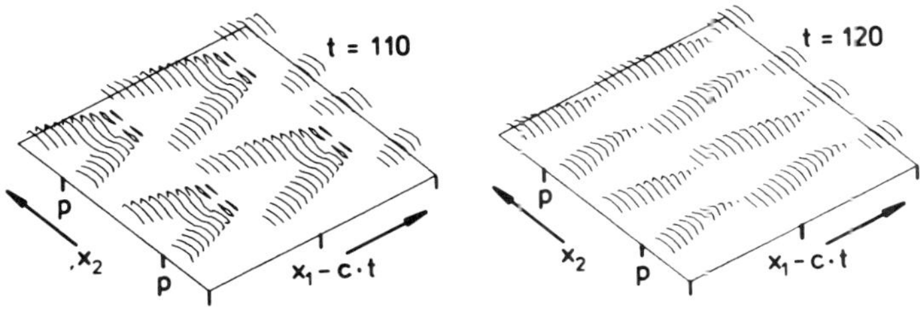

Fig. 9 Decay of constant-shear surface $\partial/\partial x_2\,(U+u_1) = 1.65$ after control applied during $70 \leq t \leq 80$.

References

[1] S.A. Orszag, A.T. Patera, JFM 128 (1983) 347-385

[2] Th. Herbert, AIAA paper 83-1759 (1983) and AIAA paper 84-0009 (1984)

[3] Th. Herbert, Secondary instability of shear flows, in AGARD Report No. 709 (1984)

[4] W.S. Saric, V.V. Kozlov, V.Ya. Levchenko, AIAA paper 84-0007 (1984)

[5] M. Nishioka, M. Asai, S. Iida, Proc. IUTAM Symposium on Laminar-Turbulent Transition, Stuttgart, 1979 (Springer Verlag, 1980) 37-46

[6] M. Nishioka, M. Asai, Proc. IUTAM Symposium on Turbulence and Chaotic Phenomena, Kyoto, 1983

[7] V.V. Kozlov, M.P. Ramasanov, Fluid Dynamics 18 (1983) 30-33

[8] Th. Herbert, Fluid Dyn. Trans. 11 (1983) 77-126

[9] L. Kleiser, Numerische Simulationen zum laminar-turbulenten Umschlagsprozeß der ebenen Poiseuille-Strömung. Dissertation, Karlsruhe, 1982 (Report KfK 3271, Kernforschungszentrum Karlsruhe)

[10] L. Kleiser, Springer Lecture Notes in Physics 170 (1982) 280-285

[11] L. Kleiser, U. Schumann, in R.G. Voigt et al. (ed.), Spectral Methods for Partial Differential Equations, SIAM, Philadelphia 1984, 141-163

[12] L. Kleiser, in H.L. Jordan et al. (ed.), Nonlinear Dynamics in Transcritical Flows, Springer Lecture Notes in Engineering, to appear

[13] A. Wray, M.Y. Hussaini, AIAA paper 80-0275 (1980)

[14] F.R. Hama, Phys. Fluids 5 (1962) 644-650

[15] F.X. Wortmann, Springer Lecture Notes in Physics 148 (1981) 268-279

[16] D.R. Williams, H. Fasel, F.R. Hama, to appear in JFM

[17] R.W. Milling, Phys. Fluids 24 (1981) 979-981

[18] H.W. Liepmann, D.M. Nosenchuck, JFM 118 (1982) 201-204

[19] A.S.W. Thomas, JFM 137 (1983) 233-250

[20] E. Laurien, L. Kleiser, to appear in ZAMM

Instability and Turbulent Bursting in the Boundary Layer

MARTIN LESSEN, PENG-HONG KOH

Department of Mechanical Engineering
University of Rochester
Rochester, NY 14627 USA

Defense Science Organization
MINDEF, Tanglin P.O. Box 822
Singapore 9124

ABSTRACT

The stability of the Blasius boundary layer with respect to a finite amplitude Tollmien-Schlichting wave is studied. The stability with respect to a non-linear, convectively driven, cross streamwise periodic disturbance of the resulting flow is then studied and finally, the stability of the resulting localized shearing layers is studied. It is found that at large Reynolds number, the primary disturbance amplitude is about 8% and the secondary disturbance amplitude is about 27% of the free stream velocity.

INTRODUCTION

Hydrodynamic stability and transition to turbulence is a central problem in fluid mechanics. The mechanism of the secondary instability adding three-dimensionality to the primary two dimensional Tollmien-Schlichting disturbance has been a matter of contention since the investigation of a convectively driven secondary disturbance by Goertler and Witting (1958) and interacting oblique Tollmien-Schlichting waves by Benney & Lin (1960). The Goertler-Witting model predicted streamwise oriented vortices with maximum amplitude located between the Kelvin cat's eyes of the primary disturbance while the Benney-Lin model predicted maximum secondary disturbance amplitude at or near the crests of the cat's eyes. Stuart (1962) took issue with the assumption of Benney & Lin that the oblique and primary waves had the same phase velocity in the flow direction and his findings were further supported by Klebanoff, Tidstrom and Sargent (1962) in their experimental measurements which indicated a modulation of the secondary flow structure which could have resulted from a difference of phase velocities. The Klebanoff et al experiment also found that the secondary disturbance had its maximum amplitude near the crests of the primary Tollmien-Schlichting waves.

Based on the foregoing, it seems natural that the turbulent bursting phenomenon is a regular one and this was found by Kline, Reynolds, Schraub and Rundstadler (1967). In this paper, the breakdown of two dimensionality of the primary Tollmien-Schlichting disturbance is studied approximately from the point of view of an improved Goertler-Witting model which results in distortion of the flow in the form of intense shearing layers that cause local wave breaking and mixing. It is this third stage which is denoted as transition to turbulence. For the case of unbounded, inviscidly unstable flows such as jets, wakes, and shearing layers, the initial instability is already this third stage for which the postulate of marginal instability describes the state of affairs as in Lessen and Singh (1974) Lessen and Paillet (1976), Lessen, Barcilon and Butler (1977) and Lessen (1978).

The present analysis is restricted to Blasius boundary layer flow and assumes that the primary Tollmien-Schlichting instability, the secondary Goertler-Witting instability and the tertiary shearing layer instability occur serially for simplicity of analysis rather than the real life phenomenon of simultaneity. It is probable that the wave interaction model of Benney-Lin and successors plays a role in initially establishing spanwise periodicity to the developing instability but that the centrifugally driven model seems to provide the most energy at large amplitude to the secondary instability.

PRIMARY INSTABILITY

For a two dimensional incompressible flow, the dimensionless vorticity equation in terms of the stream function ψ where $u = \psi_y$ and $v = -\psi_x$ is

$$\nabla^2 \psi_t + \psi_y \nabla^2 \psi_x - \psi_x \nabla^2 \psi_y = \frac{1}{R} \nabla^2 \nabla^2 \psi$$

where $R = U_\infty \delta/\nu$, U_∞ is free stream velocity, ν is kinematic viscosity,

$\delta = \sqrt{(\nu x^*/U_\infty)}$, x^* is dimensional distance from boundary layer leading edge.

Following Landau (1959), Stuart (1960) and Watson (1960) we consider a solution of the form

$$\psi(x,y,t) = \sum_{m=0}^{\infty} \sum_{n=-\infty}^{\infty} |A|^{|n| + 2m} \phi_{nm}(y) \exp in\alpha(x-c_r t)$$

where $A = A(t)$; c_r is the real disturbance phase velocity;

$\dfrac{dA}{dt} = A \sum_{m=0}^{\infty} |A|^{2m} a_m$ or $\dfrac{d|A|}{dt} = \sum_{m=0}^{\infty} |A|^{2m+1} a_{mr}$; $a_{mr} = Rl\ a_m$;

and $\phi_{nm} = \tilde{\phi}_{-nm}$.

If terms of periodicity n and order $|A|^{|n| + 2m}$ are gathered, the single equation for vorticity is represented by a doubly infinite set of ordinary differential equations, the lowest order of which deals with the mean flow. Due to the form of the solution, the mean flow it represents is an identically parallel flow. Since we consider the stability of a quasi-parallel flow, namely Blasius flow, we must substitute more accurate equations for the mean flow. The method is similar to that of Lessen and Gangwani (1976) extended for a time dependent disturbance yielding

$$2F''' + FF'' - 2R\frac{\partial F'}{\partial t} = 2R\frac{\partial}{\partial \eta}\overline{(u'v')}$$

where the boundary layer stream function $\Psi = \sqrt{(\nu x^* U_\infty)}\, F(\eta)$; $\eta = \dfrac{y^*}{\delta}$; y^* is dimensional coordinate in the direction of boundary layer thickness, and u', v' are components of the disturbance, $\Psi = \delta\, U_\infty\, F(\eta)$. The dimensionless coordinate y corresponds to the Blasius variable η and $F(\eta)$ is a dimensionless stream function. F, u and v may be expanded in powers of $|A|$ such that

$$F(\eta,t) = F_o(\eta) + |A|^2 F_1(\eta) + |A|^4 F_2(\eta) + ---,$$

the zeroth order of F in $|A|$ yielding the Blasius equation and the higher orders of F available recursively using the various levels of approximation of u and v . The boundary conditions on F_m are

$$F_m(0) = F'_m(0) = 0 \text{ and } F'_o(\infty) = 1,\ F'_m(\infty) = 0 \text{ for } m \geq 1.$$

The mean flow solutions so obtained can be inserted in the separated, recursive equations for ϕ_{nm} by recognizing that $\phi_{om} = F_m$.

In solving the equations for ϕ_{nm}, a non-uniqueness arises that can be removed as per Herbert (1976) by requiring that $\phi_{1m}(y_o) = 1$ for m = 0 and $\phi_{1m}(y_o) = 0$ for m $\geq$ to 1 where y_o is the point of normalization, which then determines a_m. The saturation amplitude (for fixed α, R) is then obtained by setting $\frac{d|A|}{dt} = 0$.

SECONDARY INSTABILITY

Utilizing the results of the previous section, the total flow field stream function (cat's eyes) in a coordinate frame moving at the phase velocity of the primary disturbance is available (see fig 4). It can be seen that the regions of large streamline curvature between and on the flank of the cat's eyes both above and below the critical layer are potentially regions subject to a Goertler and Witting (centrifugally driven) secondary instability. The Navier-Stokes equations with velocity and streamline curvature distribution are then written in this frame and it is assumed (for an approximate solution for large cross-stream disturbance wave number) that the local velocity and curvature distributions are not functions of x. Of course, the foregoing is a very crude approximation, but it was found by Goertler and Witting that it underestimated instability in the linear disturbance case. The system of equations is amenable to a solution of the form

$$U(y,z,t);\ V(y,z,t);\ P(y,z,t) = \sum_{m=0}^{\infty} \sum_{n=0}^{\infty} |A_T|^{n+2m} \; U_{nm}(y); V_{nm}(y); P_{nm}(y) \operatorname{Cos} n\alpha_T z$$

$$W(y,z,t) = \sum_{m=0}^{\infty} \sum_{n=0}^{\infty} |A_T|^{n+2m} \; W_{nm}(y) \operatorname{Sin} n\alpha_T z$$

where $V_{om} = W_{om} = 0$; U_{oo}, P_{oo} is the velocity and pressure distribution of the primary cat's eyes; U, V, W, P are the three dimensional flow field velocity components and pressure including the secondary disturbances; z is the dimensionless space coordinate transverse to the primary flow field;; α_T is the secondary disturbance wave number (in the z direction); and A_T is the secondary disturbance amplitude. Again, A_T is given by a Landau equation of the form $\dfrac{d|A_T|}{dt} = \sum_{m=0}^{\infty} |A_T|^{2m+1} a_{mr}$. Gathering terms of periodicity n and order $|A_T|^{n+2m}$, a solution in the same format as that for the primary disturbance is obtainable with application of the Herbert normalization condition.

DISTURBANCE KINETIC ENERGY EQUATION

The kinetic energy equation for a disturbance was discussed by Stuart(1958). Utilizing his method and including in addition the energy transfer between harmonics, the equation of disturbance kinetic energy for the primary instability may be written for the fundamental as

$$\frac{\partial E_1}{\partial t} = \int_{y=0}^{\infty} \int_{x=x_1}^{x_1+2\pi/\alpha} \frac{1}{2} \frac{\partial}{\partial t} (u_1^2 + v_1^2) \, dx \, dy$$

$$= - \iint u_1 v_1 \frac{\partial \bar{u}}{\partial y} \, dxdy - \iint \{(u_1^2 - v_1^2) \frac{\partial u_2}{\partial x} + u_1 v_1 (\frac{\partial v_2}{\partial x} + \frac{\partial u_2}{\partial y})\} \, dx \, dy$$

$$- \mu \iint (\frac{\partial v_1}{\partial x} - \frac{\partial u_1}{\partial y})^2 \, dx \, dy$$

where $u = \bar{u} + u_1 + u_2 \ldots$; $v = v_1 + v_2 \ldots$; $\bar{u}$ is the modified mean velocity profile and the subscripts 1 and 2 denote the fundamental and the second harmonic disturbances respectively. At fixed α,R the amplitude for

$\frac{\partial E_1}{\partial t} = 0$ is obtained as a check on the asymptotic method.

TERTIARY INSTABILITY

The secondary disturbance causes mass and momentum transport between regions of low and high velocity in the boundary layer. As a result, there exist local regions of shearing layers where, for instance, the low-velocity fluid contacts the high velocity fluid. Such shearing layers (points of inflection in the local velocity distribution) exist above and below the critical layer but the ones above the critical layer are more pronounced. From the scale of the upper shearing layers, we estimate the dominant wave number of the shearing layer primary disturbance.

RESULTS AND DISCUSSION

Including all second order effects it was found that the primary (Tollmien-Schlichting) instability is slightly sub-critical see fig 1,2. At R=580, the variation of saturation amplitude $|A|$ with wave number α is given in Fig. 3 using the asymptotic and energy methods which compare favorably except at the point of maximum amplitude. At this point, we feel that the energy method is more accurate. The distorted Kelvin cat's eye is shown in fig 4 and the secondary, centrifugally driven instabilities were examined at sections 1 for the full velocity and streamline curvature profiles above and below the critical layer and at sections 2 and 3 for the velocity and streamline curvature profiles above the critical layer with slip boundary conditions at the edge of the cat's eye. The results of calculations are given in Fig. 5,6. The modified mean velocity distribution due to the secondary disturbance is given in fig. 7,8. The instantaneous velocity distribution at the local areas of high shear are given in fig. 9.

The sub-critical nature of the primary instability explains the difficulty in reproducing experimentally the theoretical neutral stability curve for infinitesimal disturbances particularly at the minimum critical Reynolds number and the upper branch. The use of slip boundary conditions at the edge of the cat's eye for the secondary disturbance gave results that were close to those obtained by integrating through to the wall between the cat's eyes so that the use of the slip condition was felt to give reliable results along the flank of the cat's eye. The modified mean profile with the 27% secondary disturbance looks qualitatively like that obtained by Klebanoff et al but the Klebanoff primary disturbance was forced and the phase of his measurements could not be determined. The progression of the mean profile from the Blasius to that of fully developed turbulence is shown in fig. 8 where the 3 spike and 5 spike stages as in Nishioka (1981) represent the development of the tertiary disturbance in local regions between the longitudinal Goertler-Witting vortices. In the present model, as the Reynolds number increases, the primary and secondary disturbances go to a higher amplitude, the local shear layers become sharper and the length scale of the tertiary disturbance smaller, so that the wave length of the tertiary disturbance relative to the primary disturbance wave length becomes smaller therefore increasing the number of spikes in the wavelength. In our investigation at our Reynolds number, the tertiary wave length was about 10% of the primary and since the local shearing layer was only about 50% of the primary wave length, the situation was at the 5-spike stage. Had the corresponding Reynolds stresses been calculated the mean profile would have been quite close to that observed by Nishioka et al. The instantaneous profiles agreed qualitavely with those observed by Matsui and Okude (1984) who also seemed to have observed longitudinal structures below the critical layer as this theory also predicts.

ACKNOWLEDGEMENT

This work received partial support from the Office of Naval Research and the calculations were supported by the University of Rochester.

REFERENCES

BENNEY, D. J. & LIN, C. C. , 1960, Phys. Fluids, 3, 656.

GOERTLER, H. & WITTING, H., 1958, IUTAM Symposium "Boundary Layer Research", Berlin, 110.

HERBERT, T., 1976a, Springer Lecture Notes in Physics, 59, 235

HERBERT, T., 1976b, Arch. Mech. Arch. Mech. Stosowanej, 28, 1039.

KLEBANOFF, P. S., TIDSTROM, K. D. & SARGENT, L. M., 1962 J. Fluid Mech., 12,1.

KLINE, S. J., REYNOLDS, W. C., SCHRAUB, F. A. & RUNDSTADLER, P. W., 1967, J. Fluid Mech., 30, 741.

LANDAU & LIFSHITZ, 1959, Fluid Mechanics, London, Pergamon Press

LESSEN, M. & SINGH, P. J., 1974, Phys. Fluids, 17, 1329.

LESSEN, M. & GANGWANI, S. T., 1976, Phys, Fluids, 19, 510.

LESSEN, M. & PAILLET, F. L., 1976, Phys. Fluids, 19, 510.

LESSEN, M. BARCILON, A. & BUTLER, T. E., 1977, J. Fluid Mech., 82, 449.

LESSEN, M., 1978, J. Fluid Mech., 88, 535.

MATSUI, T. E. OKUDE, M., 1984, Proc. of 2nd IUTAM Symposium on Laminar-Turbulent Transition, Springer, Berlin.

NISHIOKA, M.; ASAI, M.; IIDA, S., 1981, Proc. of Symp. on Transition and Turbulence, New York: Academic Press.

STUART, J.T., 1958, J. Fluid Mech., 4, 1.

STUART, J. T., 1960, J. Fluid Mech., 9, 353.

STUART, J. T., 1962, Advances in Aero, Sciences, Pergamon, Vol. 3. 121.

WATSON, J., 1960, J. Fluid Mech., 9, 371.

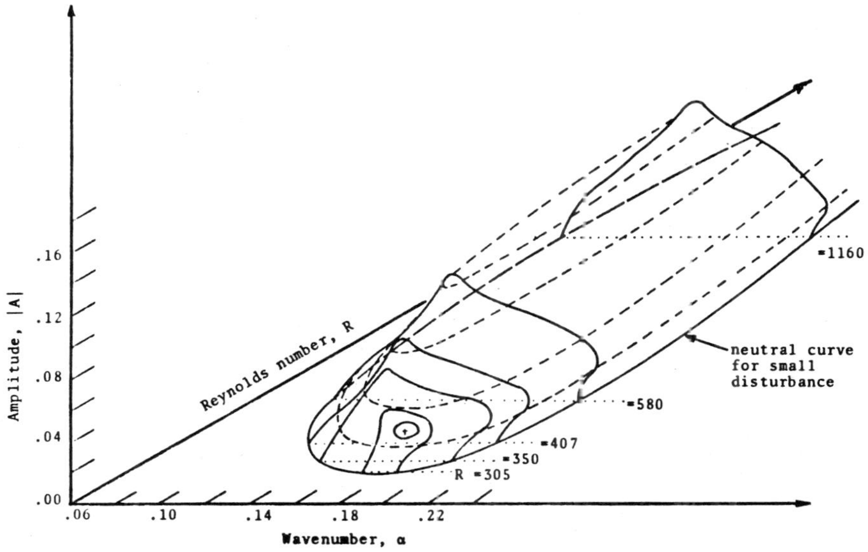

Fig. 1 The Neutral Surface for Blasius Profile

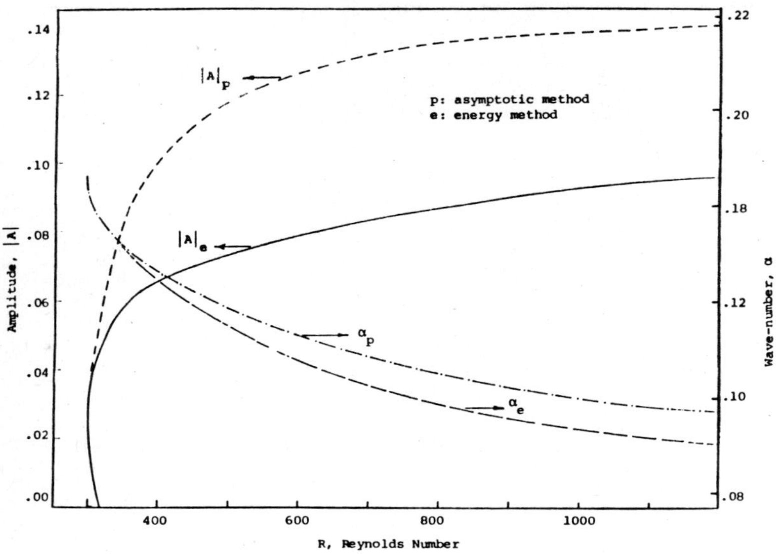

Fig. 2 Variations of Amplitude and Wave-number of the most amplified disturbance with Reynolds Number.

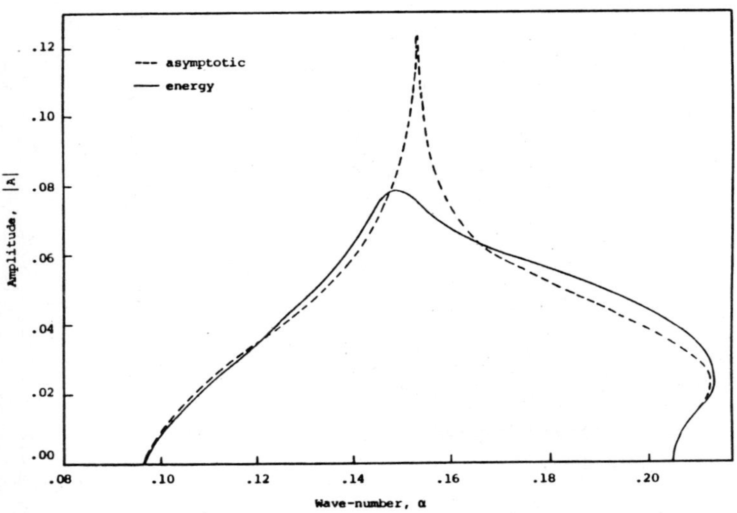

Fig. 3 The saturation amplitude of the T-S waves at R = 580

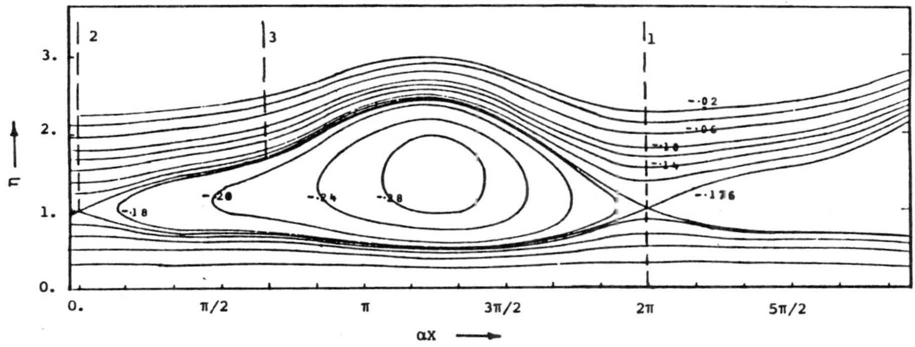

R = 580, α = .147, C_r = .390, |A| = .077, y_σ = 1.00

Fig. 4 The distorted Kelvin cats eye

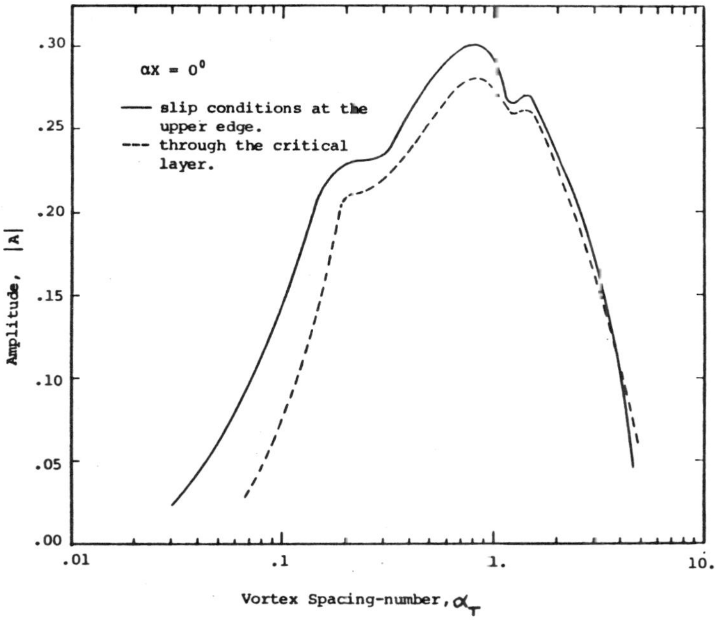

Fig. 5 The saturation amplitude of the secondary instability at the valley of the cats eye.

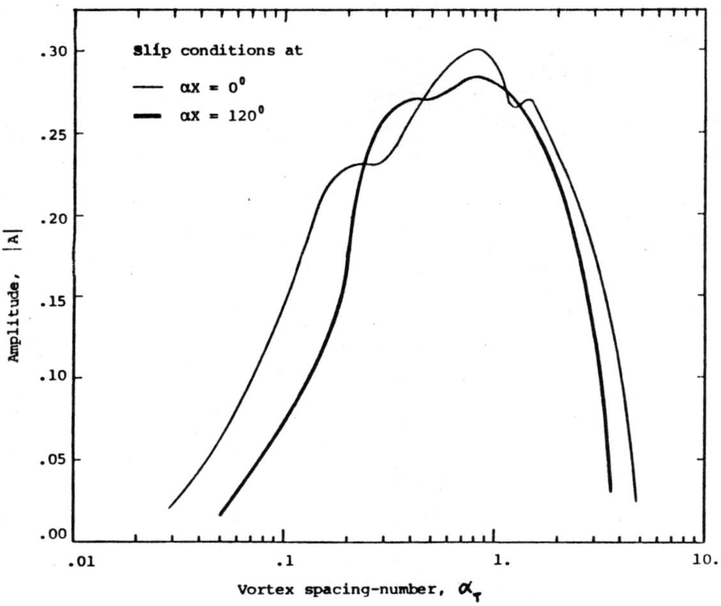

Fig. 6 Comparison of the saturation amplitude of the secondary instability at two locations.

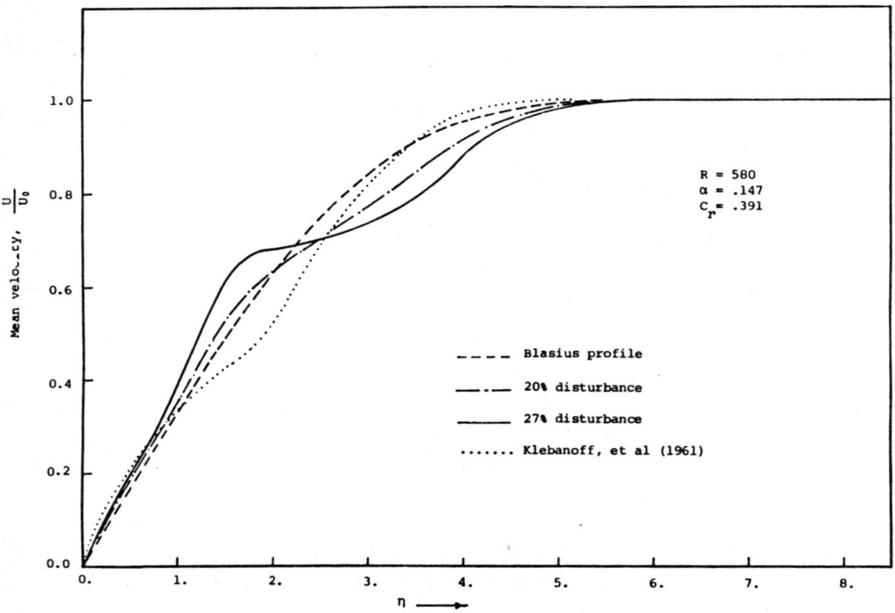

Fig. 7 The modified mean velocity distribution at the most destabilising region (largest curvature) of the secondary disturbance.

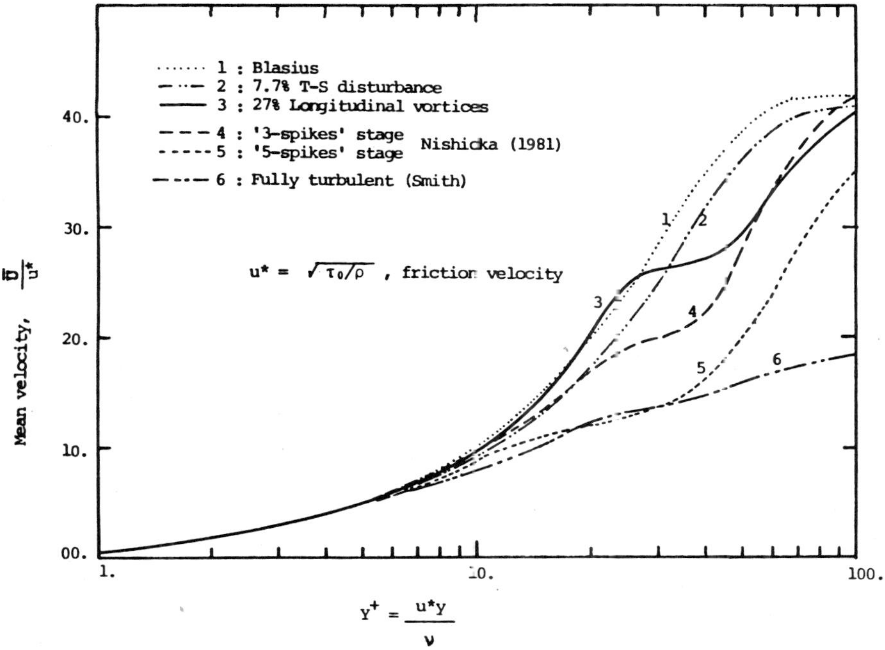

Fig. 8. Mean velocity profiles at the different stages of transition.

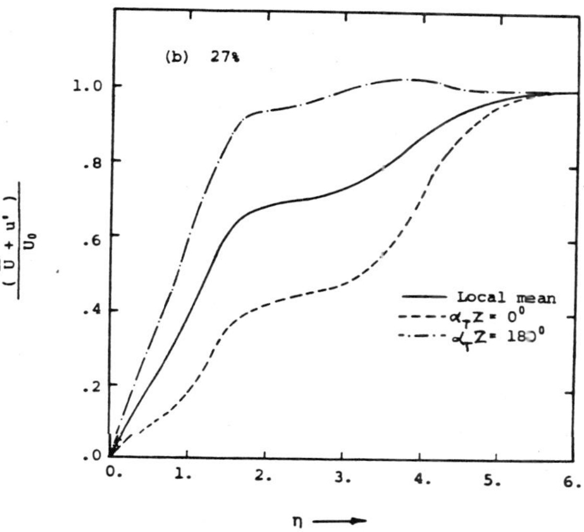

Fig. 9 The instantaneous velocity distribution prior to the turbulent burst.

Secondary Flows in Boundary Layers and Jets

N. A. ZHELTUKHIN, N. M. TEREKHOVA

Institute of Theoretical and Applied Mechanics
Siberian Branch of the USSR Academy of Sciences
Novosibirsk 630090, USSR

In a gas or fluid flow having lost its stability parallel with quickly changing components of the motion (disturbances) there are also developing relatively long-living large-scale vortex formations - coherent structures that are characterized by shape, dimensions and intensity. The orientation of the vortex axis to a main flow appears to be a characteristic of paramount importance of such formations. Thus, if in some region of the flow the coordinates x, y and z are chosen in such a way that the flow has a shape of a simple gradient flow, then it is natural to call the vortices directed along these axes longitudinal, transverse and **lateral** . The interaction of such vortices with a main flow is different. A transverse vortex becomes strongly deformed by a main flow and has an evident tendency to a fast transformation into an inclined one close to longitudinal one. It is difficult to observe it and complicated to compute. The longitudinal and **lateral** vortices become deformed by a main flow relatively slow, they are successfully recorded by the up-to-date means of diagnostics and in the theoretical developments the mechanisms of their formation are studied. It has been elucidated that they may arise and develop due to the nonlinear interaction of different kind fluctuations and wave motions taking place in the unstable flow. The carried out estimates and calculations for a number of characteristic flows (in channels, boundary layers and jet flows) have shown a reasonable nature and quantitative confirmation of the mechanisms suggested.

At present an important question about an opposite influence of the considered large-scale secondary flows (coherent structures) on a main flow and its characteristics remains scantly explored. The present study concerns with this question. We have shown how and to what extent the longitudinal vortex formations influence the velocity diagram of a main flow in the boundary layer, in subsonic and supersonic jets.

I. Boundary Layer

Consider the secondary regimes arising in the boundary layer of a viscous incompressible fluid with the velocity profile

$$\Phi_{\eta\eta\eta} + \Phi_{\eta\eta}\Phi = 0\,, \quad \eta = (\overline{U}_0/2\vartheta x)^{1/2}, \quad U_0 = \Phi_\eta$$

In a region of the nonlinear development of disturbances, as is now well-known, there arises an additive eigenwavefield in the form of the Tollmien-Schlichting waves triplet [1, 2]

$$[u', v'](x,y,z,t) = 2æ_t [u,v]_t (y) e^{\Omega_1} \cos\beta z + æ_g [u,v]_g (y) e^{\Omega_2} \quad (1)$$
$$w'(x,y,z,t) = 2æ_t i\, w_t(y) e^{\Omega_1} \sin\beta z$$

Here the subscript "t" denotes the propagating oblique (at the angle θ = arg tg β/α to the axis x) waves - "crossing", the total action of which is equivalent to that of a standing wave, the subscript "g" denotes a plane two-dimensional wave with amplitudes $æ_t$ and $æ_g$ and phases $\Omega_{1,2} = i\alpha_{1,2}(x - Ct)$. The availability of such a kind eigendisturbances leads to a qualitative change of the flow structure. In the transversal direction in the flow there are revealed the minima of mean velocity in places of the wave peaks and the maxima in places of the pits. There arise the longitudinal large-scale (Benney-Lin) vortices. This secondary quasistationary regime $U(\eta,z)$, $V(\eta,z)$, $W(\eta,z)$ is described by Reynolds system

$$VU_\eta + WU_z - \frac{1}{R}(U_{\eta\eta} + U_{zz}) = -\langle u'v'\rangle_\eta - \langle u'w'\rangle_z - U_{0\eta\eta} \quad (2a)$$
$$VV_\eta + WV_z - \frac{1}{R}(V_{\eta\eta} + V_{zz}) + P_\eta = -\langle v'^2\rangle_\eta - \langle v'w'\rangle_z$$
$$VW_\eta + WW_z - \frac{1}{R}(W_{\eta\eta} + W_{zz}) + P_z = -\langle v'w'\rangle_\eta - \langle w'^2\rangle_z \quad (2b)$$
$$V_\eta + W_z = 0$$

and satisfies the boundary no-slip and no-leakage conditions on the wall as well as the boundedness on the external boundary and periodicity with respect to the variable z.
The closure of (2) is carried out within the framework of mono-harmonic approximation [2] for waves (1). In the right-hand sides of Eq.(2) there are Reynolds stresses, second order momenta by which the energy redistribution of the main flow and disturbances occurs. A system (2) allows the partition and it is possible to find the longitudinal vortices with components V, W separately from the epures of U. A linearized problem (2b) is considered as the first approximation and it is possible to confine oneself to this problem if the vorticity $\omega^x = V_z - W_\eta$ is small. In this case ($V=\psi_z, W=-\psi_\eta$) the equation

$$\frac{1}{R}\left[\partial^2/\partial\eta^2 - \partial^2/\partial z^2\right]^2\psi = F, \quad F = -\varkappa_t^2 \mathcal{F}_t(\eta)\sin 2\beta z - \varkappa_t \varkappa_g \mathcal{F}_{gt}(\eta)\sin\beta z \qquad (3)$$

describes the vortex in which the generation action of waves (1) is balanced by viscosity. Due to the linearity of Eq.(3) the problem reduces to a consecutive determination of the induced by crossing vortex ψ_t and of a vortex conditioned by the nonlinear engagement of two- and three-dimensional waves ψ_{gt}. Figure 1 illustrates some typical vortex configurations for different relationships of $\varkappa_t$ and $\varkappa_g$. General regularities are following: the growth of $\theta, R, \varkappa_t$ causes the increase in the vorticity ω^x, moreover, the growth of θ contributes to the vortex localization in the near-wall region. A more complex pixture of the formation of a low-intensive additional near-wall vortex with contrawise rotation (Fig. 2) can be observed for some range of $\varkappa_g/\varkappa_t$ (f.e. for R_c and $\theta \sim \pi/6$ such a regime is realized at $10 \leqslant \varkappa_g/\varkappa_t \leqslant 20$).
The regard for convective terms in the transfer equations derived by integrating a complete system (2b) with the use of the finite-difference method [3] leads to additional deformation of coherent structure with respect to η, z and to decrease in the vortex intensity, that is clearly depicted in Fig. 3, obtained for the flow regime R_c, $\theta \sim \pi/6$, $\varkappa_t \gg \varkappa_g$, $\varkappa_t = 0.02$. The distortion of epures of the longitudinal velocities U answering the mass and impulse redistribution in the regime of secondary vortex motion corresponds to the following physical process: in regions where the vortices carry

away the fluid mass from the near-wall low-velocity layers to a region of external field, there form the local thinly filled distributions of U , on the contrary, in places of its contribution from a potential core the filling of the near-wall sections increases. Figure 4 illustrates such a redistribution derived from (2a) for the regime R_c , $\varkappa_t$ = 0.02. Such a process may lead to the following: the thinly filled epures of $U(\eta, z)$ in the near-wall region obtain a bend point that may be a reason of a local growth of oscillations in these places of the space and lead to the secondary instability of the flow. The study of the main characteristics of the total flow $\langle U(\eta)\rangle = \frac{1}{T}\int_0^T U(\eta, z)\,dz$ shows that for the secondary flow there arises the filling of the profile in the near-wall region where the gradients sharply change and, hence, the connected with them important flow characteristics (Fig. 5). The study of a temporal instability spectrum for Orr-Sommerfeld equation is indicative of the excitation of the third spectral mode and of the possibility of the unstable state appearance with the phase velosity of an order of the flow velocity.

2. Subsonic Laminar Jet

The flow uniform with respect to the coordinate z which in the plane x, y is approximated by the universal dependency U_0 = $1-\mathrm{th}^2\xi$ is known in hydrodynamics as a plane laminar jet. Such a flow belongs to a class of inviscidly unstable. In view of the profile symmetry with respect to the self-similar transverse coordinate $\xi = \xi(x,y)$ it is possible to carry out a separate study of antisymmetric ($R_c \sim 4$) and symmetric ($R_c \sim 88$) disturbances. The generation of vorticity is considered within the framework of linear approximation ("viscous" vortex). It turned out that the vortex shape strongly depends on the Reynolds numbers and angles θ . Thus, for antisymmetric disturbances (Fig. 6) with the growth of R at first there occurs the localization of vortices in the flow region and then the origin in the near-axis region of new low-intensive ones with counterrotation relative to the initial one. A general vortex intensity is extremely small. The vortex structures generated by symmetrical modes are shown in Fig. 7 for R = 88.29 and different θ . With the growth of θ a solitary induced vortex (a) at first displaces to a near-axis region (b) and then disappears completely (c), and in the flow field there occurs the confluence of two vortices originating in the external region.

A deformation of a total profile is characterized by the mass transfer from a near-axis region to a region of the external field (Fig. 8).

It has proved to be that the three-dimensionality of disturbances may play a definite role only at conditions of moderately high Reynolds numbers ($R \sim 40$), and in the region of the stability loss the main peculiarities of the disturbed flow are described by a plane interaction that in essence distinguishes such a jet flow of an incompressible viscous fluid from the boundary layer on a solid surface.

3. Supersonic Gas Jet

A round turbulent jet of an inviscid compressible fluid in an initial region of its development is considered. A jet velocity profile has only a longitudinal component U and is approximated by the universal Schlichting formula $U_0(\tau)=1-(1-\eta^{3/2})^2$, $\eta=\eta(\tau)$. A large-scale wave structure, a cylindrical Tollmien-Schlichting wave, is superimposed on the equilibrium flow field at the moment $t = 0$, the modes $n = 0$ being axisymmetrical and $n = \pm 1, \pm 2$ being helical oscillations. Both a solitary wave and a pair of waves exciting the oscillations of a flapping type, i.e. the superposition of right- and lefttwisted helical oscillations are considered. The disturbances of the mentioned type induce in the flow the Reynolds stresses (solitary waves - homogeneous ones, pair waves - periodical ones with respect to the angle φ) of very complex shape on the whole, and the number of their components increases up to 13.

The evolution of the mean flow is studied on the basis of numerical integration of averaged equations of motion (Euler equations) for an inviscid compressible non-heat-conducting gas (an explicit second-order MacCormack scheme). We suppose that the secondary flow is independent of the longitudinal coordinate z . It turned out that the oscillations of the flapping type cause the additional radial-tangential transfer of the gas mass which may be interpreted as existence of infinite with respect to the flow secondary vortices with the axis along the jet propagation direction. Figure 9 demonstrates the transverse cross-sections of such vortices at the moment t = 0.03 for n = ±1 and n = ±2. The increase in the spectral number of the wave mode increases the symmetry order. The presence of such secondary flow naturally leads to the change of the flow characteristics. A change of the longitudinal velocity U is the most demonstrative. It may be both synchronized and counterphase (Fig. 10) and lead to deformations of the jet flow field. The action of solitary waves on the flow is homogeneous with respect to the angle φ and by two orders less than the action of a wave pair. A similar example for n = 0,1,2 is shown in Fig. 11.

All the above-stated testifies to the significance of the mechanism of mixing of a main flow by secondary flows by the nonlinear interaction of Tollmien-Schlichting waves. A general scheme of interaction of disturbances and averaged flow is thought to be the following: the Tollmien-Schlichting waves take away the energy from a main flow (generate the secondary flows, in particular, longitudinal vortices). Longitudinal vortex flows deform the velocity diagram of the main flow, thus creating the main features inherent in the velocity diagrams of pre-turbulent flows.

References

1. Craik, A.D.D.: Nonlinear resonant instability in boundary layers. - J. Fluid Mech., 1971, vol.50, pt.2.

2. Goldshtik, M.A.; Shtern, V.N.: Hydrodynamic Stability and Turbulence. Novosibirsk, Nauka, 1977 (in Russian).

3. Dzhakupov, K.B.: On some numerical methods for solving Navier-Stokes equations of a viscous incompressible fluid. In: Proc. All-Union Seminar on Numer. Meths of Viscous Fluid. Novosibirsk, Nauka, 1969 (in Russian).

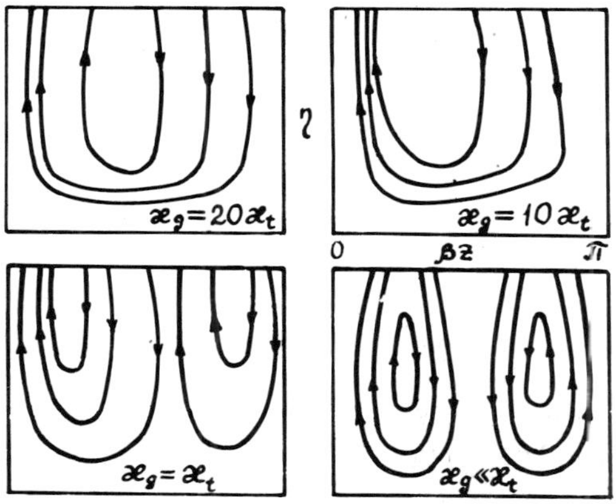

Fig. 1. Vortex configurations for different relationships of $\varkappa_g$ and $\varkappa_t$

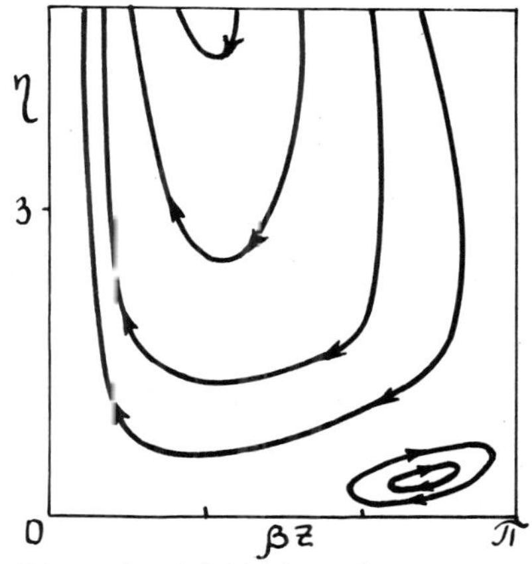

Fig. 2. Additional near-wall vortex (R_c, $\theta \sim \pi/6$, $\frac{\varkappa_g}{\varkappa_t}=12$)

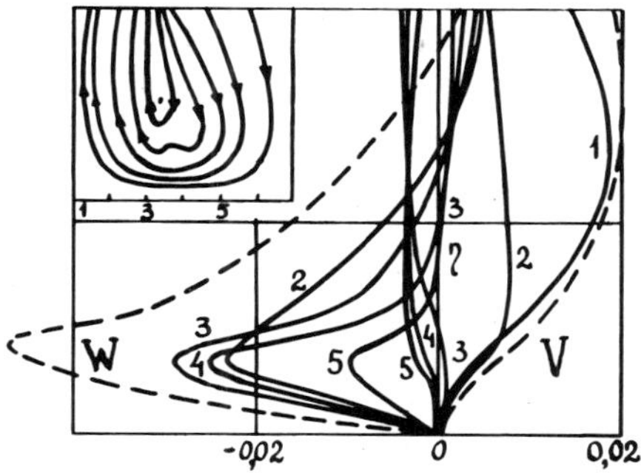

Fig. 3. Distributions V, W derived by integrating (2b), – – – V_{max}, W_{max} from (3)

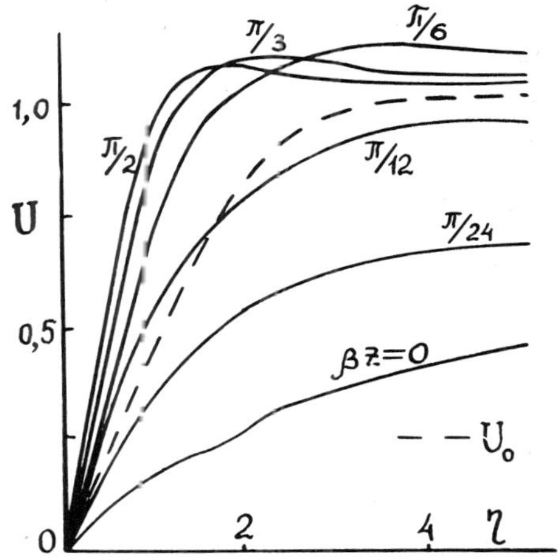

Fig. 4. Epures of longitudinal velocity $U(\eta, z)$

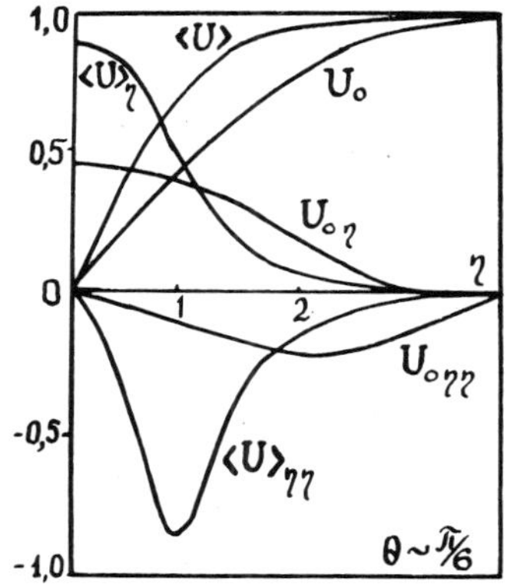

Fig. 5. Main characteristics of the total flow (R_c, $\varkappa_t = 0,02$)

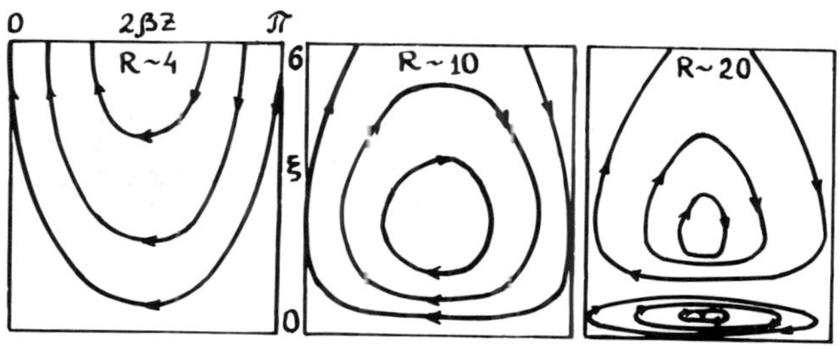

Fig. 6. Vortex configurations for antisymmetric disturbances ($\theta \sim \pi/5$, $\varkappa_t = 0,05$)

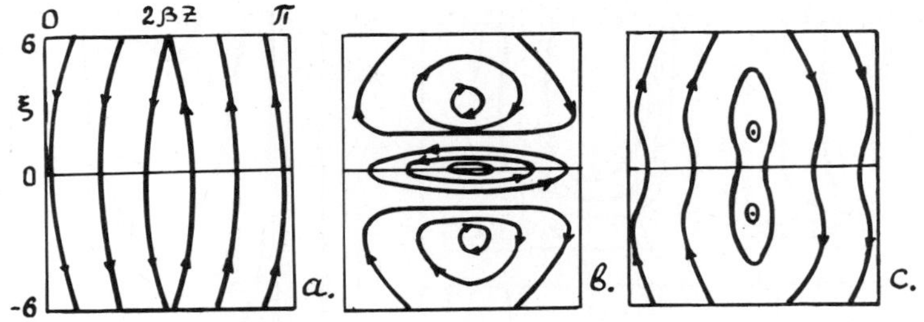

Fig. 7. Vortex configurations for symmetric disturbances $\varkappa_t$=0,02 (a - θ~π/15, b - ~π/6, c - ~ π/4)

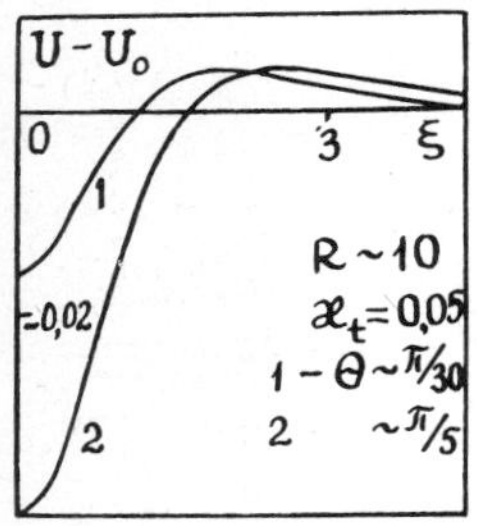

Fig.8. Derormation of a total profile for a-s disturbances

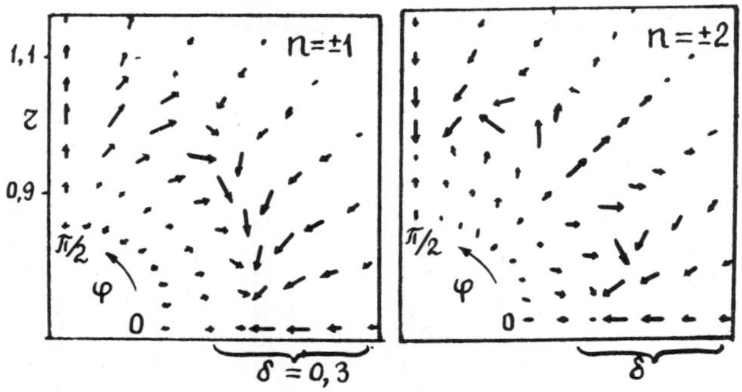

Fig.9. Vortex configurations for n=±1,±2 (M=1,5 , St =0,25)

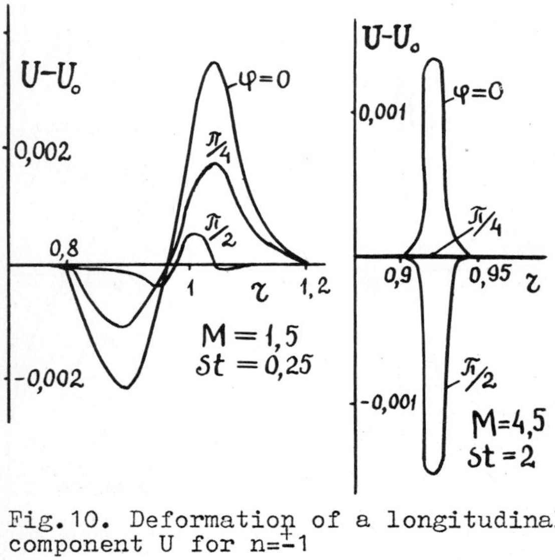

Fig.10. Deformation of a longitudinal component U for n=±1

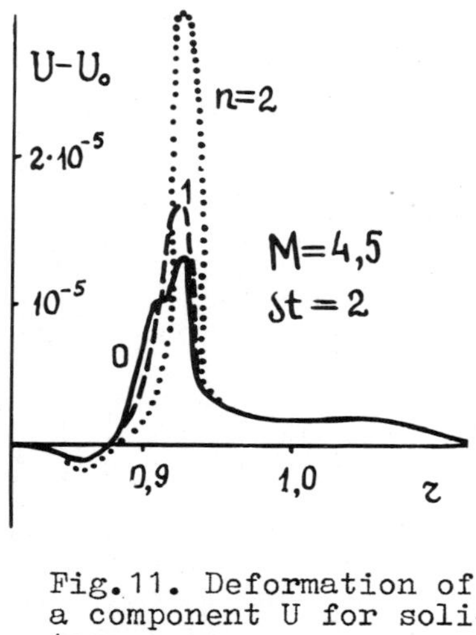

Fig.11. Deformation of a component U for solitary waves

On Nature of K-Breakdown of a Laminar Boundary Layer. New Experimental Data

YU.S.KACHANOV, V.V.KOZLOV, V.YA.LEVCHENKO, M.P.RAMAZANOV

Institute of Theoretical and Applied Mechanics,
Siberian Division of the USSR Academy of Sciences,
Novosibirsk, USSR

Introduction

The detailed analysis of nonlinear breakdown of a laminar boundary layer had been carried out in the fundamental experimental work [1] . The results of this work (a strong amplification of three-dimensionality, a sudden appearance of "spikes" and high frequency oscillations etc.) are discussed and analyzed with the unremitting interest. The work [1] remains a base for theoretical models up to the present time. For a long time the general opinion was that the discovered in [1] sequence of nonlinear processes leading to breakdown of a laminar regime has been the inherent property of the boundary layer flow. However, a new type of a laminar-turbulent transition in a boundary layer which differs from classical K-breakdown [1] in principle, had been discovered in 1976, [2] . The main features of this type of transition were the generation of the subharmonic oscillation and a packet of low frequency disturbances with subsequent nonlinear interaction of different modes of disturbances. This subharmonic type of transition was studied in subsequent experiments [3]-[6] . The theory of subharmonic activity of distrubances is developed in [7]-[8] . It was established in [3] that the subharmonic resonant three-wave interaction is the main mechanism leading to this way of transition. Possible causes of the realization of either type of transition are also discussed in [3] . A hypothesis on a competition of resonant mechanisms was formed. However, the role of resonant processes in K-regime is not clear because a dearth of spectral analysis is present in [1] and in subsequent studies of K-regime. This was the reason to carry out the present experiments. Full results of this study are given in [9] .

Procedures. The Choice of a Regime. Initial Disturbances

The experiments were conducted in the ITAM low-turbulence wind tunnel at freestream speed 9.18 m/s and turbulence level less than 0.02%. The boundary layer of a flat plate was studied at practically zero pressure gradient (see [2]-[3]). Sinusoidal disturbances (Tollmien-Schlichting waves type) were introduced into the boundary layer by a vibrating ribbon located 250 mm downstream of the leading edge. The spacers 0.045 mm thick in the form of a square with the period along the span coordinate Z λ_1 = 25 mm ($2\pi/\lambda_1=\beta_1$, $\hat{Z}=\beta_1 Z$) were placed on the plate surface under the vibrating ribbon.

The signals were registered using a DISA 55 D01 hot-wire anemometer with the linearizer. The experiments were conducted mainly in the regime of the fundamental wave frequency f_1=96.4 Hz (the principle regime), downstream distances being between X= =300 mm and X=500 mm. These values corresponded to a frequency parameter $F = 2\pi f_1 \nu / U_\infty^2 = (110 \pm 2)\cdot 10^{-6}$ and Reynolds number $R = (U_\infty X/\nu)^{1/2}$ between R = 422±5 and R = 545±6.

The K-regime of breakdown appears at sufficiently large amplitudes of a ribbon vibration. All stages of transition at K-regime are displaced while changing initial amplitudes of fluctuations. When "spikes" appeared ar X=430-450 mm, this regime was chosen as the principle one. Such a choice permitted to obtain the breakdown processes from almost two-dimensional harmonic wave up to randomization of the flow (at X $\gtrsim$600 mm).

The spectrum of initial (X=300 mm) disturbances (see [9]) consisted only of the fundamental frequency f_1 and its higher harmonics $f_n = nf_1$. Amplitudes (maximum in the y-profiles) of these oscillations were $A_1 \approx 3\%$, $A_2 \approx 0.35\%$, $A_3 \approx 0.075\%$, $A_4 \approx$ 0.023%, $A_5 \approx 0.0070\%$, $A_6 \approx 0.0021\%$. Profiles of amplitudes and phases of disturbances given in Fig. 1a change slowly in form in Z direction under the influence of spacers. Profiles for the fundamental wave are typical T-S (linear) wave profiles.*

---

* The symbols ○ , △ , □ , ● , ▲ , ■ , e , ◊ and × in Fig. 1 and further correspond to harmonics $f_1, \ldots f_6, f_{10}$, integral intensity of disturbances and mean flow velocity.

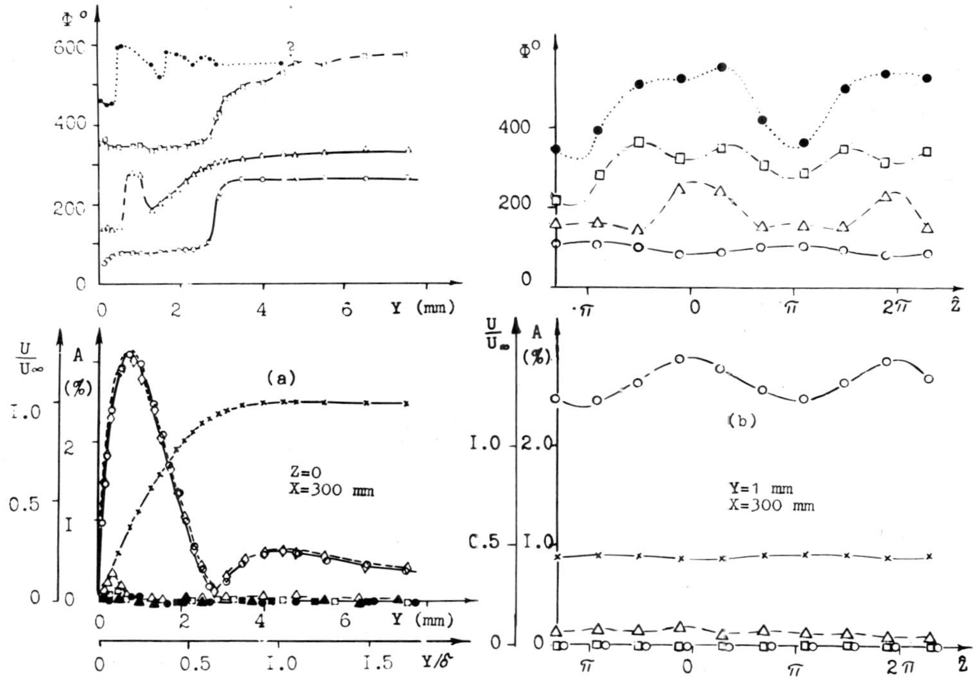

Fig.1. Initial distributions for amplitudes and phases of disturbances and for mean velocity

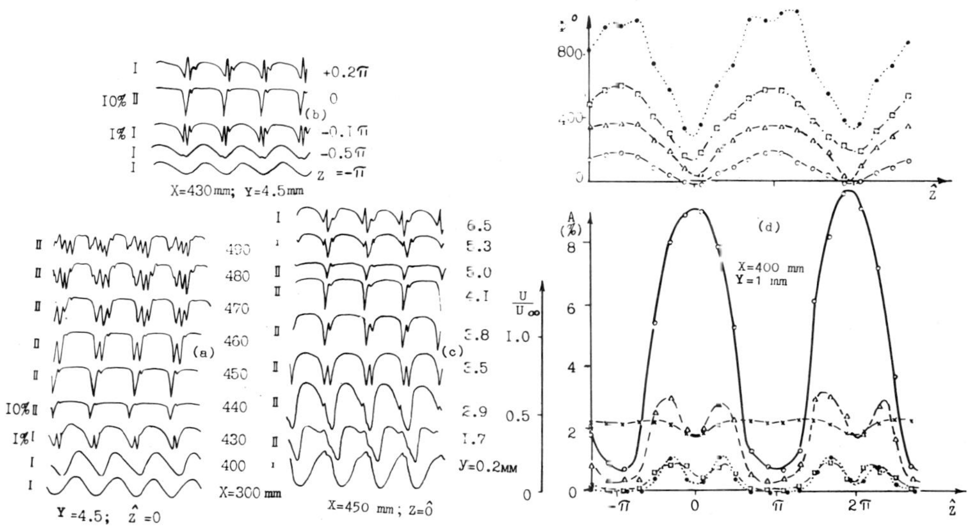

Fig.2. Oscilloscope traces (a,b,c). Growth of Z-modulation (d)

The mean velocity profiles correspond to those of Blasius. An initial z-modulation of waves and mean flow (seen in Fig. 1b) is conditioned mainly by a modulation of a local Reynolds number (boundary layer thickness).

Evolution of Oscilloscope Traces and Spectra. Growth of Three-Dimensionality

Beginning from X = 400 mm, the typical "spikes" appear suddenly at y = 4.5 mm, $\hat{Z} = 0$ ("peak") which double, treble etc.(Fig.2a). The "spikes" are localized in "y" and "z". The region of their existence is $\Delta Z \approx 1\text{-}2$ mm, e.g. $\Delta\hat{Z} \approx 15\text{-}30°$. "Spikes" have very large amplitudes (see also [9]) which are maximum just after their appearance. The amplitude of oscillations is 43% at X = 430 mm, y = 3.7 mm. Up to this region (up to X = 450 mm) the fluctuations remain strictly deterministic and periodic with the period $T_1 = 1/f_1$ at all the points!

The inherent property of K-breakdown is a quick downstream growth of initial (Fig. 1b) z-modulation of disturbances. The distributions in Fig. 2b demonstrate this fact and show the appearance of disturbances which oscillate in z with the period $n\lambda_1$ (dividing up of fluctuating vortices). A modulation of mean velocity corresponds to stationary waves which divide up also. The depth of z-modulation of disturbances at y = 4.5 mm is small (Fig.3a) and it grows slowly initially. Then sharp "spikes" with the period λ_1 in the z-distributions of harmonic amplitudes appear by jump at the place where "spikes" in

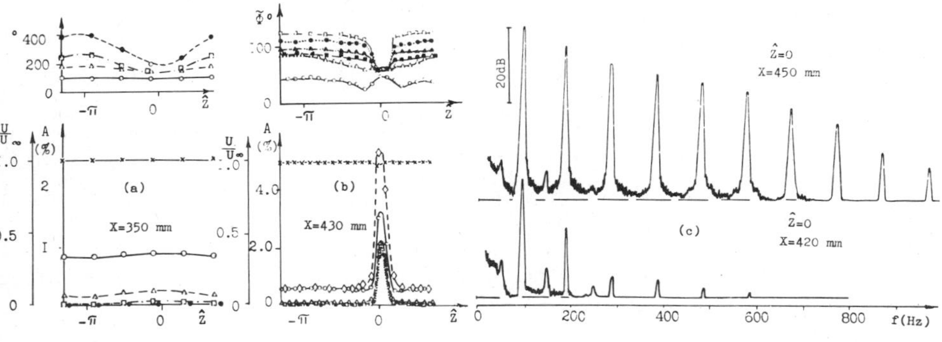

Fig.3. Appearance of "spikes" in (a,b) space and in spectra (c); y = 4.5 mm.

oscillograms appear (X=430 mm, Fig. 3b). In this place the phases of the harmonics $\tilde{\Phi}$ are synchronized within $\Delta\tilde{\Phi} = \pm 1°$ (excluding the fundamental wave), the phase being determined with the error $\pm 2°$). The mean velocity in the region where time- and space "spikes" exist, is essentially less than that in the other z-sections at the same distances from the wall (Fig. 3b).

The spectra of disturbances are a set of harmonics nf_1 both in the initial section and in the whole region of measurements up to X=450 mm (Fig. 3c). They testify to the almost strict time-periodicity of the process. The appearance of "spikes" in oscilloscope traces (Fig. 2a) at "peaks" ($\hat{Z}$=0) corresponds to sharp increasing of amplitudes of higher harmonics (Fig. 3c). This means that the process of the generation of "spikes" is essentially the process of the generation of three-dimensional higher harmonics of the fundamental wave. It is no manifestation of a high-frequency secondary instability, as it was treated in [1] and subsequent papers (see e.g. [10]). Therefore, from the spectral point of view the study of K-regime of transition at this stage ($X \leqslant 450$ mm) reduces basically to the study of harmonics $nf_1 (n = 0, 1, \ldots)$.

Profiles of Disturbances. Growth of Amplitudes and Phases

The evolution of y-profiles of mean velocity, amplitudes and phases of harmonics leads to the given in Fig. 4a* distributions in the region of formed "spikes". The fundamental wave reaches the amplitudes of $\sim 15\%$ here while a strict determinism (and periodicity) of the process is maintained! The phase profiles in the "spike" region ($y/\delta \approx 0.8 - 1.05$) demonstrate the synchronization of phases of harmonics (up to the 10th) within $\Delta\tilde{\Phi} = \pm 2.5°$, i.e. within the measurement error. The region of phase synchronization and of the existence of the "spike" moves off the wall when X increases (e.g. it places in $y \approx 0.7-1.0$ mm at X = 430 mm), the accuracy of phase synchronization increases thereat ($\Delta\tilde{\Phi} = \pm 7°$ at X = 430 mm). The mean velocity profiles

* The segments with one, two, three lines in Fig. 4a and further indicate approximately the regions of the existence of one, two, three "spikes".

have inflection points, the upper of which lies slightly lower than the "spike" region. It can be noted that the "spike" region is placed always on the outer steep slope of distributions of all higher harmonics.

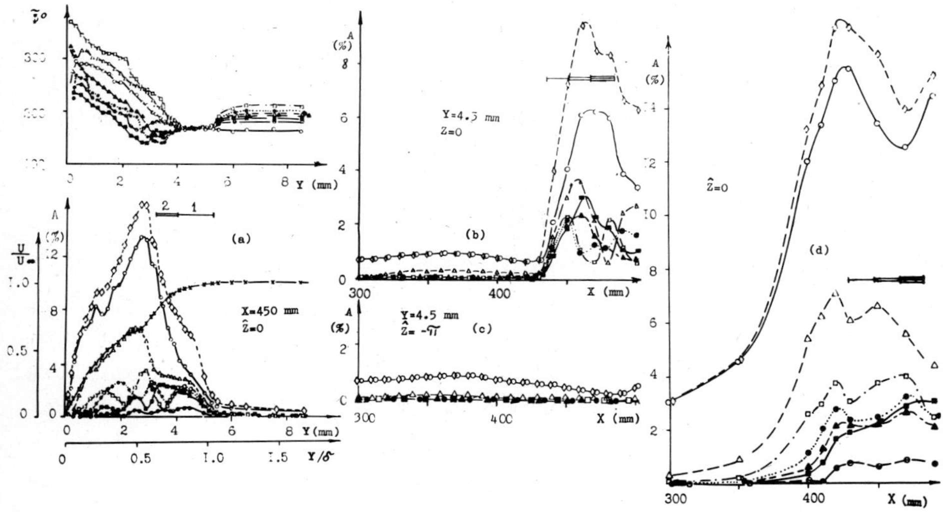

Fig.4. y-profiles of disturbances and mean velocity (a). Growth curves of harmonic amplitudes (b,c,d).

The amplification of harmonic amplitudes along the line y = const = 4.5 mm has a strong splash in the "peak" ($\hat{Z}$ = 0) in the region of formation of "spike" when harmonic amplitudes decrease in "valley" and other regions of the flow (Fig.4b,c). The jump of intensity of all harmonics by an order of magnitude and more resembles a burst and makes one think right away of the beginning of a **violent** catastrophic breakdown of the flow [1]. However, the growth curves of maxima of harmonics amplitudes (Fig. 4d) show their monotonous, strictly hierarchical amplification up to the formation of a "spike" ($300 \leqslant X \leqslant 420$ mm). On the contrary, the amplitudes of harmonics forming the "spike" remain constant or decrease in the place of the "spike" formation and downstream! There is nothing resembling a burst or a violent splash of high-frequency fluctuations. The jump in Fig. 4b at y = 4.5 mm is explained by the observed displacement of maxima in the amplitude profiles away from the wall. It is on the outer steep slope of the profiles where the "spike" appears (see Fig. 4a).

The curves in Fig. 5a demonstrate clearly the harmonic phase synchronization in the "peak" in the region of the "spike" formation.

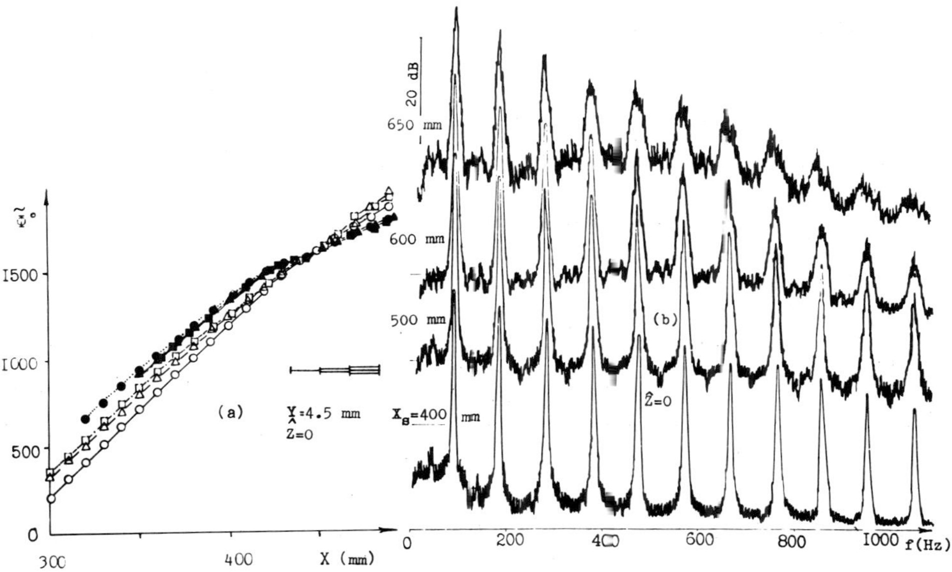

Fig.5. Growth of phases (a). Randomization of "spike" spectra (b).

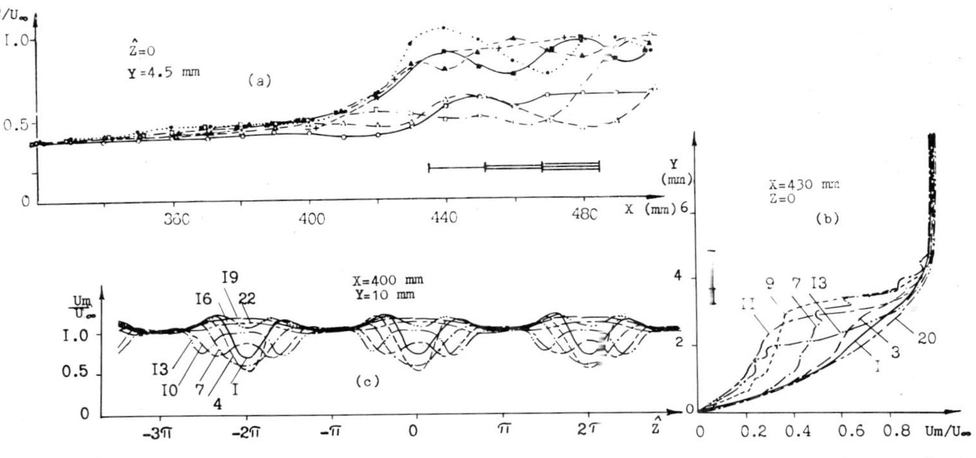

Fig.6. Phase speeds of harmonics and group velocity of "spike" (crosses) (a). Instantaneous velocity profiles (b,c)

The phase speeds of harmonics (up to the 10th) increase quickly at $X > 400$ mm (Fig. 6a) from $C/U_\infty \approx 0.4$ up to $C/U_\infty \approx 0.8 - 0.9$ (besides f_2, f_3). The phase speed of the fundamental wave (and harmonics f_2, f_3) remains in the range $C/U_\infty \approx 0.5$-0.6. The group velocity of the "spikes" increases together with the phase speeds of higher harmonics and essentially passes ahead the speed of the fundamental wave at the stage before the "spike" formation (Fig. 6a). This fact contradicts the well-known conception (and results of calculations) of Landahl [11] (for more details of analysis see [9]). The oscillations of the instantaneous y-profiles of velocities in the "peak" determined for 24 moments of the period of the fundamental wave, grow from the smooth quasi-sinusoidal ones at X = 300 mm (see [9]) to the splashing out to the region of $y/\delta \approx 0.6$-0.7 which is typical of the region of the "spike". Many authors relate the formation of "spikes" to a local high-frequency secondary instability of the flow in the layer of a strong shear with the inflection point. However, the profiles given in Fig. 6b simply represent a section (from the point of view $U_i(y)$ at t = const) of the array of the experimental data, the oscilloscope traces with "spikes" (Fig. 2a) being another section of this ($U_i(t)$ at y = const) array. Therefore, the consideration of the instantaneous profiles in Fig. 6b as the reason of the secondary instability has nothing to do with the reasons of the formation of "spikes".

The instantaneous z-profiles at y = 1 mm up to the region of formed "spikes" (to X = 450 mm) have wavy oscillations (Fig.6c). The waves appear at "peaks" (as if their sources were placed here) and propagate in the direction of the valleys. After this they interfere cancelling each other, and the cycle is repeated again.

On Randomization of "Spikes". Modulation and Intermittency

In previous studies, beginning with [1] , the "spike" had been identified with random splashes of high-frequency fluctuations, leading to the appearance of the intermittency and bounded up with the beginning of the breakdown of the deterministic oscillatory regime. What was the reason for such an opinion?

As it was mentioned, a movement of all stages of the breakdown in X-direction took place at a small change of the vibrating ribbon amplitude. Other changes were observed too. The gradual increase in non-periodicity of the process was observed at the decrease in the initial amplitude of oscillations and increase in the coordinate of the "spike" formation X_s. The amplitude (and phase) of the "spikes" begins to oscillate (see [9]). An ever-growing share of components undivisible with f_1 appears in spectra of the oscilloscope traces with "spikes" (Fig. 5b). The harmonics with large numbers n are broadened and spreaded on the spectrum. The coherency of oscillations on frequencies $f_n = nf_1$ disappears. Thus, all the 15 harmonics studied are coherent at X_s = 400 mm, only 10 at X_s = 550 mm and 5 at X_s = 650 mm (at 4 Hz bandwidth). The comparison of the results of this study with the ones of [1] (see [9]) leads to the conclusion that the experiments in [1] were carried out under less controlled conditions (larger velocities, distances and Reynolds numbers, smaller initial amplitudes, etc.). The results of this study show that the amplitudes of harmonics (and the "spike") in the region of "spike" formation are changing 2-3 times at displacement in z only by $\sim$1.2 mm (5% of period λ_1), in y by $\sim$0.2 mm (4% of δ), in X by $\sim$10 mm (1% of the distance from the leading edge), in amplitude of the current through the ribbon by $\sim$ 0.05 v (1.5% of amplitude). Analogous gradients took place at small variations of free-stream velocity too. The existence of such enormous gradients at less controlled conditions leads to (or only makes an impression of) a smaller periodicity and a larger stochasticity of the process.

The resonant mechanisms for amplification of non-controlled random initial oscillations like subharmonic resonances [7,8] may be another reason of unperiodicity and stochasticity of the process at the "spike" stage at small initial amplitudes and large X. (For the discussion of this question see [3] , [6], [9]). The available data corroborate the correctness of this hypothesis.

The experiments carried out with periodical time-modulation of oscillations of the ribbon (by means of the superposition of two frequencies f_{11} = 96.4 Hz and f_{12} = 81.4 Hz) showed that at X = 450 mm an interchange of parts of oscilloscope traces without "spikes", with one, one and half, two "spikes" takes place. It has a quasi-stationary character and corresponds to the consecutive replacement of stationary conditions which were studied separately in [9] . The subharmonic three-wave resonances have analogical properties [3] . Such a similarity of the properties serve as a telling argument in favour of the hypothesis [3] that "spikes" arise as a result of a set of resonances between harmonics of the fundamental wave and oblique ones. The clear intermittency with the period of beats $\Delta f = f_{11} - f_{12}$ was observed downstream (at X = 520 mm) like in [12] but in the conditions of K-regime.

Frequency-Wavelength Analysis of Oscillations

The evolution of amplitude frequency-wavelength spectra of harmonics (periodogramms) shows the quick amplification of three-dimensional components (Fig. 7). The numbers of the most amplified harmonics (m) with wavenumbers $\beta_m = \pm m\beta_1$, their amplification factors $\varkappa$ * and corresponding coefficients $\varkappa_0$ for plane waves (β = 0) are given in the Table.

Stage		f_1			f_2			f_3		
		m	$\varkappa$	$\varkappa_0$	m	$\varkappa$	$\varkappa_0$	m	$\varkappa$	$\varkappa_0$
1	X=300-350 mm	1-2	2.5	1.20	1-2	3.5	2.28	2-3	15	1.40
2	X=350-400 mm	3-4	23	1.02	3-5	26	1.37	5-6	33	5.36
3	X=400-450 mm	4-5	14	0.85	5-7	4.5	2.37	7-9	11	1.49

When X (and n as a rule) increases, the more three-dimensional harmonics amplify (m increases). Their amplification factors $\varkappa$ can be more than those for plane waves ($\varkappa_0$) by an order of magnitude or more at the second stage. The initial periodicity β_1 (m = 1) given by spacers doesn't manifest itself in the dynamics of three-dimensionality development. This question

* $\varkappa$ is the ratio of frequency-wavelength spectra at two neighbour sections.

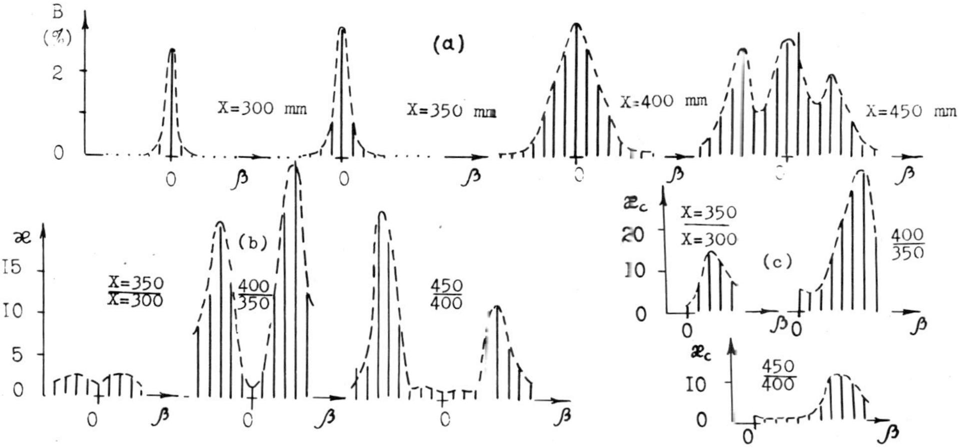

Fig.7. Frequency-wavelength spectra (a) and their ratios (b) for f_1. Ratios of symmetrized spectra for f_3 (c).

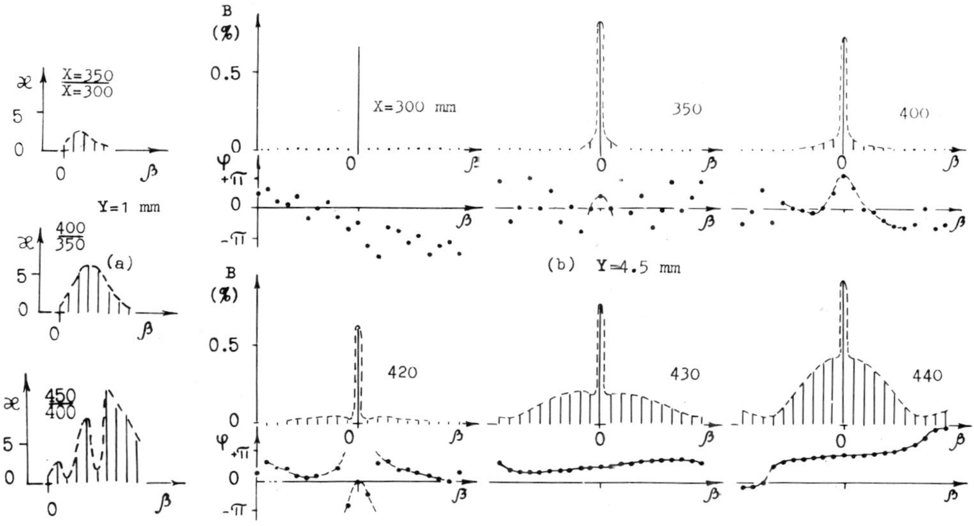

Fig.8. Ratios of frequency-wavelength spectra for mean velocity (a) and complex spectra for f_1 (b)

is discussed in detail in [9] . The wavelength spectra for the mean flow are developing like the fluctuation spectra, although the plane component ($\beta = 0$) is prevailing over others. The most amplified components at the first two stages (m = 1-2 and m =

2-4) almost coincide with the corresponding values for the first harmonic in the Table (Fig. 8a).

The complex spectra for the harmonic f_1 at y = 4.5 mm demonstrate the synchronization of the phases of frequency-wavelength harmonics in the region of the "spike" formation (Fig. 8b). The more high-frequency harmonics have the analogical spectra. The whole complex of results on frequency - and frequency wavelength analysis of oscillations showed that <u>the development of disturbances at the whole stage of formation of three-dimensionality up to formation of "spikes" is (from the spectral point of view) the process of formation of a deterministic system, peculiar "lattice", of the space-time harmonics with the steps f_1 on frequency and β_1 on z-component of the wavenumber. This "lattice" contains almost the whole energy of fluctuations and mean flow</u>.

Conclusion

The process of the formation of "spikes" in K-breakdown is not connected with a local high-frequency secondary instability of the flow and has the nature of burst only on the face of it. It takes place as a result of the strictly deterministic, gradual, hierarchical amplification of higher harmonics of the fundamental wave with the synchronization of their phases. Simultaneously, the rapid growth and phase synchronization of three-dimensional frequency-wavelength harmonics with definite wavenumbers β take place, their amplification rates being dozens of times as much as the ones for plane waves. As a result, the strictly deterministic system (lattice) of space-time harmonics which is synchronized in phase, is formed to the moment of the "spike" formation. The obtained results are in a good line with the hypothesis [3] that the main difference of K-regime of transition from the one discovered in [2] lies in the fact that the resonances of deterministic waves nf_1 with waves $2nf_1$ forestall the subharmonic resonance. Because of the discovered strict determinism of "spikes" and the process of their generation the question on the ways of randomization of the flow in K-regime proves to be open.

References

1. Klebanoff, P.S.; Tidstrom, K.D.; Sargent, L.M.: J. Fluid Mech. 12 (1982) 1-41.

2. Kachanov, Yu.S.; Kozlov, V.V. & Levchenko, V.Ya. Izv. Acad. Nauk SSSR, Mekh. Zhidk. i Gaza 3 (1977) 49-58 (in Russian) (see also Fluid Dyn. 12 (1978) 383).

3. Kachanov, Yu.S. & Levchenko, V.Ya. Preprint N°10-82, Novosibirsk: ITPM SO AN SSSR, 1982 (in Russian) (see also J. Fluid Mech. 138 (1984) 209-247.

4. Thomas, A.S.W. & Saric, W.S. Bull. Am. Phys. Soc. 26 (1981) 1252.

5. Saric, W.S.; Carter, J.D. & Reynolds, G.A. Bull. Am. Phys. Soc. 26 (1981) 1252.

6. Kozlov, V.V.; Levchenko, V.Ya. & Saric, W.S. Preprint N° 10-83, Novosibirsk: ITPM SO AN SSSR, 1983 (in Russian).

7. Craik, A.D. J.Fluid Mech. 50 (1971) 393-413.

8. Herbert, T. AIAA-Paper 84-0009 (1984).

9. Kachanov, Yu.S.; Kozlov, V.V.; Levchenko, V.Ya. & Ramazanov, M.P. Preprint N°9-84, Novosibirsk:ITPM SO AN SSSR, 1984 (in Russian).

10. Greenspan, H.P. & Benney, D.J. Fluid Mech. 15 (1963) 133-153.

11. Landahl, M.T. J.Fluid Mech. 56 (1972) 775-802.

12. Kachanov, Yu.S.; Kozlov, V.V. & Levchenko, V.Ya. Preprint N°16, Novosibirsk: ITPM SO AN SSSR, 1978 (in Russian) (see also In: Laminar-Turbulent Transition. Ed. R.Eppler & H.Fasel). Springer (1980) 135-152.

On the Issue of the Nonlinear Theory of the Instability Waves

V. N. ZHIGULEV, N. V. SIDORENKO

The Moscow Institute of Physics and Technology, USSR

Summary

The method of matching asymptotic expansions is used to investigate the nonlinear nonstationary waves with nonlinear critical layer. The structure of the nonlinear nonstationary critical layer is studied. The time function of the wave amplitude has been obtained.

The subject of the theory of the nonlinear development of the disturbances in the boundary layer remains to be the key subject for understanding the mechanism of the boundary layer transition to turbulence. It was Lin C.C. who noted, that if a nondimensional amplitude of the disturbance ε is greater than $\overset{*}{\varepsilon}=(\alpha Re)^{-2/3}$, then the nonlinear effects in the critical layer (C.L.), that is, where the phase velocity of the disturbance equals the basic flow velocity, become basic. Waves with nonlinear C.L. were first considered in [1]-[3].

Setting of the Problem

The development of the two-dimensional disturbance with nondimensional amplitude ε is considered in the stationary parallel flow of the incompressible fluid of the boundary layer type with the undisturbed velocity profile $U(y)$. In dimensionless variables in the frame of reference of a wave moving at the phase velocity C the equation for stream function φ has the following form:

$$\frac{1}{\varepsilon}\frac{\partial}{\partial t}\varepsilon\Delta\varphi+(U-C)\Delta\varphi_x-U''\varphi_x+\varepsilon(\varphi_y\Delta\varphi_x-\varphi_x\Delta\varphi_y)=Re^{-1}\Delta^2\varphi, \quad (1)$$

$$\varphi=\varphi_y = 0 \text{ at } y = 0; \varphi \to 0 \text{ at } y \to \infty$$

C is real here. Time dependence of C is neglected in the main approximations. ε is the wave amplitude in the C.L. We shall consider the disturbance to be periodic in x-axis and having the period $2\pi/\alpha$. Further we assume that $\varepsilon \ll 1$, $Re^{-1} \ll 1$, $\omega_i/U'_c = \frac{1}{\varepsilon}\frac{d\varepsilon}{dt}/U'_c \ll 1$ is satisfied with the subscript pointing to the C.L. We limit ourselves to the consideration of the harmonic in x-axis wave. In this case the stream function satisfies the Rayleigh equation in the main approximation everywhere except the C.L. and the viscous wall sublayer:

$$(U-C)(\varphi''-\alpha^2\varphi)-U''\varphi=0. \tag{2}$$

In the present study it is assumed that both the C.L. and the viscous wall sublayer separate asymptotically and there is a region between them for which equation (2) is true. In the C.L. equation (2) is inapplicable and in order to connect the solutions of both sides of the C.L., it is necessary to solve the equation (1). The C.L. will be regarded as a strong nonlinear one everywhere in the present study, i.e. $\gamma=\varepsilon^{3/2}\alpha Re \gg 1$. In the C.L. the relative value of the viscous terms is γ^{-1}, and that of nonstationary terms $b=\omega_i\varepsilon^{-1/2}/\alpha U'_c$. We assume further the nonstationarity to be small in the C.L., i.e. $b \ll 1$. As shown in [1], the flow structure in the C.L. is determined uniquely from the periodicity condition of the viscous correction γ^{-1} to the basic solution. Nonstationary terms will determine the structure of the C.L. in the main approximation when they are of the order of the viscous ones. Thus the nonlinear parameter $a=\omega_i\varepsilon Re/U'_c$ corresponds to the influence of nonstationary terms on the structure of the C.L. At $a \ll 1$ the C.L. is quasi-stationary and at $\gamma \gg a \gg 1$ we arrive at the setting [3].
The relationships among ε, ω_i, C, α, Re are defined from the eigenvalue problem for the equation (2), with the formulae for the C.L. crossover determined by the C.L. structure. They are considered, in a general way, for example in [2], [4]. In the main approximation this yields the relationships

$$C=C(\alpha); \quad \Phi\sqrt{2\alpha ReC}\,(-U_c'')/U_c'=1, \tag{3}$$

where Φ is the change of the phase of the logarithm in a singular Frobenius solution of the equation (2). As it is seen below (12) contribution of the nonstationary C.L. effect to the magnitude of the phase jump Φ and consequently to the stream function φ, has the order of b. As to the contribution of the nonstationary terms to the stream function outside the C.L. it is of the order of $\omega_i/\alpha U_c' \ll b$. Hence the nonstationary terms outside the C.L. can be neglected in the main approximation, that justifies the use of the results (3) in the nonstationary case.

The Structure of the Critical Layer

Here we use the C.L. variables $Y=(y-y_0)/\sqrt{\varepsilon(t)}$, $X=\alpha x$ and the entire stream function

$$\Psi=\int_0^y (U-C)\,dy+\varepsilon\varphi. \tag{4}$$

Let's assume that the stream function Ψ does not depend on time explicitly, but only through the parameters ε, χ, a, i.e. we shall consider a class of "self-similar" solutions. As it is shown below, the expression for Φ has the form

$$\Phi=\text{const}\; f(a)/\chi+O(\chi^{-2}). \tag{5}$$

Then we ignore the term $O(\chi^{-2})$. According to (3) and from the foregoing assumptions the value Φ is time-independent, so that $f(a)/\chi=\text{const}$. Taking it into account, we obtain

$$\frac{a}{\omega_i}\frac{\partial}{\partial t}=a\varepsilon\frac{\partial}{\partial\varepsilon}+\frac{3}{2}a\chi\frac{\partial}{\partial\chi}+\frac{3}{2}a\frac{f(a)}{f'(a)}\frac{\partial}{\partial a}. \tag{6}$$

Deriving the phase jump Φ from the equation for Ψ and comparing it with (5), we find the relationship for f(a) from which f(a) is determined. In practice, this procedure is carried out with the aid of the asymptotic power series expansions in f(a):

$$f(a)=1+\sum_{n=1}^{\infty} a^n f_n \quad \text{or} \quad f(a)= a\sum_{n=0}^{\infty} a^{-n/3} g_n \tag{7}$$

at $a \ll 1$ and $a \gg 1$ respectively and calculations of f_n and g_n in succession.

The C.L. stream function is represented as a power series expansion in ε [1]:

$$\Psi= \Psi^{(0)}+\varepsilon^{3/2}\ln\varepsilon\Psi^{(1)}+\varepsilon^{3/2}\Psi^{(2)}+\ldots \tag{8}$$

For both $\Psi^{(0)}$ and $\Psi^{(1)}_{YY}$ we obtain

$$\Psi^{(0)}=U_c'Y^2/2 + \cos X; \quad \Psi^{(1)}_{YY}=0. \tag{9}$$

The formula for the phase jump Φ is obtained from that for $\Psi^{(2)}$. The equation for $\Psi^{(2)}$ has the form

$$\Psi^{(0)}_Y \Psi^{(2)}_{YYX}- \Psi^{(0)}_X \Psi^{(2)}_{YYY}=\chi^{-1}\Big[\Psi^{(2)}_{YYYY}- \frac{a}{2}U_c'Y\Psi^{(2)}_{YYY}- \\ - \frac{a}{2}U_c'\Psi^{(2)}_{YY}- \frac{3}{2}aU_c'\frac{f(a)}{f'(a)}\frac{\partial}{\partial a}\Psi^{(2)}_{YY}- \frac{3}{2}a\chi U_c'\frac{\partial}{\partial\chi}\Psi^{(2)}_{YY}\Big]. \tag{10}$$

The phase jump Φ can be found from the relationship

$$-\pi\Phi U_c''/U_c'= \int_0^{2\pi}\Psi^{(2)}_Y \sin X\, dX \Big|_{-\infty}^{+\infty}, \tag{11}$$

further on, from (10) we obtain

$$-\pi\Phi U_c''/U_c'=\chi^{-1}\Big(\int_0^{2\pi}\Big(\Psi^{(2)}_{YY}+ \frac{a}{2}U_c'(Y\Psi^{(2)}_Y-3\Psi^{(2)})- \\ - \frac{3}{2}aU_c'\frac{f(a)}{f'(a)}\frac{\partial}{\partial a}\Psi^{(2)}- \frac{3}{2}aU_c'\chi\frac{\partial}{\partial\chi}\Psi^{(2)}\Big)dX +I_1+I_2Y\Big)\Big|_{-\infty}^{+\infty} \tag{12}$$

Here I_1 and I_2 are the constants to be determined. It is seen from (12) that the expression for Φ does have the form (5). We seek for $\Psi^{(2)}_{YY}$ in the form $\Psi^{(2)}_{YY}=\Omega+\chi^{-1}\Omega_1+\ldots$. For Ω and Ω_1 we obtain the equations

$$\frac{D(\Psi^{(0)},\Omega)}{D(Y,X)}=0; \tag{13}$$

$$\frac{D(\Psi^{(0)}, \Omega_1)}{D(Y, X)} = \Omega_{YY} + \frac{1}{2}aU_c'Y\Omega_Y - \frac{1}{2}aU_c'\Omega - \frac{3}{2}aU_c' \frac{f(a)}{f'(a)}\frac{\partial\Omega}{\partial a}. \quad (14)$$

From (13) we obtain $\Omega = (\Psi^{(0)}, a)$. To find this relationship from the condition of periodicity Ω_1 in X, we obtain:

$$\Omega'' + \Omega'\left(\frac{a}{2} + \frac{V'}{V}\right) - \frac{a}{2}\frac{V'}{V}\Omega = \frac{3}{2}\frac{V'}{V} a\frac{f(a)}{f'(a)}\frac{\partial}{\partial a}\Omega \quad (15)$$

where $V = \operatorname{sign} Y \oint\sqrt{2(\Psi^{(0)} - \cos X)/U_c'}\, dX$. At $\Psi^{(0)} \to \infty$ the asymptotics of Ω can be found from the asymptotic expansion of the external solution and has the form

$$\Omega \to U_c'' \operatorname{sign} Y \sqrt{2\Psi^{(0)}/U_c'} + H_{\pm}(a) + C(\Psi^{(0)-1/2}). \quad (16)$$

$H_{\pm}(a)$ causes the distortion of the averaged flow by a quantity $\sim\varepsilon^{1/2}$ [2]. In expanding Ω in ε-series $\Omega = G_0 + a\, G_1 + \dots$ at $a \ll 1$, for the terms of the expansion we obtain the equations

$$G_0'' + G_0'V'/V = 0; \quad (17)$$

$$G_1' + G_1'V'/V - 3/2 f_1\; V'/V\; G_1 = \frac{1}{2}(G_0V'/V - G_0'). \quad (18)$$

Equation (17) has been solved in [2]. Equation (18) at the corresponding boundary conditions resulting from (16) is solved numerically at different values of f_1. $\Psi^{(2)}$ is defined from Ω found previously, then it is substituted in (12). After comparing the latter with (5) we obtain the transcendental equation for f_1. This equation from which we obtain $f_1 = 0{,}55$ has been solved numerically (f_1 values were calculated by Erofeev E. A.). Thus we obtain

$$\Phi = 5{,}52\chi^{-1}(1 + 0{,}55a + O(a^2)). \quad (19)$$

Higher-order terms of this equation can be found in the same way. From (3) and (19) the equation for $\varepsilon(t)$ can be found:

$$\frac{d\varepsilon}{dt} = 0{,}23U_c'^2\varepsilon^{3/2}\alpha^{1/2}(\alpha Re)^{-1/2}/(-U_c'') - 1{,}32U_c'/Re. \quad (20)$$

It is seen from (20) that the nonlinear neutral curve obtained in [2] is unstable.
In the case $a \gg 1$ Φ is equal to [4]:

$$\Phi = \frac{64}{3\pi}\omega_i \varepsilon^{-1/2}/\alpha U_c' . \tag{21}$$

From both (2) and (3) one can find the time dependence of ε

$$\varepsilon(t) = \varepsilon_0 (1 - t/t_0)^{-2}, \tag{22}$$

where $t_0 = \frac{128}{3\pi}\sqrt{2ReC/\alpha\,\varepsilon_0}\;(-U_c'')/U_c'^2$.

References

1. Benney, D.G.; Bergeron, R.F.: A new class of nonlinear waves in parallel flows. Stud. Appl. Math. 48 (1969) 181-204.

2. Haberman, R.: Critical layers in parallel flows. Stud. Appl. Math. 51 (1972) 139-161.

3. Robinson, J.L.: The inviscid nonlinear instability of parallel shear flows. J. Fluid Mech. 63 (1974) 724-752.

4. Zhigulev, V.N.: In: Mekhanika turbulentnykh potokov. Moskva: Nauka 1973 109-133 (in Russian).

The Critical Layer and Nonlinear Waves in the Wall Flows

V.P.REUTOV

Institute of Applied Physics, USSR Academy of Sciences
Gorky, USSR

Abstract
Essentially nonlinear waves in the wall flows are considered. The characteristic equation for stationary waves corresponding to the saturation of two-dimensional instability is obtained. The mechanism of the formation of Λ-shaped vortices that are usually observed by the visualization of the wall flows in the transition region is discussed.

Introduction

The numerous experimental investigations show that a laminar-turbulent transition in the wall flows(a boundary layer, Poiseuille flow etc.)is closely associated with the formation of essentially nonlinear wave structures(see,for ex. [1-4]). However, at present, there are no theoretical models that would interprete the physical mechanisms leading to the formation of these structures. It is known that the wall flows show a weak instability. This fact allows to use the clear representation about a nonlinearity localization in a rather thin critical layer(CL) for the analysis of the nonlinear wave processes. Though an essential progress is made in the theory of nonlinear CL [4-7] its possibilities have not been used completely. New results of the investigation of two-dimensional and three dimensional wave structures containing the nonlinear CL will be reported below.

Two-dimensional Nonlinear Waves

Consider the stationary two-dimensional waves in the wall flow with a velocity profile $\bar{u}(y)$. It is well known that the search for inviscid solutions of linearized equations detects the singularity at a critical point $y=y_c$ that corresponds to the coincidence of a flow velocity with a phase velocity of the wave [4,8,9]. The problem of the inviscid solution continuation into the nonlinear CL was solved in papers [5,6]. It was shown that when the amplitude of a harmonic perturbation in the outer (with respect to CL) regions of the flow is small($\sim \varepsilon$), the nonlinearity may be localized in a small($\sim \varepsilon^{1/2}$) vicinity of a critical point. Also the parameters of a linear problem solution for the outer regions determine essentially nonlinear motion in CL. The clue equation in the problem of matching inviscid solutions in a critical point is the equation for the vorticity perturbation in the nonlinear CL. Representing a CL-vorticity in the form $\omega_z = -\bar{u}_c' + \varepsilon^{1/2}(-\bar{u}_c'' Y + \Omega)$ we obtain the following equation for Ω [6,7] :

$$\bar{u}_c' Y \frac{\partial \Omega}{\partial X} + V_c \sin \alpha X \frac{\partial \Omega}{\partial Y} - \lambda \frac{\partial^2 \Omega}{\partial Y^2} = \bar{u}_c'' V_c \sin \alpha X , \tag{1}$$

where X is a longitudinal coordinate in the frame of reference of the wave; $Y = (y - y_c)/\varepsilon^{1/2}$ is the inner variable of CL; α is the wave number, V_c is the amplitude of harmonic oscillations of transverse velocity when $y \to y_c$ (in units of ε); $\lambda = 1/R\varepsilon^{3/2}$. A principle of matching of inviscid solution determined above and beneath the CL is inferred with the aid of solution of eq.(1) having inviscid asymptotics:

$$\Omega \to \text{const}_{\pm} - \frac{\bar{u}_c''}{\bar{u}_c'} \left(\frac{V_c}{\alpha} \right) \frac{1}{Y} \cos \alpha X + \cdots , \quad Y \to \pm \infty \tag{2}$$

Similar to the linear theory the round of the critical point may be described by the phase jump of logarithm Φ [6]. However in the case of nonlinear stationary CL Φ is dependent on the normalized wave amplitude $S = d_n^2/d_\ell^2$ where $d_\ell = (\alpha \bar{u}_c' R)^{-1/3}$, $d_n = (\varepsilon V_c/\alpha \bar{u}_c')^{1/2}$ are the scales of viscid and nonlinear CL. A simple analytic approximation of this dependence was proposed in the paper [7].

The equation (1) is valid for a wall flow only when the distance between nonlinear CL and the wall is large enough: $y_c/\max(d_\ell, d_n) \gg 1$. It also follows from this condition that the viscid perturbations due to sticking of the fluid to the wall damp strongly during their way to CL. It was assumed in papers [6,7] that at the damping length of viscid perturbations the velocity $\bar{u}(y)$ can be replaced by $\bar{u}(0) = 0$. Nevertheless this assumption allows to get only the asymptotics of the upper branch of the neutral curve. It is known from the linear stability theory that at the moderate large Reynolds numbers R corresponding to the main part of the neutral curve loop the transformation of the viscous vorticity perturbations in CL should be taken into account to determine correctly the eigenvalues of the boundary problem [9]. For describing this process in the case of a nonlinear CL we represent the solution of eq. (1) in the form $\Omega=\Omega_1+\Omega_2$ where Ω_1 is the solution of eq.(1) with the asymptotics(2), the component Ω_2 satisfies the homogeneous equation following from (1) and turns into the viscous perturbations when $|Y|\to\infty$. The equation for Ω_2 may be used at all the distance from the CL to the wall $y=0$. Naturally, the equation (1) is one of vorticity evolution in the fixed (independent of Ω) velocity field. It is clear that the condition of location of the CL far from the wall secures the smallness of viscous additions to this velocity field. When $V_c=0$ such an equation for complex amplitude of harmonic viscous perturbations follows from the equation for Ω_2 as one to be known in the linear theory [8,9]. With $V_c\neq 0$ the contribution of the term $\sim V_c$ to eq. for Ω_2 decreases with $|Y|\to\infty$ as $1/\sqrt{\lambda Y}$. Thus, the nonlinearity is localized in CL also in the case of viscid perturbations. The characteristic equation for stationary waves can be obtained by the generalization of Tollmien method [8]. It includes the wave amplitude

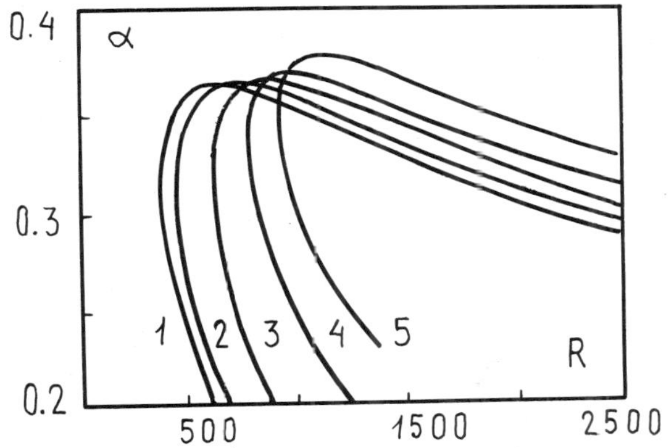

Fig.1. The isolines $S(\alpha,R)$ =const for Blasius flow (Reynolds number R is determined by using the displacement thickness). The curves 1-5 correspond to S = 0; 0.4; 1; 1.5; 2.

and has a form $f(\alpha, c, z_c, S) = 0$ where $z_c = y_c / d_\ell$. Fig.1 shows the results of its numerical solution for the Blasius flow. The line S=0 coincides with a neutral curve for the case of infinitely small perturbations obtained by Tollmien [8]. The isolines intersection indicates the hard excitation of self-oscillations. The formation of the stationary two-dimensional wave was observed experimentally at the first station of the boundary-layer transition[1]. As follows from given analysis the saturation of two-dimensional instability (of the Tollmien -Schlichting mechanism) is associated with the CL nonlinearity effect both on the effectiveness of the resonant "wave-flow" interaction and on the effectiveness of the viscous mechanism of the wave destabilization.

The Formation of Λ -shaped Vortices

Consider the nonlinear CL when one two-dimensional and two slanting waves are present in the outer flow regions. We suppose that the wave vectors of these waves are equal to $(\alpha, 0)$ and (α, β) and their downstream phase velocities are identical. A similar wave system is formed, for example, by the development of a parametric instability [1]. The flow field arising in this case was investigated in paper [10]by using the perturbation method. The time of development of a nonlinear CL is considered below. The wave amplitudes are assumed to be constant and fluid viscosity is not taken into account.

We introduce a coordinate Y (see above) and "slow" time $\tau = \varepsilon^{1/2} t$ Then the velocity and pressure fields in CL are presented as the series $u = \varepsilon^{1/2} u^{(0)} + \cdots$, $w = \varepsilon^{1/2} w^{(0)} + \cdots$, $v = \varepsilon v^{(0)} + \cdots$ and $p = const + \varepsilon p^{(0)} \ldots$. The asymptotics of $u^{(0)}$, $w^{(0)}$, $v^{(0)}$ and $p^{(0)}$ at $|Y| \to \infty$ are determined by the primary flow and three wave structure in its outer regions. The analysis shows that $p^{(0)}$ does not depend on Y and coincides with the limit of the outer pressure distribution at $y \to y_c$. In the case of three-wave structure we obtain $p^{(0)} = A\cos(\alpha x - \beta z) + A\cos(\alpha x + \beta z) + B\cos(\alpha x + \Delta)$ where $A, B = const > 0$, $\Delta = const$. Using the normalized variables we can get the following equations for the coordinates x, z and the horizontal velocity components of the individual fluid particles:

$$\frac{dx}{d\tau} = u, \quad \frac{du}{d\tau} = -\frac{\partial p}{\partial x}, \quad \frac{dz}{d\tau} = w, \quad \frac{dw}{d\tau} = -\frac{\partial p}{\partial z}; \qquad (3)$$

$$p = q\cos\gamma z\cos x + (1-q)\cos(x+\Delta)$$

where $q = 2A/(2A+B)$, $\gamma = \beta/\alpha$. Thus, the horizontal motion of fluid particles in CL is autonomous and similar to the motion of the material points in the force field with the potential $p(x,z)$. Let us consider a line of fluid particles that coincide at the initial moment t=0 with some vortex line of unperturbed flow: $x(0)=const$, $u(0)=const$, $w(0)=0$, ($u(0)=0$ corresponds to the coincidence layer). Integrating the equation of motion (3) and using the law of vortex conservation [11] we can find the distribution of a horizontal vorticity component along each vortex line at time $t>0$: $\omega_h(t)=\omega_z(0)\delta\ell_h(t)/\delta\ell_h(0)$ where $\delta\ell_h$ is the projection of a fluid element on the plane (x,z).

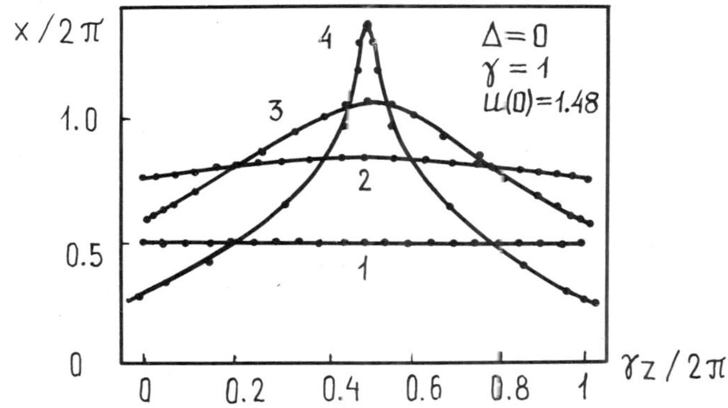

Fig.2. The formation of Λ -shaped vortex line in CL. The curves 1-4 show the positions of vortex line over a time $\Delta\tau$ =1.72.

The conservation law of energy takes place for each particle: $E=(1/2)u^2+(1/2)w^2+p(x,z)=const$. It shows that particle moves finitely or infinitely along the x-axis depending on its initial conditions. Since potential p is inhomogeneous along spanwise co-ordinate Z, the free and trapped particles may be placed on the same vortex line when $u(0)$ is not very large. In this case the particles may be scattered in the x-direction that gives rise to a big streamwise vorticity $|\omega_x|>|\omega_z(0)|=$ $=\bar{U}'_c$ – Fig.2 . We assume that the similar deformation of totality of vortex lines of unperturbed flow tends to the formation of Λ - vortices. In the natural flow the growth of the vorticity gradients is limited by the viscosity effect. It appears that such limitation tends to the formation of a stable vortex structure that will be similar to a vortex structure in the ideal fluid.

References

1. Kachanov, Yu.S.; Kozlov, V.V.; Levchenko, V.Ya.: Transition to turbulence in a boundary layer. Novosibirsk: Nauka, 1982.
2. Knapp, C.F.; Roache, P.J. : AIAA J. 6 (1968) N.1.
3. Kozlov, V.V.; Ramazanov, M.P.; : Izv. Akad. Nauk SSSR, MZhG. 1 (1982).
4. Maslowe, S.A.: In: Hydrodynamic Instabilities and the Transition to Turbulence (eds. H.L.Swinney, J.P.Gollub). Springer. 1981.
5. Benney, D.J.; Bergeron, R.F. : Stud. Appl. Math. 48 (1969) N.3.
6. Huberman, R. : Stud. Appl. Math. 51 (1972) N.2.
7. Reutov, V.P. : PMTF 4 (1982) 43-54.
8. Schlichting, H. : Boundary layer theory. 6th Ed. McGraw - Hill 1968.
9. Betchov, R.; Criminalle, W.O. : Stability of parallel flows. New York : Academic Press 1967.
10. Benney, D.J. : Phys. Fluids 7 (1964) N.3.
11. Batchelor, G.K. : An introduction to fluid dynamics. Cambridge University Press 1967.

A Turbulence Model for the Numerical Simulation of Transitional Boundary Layers

G. VANCOILLIE

Research assistant of the National Fund for Scientific Research
Department of Machinery
State University at Gent, GENT, BELGIUM

SUMMARY

Conditioned averages are used for a detailed simulation of the intermittent flow in the transition zone of a boundary layer. Conditioned continuity, momentum and turbulence equations are constructed. The equations contain source terms which express interactions between turbulent spots and laminar fluid. These terms are modelled with some simplifying assumptions about turbulent spot growth. The finite difference calculations are compared with different experiments.

MODEL EQUATIONS

A two-dimensional boundary layer in incompressible steady flow is considered. It is assumed that turbulent spots are formed in the transition region. The intermittency function I(x,y,z,t) has the value 1 inside and 0 outside the spots. The x,y,z-coordinates are streamwise, normal to the wall and transverse. The y-dependence of I is neglected, which implies that the spots are modelled with a cylindrical interface, normal to the wall.

The time average of I is the intermittency factor $\gamma(x)$. It is the fraction of the time that the flow is turbulent. The total derivative of I is zero everywhere except at the interface between laminar and turbulent fluid

$$\frac{dI}{dt} = \frac{\partial I}{\partial t} + u\,\frac{\partial I}{\partial x} + v\,\frac{\partial I}{\partial y} + w\,\frac{\partial I}{\partial z} \tag{1}$$

The global averaged velocity components are $\bar{u}$, $\bar{v}$ and $\bar{w}$. With the intermittency function, conditioned (or zone) averages are defined for the turbulent and laminar fluid parts

$$u_t = \overline{u.I}/\gamma \qquad\qquad u_l = \overline{u(1-I)}/(1-\gamma)$$

All flow properties can be decomposed in a mean and a fluctuating part in three different ways : $u = \overline{u} + u' = u_t + u_t' = u_l + u_l'$. The fluctuations u_t' and u_l' are only defined in the turbulent, respectively laminar zone. A global average can be expressed as : $\overline{u} = \gamma u_t + (1-\gamma)u_l$.

Continuity equations. The model equations are derived from Navier-Stokes equations. The unconditioned continuity equation is multiplied by I, combined with (1) and time averaged, to yield the turbulent zone averaged continuity equation. The laminar zone averaged continuity equation is obtained similarly

$$\frac{\partial \gamma u_t}{\partial x} + \frac{\partial \gamma v_t}{\partial y} = \overline{\frac{dI}{dt}} \quad (2) \qquad \frac{\partial (1-\gamma)u_l}{\partial x} + \frac{\partial (1-\gamma)v_l}{\partial y} = -\overline{\frac{dI}{dt}} \quad (3)$$

The transverse velocities w_t and w_l are zero. The source terms on the right hand sides express the mass transport through the interface.

Momentum equations. The unconditioned momentum equation for u is multiplied by I, substracted from equation (1) multiplied by u_t', and averaged. Some terms are zero because of the symmetry of turbulent spots. Others are neglected after an order of magnitude analysis. The resulting turbulent zone averaged momentum equation is :

$$\gamma u_t \frac{\partial u_t}{\partial x} + \gamma v_t \frac{\partial u_t}{\partial y} = \gamma \frac{\partial}{\partial y}\left(\nu \frac{\partial u_t}{\partial y} - (\overline{u_t'v_t'})_t\right)$$
$$- \gamma \frac{1}{\rho}\frac{\partial p_t}{\partial x} + \frac{1}{\rho}\overline{p_t' \frac{\partial I}{\partial x}} + \overline{u_t' \frac{dI}{dt}} - \nu \overline{\frac{\partial u_t'}{\partial x}\frac{\partial I}{\partial x}} - \nu \overline{\frac{\partial u_t'}{\partial z}\frac{\partial I}{\partial z}} \quad (4)$$

The first four terms are identical to those of the classical momentum equation for turbulent flow, but they are written in conditioned variables. The other terms, containing derivatives of I, are point averages. They express the interaction between the turbulent spots and the surrounding laminar flow at the interface. A laminar zone averaged momentum equation is constructed similarly.

Turbulent kinetic energy equation. A transport equation for the conditionally averaged turbulent kinetic energy k_t is constructed by multiplying the momentum equations for u, v and w by u_t', v_t' and w_t', multiplying this sum by I and time averaging.

$$\gamma u_t \frac{\partial k_t}{\partial x} + \gamma v_t \frac{\partial k_t}{\partial y} = \gamma \frac{\partial}{\partial y}\left(\nu \frac{\partial k_t}{\partial y} - (\overline{v_t'k_t'})_t - \frac{1}{\rho}(\overline{v_t'p_t'})_t\right)$$

$$- \gamma(\overline{u_t'v_t'})_t \frac{\partial u_t}{\partial y} - \nu \overline{I \sum_{j,k} (\frac{\partial u_{jt}'}{\partial x_k})^2} + \overline{k_t' \frac{dI}{dt}} - k_t \overline{\frac{dI}{dt}}$$

$$- \nu \overline{\frac{\partial k_t'}{\partial x} \frac{\partial I}{\partial x}} - \nu \overline{\frac{\partial k_t'}{\partial z} \frac{\partial I}{\partial z}} + \frac{1}{\rho} \overline{u_t'p_t' \frac{\partial I}{\partial x}} + \frac{1}{\rho} \overline{w_t'p_t' \frac{\partial I}{\partial z}}$$

The instantaneous value of the turbulent kinetic energy is : $k_t' = (u_t'^2 + v_t'^2 + w_t'^2)/2$. Convection, diffusion, production and dissipation terms, identical to those of the unconditioned k-equation, are found. The other terms are source terms.

CLOSURE OF THE EQUATIONS

A lot of research on turbulent spots (mostly artificially generated) is reported in the literature. It is observed that all spots have the same growth pattern. Some distance downstream from their point of origin, they grow linearly in all directions parallel to the plate. In x,z,t space, the growth is represented by a cone. All source terms are proportional to the intermittency frequency f_γ. This is the average number of spots that pass a point per unit time. For non-overlapping spots, the intermittency frequency is proportional to $d\gamma/dx$. If the spots overlap, the above relation is no longer exact, but can be used as a good approximation. The intermittency factor γ is related to the probability distribution of the spot generation. The spots originate in a restricted region (the transition point x_t). Dhawan and Narasimha [1] proposed the following expression for γ :
$\gamma = 1 - \exp(-0.411(x-x_t)^2/\lambda^2)$ with $\lambda = x(\gamma{=}0.75) - x(\gamma{=}0.25)$

The source term of the turbulent zone averaged continuity equation is $\overline{dI/dt}$. As the spot always entrains laminar fluid, dI/dt is positive at the interface. At a fixed location in space, dI/dt is a train of Dirac-δ pulses of unequal strength. The source term depends on the shape of the spot, the propagation velocity at the interface, and the velocity of the fluid crossing the interface. This can be modelled by : $\overline{dI/dt} = u_m\, d\gamma/dx$, where u_m is some average value of the flow velocity at the interface. Disregarding viscous and pressure forces in the momentum equations, only the source term due to momentum transport through the interface remains. The entrainment of fluid by the spots changes the momentum of the turbulent zone, but does not influence the fluid that remains laminar. Therefore the continuity and momentum equations for the laminar zone (without viscous and pressure interaction) should be equal to unconditioned continuity and momentum equations. This is only possible if $u_m = u_l$. With this assumption, the transport term in the conditioned momentum equation can also be evaluated. The viscous and pressure interactions are caused by the velocity difference between laminar and turbulent fluid. They are modelled in one term, with a new model constant c_v. The source terms of the k_t-equation express the conversion of the work of the Reynolds-stresses at the interface into turbulent kinetic energy. This production term is proportional to the square of the velocity gradient at the interface. A model constant c_k is introduced. The equation for the dissipation rate ε_t is similar to the k_t equation. A model function $c_\varepsilon(\gamma)$ is used in the source term. The model constants and low Reynoldsnumber terms of the k-ε model of Jones and Launder [2] are also included in the final model which is given below.

$$\frac{\partial u_l}{\partial x} + \frac{\partial v_l}{\partial y} = 0 \qquad\qquad \frac{\partial u_t}{\partial x} + \frac{\partial v_t}{\partial y} = \frac{1}{\gamma}(u_l - u_t)\frac{d\gamma}{dx}$$

$$u_l \frac{\partial u_l}{\partial x} + v_l \frac{\partial u_l}{\partial y} = u_p \frac{du_p}{dx} + \frac{\partial}{\partial y}\left(\nu \frac{\partial u_l}{\partial y}\right) - c_v \nu \frac{u_l - u_t}{\delta_1} \frac{1}{1-\gamma} \frac{d\gamma}{dx}$$

$$u_t \frac{\partial u_t}{\partial x} + v_t \frac{\partial u_t}{\partial y} = u_p \frac{du_p}{dx} + \frac{\partial}{\partial y}\left[(\nu+\nu_t)\frac{\partial u_t}{\partial y}\right] + \frac{(u_l-u_t)u_l}{\gamma}\frac{d\gamma}{dx} + c_v \nu \frac{u_l - u_t}{\delta_1 \gamma}\frac{d\gamma}{dx}$$

$$u_t \frac{\partial k_t}{\partial x} + v_t \frac{\partial k_t}{\partial y} = \frac{\partial}{\partial y}\left[\left(\nu + \frac{\nu_t}{\sigma_k}\right)\frac{\partial k_t}{\partial y}\right] + \nu_t\left(\frac{\partial u_t}{\partial y}\right)^2 - \varepsilon_t - 2\nu\left(\frac{\partial k_t^{1/2}}{\partial y}\right)^2 + c_k \nu_t \frac{(u_l-u_t)^2}{\delta_1 \gamma}\frac{d\gamma}{dx}$$

$$u_t \frac{\partial \varepsilon_t}{\partial x} + v_t \frac{\partial \varepsilon_t}{\partial y} = \frac{\partial}{\partial y}\left[\left(\nu + \frac{\nu_t}{\sigma_\varepsilon}\right)\frac{\partial \varepsilon_t}{\partial y}\right] + c_1 \frac{\varepsilon_t}{k_t}\nu_t\left(\frac{\partial u_t}{\partial y}\right)^2 - c_2 f_2 \frac{\varepsilon_t^2}{k_t} + 2\nu\nu_t\left(\frac{\partial^2 u_t}{\partial y^2}\right)^2$$

$$+ c_\varepsilon \frac{\varepsilon_t}{k_t}\nu_t \frac{(u_l-u_t)^2}{\delta_1\gamma}\frac{d\gamma}{dx}$$

The eddy viscosity is : $\nu_t = \nu c_\mu f_\mu R$, with the turbulence Reynolds number $R = k_t^2/\nu\varepsilon_t$.

The model constants and functions are :

$c_1 = 1.44$ $\quad c_\mu = 0.09$ $\quad f_\mu = \exp(-3.4/(1 + R/50)^2)$

$c_2 = 1.92$ $\quad c_v = 300.$ $\quad f_2 = 1 - 0.3 \exp(R^2)$

$c_k = 1.$ $\quad c_k = 60.$ $\quad c_\varepsilon = 200.(0.7 + 0.3 \cos\gamma\pi)$

$c_\varepsilon = 1.3$

RESULTS

The finite difference calculations are started in the laminar boundary layer. Comparison between the calculations and the experiments of Schubauer and Klebanoff [3] have been used to determine the new model constants. The turbulence intensity is calculated as $\overline{u'^2} = \gamma k_t + \gamma(1-\gamma)(u_t-u_1)^2$. The velocity profiles show good agreement. The turbulence intensity is also well predicted. Similar agreement is obtained for the transition region in zero pressure gradient measured by Arnal and Juillen [4]. For adverse pressure gradient, the predicted turbulence intensity is too low. This can be explained by the fact that the transition in adverse pressure gradient is less dominated by turbulent spots. The transition region in favourable pressure gradient is compared with experiments of Blair and Werle [5]. Some results are shown in figures 1 to 6.

REFERENCES

1. Dhawan, S.; Narasimha, R. : Some properties of boundary layer flow during the transition from laminar to turbulent motion. J. Fluid. Mech., Vol. 3, pp. 418-436 (1958).

2. Jones, W.P.; Launder, B.E. : The calculation of low-Reynolds-number phenomena with a two-equation model of turbulence. Int. J. Heat Mass Transfer, Vol. 16, pp. 1119-1130 (1973).

3. Schubauer, G.B.; Klebanoff, P.S. : Contributions on the mechanics of boundary layer transition. NACA report 1289 (1955).

4. Arnal, D.; Juillen, J.C. : Etude expérimentale et théorique de la transition de la couche limite. La recherche aérospatiale n° 2, pp. 75-88 (1977).

5. Blair, M.F.; Werle, M.J. : Combined influence of free-stream turbulence and favorable pressure gradients on boundary layer transition and heat transfer. UTRC report 81-914388-17 (1981).

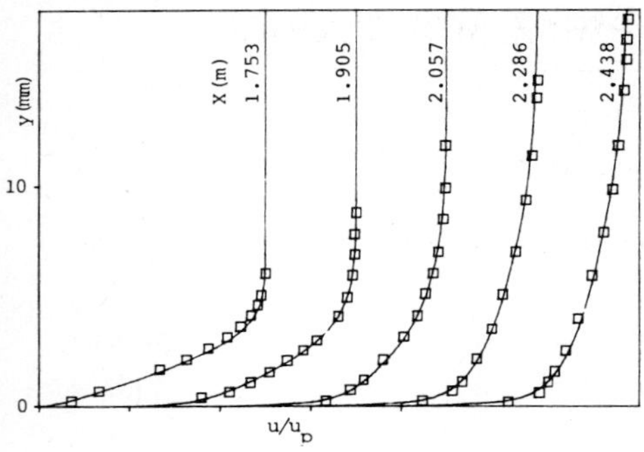

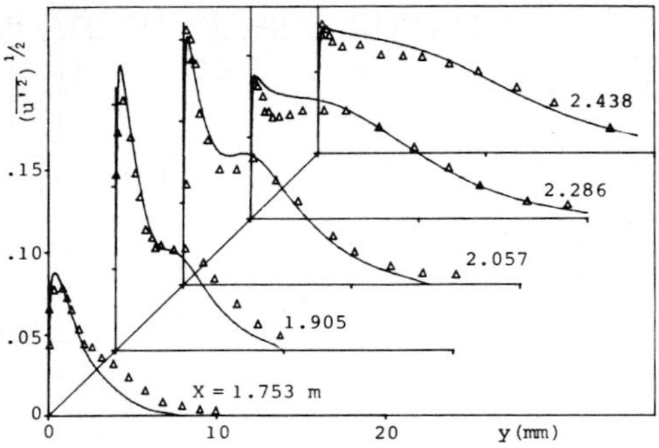

Fig. 1 : Velocity profiles

Fig. 2 : Turbulence intensity

Experiment of Schubauer and Klebanoff [3] : $dp/dx = 0$; $T_u = 0.03$ %

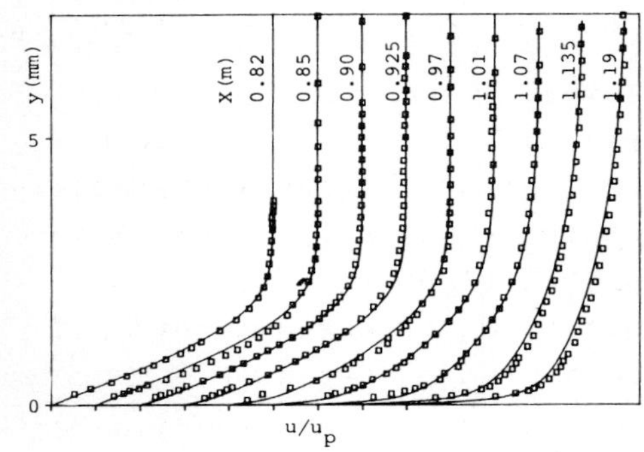

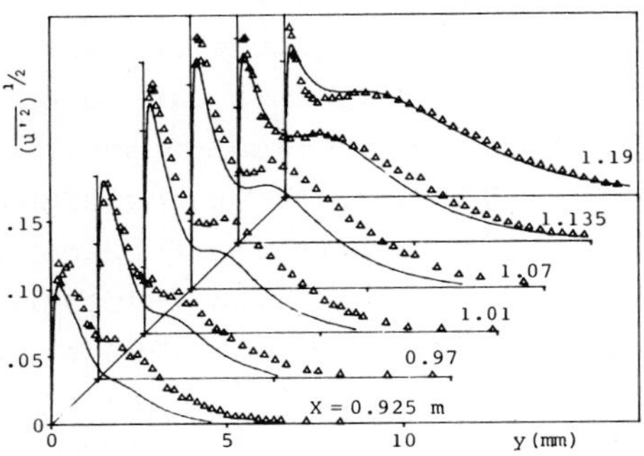

Fig. 3 : Velocity profiles

Fig. 4 : Turbulence intensity

Experiment of Arnal and Juillen [4] : $dp/dx = 0$; $T_u = 0.25$ %

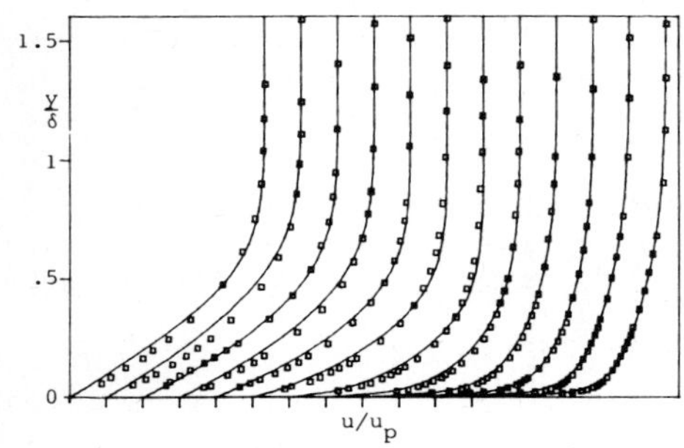

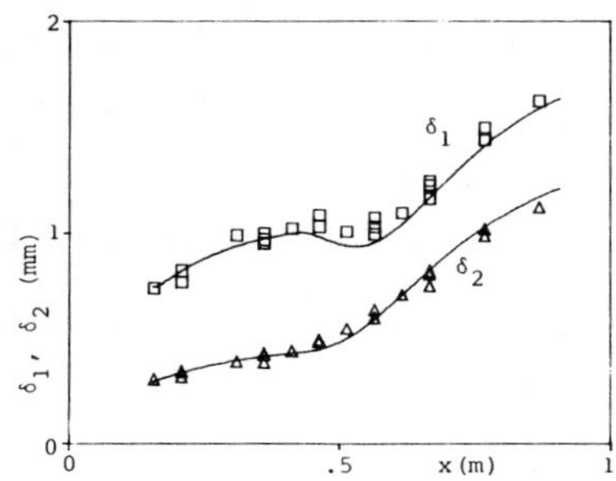

Fig. 5 : Velocity profiles

Fig. 6 : Displacement and momentum thickness

Experiment of Blair and Werle [5] : $dp/dx < 0$; $T_u = 1$ %

Characteristic Fields of the Natural Transitional Boundary Layer

M. HACKESCHMIDT, M. RÖSSLER, H.-D. HILBRICH

"Friedrich List" University of Transport and Conmunications, Dresden, GDR

Summary

As a result of theoretical and experimental investigations characteristic correlations between momentum thickness Reynolds number and free stream turbulence level with the intermittency factor as a parameter of the flat plate without pressure gradient are obtained and furthermore a characteristic of the pressure gradient influence on the laminar-turbulent transition.

The Start of Transition

Varied measuring results served to design fields of characteristics of the laminar-turbulent transition on the flat plate with/without pressure gradient. The values were obtained by means of hot wire anemometry, heat transfer measurements and according to the soot-oil-petroleum erosion method and considerations on the similarity theory. Accordingly the transition start S may be determined as a location where the first longitudinal vortex pairs near the wall separate. The last longitudinal vortex pairs only separate from the wall after the first turbulence spots have already left their prints on the wall. The start of transition is defined by the intermittency factor $\gamma = 0$ [1].

Different Turbulence Level Regions

The dimensions of the longitudinal vortex pairs formed in the unstable laminar boundary layer are dependent on the type of excitation. Accordingly different turbulence level regions may be differentiated in the field transition characteristics.

(i) In the turbulence level region SE transition is exclusively initiated by self-excitation in the unstable laminar boundary layer,

(ii) while in region FE it is initiated by external flow turbulence (foreign excitation).

(iii) Region ME is considered to be a transition region.

(iv) In region TE external flow turbulence is greater than that of the boundary layer and there is no pressure gradient influence. In this region one has to reckon with a distinct influence of turbulence anisotropy (k [1]).

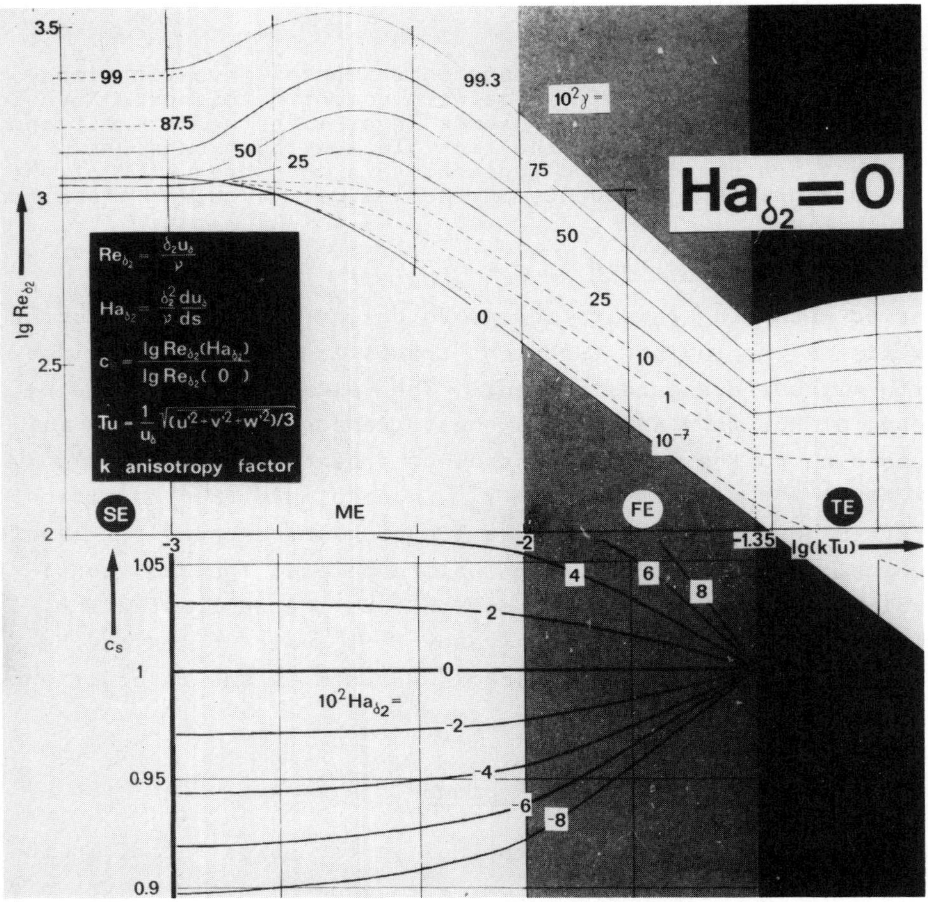

Figure above: Relationship between momentum thickness Reynolds number and free stream turbulence level for transitional boundary layer with zero pressure gradient with the intermittency factor as parameter

below: Effect of pressure gradient $c_s = c(\gamma = 5\%)$

The Intermittency Factor

Presupposing foreign excitation and increasing Reynolds number the intermittency factor will increase at the beginning, i.e. when bursting vortex loops are formed, only in a very slight manner. As is generally known it is only in the middle of the transition region that it increases rapidly. In case of self-excitation, however, the development of the intermittency factor is balanced due to hairpin vortex formation.

The intermittency factor appears as parameter between the lines of transition start S ($\gamma = 0$) and of transition end E ($\gamma \geqq 0.993$) in the Fig. being dependent on the Reynolds number $Re_s = u_\delta s/\nu$. It may be written in the standardized form as follows:

$$\gamma = \frac{\overline{F} - F_1}{\overline{F_t} - f_1} = 1 - \exp\left[-5\left(\frac{Re_s - Re_{sE}}{Re_{sS} - Re_{sE}}\right)^{n(kTu)}\right] \qquad (1)$$

where F are any shape factors of the transitional boundary layer. The subscript t applies to the turbulence spot region and the subscript l to the laminar flow surrounding.

In the formula (1) the changed turbulent velocity profile shape resulting from the external flow turbulence in the region where turbulent spots are observed is taken into account by means of coefficients (k_1 and k_2 in [1]) being themselves dependent, in the turbulence region of mixed excitation ME, on the turbulence level.

The Incidence of Pressure Gradient

The pressure gradient influence may be shown separately (see Figure below). As dependent quantity the exponent c appears by which the momentum thickness Reynolds number may be reduced, in case of a flow of non-zero pressure gradient, to that which represents the flow state of zero pressure gradient. For the latter the following correlation was found

$$\gamma \approx (c - c_S)/(1 - c_S) \qquad (2)$$

Reference

1. Hackeschmidt, M.: Über den Einfluß des Druckgradienten auf ... Maschinenbautechnik Berlin 33 (1984), H.5 214-220

Causes of Vortex Bursting in Transitional Boundary Layers

M. HACKESCHMIDT

"Friedrich List" University of Transport and Communications, Dresden, GDR

Summary

The cause of vortex bursts in transitional boundary layers may be explained by secondary flow formation at the turbulence spot flanks due to a lateral deflection of the main flow in the viscous sublayer and by the formation of counter-moving secondary flows in case of a changed curvature of the main flow path.

Hypothesis

Information on the flow type observed directly on the wall overflown by the fluid may be obtained by means of snapshots demonstrating the erosion of a soot-oil-petroleum layer. In Figure 1 a turbulence spot on a flat plate without pressure gradient is shown in the region of laminar-turbulent transition with an intermittency factor of approximately $\gamma = 5\%$ and a local free stream turbulence level of about 2.4%. Compared with its surrounding it moves in the main flow direction with reduced velocity.

In this case the surrounding fluid at both sides avoids the spot; the fluid is laterally deflected at point D_1, whereby secondary flows ω'_{s1ind} are induced at the rear-end flanks of the spot. The surrounding fluid is deflected for the second time at point D_2, i.e. compared with the first deflection in the rear area it is deflected in opposite direction. By that a new secondary flow ω'_{s2ind} is induced in the viscous sublayer which as compared with the "older" vortex rotates in opposite direction and forces it out of its path. The diagonal vortices ω'_{sz} are formed generating in the turbulence spot centre (ω_n) velocity components against the main flow direc-

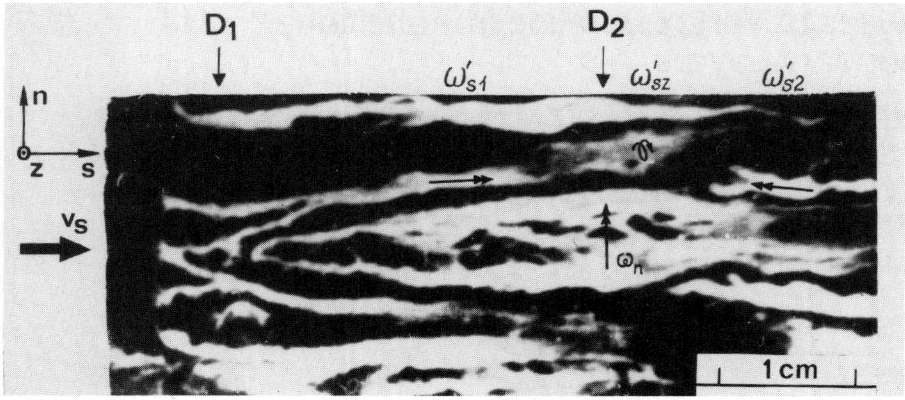

Figure 1. Erosion of a soot-oil-petroleum layer on the flat plate without pressure gradient in the region of laminar-turbulent transition according to M. Rößler

tion s. Consequently the diagonal vortices ω'_{sz} must move more rapidly than the whole turbulent spot, a fact explaining their eruptive or burst-like behaviour.

According to this model presentation each turbulence spot generates at its flanks two bursting diagonal vortices which generate, on their part, a turbulence spot each. As a logical consequence of this theoretical model the number of burst processes escalates in the downstream area within the laminar-turbulent transition region. A restriction of this process is

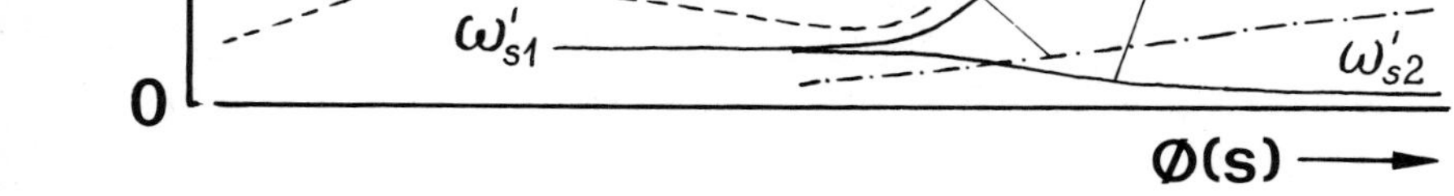

Figure 2. Burst of the "older" longitudinal vortex caused by induction of a counter-moving "younger" longitudinal vortex. δ boundary layer thickness

seen in the more or less centrical collisions of longitudinal vortex threads with counter-moving secondary flows, which might otherwise be formed.

Mathematical Models

To verify the hypothesis of secondary flows forming at the turbulence spot flanks first of all a very simplified mathematical model must be taken, i.e. a guiding model. This means that in the vorticity equation at first only those terms must be considered causing, due to a lateral deflection of the main flow, induction of secondary vorticity [1].

While this guiding model only proceeds from the collateral boundary layer, i.e. that of the main flow without secondary flow interaction, to calculate the non-friction secondary flow, the modified model also considers the shearing action of the secondary flow (bilateral shear flow). Moreover there is an interaction between the secondary flow and the main flow.

Results of Numerical Calculations

The result of the calculations is shown in Figure 2. When comparing the Figures 1 and 2 it must be borne in mind that the former represents the absolute system and the latter must be regarded from the turbulence spot, i.e. in the relative system.

Although the very simple secondary flow model was designed for the section near the wall it is principally able to describe the diagonal vortex ω'_{sz} in a correct manner.

As is shown in Figure 2 the bilateral boundary layer changes the result of the collateral one. In the z-s plane the "older" vortex "rises" more slowly than in case of an ideal secondary flow; after the "older" vortex having burst, the centre of the "younger" one does no longer move off the wall, but towards it.

Reference

1. Hackeschmidt, M.; Hilbrich, H.-D.; Rößler, M.; Krüger, N.: Secondary Flows in Curved Channels of Any Sectional Change. Paper presented to the "Hydrosoft" Conference in Portorož, Yugoslavia, Sept.,10 to 14, 1984. CMC Publications Ashurst Lodge, Ashurst, Southampton SO4 2 AA, England

Experimental Investigation of Statistical Characteristics of Velocity Fluctuations in Natural Laminar-Turbulent Transition

V. A. TETIANKO

Krylov Shipbuilding Research Institute
Leningrad, USSR

The problem of laminar-turbulent transition draws recently an increasing authors' attention. A lot of experimental studies have been mainly concerned with an artificial laminar-turbulent transition, special attention has been payed to investigation of detailed structure of the artificial turbulent spots. The main part of our investigations was concerned with the natural laminar-turbulent transition. Our early investigations showed the existence of a number of zones which are by the thickness and lengthways of the boundary layer associated with the nature of the velocity fluctuations [I] . This work contains some results obtained during two cycles of investigations performed in different years.

Laminar-turbulent transition in boundary layer on a flat plate was investigated in the low-turbulence wind tunnel T-324 of the Institute of Theoretical and Applied Mechanics.Of the greatest interest was the region where the transition process appears to arise from the instability which leads to formation of the turbulent spots and their growth in a laminar boundary layer with zero gradient.Velocity measurements were made using single hot-wire anemometer DISA 55MIO with linearizer DISA 55DIO.The longitudinal component of mean velocity u and rms fluctuations $\sigma = (\overline{u'^2})^{I/2}$ were measured during the experiment.At the same time the signal from the anemometer was driven up to two Signal Conditioners 55D26 and further to two RMS Voltmeters 55D35.This gave possibility to investigate the velocity fluctuations in different frequency bands during the experiment.The principal results were obtained in I Hz - IO kHz and I kHz - IO kHz bands.The rms velocity fluctuations made dimensionless by the velocity on the outer edge of the boundary layer show that LF and HF fluctuations develop differently in the region of transition.The maximum level of rms high-frequency fluctuations displaces from the surface with the development of the laminar-turbulent transition, but at the same time the maximum level of rms velocity fluctuations approaches, when the low-frequency fluctuations are taken into account, the surface (Figs.I2,I3 from [2]).This result is determined by the character of the turbulent spots' growth.

For a more detailed analysis of the velocity fluctuations the output from the hot-wire electronics was high-pass filtered at I Hz and low-pass filtered at IO kHz and the fluctuating component was recorded by the Portable instrumentation recorder MP-552I in frequency modulation with a band width of IO kHz.The tape recordings were later replayed and the detailed analysis of velocity fluctuations was made.Using a specially designed and built apparatus the intermittency factor γ was obtained for a number of cross sections along the transition zone (Fig.9 of [2]).

To obtain the probability density functions of velocity fluctuations the tape recordings were replayed and analysed on the statistical characteristics analyser Histomat-S.Probability density functions were built on IOO-I24 intervals with using 10^6 events, the total time of processing being 90 sec.for each distribution. Using a standard experimental data processing method defined were: $E[V]$, $E[V^2]$, $D=E[V^2]-(E[V])^2$, $\sigma=D^{1/2}$, skewness factor $\nu = E[\mathring{V}^3]/D^{3/2}$.The obtained probability density functions of velocity fluctuations $f(u')=\sigma\cdot P$ for five investigated cross sections of laminar-turbulent transition are shown in the Figs.26-30 of[2] . It does not seem possible to reproduce the indicated results in this paper. This probability density functions characterize qualitative and quantitative development of the two processes: ejection of low-velocity fluid from near-surface region of the boundary layer and inward motion of high velocity fluid to the surface. As soon as the intermittency factor is a good characteristic for the reorganisation of the fluid's motion in transition,the statistical characteristics of velocity fluctuations are regarded as functions of maximum value of the intermittency factor

$\gamma^*(X)$ for each cross section.

In the region of the beginning of the turbulent spots formation ($\gamma^* \leqslant 0.15$) the moda of probability density functions maintains negative constant value till $Y/\delta^*=0.4 \div 0.5$ and it changes rapidly to zero for the greater Y/δ^*. At the distance from the surface of about three displacement thicknesses the Gaussian curve of exponential distribution was observed. From the experimental data for $0.2 < \gamma^* < 0.6$ we can say that the moda through the thickness of the boundary layer has a clear-cut minimum at $Y/\delta^*=0.4$, that the absolute value increases with the increase of γ^* (Fig.I). The least squares analysis of the experimental data showed that function $M^*=-0.07+0.76\ \ln(1-\gamma^*)$ describes the minimum values of the moda adequately.

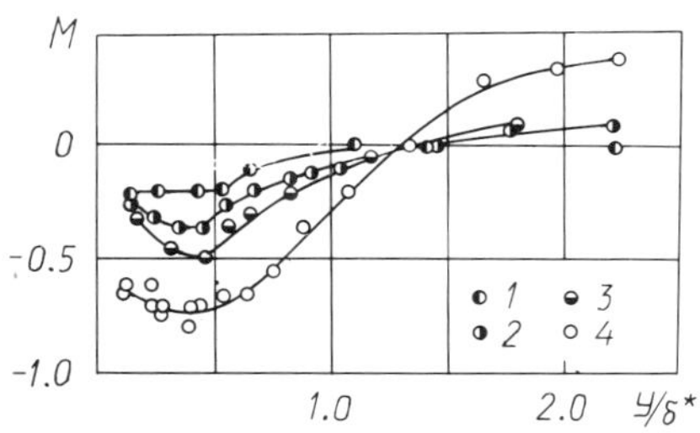

Fig.I. Variation of the moda of the velocity fluctuations distribution in transition. γ^*: I-0.I5; 2-0.32; 3-0.44; 4-0.58.

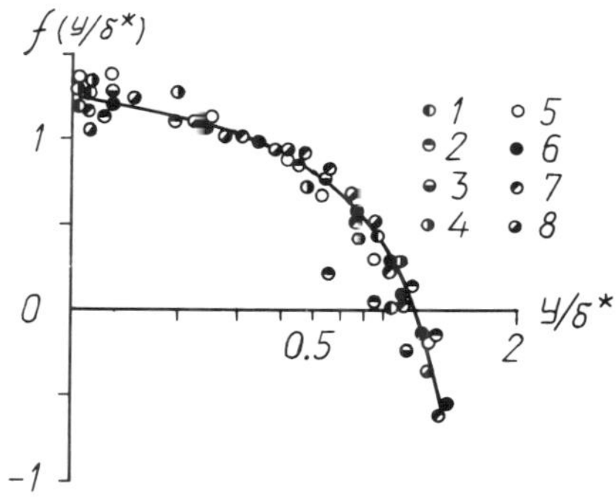

Fig.2. Normalized skewness factor. γ^*: I-0.08; 2-0.20; 3-0.45; 4-0.65; 5-0.I5; 6-0.32; 7-0.44; 8-0.58.

In the near-wall region the bimodal probability density functions are observed which are connected with definite values of the intermittency factor. This is in a good agreement with the results of the investigation on the natural transition in two-dimensial duct flow [3]. The low-frequency velocity fluctuations connected with the appearance and passing of the turbulent spots play the principal role in the formation of bimodal probability density function. This can be confirmed by comparing the probability density functions of velocity fluctuations obtained for just the same tape recording at $\gamma^*=0.58$, but using different high-pass and low-pass filters. The character of probability density functions for I Hz - IO kHz and for I Hz - I kHz bands is identical; there is a clear-cut bimodality (Figs.30,33 of [2]). The symmetrical probability density functions with zero moda were obtained for the velocity fluctuations between I kHz and IO kHz (Fig.34 from [2]).

The analysis of the skewness factor ν in transition showed:

(a) $\nu=0$ for $Y=\delta^*$ when $0.1 \leqslant \gamma^* \leqslant 0.8$ (Y- the distance from the surface, δ^*- the displacement thickness, γ^*- the maximum value of the intermittency factor for the investigated section);

(b) similar character of variation of ν depending on Y/δ^* when $0.1 \leqslant Y/\delta^* \leqslant 1.2$.

It was supposed that the behaviour of the skewness factor ν in the mentioned region could be represented as

$$\nu(\gamma^*;\ Y/\delta^*)=f_I(\gamma^*;\ Y/\delta^*=\text{const})\cdot f(Y/\delta^*)$$

With that assumption the experimental values of the skewness factor obtained for the different intermittency factors were divided by the skewness factor for the respective γ^* at the distance from the surface $Y/\delta^*=0.3$, which corresponds to the region of maximum velocity fluctuations at the end of transition. As a result of such processing we obtained a narrow region for all the experimental data (Fig.2) which can be described by function

$$f(Y/\delta^*)=2-0.709\ \exp\ (1.088\ Y/\delta^*),$$

plotted as a solid curve in the same figure. At the same time the function f_I which defines the value of the skewness factor ν at $Y/\delta^*=0.3$ depending upon the maximum intermittency factor γ^* can be presented by

$$f_I(\gamma^*;\ Y/\delta^*=0.3) = 2.20 - 3.16\gamma^* + 3.05\ \exp\ (-3.8\gamma^*).$$

The experimental data on the skewness factor obtained during the second cycle of investigations are shown in the Fig.3. The solid lines were plotted by using the above mentioned functions $f(Y/\delta^*)$

and $f_I(\gamma^*; Y/\delta^*=0.3)$. There is a good agreement between the experimental data and these curves. The analysis of ν for $\gamma^*=0.58$ showed that the skewness factor of the velocity fluctuations obtained for the I Hz - IO kHz and I Hz - I kHz bands coincide for $Y/\delta^* \leqslant I$, but the skewness factor for the I kHz - IO kHz band has a different character (Fig.4). The further analysis showed that the low-frequency velocity fluctuations connected with the appearance and passing of the turbulent spots define the skewness factor.

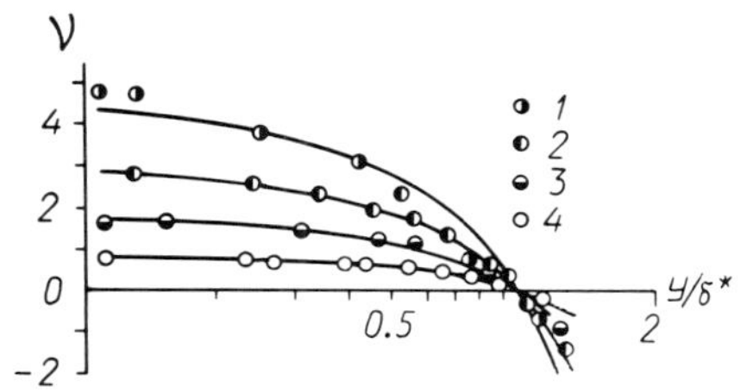

Fig.3. Variation of the skewness factor in transition.
γ^*: I-0.I5; 2-0.32; 3-0.44; 4-0.58.

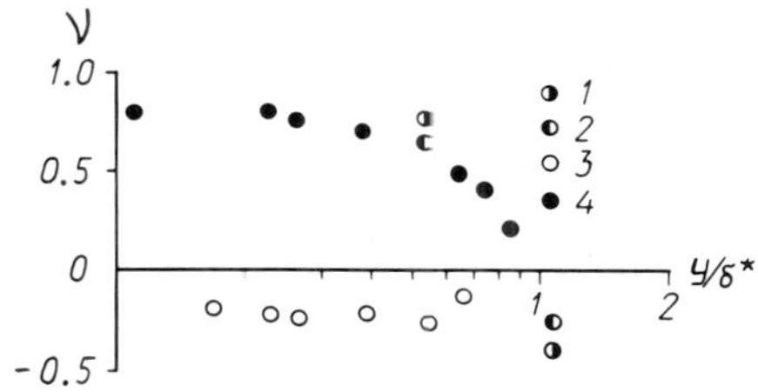

Fig.4. Dependence of the skewness factor on the frequency band at $\gamma^*=0.58$: I-I Hz÷IOkHz, 2- I Hz÷I kHz, 3- IkHz÷IO kHz, 4- coinciding points I and 2.

The results of the investigation on the natural laminar-turbulent transition on a flat plate show the existence of some statistical functions in the near-wall part of the boundary layer.

There is a difference between the regions of maximum growth for low-frequency velocity fluctuations and for high-frequency fluctuations. The experimental data show the minimum value of moda of the distribution of the velocity fluctuations at $Y/\delta^*=0.4$ and its absolute value increases as the intermittency factor increases.

Analysis of the skewness factor ν behaviour showed:
(a) $\nu=0$ for $Y=\delta^*$ when $0.I \leqslant \gamma^* \leqslant 0.8$;

(b) similar character variation of γ depending on Y/δ^*.

The mathematical processing of the experimental data gave simple functions for calculating the skewness factor in the near-wall region of transition.

References

I. Tetianko,V.A.; Shtatnov,Yu.V.: Issledovanie kharakteristic laminarno-turbulentnogo perekhoda.-Pristennoe turbulentnoe techenie. Trudi XVIII Sib. teplophiz.seminara. Red.S.S Kutateladze. Novosibirsk, 1975, ch.2, s.209-223.

2. Tetianko,V.A.:Experimentalnoe issledovanie statisticheskikh kharakteristik pulsatsii skorosti pri perekhode laminarnogo pogranichnogo sloya v turbulentniy. Preprint 70-81. Institut teplophiziki, Novosibirsk, 1981, 41 s.

3. Rajagopalan,S.; Antonia,R.A.: Investigation of Natural Transition in the Inlet Region of a Two-Dimensional Duct Flow.- Phys.Fluids 23(IO), I980, pp.I938-I948.

A Non-Linear Transfer Function Description of Wave Growth in a Boundary Layer

M.GASTER

NMI Ltd,
Teddington, Middlesex, ENGLAND

Linear and quadratic transfer functions relating the velocity fluctuations at a downstream location to an excitation of the flow at some upstream point have been determined by experiment. These functions have then been used to reconstruct the responses arising from various excitations and the results compared with measured hot-wire signals.

1. INTRODUCTION

I am interested in the mechanics of the laminar/turbulent transition processes that occur in boundary layers and other related flows within the class of "open systems". Experimenters studying these types of flow generally spend a great deal of time and effort trying to eliminate, or at least minimise, all forms of extraneous excitation from the environment. The level of turbulence and of acoustic noise in the on-coming flow can be reduced to very low values by careful design of the wind tunnel or flow rig. In some sense this has been self defeating as the signals remaining after all this effort may then be too small to measure precisely. It then becomes necessary to artificially excite the flow by some means so as to generate disturbances that are large enough to study. Mostly two-dimensional periodic wave trains are generated, as these can be compared directly with linearised theory, but other more complex forms of excitation are now sometimes also used. If one is only concerned with the verification of linear theory then the measurement of the evolution of individual Fourier components is sufficient. But there is a need to the study the non-linear phase of evolution involving the interaction of different wave components and then it is essential to consider more complex forms of excitation.

In my own series of experiments at the NMI Ltd I have concentrated on one of the simplest forms of excitation from a point-source on the centre-line of a flat plate. The perturbations downstream of this source have been measured for a variety of different excitations: sine waves, pairs of sine waves, discrete pulses, white noise and for various combinations of these wave forms. In each case the non-linear features that appeared seemed to be unrelated to one another. If each type of excitation really does produce a different non-linear pattern of behaviour then we have enough experimental topics for Ph.D students for the next hundred years or so!. I believe that we need to find more general ways of describing the outcome of such diverse forms of excitation, and this short paper attempts to formulate a way of approaching the problem.

Consider the experiment depicted on figure 1, where the response to some form of input excitation is detected by a hot-wire anemometer at a fixed downstream position. We want to determine - by experiment in this case - a description of the "Black-Box", which may be thought of as a non-linear operator of some sort, capable of forming an output waveform from a given input excitation. In this type of experiment the hot-wire signal will be directly related to the input excitation (apart from any residual noise in the system) through the character of the "Black-Box". At sufficiently low amplitudes the "Black-Box" can be expected to have linear characteristics and a simple linear transfer function, $T_1(Q)$, can then be used to define the phase and gain of each Fourier component:

$$T_1(Q) = \frac{OUT(Q)}{IN(Q)}$$

where IN(Q) and OUT(Q) are respectively the Fourier coefficients of the excitation and the hot-wire response signal at frequency Q. Once defined, either by experiment, or possibly by computation, this transfer function should be capable of describing all linear responses of the flow. The modulated signals arising from white noise excitation, or the wavepackets from isolated pulses can all be described by a linear transfer function. At somewhat larger levels of excitation quadratic, or "wave-wave", interactions arise. This weak form of non-linearity can be defined by the bispectral transfer function, $T_2(N,M)$, which specifies the response at the sum and difference signal frequencies arising from the interaction of two periodic waves of frequency N and M.

$$T_{\pm 2}(Q) = \frac{OUT(Q)}{IN(N) * IN(M)} \quad , \quad Q = M \pm N$$

In principle transfer functions of triple or higher degree can also be defined, but here we will only attempt to describe the first stages of the non-linear evolution through the bispectral transfer function.

2. EXPERIMENTAL SET-UP

The transfer functions T_1 & T_2 were determined from measurements of the hot-wire response at one downstream point in the flow arising from excitations at an upstream location. Excitation at a point on the centre-line of a flat plate some 20 cm from the leading edge was provided by a buried microphone acting through a small hole in the plate's surface. The measuring station was 70 cm downstream of this, or at 90 cm from the leading edge. The electrical signal used to drive the microphone was generated by a micro-computer via a 12 bit digital-to-analogue converter. The excitation signal consisted of pairs of sine waves of different frequencies summed together to form a record of 1024 values. The signal frequencies were defined so that an integral number of cycles in the range 1-32 were formed over the 1024 samples. This waveform was clocked out to the D/A converter at 16 kHz repeatedly some 256 times. The hot-wire response to this excitation was simultaneously recorded in digital form as an ensemble average. The BBC micro-computer was also interfaced to a digital 9-track tape recorder so that both input and output data could be stored for further processing on a main-frame or mini-computer to form the cross-spectral and cross-bispectral responses. The wind tunnel was run a speed of 18.4 m/s and the T/S waves were in the range 100 to 300 Hz.

3. RESULTS

The response of the boundary layer flow as registered by the hot-wire anemometer for the case when the exciter was driven by single sine waves of various frequencies at different amplitudes is shown on figure 2. The magnitude of the hot-wire signal, OUT(Q), versus the excitation, IN(Q), shows a reasonably linear behaviour at low excitation levels. Examples of the quadratic sum and difference responses to pairs of sine waves of different amplitudes are shown on figures 3 & 4. Again straight lines can be drawn through both sets of data points and the origin showing that the quantities T_1 and T_2 are independent of exciter amplitude, at least over the range of this experiment. Actual frequencies driving the microphone are given by the index multiplied by 16 Hz.

The BBC micro-computer was programmed to select all pairs of frequencies in the index range 1-32, or rather one half of this number because of symmetry. For each combination of waves the anemometer response was recorded for some 16 seconds and formed into an ensemble average before being written to magnetic tape for later analysis on a large machine. This process took roughly 5 hours to complete for some 512 cases. The resulting linear behaviour is depicted on the perspective display on figure 5 which shows the modulus of the transfer function $T_1(N)$ over the N & M surface - data was only acquired over one half of the plane, but it has been reflected in the axis of symmetry along the diagonal. The response shows that $T_1(N)$ is independent of the value of M, as expected for a linear system. The shape of this response results from the selective amplification of Tollmien-Schlicting waves. The sum and difference bispectral responses are shown on figures 5 & 6 respectively, again using the symmetry condition about the diagonal. In these displays the surfaces are less smooth than in the linear case and there is more variability, nevertheless there are clearly zones where there are significant non-linear effects given by values of $T_2'(N,M)$.

4. SYNTHESIS

The experimentally measured values of both $T_1(N)$ and $T_2(N,M)$ define the linear and weakly non-linear behaviour expected for any form of excitation of appropriate amplitude. We can therefore use these quantities to "predict" the outcome of an arbitrary excitation. For example, consider the response from white noise excitation of different amplitudes. Figure 8 shows a set of four power spectra generated by a white noise source at various levels. At low excitation, (A), the hot-wire signal consists mainly of the parabolic response of the T/S waves together with a certain amount of additional low frequency background noise from the free-stream. Increasing levels of excitation, (B)-(C)-(D), show that the primary T/S response increases in step with the level of drive, but there is also additional power at roughly one half to three quarters of the main peak and a slight hint of some power at the second harmonic of the dominant peak. Figure 9 contains the reconstructed power spectra computed from the measured transfer functions at different excitation levels. It can clearly be seen that the larger amplitudes provide a slight bulge to the spectrum below the peak as well as some second harmonic. I must confess that the experimental spectra shown on figure 8 were recorded some 10 cm downstream of the present work and the comparisons should be treated as only qualitative.

A second illustration is provided by the wavepacket created by a discrete pulse. Figure 10 contains the measured hot-wire signals obtained at three different levels of excitation scaled by the amplitude of the driving pulse. Small pulses of either sign produced virtually identical records, shown by (B), whereas larger excitations generated the distortions shown on (A) & (C) for positive and negative signs respectively. These deviations from the low amplitude linear result are different for positive and negative pulses. The signature of the reconstructed wavepacket for a low amplitude pulse and for large positive and negative pulses are shown on figure 11. Although the reconstructed records are not identical to those measured directly there are a number of quite similar features. The peak-to-peak amplitudes are increased by a large positive pulse and reduced by a negative one, and the wave distortions towards the tail of the packets are fairly similar in experiment and reconstruction.

5. CONCLUDING REMARKS

A first attempt has been made to measure a non-linear transfer function linking the hot-wire response to a given input excitation. The approach indicates that the initial non-linear evolution of the signals generated by white noise and by isolated pulses are first influenced by quadratic wave-wave interactions as described by the bispectral transfer function. In particular, the initially rather surprising finding that the sign of the pulse affects the non-linear behaviour of a wavepacket has been confirmed by this simple model.

The rather crude algebraic relationship considered initially is not really appropriate to the boundary layer problem which behaves more like a distributed system consisting of a large number of elemental "Black-Boxes". Each of the transfer functions operating across slices of the flow will have slowly changing characteristics to account for variations in Reynolds number and boundary layer scale. Some of the important characteristics of a distributed system can, however, be obtained by treating the distributed transfer functions as fractions of T_1 and T_2 over the spatial range of the experiment. Then the relationship between input and response is given by the set of first order differential equations:-

$$\frac{\partial}{\partial x} SIG(Q,x) = SIG(Q,x) * \frac{\partial}{\partial x} T_1(Q,x) + \sum_{M \neq N} SIG(N,x) * SIG(M,x) * \frac{\partial T_{\pm 2}}{\partial x}(Q,x)$$

The solution of this set of equations for the 32 coefficients used here is a straightforward task that can be carried out for any arbitrary input excitation. Not only will the result of such a calculation provide the linear and quadratic responses from which the raw coefficients were evaluated, but certain cubic and higher degree elements should also be generated. It is these components that contribute to the large broad band motion often observed far downstream in experiment. These disturbances will almost certainly be three-dimensional and may be linked with known forms of non-linear interaction, but such considerations do not affect the ideas expressed here. It is expected that a more refined experiment will be able to provide quantitative correlations with direct measurements for a wide range of excitations, and will therefore help to bring out the important non linear interactions.

6. REFERENCE

1. Miksad, R.W.; Kim, Y.C.; Beall, J.M. and Powers, E.J.: Bispectrum and Non-Linear Wave Coupling, Physics of Fluids, 23 (1980) 258-263.

This work is supported by the USAF under contract F49620-83-C-0146DER, and by the Procurement Executive Ministry of Defence, ARE.

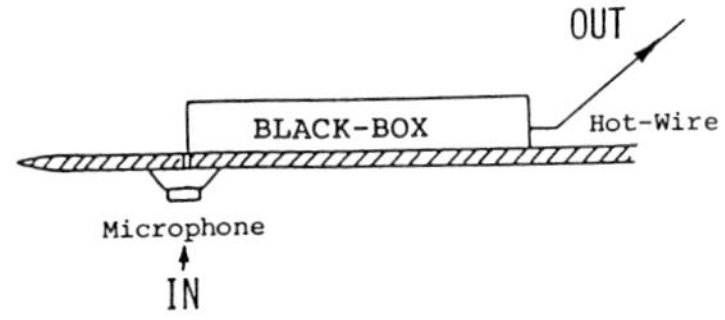

Fig.1 Experimental Set-up

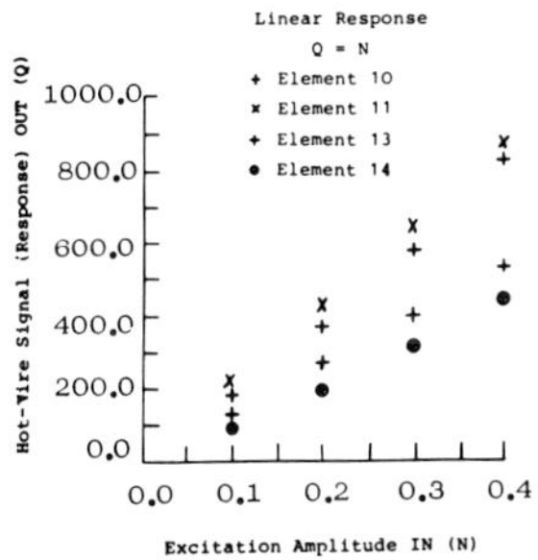

Fig.2. Response to Sine Wave

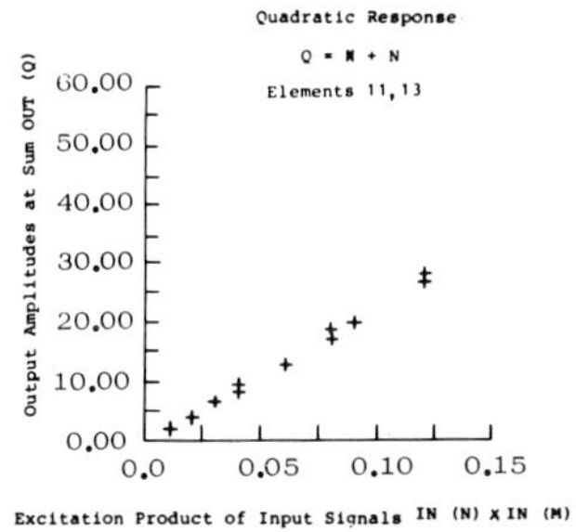

Fig.3. Response to Pairs of Waves

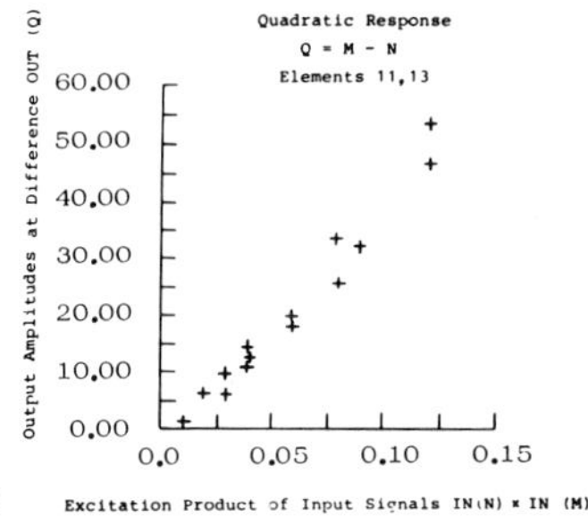

Fig.4. Response to Pairs of Wave

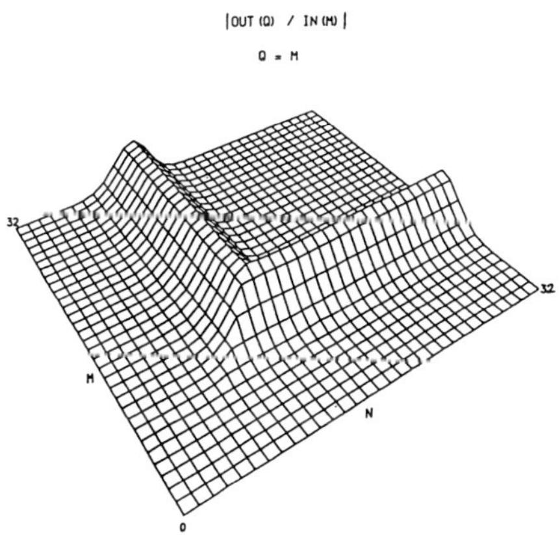

Fig.5 Linear Response

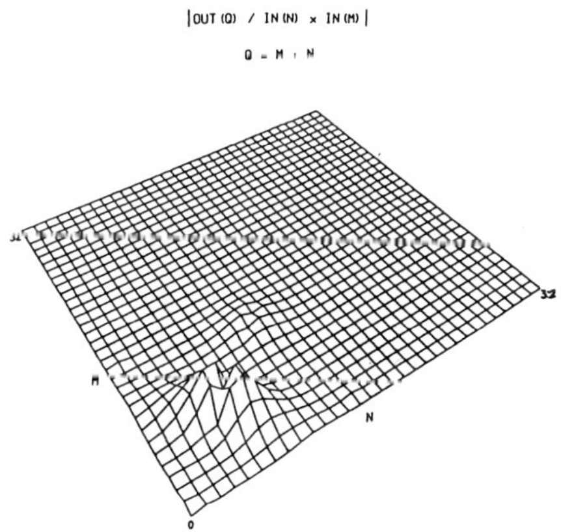

Fig.6 Quadratic Summation Response

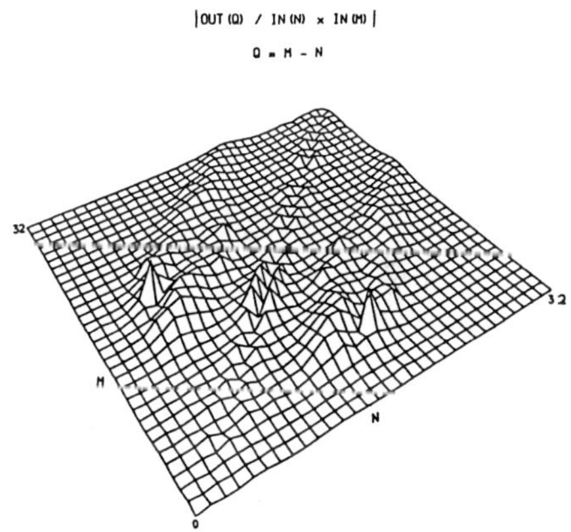

Fig.7 Quadratic Difference Response

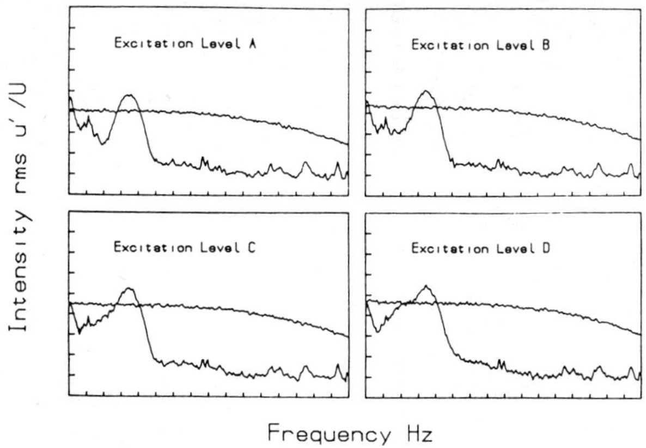

Fig.8 Measured Response to White Noise

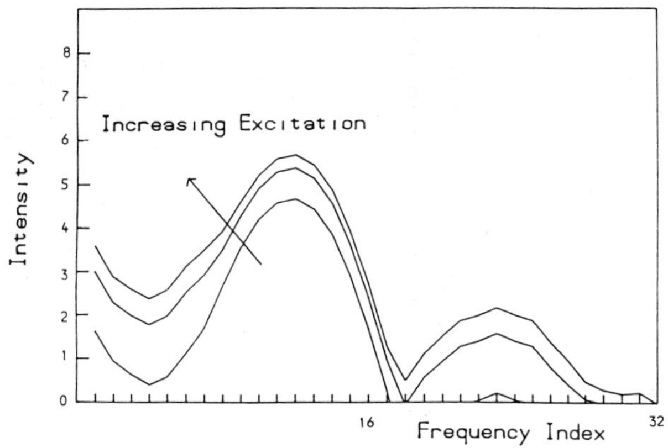

Fig.9 Reconstructed Response to White Noise

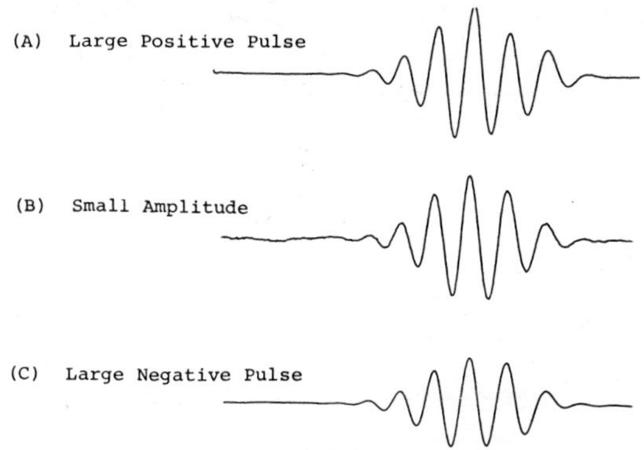

Fig.10 Measured Wave Packet

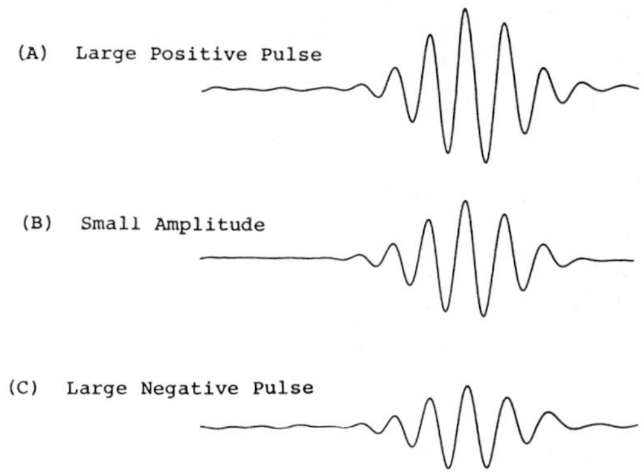

Fig.11 Reconstructed Wave Packet

Development of Spatial Wave Packets in Boundary Layer

Yu. S. KACHANOV

Institute of Theoretical and Applied Mechanics
Siberian Division of the USSR Academy of Sciences,
630090 Novosibirsk, USSR

Summary

The main results of the experimental investigation on an evolution of localized in space and harmonic in time linear packets of Blasius boundary layer waves are presented. A complex frequency-wavelength Fourier analysis of the data permits to study stability characteristics (including dispersive ones) for a fan of 3-D waves in a wide range of propagation angles.

Introduction

The pulsations observed in "natural" transition experiments represent quasi-harmonic in time oscillations corresponding to a rather narrow Tollmien-Schlichting wave (T-S wave) packet selectively amplified by an unstable boundary layer from an initial pulsation spectrum [1] . In space there takes place a complicated dependence of amplitudes and phases of frequency spectrum components on the spanwise coordinate z. In references [2]-[5] and others different experimental transition models were studied when introducing in the flow controlled determined T-S waves simulating a "natural" frequency-wavelength pulsation spectrum.

In the model which has been investigated in references [5] - [7] the attention was focussed on the pulsation splashes localized in space and in time observed in a "natural" Tollmien-Schlichting wave with negligibly small oscillations between splashes. The study of such transition model seems to be of much prospect [8] . In references [6] - [7] for the first time frequency-wavelength pulsation spectra were obtained which can be compared (and this was done in [7]) with the calculations according to a linear stability theory. Unfortunately, these are nothing but amplitude spectra containing no phase information which did not permit the authors [6] to determine x-components of wave vectors (α -components) and, consequently, to determine the propagation angles of disturbances. The second drawback of the works [6], [7] is the absence of the information on the y-profiles of disturbances and, possibly, an unsuccessful choice of a fixed coordinate for all the measurements beyond the outer edge of the boundary layer. In the present study a complete complex Fourier analysis of disturbances was carried out. It enables one to obtain a fan of oblique waves with known wave vectors, phase velocities, etc. The structure of eigenfunctions of oblique waves was also studied.

The experiments were conducted in the subsonic low-turbulence wind tunnel at the Institute of Theoretical and Applied Mechanics, at freestream velocity u_∞ = 9.18 m/s and turbulence level less than 0.02%. The disturbances were sinusoidal in time and localized along z and were introduced into the laminar boundary layer on a flat plate mounted at a zero angle of attack (the technique is described in [8]). The signal frequency was f_1 = 81.4 Hz in a principle regime and in two others f_2 = 55 Hz, f_3 = 120 Hz, which correspond to frequency parameters $F = 2\pi f \nu / u_\infty^2$: $F_1 = 94\cdot 10^{-6}$, $F_2 = 64\cdot 10^{-6}$, $F_3 = 139\cdot 10^{-6}$. The T-S wave packet evolution was registered by the equipment based on a hot-wire anemometer with the linearizer. A frequency-wavelength analysis of disturbances was conducted with the help of a computer. Apart from these measurements the study of plane and oblique waves (harmonic in space) was carried out. These waves were introduced into the boundary layer by means of a vibrating ribbon stretched over the plate surface at different angles relative to the mean flow (0°, 20°, 42°).

Evolution of Wave Packets

Isolines of the pulsation intensity and phases (with a half a period interval) downstream the source showing the spreading of the spatial wave packet is presented in Fig. 1. The measurements were carried out near the interior pulsations maximum at u = const = 3.82 m/s (i.e. y/δ = 0.25) in a principal re-

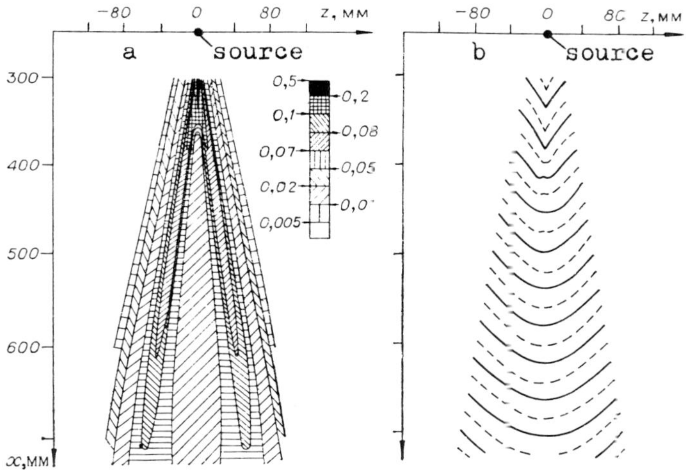

Fig.1. Isolines of amplitudes (a) and phases (b) of disturbances

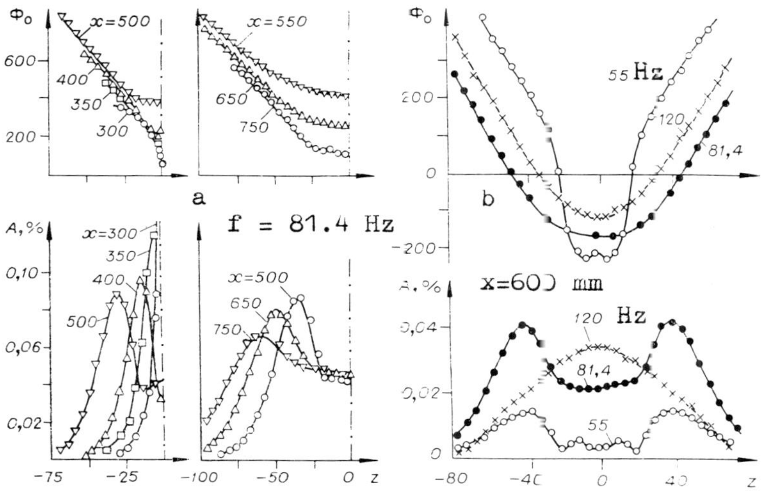

Fig.2. Z-distributions of amplitudes and phases of disturbances

gime (f_1 = 81.4 Hz). The sections of two-dimensional distributions presented in Fig. 1 at some fixed values of the x-coordinate are given in Fig. 2a. From the distributions presented here it is extremely difficult to distinguish the features showing some generalized objective properties of the spatial wave packet evolution disregarding the subjective properties of the source and the values of the parameters Re, F. The distributions of amplitudes and phases of the disturbances along z obtained for three different frequency values demonstrate this fact other conditions being equal (Fig. 2b). The measurement of the distributions in the region of the external maximum ($y \approx 4.5$ mm) showed that they are "smoother" (see [8]) and, consequently, far from the wall distributions A(z) and $\Phi(z)$ contain less quantity of pulsations with high wavenumber β . The x-distributions of amplitudes of the T-S wave packet obtained along the central line z = 0 for disturbances frequency f_1 = 81.4 Hz (Fig. 3a) show that the amplification rate of disturbance amplitude along the external maximum (y = 4.5 mm) is higher than in the region of the internal one (y/δ = 0.25) which is explained by an above-mentioned more rapid attenuation of three-dimensionality near the outer edge of a boundary layer and, consequently, by approaching the amplification rate of the spatial wave packet to a plane wave amplification. The amplification curves of the wave phases (Fig. 3b) for all three cases are quite linear and at $x > 450$ mm their amplification rates $\Delta\varphi/\Delta x$ are practically equal. This circumstance testifies to a weak dependence $\alpha = \alpha(\beta)$ in an energy-containing part of a frequency-wavelength spectrum. The evolu-

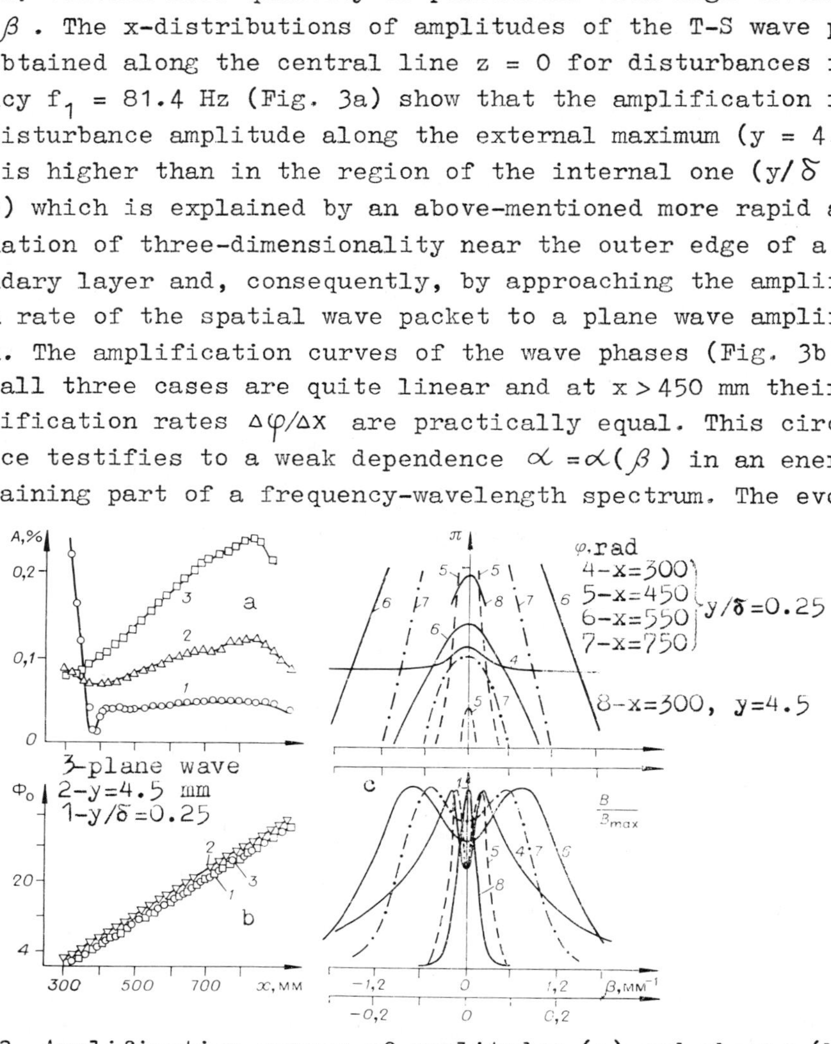

Fig.3. Amplification curves of amplitudes (a) and phases (b). Wavelength spectra (c).

tion of disturbances is characterized by a complicated rearrangement of amplitude and phase pulsation of y-profiles showing an interference of waves with different wave vectors. A coincidence of amplitude and phase pulsations distributions along y and z [8] measured at different initial amplitudes (normalized) showed that the superposition principle was present and that the process of the disturbance evolution was linear.

Frequency-Wavelength Spectrum Evolution

Amplitude wavelength spectra are characterized by two maxima showing that the disturbances with corresponding wavenumbers prevail in the spectra, see Fig. 3c. (Though it doesn't mean that the disturbances with indicated wavenumbers possess prevailing amplification rates.) The phase spectra (Fig. 3d) undergo great **changes** from a nearly plane one at x = 300 mm to the spectra with a rapid phase decrease while the absolute

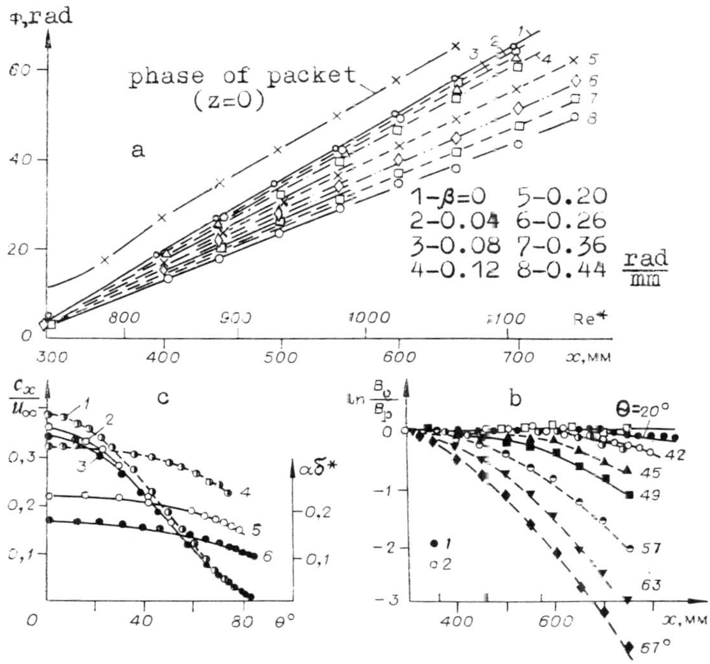

Fig.4. Amplification curves of phases (a) and amplitudes (b) of fan of waves. Dispersive dependences (c): 1,2,3 - C_x/u_∞ and 4,5,6 - $\alpha\delta^*$ for F = 139, 94, $64\cdot10^{-6}$.

value of β is growing. The linear character of the shown in Fig. 4a dependencies $\varphi(x)$ for some wavenumbers β as well as a rather slow downstream change of wave amplitudes permit to draw a conclusion that for each fixed z-component of the wave vector β the wavelength spectrum relative to α (x-component of the wavevector) in all the x-range has the form of a very narrow strip so that it can be practically identified with a harmonic which has a definite wavenumber $\alpha = \alpha(\beta)$ and is equal to a tangent of the inclination angle of the curves in Fig. 4a. In Fig. 4c there are presented dependencies of the nondimensional wavenumber α and of x-component of the phase velocity $C_x = \omega\alpha/(\alpha^2 + \beta^2)$ on the angle $\theta = \arctan(\beta/\alpha)$ for different disturbance frequencies. The corresponding dispersion curves $C_x(f)$ drawn up for different values of θ are presented in Fig. 5b.

In Fig. 4b there are presented amplification curves of relative amplitudes of 3-D waves $\ln(B_0/B_p)$ (where B_0 and B_p are the amplitudes of an oblique and a plane waves) versus θ. The amplification rate lags of all 3-D waves are quite obvious in all x-range. The evolution of the wave amplitude spectrum nor-

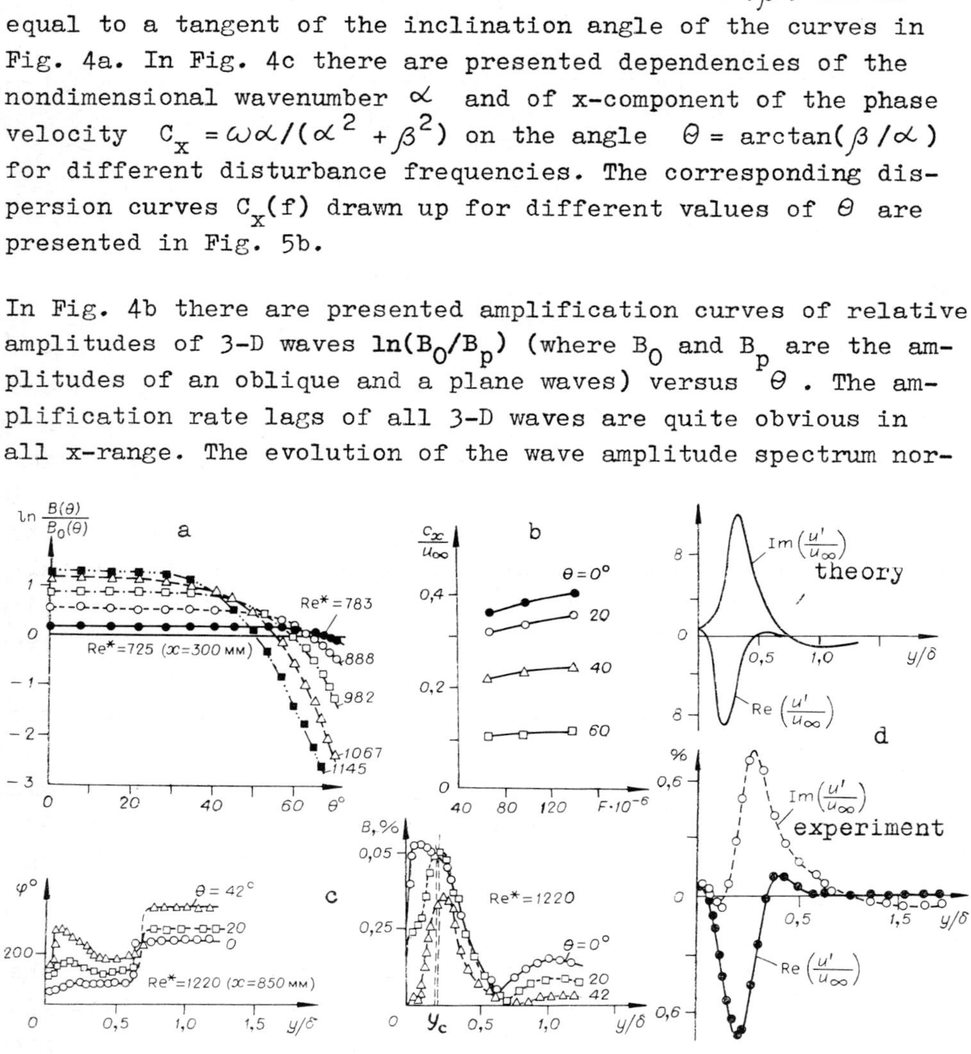

Fig.5. Normalized wavelength spectra (a), dispersive curves (b) and y-profiles (c,d) for 3-D waves.

malized with respect to the initial spectrum at R_{e*} = 725 (x = 390) is shown in Fig. 5a. The normalization enables to exclude from the spectrum such a subjective information as the initial spectral composition which depends entirely on the properties of the specific disturbance sources and to single out the laws conditioned by the stability of the boundary layer towards the disturbances propagating at different angles relative to the flow.

The Fourier analysis of the disturbances was also carried out at the level of the external maximum of the pulsations used in the measurements [6] . The corresponding wavelength spectra of the pulsations given in Fig. 3c (see also [8]) show the constriction of spectra along β at this distance up to the wall in comparison with the spectra at y/δ = 0.25.

Evolution of Harmonic Spatial Waves

The fronts of T-S waves for the angles Θ = 20° and 42° obtained from the phase distributions of disturbances presented straight lines inclined to the flow at the angle $\pi/2 - \Theta$ in the range $400 \leqslant x \leqslant 700$ mm, $-125 \leqslant z \leqslant +125$ mm (see [8]). The measurements of the wave phase dependence $\varphi = \mho(x)$ for angles Θ = 20° and 42° showed a complete coincidence of the obtained results with the data of the previous point (Fig. 4a). Thus, the values of the x-component of the wavenumber constituted α = 0.150 and α = 0.141 for the angles Θ = 20° and 42° as at the previous point. The measurements were carried out at the level y/δ = 0.25. The amplification curves of the disturbances presented in Fig. 4b (points 1,2) also are in agreement with the earlier data and define them more exactly in the range of small angles. The velocity growth lag for the wave having Θ = 20° from the plane one (Θ = 0°) is in good agreement with the calculation results (see [8]).

Of great interest are the amplitude and phase profiles of the pulsation some of which are presented in Fig. 5c for Re*=1220 (x = 850 mm). With the growth of the angle Θ there is observed a considerable decrease in the amplitude of pulsations near the wall and a corresponding constriction and displacement of the maximum towards the outer edge of the boundary layer. A relative stability of position of the maximum of oblique waves together with the usually observed shift of the plane wave maximum leads to a growing discrepancy of maximum position at x-increasing. From this it may be concluded that a too "lazy" lag in the growth of harmonics propagating at small angles ("shelves" at $\theta < 30°$ in spectra, Fig. 5a), is connected with the choice of the distance between a probe and the wall (y/δ = 0.25). The phase distributions in the case of oblique waves acquire a form of a "dipper" with a typical phase increasing at $y/\delta < 0.4$. But the phase jump at the outer edge and its constant value after the jump remain an invariable property of the phase profiles. And, finally, it strikes the eye that the amplitude of pulsations in the region of the external maximum ($y/\delta = 1.1$) rapidly decreases with increasing θ. This fact was repeatedly observed earlier and testifies to the "smoothing down" of the wave spectra when conducting the measurements in the region of the external maximum. In Fig. 5d there are presented the distributions of the real and imaginary part of the eigenfunctions (theory [9]) of the 3-D wave inclined at the angle $\theta = 37.4°$ ($Re^* = 882$, $\alpha\delta^* = 0.25$). Except for a small region near the wall the profiles obtained from the experiment ($Re^* = 1220$, $\alpha\delta^* = 0.2$, $\theta = 42°$) coincide qualitatively with the theoretical ones.

References

1. Schubauer, G.B.; Scramsted, H.K. NACA Rep. 909 (1948).

2. Kachanov, Yu.S.; Kozlov, V.V.; Levchenko, V.Ya. Izv. Akad. Nauk SSSR, Mekh. Zhidk. i Gaza 3 (1977) 49-58 (in Russian). (See also Fluid Dyn. 12 (1978) 383).

3. Klebanoff, P.S.; Tidstrom, K.D.; Sargent, L.M. J. Fluid Mech. 12 (1962) 1-41.

4. Kachanov, Yu.S.; Kozlov, V.V.; Levchenko, V.Ya. In: Laminar-Turbulent Transition. Berlin, Springer-Verlag, 1980 (See also Preprints N 16, Novosibirsk, ITPM SO AN SSSR, 1978 (in Russian)).

5. Gaster, M. In: Twelfth Naval Hydrodynamics Symposium, Washington 1978.

6. Gaster, M.; Grant, I. Proc. Roy. Soc. Lond. A.347 (1975) 253-269.

7. Gaster, M. Proc. Roy. Soc. Lond. A.347 (1975) 271-289.

8. Gilyov, V.M.; Kachanov, Yu.S.; Kozlov, V.V. Preprint N 34, Novosibirsk, ITPM SO AN SSSR, 1981 (in Russian).

9. Usher, J.R.; Craik, A.D.D. (Appendix by F.Hendriks) J. Fluid Mech. 70 (1975) 437-461.

Instability Wave Patterns from Harmonic Point and Line Sources in Laminar Boundary Layers

LESLIE M. MACK

Jet Propulsion Laboratory
California Institute of Technology
Pasadena, California 91109
U.S.A.

Summary

This paper presents the results of recent calculations of wave patterns produced by point and line harmonic sources in laminar boundary layers. It is shown that quasi-parallel stability theory can produce results that are in good agreement with experimental measurements in both the Blasius and rotating-disk boundary layers.

Introduction

Instability waves are unlikely to be produced as single normal modes, but as a spectrum of phase-related normal modes of many frequencies and orientations which is then filtered by the amplification and damping mechanism of the boundary layer. The simplest problem of this type is a harmonic point source, where the normal modes produced by the source have the frequency of the source and all possible orientations. A harmonic point source in a Blasius boundary layer was the subject of a detailed experiment by Gilev, Kachanov and Kozlov [1], which revealed that the wave pattern formed downstream of the source is of considerable complexity. Other measurements, as well as calculations based on the quasi-parallel stability theory, were later carried out by Mack and Kendall [2] (see also Chapter 7 of [3]). As in previous work by Gaster [4,5] on wave packets, the calculations of [2] made use of both the direct numerical evaluation of the Fourier integral over spanwise wavenumber, and on the method of steepest descent as modified by Gaster for a growing boundary layer [5]. The experimental results of Wilkinson and Malik [6] for the rotating-disk boundary layer, which demonstrated that the steady waves responsible for the spiral streaks that appear in flow-visualization experiments emanate from random point sources on the disk surface such as dust particles, were the inspiration for extending similar calculations to this flow also [3,7].

Once the point-source solution is known, line-source solutions can be obtained by superposition. A number of these results were given in [8] for finite-length spanwise and oblique line sources in a Blasius boundary layer. The former is of interest because it provides a simulation of a vibrating ribbon. Some additional information on an oblique line source is given here.

Harmonic point source - Blasius boundary layer

With the assumption that a point source produces a uniform orientation spectrum with respect to the spanwise wavenumber β, the downstream wave pattern can be obtained by numerically evaluating the Fourier integral

$$u(x,z,F) = (1/2\pi)\int_{-\infty}^{\infty} \exp[i\chi(\beta;x,z)]d\beta. \tag{1}$$

We have ignored the eigenfunctions, so that the quantity u is independent of both the flow variable and the distance from the wall (y), and is equivalent to the usual complex amplitude ratio of the quasi-parallel stability theory. All quantities have been made dimensionless with respect to the freestream velocity U_1^* and the inverse unit Reynolds number ν^*/U_1^* as velocity and length scales, respectively (asterisks refer to dimensional quantities). The (circular) frequency of the source is ω^*, and the dimensionless frequency F is $\omega^*\nu^*/U_1^{*2}$. With the source at $x=x_s$, $z=0$, the complex phase is

$$\chi(\beta;x,z) = \int_{x_s}^{x} \alpha dx + \beta z , \tag{2}$$

where α is the compex wavenumber in the streamwise (x) direction. The phase is evaluated by the quasi-parallel spatial stability theory with both the frequency and the spanwise wavenumber real and constant. The latter condition is the irrotationality condition of kinematic wave theory.

Gilev et al. [1] measured the constant-phase lines at a constant value of the Blasius similarity variable. They found the lines to be V-shaped near the source and arc shaped far from it, with a transition region in between characterized by the appearance of a dimple near the centerline. Figure 1 shows the calculated constant-phase lines for a dimensionless frequency of $F=0.60\times10^{-4}$, and the source located at $R_s=485$, where $R=x^{1/2}$ $(x=U_1^*x^*/\nu^*)$.

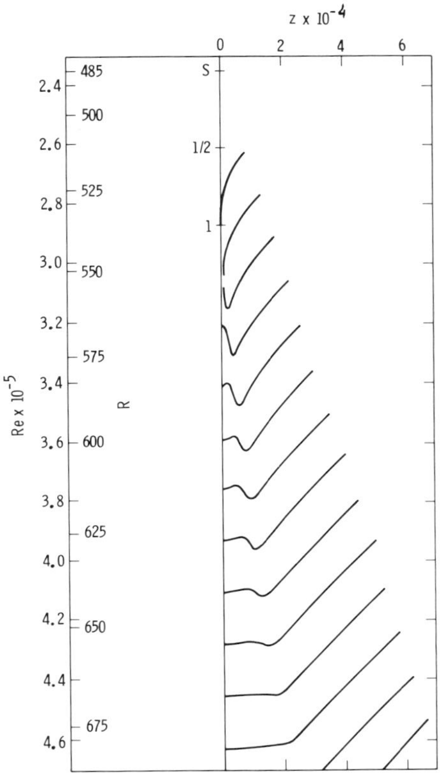

Fig. 1. Constant-phase lines, one-half wavelength apart, of wave pattern produced by $F=0.60\times10^{-4}$ point source located at $R_s=485$ in Blasius boundary layer.

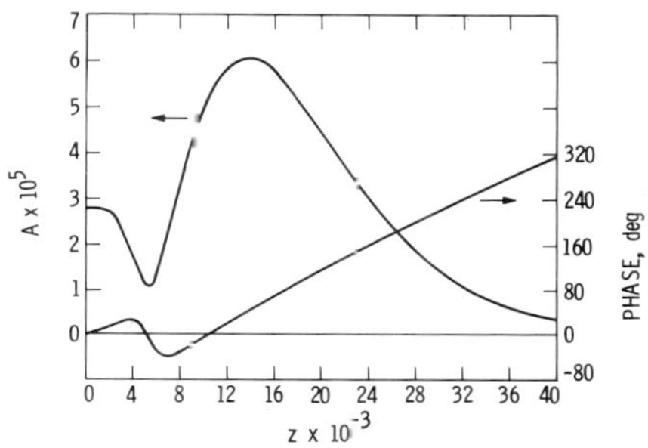

Fig. 2. Spanwise distribution of amplitude and phase at R=600 for $F=0.60\times10^{-4}$ point source located at $R_s=485$ in Blasius boundary layer.

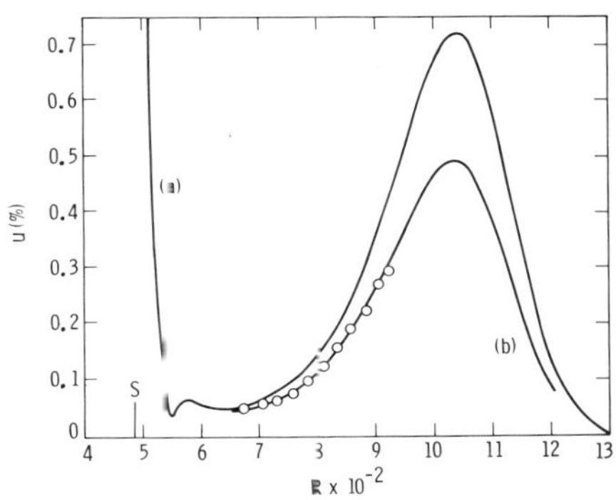

Fig. 3. Centerline amplitude distribution behind $F=0.60\times10^{-4}$ point source located at $R_s=485$ in Blasius boundary layer; (a) Eq. (1); (b) with $x^{-1/4}$ correction; o, JPL experiment.

The constant phase lines are one-half wavelength apart, and the first one shown is one wavelength from the source. The sequence of events moving downstream from the source is similar, but not identical, to the measurements of [1] for a different frequency and source location. The arc-shaped constant-phase lines measured in [1] far from the source appear in this example for R>750.

It can be observed from Fig. 1 that along and very close to the centerline the wavelength count is one less than for $z\geq1000$. This is because the phase, which increases monotonically with z for R<555, increases only slightly for $R \geq 555$ before it decreases sharply, reaches a minimum, and then increases monotonically with z. This change in behavior occurs just below R=555, where there is a point of zero amplitude near z=1000. Two constant-phase lines that are one-half wavelength apart actually join at this point. In effect, a phase discontinuity of 2π radians exists on the centerline at R=555.

Typical spanwise distributions of amplitude and phase (referenced to the centerline phase) are shown in Fig. 2 at R=600. The off-centerline location of the maximum amplitude is established close to the source and persists to about R=825. The centerline amplitude distribution, suitably normalized, is shown in Fig. 3 and compared to measurements of Mack and Kendall [2] of the peak velocity fluctuation (as percent of U_1^*) in the boundary layer. When the calculated amplitudes are corrected by the factor $x^{-1/4}$ used by Gaster [4] to account for the growth of the boundary layer, the agreement between theory and experiment is excellent. The method of steepest descent, as modified by Gaster [5] for a growing boundary layer, gives an explicit formula for the centerline amplitude ratio which, for R>650, yields the same result shown in Fig. 3. For z>0, this method is less successful until the off centerline peaks have disappeared, when it then gives nearly the same results as does numerical integration.

Harmonic point source - Rotating-disk boundary layer

The steady wave pattern produced by a point source of zero frequency on a disk rotating at angular velocity Ω^* can be calculated by the same numerical-integration method used for the Blasius boundary layer. An important difference, however, is that the sixth-order stability equations of Malik, Wilkinson and Orszag [9], which include the effects of Coriolis force, streamline curvature and axial mean velocity, must be used in place of the usual fourth-order system.

We show two principal results that may be compared directly with the measurements of [6]. For zero frequency and the source located at R=250, where $R=r^*(\Omega^*/\nu^*)^{1/2}$ and r^* is the radius, the wave forms normalized by the maximum amplitude at each radius and the constant-phase lines of the wave pattern are given in Fig. 4. The corresponding amplitude distributions along the constant-phase lines are given in Fig. 5. The

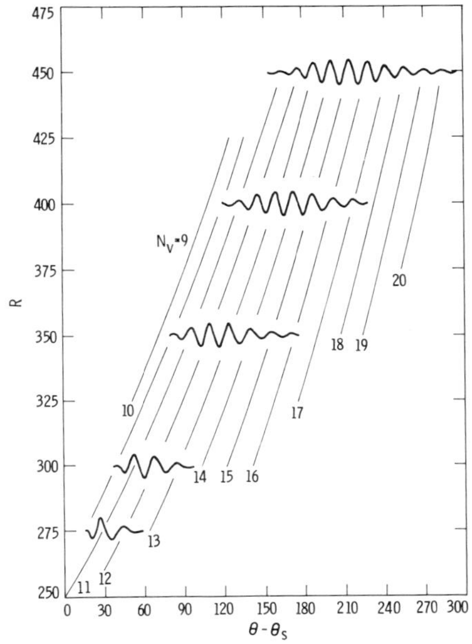

Fig. 4. Normalized wave forms and constant-phase lines of wave pattern produced by zero-frequency point source located at R_s=250 in rotating-disk boundary layer.

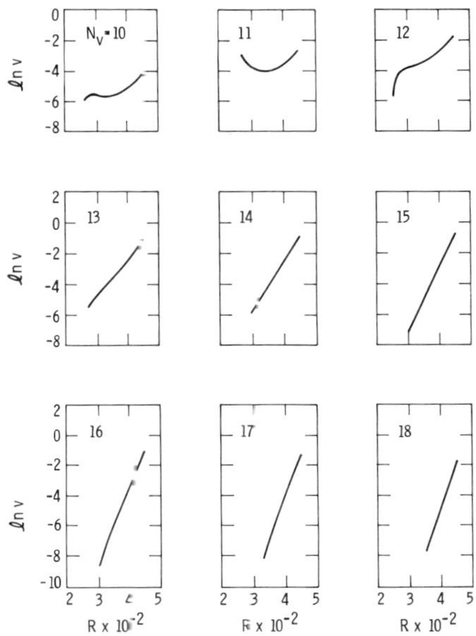

Fig. 5. Amplitude distributions along nine constant-phase lines behind zero-frequency point source located at R_s=250 in rotating-disk boundary layer.

numbering of the constant-phase lines, which correspond to the spiral streaks in flow-visualization experiments, is the same as in [6]. In Fig. 5, the maximum amplitude at R=400 has been set equal to the maximum experimental amplitude (in m/sec) at R=397. These results, which are a corrected version of those in Chapter 12 of [3] where three of the four axial mean-velocity terms were inadvertently omitted, are presented in much more detail in [7]. The essential point to note is that in virtually all respects the present approach gives results that are remarkably close to the measurements of Wilkinson and Malik [6]. This agreement demonstrates that the steady spiral streaks that occur on a rotating disk are a consequence of the superposition of normal modes of all wavenumbers, both amplified and clamped.

It is necessary to point out that there are two solutions at each azimuthal wavenumber in the Reynolds number range of Figs. 4 and 5. The family of solutions with the larger amplification rates has been chosen for the numerical calculations. For R>450, the short wavelength amplified part of the first solution transfers to the second solution, and the consequences of this change have not yet been worked out.

Oblique line sources

Once the point-source solution is known, the wave patterns from line sources of arbitrary shape can be built up by the superposition of closely spaced point sources. Several line-source problems were solved in this manner by Mack [8]. For example, a spanwise source of finite length simulates a vibrating ribbon, and the calculations of [8] show that the influence of the source tips propagates at an angle of about 16° to the freestream direction.

Another problem of interest is the wave pattern formed by an oblique line source in a growing boundary layer. This technique has been used experimentally to produce oblique waves, e.g., by Kendall [10] and by Gilev et al. [1], but it has never been clear to what extent the growing boundary layer causes a departure from the oblique normal mode that would occur in a truly parallel flow.

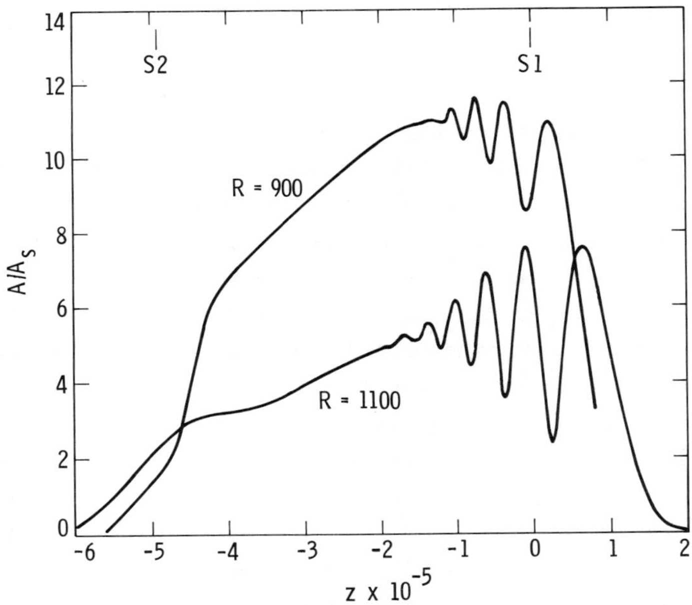

Fig. 6. Spanwise amplitude distribution at two Reynolds numbers behind 30° line source with $F=0.60\times10^{-4}$; upstream tip (S1) at R=450, z=0; downstream tip (S2) at R=696.7, $z=4.90\times10^{5}$.

Typical results for the spanwise amplitude distribution are shown in Fig. 6 at two downstream locations for a line source of frequency $F=0.60 \times 10^{-4}$ at an angle of 30° with respect to the z axis in a Blasius boundary layer. The source, which has a spanwise length of 4.9×10^{5} in units of z, has its upstream tip at R=450, z=0 and was represented by 99 equally spaced point sources. There is indeed a spanwise region at each R of nearly constant wave angle (to less than 0.1°), but the wave angle is slightly greater than the angle of the source and increases with increasing x. In these regions of nearly constant wave angle, the amplification rate and wavenumber components as calculated from the amplitude and phase at neighboring stations, all vary slightly in the spanwise direction. However, along a streamwise path at a constant value of z, the wave development is close to that of a normal mode with its initial wave angle equal to that of the source and which satisfies the simplified irrotationality condition of constant spanwise wavenumber that is appropriate for a wave motion independent of z in a two-dimensional boundary layer. The small departure from normal-mode behavior is consistent with the general irrotationality condition on the complex wavenumber vector, and the small spanwise variations of the wavenumber and amplification rate.

Concluding Remarks

The quasi-parallel stability theory has been shown to be a useful tool for point-source problems in both Blasius and rotating-disk boundary layers. Even though the amplitude and phase given by this theory have no obvious connection with any physically measureable quantity, some results in good accord with consistently measured experimental quantities have been calculated. Remarkably close agreement of waveforms and constant phase lines was obtained for the rotating-disk boundary layer with the hot-wire measurements of [6] made at a constant distanee from the disk surface. For this flow, it is not known what the distributions of amplitude and phase through the boundary layer actually are. However, for Blasius flow, the measurements of Gilev et al. [1] showed that on the centerline near the source there is a large phase change through the boundary layer, and the amplitude distribution has double peaks. Both the amplitude and phase distributions change drastically with respect to z, even far from the source. Although these circumstances seem unfavorable, excellent agreement was obtained for the centerline amplitude distribution, and the calculated constant-phase lines and spanwise amplitude have many points of similarity with the measurements. To go beyond the present approach and include eigenfunctions in the calculations is a formidable task, and requires some form of the non-parallel theory.

Acknowledgment

The work described in this paper was carried out at the Jet Propulsion Laboratory, California Institute of Technology, under contract with the National Aeronautics and Space Administration (NASA). Support from the Fluid and Thermal Physics Branch of the Aerospace Research Division, Office of Aeronautics and Space Technology, NASA, is gratefully acknowledged.

References

1. Gilev, V.M.; Kachanov, Y.S. and Kozlov, V.V.: Development of Spatial Wave Packets in a Boundary Layer, Preprint No. 34-81, Institute of Theoretical and Applied Math., USSR Acad. Sci., Novosibirsk 1981.

2. Mack, L.M. and Kendall, J.M.: Wave Pattern Produced by a Localized Harmonic Source in a Blasius Boundary Layer, presented at AIAA Aerospace Meeting, Reno, NV, 1983, as AIAA Paper 83-0046 (to be published).

3. Mack, L.M.: Boundary-Layer Linear Stability Theory, Lecture Notes of VKI/AGARD Special Course on Stability and Transition, Chapters 7 and 12, Von Karman Institute of Fluid Dynamics, Rhode-St-Genese, Belgium, 1984 (to be published as AGARD Report).

4. Gaster, M.: A Theoretical Model of a Wave Packet in the Boundary Layer on a Flat Plate, Proc. Roy. Soc. A, Vol. 347, pp. 271-289, 1975.

5. Gaster, M.: The Development of a Two-Dimensional Wave Packet in a Growing Boundary Layer, Proc. Roy. Soc. A., Vol. 384, pp. 317-332, 1982.

6. Wilkinson, S.P. and Malik, M.R.: Stability Experiments in Rotating-Disk Flow, AIAA Paper No. 83-1760, 1983.

7. Mack, L.M.: The Wave Pattern Produced by a Point Source on a Rotating Disk, to appear.

8. Mack, L.M.: Line Sources of Instability Waves in a Blasius Boundary Layer, AIAA Paper No. 84-0168, 1984.

9. Malik, M.R.; Wilkinson, S.P. and Orszag, S.A.: Instability and Transition in Rotating Disk Flow, AIAA J., Vol. 19, pp. 1131-1138, 1981.

10. Kendall, J.M.: Supersonic Boundary Layer Stability Experiments, in Proceedings of Boundary Layer Transition Study Group Meeting (W.D. McCauley, ed.), Aerospace Corp. Report No. BSD-TR-67-213, Vol. II, 1967.

The Growth of Instability Waves in a Slightly Nonuniform Medium

Marten T. Landahl

Massachusetts Institute of Technology, Cambridge, Mass., and Royal Institute of Technology, Stockholm, Sweden

Summary

Kinematic wave theory, extended to include weak dissipation, higher-order amplitude dispersion, and weak nonlinearity, is used to calculate the long-time evolution of instability waves in a three-dimensional boundary layer. An unstable wave on a convected inhomogeneity reaches criticality when propagating at the velocity of the inhomogeneity; it then becomes absolutely unstable in that moving reference frame. A trapped such a wave is found to be governed by the Ginzburg-Landau equation within an inner region of size $\varepsilon^{-1/2}$ times the wave lengt where $\varepsilon \ll 1$ measures the non-uniformity of the wave system or the background medium.

1. Introduction

Of interest in transition to turbulence is how an initially small wave-like disturbance of a given wave number or frequency evolves during its downstream travel. A number of different suggestions have been made for how to extend the results for an infinite, uniform wave train propagating through a uniform medium to wave trains and wave packets propagating in a medium with properties that vary slowly with downstream distance (see [1] and the references cited therein). The effects of variations in both space and time is of interest also in the study of the final stage of transition at which the instability waves break down into small-scale, chaotic motion.

A natural starting point for such investigations is kinematic wave theory. However, doubts have been raised [7] regarding how it should be properly extended to modal waves with dissipation, however, in particular for the long-time evolution of wave packets propagating over distances many times their own length.

Here model equations are proposed which incorporate the effects over long times of weak dissipation, weak nonlinearity and dispersion due to large wave envelope curvature.

2. Model equations

Kinematic wave theory in its original form describes the evolution of a conservative wave group with slow variations of amplitude and wave number by the system (see [2])

$$\omega = \Omega(k;x,t) \ , \tag{1}$$

$$\partial k/\partial t + \partial \Omega/\partial x = 0 \ , \tag{2}$$

$$\partial A/\partial t + \partial B/\partial x = 0 \ , \tag{3}$$

where k is the wave number, ω the frequency, c is the group velocity

$$c = \Omega_k \ , \tag{4}$$

A the wave action density (A a^2), and B the wave action flux, in the linear case given by

$$B = cA \ , \tag{5}$$

(subscripts denoting partial derivativesin the extended space x,t,k [3]). The equations are here written for a one-dimensional wave group but are easily extended to higher dimensions by replacing the partial derivatives with the appropriate vector quantities. It is also valid for modal waves with the average Lagrangian, and hence the wave action density, determined as an integral over cross space [3].

These evolution equations determine the variation of wave number and amplitude with respect to the slow space and time variables $X = \varepsilon x$, $T = \varepsilon t$, where ε is a small quantity measuring the nonuniformity of the wave train and the background medium (x,t are dimensionless with the centered wave length $\lambda_c = 2\pi/k$ and centered frequency ω_c used as reference lenth and time, respectively). The kinematic wave theory may be regarded as an asymptotic theory in ε valid within $X = O(1)$, $T = O(1)$ with errors of order ε^2.

The major change needed to extend the kinematic wave theory to hold also for the long-time evolution ($T = O(\varepsilon^{-1})$) of waves with weak dissipation and nonlinearity is to replace the dispersion relation (1) with

$$\omega + i\mu = \Omega + f_2 a^2 - a_{xx}\Omega_{kk}/2a \quad , \tag{6}$$

where μ is the growth rate, assumed small, so that $|\mu|/|\omega| = O(\varepsilon) << 1$ and Ω and f_2 are complex functions with small imaginary parts. The last term in (6), first introduced by Chu and Mei [4], corrects for the effects of rapid variations of the amplitude $a(x,t)$. Since the wave action density is proportional to a^2 these terms may also be expressed directly in terms of A, if one neglects the variation of A with k. The second modification needed is to select a complex wave number $k^c = k_r^c + i\,k_i^c$ such that the group velocity becomes purely real, i.e.,

$$\Omega_k^i(k^c) = 0 \ . \tag{7}$$

The conservation equation for the wave action density, (3), is then replaced by [1]

$$\partial A/\partial t + \partial(cA)/\partial x = 2\mu_e A \ , \tag{8}$$

where μ_e is the effective growth rate,

$$\mu_e = \mu(k^c) - k_i^c \ , \tag{9}$$

all dispersion quantities to be evaluated at the complex wave number k^c. The aplication to a linear wave packet in a homogeneous medium [1] suggests that the use of a complex wave number is only necessary for the long-time evolution of waves with small dispersivity such that $|\Omega_{kk}^r|/|\Omega_{kk}^i| = O(1)$, the group velocity then being determined from the real part of the linear dispersion relation.

The system (7) -- (9), which will be considered in the following, contains as special cases most of the model equations used as bases for earlier linear and nonlinear stability investigations.

3. Evolution of three-dimensional boundary layer instability wave train.

In studies of boundary layer stability one usually considers the growth of a steady three-dimensional Tollmien-Schlichting wave train to investigate which wave frequencies may be subject to large amplifications and thereby be susceptible to nonlinear effects leading to transition. For a complete such investigation one needs to follow wave trains of different initial orientation and (fixed) frequency over large downstream distances. This requires that higher-order terms having cumulative effects be retained, which is generally the lowest-order derivative terms. The use of a complex wave number will handle those appearing because of $\Omega_k^i \neq 0$. Hence, the proper model equations to use for the long-time evolution of a wave train are (7)--(9) with

$$\mu_e = \Omega^i(k^c) - ck_i^c \ . \tag{10}$$

For a steady wave train with $d/dt = \underline{c}\cdot\nabla$ this is accomplished by a straightforward quadrature. The final result for the wave action density is given by (correcting misprints in [1])

$$A/A_o = J^{-1}\exp[2\int_0^x \mu_e\underline{c}.d\underline{x}/c^2)] \ , \tag{11}$$

where

$$\ln J = |\int_0^t \nabla.\underline{c}dt| \ . \tag{12}$$

As an illustration we reproduce some results for Tollmien-Schlichting instability waves in the nose region of a swept wing obtained by Nordstrom and Landahl [5]. Since the boundary layer is independent of the spanwise coordinate z it follows that the spanwise wave number component is constant along the ray (Snell's law). From (12) one then finds directly

$$J = c_1/(c_1)_{x=0} \ , \tag{13}$$

where c_1 is the x-component of the group velocity. Some results are shown in Figs. 1 and 2.

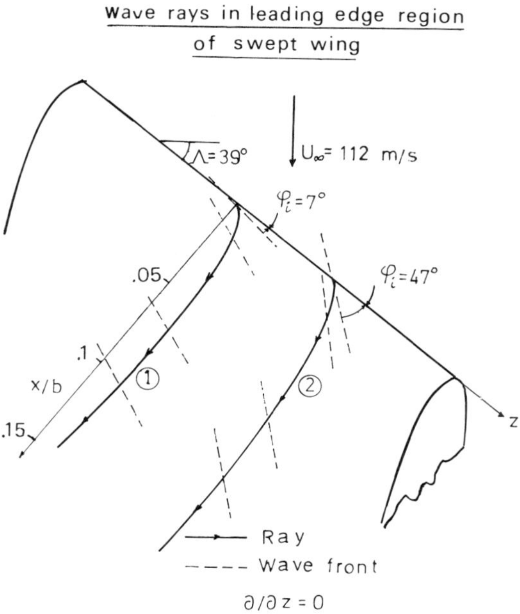

Fig. 1. Instability wave rays along the underside of the leading-edge region of a swept wing for two initial wave angles 7° and 47°. Airfoil NACA 64A010, Mach no 0.333, altitude 1 km angle of attack $5,1^{\circ}$ (From [5]).

The calculations were for instability waves of a frequency 5 kHz on the lower surface of the leading edge region of a swept wing with a NACA 64A010 airfoil and chord 1.45 m flying at a Mach number of 0.333 at an altitude of 1 km and angle of attack $\alpha = 5.1^{\circ}$ (This was chosen to correspond to an actual flight test case for a Saab A32 Lansen aircraft). The calculated ray paths for two initial wave front angles 7° and 47° are shown in Fig 1 and the corresponding wave action density growth in Fig.2. As seen from Fig.1, for these initial wave angles the wave rays tend to turn into a direction approximately normal to the leading edge. Fig.2 shows that the instability waves grow rapidly with distance from the leading edge, especially the wave with 7° initial wave angle.

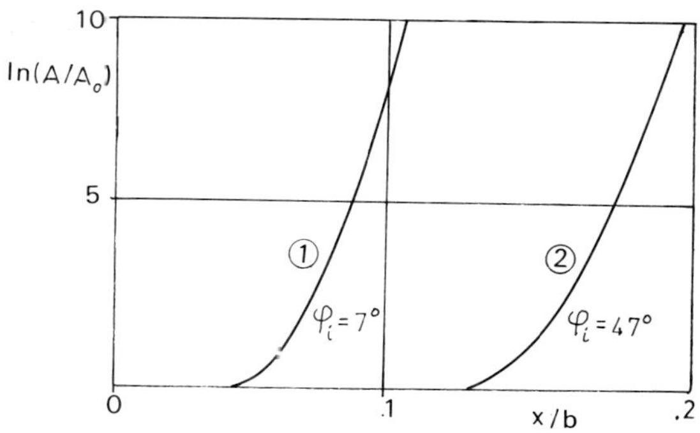

Fig. 2. Variation of wave action density along the two rays with initial wave angles of 7° and 47°, respectively, as function of distance from the leading edge (From [5]).

4. Critical region behavior.

In analyzing the behavior of short-wave disturbances propagating through a background velocity field caused by the presence of a long wave with a phase velocity c_o, Landahl [6] found that an unstable short wave would reach a critical condition for $c = c_o$ involving either (space-time) focusing (complete compression of the wave group) or focusing and trapping (the wave becoming stuck on the primary disturbance). However, the analysis was based on the lower-order linear kinematic wave theory, extended for weak dissipation, the validity of which was challenged by Stewartson [7]. Here we shall derive equations for the inner focal region starting from (7) -- (9) with both nonlinearity and amplitude dispersion incorporated.

Starting from the system (7) -- (9) (for propagation in one dimension) we look for an inner expansion around the critical point of the form

$$k - k_C = \varepsilon^{\alpha}\,\bar{K}\,, \tag{14}$$

$$\omega - \Omega^r(k_C) = \varepsilon^{\beta}\,\bar{\Omega}\,, \tag{15}$$

$$x - c_o t - x_C = \varepsilon^{\gamma}\,\bar{X}\,, \tag{16}$$

$$t - t_C = \varepsilon^{\delta}\,\bar{T}\,, \tag{17}$$

$$a = \varepsilon^{\zeta}\,\bar{a}\,, \tag{18}$$

where subscript C denotes values at the critical condition where the group velocity vanishes relative to the convected frame of reference $x - c_o t$.

It is necessary to consider two different cases, namely

i) (focusing only:) $\Omega_{x,C} = O(\varepsilon) \neq 0$ and $\Omega_{kx,C} = O(\varepsilon) \neq 0$

ii) (focusing and trapping) $\Omega_{x,C} = 0$, $\Omega_{kx,C} = 0$.

In case i) the proper stretching to obtain balance between all important terms is found to be

$$\alpha = 1/3 = -\beta = \zeta, \quad \gamma = -2/3 = \delta \quad . \tag{19}$$

We introduce the complex function

$$\psi = a(\bar{X},\bar{T}) \exp[i\bar{\Theta}(\bar{X},\bar{T})] , \tag{20}$$

where $\bar{\Theta}$ is a slowly varying phase defined by

$$\bar{\Theta}_{\bar{X}} = \bar{K}, \ \bar{\Theta}_{\bar{T}} = - \bar{\Omega} \tag{21}$$

and expand around the critical condition. To lowest order in $\varepsilon^{1/3}$, ψ is found to satisfy

$$\psi_{\bar{T}} - i\Omega_{kkC}\psi_{\bar{X}\bar{X}}/2 + i(f_2|\psi|^2 + \bar{X}\Omega^r_{XC})\psi = 0 , \tag{22}$$

where $X = \varepsilon x$.

In case ii) he proper stretching is

$$\alpha = 1/2 , \ \beta = 1 , \ \gamma = -1/2 , \ \delta = 1, \ \zeta = 1/2 \tag{23}$$

and ψ satisfies

$$\psi_{\bar{T}} - i\Omega_{kk}\psi_{\bar{X}\bar{X}}/2 + if_2|\psi|^2\psi - \Omega^i\psi = 0 . \tag{24}$$

4. Discussion

Since in case i) the wave spends only a time of the order $\varepsilon^{-2/3}$ near the focus, the total exponential amplifiction is only of the order $\varepsilon^{1/3}$ and hence negligible to lowest-order. Furthermore, amplitude dispersion will generally be expected to act so as to limit the amplitude. Thus, focusing alone is not likely to lead to large amplitudes.

In case ii) with both focusing and trapping the governing equation (26) may be identified as the Ginzburg-Landau equation which has been studied by many as a master equation for hydrodynamic stability problems. Hocking et al. [9] found that in the case of $\Omega^i>0$ and $f^r_2\Omega_{kk} > 0$, self focusing of the wave train may occur leading to infinite amplitudes at the center of the wave group. However, a more relevant result in this context is the recent findings by Moon et al. [10] that this equation could have solutions with chaotic behavior.

Acknowledgement

This research was sponsored in part by the U.S. Air Force under Contract F49-49620-83-C-0019.

References

1. Landahl, M.T.: The application of kinematic wave theory to wave trains and packets with small dissipation. Phys. Fluids 25 (1982) 1512-1516

2. Whitham, G.B. Linear and nonlinear waves. New York, London, Sydney, Toronto: John Wiley & Sons 1974.

3. Hayes, W.D.: Conservation of action and modal wave action. Proc. Roy. Soc. A320 (1970) 187-208.

4. Chu, V.H.; Mei, C.C.: On slowly varying Stokes waves. j. Fluid Mech. 41 (1970) 873-887.

5. Nordstrom, J.; Landahl, M.T.: The growth of instability waves in a three-dimensional boundary layer (to be published)

6. Landahl, M.T.: Wave mechanics of breakdown. J. Fluid Mech. 56 (1972) 775-802.

7. Stewartson, K.: Some aspects of nonlinear stability theory. Polish Academy of Sciences, Fluid Dynamic Transaction 7 (1977) 101-128.

8. Bers, A.: Linear waves and instabilities. In Physique des Plasmas (ed. C. DeWitt & J. Peyraud) New York: Gordon and Breach 1975.

9. Hocking, L.M.; Stewartson, K. Stuart, J.T.: A nonlinear instability burst in plane Poiseullle flow. J. Fluid Mech. 51 (1972) 705-735.

10. Moon, H.T.; Huerre, P.; Redekopp, L.G.: Transition to chaos in the Ginzgurg-Landau equation. Physica 7D (1983) 135-150.

Subtransitions in the Transition Zone

R. NARASIMHA

Department of Aerospace Engineering
Indian Institute of Science
Bangalore 560 012

SUMMARY

A general expression is derived for the transitional intermittency in boundary layers on arbitrary surfaces, assuming the hypothesis of concentrated breakdown. An analysis of experimental data based on this expression shows that when there are rapid and substantial changes in the geometry of the spot dependence cone, the transition zone reveals corresponding changes that may be best viewed as subtransitions. Examples offered include: plane boundary layers subjected to pressure gradients varying over the transition zone, and flow past axisymmetric bodies.

1. INTRODUCTION

The transition zone of this paper is the region that begins (in the boundary layer) with the appearance of turbulent spots and ends through an asymptotic approach to the fully turbulent flow far downstream. The key flow variable in this zone is the intermittency. We analyse the results of a series of experiments that show how certain peculiarities appear in the intermittency distribution in the presence of pressure gradients and transverse curvature. A theoretical framework for interpreting these distributions is first set up.

2. A GENERAL TRANSITIONAL INTERMITTENCY DISTRIBUTION

Emmons [1] first provided a statistical analysis relating the intermittency γ to the rate of generation of turbulent spots. Consider xyt space, where x is a coordinate along the flow, y is transverse and t is the time. (The z-coordinate, normal to the surface, plays no role, as we are interested in the **transitional** intermittency at $z = 0$, e.g. in the wall stress; see [2] for comments on the relation to "edge" intermittency.) A spot generated at $P_o(x_o y_o t_o)$ will in general sweep out a volume in xyt space, called the "propagation cone" from P_o. The set of all points in xyt space from which the propagation cones cover P_o constitute the "dependence cone" for P_o. A function $g(x,y,t)$ is also introduced such that $g\,dV$, where $dV = dxdydt$, is the probability of formation of a spot in dV.

Emmons's relation between γ and g can be written down almost immediately by postulating that spot formation obeys a Poisson process. This is ensured if one accepts the hypotheses of "orderliness" and "absence of after-effects" [3] that are familiar in the theory of queues, provided the latter (often also called "independence") is widened to include the absence of any effect on the **propagation** of a spot due to other spots that may be present in the flow. (The **physical** validity of these **mathematical** assumptions has been discussed elsewhere [2].) It then follows that the probability of turbulence γ at any point $P(x,y)$ is just the probability that at least one spot is formed in the dependence cone $R(P)$ for P; by a generalization of a standard result for the idle period in one-dimensional queues [3] to three-dimensional xyt space, we then get

$$\gamma(P) = 1 - \exp\left[- \int_{R(P)} g(x', y', t')\, dx'\, dy'\, dt'\right]. \qquad (1)$$

Emmons's assumptions on g can not describe measured γ, but the following "hypothesis of concentrated breakdown" [4] has proved successful in plane boundary layer flow :

> turbulent spots are formed at a preferred streamwise location randomly in time and in the transverse coordinate. (2)

As there is considerable evidence in support of this hypothesis [2], we will now assume it is valid in more general cases as well, and put

$$g(x,y,t) = n(y)\,\delta\,[x - x_t(y)] \qquad (3)$$

in stationary flow, $x = x_t(y)$ being the transition onset curve on the surface. Putting this in (1) we get

$$\gamma(P) = 1 - \exp\left[- \int_{R_t(P)} n \, dA_t \right] \tag{4}$$

where the integration is now carried out over what we shall call "the dependence area for P", $R_t(P)$, which is the intersection of the dependence cone with the surface $x = x_t$. If the spot formation or flow breakdown rate n (per unit time, unit length) is independent of y, (4) reduces to

$$\gamma(x) = 1 - \exp\left[- nA_t(x) \right] ; \tag{5}$$

and the problem of finding the **form** of $\gamma(x)$ is reduced to that of finding $A_t(x)$, Conversely, if γ is measured, we can infer $A_t(x)$ from

$$F^2(\gamma) == - \ln(1-\gamma) = nA_t(x) . \tag{6}$$

Armed with the general intermittency distributions (4) and (5), we are in a position to relate spot characteristics and intermittency. In particular, when the volume R is a true cone with straight generators, A_t is proportional to $(x-x_t)^j$ where j is the number of dimensions. In constant-pressure flow at velocity U on a flat plate, j = 2 and we obtain Narasimha's 2D result (4)

$$\gamma = 1 - \exp[-n\sigma(x-x_t)^2/U] ; \tag{7}$$

in slug-flow in a pipe [5], or in axial flow past a circular cylinder [6] j = 1, and we have the 1D distribution

$$\gamma = 1 - \exp\left[- n_1 \sigma_1 (x-x_t)/U \right] . \tag{8}$$

Here σ, σ_1 are non-dimensional spot propagation parameters that we shall call "dependence area factors", and n, n_1 the appropriate breakdown rates.

3. SUBTRANSITION IN PRESSURE GRADIENT FLOW

Some recent studies have shown the strong effect that pressure gradients can have on spot propagation. Wygnanski [9] reports a slow spread-rate (half-angle = 5°) in a flow with Falkner-Skan parameter $\beta = 0.12$; furthermore the velocity of propagation does not increase even when the free-stream accelerates downstream. Narasimha et al. [10] report several experiments in which favourable pressure gradients inhibit spot growth: they suggest the explanation that the flow is rendered subcritical by the favourable gradient, and find that growth resumes at normal rates once the flow becomes super-critical again. Their analysis showed that the Wygnanski flow remained subcritical throughout. These experiments demonstrate that a favourable pressure gradient can make the propagation and dependence cones substantially smaller, and induce rapid changes in their geometry when critical conditions are attained.

A comprehensive set of new intermittency measurements in flows subjected to pressure gradient have recently been reported [11]. An analysis of these data is best done using the variable $F(\gamma)$ of (6), which should be linear in x if (7) is true. In fact, it had been shown many years ago, in an isolated experiment [8], that a favourable gradient imposed near the onset of transition could result in an $F(\gamma)$ plot with a kink. To study this phenomenon better, a set-up was devised [11] where, by judicious use of tunnel wall liners and transition-fixing devices, the location of pressure gradient and transition onset could be varied almost independently.

Figure 1 shows the results in flow DFU3 of [11]: the free-stream velocity distribution U(x), the intermittency γ (on the $F(\gamma)$ scale that makes (7) linear), the momentum thickness Reynolds number Re_θ and the shape factor H are all shown. The $F(\gamma)$ slope suddenly goes up around $x \simeq 120$ cm, where the pressure gradient diminishes to zero. Around the same point there is a sharp increase in Re_θ, and the beginning of a drop in H that is just noticeable. What is striking is that all these features occur at the same station, indicating a sudden change in the nature of the flow that is best viewed as a "subtransition" - from a subcritical to a supercritical state, as in the spot experiments of [10]. Figure 2 shows another instance, flow DFU1, in which the presure gradient is milder, the subtransition being less pronounced but nevertheless definitely present.

When the flow goes from a zero to an adverse gradient, as in DAD1 (Figure 3), the kink in $F(\gamma)$ is just as clearly noticeable; however it does not show up as strongly in Re_Θ, as there was no favourable gradient in this case to suppress boundary layer growth in the upstream half of the transition zone.

Flow DFU2 shows similar behaviour in γ, but is not shown here as other flow parameters were not measured.

If the pressure gradient is very mild [7] or occurs in the downstream half of the transition zone where its stabilizing effect on a nearly full-time turbulent flow would be less, $F(\gamma)$ shows no break, as seen from a comparison of the flows NFD1 and NFU1 (Figure 4).

4. SUBTRANSITION IN AXISYMMETRIC FLOW

We now examine briefly a different class of sub transitions that can occur on three-dimensional surfaces of finite lateral extent. Rao [6] has reported intermittency measurements in flow past a circular cylinder with axis aligned to the flow, and shown that the observed distributions tend to follow the 2D law (7) in an initial or I region and the 1D law (8) in a later or L region. The explanation briefly is that a spot created on the cylindrical surface first propagates as in plane flow on the developed surface; at a certain station x^* the spot will wrap around the cylinder and propagate eventually as a one-dimensional "sleeve".

Rao also reported failure of an attempt at a mixed (1D + 2D) theory, taking the cut-off portion of the dependence volume to be the same fraction as that of a right circular cone. However, in the L-region turned out to obey effectively the 1D law with origin at the end of the I region.

A rational approach to the problem must start with a determination of the form of A_t (x). To help visualization, let us imagine the spot is an isosceles triangle in shape. From any point P on the cylinder, the cones we are considering will thus start out as pyramids (as in flat plate flow). After the critical point x^*, there can be no change in the maximum cross-flow width of the propagation cone (section A,B, Figure 5). Consequently the spot shape on the developed surface will be a cropped triangle, and the pyramid becomes pentagonal in section, with a base of constant width and increasing length, capped by an arrow shaped head (section B) According to Rao's measurements velocities change little at x^* if the Reynolds number $Re_\delta \gtrsim 5000$; so the corners of the pyramid have the same slope in the xt plane (but not in the xy plane) on either side of x^*.

The geometry of the dependence cone can also be described in similar terms as long as x_t remains on the cylindrical surface (if transition occurred on the nose there is a slightly more complicated situation that will be discussed elsewhere). If we assume that the arrow-head mentioned above remains similar, a little consideration shows that

$$A_t = \sigma(x-x_t)^2/U, \quad x \leq x^*, \tag{9a}$$

$$= [\sigma\,(x^*-x_t)^2 + 2\pi a\sigma'(x-x^*)]/U, \quad x \geq x^*. \tag{9b}$$

Here (9a) is the 2D result; the first term in (9b) is the contribution from the "head" of the sleeve, and the second from the body (of a radius a,say), with a factor that accounts for the ratio of the propagation velocity to U. Noting that n_1= 2 an, it immediately follows from (5) that

$$\gamma(x) = 1 - \exp[-n\sigma(x-x_t)^2/U], \quad x_t \leq x \leq x^*, \tag{10a}$$

$$= 1 - (1-\gamma^*)\exp[-\,2\pi a n\sigma'(x-x^*)/U], \quad x^* \leq x, \tag{10b}$$

where $\gamma^* = \gamma(x^*)$, from (10a) or (10b). Figure 6 compares experimental data from [6] with (10a,b); the agreement is good, and the subtransition at x^* is clearly seen. Note that (10b) is the 1D distribution with origin at x^*, except for the factor $(1-\gamma^*)$. When γ^* is relatively low (0.2 in Figure 6), the rise is given by (10b) is so rapid near the origin that there is little room for adjusting its location.

The mixed nature of (10b) is slightly better seen when it is written in the equivalent from

$$\gamma = \underset{(2D)}{\gamma^*} + \underset{(1D)}{(1-\gamma^*)}\,(1-\exp[-(2\pi\, an\sigma'/U)(x-x^*)]),\quad x^* \leq x. \qquad (10c)$$

An incidental but important consequence of the present model needs to be noted for cone flows. The developed surface of a cone of half-angle δ is a sector of half-angle $\beta = \pi \sin\delta$. A spot envelope that is a wedge of half-angle α wraps around the cone only if $\beta < \alpha$, i.e. if $\delta < \sin^{-1}(\alpha/\pi)$, $\doteq$ 3.2 deg. for α= 10 deg. This leads us to the prediction that

> as far as the transition zone is concerned, cones with half-angle less than about 3-3.5 deg. are cylinder-like, those with greater half-angles plate-like. (11)

As a corollary, the 2D-1D subtransition can be expected on slender cones, but not on wider ones or (a fortiori) fat, blunt noses.

These predictions still need to be tested.

Finally we consider how γ^* can be predicted if x_t is known. Defining a zone-length parameter λ [4] as the distance between γ = 0.25 and 0.75, we can show from (7) that

$$n\sigma = 0.412\, U/\lambda^2 . \qquad (12)$$

Now Dhawan & Narasimha [12] have collected data on λ and proposed the rough correlation $Re_\lambda \simeq 5\, Re_t^{0.8}$. Narasimha (13) noticed that a change in the exponent from 0.8 to 0.75 would still be consistent with the Dhawan-Narasimha data (Figure 7: where the proposed new correlation

$$Re_\lambda \simeq 9\, Re_t^{3/4} \tag{13}$$

is also shown), and would lead to the significant conclusion that n depends primarily on the local boundary layer thickness. It is easy to show from (12), (13) and the Blasius relation that the appropriate non-dimensional parameter for the spot formation rate is $n\sigma\theta_t^3/\nu$, and to determine an approximate value for it: we get (see Figure 7)

$$N == n\sigma\theta_t^3/\nu \doteq 10^{-3}, \tag{14}$$

where θ_t is the momentum thickness at x_t. If we took $\sigma = 0.25$ [13], and the Blasius boundary layer thickness as $\delta \doteq 5xRe_x^{-1/2}$, then $\delta/\theta = 7.53$ and (14) gives $n\delta_t^3/\nu \doteq 2$. This clearly suggests that spot formation scales primarily with the boundary layer thickness and the viscous diffusion time δ_t^2/ν - a physically appealing conclusion.

Using (4) in (10a), and noting that $x^* - x_t = \pi\, a \cot\alpha$,

$$\gamma^* = 1 - \exp[-(N/Re_{\theta t})(\pi\, a \cot\alpha/\theta_t)^2]. \tag{15}$$

Using $\alpha = 10$ deg., $N = 10^{-3}$ this gives $\gamma^* = 0.22$ in flow R CL2, compared to the measured value of 0.2 - agreement that is almost too good to be true !

I thank Mr.J. Dey for help with some calculations.

REFERENCES

1. Emmons, H.W.: J. Aero. Sci. 18 (1951) 490-498.

2. Narasimha, R.: Proc. 2nd Asian Cong. Fluid Mech. Beijing, Oct. 1983. Also Report 83 FM 5, Dept. Aero. Engg. Ind. Inst. Sci.

3. Khintchine, A.Y.: Mathematical methods in the theory of queueing. Charles Griffin, London 1960.

4. Narasimha, R.: J. Aero. Sci. 24 (1957) 711-712.

5. Pantulu, P.V.: MSc thesis, Ind. Inst. Sci. (1962).

6. Rao, G.N.V.: ZaMP 25 (1974) 53-75.

7. Abu-Ghannam, J.B; Shaw, R.I.: J. Mech. Engg. Sci. 22 (1980) 213-228.

8. Narasimha, R.: AIISc thesis, Dept. Aero. Engg., Ind.Inst. Sci. 1958.

9. Wygnanski, I.: Proc. Symp. Structure of turbulence and mixing, Madrid, Springer. 1980.

10. Narasimha, R.; Badri Narayanan,M.A.; Subramanian, C.: To appear in AIAA J. Also Report 82 FM 12, Dept. Aero. Engg., Ind. Inst. Sci. 1982.

11. Narasimha, R.; Devasia, K.J.; Gururani, G.; Badri Narayanan, M.A.: To appear in Experiments in Fluids. Also Report 84 FM 2, Dept. Aero. Engg., Ind. Inst. Sci. 1984.

12. Dhawan, S.; Narasimha, R.: J. Fluid Mech.3 (1958) 418-436.

13. Narasimha, R.: Report 78 FM 10, Dept. Aero. Engg., Ind. Inst. Sci. 1978.

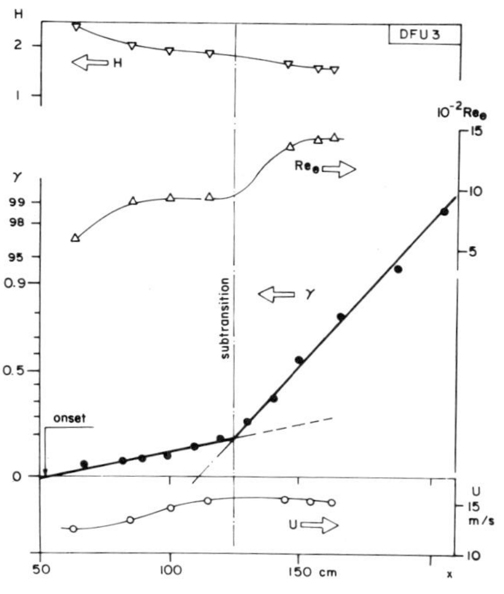

Figure 1: Flow parameters in transition zone, flow DFU3 of [11].

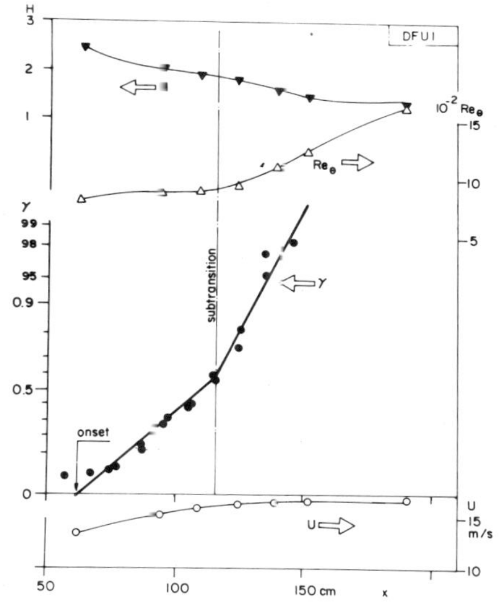

Figure 2: Flow parameters in transition zone, flow DFU1 of [11].

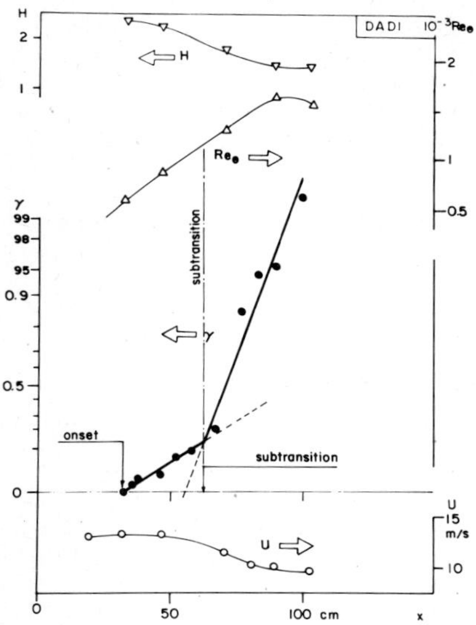

Figure 3: Flow parameters in transition zone, flow DAD1 of [11]

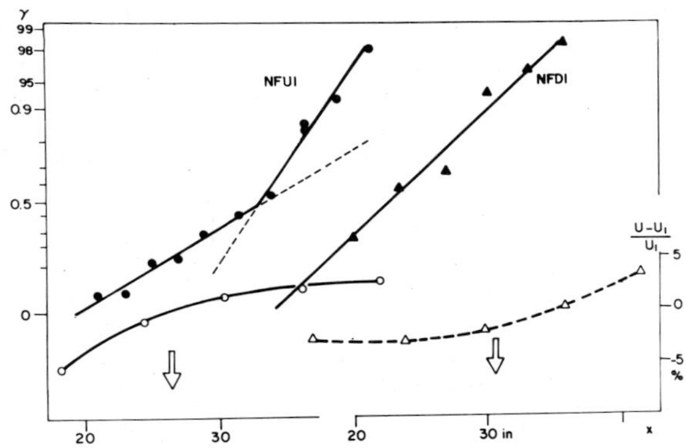

Figure 4: Intermittency distribution in two flows of [8], showing that moving the same pressure gradient downstream in the transition zone (NFD1) removes the subtransition noticed when the pressure gradient is upstream (NFU1).

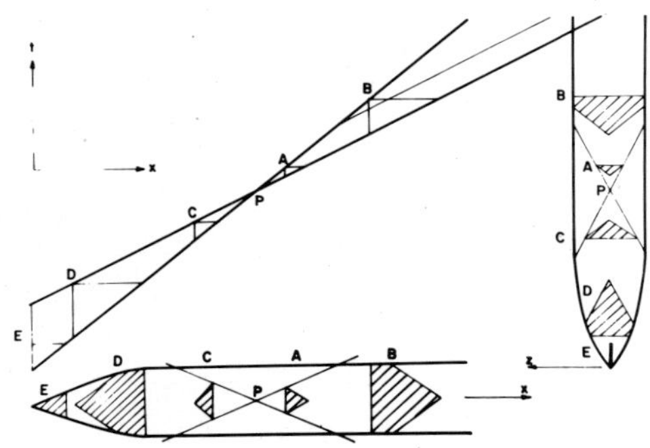

Figure 5: Idealized propagation and dependence cones in axial flow past a circular cylinder with an ogive-shaped nose. Note how the dependence cone is squeezed to a line along the time axis at the stagnation point.

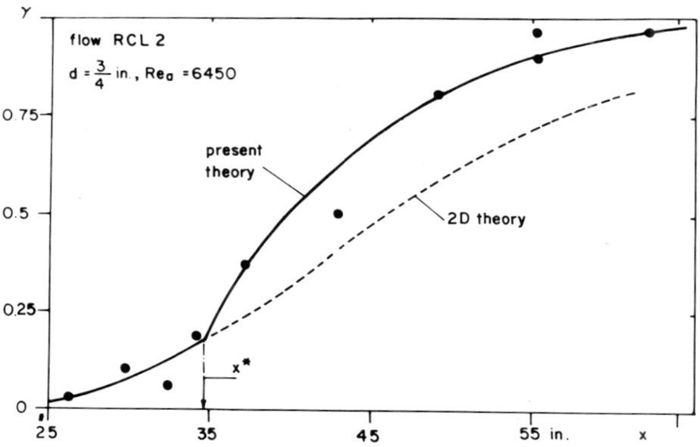

Figure 6: Intermittency distribution in flow RCL2 of [6], compared with the present theory (full line).

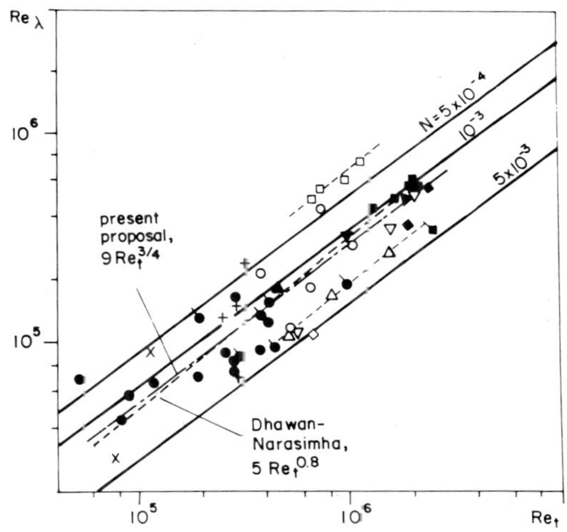

Figure 7: Experimental data on extent of transition zone, showing two proposed correlations and lines of constant breakdown parameter $N \simeq 0.12\ Re_t^{3/2}\ Re_\lambda^{-2}$. The scatter in the diagram is in part due to widely different definitions adopted for the zone length; in some cases where values appear numerically different, the **trend** of Re_λ with Re_t is the same in a given series of experiments (as shown by short dashed lines).

Control of Water Boundary Layer Stability and Transition by Surface Temperature Distribution

Carl Gazley, Jr.* and A. R. Wazzan

Department of Chemical Engineering
University of California, Los Angeles, CA 90024

Summary

The strong variation of water viscosity with temperature makes surface heating a powerful mechanism for boundary-layer control. Previous numerical investigations by the present authors and others have examined the effects of heating, with constant surface temperature, on the stability and transition of water boundary layers. These computations indicated that the Reynolds numbers corresponding to neutral stability and predicted transition are both increased significantly by relatively moderate uniform temperature differences between surface and free stream.

Preliminary studies showed that even greater stability and more efficient heat utilization could be achieved by tailoring the surface temperature distribution. This paper describes an investigation of both the neutral stability and predicted transition for an intriguing case: flow over a flat plate with $\Delta T \simeq x^{\frac{1}{4}}$. This surface temperature distribution is particularly interesting because the minimum critical Reynolds number increases along the flow in the same way the Reynolds number of the developing boundary layer increases--i.e., the stability keeps pace with the boundary layer growth and the flow can be kept outside the region of disturbance amplification.

The results of this study indicate that very large increases in the Reynolds numbers of neutral stability and predicted transition can be achieved by relatively modest surface temperatures and that under these conditions of large stability the predicted transition Reynolds number is only about twice the critical value.

Introduction

The effects of variable fluid properties are known to modify appreciably some of the characteristics of laminar boundary layers. In the case of liquid water at moderate temperature, the specific heat and thermal conductivity are relatively constant. It is primarily the strong variation of viscosity (and

---

*Also consultant to the Rand Corporation Santa Monica, CA

consequently Prandtl number) with temperature that modifies the velocity profile in a heated or cooled laminar boundary layer in water.

Studies by Hanna and Myers [1], Seban [2], and Poots and Raggett [3,4] have shown that although the velocity and temperature profiles in the laminar water boundary layer are strongly affected by the viscosity variation, the skin friction and heat transfer are influenced only slightly. It is these internal profiles, that influence the stability of the laminar boundary layer. Analytical and numerical studies by Hauptmann [5] and by Wazzan et al. [6-8] indicated very strong effects of surface heating on the stability of the laminar boundary layer.

The variation of liquid viscosity with temperature is frequently approximated by an expression of the form

$$\mu \simeq \frac{1}{a+bT} \tag{1}$$

By choice of temperature scale, this expression can be simplified to

$$\mu \simeq \frac{1}{b'T} \tag{2}$$

In the case of water, this is a useful (and perhaps amusing) approximation over an interesting range of temperature (5-60°C, 40-140°F) when T is expressed in degrees Fahrnheit!

This approximation allows one, for example, to see rather easily the relation between velocity and temperature profiles near the wall: a reasonable assumption is a linear temperature distribution (since thermal conductivity is near constant); then by (2) the velocity is a quadratic function of distance from the wall.

The analytical investigations of both Hanna and Myers [1] and Hauptmann [5] found that the boundary-layer properties are dependent on a parameter of the form

$$\left[\frac{-d\,\ell n\,\mu}{d\,\ell n\,T}\right]_w \frac{T_w-T_e}{T_w}\,Pr_w^{1/3} \tag{3}$$

Liepmann et al. [9] recently pointed out that heating of a liquid and suction are roughly equivalent and that a parameter similar to the above had a suction counterpart

$$\left[\frac{-d\ \ell n\ \mu}{d\ \ell n\ T}\right]_w \frac{T_w - T_e}{T_w} Pr_w^{1/3} \approx \frac{-v_w}{u_e}\left(\frac{u_e x \rho}{\mu}\right)^{\frac{1}{2}} \qquad (4)$$

They point out that both heating and suction are stabilizing and are, in turn, roughly equivalent to a favorable (i.e., negative) pressure gradient.

It may be noted that the approximation for viscosity variation with temperature given in (2) above allows the parameter (3) to be simplified to

$$\frac{T_w - T_e}{T_w} Pr_w^{1/3} \qquad (T \text{ in } °F)$$

since $d\ \ell n\ \mu / d\ \ell n\ T = -1$.

The previous studies of boundary-layer stability for various constant surface temperatures [5,6] suggested that it would be useful to consider the effects of surface temperature variation. Preliminary stability computations for surface-temperature distributions of the form

$$(T_w - T_e) \sim x^n \quad \text{where} \quad \tfrac{1}{4} < n < 2$$

have been presented elsewhere [10]. The purpose of this paper is to describe results of computations for both neutral stability and predicted transition for a particular interesting case of this type:

$$\Delta T \sim x^{\frac{1}{4}}$$

was chosen because the minimum critical Reynolds number increases approximately linearly with flow distance.

Computations

The choice of this particular case came about for practical reasons. In the use of heating for laminar boundary-layer stabilization, it is desirable to distribute the heat in the most efficient manner. For example, the simple case of constant surface temperature provides more heat than necessary up to the point where the boundary layer Reynolds number becomes equal to the minimum critical Reynolds number. Upstream of this point the constant critical Reynolds number exceeds the growing Reynolds number of the developing boundary layer.

A useful alternative criterion is suggested: that the critical Reynolds number be kept equal to, or somewhat larger than, the Reynolds number of the developing boundary layer--i.e., the proportionality

$$Re_{x(crit)} \sim Re_x$$

How such a proportionality might be achieved was gleaned from the results for suction and for heating with constant surface temperature. Figure 1 shows the stability results for suction [11] and for heating with constant surface temperature [6]. In this figure, the critical Reynolds number is shown as a function of the respective boundary-layer parameters described above. It is seen that the critical Reynolds number increases strongly; in fact, approximately with the fourth power of these parameters. Thus, for heating

$$Re_{x(crit)} \sim (\Delta T)^4$$

approximately, or

$$\Delta T \sim x^{1/4}$$

This is, then, a temperature distribution which should yield an approximately linear increase of $Re_{x(crit)}$ with x.

Computations were restricted to the zero pressure gradient flow over a flat plate with flow distances up to 100 ft and temperature differences varying with the one-fourth power of distance as shown in the bottom portion of Figure 2. Most of the computations were made for an ambient temperature of 67°F; however, one stability computation was made with an ambient temperature of 42°F to demonstrate the strong effects of viscosity variation with temperature.

The mean flow profiles, the neutral stability characteristics of the boundary layer with the resulting minimum critical Reynolds number, and the assessment of transition were all computed by methods previously described [6-8]. In brief, the neutral stability was computed using the Orr-Sommerfeld criteria to evaluate a minimum critical Reynolds number--i.e., the number below which all small disturbances are damped. The transition prediction was computed using the Van Ingen [12]/ Smith [13] criterion: a Reynolds number corresponding to a disburbance growth factor of e^9.

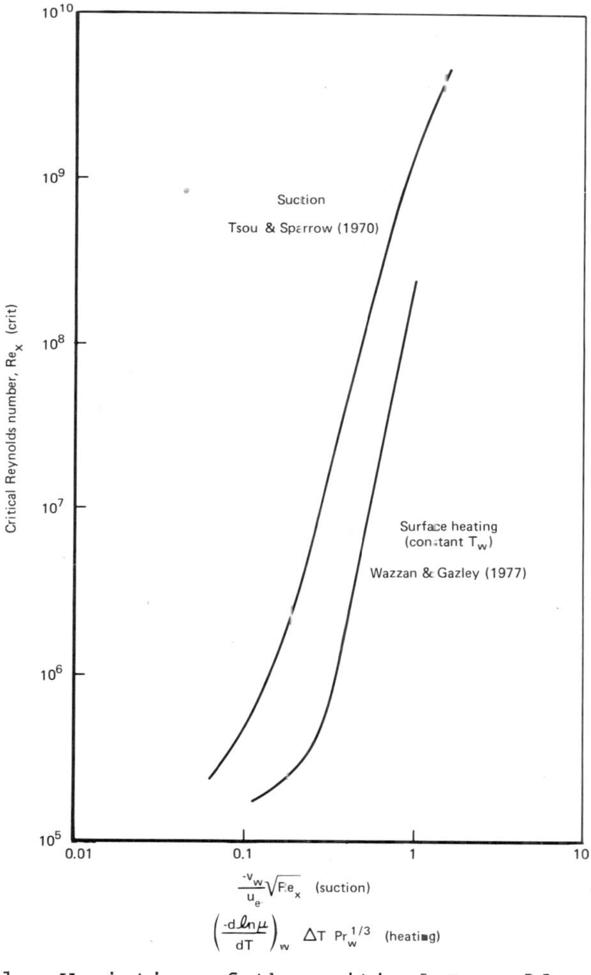

Figure 1. Variation of the critical Reynolds number with generalized suction and heating parameters.

Results

The neutral stability results are shown in Figure 2 with the minimum critical Reynolds number as a fuction of distance along the flat plate for the three cases described above. For the more moderate heating (Case 2), the minimum critical Reynolds number increases linearly along the plate--as surmised above.

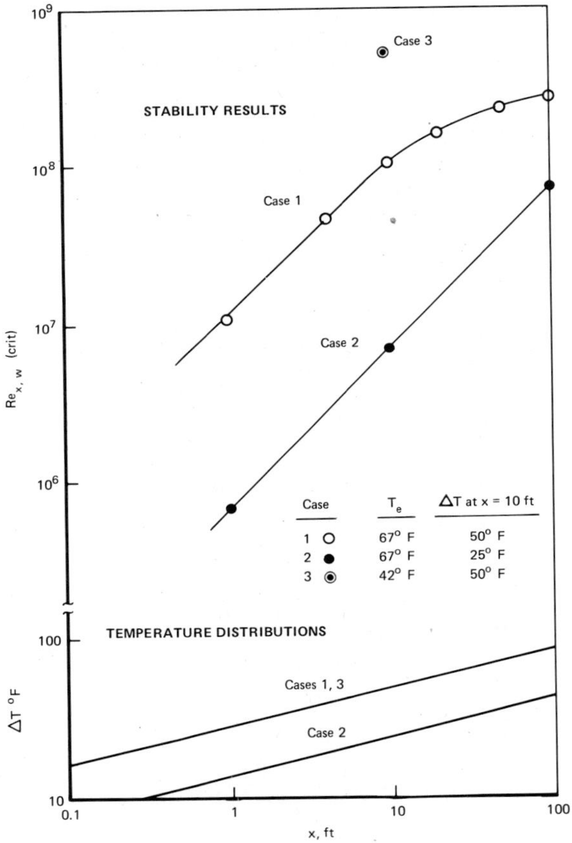

Figure 2. The effect of surface temperature distribution on the critical Reynolds number.

Increased heating (twice the temperature difference), with the same type of temperature distribution and the same ambient temperature (Case 1), yields critical Reynolds numbers higher by an order of magnitude. Here also the critical Reynolds number increases linearly with distance along the plate up to about 10 ft. At larger distances and higher surface temperatures, the rate of increase is less rapid than linear--perhaps due to the somewhat reduced dependence of viscosity on temperature in the higher temperature range.

The result for the lower ambient temperature (Case 3) give an even higher critical Reynolds number. At this lower temperature level, the rate of change of viscosity with temperature is much higher and this factor is presumably responsible for the greater stability.

These same results are shown in Figure 3 in terms of the parameter.

$$\left[\frac{-d\,\ell n\,\mu}{dT}\right]_w \Delta T\, Pr_w^{1/3}$$

In this formulation the stability results collapse into a single curve. The analytical results of Hauptmann [5] for constant surface temperatures is seen to predict the initial trend but to underestimate the effect of larger temperature differences. Previous computations for the effects of various constant surface temperatures on stability [6] are slightly above these values at lower temperature differences and somewhat below at the higher temperature differences. See also Gazley [10] for a preliminary comparison of the stability results for constant and for variable surface temperatures.

The predicted transition Reynolds numbers for Cases 1 and 2 are also shown in Figure 3. This method of plotting also collapses the results into a single curve.

The same results are shown in Figure 4 in terms of the approximation:

$$\frac{\Delta T}{T_w}\, Pr_w^{1/3}$$

for water at moderate temperatures (see above) where T is expressed in degrees Fahrenheit.

A significant feature of the results shown in Figures 3 and 4 is the apparent very rapid growth of disturbances as the minimum critical Reynolds number increases. For a flat plate flow with no heating (or constant properties), the predicted transition distance is about 30 times the distance along the place at which disturbances begin to grow. In the more stable flows, the predicted transition occurs at only about twice the distance at which disturbances begin to grow. The same was found by Wazzan and Gazley [6] for heated wedge flows with constant surface temperature. Thus, the heated boundary layers are much more stable, but when they become unstable the disturbances grow rapidly. Even, then, however, transition can hardly be considered a precipitous process--since the growth of disturbances to transition occupies at least as long a flow distance as does the damping of disturbances.

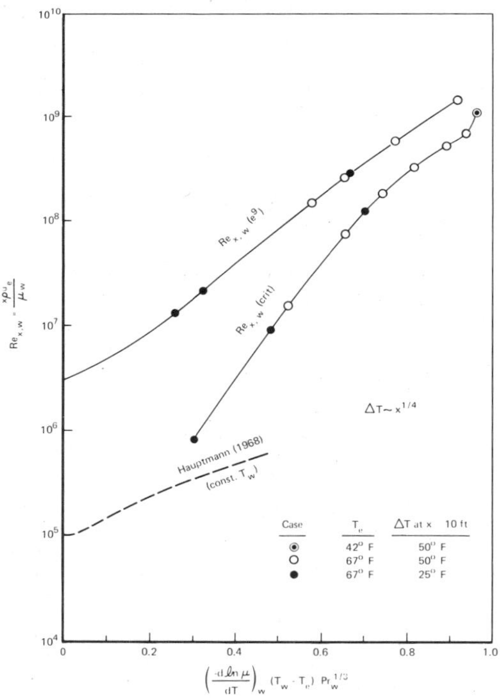

Figure 3. Variation of the critical and computed transition Reynolds numbers with a generalized heating parameter.

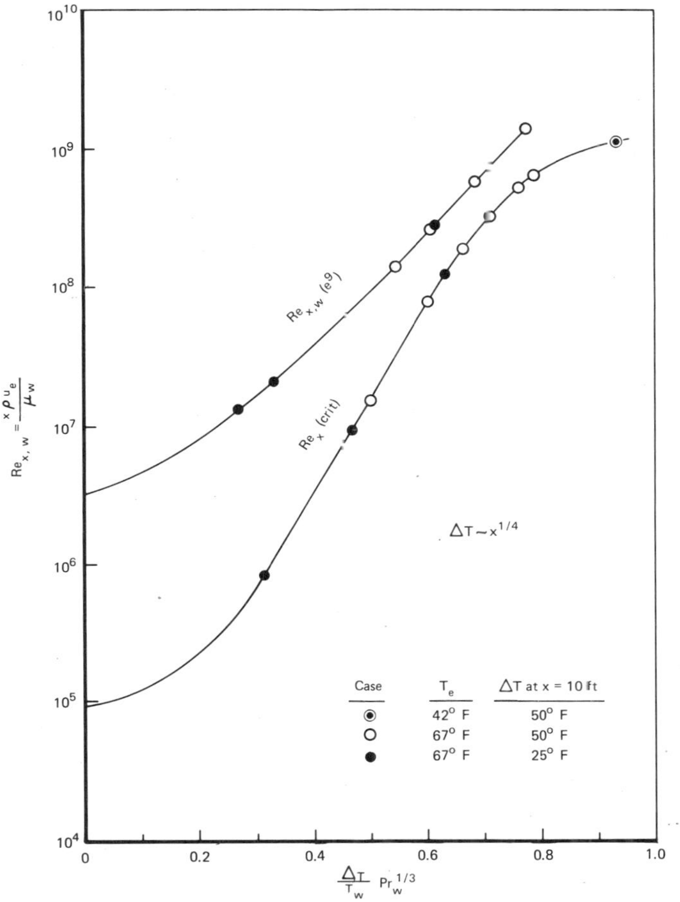

Figure 4. Variation of the critical and computed transition Reynolds numbers with a generalized temperature difference parameter.

References

1. Hanna, O. T. and J. E. Myers; Laminar boundary layer flow and heat transfer past a flat plate for a liquid for variable viscosity. J. Am. Inst. Chem. Engrs. 7 (1961) 437-441.
2. Seban, R. A.: The laminar boundary layer of a liquid with variable viscosity. In Boelter Heat Anniversary Volume: Thermodynamics, and Education (ed. H. A. Johnson), New York: McGraw-Hill 1964.

3. Poots, G. and G. F. Raggett: Theoretical results for variable property, laminar boundary layers in water. Intl. J. Heat Mass Transfer 10 (1967) 597-610.

4. Poots, G. and G. F. Raggett: Theoretical results for variable property, laminar boundary layers in water with adverse pressure gradients. Intl. J. Heat Mass Transfer 11 (1968) 1513-1534.

5. Hauptmann, E. G.: The influence of temperature dependent viscosity on laminar boundary layer stability. Intl. J. Heat Mass Transfer 11 (1968) 1049-1052.

6. Wazzan, A. R. and Carl Gazley, Jr.: The combined effects of pressure gradient and heating on the stability and transition of water boundary layers. Proc. Second Intl. Conf. Drag Reduction, Paper E-3, Cambridge University, England, August 1977. (Also Rand Corp. Report R-2175-ARPA, March 1978).

7. Wazzan, A. R., T. T. Okamura and A. M. O. Smith: The stability of water flow over heated and cooled flat plates. J. Heat Transfer 90 (1968) 109-114.

8. Wazzan, A. R., T. T. Okamura and A. M. O. Smith: The stability and transition of heated and cooled incompressible boundary layer. Proc. Fourth Intl. Heat Rans. Conf. Paris Versailles, Vol. II, Paper FC1.4 1970.

9. Liepmann, H. W., G. L. Brown and D. M. Nosenchuck: Control of laminar-instability waves using a new technique. J. Fl. Mech. 118 (1982) 187-200.

10. Gazley, Carl, Jr.: Effects of surface temperature distribution on the stability of laminar boundary layers in water. In Studies in Heat Transfer: A Festschrift for E. R. G. Eckert (ed. J. P. Hartnett, T. F. Irvine, Jr., E. Pfender and E. M. Sparrow). New York: Hemisphere 1979.

11. Tsou, F. K. and E. M. Sparrow: Hydrodynamic stability of boundary layers with surface mass transfer. Appl. Sci. Res. 22 (1960) 273-286.

12. Van Ingen, J. L.: A suggested semi-empirical method for the calculation of the boundary-layer transition region. Dept. Aero. Engr., Inst. of Techn., Delft, Netherlands, Report V.T.H. 74 1956.

13. Smith, A. M. O.: Transition, pressure gradient, and stability theory. Proc. 9th Intl. Congr. Appl. Mech., Brussels Vol. 4, p. 234 1957.

Secondary Flows in the Channels and Pipes, their Connection to and Comparison with Turbulent Flows

B.L. ROZHDESTVENSKY
Keldysh Institute of Applied Mathematics
Academy of Sciences of the USSR,
Moscow, 125047, USSR

Summary

Two- and three-dimensional nonstationary viscous fluid flows in the plane channels and pipes are considered. By means of efficient computational algorithms for direct integration of the incompressible Navier-Stokes equations the classes of two- and three-dimensional nonstationary flows with the stationary integral characteristics are found. They are called the secondary flows. 2-D secondary flows have only qualitative similarity to the turbulent flows observed in experiments. 3-D secondary flows agree rather well even quantitatively with the turbulent flows.

1. The main purpose of this work was to find a class of time-dependent flows in a plane channel and in a pipe with stationary integral characteristics. We accept that a flow in an infinite channel (pipe) is completely described by the incompressible Navier-Stokes equations, the initial velocity field $v_o(x)$ and an additional (external) condition. As the external condition the requirement of the constancy of the flow rate or of the space-averaged pressure gradient is usually applied. It is convenient to systematize the nonstationary flows in channels by two different Reynolds numbers R_Q, R_p, where R_Q is calculated by the flow rate and R_p by the mean pressure gradient modulus [1] . A nonstationary flow with the constant Reynolds numbers R_Q and $R_p (R_p > R_Q)$ will be called the secondary flow. Note that for a stationary Poiseuille flow $R_Q = R_p = R$.

In our computations the flows with the fixed mean pressure gradient $\nabla p = -2\nu \cdot \vec{i} = -2\vec{i}/R_p$ were simulated, where $\vec{i} = \{1,0,0\}$, ν is the kinematic viscosity and the density of fluid is assumed to be constant $\rho_0=1$. Later on we shall consider the case of a plane channel and restrict ourselves to a few remarks on the case of the pipes.

2. We investigate only the periodic in the homogeneous spatial coordinates x,y flows of incompressible viscous fluid in an infinite plane channel $K \equiv [\vec{x} = (x,y,z),\ x,y < \infty,\ |z| \leq 1]$, which is a simplest problem in infinite channels. It reduces to determinating the solution of the Navier-Stokes equations $\vec{v}(\vec{x},t) = \{u(\vec{x}, t), v(\vec{x}, t), w(\vec{x}, t)\}$ that satisfies the no--slip condition on the channel walls $z = 1$ and the conditions of periodicity in the x,y variables with the periods $X=2\pi/\alpha_0$, $Y=2\pi/\beta_0$, respectively.

For obtaining the periodic (in x,y) secondary flows we use the method of relaxation. Imposing a certain initial velocity field $\vec{v}_0(\vec{x})$ (which differs from the velocity in a Poiseuille flow) we compute the evolution of a corresponding flow until $R_Q(t)$ becomes constant, i.e. the flow becomes a secondary or Poiseuille flow. The mean time of the passage of 2-D and 3-D flows to a secondary regime is $T=10^3$ to 10^4 for $R_p=10^3$ to 10^4. The process of relaxation to the secondary flow was controlled by the stabilization of its principal integral characteristics such as the flow rate Q(t), the total energy of disturbances E(t) and others. At the final stage of all our computations of secondary flows the flow rate is practically constant. Its variation in the end of all runs does not exceed 0,1% and 1.0% for 2-D and 3-D cases respectively.

3. The searches for secondary flows were performed for plane channels, pipes and annular pipes. For a plane channel in the range R_p=2000 to 12000 more than forty 2-D and ten 3-D runs were performed. The most part of them was successful, i.e. in these runs some secondary flows were obtained. The 2-D and 3-D secondary flows were obtained also in the annular pipes; in a circular pipe a few 3-D secondary flows were found, but 2-D secondary flows apparently do not exist.

For $R_p \in [2750, 3250]$ the 2-D secondary flows in the plane channel only with large intervals of pericdicity $X \gtrsim 6\pi$ were obtained, for $R_p \geqslant 3250$ the 2-D secondary flows with smaller $X \lesssim 2\pi$ were also found. Thus, the instability of the Poiseuille flow to 2-D finite-amplitude disturbances was established by our computations for all the Reynolds numbers $R_p \geqslant 2750$ ($R_Q \geqslant 2612$). It was discovered that for any $R_p \geqslant 2750$ there exists an infinite set of 2-D secondary flows with different $X(\alpha_o)$. Among them there exists the limit flow in which the mininal flow rate (at fixed R_p) and the maximal drag (at fixed R_Q) are realized. For some Reynolds numbers the 2-D secondary flows were obtained which are sufficiently close to the limit flows.

Each secondary flow may be depicted by a point in the plane of variables R_Q, R_p or by a corresponding point in the R_Q, c_f - plane, where $c_f = 9 \cdot R_p / R_Q^2$ is the drag coefficient (in a plane channel). In Fig. 1 all limit 2-D secondary flows we obtained are depicted in the R_Q, c_f - plane as well as the experimental results on the drag coefficient. It may be seen from this figure that the drag coefficient in all the 2-D secondary flow differs substantially from experimental data. Note, however, some similarity in the drag coefficient dependence on Reynolds number R_Q for 2-D limit secondary flows with this dependence obtained in experiments [2] . Such qualitative similarity of 2-D secondary flows with the turbulent ones is observed also in some other aspects (the flattening of the mean velocity profiles, the substantial reduction of the flow rate, etc.).

Note that some control runs with a better resolutions in space and time and with a different numerical technique confirm rather good accuracy of computations of 2-D secondary flows.

4. Next part of our work was devoted to numerical simulation of 3-D secondary flows. The first 3-D secondary flow was computed at $R_p = 5000$, $\alpha_o = 1{,}25$, $\beta_o = 2$. As the initial condition in this run the velocity field of 2-D secondary flow with the same R_p and α_o was taken on which small 3-D disturbances were imposed. This flow passed to a steady regime with $R_Q = 2037$. Note that the reduction of the flow rate proportional to $R_p - R_Q$ increased more than by a factor of two as compared to 2-D se-

condary flow. After that the 3-D secondary flows at R_p=7500, 3500, 2800, 2100 were obtained. The minimal Reynolds number is R_p=2100 (R_Q=1313). Thus, the instability of the Poiseuille flow to the 3-D finite-amplitude disturbances is established at $R_p \geq 2100$ ($R_Q \geq 1313$). This result rather well agrees with the experiments in which the transition to turbulence in a plane channel was observed at $R_Q \gtrsim 1000$ (see 2,3).

In Fig. 1 are plotted the points corresponding to all the computed 3-D secondary flows in a plane channel. For all of them the drag coefficient is sufficiently close to the experimental data. The comparison of mean flow velocity in the 3-D secondary flows with the experimental ones is presented in Fig.2. We may say that beyond all our expectations agreement between mean velocity profiles is very good. For contrast the mean velocity profile in the 2-D limit secondary flow at R_Q=2952 (R_p=3500) is also plotted in Fig. 2. In Fig. 3 a mean velocity profile in the 3-D secondary flow at R_Q=2037 (R_p=5000, α_0= =1.25, β_0=2.0) is compared with the experimental[2]at R_Q=2070. There $z^+=(1-z)u_\tau/\nu$, $U^+=U/u_\tau$; (1)-$U^+=z^+$, (2)-$U^+=5.5 \log_{10}z^+$ + +5.45. We may see rather good quantitative agreement of mean velocity profiles both near the walls and in the middle of the channel.

The comparisons presented in Figs. 1-3 show rather good agreement of basic parameters of turbulent flows with those in the 3-D secondary flows. Note that some other characteristics of turbulent flows (r.m.s. pulsation velocity component, the distribution of Reynolds stresses) are reproduced in the 3-D secondary flows with good accuracy too. Analogous conclusions may be made about the 3-D secondary flows in the pipes and annular pipes. For example see Fig. 4 where the mean velocity profile of the 3-D secondary flow computed at r_1/r_2=1/3, R_Q= =6557 is compared with the experimental one by Knudsen & Katz [4] at r_1/r_2=0.28, R_Q=7059.

A more detailed description of the results of our computations and their comparisons with experiments and other works may be found in [1,5,6,7] .

5. The solution of the problem on a viscous fluid flow in the plane channel with a fixed mean pressure gradient is presented in the form $\vec{v}(\vec{x},t)=(1-z^2)\vec{i} + \vec{v}'(\vec{x},t)$; $p(\vec{x},t)=-2\nu x + p'(x,t)$ (in what follows the primes above v and p are omitted). These deviations are searched for in the form

$$\begin{Bmatrix} \vec{v}(\vec{x},t) \\ \Pi(\vec{x},t) \end{Bmatrix} = \sum_{m,n,p} \begin{Bmatrix} \vec{v}_{mnp}(t) \\ \Pi_{mnp}(t) \end{Bmatrix} \cdot T_p(z)\exp(im\alpha_0 x+in\beta_0 y) \tag{1}$$

This approximation assumes the periodicity of the solution in the x,y variables. Here $\Pi(\vec{x},t)=p(\vec{x},t)+\vec{v}^2(\vec{x},t)/2$; $T_p(z)$ are the Chebyshev polynomials and $|m| \leq M$, $|n| \leq N$, $0 \leq p \leq P$. Note that Chebyshev polinomials in such a problem were used by Herbert, Orszag & Kells, Moin & Kim and some others. The integration in time of the Navier-Stokes equations approximated in terms of $\vec{v}_{mnp}$ and Π_{mnp} is carried out by the fractional steps method.

At the first fractional step the solution $\vec{v}^{k+1/3}$ of the explicit difference scheme

$$\begin{aligned} 2(\vec{v}^{k+1/3}_{mnl} - \vec{v}^{k}_{mnl})/\tau &= 3\vec{F}^{k}_{mnl} - \vec{F}^{k-1}_{mnl} - im\alpha_0 U_l \vec{v}^{k}_{mnl} - \\ &- \vec{i}U'_l(5w^{k}_{mnl} - 2w^{k-1}_{mnl}) \end{aligned} \tag{2}$$

is determined. Here $\vec{F}=[\vec{v},\vec{\omega}]$, $\vec{\omega}=[\nabla,\vec{v}]$, the subscript l corresponds to the points $z_l=\cos(\pi l/P)$, $U_0(z)=1-z^2$ and $U_l=U_0(z_l)$, $U'_l = U'_0(z_l)$.

At the second fractional step the solution $\vec{v}^{k+2/3}$ of following purely implicit difference scheme

$$\begin{aligned} (\vec{v}^{k+2/3} - \vec{v}^{k+1/3})/\tau &= -\nabla\Pi^{k+2/3} + \nu\nabla^2\vec{v}^{k+2/3}, \\ \vec{v}^{k+2/3} = 0, \quad & \vec{v}^{k+2/3}(x,y,z=\pm 1) = 0 \end{aligned} \tag{3}$$

is constructed. The problem (3) is effectively solved in the representation of solution being spectral in all spatial variables. In the z-variable the method by Petrov (it is also called the spectral tau-method) is employed.

It should be noted that this method possesses a high acccuracy. However, the application of this method results in spurious roots which make the spatial approximation of the original evolution differential problem absolutely unstable. In the case of the purely implicit difference scheme (3) this shortcoming is manifested in a quite original manner: the scheme (3) becomes conditionally stable with the limitation on the time step τ from below ($\tau > \tau_o \sim 2/(\nu P^4) \rightarrow 0$ as $P \rightarrow \infty$). In our computations $R_p = 1/\nu \leq 1.2 \cdot 10^4$ and $P \geqslant 32$, consequently $\tau_o \lesssim 0.02$, and at $\tau = 0.1$, 0.05 which we have used the scheme (3) possesses a high reliability and accuracy.

The final third fractional step

$$2(\vec{v}^{k+1} - \vec{v}^{k+2/3})/\tau = -im\,\alpha_o U_1 \vec{v}^{k+1}_{mnl} + \vec{i} U_1' w^{k+1}_{mnl} \qquad (4)$$

uses the mixed spectral-physical representation of the solution $\vec{v}$ like at the first fractional step and has the purpose to improve the approximation of the linear terms in the Navier-Stokes equation. It is not difficult to show that the solution $\vec{v}^{k+1}(x)$ satisfies the incompressibility condition and equals to zero at the walls. The third fractional step concludes the computation of full time step. Note that we have constructed an analogous scheme also for the second problem formulation when flow rate is fixed.

6. The most part of 2-D secondary flows was computed with rather good accuracy. The numbers 2M and P reached up to 64 which was quite enough for correct description of the 2-D secondary flows at moderately large Reynolds numbers $R_p \lesssim 10^4$ and not very small $\alpha_o \gtrsim 0.3$.

In all our 3-D computations M=N=4, P=32 which corresponds to nearly 8×10^3 degrees of freedom of the representation (1) of the velocity field in our numerical method. The small numbers

M and N as well as small number of runs performed are related to a substantial increase in computer time needed with increasing M and N. These circumstances do not allow us to make justified conclusions on the limit secondary flows with maximal drag like in the 2-D case. In spite of the fact that the accuracy of 3-D computations is not so ideal as in the 2-D case due to the smallness of numbers M and N the agreement between the integral parameters of the computed 3-D secondary flows and of turbulent flows is surprisingly good. It gives the reason to believe that the basic characteristics of turbulent flows may be computed theoretically with good accuracy on the basis of the Navier-Stokes equations.

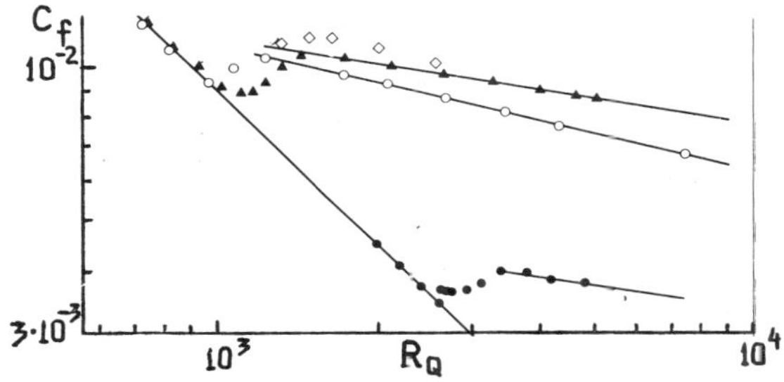

Fig. 1. Drag coefficient c_f versus R_Q. Experiments: ▲ - [2], ○ - [3]. Our computations: ◇ - 3-D secondary flows, ● - 2D limit secondary flows.

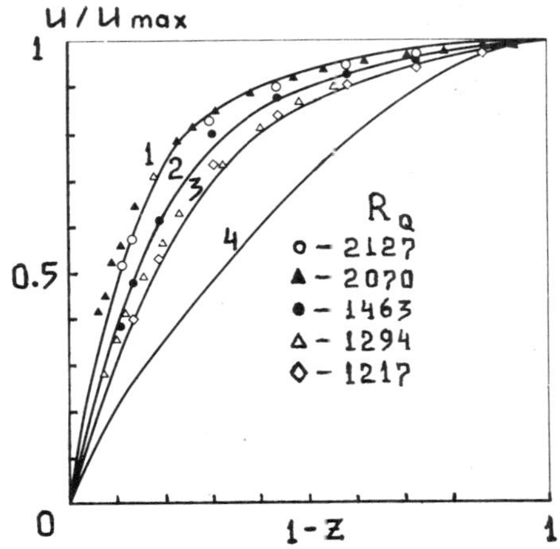

Fig. 2. Normalized mean velocity profiles. Our computations: 1-3 - 3D secondary flows at R_Q=2037, 1498, 1313; 4 - 2-D secondary flow at R_Q=2952. Experiments: ▲,△ - [2]; ○,●,◇ - [3].

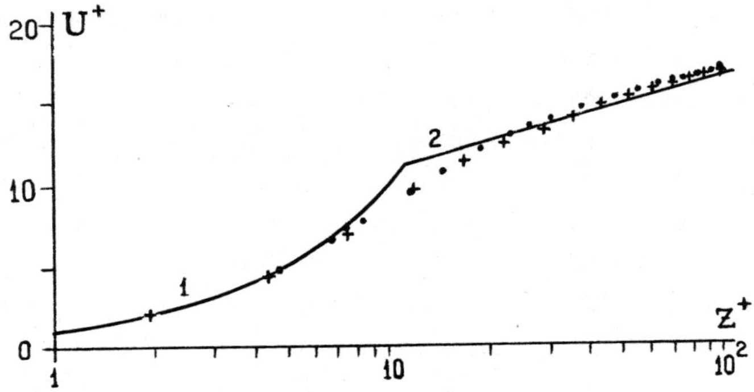

Fig. 3. Mean velocity profile versus $z^+=(1-z)u_\tau/\nu$. Experiment [2] at R_Q=2078 - •; 3-D secondary flow at R_Q=2037 - +.

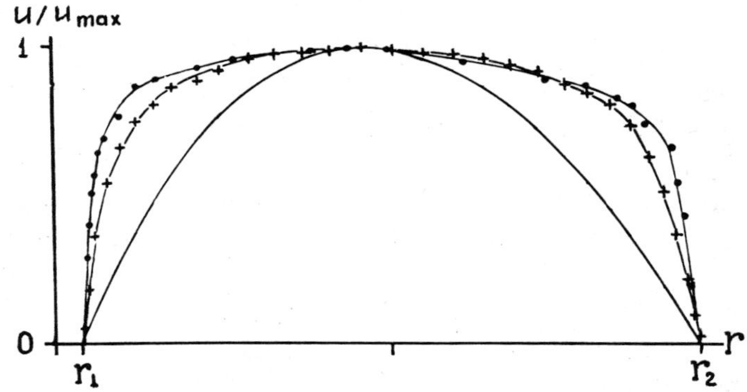

Fig. 4. Mean velocity profiles in annular pipes. Experiment [4] - •, 3-D secondary flow - +, inner curve-laminar flow. For other details see text.

References

1. Rozhdestvensky, B.L.; Simakin, I.N.: Nonstationary secondary flows in a plane channel and stability of the Poiseuille flow with respect to finite disturbances. Dokl. Acad. Nauk SSSR, 266, (1982) No.6, 1337-1340, (Russian).

2. Patel, V.G.; Head, M.R.: Some observation on skin friction and velocity profiles in fully developed pipe and channel flows. J.Fluid Mech. 38 (1969), No.2, 308-311.

3. Whan, G.A.; Rothfus, R.R.: Characteristics of transition flow between parallel plates. A.I.Ch.E.J. 5 (1959), No.2, 204-209.

4. Knudsen, J.G.; Katz, D.L.: Velocity Profiles in Annuli. Proc. Midwestern Conf. on Fluid Dynamics, 1951, 175-203, Ann. Arbor, Michigan.

5. Rozhdestvensky, B.L.; Simakin, I.N.: Stability of the Poiseuille flow with respect to finite disturbances. Chisl. Metody Mekh. Sploshnoy Sredy, Novosibirsk 13 (1982), No. 4 , 89-128, (Russian).

6. Rozhdestvensky, B.L.; Simakin I.N.: Two- and three-dimensional secondary flows in a plane channel, their relationship and comparison with turbulent flows. Dokl. Akad. Nauk SSSR, 273 (1983), No. 3, 553-558, (Russian).

7. Rozhdestvensky, B.L.; Simakin I.N.: Secondary Flows in the Plane Channel,Their Relationship and Comparison with Turbulent Flows, J.F.M., 1984, to be published.

Three-Dimensional Wave-Disturbances in Plane Poiseuille Flow

MICHIO NISHIOKA and MASAHITO ASAI

College of Engineering
University of Osaka Prefecture
Sakai, Osaka, Japan

Summary

To clarify the initial 3D wave development in ribbon-induced transition of plane Poiseuille flow, double Fourier analyses, with respect to frequency and spanwise wavenumber, of u-velocity fields are made. The results indicate the validity of the concept of secondary instability, i.e. a linear instability of the near-equilibrium periodic flow (with 2D T-S wave) with respect to phase-locked 3D wave-disturbances, which has been proposed by Orszag & Patera and Herbert recently.

Introduction

For boundary-layer transition initially governed by 2D T-S wave, as shown by the experiment of Klebanoff et al (1), the prerequisite for the eventual breakdown into turbulence (or at least into the hairpin eddies) is the occurrence and growth of three-dimensionality leading to a spanwise-periodic wave motion, i.e. the so-called peak-valley structure. Many experiments were made on the phenomenon but could not throughly reveal the actual mechanism for the onset and the subsequent growth: See Anders & Blackwelder (2). As for the theoretical attempt, Benney & Lin (3) modeled the so-called longitudinal vortices associated with the peak-valley structure (that cause the local high-shear layers to form at peak positions (4)) through considering the second-order effects of the interaction between the primary 2D wave and a pair of oblique waves (of the same streamwise wavenumber) with the spanwise periodicity: Klebanoff et al supported this model. Craik (5) proposed a resonant wave-triad: The resonance between the fundamental 2D and a pair of subharmonic oblique modes. Kachanov & Levchenko (6) found Craik's mechanism to operate in experiments depending on the disturbance environment, and Saric & Thomas (7) visualized the subharmonic modes developing into a staggered peak-valley structure, which is quite different from the Klebanoff-type orderd peak-valley. Quite recently Orszag &Patera (8) and Herbert (9) proposed an instability mechanism. They studied a linear secondary instability of the basic flow modified by finite amplitude T-S wave with respect to 3D disturbances. The main purpose of the present study is to examine the validity of the concept of the secondary instability experimentally in plane Poiseuille flow.

Ribbon-induced transition in plane Poiseuille flow

Using a 6 m long wind-channel of rectangular cross-section, whose half-depth h is 7.3 mm and aspect ratio is 27.4, we have been studying the ribbon-induced transition at sub- and super-critical Reynolds numbers. The results have been published to clarify the linear instability, the non-linear subcritical instability put forward by Stuart (10), the peak-valley development, the high-frequency secondary instability (of the high-shear layer) into hairpin eddies, and the final evolution into wall-turbulence (11~15). Here, we go back to the problem of the peak-valley splitting to study the underlying mechanism. Let us briefly review what we have learned from an experiment (12, 13) made at a subcritical Reynolds number R = 5000 and at a ribbon-frequency f = 72 Hz ($2\pi fh/U_c = 0.337$, $2\pi h/\lambda_{TS} = 1.16$), where R is defined on the centre-line velocity U_c and h. First of all it should be mentioned that in our channel U_c slightly varies in the spanwise direction with wavelengths of 24 ~ 26 mm. Therefore the vortical wave induced by a vibrating ribbon can not be perfectly two-dimensional. Indeed, the phase of u-fluctuation at and around the critical layer exhibits spanwise variations which are almost periodic with the wavelengths imposed by the corresponding distortion in the basic parabolic-flow,though remain small in magnitude until the peak-valley splitting occurs. In spite of the subcritical Reynolds number (about 86 % of the critical R for the linear instability), the excited T-S wave can grow when the initial amplitude exceeds a threshold, about $0.01U_c$. As the wave grows, the initial spanwise distortion of the wave front is intensified to develop into the peak-valley structure. In the beginning, the intensification proceeds near the linear critical layer only.

No large distortion undergoes near the centre-line until the high-shear layer breaks down into hairpin eddies.

The observed threshold behaviour of the initial wave growth is in reasonable agreement with the prediction from the weakly non-linear theory for 2D waves. More importantly, the amplification of the initial three-dimensionality clearly indicates that the periodic flow near the subcritical equilibrium is highly sensitive to 3D disturbances. On the basis of Stuart's non-linear theory (16) including 3D disturbances, Itoh (17) studied 3D effects and found similar threshold behaviours in plane Poiseuille flow, though the parameter range in his calculation was different from ours. Subsequently, Kleiser (18) succeeded in reproducing our experiment through numerical simulation and also clearly demonstrated the threshold behaviour in the growth of the initial small-amplitude 3D disturbances. This encouraged Herbert to formulate his theory of the secondary instability.

The mechanism of the secondary instability is a kind of parametric resonance. Solving the eigenvalue problem for the case of plane Poiseuille flow, Herbert found two different 3D modes with the fundamental and subharmonic frequencies, and thought that these are respectively causes of the ordered peak-valley studied by the present authors and staggered peak-valley visualized by Kozlov & Ramazanov (19). However, experimental information on the structure of 3D disturbances is rather limited. In particular for the case of the ordered peak-valley, the 3D waves have not been singled out. Generally speaking, to identify an instability we should know the structure of the excited wave. So in the present investigation we try to single out 3D waves in our plane Poiseuille flow. This is rather painstaking because the dominant 2D and 3D waves have the same frequency and common instruments for filtering are useless.

Experimental procedure

The experiment reported here is carried out in the same wind-channel under the same flow conditions as our previous works: R = 5000 and f = 72 Hz. For the measurements of the streamwise velocity a hot-wire probe (with sensitive length 1.5 mm) of a constant-temperature system is fixed, just like as in the previous experiment, at 48h downstream from a vibrating ribbon and can move continuously in the normal-to-wall (y-) and spanwise (z-) directions only. The wave growth in x is simulated at the observation position by varying the amplitude of the ribbon oscillation. As the amplitude increases, successive stages of the wave development are realized as if they evolved in x. The velocity signal is separated into the time-mean and fluctuating components U and u. These signals together with the input to the ribbon are stored on a multi-channel analogue data recorder after pre-amplification, and converted into 12 bit-digital form and then processed and analyzed by a micro-computer (DATA GENERAL ECLIPSE S/20): The maximum conversion rate available is 33 kHz. The computer processing includes the linearization of non-linear response of hot-wire, the periodic sampling and ensemble averaging for wave-form of u-fluctuation, Fourier analysis of the averaged wave-forms and the graphic representation of these results.

Although the background turbulence is less than 0.05 % of U_c, this can no doubt contaminate the periodic velocity signal of our interest and thus limits the accuracy of measurements. So the periodic sampling is made to eliminate the possible contamination which is not related to the ribbon-induced periodic wave. The number of samples is fixed at 256. As a matter of course we ascertained the reproducibility of the phenomena observed in the previous experiment (12). We also noticed minute but non-negligible differences in y-distributions of r.m.s. u-fluctuation u' from the previous ones. So we examined the cause and found two main factors. One is related to the quality of linearization of hot-wire response in the wind speed range less than 3.5 m/s (in particular less than 1 m/s where the heat-loss deviates from the King's law): It turned out that the piecewise linear fitting (to the calibration curve) by means of Transister-chain circuit used in (12) probably had about 5 % error (in each linear fitting) in the low speed range. This was already solved in the latter experiments by using an Integrated Circuit and the linearization is further qualified by the use of computer in the present experiment. The other is related to slits (in the channel wall) provided for the purpose of inserting a probe into the test section. Those not in use are filled up with cover blocks flush with the channel inner wall. However, every time we open and close those, the alignment may change slightly and this may have some effects on the wave development. In fact we found this also to be responsible for the small differences mentioned.

We tried two methods to single out 3D wave-disturbances. These are rough and detailed double Fourier analyses with respect to the frequency and the spanwise wavenumber.

Results and discussion

Because of the limmited space available here, what we can show is only a part of the new results. Fig.1 gives a 3D plotting of u at y/h = 0.185 near the critical layer of the ribbon-induced small-amplitude (0.6 % of U_c) T-S wave, comparing with z-distributions of local U at y/h = 0.1 and 1.0: Time t runs backward so as to provide a near simulation of a wave coming out of the paper. This comparison indicates that the initial wave distortion is imposed by the corresponding distortion in the basic parabolic flow. As mentioned earlier, when the amplitude exceeds $0.01U_c$, the wave grows downstream, and the initial three-dimensionality is intensified: The growth is enhanced at a spanwise position noted 'peak' and reduced at 'valley' as illustrated in Fig.2. To name various stages of the wave development we use the non-dimensional y-maximum of u' at a peak z = 7.5 cm, u'_m/U_c. Fig.3(a) compares the ensemble-averaged wave-forms of u at the neighbouring peak and valley at the 2 % stage, and shows significant differences around y/h = 0.185. It is quite interesting to note that the phase averaging of the wave-forms at the paired peak and valley at each y/h agrees almost perfectly with the corresponding one at the spanwise mid position, as demonstrated in Fig.3(b). We may interpret this feature as indicating that phase-locked spanwise-periodic 3D wave-disturbances are 'linearly' superposed on the 2D T-S wave, and the dominant 3D component has the spanwise wavelength twice the distance between the peak and valley. If this is the case, the 3D can be singled out by simple subtraction as displayed in Fig.3(c). The wave-forms contain higher harmonics, so the fundamental represented by dots is filtered out by Fourier analysis. What we have done thus far is a kind of crude double Fourier analysis and the results encourage us to go to the detailed analysis with respect to the frequency and spanwise wavenumber, at various y/h positions, as described below.

The spanwise extent covered by double Fourier analysis is $2\lambda_z$ so that the fundamental spanwise wavenumber is $\alpha_z/2$, where λ_z and α_z are the distance between the neighbouring peaks and the corresponding wavenumber. We took 10 or 25 points equally spaced over $2\lambda_z$ and 56 equal-interval points for the period of T-S wave. The double Fourier coefficient C(m,n), each Fourier mode $u_{m,n}$ and a pair of oblique modes $u^{(m,n)}$ $(= u_{m,n} + u_{m,-n})$ are given by the following equations:

$$C(m,n) = \sum_{I=0}^{M-1} \sum_{J=-N/2+1}^{N/2} u(I,J)\exp\{-i(2\pi mI/M + 2\pi nJ/N)\}$$

$$u_{m,n} = |C(m,n)|\cos(m\omega t + n\alpha_z z + \delta^{(m,n)})$$

$$u^{(m,n)} = \{|C(m,n)| + |C(m,-n)|\}\cos(m\omega t + \theta^{(m,n)})\cos\{n\alpha_z(z-z_0)\}$$

$$+ \{|C(m,n)| - |C(m,-n)|\}\sin(m\omega t + \theta^{(m,n)})\sin\{n\alpha_z(z-z_0)\}$$

where, $\omega = 2\pi f$, $\alpha_z = 2\pi/\lambda_z$, $m = 1,2,3,\cdots$, $n = 0,\pm 1/2,\pm 1,\pm 3/2,\cdots$. Fig.4 shows the magnitude of coefficients for the fundamental frequency modes at three selected stages. At the 1.8 % stage, the most energetic is of course 2D mode. The dominant 3D modes are a pair of α_z and $-\alpha_z$ modes with nearly equal magnitude. At the 5.5 % stage the pair of α_z modes get comparable to 2D mode. The other pairs get large too and nearly even in magnitude. Therefore, the sin term in the above equation becomes negligible compared with the cos term.

As an example, Fig.5 compares 3D plottings of the fundamental 2D and α_z mode $u^{(1,1)}$ at the 3 % stage, which dominate the peak-valley structure. The relative phase relation is such that the wave growth is enhanced at peaks and reduced at valleys. The 3D mode no doubt develops into a secondary-flow like structure with a pair of counter-rotating streamwise vorticity fields, which soon causes local high-shear layer to form at the peak. The instantaneous maximum amplitude of the 3D mode is 0.8 % and 4.6 % at the 1.8 % and the 5.5 % stages respectively, indicating a rapid growth. Fig.6 compares the amplitude and phase distributions of the fundamental α_z mode $u^{(1,1)}$ with Herbert's eigenfunction and those calculated by the present authors on the basis of the Benney-Lin model including the velocity components parallel to the oblique wave fronts. Herbert s solution comes closer to the experiment than the Benney-Lin model. This is true at the 5.5 % stage too. Moreover, the amplitude and phase distributions of $u^{(1,1/2)}$, $u^{(1,1)}$, $u^{(1,2/3)}$, $u^{(1,2)}$, $u^{(2,1)}$, $u^{(2,2)}$ at the 5.5 % stage are all similar in shape, though not the same. The reason is probably that the 3D structure is mainly determined by the velocity components parallel to the oblique wave fronts emphasized by Craik (20). Fig.7 shows the structure of 2D wave at the 1.8 % and the 5.5 % stages and compares with the corresponding results of the linearised stability theory (21). The experimental points for amplitude are plotted in such a way that they lie on the theoretical curve away from the wall, where all the 3D modes almost vanish. We see a dip in the amplitude distribution around the linear critical layer, near which all the 3D modes attain their maximum amplitudes. The dip gets larger as the wave grows as if the 2D mode gave energy to 3D modes. Whatever the causes are, this feature indicates non-linear nature of the peak-valley development, which is not considered in Herbert's secondary instability.

Concluding remarks

The important findings of the present experimental study are summarized as follows. The wave-disturbances which govern the initial stage of the ordered peak-valley development can be decomposed into the 2D T-S mode and various 3D Fourier modes. The 2D mode gradually changes the structure around the critical layer as the 3D modes grow as if it were taken energy by the 3D modes. The 3D modes observed are all similar in shape though not the same. Concerning the fundamental 3D mode ($u^{(1,1)}$), a satisfactory agreement is obtained with the corresponding Herbert's eigensolution for the secondary parametric instability. This indicates the validity of the concept of the secondary parametric instability, i.e. an instability of the near equilibrium periodic flow with respect to 3D disturbances. However, we can not forget the above-mentioned deformation of the 2D T-S wave, which indicates non-linear nature of the 3D wave development.

The authors wish to express their sincere gratitude to Prof. Th. Herbert for kindly providing the results of his stability calculations, and to Profs. S.Iida, I.Tani and H.Sato for their continual encouragement . This work was in part supported by Grant-in-Aid for Special Project Research (No. 57109004), The Ministry of Education, Science and Culture.

References

1. Klebanoff, P.S., Tidstrom, K.D. & Sargent, L.M.: J. Fluid Mech. 12 (1962) 1-34.
2. Anders, J.B. & Blackwelder, R.F.: Proc. IUTAM Symposium on Laminar-Turbulent Transition. eds. Eppler, R. & Fasel, H. (Springer-Verlag, 1980) 110-119.
3. Benney,D.J. & Lin, C.C.: Phys. Fluids 3 (1960) 656-657.
4. Stuart, J.T.: AGARD Report 514 (1965).
5. Craik, A.D.D.: J. Fluid Mech. 50 (1971) 393-413.
6. Kachanov, Yu.S. & Levchenko, V.Ya.: J. Fluid Mech. 138 (1984) 209-247.
7. Saric, W.S. & Thomas, A.S.W.: Proc. IUTAM Symposium on Turbulence and Chaotic Phenomena in Fluids. Kyoto (Sep. 1983).
8. Orszag, S.A. & Patera, A.T.: J. Fluid Mech. 128 (1983) 347-385.
9. Herbert, Th.: Proc. IUTAM Symposium on Turbulence and Chaotic Phenomena in Fluids. Kyoto (Sep. 1983).
10. Stuart, J.T.: J. Fluid Mech. 9 (1960) 353-370.
11. Nishioka, M., Iida, S. & Ichikawa, Y.: J.Fluid Mech. 72 (1975) 731-752.
12. Nishioka, M., Iida, S. & Kanbayashi, Y.: Proc. 10th Turbulence Symposium Inst. Space Aeron. Sci., Tokyo Univ. (1978) 55-62.
13. Nishioka, M., Asai, M. & Iida, S.: Proc. IUTAM Symposium on Laminar-Turbulent Transition. eds. Eppler, R. & Fasel, H. (Springer-Verlag, 1980) 37-46.
14. Nishioka, M., Asai, M. & Iida, S.: Proc. Symposium on Transition and Turbulence. ed. Meyer, R.E. (Academic Press, 1981) 113-126.
15. Nishioka, M. & Asai, M.: Proc. IUTAM Symposium on Turbulence and Chaotic Phenomena in Fluids. Kyoto (Sep, 1983).
16. Stuart, J.T.: Adv. Aeron. Sci. 3 (1962) 121-142.
17. Itoh, N.: Trans. Japan Soc. Aeron. Sci. 23 (1980) 91-103.
18. Kleiser, L.: Numerishe Simulationen zum Laminar-Turbulenten Umshlagsprozeß der ebenen Poiseuille Stromung. Dissertation, Universitat Karlsruhe (1982).
19. Kozlov, V.V. & Ramazanov, M.P.: I.T.A.M., USSR AN Novosibirsk, Preprint 4-82 (1982).
20. Craik, A.D.D.: J. Fluid Mech. 99 (1980) 247-265.
21. Itoh, N.: Trans. Japan Soc. Aeron. Sci. 17 (1974) 65-75.

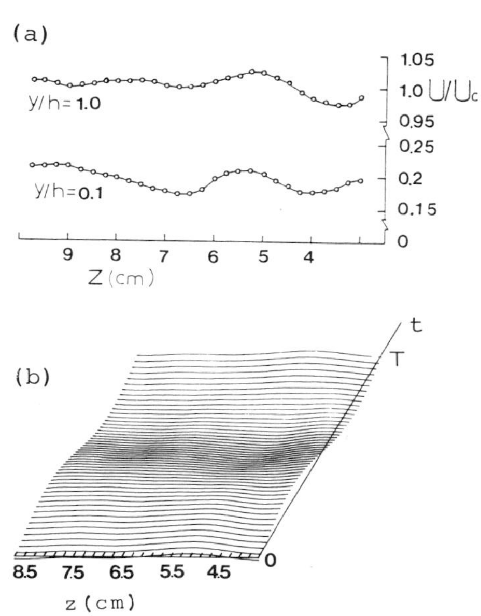

Fig.1. Initial three-dimensionality: (a) basic flow(the scale U_c is the value at z=7.5 cm), (b) u-fluctuation of small amplitude T-S wave(u'_m/U_c=0.6%) at y/h=0.185.

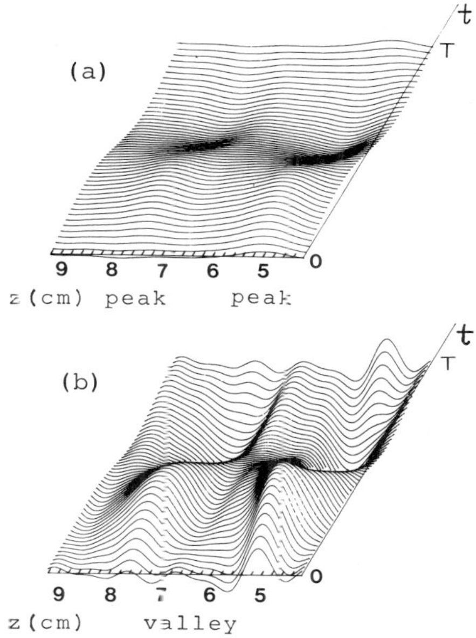

Fig.2. Three-dimensional development of u-fluctuation at y/h=0.185: u'_m/U_c =1.4% in (a) and 3.0% in (b).

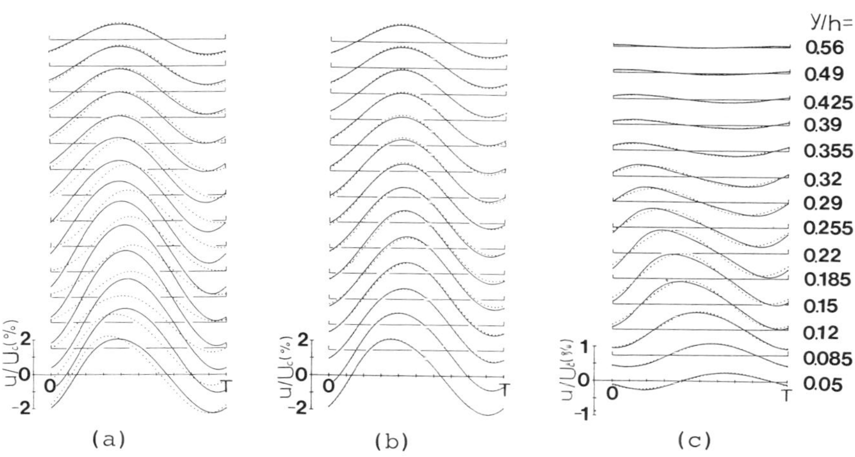

Fig.3. Ensemble-averaged wave-forms of u-fluctuations at u'_m/U_c= 2.0%. (a) —— peak, ··· valley. (b) —— phase-averaged, ··· mid-position between peak and valley. (c) 3D component (··· fundamental frequency).

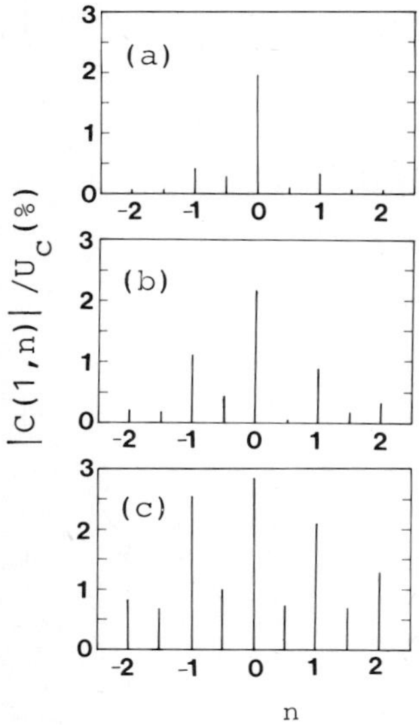

Fig.4. Magnitude of Fourier coefficients for fundamental frequency modes, $|C(1,n)|$: $(u'_m/U_c,\ y/h)=(1.8\%,0.185)$ in (a), $(3.0\%,0.185)$ in (b) and $(5.5\%,0.215)$ in (c).

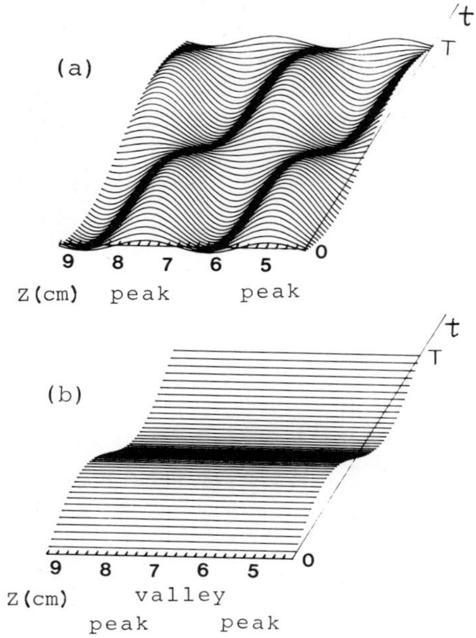

Fig.5. Dominant 3D and 2D modes($u^{(1,1)}$ in (a) and $u_{1,0}$ in(b)) at $u'_m/U_c=3.0\%$, at $y/h=0.185$.

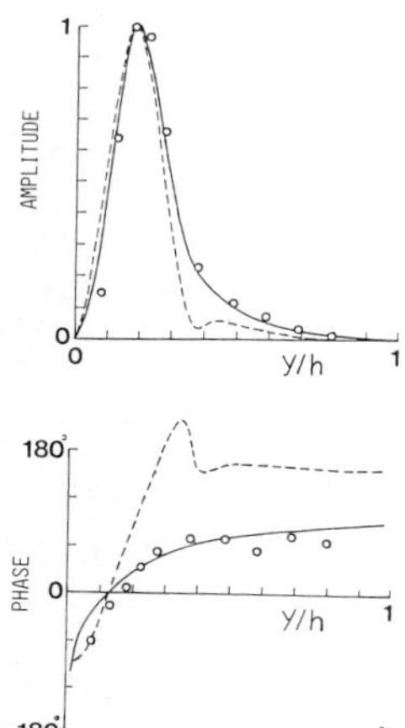

		$\omega h/U_c$	$\alpha_x h$	$\alpha_z h$
o	Experiment	0.337	1.16	1.84
---	Benney-Lin	0.364	1.12	2.00
——	Herbert	0.332	1.12	2.00

Fig.6. Comparison of fundamental 3D mode $u^{(1,1)}$ at $u'_m/U_c=1.8\%$ with Herbert's and Benney-Lin models.

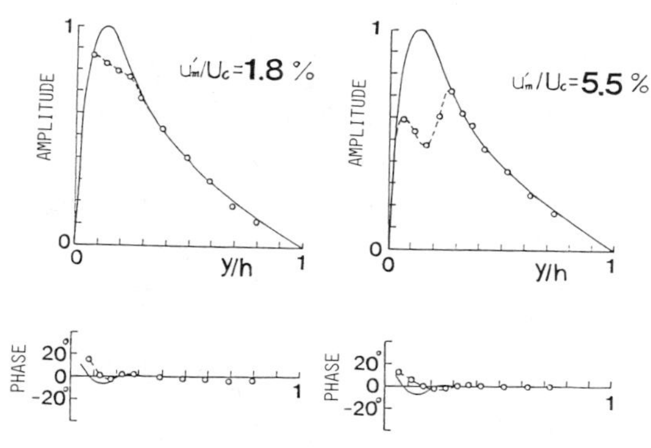

Fig.7. Structure of fundamental 2D mode $u_{1,0}$:
o experiment, —— result of linearised stability theory.

Development of Finite-Amplitude Disturbances in Poiseuille Flow

M. P. RAMAZANOV

Institute of Theoretical and Applied Mechanics,
USSR Academy of Sciences, Novosibirsk, USSR

Summary

The process of three-dimensional distortion of previously two-dimensional disturbances was investigated in a rectangular channel. For the first time the three-dimensional structures at the breakdown station of the two-dimensional wave were studied by flow visualization. It was shown that the structures have forms identical to Λ-shaped vortices in a boundary layer of the flat plate. The spanwise spacing of the Λ-shaped vortices is quite independent of the mean flow velocity and of the frequency of artificial disturbances.

Introduction

The flow in a plane channel is usually considered for modelling and investigating the physical processes at a laminar-turbulent transition. This is related to the exact solution of Navier-Stokes equations for the flow and to the simpler type of equations for disturbances. Therefore many authors use this flow for the problem of three-dimensional growth of the disturbances (see Herbert [1], Goldshtic, Lifshitz & Shtern [2]). The experimental investigations of Nishioka et al. [3, 4], Kozlov & Ramazanov [5] for the finite amplitude at subcritical and supercritical Reynolds numbers indicated the presence of the spanwise periodic three-dimensional structures in the flow that are similar to that defined by Klebanoff, Tidstrom & Sargent [6] in flat plate boundary layer.

Experimental apparatus

The channel flow is studied in the installation described in [5]. Fig.1 shows a schematic diagram of the installation. The channel is made of plexiglass and is 4500 mm long, 15 mm wide

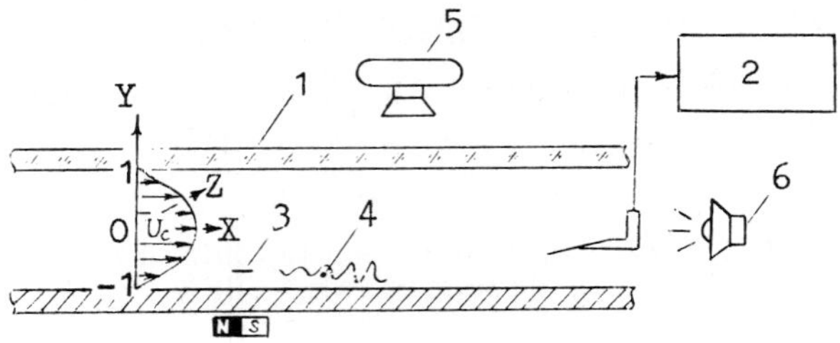

Fig.1. Schematic diagram of experiment

1 - channel

2 - hot-wire anemometer

3 - ribbon

4 - smoke wire

5 - photographic camera

6 - photo-flash

and 400 mm high. This channel is placed in the test section of the low-turbulence wind tunnel T-324 of the Institute of Theoretical and Applied Mechanics. The level of turbulence in the channel is reduced to approximately 0.1% by means of a large-ratio (3:80) contraction at the inlet and low turbulence level of free stream. Disturbances are introduced into the flow by a metal ribbon which is 0.05 mm thick, 3 mm wide and the distance from the inlet X_o=3600 mm. This ribbon is placed in a constant magnetic field near one of the channel walls at the distance of Y=0.85 mm. Sinusoidal current passes through the ribbon with the frequency given by a wave generator. The frequency is checked by a frequency meter. The Reynolds number is based on the maximum velocity U_c in the channel and the channel half-width h. With the channel center at Y=0, wall dimensionless coordinates are 1 and -1 respectively. As it was shown in paper of Kozlov & Ramazanov [7], plane Poiseuille flow can be produced in the whole measured region in the absence of artificial finite amplitude disturbances.

The structure of the channel flow is visualized by the smoke wire which is located close to the maximum of the Y - distribution of disturbance velocity, at a position 2.2 mm (Y/h=0.29) from the wall, that approximately corresponds to the critical layer. As it was shown by Hama [8], streaklines introduced into the flow at a critical layer, distinguish more accurately a wave structure of the flow field, and the distance between the smoke accumulations coincides very well with the wavelength.

Results and discussion

Fig.2 shows the three-dimensional distortion process of the initially two-dimensional wave front at the frequency of f= =73 Hz and Re=3850. The flow passes from left to right, $X-X_o$ is a distance from the vibrating ribbon, Z=0 corresponds to middle of the channel. The bright stripes are observed at the left side of each photograph. These stripes are the smoke accumulations which are sufficiently uniform in Z-direction. As it is shown by Hama [8], streaklines which were propagated in the flow with the wave velocity, were accumulated gradually in the regions of local decrease and began to turn back, forming a clearly visible wide strip. As the amplitude of the disturbance reached a value of 1%, the double longitudinal vortices began to appear in the flow, the streaklines were accumulated in the region of the maximum vorticity forming two sloped lines in the form of the Greek letter Λ. This Λ-shaped vortex may be clearly observed at the top photograph in the region of

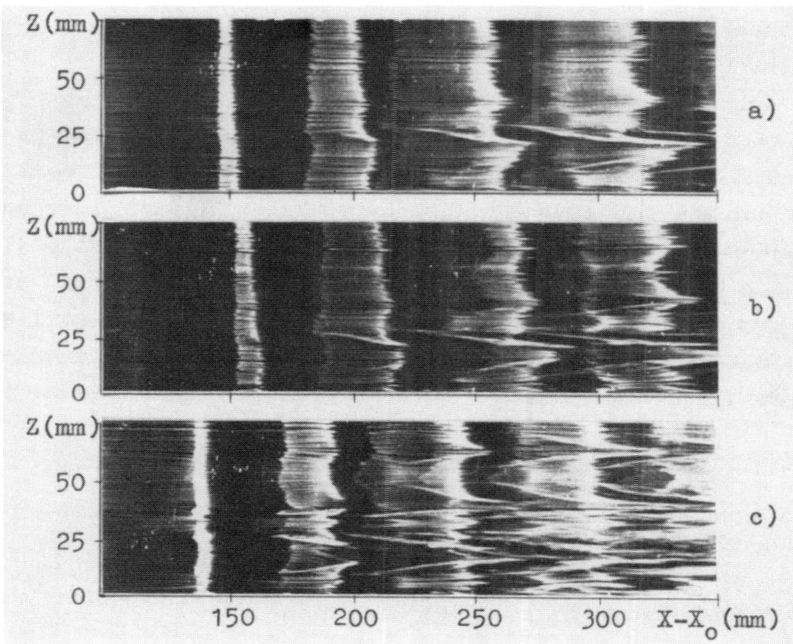

Fig.2. Development of three-dimensionality with increasing of disturbances amplitudes. Re=3850, f=73 Hz
a) u'_{max}/U_c=1.0%; b) u'_{max}/U_c=1.26%; c) u'_{max}/U_c=1.36%

$X-X_0$=250÷300 mm. As long as the amplitude increases, the Λ-shaped vortex moves to the left (Fig.2 b) and c)). A chain of such vortex lines arranges strictly one after another, the second and then the third chain of the Λ-shaped vortices appears beside it. So, the region of appearance of these Λ-shaped vortices moves upstream with increasing amplitude and also spreads from the center of the channel to the periphery. Non-uniformity of the vortex appearence in the cross-section direction is associated with that the vibrating ribbon is fixed at the edges. Therefore, the amplitude of the ribbon has a maximum at the center and decreases towards the edges. Moreover, the disturbances are introduced near only one of the walls, so that the amplitude maxima are different at the two walls. This asymmetry was also observed by Nishioka, Iida & Ichikawa [3] and caused the "spikes" to emerge in the wave form of the fluctuation at only one of the channel walls, namely, that with the larger disturbance amplitude. At the same time, the wave front

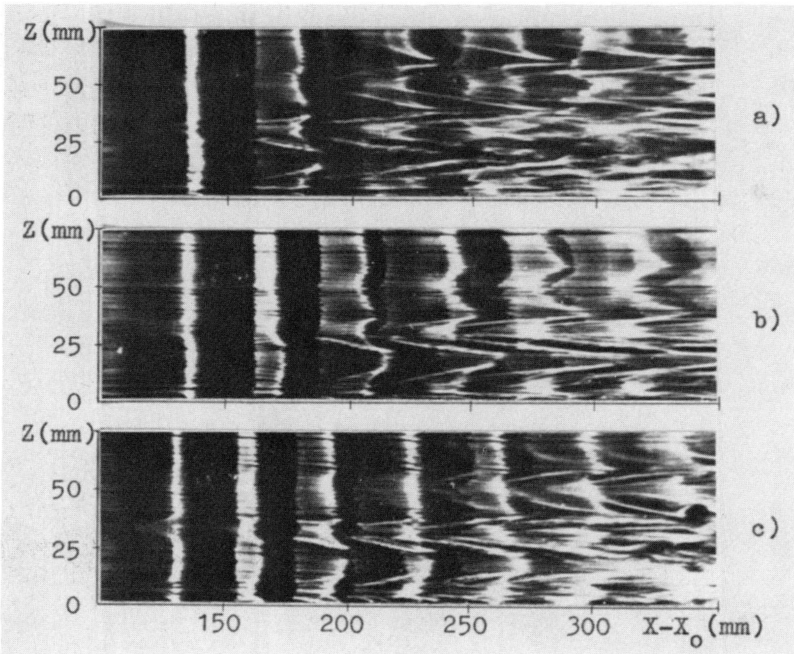

Fig.3. Three-dimensional deformation of two-dimensional waves with different frequencies at Re=3850
a) f=90 Hz; b)f=110 Hz; c) f=120 Hz

at the other wall had not been distorted. Fig.3 shows the Λ-shaped vortex formation for different frequencies. It is worthy of note that in spite of **considerable** changes of the wavelength the spanwise spacing of the Λ-shaped vortices remains constant and is approximately 23 mm. It does not change as the amplitude increases to a certain magnitude and Fig.2 confirms it. Also, the mean flow velocity does not change the arrangement or spacing of the Λ- shaped vortices. Apparently, the wavelength along the axis Z is a certain invariant that depends on the channel width but is independent of the flow parameters. A channel with similar size was used by Nishioka, Iida & Kanbayashi [4] and the disturbance irregularity in the Z-direction had the same wavelength (see Fig.4).

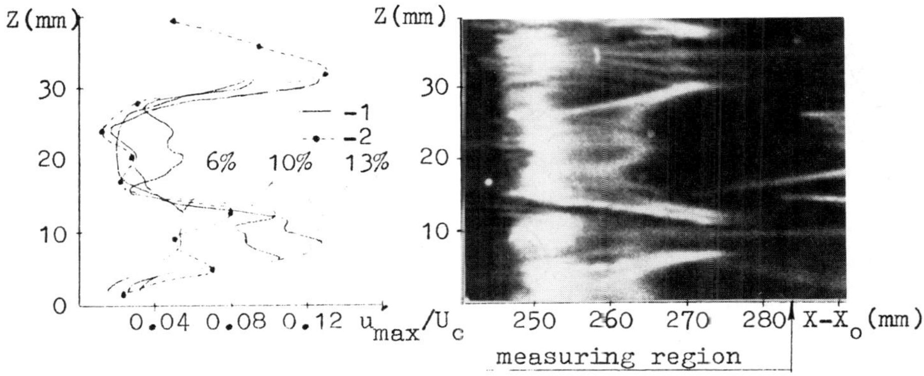

Fig.4. Variation of disturbance amplitude against Z coordinate 1-data from paper of Nishioka et al. [4], 2-this study, f=72 Hz

When the fundamental wave was introduced into the flow with a little signal of subharmonic, the resonant interaction was observed near the wall with maximal disturbances. At this time spikes of subharmonic period **arose** in the spectrum of hot-wire anemometer signal (Fig.5). The interaction of two waves is highly dependent of their phase difference and starts earlier in the case of zero phase difference (Fig.6). This phenomenon says for a resonant nature of interaction and had been observed for the first time by Kachanov & Levchenko [9], Saric, Kozlov & Levchenko [10] in flat plate boundary layer.

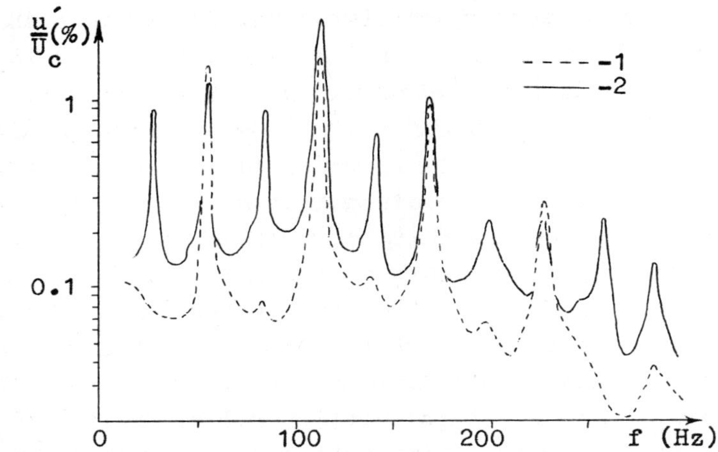

Fig.5. Spectra of disturbances, Re=3850; $X-X_o$=265 mm; Z=23 mm; Y=2 mm. 1- only f, 2- f+f/2

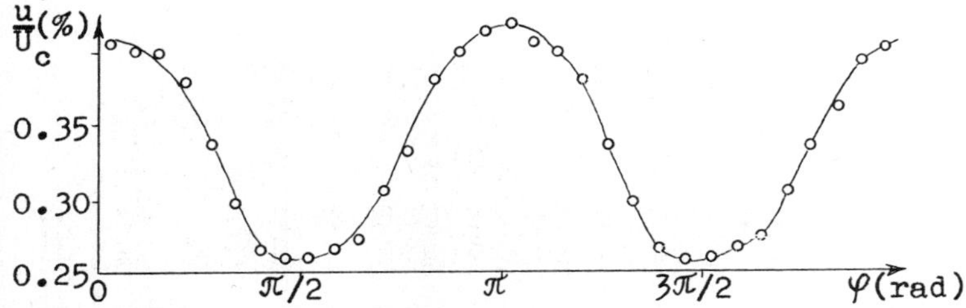

Fig.6. Subharmonic amplitude against phase shift between f and f/2. Re=3850; $X-X_o$=170 mm; Y=2 mm

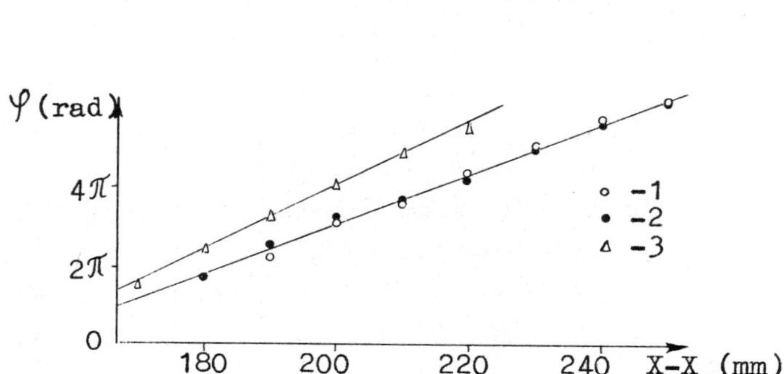

Fig.7. Phase sinchronization of fundamental (1) and subharmonic (2) at resonant interaction; (3)- two-dimensional subharmonic wave

The distribution of subharmonic amplitude is periodic in Z-direction. In the regions of an amplitude minimum the subharmonic phase is rapidly inverted that says for the presence of two oblique subharmonic waves in the flow. The resonant interaction of the waves **arise** from spatial growth of subharmonic wave on the background of decreasing fundamental wave. In this case the subharmonic phase speed is equal to that of the fundamental one (Fig.7) that is slightly less than in case of the development of only subharmonic wave. The flow reconstruction is observed in the case of resonant interaction (Fig.8c) as compared to the Klebanoff-type transition (Fig.8a, b). In K-transition Λ -shaped vortices aline one after another but in the case of resonant interaction Λ -vortices are thatched.

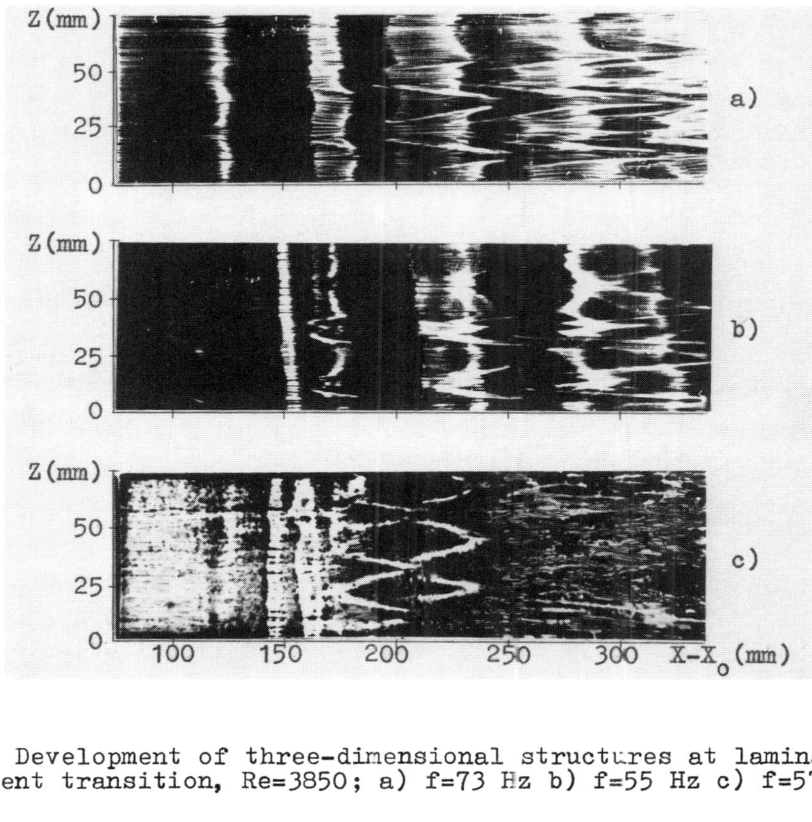

Fig.8. Development of three-dimensional structures at laminar-turbulent transition, Re=3850; a) f=73 Hz b) f=55 Hz c) f=57 Hz

Conclusion

Finally, the following conclusion is arrived at:
The transition to turbulence happens in a spatial development of the hydrodynamic stability waves like Tollmien-Schlichting waves, via three-dimensional distortion, Λ-shaped vortices formation and their breakdown into turbulence, i.e. in the same way as in the experiment of Klebanoff et al. [6].

The spanwise size of a Λ-shaped vortex seems to be invariant, independent of frequency of the disturbances, parameters of the flow but apparently dependent on the channel width.

The presence of subharmonic disturbances in the flow leads to their resonant interaction with the fundamental wave, and the new type of transition to turbulence is realised in a plane Poiseuille flow. It is clearly seen from the photographs illustrating an exchange of three-dimensional structures arrangement.

References

1. Herbert, Th.: AIAA Paper No. 83-1759 (1983).

2. Goldshtic, M.A.; Lifshitz, A.M.; Shtern, V.N.: Doklady Acad. Sci. of USSR v.273 No.1 (1983) 75-78.

3. Nishioka, M.; Iida, S.; Ichikawa, Y.: J. Fluid Mech. v.72 (1975) 731-752.

4. Nishioka, M.; Iida, S.; Kanbayashi, S.: 10-th Turbulence Symp., Inst. Space Aeron. Sci., Tokyo Univ. (1978) 55-62.

5. Kozlov, V.V.; Ramazanov, M.P.: Preprint 21, Inst. Theor. Appl. Mech., AN SSSR, SO (1980).

6. Klebanoff, P.S.; Tidstrom, K.D.; Sargent, L.M.: J. Fluid Mech. v.12 (1962) 1-34.

7. Kozlov, V.V.; Ramazanov, M.P.: Izv. SO AN SSSR Sec. Tech. Sci. v.8 (1981) 45-48.

8. Hama, F.R.: Phys. of Fluids, v.5 No.6 (1962) 644-650.

9. Kachanov, Yu.S.; Levchenko, V.Ya.: J. Fluid Mech. v.138 (1984) 209-247.

10. Saric, W.S.; Kozlov, V.V.; Levchenko, V.Ya.: AIAA Paper No. 84-0007 (1984).

Threshold Regimes in the Plane Channel Flow

M. A. GOLDSHTIK, A. M. LIFSHITS, V. N. SHTERN

Institute of Thermophysics, 630090, Novosibirsk, USSR.

Abstract. Steady self-oscillations in the Poiseuille flow are calculated by using the approach concerning the resonance interaction between three waves: the plane wave with a certain fundamental frequency and three-dimensional half-frequency symmetrical couple. The problem is reduced to the eigenvalue problem for the system of nonlinear differential equations and is solved by the Newton method. The relation between the pulsation energy of the threshold regimes and the Reynolds number in the subcritical region has been obtained. The calculated results are in agreement with the experimental data for the transition in a plane channel.

Introduction.

The laminar-turbulent transition in pipes and channels is known to occur subcritically and hysteresis takes place. Both regimes exist and are stable at some Reynolds numbers. A finite disturbance exceeding a certain threshold value is necessary for the transition. One of the main problems in the nonlinear theory of hydrodynamic stability is to determine the magnitude of this threshold depending on the Reynolds number.

The threshold intensity turns into zero when the Reynolds number attains the value R_1 determined by the linear stability theory. For the plane Poiseuille flow $R_1=3848$ [1] (the average velocity and the channel half-width are used here as scales), the turbulent regime may, on the other hand, exist at $R>R_\pi>700$. The analysis of the self-oscillating solutions for the Navier-Stokes equations may be useful to theoretically estimate the amplitude of critical disturbances in the range $R_\pi<R<R_1$ and its lowest boundary. Two-dimensional self-oscillating solutions branch from the linear-stability neutral curve into the region of the lower Reynolds numbers [2]. But as is shown by calculation [1,3], they are not sufficient to estimate the threshold because they

exist only at $R>R_a \approx 1650$ and don't cover all the hysteresis interval. These solutions which may be referred to as fundamental periodic oscillations bifurcate in their turn, and three-dimensional self-oscillating solutions of double period arise.
The time and space scales of the fundamental self-oscillations may be characterized by parameters (ω, α) which, accordingly, mean the frequency and the longitudinal wave number. When the amplitude increases the oblique waves become unstable first of all. Their parameters are $(\omega/2, \alpha/2, \pm\beta)$, where β is a transversal wave number, its value at the bifurcation point can be determined. As has been found in [1], these solutions may branch directly from the linear-stability neutral curve. This instability was investigated in [4,5] for the case of finite amplitudes. Such instability is sometimes referred to as parametric resonance and the three above mentioned waves as a symmetrical triplet. Apart from this subharmonic bifurcation due to the double oscillation period, there is instability because of the three-dimensional oscillations of the same period [4-6] which are in a resonance connection with the multiple wave of the fundamental oscillation. The up-to-date view concerning the contribution of the triplet interaction to the formation of structures in the transition area is reported in review [7].
The recent work considers steady subharmonic finite amplitude self-oscillations as an approximation of the symmetrical triplet (multiple waves are ignored). It is the development of the work by the authors [8] where the area of the existence of such solutions was shown to be limited below by the Reynolds number R_t which is in satisfactory agreement with the lower boundary R_π for the existence of the turbulent regimes. Here the critical amplitude dependence on the Reynold number is calculated and facts about the subharmonic manifold structure when the Reynolds numbers are close to the minimum value R_t are presented. The threshold is determined as the minimum amplitude of the self-oscillations.

Basic equations.

The general scheme for the analysis of hydrodynamic stability was proposed in [1]. It is based on the consecutive consideration of the basic interactions between the disturbance and is referred to as an escalator scheme. At the first monoharmonic stage, the pulsation velocity is approximated by one wave and the problem is reduced to the joint solutions of the Orr-Sommerfeld and Reynolds equations. The fundamental periodic regimes that branch from the linear stability neutral curve are well described by this approximation. Considering multiple harmonics doesn't result in any essential changes [5,6].

The three-waves stage is as follows. The pulsation velocity is presented in the form

$$(v_x,v_y,v_z)=\sum_{j=1}^{3}\{[u_j(y),v_j(y),w_j(y)]\times$$
$$\times\exp[i\alpha_j(x-X_jt)+i\beta_jz]+\text{complex conjugate}\}$$

where X_j are the phase velocities, $\alpha_jX_j=\omega_j$ are the frequencies. So that the disturbances could interact it is necessary that $\Sigma\alpha_j=0$, $\Sigma\beta_j=0$, $\Sigma\alpha_jX_j=0$. For the symmetric triplet, these relations may be reduced to the simplest form

$$\alpha_2=\alpha_3=\alpha,\quad \alpha_1=-2\alpha,\quad \beta_2=-\beta_3=\beta,\quad \beta_1=0 \qquad (1)$$
$$X_j=X,\quad J=\overline{1,3}$$

After projecting the Navier-Stokes equations on the basis and zero harmonics, one can obtain, for the general-form triplet, the following system

$$L_ju_j-U'v_j-i\alpha_jp_j=[2i\alpha_ju_m^*u_l^*+(u_m^*v_l^*+u_l^*v_m^*)'+i\beta_j(w_m^*u_l^*+w_l^*u_m^*)] \qquad (2.1)$$

$$L_jv_j-p_j'=[i\alpha_j(u_m^*v_l^*+u_l^*v_m^*)+2(v_m^*v_l^*)'+i\beta_j(w_m^*v_l^*+w_l^*v_m^*)] \qquad (2.2)$$

$$L_jw_j-i\beta_jp_j=[i\alpha_j(u_m^*w_l^*+u_l^*w_m^*)+(v_m^*w_l^*+v_l^*w_m^*)'+2i\beta_jw_m^*w_l^*] \qquad (2.3)$$

$$i\alpha_ju_j+v_j'+i\beta_jw_j=0 \qquad (2.4)$$

$$U_d''=R_{p_j}\Sigma\tau_j \qquad (2.5)$$

$$L_j=[1/R_p(d^2/dy^2-k_j^2)-i\alpha_j(U-X_j)],\quad k_j^2=\alpha_j^2+\beta_j^2,$$

$$\tau_j=u_jv_j^*+u_j^*v_j$$

Here τ_j is the contribution of the harmonic denoted by the subscipt j to the Reynolds stresses. The total velocity $U=U_l+U_d$, $U_l=3/2(1-y^2)$; j,m,l are cyclic permutations of 1,2,3; the Reynolds number R_p is calculated via the laminar-regime average-velocity which is used as the velocity scale. Hence $R=R_p\times U_a$ where U_a is the nondimensional secondary-regime average-velocity at the same pressure gradient. If the triplet is symmetrical, system (2) may be essentialy simplified, since, besides (1), $u_2=u_3$, $v_2=v_3$, $w_2=-w_3$, $w_1=0$. System (2) and the no-slip boundary conditions are together the eigenvalue problem.
In the symmetric triplet case any three parameters of the set $\{\alpha,\beta,R_p,X\}$ may be taken as arbitrary values within a certain range and one of them is searched as an eigenvalue.
There are several degeneration levels in this problem. Firstly, it is the trivial solution, secondly, it is the monoharmonical level when only one harmonic has the non-zero amplitude. And, finally, the case when multiple harmonics are simply taken in consideration. However, if all the harmonics are close to the neutral one, then the right-hand sides of (2.1-2.3) act as a driving force, at the resonance frequency. Just such triplets corresponding to the subharmonic finite-amlitude self-oscilations will be considered later on.

Results and discussion.

The three-dimensional manifold of the solutions touches the linear-theory neutral curve at the only point 1 corresponding to

$$R=7776,\ \alpha=0.544,\ X=0.359,\ \beta=0.71. \quad (9)$$

and the amplitudes of the harmonics being the linear-problem eigenfunctions are symmetrical. It is easy to be convinced that the solution of such symmetry for system (2) is impossible if the intensities are finite. The solution will be asymmetrical including the average-velocity profile. Although the turbulent channel flow is generally assumed to be symmetrical, the velocity field may during some time intervals be in the vicinity of the asymmetrical solutions. The asymmetry of the average velocity profile at moderate Reynolds numbers is small. The lower parts of curves 1 in fig. 1,2 correspond to the regimes of the minimum intensity $I=1/2\int_{-1}^{\alpha}\sqrt{2E}\,dy/U_a$ at fixed R where E is the total pulsation energy. Curves 2 represent the analogous relations for the monoharmonic model. The circles in fig.1 have been plotted according to the data in [9] , being the threshold values of $\sqrt{3}(\overline{Vu'}/U_a)$ measured at the inlet of the channel. In fig.2 the circles have been plotted according to the data in [10], curve 3 represents the laminar friction law.

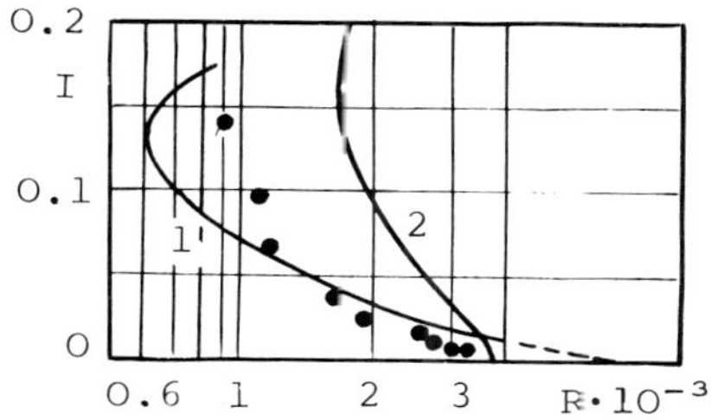

Fig.1. Intensity of threshold regimes.

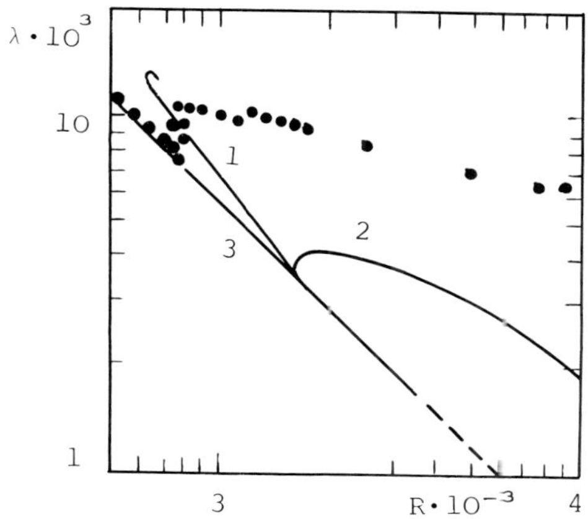

Fig.2. Friction factor λ vs. Reynolds number R.

Fig.3,4 give some information about the shape of the neutral manifold near its nose. In fig.3 the cross-section passing through the nose and corresponding to the fixed values of certain "intensities" of both harmonics (special "intensities" were used for the convenience of our calculations) is presented. In fig.4 the cross-sections $\alpha = 0.48$ (dashed line) and $\beta = 1.84$ at $R_p = 10$ are presented. The parameters on the neutral-surface nose, i.e. of the critical regime, are as follows

$$R_p = 752.1 \quad U_a = 0.83, \quad \alpha = 0.56, \quad \beta = 1.12, \qquad (10)$$
$$X = 0.6, \quad I = 11\%, \quad (R = R_t = 624)$$

Thus the critical value of R for the steady solution has tur-

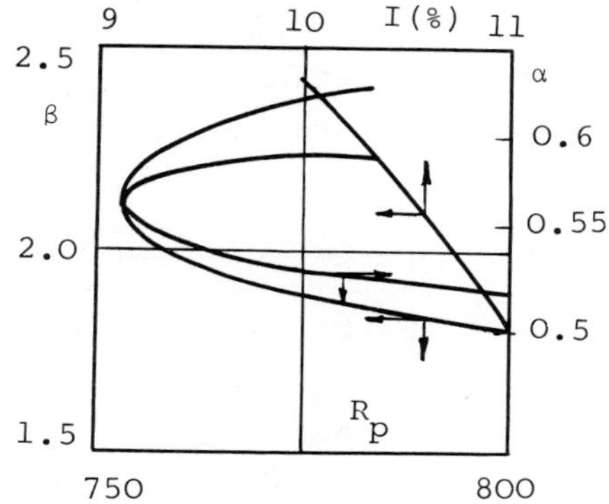

Fig.3. Cross-section passing through the neutral surface nose.

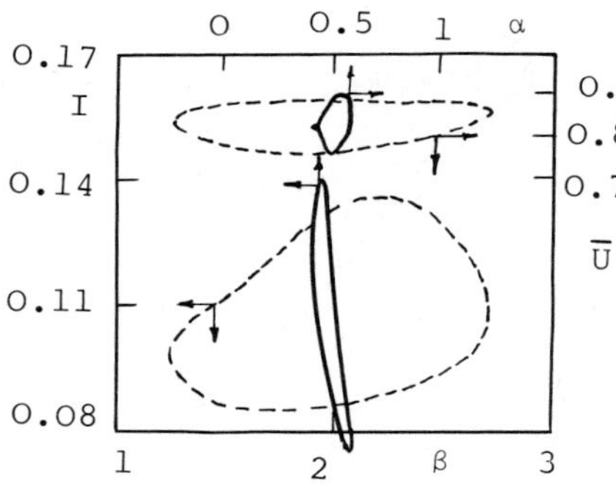

Fig.4. Cross-sections α=const and β=const (solid line) at $R_p=10^3$.

ned out to be somewhat smaller than that registered in the experiment. The gap may be due to the approximation of the approach. But it should be noted that the theoretical area is always wider than the experimental one when hysteresis takes place. And, moreover, it is possible that in a certain range of Reynolds numbers pulsating solutions exist, but are unstable. In the experiments on the transition in near-wall shear flows the so-called Λ-vortices were registered by visualization, at a certain stage of the disturbances development [4,7]. They can be observed when three-dimensional perturbations develop, which arise due to the secondary instability of the Tollmien-

Schlichting waves. Now, if the oscillations of the ribbon are sum of the frequencies f and f/2 then the Λ-vortices are located in staggered rows. In fig.3 the results calculated according to a procedure analogous to the visualization are given for solution (10). To better identify the Λ-formations located in staggered rows their schematic axes are given as solid lines. Further downstream the particles intensively get mixed, the Λ-formations lose their individuality and one may get the impression that the stochastization is beginning to take place, though the velocity field is absolutely periodic. This picture coincides with the experimental one with an accuracy of the scales.

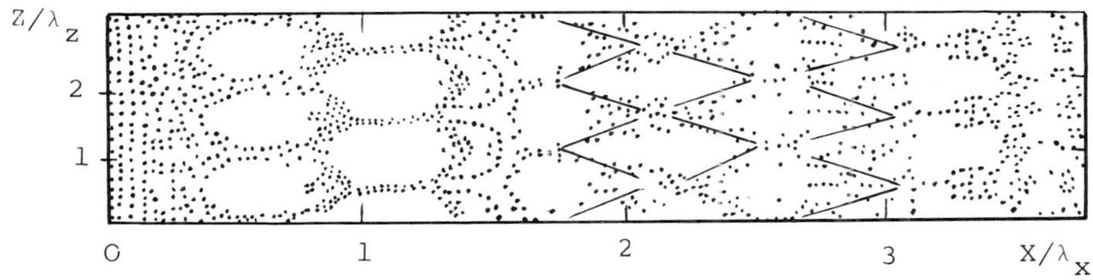

Fig.5. Λ-formation arising by numerical flow "visualisation".

Conclusion

The presented calculated results testify to that three-dimensional subharmonic selfoscilation exist nearly in the same subcritical range of the Reynolds numbers as the turbulent regimes. The energy minimum of these oscilations satisfactorily conforms with the experimental data for the threshold perturbation magnitudes. It should be noted that the used here appoach is more effective to determine the threshold of the transition to turbulence than the direct Cauchy problem for the solution of the Navier-Stokes equations. In spite of the impressing achievements of recent years concerning the solution of the Cauchy problem [11, 12], this approach can't be used to calculate the separatrix surface bounding the attraction regions of the laminar and turbulent regimes. The obtained steady triplet solutions are, on the whole, in agreement with the observed staggered arrangement of the Λ-structures in the transition area whereby, they describe not only the structures themselves, but also the initial stage of their destruction.

References.

1. Goldshtik, M.A.; Shtern, V.N.: Hydrodynamic stability and turbulence. Novosibirsk; Nauka, 1977, 366 p.
2. Andreichikov, J.P.; Youdovich, V.I.: About self-oscillating regimes bifurcating from Poiseuille flow in a plane channel. - Soviet Doklady, 1972, v.202, N4, p.791-794.
3. Herbert, Th.: Finite amplitude stability of plane parallel flows, AGARD Conf. Proc., 1977, CP-224, paper N3.
4. Kozlov, V.V.; Ramazanov, M.B.: The development of finite - amplitude disturbances in Poiseuille flow - Soviet Fluid Mechanics, 1983, N1, p. 43-47.
5. Herbert, Th.: Secondary instability of plane channel flow to subharmonic three-dimensional disturbances, Phys. Fluids. 26(4), April 1983, p.871-874.
6. Orszag, S.A.; Patera, A.T.: Secondary instability of wall-bounded shear flows, J.Fluid Mech., 1983, v.128, p.347-385.
7. Kachanov, Yu.S.; Levchenko, V.Ya.: The resonant interaction of disturbances at laminar-turbulent transition in a boundary layer, J.Fluid Mech., 1984, v.138, p.209-247.
8. Goldshtik, M.A.; Lifshits, A.M.; Shtern, V.N.: Transition Reynolds number for two-dimensional channel flow, - Soviet Doklady, 1983, v.273, N1, p.75-78.
9. Karnitz, M.A.; Potter, H.C.; Smith, M.C.: An experimental investigation of transition of a plane Poiseuille flow - Trans. ASME. J., 1974, v.96, N4, p.384-388.
10. Whan, G.A.; Rothfus, R.R.: Characteristics of Transition Flow between Parallel Plates - A.I.Ch.J., 1959, v.5, N2, pp.204-208.
11. Orszag, S.A.; Kells, L.C.: Transition to turbulence in plane Poiseuille and plane Couette flow - J. Fluid Mech., 1980, v.96, N1, p.159-205.
12. Rozhdestvenskii, B.L.; Simakin, I.N.: Two- and three-dimensional secondary flows in plane channel, their relationship and comparison with turbulent flow - Soviet Doklady, 1983, v.273, N 3, p.553-557.

Stability of Finite-Amplitude Autooscillations in Poiseuille Flow

N. N. YANENKO, B. YU. SCOBELEV, A. ZHARILEASIMOV

Institute of Theoretical and Applied Mechanics
USSR Academy of Sciences, Novosibirsk 630090

The conditions of existence of analytical invariant manifolds for Navier-Stokes equations are derived. A method of constructing the invariant manifolds in specific problems is suggested. The bifurcations of subcritical autooscillating regimes in a plane Poiseuille flow are studied.

Consider the flow of a viscous incompressible fluid in a bounded domain Ω with piecewise-smooth boundary $\partial\Omega$. The flow velocity $v(x,t)$ and pressure $p(x,t)$ are determined by a system of Navier-Stokes equations

$$\frac{\partial v}{\partial t} + v\cdot\nabla v = -\nabla p + \nu\Delta v + f \quad ; \quad \nabla\cdot v = 0 \tag{1}$$

Assume that at the given boundary conditions there exists a stationary solution of the system (1) (v_0, p_0) and we shall search for the solution of the form

$$v(x,t) = v_0(x) + u(x,t) \quad , \quad p(x,t) = p_0(x) + q(x,t)$$

Introduce the Hilbert spaces

$$H = \{u \in [L_2(\Omega)]^3 \; ; \; \nabla\cdot u = 0 \quad , \quad u\cdot n|_{\partial\Omega} = 0\}$$

$$K = \{u \in [H^1(\Omega)]^3 \; ; \; \nabla\cdot u = 0 \quad , \quad u\cdot n|_{\partial\Omega} = 0\}$$

$$D = \{u \in [H^2(\Omega)]^3 \; ; \; \nabla\cdot u = 0 \quad , \quad u|_{\partial\Omega} = 0\}$$

where $H^s(\Omega)$ is the Sobolev's space.

Then (see, f.e. [1]) a problem of finding (u, q) reduces to a differential equation with unbounded operators in the Hilbert space H

$$\frac{du}{dt} = -L_\nu u + N(v; u) \tag{2}$$

where L_ν is the linear operator with the domain D

$$L_\nu = -\Pi(\nu\Delta u - v_0\cdot\nabla u - u\cdot\nabla v_0)$$

(Π is the orthogonal projector on H in $[L_2(\Omega)]^3$).

The nonlinear operator N has the domain K and in the present case does not depend on the parameter ν

$$N(u) = -\Pi u \cdot \nabla u$$

Consider a general case of Eq.(2) in some Hilbert space. Assume that the linear L_ν and nonlinear N operators are closed and satisfy the following conditions:

I. $N(\nu; 0) = 0 \quad , \nu \in R^m \quad , \quad N_u(\nu; 0) = 0$.

II. A domain of the operator L_ν - $D(L)$ does not depend on ν .

III. The operator L_ν is m-sectorial. Hence it follows that $(-L_\nu)$ is the generating operator of the analytical semigroup and that there exists the operator $(-L_\nu - a_\nu)^\alpha$, $0<\alpha<1$, $D[(-L_\nu - a_\nu)^\alpha] \supset D(L)$ (a_ν is the vertex of sector inside which the eigenvalues of the operator L_ν lie).

IV. The operator $N(\nu, x)$ has a domain $D(N) \supset D(L)$ independent of ν and is the mapping of $D(L)$ in $D[(-L_\nu - a_\nu)^\alpha]$ at some α . This mapping possesses the Lipschitz-continuous first derivative.

V. A spectrum of the $(-L_\nu)$ - σ permits the partition $\sigma = \sigma_1 \cup \sigma_2$ where σ_1 is a bounded portion of σ .

VI. $\quad \varkappa_\nu > 0 \quad , \varkappa_\nu > q_\nu$

where $\quad \varkappa_\nu = -\sup_{\varrho \in \sigma_2} Re\,\varrho \quad , \quad q_\nu = -\inf_{\varrho \in \sigma_1} Re\,\varrho$

We shall study classical solutions of the initial problem for Eq.(2)

$$u(0) = u_0 \in D(L) \quad ; \quad u \in C^0(0,\infty; D(L)) \cap C^1(0,\infty; H) \tag{3}$$

By virtue of the condition V there exists the decomposition of the operator L_ν with respect to a direct sum of the orthogonal subspaces $H = P_\nu H \oplus (I-P_\nu)H$ where P_ν is such a projector that the spectrum of the operator $(-P_\nu L_\nu)$ equals σ_1 . Any vector $u \in H$ can be represented in the form of sum $u = y + z$, $y \in P_\nu H$, $z \in (I-P_\nu)H$.

In what follows a notion of invariant manifold will be of importance. We shall call an invariant manifold of Eq.(2) a set $\mathcal{M} \subset H$ of the form

$$\mathcal{M} = \{y, \tilde{z}(y)\} \quad ; y \in P_\nu D(L) \, , \, \tilde{z}(y) \in (I-P_\nu) D(L)$$

such that if $u_0 \in \mathcal{M}$, then the whole trajectory $u(t)$, $0 \le t < \infty$ belongs to $\mathcal{M}$.

Then there holds the following

T h e o r e m 1. Let the conditions I - VI be satisfied. Then there exists such a number $\beta > 0$ that the following statements hold:

1. Eq.(2) has an invariant manifold

$$\mathcal{M} = \{y, \tilde{z}(y)\} \quad , \tilde{z}(0) = 0 \ , \ \|y\|_d \leq \rho$$

$$\rho + \rho^{1/\alpha} < \beta(\varkappa_\nu - q_\nu)$$

The mapping $\tilde{z} : P_\nu D(L) \to (I - P_\nu) D(L)$ satisfies the Lipschitz condition and $\|\tilde{z}\|_d \leq \rho$

$$(\|x\|_d = \|x\| + \|L_\nu x\|)$$

2. Any global solution of the problem (2) (3) with the property

$$\|P_\nu u(t)\|_d \leq \rho \quad , t \in [0, \infty)$$

is attracted by the manifold $\mathcal{M}$, i.e. there may be found such a number $\gamma > 0$ that

$$\|\tilde{z}(P_\nu u(t)) - (I - P_\nu) u(t)\|_d \leq const \|\tilde{z}(y_0) - z_0\|_d \, e^{-\gamma t} \ , \ t \to \infty$$

where

$$y_0 = P_\nu u_0 \quad , \ z_0 = (I - P_\nu) u_0$$

From the theorem I it follows that the study of trajectories of the evolutionary equation (2) with unbounded operator may be reduced to the analysis of solutions of the evolutionary equations with bounded operators $L_1 = P_\nu L_\nu$ and $N_1 = P_\nu N$. If the spectrum σ_1 consists of a finite number of isolated eigenvalues, then the problem reduces to the study of the finite-dimensional dynamic system. For finding the manifold of $\mathcal{M}$ we shall formulate the additional conditions on the operators L_ν and N.

VII. The nonlinear operator $N(\nu ; x)$ is an analytical operator acting from $\{\nu ; D(N)\}$ in H .

VIII. The vector $L_\nu x$ depends analytically on ν at any $x \in D(L)$.

IX. The spectrum σ_1 consists of n pairs of simple isolated eigenvalues $\{\lambda_i, \bar{\lambda}_i\}$

$$\lambda_i(\nu) = \gamma_i(\nu) + i\omega_i(\nu) \quad , \ \omega_i(\nu) > 0 \quad , \quad i = 1, 2, \dots, n$$

If at some natural $k > 0$ there holds the equality

$$k\omega_1(\nu_0) = \omega_{i_0}(\nu_0) \quad , \ \nu_0 \in R^m$$

then

$$\gamma_{i_0}(\nu_0) \neq 0$$

Denote by φ_i the eigenfunctions of the operator $(-L_\nu)$ and by ψ_i the eigenfunctions of the conjugate operator $(-L_\nu^*)$. Let $P^{(i)}$ be a projector on the eigenspace $(-L_\nu)$ answering a pair of eigenvalues $(\lambda_i, \bar{\lambda}_i)$. Then

$$P^{(i)} x = (x, \psi_i)\varphi_i + (x, \bar{\psi}_i)\bar{\varphi}_i \equiv y_i \qquad (\varphi_i, \psi_i) = 1$$

The equation (2) is equivalent to the following system

$$\frac{dy_i}{dt} = -P^{(i)} L_v y_i + P^{(i)} N(v; y_1 + y_2 + \dots + y_n + z) \; ; i = 1, 2, \dots, n$$
$$\frac{dz}{dt} = -Q L_v z + Q N(v; y_1 + y_2 + \dots + y_n + z) \; ; z \in QD(L) \tag{4}$$
$$Q = I - \sum_{i=1}^{n} P^{(i)}$$

Introduce in the subspaces $P^{(i)}H$ the systems of polar coordinates and we shall search for the solutions of Eq.(4) in the form

$$y_i = Re(r_i e^{i\theta_i} \varphi_i) + \tilde{y}_i(r_1, \dots, r_n; \theta_1, \dots, \theta_n)$$
$$\int_0^{2\pi} e^{-i\theta_i} (\tilde{y}_i, \psi_i) d\theta_i = 0 \tag{5}$$
$$z = \tilde{z}(r_1, \dots, r_n; \theta_1, \dots, \theta_n)$$

Substituting (5) into (4), we can derive a finite-dimensional autonomous system of equations for r_i, θ_i

$$\frac{dr_i}{dt} = (\gamma_i + b^{(i)}) r_i$$
$$\frac{d\theta_i}{dt} = \omega_i + c^{(i)} \qquad ; \; i = 1, 2, \dots, n \tag{6}$$
$$b^{(i)} = Re \ll N, \psi_i \gg \;, \; c^{(i)} = Im \ll N, \psi_i \gg \; ; \ll N, \psi_i \gg = \frac{1}{2\pi r_i} \int_0^{2\pi} e^{-i\theta_i} (N, \psi_i) d\theta_i$$

In what follows we shall construct the invariant manifold from integral curves of Eq.(2). For this purpose we make use of the first integrals of the system (6) which establish the relationship between two points of the integral curve of this system $(r^{(1)}, \theta^{(1)})$, $(r^{(2)}, \theta^{(2)})$

$$r_i^{(2)} = \hat{r}_i(\theta_1^{(2)}; r_1^{(1)}, \dots, r_n^{(1)}, \theta_1^{(1)}, \dots, \theta_n^{(1)}) \quad , \quad i = 1, 2, \dots, n \tag{7}$$
$$\theta_i^{(2)} = \hat{\theta}_i(\theta_1^{(2)}; r_1^{(1)}, \dots, r_n^{(1)}, \theta_1^{(1)}, \dots, \theta_n^{(1)}) \quad , \quad i = 2, \dots, n$$

If $\omega_1 + c^{(1)} \neq 0$, then the functions $\hat{r}_i$, $\hat{\theta}_i$ are found from equations

$$\frac{d\hat{r}_i}{d\theta_1} = \frac{(\gamma_i + b^{(i)}) \hat{r}_i}{\omega_1 + c^{(1)}} \qquad ; \; i = 1, 2, \dots, n$$
$$\frac{d\hat{\theta}_i}{d\theta_1} = \frac{\omega_i + c^{(i)}}{\omega_1 + c^{(1)}} \qquad ; \; i = 2, \dots, n \tag{8}$$

The equation

$$\frac{d\theta_1}{dt} = \omega_1 + c^{(1)} \tag{9}$$

determines the dependence of solution of the system (8) on time. When deriving the equations for $\tilde{z}$ and $\tilde{y}$ we shall assume that in formulas (7) the point $(r^{(2)}, \theta^{(2)})$ is fixed and the point ($z^{(1)}$, $\theta^{(1)}$) changes along the integral curve. As a result we obtain

$$\omega_1 \frac{\partial \tilde{y}_i}{\partial \theta_1} + P^{(i)} L_\nu \tilde{y}_i = - c^{(1)} \frac{\partial \tilde{y}_i}{\partial \theta_1} + P^{(i)} N - Re(\langle\langle N, \psi_i \rangle\rangle z_i e^{i\theta_i} \varphi_i)$$
$$\omega_1 \frac{\partial \tilde{z}}{\partial \theta_1} + Q L_\nu \tilde{z} = - c^{(1)} \frac{\partial \tilde{z}}{\partial \theta_1} + Q N \qquad (10)$$

T h e o r e m 2. Let the conditions I-IX be satisfied. Then there may be found such $\rho_0 > 0$ that at $|z| < \rho_0$ ($|z| = z_1 + \dots + z_n$) the equation (1) has a unique 2n-dimensional invariant manifold $\mathcal{M}_{2n}$.

$$\mathcal{M}_{2n} = \{ Re(z_1 e^{i\theta_1} \varphi_1) + \tilde{y}_1(z;\theta), \dots, Re(z_n e^{i\theta_n} \varphi_n) + \tilde{y}_n(z;\theta), \tilde{z}(z;\theta) \}$$
$$\tilde{z} \in QD(L), \quad \tilde{z}(0;\theta) = 0$$

The functions $\tilde{z}(z;\theta)$, $\tilde{y}(z;\theta)$ satisfy the equations (10), analytically depend on z_i $(i=1,\dots,n)$ and on the parameter ν and are continuously differentiable 2π-periodical functions of θ_i $(i=1,\dots,n)$. The phase trajectories of Eq.(2) on the manifold are determined by the system (8) where $b^{(i)}(z;\theta_1,\dots,\theta_{i-1},\theta_{i+1},\dots,\theta_n)$, $c^{(i)}(z;\theta_1,\dots,\theta_{i-1},\theta_{i+1},\dots,\theta_n)$ are analytical functions of z_i and ν and are continuously differentiable, 2π -periodical functions of θ_i. There may be found such $\beta_n > 0$ that at $|z| + |z|^{1/\alpha} < \beta_n(\alpha_\nu - q_\nu)$ the manifold $\mathcal{M}_{2n}$ is attracting.

From the Theorem 2 it follows that the solution of Eq.(2) on the manifold $\mathcal{M}_{2n}$ has the form

$$u = \sum_{i=1}^{n} Re(z_i e^{i\theta_i} \varphi_i) + \tilde{\tilde{z}} \quad ; \tilde{\tilde{z}} = \sum_{i=1}^{n} \tilde{y}_i + \tilde{z} \; ; \tilde{\tilde{z}} = \sum_{|S|=2}^{\infty} z_S z^S \qquad (11)$$
$$z^S = z_1^{s_1} \dots z_n^{s_n}$$

Assuming also

$$b^{(i)} = \sum_{|S|=1}^{\infty} b_S^{(i)} z^S \quad , \quad c^{(i)} = \sum_{|S|=1}^{\infty} c_S^{(i)} z^S \qquad (12)$$

we obtain a recurrent system of linear equations for z_S

$$\omega_1 \frac{\partial z_S}{\partial \theta_1} + L_\nu z_S = - Re \sum_{i=1}^{n} (b_{S_i}^{(i)} + c_{S_i}^{(i)}) e^{i\theta_i} \varphi_i - \sum_{K+P=S} c_K^{(1)} \frac{\partial z_P}{\partial \theta_1} + N_S \qquad (13)$$
$$N = \sum_{|S|=2}^{\infty} N_S z^S \quad , \quad S_i = (s_1, \dots, s_i - 1, \dots, s_n)$$

The periodical functions $b_S^{(i)}$, $c_S^{(i)}$ are unambiguously determined by the condition

$$\int_0^{2\pi} e^{-i\theta_i} (z_S, \psi_i) \, d\theta_i = 0$$

It is possible to show that for Navier-Stokes equations in a bounded domain the conditions I-VIII are satisfied, for α in the condition IV the estimate $\alpha < 1/4$ is valid. In the case of spatial periodicity of the flow we can take a periodicity cell as Ω. The implementation of the condition IX depends on a specific problem. For example, it was shown in [2] that for a plane Poiseuille flow the neutral eigenvalue is almost everywhere simple on the neutral curve.

Consider the two-dimensional invariant manifold for disturbances of a plane Poiseuille flow. Take a pair of the first eigenvalues of the corresponding Orr-Sommefeld problem $(\lambda_1, \bar{\lambda}_1)$ (the ordering is carried out with respect to γ_i) as σ_1. It is easy to show that the coefficients in expansions (12) are numbers and different from zero are only the coefficients with even indices. For numerical computations it is convenient to go over to the equation for stream function. Then the system (13) reduces to a system of boundary-value problems for ordinary differential equations for the determination of harmonics of functions $z_S - z_{SK}$, $|K| \leq S$. The fixed points of the equation for the phase trajectories

$$\frac{d\tau}{d\theta} = \frac{(\gamma_1 + b(\tau))\tau}{\omega_1 + c(\tau)}$$

determine the amplitudes of limiting cycles. It is known that for the Poiseuille flow there takes place a subcritical bifurcation. In [3,4] it was shown on the basis of the asymptotic method that the amplitude curve in the (τ, R) plane, R being the Reynolds number, has a special point - a turning point.

Consider a subcritical domain of Reynolds numbers and disturbance wavenumber α values. Introduce the notations

$$\nu = (\alpha, R) \;;\; \tau^2 = p \;;\; \xi_0 = -\gamma_1 \;;\; \xi_\ell = b_{2\ell} \; (\ell = 2m-1);\; \xi_\ell = -b_{2\ell} \; (\ell = 2m),\; m = 1,2,\ldots$$

Write down the equation for amplitudes of limiting cycles in the form

$$-\xi_0(\nu) + f_1(\nu,p)\,p - f_2(\nu,p)\,p^2 = 0 \tag{14}$$

where

$$f_1(\nu,p) = \xi_1(\nu) + \sum_{K=1}^{\infty} \xi_{2K+1}(\nu)\,p^{2K} \;;\; f_2(\nu,p) = \xi_2(\nu) + \sum_{K=1}^{\infty} \xi_{2K+2}(\nu)\,p^{2K}$$

From (14) we derive equations for two pieces of the amplitude surface

$$p = \frac{f_1(\nu,p) + \Phi^{1/2}(\nu,p)}{2f_2(\nu,p)} \tag{15}$$

$$p = \frac{f_1(\nu,p) - \Phi^{1/2}(\nu,p)}{2f_2(\nu,p)} \quad ; \quad \Phi = f_1^2 - 4\ell_0 f_2 \tag{16}$$

At the fixed $\alpha = \alpha_0$ the amplitude curve has a turning point at $R = R_H$ where $\Phi = 0$. At small p

$$f_1 \approx b_2(\alpha, R), \; f_2 \approx -b_4(\alpha, R), \; \Phi \approx b_2^2 - 4\gamma_1 b_4$$

In the paper the numerical calculation of coefficients b_2, b_4 and approximate values of R_H is carried out. Table I represents the results; z_H is the critical value of the amplitude, R_0 is the linear neutral value of R.

Table I

α	1,10	1.08	1.06	1.04	1.02	1.0	0.98
R_H	4960	4896	4905	4968	5068	5197	5301
$z_H 10^2$	0.63	0.71	0.79	0.89	1.03	1.28	1.84
R_0	8417	6359	5945	5849	5774	5880	5898

The computations have shown that at $0.94 \leq \alpha \leq 0.96$ the coefficient b_4 changes its sign in the vicinity of the turning point. In this connection on the basis of equations (15),(16) we can show that the amplitude curve may have a second turning point. This possibility is depicted in Fig. 1 by dotted line. Then in the (α, R) plane there may exist a point where two turning points mix together. The cusp of the amplitude surface depicted in Fig. 2 corresponds to this point.

The calculation of four-dimensional invariant manifold with the accuracy up to the second order term with respect to z_1, z_2 has been carried out. The first eigenvalues for odd eigenvalues of Orr-Sommerfeld equations were taken as the second pair of eigenvalues for σ_1. Numerical computations of Eq.(8) have shown that at $\alpha = 1$, $R = 5000$ and $z_1 < 0.9 \cdot 10^{-2}$ the two-dimensional manifold is stable with respect to four-dimensional disturbances.

In calculations the Poiseuille flow velocity profile was taken

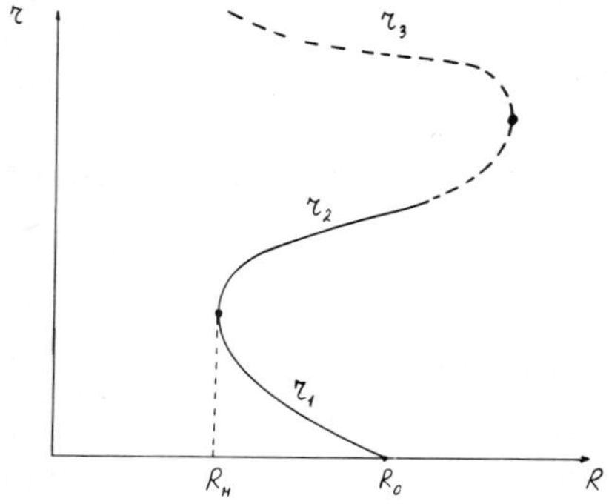

Fig.1. τ_1 , τ_3 are unstable and τ_2 stable limit cycles

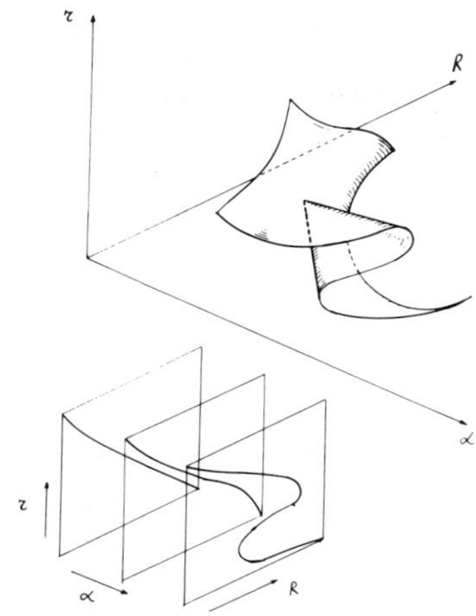

Fig.2

in the form $v_0 = y(2-y)$. The eigenfunctions of the Orr-Sommerfeld equation $\varphi(y)$ were normalized by conditions $\varphi(1)=1$ for even functions and $\varphi'(1)=1$ for odd ones. The integration of equations was carried out by the orthogonalization method. Each equation was integrated on its nonuniform mesh which condensed in a region of critical layer and near wall.

References

1. Iooss, G.: Bifurcation of a periodic solution of the Navier-Stokes equations into an invariant torus. Arch. Rat. Mech. Anal. 58 (1975) 35-56.

2. Scobelev, B.Yu.: Secondary flows in a plane channel. Zhurn. Prikl. Math. Mekh. 42 (1978) 361-367.

3. Struminsky, V.V.; Scobelev, B.Yu.: Nonlinear neutral curve for Poiseuille flow. DAN SSSR, 252 (1980), 566-570

4. Scobelev, B.Yu.; Molorodov, Yu.I.: Subcritical autooscillations and nonlinear neutral curve for Poiseuille flow. Comp. Math. with Appls., 6 (1980) 123-133.

On Linear Three-Dimensional Stability of Pipe Poiseuille Flow

M. WADIH

Institut de Mécanique des Fluides de Marseille (France)

Summary

Theoretical investigations on the stability of pipe Poiseuille flow have shown the existence of two disturbances family modes : center modes for which $c_r \to 1$, and wall modes for which $c_r \to 0$ when $\alpha R \to \infty$ (Pekeris, [5] ; Corcos & Sellars, [1] , and Gill, [4]). In Davey and Drazin's calculations [2] , another family of modes appears : $c_r \to 2/3$, which we may call " mean modes". A theoretical explanation was still lacking for these modes.

The object of this paper is to show theoretically the existence of this latter family and, more generally, to establish the eigenvalue relation for the three-dimensional case, since Squire's theorem does not hold in cylindrical geometry.

1. EIGENVALUE PROBLEM

In the adimensional cylindrical components (x,r,θ), we define the Reynolds number as

$$R = \frac{U_{max} \cdot a}{\nu}$$

where U_{max} is the basic flow velocity on the pipe axis, a, the pipe radius and ν the fluid kinematic viscosity. The cylindrical components of the basic flow velocity are

$\vec{U}\ (1 - r^2,\ 0\ ,\ 0)$, and the disturbances are of the form

$$\vec{V} = \vec{V}\ (r)\ e^{i\ (\alpha x\ -\ \beta t\ +\ n\theta)}$$

Then, the Navier-Stokes equations are written in linear theory

$$L_{n^2+1}\ v_r - i\ \frac{2n}{r^2}\ v_\theta - R\ Dp = 0 \qquad (1\text{-}1)$$

$$L_{n^2+1}\ v_\theta + i\ \frac{2n}{r^2}\ v_r - i\ \frac{n}{r}\ Rp = 0 \qquad (1\text{-}2)$$

$$L_{n^2} v_x + 2 Rrv_r - i\alpha Rp = 0 \tag{1-3}$$

$$\frac{D(rv_r)}{r} + \frac{in}{r} v_\theta + i\alpha v_x = 0 \tag{1-4}$$

where $D = \frac{d}{dr}$

$$L_q = D^2 + \frac{D}{r} - [\frac{q}{r^2} + \lambda - \gamma r^2] \tag{1-5}$$

with $\lambda = \alpha^2 - i(\beta - \alpha) R$ (1-6)

and $\gamma = i\alpha R$ (1-7)

v_r, v_x and v_θ being respectively radial, longitudinal and azimuthal components of $\vec{V}(r)$.

The boundary conditions are

$$\vec{V}(1) = 0 \quad \text{and} \quad \vec{V}(0) \text{ finite} \tag{1-8}$$

The elimination of p between (1-1) and (1-2) gives a first equation for v_r and v_θ. If we neglect v_x compared to $D(r v_r)$ because $r v_r$ varies rapidly in (1.4), we obtain a second equation for v_r and v_θ. Also, with this system, we can suppose that v_r and v_θ are known in a first approximation independantly of v_x.

By eliminating p between equations (1-1) and (1-3), we obtain an inhomogeneous equation

$$L_{n^2+1} f = h \tag{1-9}$$

with $f = -D v_x + i\alpha v_r$ (1-10)

$$h = (\frac{2n\alpha}{r^2} + 2i n R - \frac{2n^3}{\alpha r^4}) v_\theta - i \frac{2n^2}{\alpha r^4} D(r v_r) \tag{1-11}$$

The solution of (1-9) is of the form

$$f = f_g + f_o$$

where f_g is the general solution of

$$L_{n^2+1} f_g = C \tag{1-12}$$

and f_o a particular solution of (1-9).

Writing f_g in the form

$$f_g = \exp(z) \sum_{m=0}^{\infty} k_{2m} r^{2m+\beta} ; \quad k_{2m} = 0 \quad \text{if} \quad m < 0 \tag{1-13}$$

where
$$z = \frac{\delta}{4} r^2 \tag{1-14a}$$

and
$$\delta = 2 e^{i \frac{3\pi}{4}} (\alpha R)^{1/2} \tag{1-14b}$$

we find two values of β: $\pm\sqrt{n^2+1}$

and $f_g = \exp(z) [c_1 V_\lambda^\delta(p,r) + c_2 V_\lambda^\delta(-p,r)]$ if $p \neq \pm 1$

(1-15)

with
$$p = \sqrt{n^2 + 1} \tag{1-16}$$

and hypergeometric functions $V_\lambda^\delta(p,r)$ and $V_\lambda^\delta(-p,r)$ are defined by

$$V_\lambda^\delta(s,r) = r^s \sum_{m=0}^{\infty} \frac{\prod_{q=1}^{m} [\lambda - (2q + s - 1)\delta]}{2^{2m} . n! (s+1)(s+2)...(s+m)} r^{2m}$$

(1-17)

Then, for $p \neq \pm 1$, the general solution of (1.9) is

$$f = \exp(z) [c_1 V_\lambda^\delta(p,r) + c_2 V_\lambda^\delta(-p,r)$$
$$- V_\lambda^\delta(p,r) \int_o^r \frac{h(r) V_\lambda^\delta(-p,r)}{W} dr$$
$$+ V_\lambda^\delta(-p,r) \int_o^r \frac{h(r) V_\lambda^\delta(p,r)}{W} dr] \tag{1.18}$$

where W is the wronskian W $[\exp(z)V_\lambda^\delta(p,r), \exp(z)V_\lambda^\delta(-p,r)]$.
The case $p = \pm 1$ (n=0) corresponds to bidimensional disturbances and in general, when p is an integer, the solution $V_\lambda^\delta(p,r)$ and $V_\lambda^\delta(-p,r)$ are not linearly independent. In that case, the second independent solution is a function of the second kind. But since, our aim here is to establish the eigenvalue relation, and that the boundary conditions require the solution to be regular at r = 0, then, we have necessarity $c_2 = 0$. The following results are valid for any value of p.

2. EIGENVALUE RELATION FOR αR INFINITE

By imposing boundary conditions at r = 1, the hypergeometric function $V_\lambda^\delta(p,r)$ must be interrupted when R is infinite. Then, the eigenvalue relation for large values of R is :

$$2a = (2q - 1 + \sqrt{n^2 + 1}) \qquad q = 1 ; 2 ; \dots \tag{2.1}$$

$$\text{with} \quad a = \frac{\quad}{2} \tag{2.2}$$

If we define $c = \frac{\beta}{\alpha}$, relation (2.1) can be written

$$c = 1 - 2\,(2q-1 + \sqrt{n^2 + 1})\; e^{i\frac{\pi}{4}} (\alpha R)^{-1/2} + \frac{\alpha}{i\alpha R} \tag{2.3}$$

If, we take n = 0 in (2.3), we find the classical result for bidimensional disturbances.

For fixed values of q and n, relation (2.3) gives a first family of modes for which $c \to 1$, when R is infinite.

It can be seen, that there is a finite number of center modes, and that disturbances are damped very rapidly, when n increases. The effect of azimuthal disturbances is on the one hand, to reduce the number of center modes. Then, a theoretical study for large values of a, i.e. n or q large is well founded.

3. BEHAVIOUR OF DISTURBANCES FOR LARGE VALUES OF a

We have defined the parameter a as

$$a = \frac{1-c}{4} e^{-i\frac{\pi}{4}} (\alpha R)^{1/2} \tag{3.1}$$

The form of solution (1.18) shows that for αR infinite, the principle term is the general solution of equation (1.12), which can be written

$$f'' + \frac{f'}{r} - \left[\frac{p^2}{r^2} + \alpha^2 - \lambda^2 \varepsilon(1 - r^2 - c)\right] f = 0 \tag{3.2}$$

with $\varepsilon = \frac{1}{16a^2}$ (3.3)

and p defined by (1.16).

For small values of ε, an approximate solution can be found by a similar method as that used by Pekeris [5] for small values of αR.

$$f(r) = \sum_{s=0}^{\infty} \varepsilon^s f^{(s)}(r) \quad \text{and} \quad \varepsilon c = \sum_{s=0}^{\infty} \varepsilon^s c^{(s)} \tag{3.4}$$

We obtain for s = 0

$$c^{(0)} = -\frac{\alpha^2 + j_{p,m}^2}{\lambda^2} \tag{3.5}$$

$j_{p,m}$ being the zeros of Bessel's function J_p and the solvability condition for s = 1

$$c^{(1)} = \frac{\int_0^1 r\,(1-r^2)\,[J_p\,(j_{p,m}\,r)]^2\,dr}{\int_0^1 r\,[J_p\,(j_{p,m}\,r)]^2\,dr} \tag{3.6}$$

In particular, for higher modes, i.e. m large

$$\varepsilon c = -\frac{j_{p,m}^2 + \alpha^2}{\lambda^2} + \frac{2}{3}\varepsilon \tag{3.7}$$

Then, we have the mean modes

$$c_r \to \frac{2}{3} \quad \text{and} \quad c_i = -\frac{j_{p,m}^2 + \alpha^2}{\alpha R} \tag{3.8}$$

We expect that there is a finite number of these modes,

because when $j^2_{p,m} >> O(\alpha R)$, $|c|$ becomes infinite, and convergence condition of (3.4) may not be satisfied.

In fact, when $|c|$ is large, we suppose that $|\varepsilon| \to 0$ very rapidly, and we write

$$f'' + \frac{f'}{r} - [\frac{p^2}{r^2} + \lambda^2 (\frac{1}{\lambda} + \varepsilon r^2)] f = 0 \tag{3.9}$$

If $|\lambda| >> O(\alpha R)$, an approximate solution of (3.9) is given by

$$f = J_p (i \lambda^{1/2} r) \tag{3.10}$$

and that

$$\lambda^{1/2} = i j_{p,m} \tag{3.11}$$

then

$$c = 1 + \frac{\alpha^2 + j^2_{p,m}}{i \alpha R} \tag{3.12}$$

For the case $j^2_{p,m} >> O(\alpha R)$, we have at the limit

$$c_r \to 1 \quad \text{and} \quad c_i \to -\infty \tag{3.13}$$

4. CLASS OF WALL MODES

Using a similar method in the bidimensional case (see Gill, [4]), we can find the wall modes. The turning point r_c is defined by

$$\frac{p^2}{r_c^2} + \lambda - \gamma r_c^2 = 0 \tag{4.1}$$

r_c is real when we have $r_c^2 = 1 - c_r$ (4.2)

and

$$c_i = - \frac{p^2}{r_c^2 (\alpha R)} - \frac{\alpha}{R} \tag{4.3}$$

If $c_r \to 0$, we put $y = 1-r$ and the homogeneous equation becomes

$$f''_\eta - \eta f = 0 \tag{4.4}$$

with

$$\eta = (2\gamma)^{1/3} (y + \frac{p^2 + \lambda - \gamma}{2\gamma}) \tag{4.5}$$

Then $f(\eta) = A_i\ (\eta)$ Airy function

and

$$c = -\ 2^{2/3}\ e^{-i\frac{\pi}{6}}\ (\alpha R)^{-1/3}\ \eta_{\pm q}\ -\ i\ \frac{p^2}{\alpha R}\ -\ i\ \frac{\alpha^2}{\alpha R} \qquad (4.6)$$

$$\text{with}\quad \eta_q = (2\gamma)^{1/3}\ \left(\frac{p^2+\lambda-\gamma}{2\gamma}\right) \quad \text{root of}\quad A_i\ (\eta) = 0 \qquad (4.7)$$

At the limit, we see easily a good agreement with (4.3)

$$c_r \to 0 \qquad \text{and} \qquad c_i \to -\ \frac{p^2}{\alpha R}\ -\ \frac{\alpha}{R}$$

CONCLUSION

The pipe Poiseuille flow is stable for three-dimensional disturbances, confirming Authors calculations : Salwen & Grosch (1972) and Garg & Rouleau (1972). There are three classes of families modes (center modes i.e. $c_r \to$ [illegible] ; wall modes i.e. $c_r \to 0$ and mean modes $c_r \to \frac{2}{3}$).

REFERENCES

1. Carcos, G.M. & Sellars, J.F. 1959, J.F. Mechanics, 5, 47.
2. Davey, A. & Drazin, P.G. 1969, J.F. Mechanics, 36, 109.
3. Garg, V.K. & Rouleau, W.T. 1972, J.F. Mechanics, 54, 193.
4. Gill, A. 1965, J.F. Mechanics, 21, 145.
5. Pekeris, L.L. 1948, U.S. Proc. Nat.Ac.Sc., 34.
6. Salwen, H. & Grosch, C.E. 1972, J.F. Mechanics, 54, 93.

Formation of Vortical Particles in Inviscid Flow

F. Schultz-Grunow

Institute for Mechanics
Technical University Aachen, F.R.G.

Summary

In shear layers between fluids with different densities and more recently in bursting flames individual particles were detected. It is shown in the following that they are produced by an unstable interface. They are the very reason for the burst. The conditions for the appearance of a burst are presented.

Introduction

The assumption that turbulence produces in any case homogeneous mixing had to be revised when it was observed that in shear layers between liquids with different densities all over the layer density fluctuations occur amounting the total difference of densities /1/. The spread across the shear layer should be managed by coherent structures but, to all appearance, one cannot exclude that the particles also propel themselves by means of vorticity which they contain. Such particles can be observed in Schlieren pictures of bursting flames /2/. The burst almost seems to be determined by sideways moving particles. These photographs are of great help for an analysis on behalf of exceptionally favourable conditions.

Formation of vortical particles

The photographs were recently published in a thesis /2/ in which turbulence models were proved on their applicability to diffusion flames which burst. The Schlieren pictures which were made depict the yet not sufficiently recognized formation of particles by a perturbed interface which, as a whole, form the burst. The flame starts as thin jet of burning hydrogen surrounded by a slower air jet which furnishes the needed oxygen.

The exhaust speed of the gas is fast enough to produce homogeneous turbulent mixing in a hose inside the air jet. Fig. 1 depicts it distinctly. It belongs, as well as Fig. 2 to the cited thesis. They are published here with the kind permission of Prof. Knoche, Institute of Thermodynamics, T.U. Aachen. The temperature distribution inside the hose, in fact, is practically homogeneous (/3/, Figs. 6.13,.17,.18) with a sharp drop to the outside. At this interface waves propagate upwards with increasing amplitude. They look occasionally like standing waves obviously because of an unintended slow fluctuation of the jet velocity which temporarily equalizes the velocities at the interface. Thus one has the exceptional opportunity to observe the growth of the waves without being hidden by their roll up. One remembers Lord Rayleighs flapping flag. It is suitable as start to a nonlinear approach. The problem here is axisymmetric. Radius $r = c + r_o$ with r_o as radius of reference and axial length x are used as coordinates.

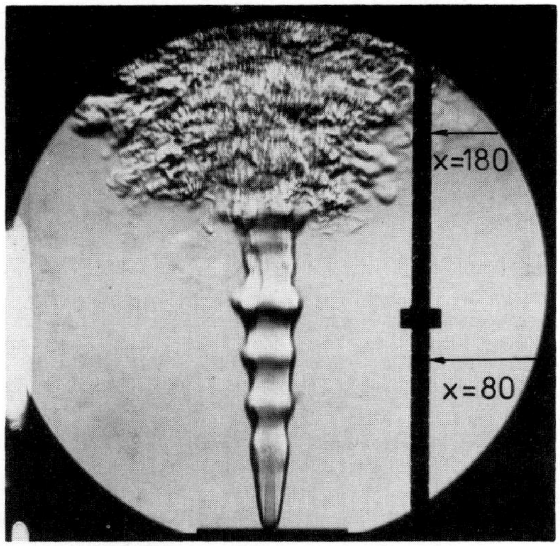

Fig. 1: Bursting flame with coherent structure, Exposure 10^{-3} s, Dimensions in mm.

The velocity potential of the perturbation is in this case

$$\Phi = a\, I_o(kr) \cos kx$$

With the wave length = 23.3 mm (s. Fig. 1) one has $k = 270\ m^{-1}$. The radial velocity is

$$v = a\, k\, I_1(kr) \tag{1}$$

The function $K_1(kr)$ is omitted. It is not applicable because of its singularity on the axis which is not realized. It would yield a wrong influence on waves with large amplitudes as considered here. $I_1(kr)$ describes the flow inside the hose and K_1 on the outside. Omitting K means that the perturbed interface proceeds in air at rest. This is acceptable since the pressure condition at the interface is satisfied exactly in first order because of the time independent potential and exactly at pointed crests which alone are here of interest.

The kinetic condition at the interface is approximately that of the components normal to the undisturbed interface:

$$\partial c/\partial t = v$$

It is exact at the troughs and crests. Hence

$$c = atk\, I_1(kr) \cos kx \qquad (2)$$

The linear increase with time agrees with Figs. 1, 2 since the jet speed is constant (/3/, Fig. 4.3). This solution assumes constant hose radius. This is the case at the top wave in Fig. 1 where combustion is completed. This wave is of special interest since it disappears and, instead, a cloud of particles appears. Fig. 1 depicts the dimensions of the top wave. The radius of the crest is $c_1 + r_o$ = 15.85 mm that of the trough $r_o - c_2$ = 11 mm. In order to determine the factor akt eq. (2) is applied to the crest and

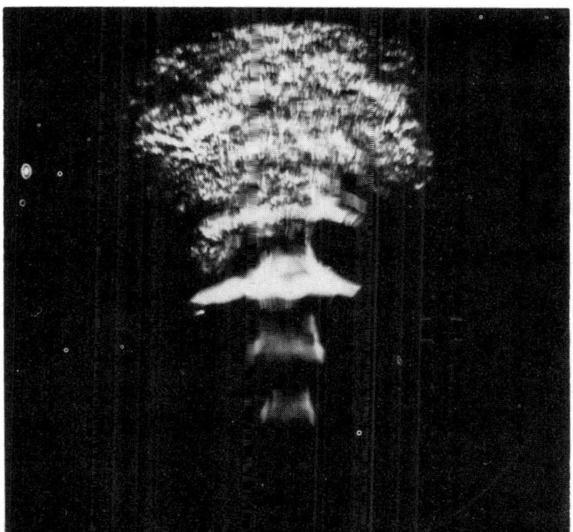

Fig. 2: Bursting flame with collapsing cusps.

the trough

$$c_1 + r_o = r_o + \text{atk } I_1(k\ (c_1+r_o))$$
$$r_o - c_2 = r_o - \text{atk } I_1(k\ (r_o-c_2))$$

Hence

$$c_1 + c_2 = \text{akt } (I_1(k\ (c_1+r_o)) + I_1(k\ (r_o-c_2))\ )$$

One finds akt = 0.295 mm. The wave now can be computed by determining x in dependence of r. It is plotted in Fig. 3 as profile (1). One sees that the formation of a cusp is imminent. A half wave length later, profile (2), the cusp is formed. It gets more and more sharp edged and finally annexes the whole wave and forms a spray, profile (3). This is in close agreement with Fig. 2.

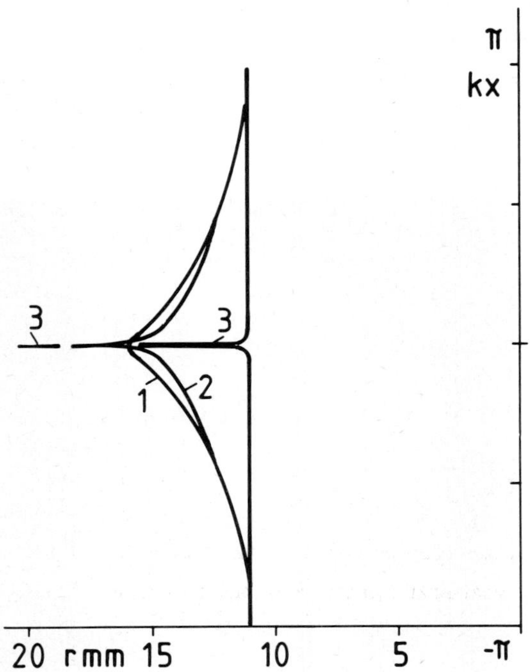

Fig. 3: Computed Wave Profiles Formation of Cusp

To dissolve the cusp in particles radial fluctuations are needed. They are furnished by strong background turbulence. Indeed, radially $v'^2 = 0.2\ m^2/s^2$ was measured (/3/ Fig. 4.7) in comparison with v = 0.4 m/s (/3/ Fig.4.6). Take a sharp edged cusp which is exhibited to a radially directed fluctuation. The cusp then is enhanced by velocities tangentially to the interface. As soon as the fluctuation stops the excess interface must vanish again. It therefore acts as sink which produces potential flow. Thus the vorticity of the excess interface is left outside the retreated interface and begins to travel as independent particle. The particle formation keeps step with the turbulent fluctuations until the wave passing the stages shown in Fig. (3) is used up. The burst now is formed. The particle speed is of the order of that of the cusp given by eq. (1). t is the travel time of the wave with 12 m/s. It is 0.008 s. For profile (3) in Fig. (3) one finds 0.8 m/s whilst Fig. (1) depicts not more than about 1 m/s. A negative fluctuation acts in the same way.

The formation of particles evidently begins as soon as a sharp edged cusp is formed. That means that the burst begins at least when the second derivative $\partial^2 c/\partial x^2$ which easily is derived from eq. (2) gets infinite at the cusp. Nevertheless one has to prove if for a given radius of the interface and a given wavelength eq. (2) after all, yields a solution. The experiments show that a slow enough roll up does not hinder the formation of particles.

In the theoretical case that an exactly axisymmetric turbulent fluctuation would exist the separated cusp would form approximately a torus. The continuity equation in torus coordinates is

$$\partial^2\Phi/\partial r^2 + \partial\Phi/r\partial r + \partial^2\Phi/R^2\partial\vartheta^2 = 0$$

with the solution

$$\Phi = a\, I_o(s\, r/R)\, \cos s\vartheta$$

Thus large peripheral waves evolve in the same way as the axial waves and form with them particles.

Much interest is devoted presently to coherent structures. One observes here successive clouds. They are relatively simple to understand. Each wave collapses at the same position. Thus the coherent structure is the representation of successive bursts. Their distance is somewhat larger than the wavelength of the perturbation on account of thermal lift.

In the plane case one has apart of the roll up process the exponential function instead of I. The asymmetry with respect to the interface now is due to gravity. The wave profiles now are trochoids. Thus again sharp edged cusps are formed.

The question arose why the burst is closely related to combustion. The influence of viscosity was quoted. The burst, however, could be described with not more than potential flow. The condition for a burst rather is the presence of an interface. Without combustion it does not exist at least with the geometric conditions of the analysed experiments. Turbulent mixing then is prevalent.

References

1. Brown, G.L., Roshko, A.: On Density Effects and Large Structure in Turbulent Mixing Layers. J. Fluid Mech. 64 (1974) 775 - 816.

2. Altgeld, H.: Laser-Doppler-Measurements in a Turbulent Hydrogen-Air Diffusion Flame, Thesis Tech. Univ. Aachen, F.R.G., 1979, Institute for Technical Thermodynamics.

3. Kolbe, W.: Modeling of a Turbulent Hydrogen-Air Diffusion Flame, Thesis Techn. Univ. Aachen, F.R.G., 1978, Institute for Technical Thermodynamics.

Numerical Simulation of Two- and Three-Dimensional Nonstationary Viscous Fluid Flows in a Plane Channel

I.N. SIMAKIN

Central AeroHydrodynamics Institute, USSR

Summary

Efficient and sufficiently accurate pseudo-spectral methods of numerical simulation of two- and three-dimensional nonstationary incompressible viscous fluid flows in an infinite plane channel with a fixed flow rate and of flows with a fixed space-averaged pressure gradient are considered. A subcritical instability of the plane Poiseuille flow with respect to two- and three-dimensional finite-amplitude disturbances periodic in homogeneous spatial coordinates is investigated using the constructed methods.

The problem of incompressible viscous fluid flow in a plane channel is reduced to the solution of the Navier-Stokes equations

$$\frac{\partial \vec{V}}{\partial t} = [\vec{V}, \vec{\omega}] - \vec{L}[\vec{V}] - \nabla \Pi + \nu \Delta \vec{V}, \qquad \Pi = p + \frac{1}{2}\vec{V}^2, \tag{1}$$

$$\nabla \vec{V} = 0, \tag{2}$$

which satisfies the no-slip condition on the channel walls $z = \pm 1$, conditions of periodicity in variables x, y with periods X, Y and one of the following external conditions $\vec{Q}(t)=\text{const}$ or $\langle \nabla P(t) \rangle = \text{const}$. Here, $\vec{V} = \{u, v, w\}$, $\vec{\omega} = [\nabla, \vec{V}]$, p are the deviations of velocity, vortex and pressure, respectively, from their values for the Poiseuille flow, $\vec{Q}(t)$ is the flow rate vector, and $\langle \nabla P(t) \rangle$ is the space-averaged pressure gradient. The linear operator

$\vec{L}[\vec{v}]$ is defined by

$$\vec{L}[\vec{V}] = \left\{ U_o \frac{\partial u}{\partial x} + U_o' \cdot w \,,\; U_o \frac{\partial v}{\partial x} \,,\; U_o \frac{\partial w}{\partial x} \right\}$$

where $U_o(z)$ is the Poiseuille flow velocity profile.
A numerical solution of the problem under study is rather time-consuming. The experience shows that the nonstationary flow simulation in the plane channel at $R \sim 10^3 - 10^4$ is associated with calculations for large time intervals $T \sim O(R)$ (see [1 - 3]). In this connection a numerical investigation of such flows is possible only by using rather efficient and stable algorithms.
To correctly simulate flows at large Reynolds numbers it is also necessary that the numerical solution of the Navier-Stokes equations should represent adequately time histories of small disturbances of the laminar flow. In other words, it is essential that the reference differential problem and the algorithm for its solution linearized on their common exact solutions have sufficiently close spectral characteristics in a spectrum zone typical of the problem of interest. Outside this characteristic zone, i.e. in that which corresponds to a strong disturbance decay, merely a qualitative conformity of their spectral characteristics suffices (see [4], [5]). For difference methods this condition results in fine meshes across the channel (more than 100 points) and a small time step, respectively. As a result, the application of these methods becomes irrational to solve the problem under discussion, when spectral methods (Galerkin, and Petrov methods) are used, about a two times lesser number of degrees of freedom across the channel $S \simeq 60$ is sufficient to obtain a required accuracy. If the choice of the expansion functions in Petrov method (linear combination of Chebyschev polynomials) is appropriate then a satisfactory space approximation is already achieved at $S \simeq 30 - 40$ [5] . Moreover, the spectral methods permit to construct algorithms enabling a numerical integration of the problem with a comparatively large time step.
Now we shall present the method of the problem solution which

describes a nonstationary incompressible viscous fluid flow in a plane channel with a fixed mean pressure gradient. The approximate solution of this problem is sought in the form

$$\begin{Bmatrix} \vec{V}(\vec{x},t) \\ \Pi(\vec{x},t) \end{Bmatrix} = \sum_{|m|\leq M} \sum_{|n|\leq N} \sum_{s=0}^{S} \begin{Bmatrix} \vec{V}_{mns}(t) \\ \Pi_{mns}(t) \end{Bmatrix} \cdot \qquad (3)$$

$$\cdot \exp\left[im\alpha_0 x + in\beta_0 y\right] \cdot T_s(z)$$

where $\alpha_0=2\pi/X$, $\beta_0=2\pi/Y$, $T_s(z)=\cos(s\cdot\arccos z)$ is the Chebyschev polynomials. It may be shown that the external condition $\langle \nabla P(t)\rangle =\text{const}$ follows from the solution representation (3).

The integration in time of problem (1), (2) is carried out by the fractional steps method. Three fractional steps are introduced. At the first fractional step the influence of the nonlinear term $\vec{F}=[\vec{v},\vec{\omega}]$ and, in part, that of the linear operator $\bar{L}[\vec{v}]$ are accounted for. The influence of the dissipative term and of $\nabla\cdot\Pi$ is taken into account at the second step. The third fractional step symmetrizes the first one to improve the approximation of the linear operator $L[\vec{v}]$. The transform and pseudospectral methods are applied to calculate the Fourier coefficients of the nonlinear term $\vec{F}$ in variables x, y and approximate in variable z in a physical space $\vec{F}$ and $\vec{L}$ (see [6], [7]).

The first fractional step is a three-timelevel explicit difference scheme; the solution is sought in the mixed spectral-physical representation (in x, y variables - in the space of Fourier coefficients and in z in the physical space - at the mesh nodes $z_l=\cos(\pi l/S)$, $l = 0 - S$):

$$\frac{\vec{V}^{k+1/3}_{mnl} - \vec{V}^{k}_{mnl}}{\tau} = \frac{3}{2}\vec{F}^{k}_{mnl} - \frac{1}{2}\vec{F}^{k-1}_{mnl} - \frac{1}{2} im\alpha_0 U_l \vec{V}^{k}_{mnl} - \qquad (4)$$

$$- \vec{i}\, U_l'\left(\frac{5}{2} w^{k}_{mnl} - w^{k-1}_{mnl}\right)$$

where $\vec{i}=\{1, 0, 0\}$, U_l and U_l' are the values of $U_0(z)$

and its derivative at the points z_l .
The solution of the next purely implicit difference scheme is determined at the second fractional step

$$\frac{\vec{V}^{k+2/3}(\vec{x}) - \vec{V}^{k+1/3}(\vec{x})}{\tau} = -\nabla \Pi^{k+2/3}(\vec{x}) + \nu \Delta \vec{V}^{k+2/3}(\vec{x}) \ ,$$

$$\nabla \vec{V}^{k+2/3}(\vec{x}) = 0 \ , \tag{5}$$

$$\vec{V}^{k+2/3}(x,y,z = \pm 1) = 0 \ ,$$

Problem (5) is solved in a spectral representation in all variables using an efficient and accurate technique (see [8]). The solution of the implicit difference scheme is obtained at the third final fractional step in the mixed spectral-physical representation (as well as at the first fractional step)

$$\frac{\vec{V}^{k+1}_{mnl} - \vec{V}^{k+2/3}_{nml}}{\tau} = -\frac{1}{2}\, i m \alpha_o U_l \vec{V}^{k+1}_{mnl} + \vec{i}\,\frac{1}{2}\, U'_l w^{k+1}_{mnl} \ . \tag{6}$$

This scheme can be solved explicitly. It is easy to show that the solution $\vec{V}^{k+1}$ satisfies the no-slip condition at the channel walls and the equation of incompressibility (2).
The third fractional step completes the calculation of the full step and we have vector $\vec{V}^{k+1}_{mnl}$ as the initial condition for the next time step. The scheme (4) - (6) can be shown to have a total approximation order $O(\tau^2)$ for the nonlinear term $\vec{F}$ and the linear operator $\vec{L}$, and an approximation order $O(\tau)$ for the influence of the dissipative term $\nu \Delta \vec{V}$ and $\nabla \Pi$.
A similar scheme is constructed for another problem formulation at a fixed flow rate vector as well (see [8]). Note that the developed algorithm may be also used to simulate finite disturbances of the plane Couette flow.
A spectral stability and accuracy analysis of the constructed shemes both for sub- and supercritical Reynolds numbers, and in a wide range of wavenumbers as well shows that at a rela-

tively small number of Chebyshev polynomials $S \simeq 30 - 40$ and at a sufficiently large time step $\tau \simeq 0,1-0,05$ the scheme are stable and have an acceptable accuracy both in space and in time. In addition, it was found that the "excessive" eigenvalues due to the three-timelevels character of the scheme at the first fractional step have a considerable negative real part as compared to the senior true eigenvalues. Hence, the disturbances corresponding to the "excessive" roots are rapidly decaying, and their influence on the true solution is insignificant. Detailed results of the spectral analysis of the constructed schemes and their comparison with other schemes are given in [5] .

Two- and three-dimensional secondary flows in a plane channel were sought using the developed procedure and some of its modifications. The results of this study may be found in [1-3].

Acknowledgement. The author wishes to express his thanks to B.L. Rozhdestvensky for his consistent concern and help.

References

1. Rozhdestvensky, B.L.; Simakin, I.N.: Nonstationary secondary flows in a plane channel and stability of the Poiseuille flow with respect to finite disturbances. Dokl. Akad. Nauk SSSR 266 (1982) 1337-1340.
2. Rozhdestvensky, B.L.; Simakin, I.N.: The stability of the Poiseuille flow with respect to finite disturbances. Chisl. Metody Mekh. Sploshnoy Sredy 13 (1982) 89-128.
3. Rozhdestvensky, B.L.; Simakin, I.N.: Two- and three-dimensional secondary flows in plane channel, their relationship and comparison with turbulent flows. Dokl. Akad. Nauk SSSR 273 (1983) 553-558.
4. Rozhdestvensky, B.L.: On the applicability of difference methods to the solution of the Navier-Stokes equations at large Reynolds numbers. Dokl. Akad. Nauk SSSR 211 (1973) 308-311.
5. Simakin, I.N.: Investigation of the stability and accuracy of the methods of numerical simulation of nonstationary flows in a plane channel. Spectral analysis. Preprint IPMatem. Akad. Nauk SSSR №49 (1983).
6. Orszag, S.A.: Numerical simulations of incompressible flows within simple boundaries. I. Galerkin (spectral) representations. Stud. Appl. Math. 50 (1971) 293-327.

7. Gottlieb, D.; Orszag, S.A.: Numerical analysis of spectral methods: theory and applications. Philadelphia: SIAM 1977, NSF-CBMS Monogr. N 26.

8. Rozhdestvensky, B.L.; Simakin, I.N.: Methods of numerical simulation of unsteady flows of incompressible viscous fluid in a plane channel. Preprint IPMatem. Akad. Nauk SSSR №191 (1982).

On a Bifurcation and Emerging of a Stochastic Solution in a Variational Problem for Poiseuille Flows

C. I. CHRISTOV+ and V. P. NARTOV++

+Dept. of Fluid Mechanics, Institute of Mechanics and Biomechanics, Bulgarian Academy of Sciences, 1090 SOFIA P.O.Box 373 BULGARIA

++Institute of Theoretical and Applied Mechanics, Siberian Division of Academy of Sciences of the USSR, Institutskaya 4/1, 630090 Novosibirsk, USSR

SUMMARY

For the pressure-driven viscous flow between two flat plates and in tubes (Poiseuille flows) the principle of minimal dissipation is applied. Minimum is sought in the class of point random functions. For kernels of stochastic integrals involved a boundary value problem is derived and appropriate numerical method is proposed. Calculations show that beyond certain critical magnitude of Reynolds number a bifurcation takes place and non-trivial stochastic solution emerges. Various characteristics of this solution are calculated and most of them are in good quantitative or qualitative agreement with the experimental data concerning turbulent Poiseuille flows. The kernels of stochastic integrals are interpreted as large eddies (coherent structures).

Introduction

Transition to turbulence in Poiseuille flows has always been one of the major problems of hydrodynamic instability. All principal theoretical approaches to instability has been checked on these flows. Reynolds called unstable those regimes for which a disturbance of increasing energy can be traced [1]. This led him to the variational problem of minimizing of the time derivative of full energy of a perturbation. After in [2] Sommerfeld turned directly to the Navier-Stokes equations when investigating the evolution of disturbances and posed the respective boundary value problem his method became the most popular approach to instability and now an extensive literature is availble (see for instance survey [3]).

Recently, the variational approach has enjoyed some revived interest. Malkus [4] introduced the principle of maximal dissipation of disturbances in turbulent flow. Goldshtik [5]

stated the principle of maximal stability of the mean-averaged profile of the turbulent flow. Christov discussed in [6] the existence of a kind of least rate of dissipation principle. It turned out that the principle of minimal dissipation predicts the lower critical Reynolds number of transition fairly well. Advantage of variational methods is the opportunity to derive somewhat simpler equations for disturbances. In turn, the results are to be thought of as bounds for the full-scale solutions and are more rough as a rule. In the present work the class of random point functions is considered when minimizing the dissipation functional which is a step ahead of the previous author's paper [7] where the Gaussian random functions are considered. The existence of random solutions is named "stochastic bifurcation".

Principle of minimal dissipation

In parallel flows there exists only one non-trivial component of the averaged velocity - the longitudinal velocity $\bar{u}_x$ and and hydrodynamic characteristics depend only on transverse coordinate y (see Fig.1), the latter being either cartesian coordinate or polar radius for channels and tubes respectively. The averaged flow is governed by the Reynolds equation:

$$\nu \frac{d\bar{u}_x}{dy} = -Ay + \overline{u'_x u'_y} \tag{1}$$

where A is the magnitude of the average pressure gradient and $\overline{u'_x u'_y}$ is the so-called Reynolds stress. The last equation is not closed. One of the ways to that is to summon the principle of minimal dissipation as an additional empirical information. The total rate of dissipation in the half of the channel is given by (for pipe flow a is the radius)

$$\int_0^a \left[\nu\left(\frac{d\bar{u}_x}{dy}\right)^2 + \varepsilon\right] y^\alpha dy = min \tag{2}$$

where ε is the density of the dissipation rate of turbulent pulsations, $\alpha=0$ refers to channel flow and $\alpha=1$ - to pipe one.

The class of random functions for minimization

Variational problem (2) with constraint (1) is still not defined unless a kind of connection between the Reynolds stress and turbulent dissipation is specified. In [6] those two are connected through semi-empirical considerations. Here we shall follow the idea of [7] where the dissipation functional is minimized over the class of Gaussian random functions. However, in the present work the class of random point functions is considered instead of that of Gaussian random functions. The random point functions are better model for the situation at the threshold of instability where the disturbances occur seldom and the random process resulted can be considered as a random point function. Let us also assume that velocity field depends on time t as on parameter. Then velocities of a disturbance can be expressed as follows (for brevity only component u_x' is displayed):

$$u_x' = \int_{-\infty}^{\infty} K_x(x-\xi;y)[g(\xi)-\gamma]\,d\xi \,, \tag{3}$$

where K_x is non-random function and $g(\xi)$ is the so-called random density function (for definition see [8],[9]) with mean γ.

Here is to be mentioned that the class of random functions is oversimplified and the results which will be obtained will bear only a distant resemblance to the real developed turbulence, but the solution to the presented problem is of profound physical meaning since it answers the question whether a random flow can bring the value of total dissipation to a level lesser than that for the laminar flow.

Now the dissipation and Reynolds stress are readily obtained:

$$\int_0^a \left\{ \nu\left(\frac{d\bar{u}_x}{dy}\right)^2 + 2\nu\gamma \int_{-\infty}^{\infty}\left[\left(\frac{\partial K_x}{\partial \xi}\right)^2 + \left(\frac{\partial K_y}{\partial y}\right)^2 + \frac{1}{2}\left(\frac{\partial K_x}{\partial y} + \frac{\partial K_y}{\partial \xi}\right)^2\right] d\xi \right\} dy, \tag{4}$$

$$\nu \frac{d\bar{u}_x}{dy} = -Ay + \gamma \int_{-\infty}^{\infty} K_x(\xi;y)\, K_y(\xi,y)\, d\xi \,. \tag{5}$$

On the other hand equation of continuity yields

$$\frac{\partial K_x}{\partial \xi} + \frac{\partial K_y}{\partial y} = 0 \,. \tag{6}$$

So far the variational problem is completed and (4)-(6) appear to be a stochastic implementation of the principle of minimal dissipation.

Euler-Lagrange equations

Introducing Lagrange factors for each of the constraints (5) and (6) the functional (4) can be recast and the problem of minimization without constraints can be considered. Without going in details we can display here only the final dimensionless form of the respective Euler-Lagrange equations

$$Re^{-1}\,\Delta\Delta\Psi = -\left[2\frac{du}{dy}\cdot\frac{\partial\Psi}{\partial y\partial\xi} + \frac{\alpha}{y}\frac{du}{dy}\cdot\frac{\partial\Psi}{\partial\xi} + \frac{d^2u}{dy^2}\cdot\frac{\partial\Psi}{\partial\xi}\right], \tag{7}$$

$$Re^{-1}\frac{du}{dy} = -y - \int_{-\infty}^{\infty}\frac{1}{y^{\alpha}}\left(\frac{\partial}{\partial y}y^{\alpha}\Psi\right)\frac{\partial\Psi}{\partial\xi}\,d\xi\ , \tag{8}$$

which are solved with the following boundary conditions

$$\frac{\partial\Psi}{\partial y} = \frac{\partial\Psi}{\partial\xi} = 0 \quad \text{at} \quad y = \pm 1\ , \quad \int_{-\infty}^{\infty}\left[\left(\frac{\partial\Psi}{\partial y}\right)^2 + \left(\frac{\partial\Psi}{\partial\xi}\right)^2\right] < +\infty\ . \tag{9}$$

In the above formulae $Re = \sqrt{Aa^3/\nu_2^{\alpha}}$ and $\Delta = \frac{\partial}{\partial y}y^{-\alpha}\frac{\partial}{\partial y}y^{\alpha} + \frac{\partial^2}{\partial\xi^2}$.

Due to its complexity the above boundary value problem is treated numerically by a method similar to that of [7] . The infinite interval is "compressed" into the finite interval [-1,1] by means of appropriate coordinate transformation.

Results and discussion

In the previous work [7] we have already presented the results for plane Poiseuille flow and here we concentrate on pipe flow. Our theoretical predictions are compared with the work of Laufer [10] , so all the experimental data plotted in figures below are taken from that work.

In Fig.2 is presented the total dissipation as a function of frictional Reynolds number. It is well seen that the total dissipation of the stochastic flow becomes lesser than that of the laminar one for Reynolds numbers greater than 15. It does mean that we have indeed obtained a solution which brings a minimum to the dissipation functional. So that it stands clear that the stochastic solution has a right of existence and its occurrence is consistent with the least-dissipation principle. This flow, however, can be employed also to predict to some extent even the specific quantitative characteristics of the turbulent flow. Fig.3 presents the evolution of Reynolds stress with Reynolds number and comparison with experiments [10]. It is seen that the agreement is quantitatively good both in the core of the flow and in the viscous sublayer. In Fig.4 is plotted the energy "component" $\sqrt{\overline{u_x'^2}}/u^*$ and compared with experiment. On Fig.5 are presented the averaged values of derivatives of longitudinal pulsation. Agreement with experiment is qualitatively good. In the end Fig.6 presents the shape of a structure which is in qualitative agreement with the structures observed in [11] for the flow in turbulent boundary layer.

References

1. Reynolds, O.; On the dynamical theory of incompressible viscous liquids and the determination of the criterion, Phil. Trans.Roy.Soc.London, 186A (1895) 123-340.
2. Sommerfeld, A.: Ein Beitrag zur hydrodynamischen Erklaerung der turbulenten Flüssigkeitsbewegungen, Atti IV Congresso Intern. dei Matematici, vol. III (1908) 116-124.
3. Christov, C.I.: Critical discussion on the stability of the plane and circular Poiseuille flows, Bulg.Acad.Sci., Theor. and Appl.Mech., 13, No.1 (1982) 59-64.
4. Malkus, W.V.R.: Outline of a theory of turbulent shear flow, J.Fluid Mech., 1, No.3 (1956) 521-539.
5. Goldchtik, M.A.: Principle of maximal stability of average turbulent flow (in Russian), Dokl.Acad.Nauk SSSR, 182 (1968) 1026-1028.
6. Khristov, Kh.I.: The principle of least energy-dissipation rate and possibility of its application to turbulent flow in a plane channel, Sov.Phys.Dokl.,24(4) (1979) 252-254.
7. Christov, C.I. and Nartov, V.P.: Bifurcation and emerging of stochastic solution in one variational problem for plane Poiseuille flow (in Russian), In: Numerical Modelling in Fluid Dynamics, Novosibirsk, 1983, pp.124-144.
8. Snyder, D.L.; Random point processes, New York, John Wiley and Sons, 1975.
9. Stratonovich, R.L.; Topics in theory of random noise, vol.I, Gordon and Breach, 1967.
10. Laufer, J.: The structure of turbulence in fully developed pipe flow, NACA Rep. 1174, 1954.
11. Zilberman, M.; Wygnanski, I. and Kaplan, R.E.; Transitional boundary layer spot in a fully turbulent enviroment, Phys. fluids, 20, No.10, pt. II (1977) S258-S271.

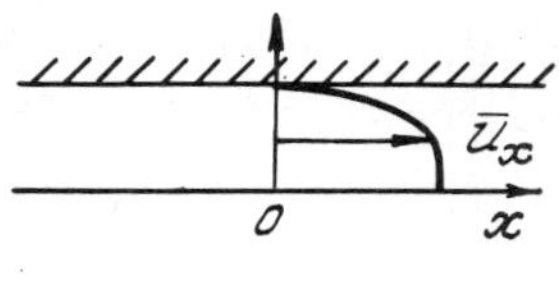

Fig. 1

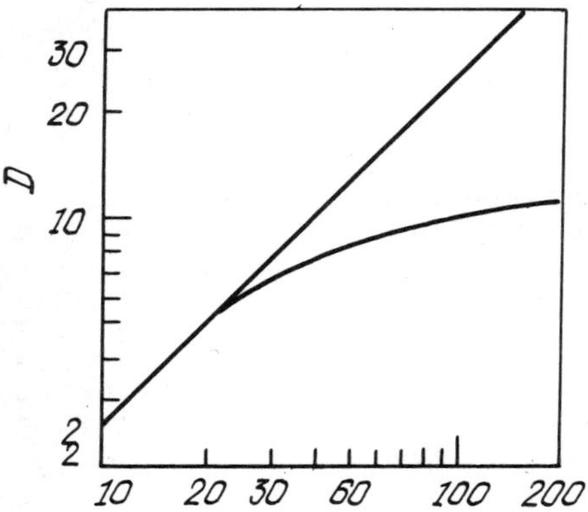

Fig. 2

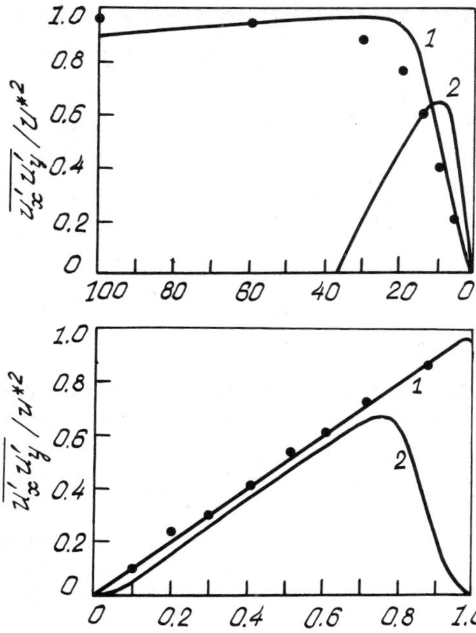

Fig. 3

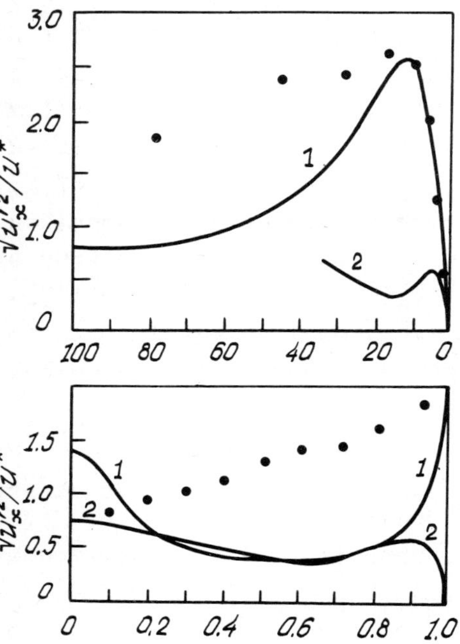

Fig. 4

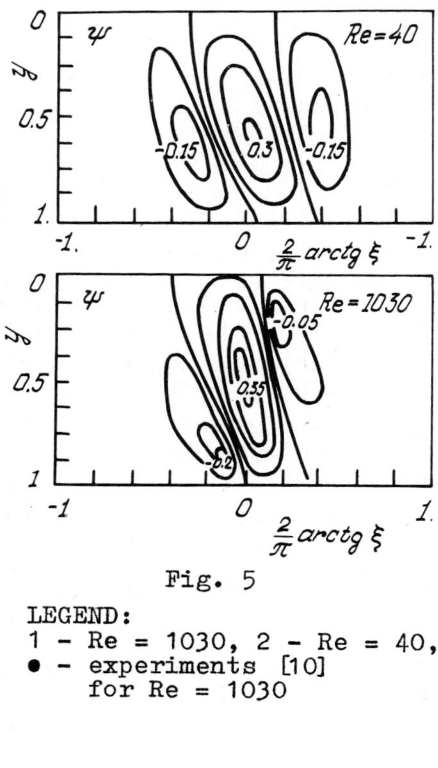

Fig. 5

LEGEND:
1 - Re = 1030, 2 - Re = 40,
● - experiments [10] for Re = 1030

On Generation and Evolution of Waves in a Laminar Boundary Layer

V. S. KOSORYGIN, V. Ya. LEVCHENKO, N. Ph. POLYAKOV

Institute of Theoretical and Applied Mechanics
Siberian Branch of the USSR Academy of Sciences
Novosibirsk 630090, USSR

Summary

Generation and development of disturbances in a flat plate laminar boundary layer was studied at the natural transition and sound excitation. Special attention was given to boundary conditions. The transformation of outer disturbances spectrum is found to take place in a neighbourhood of the leading edge. Evolution of velocity oscillations with definite frequencies was traced in the control volume. A number of features of formation and development of space-amplified waves was elicited.

Introduction

The transition to turbulence in a laminar boundary layer at low outer disturbances is known to be the consequence of flow instability. The position of transition is strongly influenced by non-stationarity of a flow. This gave rise to the problem of transformation of outer disturbances into space-amplified oscillations in a boundary layer. The classical theory of hydrodynamic stability gives no answer to the question about the formation of time-periodical oscillations in the boundary layer, so-called Tollmien-Schlichting (TS) waves. Such oscillations of velocity at the natural transition were observed for the first time by Schubauer & Scramstad. They carried out also the first model experiments on excitation and evolution of waves in a boundary layer. Some conclusions of theory were corroborated by these experiments. Subsequent experimental works can be divided into two groups: model experiments (see, for instance [1]) and experiments on transition under natural conditions. At the study of natural transition an investigator, as a rule, deals with low-level disturbances of different nature. The measurements of such conditions are connected with known difficulties. Apparently it resulted in insufficient number of works and a scatter of data. The review of some of such works and original results may be found in [2] . It should be noted that a dearth of the control for outer conditions was present in early works. This did not permit to separate out the main process/processes determining the beginning of space-amplified waves in each case. Presently one can separate vortex, acoustic and vibration disturbances which have to be necessarily fixed as the initial data of the problem of boundary layer transition. It is necessary to know their spectral structure of disturbances rather than their integral level. The sufficient attention was given to these questions only in recent works [3, 4]. A variable component of signal on hot-wire anemometer output is conditioned by vortex and acoustic pulsations of velocity, by noise of instrumentation and by oscillations of velocity because of probe vibrations. Therefore, a reduction of model vibrations and an amplification of signal/noise ratio is an important issue. After all, the pressure distribution in the neighbourhood of the leading edge demands an attention in experiments on a flat plate. Its strong influence on the position of transition was mentioned in [2, 5] . All above-mentioned considerations were taken into account in experiments which are discussed below.

Experimental conditions

The experiments were conducted in ITAM low-turbulence wind tunnel. Two flat plates having chord lengths of 2.0 m, spans of 1.0 m and thicknesses of 6 mm were used. The plates differed in sizes of leading edges (see Fig. 1a). The chosen geometry of leading edges permitted, in particular, to diminish the region of pressure gradients. The equipment of plates, traversing mechanism and apparatus was used such as in [3] . At acoustic excitation of unstable waves a loud speaker was placed downstream from the plate with $r_{l.e.}$ = 2 mm but a condenser microphone was put at x = -800 mm. An analysis of signals was done in narrow bandwidth of frequencies Δf = 3.16 Hz or 4 Hz.

The main part of experiments was carried out at freestream spead 18 m/s, as in [3] . The pressure field in the region of the leading edge ($r_{l.e.}$ = 1 mm) is given in Fig. 1b. Solid curves represent isobars and numbers are the values of pressure coefficients $C_p=2(p_i-p_\infty)/\rho U_\infty^2$. Here p_i is static pressure at i-th point of the field; p_∞ , U_∞ are pressure and velocity at x = -800 mm. The distribution of pressure on the plate surface is given in Fig. 1c as the dependence on $Re^* = U_\delta \cdot \delta^*/\nu$ where δ^* is a displacement thickness, U_δ is a velocity in the outer part of the boundary layer and ν is a kinematic viscosity. The boundaries of the region under study were determined by the pressure field: AB ($x = -100$, $0 \leqslant y \leqslant 40$), BC ($-100 \leqslant x \leqslant 1560$, $y = 40$), CD ($x = 1560$, $0 \leqslant y \leqslant 40$) and AOD ($-100 \leqslant x \leqslant 1560$, $y = 0$).

Results at natural disturbances

The spectra of the root-mean-square values of the longitudinal component of velocity fluctuations are given in Figs. 2 a-e. The lower curve represents a general noise of the measurement equipment at zero speed of the flow. It can be seen from Fig. 2a that the summary spectrum on the left boundary consisting of acoustic and vortex components is sufficiently smooth and has no discrete frequencies. Also given here are the spectra of leading edge vibration velocity and the spectrum of the fluctuation velocity calculated in acoustic approximation from measurements of pressure fluctuations. Fig. 2b demonstrates the transformation of the initial spectrum in the region of the leading edge. An amplification of the amplitude of the fluctuation velocity is observed in a rather broad bandwidth of frequencies. A packet with the central frequency f = 137 Hz is separated out in the case $r_{l.e.}$ = 1 mm. Amplitudes of excited fluctuations of velocity weakly decrease downstream up to x = 1000 mm. (Figs. 2 c, d), then an amplification of the packet with the central frequencies f = 137 Hz for the plate with $r_{l.e.}$ = 1 mm and f = 110 Hz for $r_{l.e.}$ = 2 mm was observed (Fig. 2e). The behaviour of just these main frequencies will be considered in what follows.

An analysis of vibration spectra of leading edges and spectra of pressure fluctuations showed that a variation of free stream velocity did not change the form of spectra and only led to an amplification of amplitudes of vibration and acoustic velocities. At the same time maximum amplified frequencies from the packet of velocity fluctuations in the boundary layer is related to the flow velocity practically linearly [3] . Therefore, it appears that there is no direct connection of amplified packet frequencies in the boundary layer with the leading edge vibration and acoustic disturbances in the free stream.

The lines of equal intensities of longitudinal component of the flow velocity in the region of the leading edge are shown in Fig. 3 for f = 110 Hz. Approximately the same picture was obtained numerically by Maximov (see [1]) for the transformation of vortex disturbances near the leading edge. The avolution of fluctuation velocity profiles is shown in Fig. 4. The mean velocity profiles were near to Blasius's profile. The fluctuation profiles acquire some feature of TS waves inside instability region beginning only from x = 1000 mm, and intensive growth of their amplitudes taking place after that (see [3]). The same result is obtained for the plate with $r_{l.e.}$ = 2 mm and f = 110 Hz. It should be mentioned as essential feature of obtained profiles: unlike TS profiles they have a pronounced additional maximum near the wall. Such maxima were observed in [4] and were related by the author to a heat transfer to the wall. However, as a result of our special study (variation of the overheat, big growth rates of pulsations along coordinate x) one can say with certainty that the maximum is not conditioned by the heat transfer to the wall. Growth rates for fundamental modes are presented in Figs. 5 and 9. For the plate with $r_{l.e.}$ = 1 mm the packet with central frequency being twice as much as the fundamental one appears and intensively grows near second branch of the neutral curve. The difference between experimental and theoretical values of α_i at Re^* = 2000 is connected probably with non-linear interactions of fundamental frequency with its harmonic. An intensive growth of natural fluctuations of fundamental frequency took place only inside of instability region for both plates. Our results contradict the conclusion of [6] about earlier growth of disturbances in comparison with theory. Obviously the conclusion of [6] is the consequence of extended region with adverse pressure gradient in this experiment (see also [7]).

Results at acoustic excitation

The second part of the work is devoted to the study of the transformation of longitudinal acoustic disturbances into boundary layer vortex waves. Initially, the conditions were determined as the vibration of model remained practically constant. The experiments were performed on the plate with $r_{l.e.} = 2$ mm and the results are given in Fig. 6 for one of frequencies. The decrease in the velocity is seen to lead to the narrowing of a range of the acoustic field intensity in which vibration amplitudes do not change. It makes difficult pure investigation of the transformation of acoustical disturbances into unstable waves of the boundary layer at low stream speeds. Taking into account obtained results, further experiments were carried out at free stream velocity equal to 18 m/s and a_s/a_s^o equal to 0.5 or less (the sound level equal to 96 dB or less). The natural sound level corresponded to 76 dB but the vibration velocity was constant and equal approximately to 0.06 mm/s. The spectrum on the left boundary differed from that of Fig. 2a by presence of the discrete frequency f = 110 Hz (frequency of sound). The same applies to the top boundary. The amplitude of discrete component depended here on the x-coordinate and could be represented as the superposition of two sound waves having different amplitudes and traveling to opposite directions.

A change in the RMS signal from hot-wire anemometer along the x-coordinate inside the boundary layer is shown in Fig. 7. The amplitude modulation of the signal is typical of the superposition of a wave with large phase speed and a hydrodynamic wave propagating with an own phase speed (see, e.g. [1]). In the present experiments the phase speed of the hydrodynamic wave was equal to $0.29 \div 0.3\ U_\infty$ (the linear theory gives the value of 0.307). In spite of linear theory predicting more amplitudes for hydrodynamic wave at Reynolds number less than ones corresponding to first branch of the neutral curve we could not see modulation of signal. A variation of sound amplitude did not change the space picture of the modulation and the relative modulation of the signal. It may be interpreted as the independence of the initial phase of the generated hydrodynamic wave on sound amplitude and the linearity of coupling of sound and hydrodynamic wave amplitudes.

The amplitude and phase profiles are shown in Fig. 8. At this section the hydrodynamic wave is larger than acoustic one and determines the form of the profiles completely. The profiles are the same as in Fig. 4. Scaled profiles are similar at a change of amplitudes too. Unlike the usual TS waves they have also an additional maximum near the wall. Growth rates of the hydrodynamic wave excited by the acoustic one are shown in Fig. 9. The growth rate is higher than its theoretical value up to the middle of the instability region.

The lines of equal RMS values of velocity fluctuations are shown in Fig. 10. It is seen that the intensity of the fluctuations is increased over the plate, $x > 0$, and the fluctuations are decreased before the leading edge, $x < 0$, in comparison with their values on the boundary of measurements region. A comparison with the data for natural disturbances in Fig. 3 shows that the amplification occurs in a region $x = 2\pi f \cdot x/U_{\infty} \cong O(1)$. It can be noted that the direct difference of local values of the fluctuations and of their values at the boundary is close to the values obtained for natural disturbances. However, a simple addition is incorrect of course because it is necessary to know the distribution of phase in this region. Unfortunately, the method used was not applicable in the vicinity of the leading edge.

Finally, it is interesting to compare the present experimental data with theory. A ratio of the amplitude of hydrodynamic wave on the first branch of the neutral curve to the acoustic velocity in the free stream is equal approximately to $5 \cdot 10^{-2}$ in our experiments. The numerical calculation [8] yields the values of the order $O(10^{-4})$. The experiments of [6] yielded values $O(1)$, which in our opinion had been a consequence of the extended region of adverse pressure gradient. A ratio of the maximum measured amplitude of fluctuations in the region of the leading edge to the minimum amplitude of the boundary layer wave in our experiments is about 30 for the acoustic excitation and it is

10 in case of the natural disturbances. The calculation [9] gives approximately the same value for the acoustic excitation.

Conclusion

The transformation of the spectrum of natural disturbances and their amplification takes place in the region of the flat plate leading edge. As a result of the transformation a frequency packet is formed which damps downstream weakly. The disturbances damp up to inner region of the neutral curve. The amplitude profiles receive some characteristics of TS waves inside the neutral curve only and then the amplitude begins to grow. The transformation of outer acoustic field also takes place in the region of the leading edge. In this case, it is possible to separate out the hydrodynamic wave near the first branch of neutral curve. The amplitude of hydrodynamic wave is proportional to that of the sound. The amplification rates are higher than theoretical ones at initial stages of development of the wave at sound excitation. The profiles of fluctuation velocity differ from the ones of TS waves by an additional maximum near the wall.

References

1. Kachanov, Yu.S.; Kozlov, V.V.; Levchenko, V.Ya.: Beginning of Turbulence in a Boundary Layer. Novosibirsk: Nauka 1982 (in Russian).
2. Polyakov, N.Ph.: in Development of Disturbances in Boundary Layer (ed. V.Ya.Levchenko) pp.23-67, Novosibirsk: ITPM SO AN SSSR 1979 (in Russian).
3. Kosorygin, V.S.; Levchenko, V.Ya.; Polyakov, N.Ph.: in Instability of Sub- and Supersonic Flows (ed. V.Ya.Levchenko) pp. 44-52, Novosibirsk: ITPM SO AN SSSR 1982 (in Russian).
4. Gedney, C.J.: MIT Acoustics and Vibration Lab. Rept. N° 83560-3, 1983.
5. Maximov, V.P.: in Development of Disturbances in a Boundary Layer (ed. V.Ya.Levchenko) pp. 68-75, Novosibirsk: ITPM SO AN SSSR 1979 (in Russian).
6. Shapiro, P.J.: MIT Acoustics and Vibration Lab. Rept. N° 83458-83560-1, 1977.
7. Leehey, P.; Shapiro, P.J.: in Laminar-Turbulent Transition (eds. R.Eppler & H.Fasel), pp.321-331, Springer-Verlag 1980.
8. Murdock, J.M.: in Laminar-Turbulent Transition (eds. R.Eppler & H.Fasel), pp.332-340, Springer-Verlag 1980.
9. Goldshtein, M.E.: Journal of Fluid Mech. 127 (1983) 59-81.

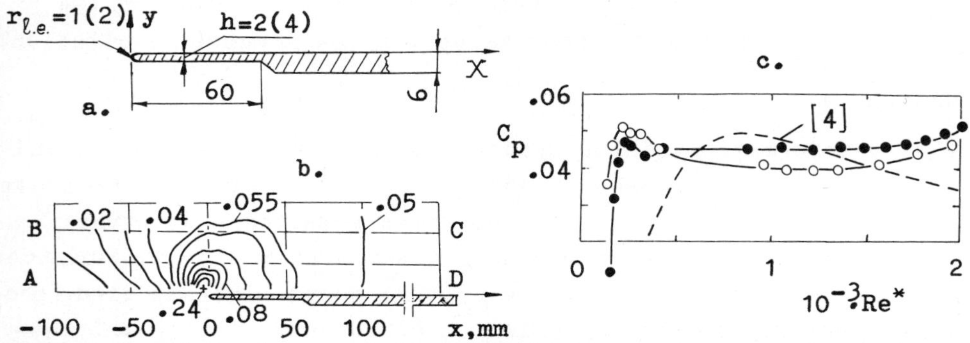

Fig.1. Leading edges and distributions of pressure:a)geometry; b)field of pressure, $r_{\ell.e.}$ = 1 mm; c)distribution of pressure, ○ -$r_{\ell.e.}$ = 1 mm, ● -$r_{\ell.e.}$ = 2 mm.

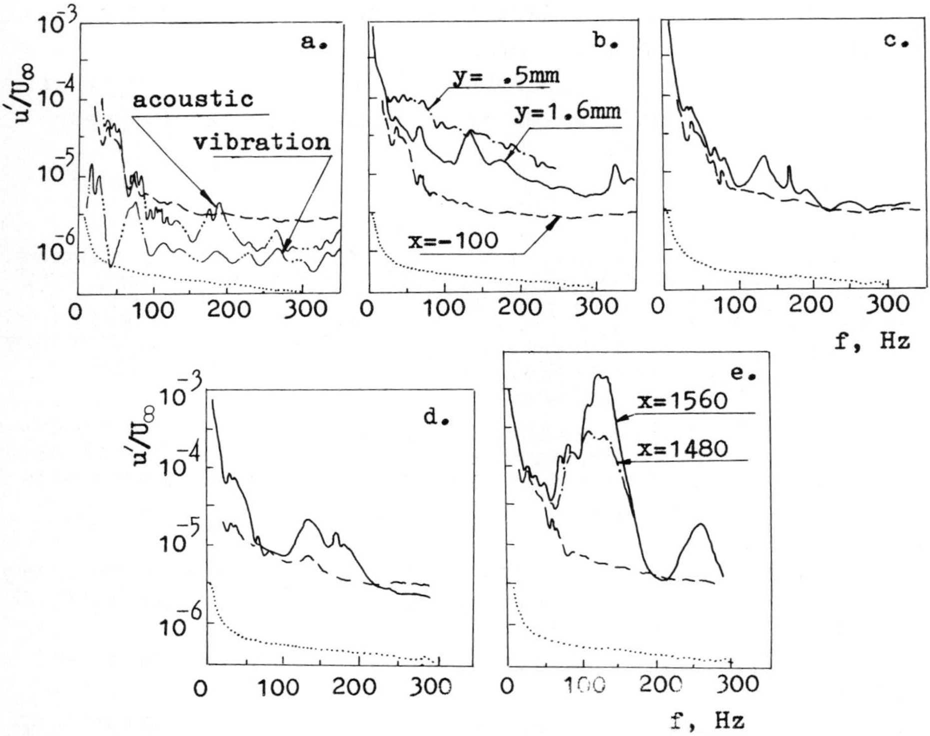

Fig.2. Evolution of spectra of velocity oscillations at U_∞= 18 m/s:a)x=-100 mm;b)x=+4 mm;c)x=650 mm;d)x=1000 mm;e)x=1560mm; .... - apparatus noise at U_∞ = 0; --- - $y_=$ 40 mm; —— - $r_{\ell.e.}$= 1 mm, $\overline{U}$ = U/U_δ = 0.4; —·—·— - $r_{\ell.e.}$ = 2 mm, $\overline{U}$ = 0.4.

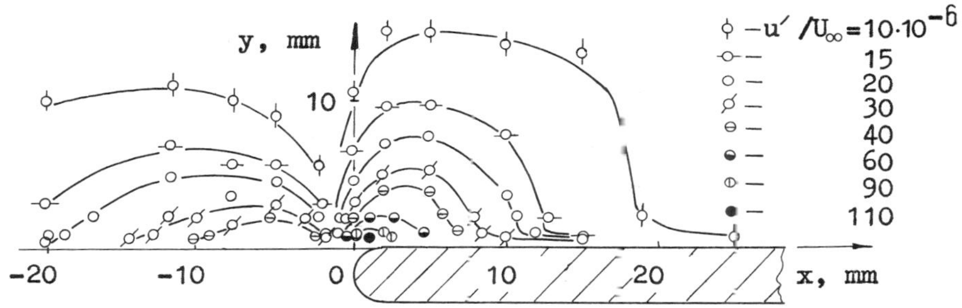

Fig.3. Field of a longitudinal component of velocity oscillations near the leading edge: $r_{\ell.e.}$ = 2 mm; f = 110 Hz; Δ f=4 Hz; $F = 2\pi f\cdot\nu/U_\delta^2 = 33.4\cdot10^{-6}$; natural sound 76 dB.

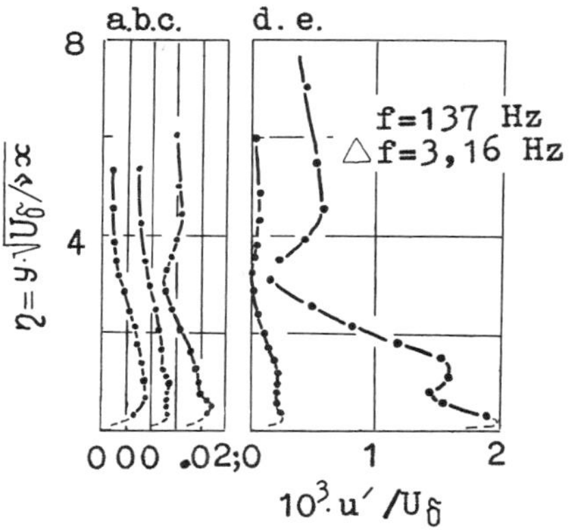

Fig.4. Formation of profiles of velocity oscillation: $r_{\ell.e.}$ = 1 mm, f=137 Hz, Δ f=3.16 Hz; a) x=.35 m, b) .60m, c) 1.0m, d) 1.33m; e) 1.56 m.

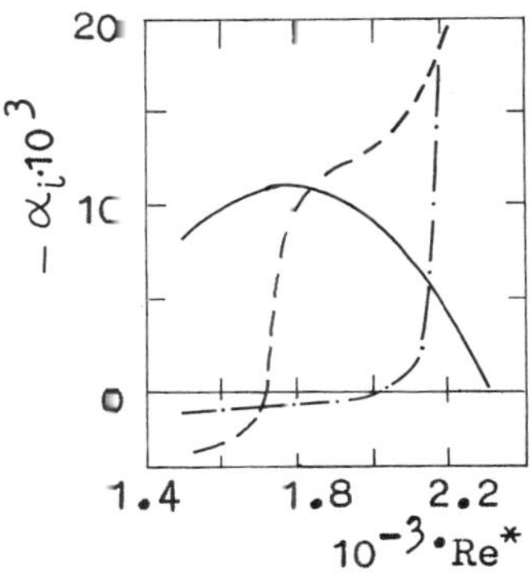

Fig.5. Growth rates of disturbances: $r_{\ell.e.}$ =1 mm; ——— - non-parallel theory; - - - - $F=43.8\cdot10^{-6}$ —·—·— 2F (natural sound 76 dB)

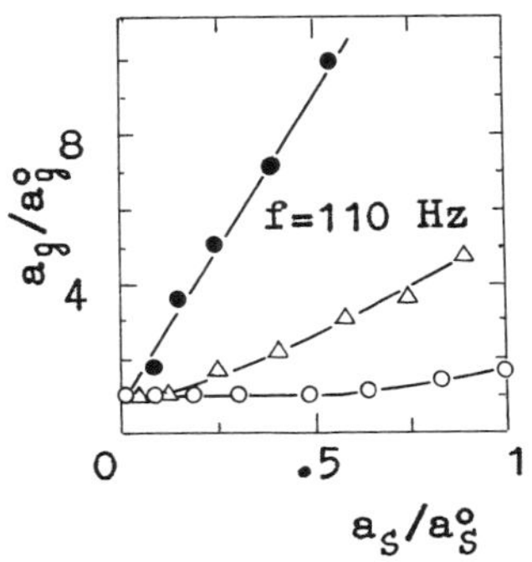

Fig.6. Vibration as a function of sound level: $r_{\ell.e.}$=2 mm, ● -5.7 m/s, △ -10.4 m/s, ○ -18.3 m/s; a_g^o - vibration by natural sound

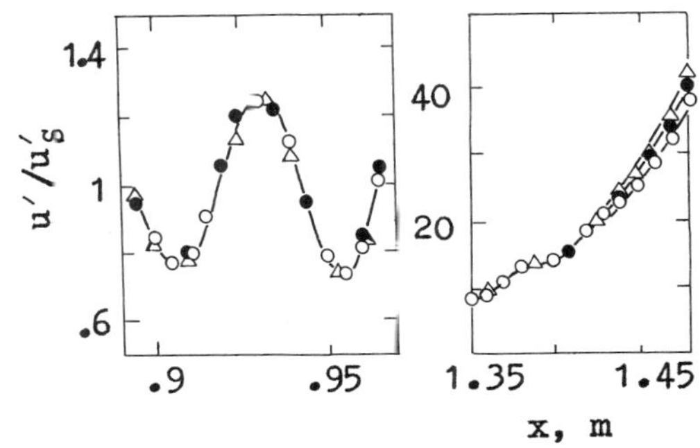

Fig.7. Normalized modulation of hot-wire anemometer signal at acoustic excitation: $r_{\ell.e.}$ = 2 mm, f = 110 Hz, △ - 84 dB, ● - 90 dB, ○ - 96 dB.

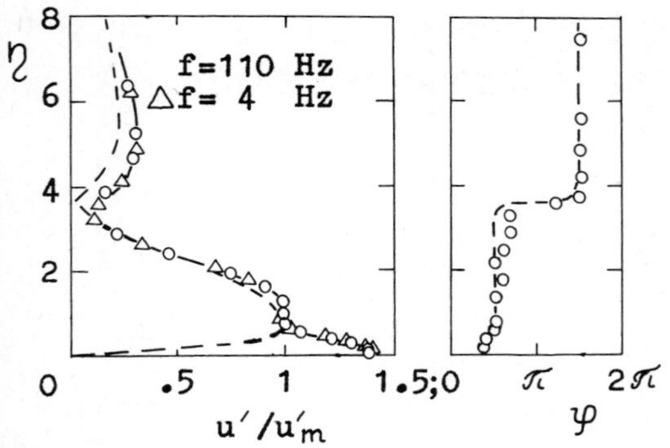

Fig.8. Normalized wave amplitudes and phases as a function of η : $r_{\ell.e.}$ =2 mm, ○ -96 db, △ -102 db, Re*= 2200, F=34.6·10^{-6} ; ----- calculation at F=31.5·10^{-6} , Re* = 2183.

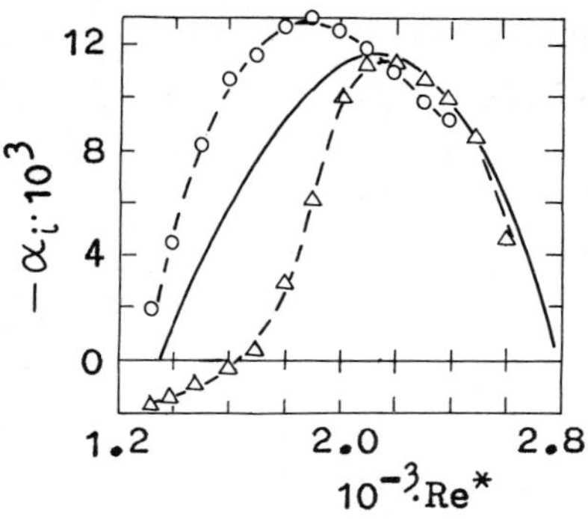

Fig.9. Growth rates of disturbances: $r_{\ell.e.}$=2mm, F=33.4·10^{-6}
——— - nonparallel theory;
--△-- natural sound 76 db;
--○-- sound level 96 db.

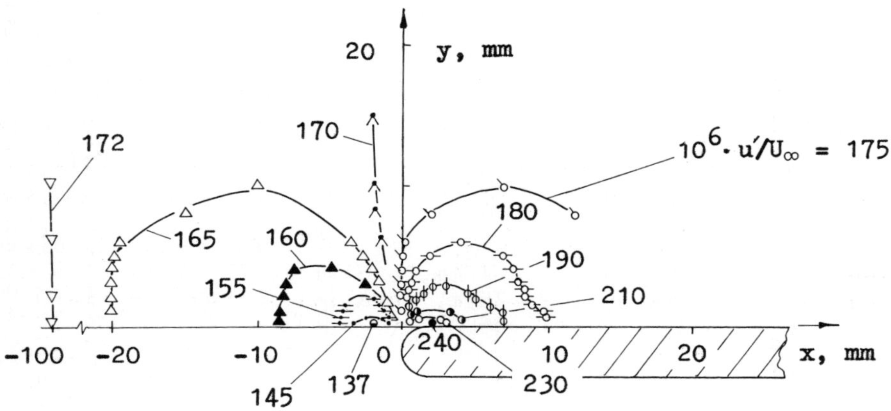

Fig.10. Field of a longitudinal component of velocity oscillations near the leading edge: $r_{\ell.e.}$= 2 mm, f=110 Hz, △f= 4 Hz, F = 33.4·10^{-6} ; sound level 96 db.

Tollmien-Schlichting Waves Excitation on the Vibrator and Laminar-Turbulent Transition Control

V. M. GILEV

Institute of Theoretical and Applied Mechanics,
USSR Academy of Sciences, Novosibirsk, USSR

Summary

The results of experimental investigation of the generation of Tollmien-Schlichting waves in the boundary layer on the vibrator are discussed. It is shown that the vibrations of surface of amplitude of a few microns cause Tollmien-Schlichting waves in the boundary layer downstream the vibrator. It was discovered that the surface vibration localized in space can become one of possible mechanisms of control over a boundary layer transition.

The question on the transition of external disturbances into Tollmien-Schlichting waves in the boundary layer has been intensively studied lately because it is closely related to the general problem of laminar-turbulent transition. Modern flying devices and aerodynamic installations are characterized by vibrating surfaces which can generate instability waves in the boundary layer. In the paper [1] the generation of Tollmien-Schlichting waves in the boundary layer of a flat plate exposed to radiation of an external acoustic field was explained by the transverse vibration of the plate nose. In theoretical papers [2, 3] dealing with Tollmien-Schlichting waves generation on the surface with a localized source of vibrations far from the leading edge of the plate the excitation of boundary layer eigenoscillations was shown to be possible. The first part of this paper deals with an experimental test of a possibility of transforming small localized surface vibrations into Tollmien-Schlichting waves and with obtaining quantitative results describing this relation. In the second part of the paper the task formulated was to show that the surface vibrations localized in space could be used for the laminar-turbulent transition control.

1. The experiment was performed in a low turbulence subsonic wind tunnel of the Institute of Theoretical and Applied Mechanics of the USSR Academy of Sciences (the integral turbulence degree with respect to the spectrum $\varepsilon \sim 0.04\%$) on an acrylic plastic plate with an ellipse-like nose which was placed parallel to the flow [4]. The forward flow velocity equalled u = =7.25 m/sec. The vibrator was built into the plate surface. For this purpose at the distance x=490 mm from its nose a through rectangular orifice 180 mm x 28 mm (fig.1) was chosen. Then it was covered by a 20 microns thin lavsan film. Around the center of its lower surface a metal ribbon 0.05 mm thick was glued. A permanent magnet was fixed below the plate (2). When sinusoid current was conducted from sound generator through the ribbon, the latter vibrated together with the film.

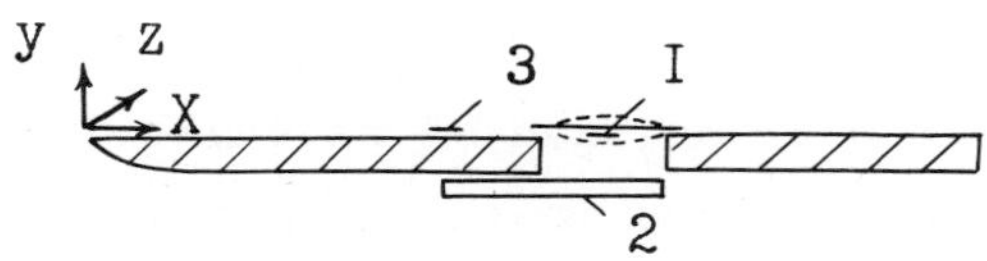

Fig.1 The scheme of the excitation of surface vibration.

The development of disturbances was controlled by means of a thermoanemometer probe. The measurements were carried out in the region of the disturbances increase for the frequency f= =45 Hz ($F=2\pi f \nu / u_\infty^2 = 81 \cdot 10^{-6}$) and f=70 Hz ($F=121 \cdot 10^{-6}$) as well as in the region of disturbances damping for the frequency f= =120 Hz ($F=225 \cdot 10^{-6}$).

In fig.2 there are given the amplitude profiles and the phases of the disturbances of the longitudinal component of the velocity u depending on the transverse coordinate y for different distances along the longitudinal coordinate x. It can be seen that upstream the vibrator the amplitude of disturbances and their phase across the boundary layer change little, which is evidently related to the fact that only pressure pulsations are present. Behind the vibrator while moving downstream the amplitude profiles and the phases of disturbances gradually acquire the form typical of the Tollmien-Schlichting waves, i.e. in the amplitude profile there are observed two maxima and a jump-like phase change by the value equal to π at the point of minimum disturbances. A comparison with the profiles obtained for the usual Tollmien-Schlichting wave which was done at $x - x_0 = 60$ mm shows their total coincidence.

In fig.3 the amplification curves of the amplitude of the disturbances from the longitudinal coordinate x for the frequency 45 Hz are presented. It is seen that at the distance $x-x_0 = 20$ mm

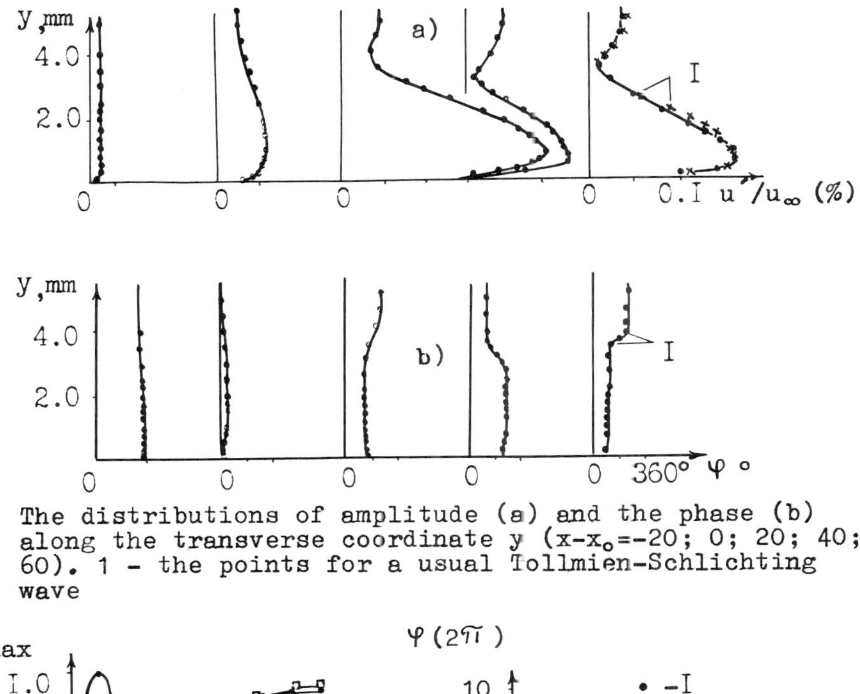

Fig.2 The distributions of amplitude (a) and the phase (b) along the transverse coordinate y ($x-x_0=-20$; 0; 20; 40; 60). 1 - the points for a usual Tollmien-Schlichting wave

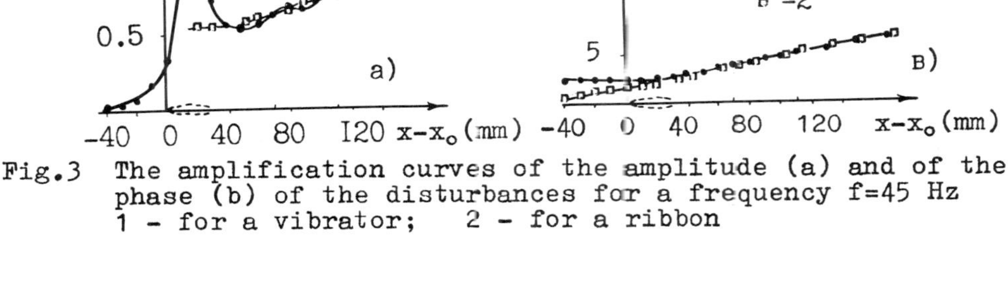

Fig.3 The amplification curves of the amplitude (a) and of the phase (b) of the disturbances for a frequency f=45 Hz 1 - for a vibrator; 2 - for a ribbon

from the beginning of the vibrator a maximum of the pulsations amplitude occurs, then while moving downstream the disturbances are suddenly damped and then increase again. For the same frequencies here are given the amplification curves of the disturbances of the phase depending on the longitudinal coordinate. It is seen that up to the region $x-x_0=20$ mm the wave phase practically does not change which is accounted for pressure pulsations emitted by the vibrator. Then the phase starts growing which indicates the presence of a Tollmien-Schlichting wave.

In fig.4 there is given a comparison of the amplification curves of the amplitude of disturbances frequencies 120 Hz, 70 Hz, 45 Hz generated by the vibrator and the ribbon. It can be seen that at the distance equal to several downstream wave lengths all the given amplification curves practically coincide, the amplitude of disturbances generated by the vibrator is quickly damped upstream. It follows from this graph that for different frequencies at the constant amplitude of vibration the surface vibrations transform into an equivalent Tollmien-Schlichting wave in the same way (under an equivalent wave here it is understood a two-dimensional Tollmien-Schlichting wave which in a region behind the vibrator coincides with the wave generated by the localized vibrator) but further downstream the

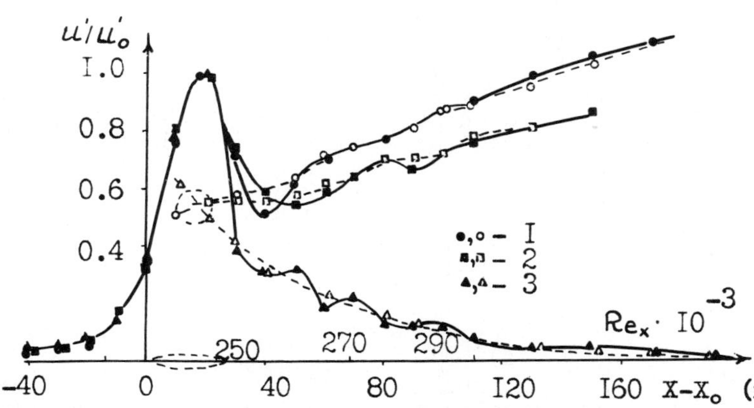

Fig.4 The comparison of the amplification curves of the disturbances for a vibrator and ribbon (dark symbols refer to the vibrator, light ones to the ribbon)
1 - f=70 Hz; 2 - f=45 Hz; 3 - f=120 Hz

disturbances developed differently according to the linear stability theory. The experimental results showed that the non-dimensional amplitude of an equivalent Tollmien-Schlichting wave (u'_{max}/u_∞) in the frequency range under study at the constant amplitude of surface oscillation practically does not depend on the vibration frequency, linearly depends on their amplitude and constitutes approximately (0.07±0.02)% at the amplitude of surface vibration being 10 microns.

2. Below it is shown that the surface vibration can be used to control the laminar-turbulent transition in a boundary layer. The scheme of the experiment remained the same as in the first section with the exception that in the boundary layer two waves could be generated simultaneously: one by the vibrating ribbon method, the other by means of a vibrator. The ribbon and vibrator were connected up to (one and) the same generator of rectangular signals, thereat the phase correlation between the signals transmitted there could discretely vary by means of a special phaseshifting device. The oscillation frequency equalled 70 Hz.

In the process of the experiment the intensities of the disturbances u' were measured versus the longitudinal coordinate x.

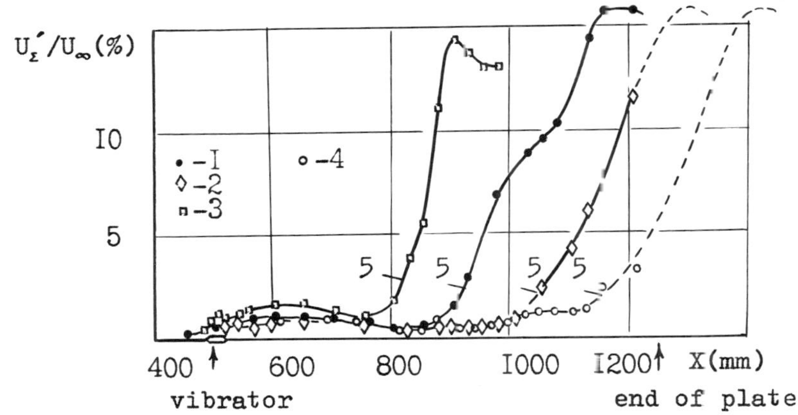

Fig.5 The integral amplification curves of the disturbances: 1 - for a ribbon; 2 - for a vibrator; 3 - wave addition; 4 - wave subtraction; 5 - the point of the transition onset

The amplification curves were built for the case when Tollmien-Schlichting waves were generated by a ribbon or a vibrator separately, as well as for the case of their superposition. In fig.5 integral amplification curves of the amplitude of the disturbances are given. It is well seen from the graph that in the case of wave compositions when both the waves are in phase (3), their additive amplitude increases, thereat the transition point (5) shifts upstream. In the case of wave subtraction when both the waves are in an antiphase (4), the additive amplitude decreases and the transition point shifts downstream respectively. In the experiment in question this displacement equalled 250 - 300 mm.

In conclusion I wish to thank Dr. Kozlov V.V. for the formulation of the problem, the help in conducting the experiment and useful discussions.

References

1. Kachanov, Yu.S.; Kozlov, V.V.; Levchenko, V.Ya.: Beginning of turbulence in a boundary layer. Novosibirsk: Nauka 1982 (in Russian).

2. Terentyev, E.D.: Linear problem on the vibrator in a subsonic boundary layer. Prikl. Mat. i Mekh. 45(1981) 1050-1055.

3. Tumin, A.M.; Feodorov, A.V.: Excitation of instability waves in a boundary layer of a vibrating surface. Zhurn. Prikl. Mekh. i Tekhn. Fiz. 3 (1983) 72-79

4. Gilev, V.M.; Kozlov, V.V.: Excitation of Tollmien-Schlichting waves in a boundary layer of the vibrator. Novosibirsk, Preprint ITPM SO AN SSSR 19(1983).

Excitation of Tollmien-Schlichting Waves by Acoustic Disturbances in a Compressible Boundary Layer

A. V. FEDOROV

Moscow Institute of Physics
and Technology, USSR

Summary

This work represents a theoretical study of the excitation of TS waves by acoustic waves on inhomogeneity resulting from the nonparallelism of the boundary-layer flow. The analysis is carried on for a compressible boundary layer. The mathematical model is based on perturbation theory method.

1. Consider the flow in a compressible boundary layer on semi-infinite flat-plate. Choose the coordinate-system with x-axis along the flow and y-axis normal to the surface of the plate. Derive characteristic scales: $L \sim 1$ in x-axis and $\delta = (\nu_\infty L/U_\infty)^{1/2}$ in y-axis; with ν_∞, U_∞ being the kinematic viscosity and the velocity of the oncoming flow, respectivly. Time t, velocity, pressure, temperature and density are nondimensionalized by δ/U_∞, U_∞, $\rho_\infty U_\infty^2$, T_∞, ρ_∞ respectively. For two-dimensional disturbances we shall introduce a vector-function

$$\vec{\Psi}(x,y,t) = \left(u, \frac{\partial u}{\partial y}, v, p, \theta, \frac{\partial \theta}{\partial y}, \frac{\partial u}{\partial x}, \frac{\partial v}{\partial x}, \frac{\partial \theta}{\partial x}\right)$$

where u, v, p, θ are the perturbations of "x" and "y" components of the velocity, pressure and temperature, respectively. Consider a fixed frequency disturbance

$$\vec{\Psi}(x,y,t) = \mathrm{Re}(\vec{A}(x,y)\, e^{-i\omega t}) \tag{1.1}$$

Upon substituting (1.1) into linearized Navier-Stokes equations for an amplitude function $\vec{A}(x,y)$ we arrive at the following system of equations

$$\frac{\partial}{\partial y}(L_o\frac{\partial \vec{A}}{\partial y}) + L_1\frac{\partial \vec{A}}{\partial y} = H_1\vec{A} + \varepsilon H_2\frac{\partial \vec{A}}{\partial x} + \varepsilon H_3\vec{A} \tag{1.2}$$

where $\varepsilon = (\nu_\infty / U_\infty L)^{-1} \ll 1$. Matrices L_o, L_1, H_1, H_2 have the dimensionality 9 x 9. Their elements are determined through the characteristics of the mean flow in a locally uniform approximation in x and depend on x as a parameter. Matrix H_3 is stipulated by the nonparallelism of the flow. For a parallel boundary layer ($H_3 = 0$) (1.2) is reduced to the system of Lees-Lin equations. The following conditions are satisfied

$$\vec{A}_1(x,0) = \vec{A}_3(x,0) = \vec{A}_5(x,0) = 0; \ |\vec{A}|<\infty, \ y \to \infty \tag{1.3}$$

$$\vec{A}_o(x_o,y) = \vec{A}_o(y) \tag{1.4}$$

As the equations (1.2) are elliptical ones, complex problem (1.2) - (1.4) is incorrect. For its regularization it is necessary that $\vec{A}_o(y)$ should not allow the solutions with an infinite increment of the downstream growth [1] . Introduce a biorthogonal system of eigen-functions of a locally-uniform approximation $\{\vec{A}_\alpha, \vec{B}_\alpha\}$ [1] :

$$\frac{\partial}{\partial y}(L_o\frac{\partial \vec{A}_\alpha}{\partial y}) + L_1\frac{\partial \vec{A}_\alpha}{\partial y} = H_1\vec{A}_\alpha + i\alpha H_2\vec{A}_\alpha$$

$$\vec{A}_{\alpha 1} = \vec{A}_{\alpha 3} = \vec{A}_{\alpha 5} = 0, \ y = 0; \quad |\vec{A}_\alpha| < \infty \quad , \ y \to \infty$$

$$\frac{\partial}{\partial y}(L_o^*\frac{\partial \vec{B}_\alpha}{\partial y}) - L_1^*\frac{\partial \vec{B}_\alpha}{\partial y} = H_1^*\vec{B}_\alpha - i\bar{\alpha} H_2^*\vec{B}_\alpha$$

$$\vec{B}_{\alpha 2} = \vec{B}_{\alpha 4} = \vec{B}_{\alpha 6} = 0, \ y = 0; \quad |\vec{B}_\alpha| < \infty \quad , \ y \to \infty$$

where "*" denotes a transposed matrix with the complex-conjugate elements; "-" denotes a complex conjugate. Eigenfunctions $\vec{A}_\alpha$ (x,y) describe the wave disturbances of the type $\vec{A}_\alpha$ (x,y) $\exp(\varphi_\alpha(x) - i\omega t)$ where $\varphi_\alpha(x) = i\varepsilon^{-1}\int_{x_o}^{x}\alpha\, dx$, α is the eigenvalue nondimensionalized by $(U_\infty/\nu_\infty 1)^{1/2}$. The orthogonal relationships are satisfied

$$\langle H_2\vec{A}_\alpha , \vec{B}_\beta\rangle = \Delta_{\alpha\beta}, \quad \langle \vec{A}, \vec{B}\rangle = \int_0^\infty \sum_j A_j \bar{B}_j dy, \qquad (1.5)$$

where $\Delta_{\alpha\beta}$ is Kronecker's simbol, if one of the eigenvalues α, β belongs to the discrete spectrum; $\Delta_{\alpha\beta} = \delta(\alpha - \beta)$ is a delta-function if α, β belong to the continuous spectrum. The orthogonality of eigenfunctions points out that there is no interaction among modes in a parallel boundary layer.

2. Consider a two-dimensional acoustic wave of the frequency ω which is incident to the plate at an angle ψ. It is scattered at the leading edge of the plate and the disturbance with the amplitude function $\vec{A}_0(y)$ is formed at the cross-section x_0 which is rather far away downstream from the leading edge. For $x > x_0$ we shall expand the disturbance amplitude in eigenfunctions

$$\vec{A}(x,y) = \sum_\alpha{}' C_\alpha(x) \vec{A}_\alpha(x,y) \exp(\varphi_\alpha(x))$$

where $\sum'$ denotes the summation in the discrete spectrum and the integration in the continuous spectrum. Using the orthogonality condition (1.5), we obtain a system of equations for $C_\alpha(x)$

$$\frac{dC_\alpha}{dx} = C_\alpha W_{\alpha,\alpha} + \sum_{\beta\neq\alpha}{}' C_\beta W_{\beta,\alpha} \exp(\varphi_\beta - \varphi_\alpha) \qquad (2.1)$$

$$C_\alpha(x_0) = \langle H_2\vec{A}_0, \vec{B}_\alpha\rangle_0$$

$$W_{\beta\alpha} = - \langle H_3\vec{A}_\beta , \vec{B}_\alpha\rangle - \langle H_2\frac{\partial \vec{A}_\beta}{\partial x}, \vec{B}_\alpha\rangle$$

where subscript "0" marks the values calculated at $x = x_0$. Integrating (2.1) for the amplitude of an excited TS wave $\vec{Q}_{TS}(x,y)$ we have

$$\vec{Q}_{TS} = C_{TS}(x_0)\vec{A}_{TS}e^{\varphi_{TS}} + \left[\int_{x_0}^{x} \sum_{\beta\neq\alpha}{}' C_\beta W_{\beta,TS}e^{\varphi_\beta - \tilde{\varphi}_{TS}}dx\right] \times$$

$$\times \vec{A}_{TS} e^{\tilde{\varphi}_{TS}} \; ; \quad \tilde{\varphi}_{TS} = \int_{x_0}^{x} (i\,\varepsilon^{-1}\alpha_{TS} + W_{TS,TS})dx \qquad (2.2)$$

where subscript "TS" refers to TS wave. The first term in (2.2) **describes the development of the TS wave excited in the vicinity of the leading edge in the region** $0 < x < x_0$. The second term describes the excitation of the TS wave on an inhomogeneity due to nonparallelism of the boundary layer flow at $x > x_0$. From the whole spectrum we shall single out only one sound wave with the wave number $\alpha = \alpha_A$. Neglecting the self action of a sound wave and the fact that it is being affected by other modes, we shall consider $C_A(x) = C_A(x_0)$. The subscript "A" is attributed to the acoustic mode. The amplitude of instability wave excited at $x > x_0$ is as follows:

$$\vec{F}_{TS}(x,y) = \left[\int_{x_0}^{x} C_A(x_0)\, W_{A,TS}(x)\, e^{\varphi_A - \tilde{\varphi}_{TS}} dx \right] \vec{A}_{TS} e^{\tilde{\varphi}_{TS}} \qquad (2.3)$$

Shown in fig 1 is the dependence in y-axis of modulus of maximum value of x-component of mass-flow rate q_m, based on vector $\vec{F}_{TS}$, on Reynolds number $R = (U_\infty x\, L/\nu_\infty)^{1/2}$ (curve 1).

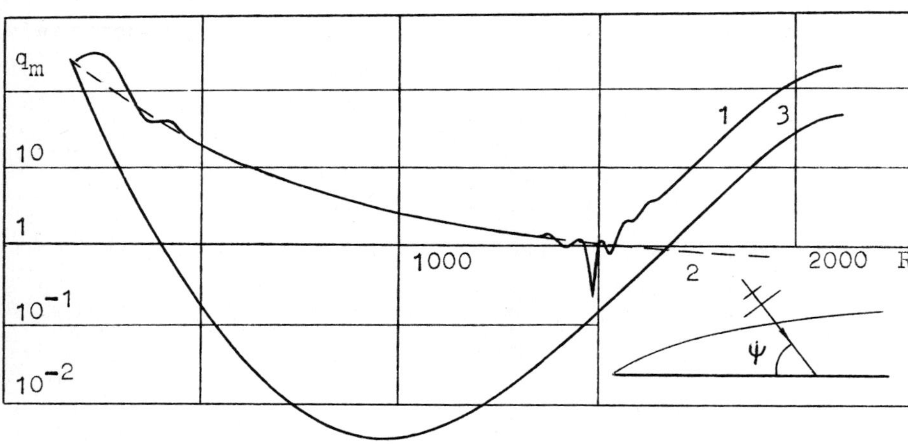

Fig. 1

Calculations have been made for Mach number 0,6; Prandtl number 0,72; total temperature 310°K. Viscosity was calculated via Sutherland's formula. Dimensionless frequency parameter $F = 2\pi f \nu_\infty / U_\infty^2 = 20 \cdot 10^{-6}$. The angle of incidence is $\psi = 86°$, the disturbance of x-component of incident wave velocity being $u_A = 1$, $C_A(x) = 1$. As seen from the graph, the excited wave slowly attenuates in the subcritical region, passes through the point of stability loss without changing its behavior and begins to grow like a free TS wave in the central part of instability region. Integrating (2.3) by parts we have

$$\vec{F}_{TS} = \varepsilon \frac{W_{A,TS}(x)}{i(\alpha_A - \tilde{\alpha}_{TS})} e^{\varphi_A} \vec{A}_{TS} - \varepsilon \frac{W_{A,TS}(x_o)}{i(\alpha_A - \tilde{\alpha}_{TS})_o} e^{\tilde{\varphi}_{TS}} \vec{A}_{TS} + \tag{2.4}$$

$$+\left\{\varepsilon \int_{x_o}^{x} \frac{d}{dx}\left[\frac{W_{A,TS}(x)}{i(\alpha_A - \tilde{\alpha}_{TS})}\right] e^{\varphi_A - \tilde{\varphi}_{TS}} dx\right\} e^{\tilde{\varphi}_{TS}} \vec{A}_{TS}$$

where $\tilde{\alpha}_{TS} = \alpha_{TS} - i\varepsilon W_{TS,TS}$ is an eigenvalue of TS wave with the nonparallelism of the boundary layer taken into acount. The first term in (2.4) describes an "intermediate" wave (curve 2) which develops downstream as acoustic wave and is distributed in y-axis as a TS wave. The intermediate wave is predominant at $R \simeq 400 \div 1400$. The second term in (2.4) describes a free TS wave (curve 3) with initial amplitude depending on the choice of the cross-section $x = x_o$. This wave quickly attenuates in the subcritical region and markedly affects integral curve 1 in a narrow zone $R \simeq 160 \div 450$ causing pulsations. Calculations showed that for initial section $R_o \lesssim 350$ the behaviour of curve 1 did not depend on the choice of R_o. The amplitude of a continuously excited TS wave is included in the 3[d] term of eq. (2.4). Its magnitude is small because of the quick oscillations of the integrand. However, growing up in instability region, it becomes predominant at $R \gtrsim 1500$. In the vicinity of point $R \simeq 1500$ the amplitude of a continuously excited TS wave is comparable with that of an intermediate wave and their superposition results in the pulsations of curve 1. The distributed excitation of the TS wave takes place in the final region of surface $R \simeq 350 \div 1500$ which has no vicinity of the leading edge. As the intermodes exchange scale $\ell \sim \varepsilon L/|\alpha_A - \alpha_{TS}|$ is much less than the inhomogeneity scale L, the mechanism of the distributed generation is weak. The distributed generation is capable of competing with the excitation near the leading edge, if only the latter is separated from the point of stability loss by a region of strong attenuation of TS waves.

In [2] Tam examined the distributed generation of a TS wave by the sound wave. In Tam's model an incident acoustic wave suddenly enters the Blasius boundary layer in the section $x = x_0$ which is rather far away from the leading edge of the plate. It is not difficult to show that Tam's solution does not describe the distributed generation, it rather describes the excitation of the TS wave at the local inhomogeneity due to a sudden "switching on" of the boundary layer at the initial section $x = x_0$. This model is suitable under the condition that the intermodes exchange scale is much less than the one at which the boundary layer allows a local self-similar solution. This condition is not satisfied in Tam's calculations.

References

1. Tumin, A. M.; Fedorov, A. V.:- J. Pricl. Mech. and Tech. Phys., 1983, N 4 (in Russian).

2. Tam, C. K. W. - J. Fluid Mech., 1981, V. 109.

Acceleration of Transition to Turbulence in a Boundary Layer due to the Combination Effect of Sound Waves

A.B. EZERSKII

Institute of Applied Physics, Academy of Sciences of the USSR
Gorky, USSR

Abstract

The amplitude of Tollmien-Schlichting waves excited under the nonlinear resonance action of two ultrasonic beams in a boundary layer is calculated. It was ascertained that when the ultrasonic intensity is of the order of 140 dB its combination effect can result in the accelerated transition from a laminar flow to a turbulent one.

Introduction

Nowadays much attention is paid to the problem of the acoustic fields action on the transition to turbulence in shear flows, in particular, in a boundary layer [1,2]. As it was found out, sound waves can transform into the flow modes due to the flow inhomogeneities, viscous shearing stress and vibrations of a model. In these cases the sound action is most efficient when its frequency coincides with that of unsteady hydrodynamic perturbations. For a subsonic flow, the propagation velocity of the hydrodynamic perturbation is less than that of the acoustic waves. Consequently, when frequencies coincide, the wavelengths differ considerably and the interaction occurs locally. Accumulated in space and time interactions can be organized by using nonlinear effects, A combination perturbation arising from a nonlinear interference of HF sound waves with $(\omega_1, \vec{k}_1)$, $(\omega_2, \vec{k}_2)$ is in resonance with hydrodynamic waves with frequency ω_T and wavenumber $\vec{k}_T$

$$\omega_T = \omega_2 - \omega_1 \qquad |\vec{k}_T|^2 = \vec{k}_T(\vec{k}_2 - \vec{k}_1) \tag{1}$$

The resonance can take place at any flow velocity because the sound waves can be directed at arbitrary angles to the flow. Precisely such a sound action on Tollmien-Schlichting (T-Sch) waves in a boundary layer is investigated in the present paper.

The Presentation of the Acoustic Waves Field

When investigating nonlinear interactions we shall assume that a flow velocity scale U is much less than the sound velocity c : $U_0/c = M \ll 1$. Owing to the fact that the frequencies and propagation velocities of sound waves and T-Sch waves differ greatly, some simplification can be made. Thus, the influence of the viscosity on the field structure of the acoustic wave can be neglected. The viscosity leading to the sound damping should be taken into account only for small distances h from rigid walls $h<(\nu/\omega_{1,2})^{1/2}$ (ν is the kinematic viscosity), but in this region hydrodynamic fields are small and the interaction is absent. If a characteristic scale of the main flow L is such that $U_0/L \ll \omega_{1,2}$ (the flow vorticity is much less than the acoustic wave frequency), the sound field can be presented through quasipotential [3] :

$$\vec{u} = \nabla\varphi - \int^{t}[\operatorname{rot}\vec{U}\,\nabla\varphi_0]\,dt$$
$$\varphi''_{tt} = c^2\Delta\varphi + c^2\int^{t}(\Delta\varphi_0\,\Delta\vec{U})\,dt \tag{2}$$

Here φ_0 is the potential of sound waves in the absence of the shear flow $\vec{U}$. For two harmonic waves falling on the acoustically rigid plate at the angles θ_1 and θ_2 (Fig.1)

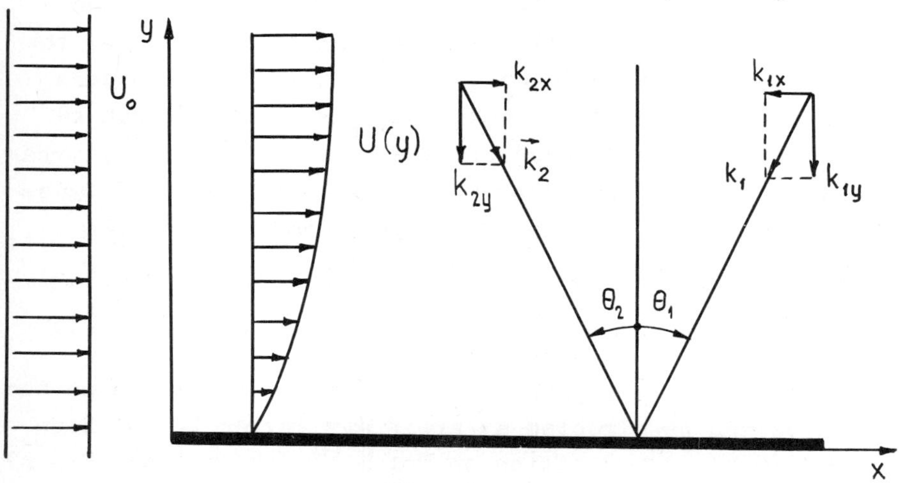

Fig. 1

the potential φ_0 can be presented in the form:

$$\varphi_0=\frac{c^2}{\omega_1}A_1\cos k_{1y}y\, e^{i(\omega_1 t+k_{1x}x)}+\frac{c^2}{\omega_2}\cos k_{2y}y\, e^{i(\omega_2 t-k_{2x}x)}+k.c. \quad (3)$$

As it is seen from (2) the wave is not purely potential due to the presence of the shear flow. It will be found below that a vortex addition is an important factor at a nonlinear action on hydrodynamic waves.

<u>Calculation of the Efficiency of Nonlinear Sound Effect on T-Sch Waves</u>

For hydrodynamic perturbation at $M \ll 1$ the fluid compressibility is not essential, but the viscosity should be taken into account when calculating the transverse structure of eigenmodes and when determining the increments.

Taking into account the simplification made we shall consider the wave excitation by external sources in a boundary layer which are due to a nonlinear interference of acoustic waves.

For the stream function of T-Sch wave ψ we have Orr-Sommerfeld equation:

$$\left(\frac{\partial}{\partial t}+U\frac{\partial}{\partial x}\right)\left(\frac{\partial^2}{\partial x^2}+\frac{\partial^2}{\partial y^2}\right)\psi-\frac{\partial^2 U}{\partial y^2}\psi-\nu\left(\frac{\partial^2}{\partial x^2}+\frac{\partial^2}{\partial y^2}\right)^2\psi=$$

$$=u\frac{\partial}{\partial x}\left(\frac{\partial v}{\partial x}-\frac{\partial u}{\partial y}\right)+v\frac{\partial}{\partial y}\left(\frac{\partial v}{\partial x}-\frac{\partial u}{\partial y}\right)+\frac{\partial U}{\partial y}\left(u\frac{\partial \rho}{\partial x}+v\frac{\partial \rho}{\partial y}\right)\frac{1}{\rho_0}+ \quad (4)$$

$$+\frac{\partial \rho}{\partial t}\left(\frac{\partial u}{\partial y}-\frac{\partial v}{\partial x}\right)\frac{1}{\rho_0}\equiv F$$

Here u, v are velocity components $\vec{u}=(u,v)$ and ρ is the fluctuation of the density in a sound field.

If we neglect the influence of the shear flow on the sound wave propagation, the right-hand side of (4) will be equal to zero. Taking into account the expression (2) for a quasipotential, in the first order of the perturbation theory in U_0/c and $U_0/L\omega_{1,2}$, we have for F :

$$\frac{\partial^2 U}{\partial y^2}\left(u\int^t\frac{\partial v}{\partial x}dt+v\int^t\frac{\partial v}{\partial y}dt\right)$$

Thus, the action of the acoustic waves is concentrated in the region of high gradients of vorticity of the main flow.

First we consider a flow of the inviscid fluid with a piece-wise linear profile of the velocity of the form:

$$U(y) = y/d\, U_0 \qquad \text{at} \qquad 0 < y < d$$

$$U(y) = U_0 \qquad \text{at} \qquad y > d$$

In this case a combination sound action is concentrated on the interface of fluids of different vorticity. Hydrodynamic waves in such a flow have the frequency:

$$\omega_T = U_0 k_T - U_0/d\, \mathrm{sh}\, k_T \exp(-k_T d)$$

Using the standard method of description of nonlinear interactions, for a slowly varying velocity amplitude of hydrodynamic waves when a HF acoustic wave propagates along the plate $\theta_2 = \pi/2$, we obtain:

$$\frac{\partial A}{\partial \tau} + \frac{\partial \omega_T}{\partial k_T}\frac{\partial A}{\partial x} = -i\sigma_T A_1^* A_2, \qquad \sigma_T = \frac{k_{1y} k_{1x}}{M k_2^2} \sin k_{1y} d \quad (6)$$

Here the time τ is scaled by d/U_0 and the coordinate x, to the thickness of the boundary layer d.

Now we give the estimation for perturbation induced in an air flow by ultrasonic waves $f_{1,2} = \omega_{1,2}/2\pi = 20\,\mathrm{kHz}$ with intensity $|A_1| = |A_2| = 3\cdot 10^{-3}$ or 143 dB with respect to the level $2\cdot 10^{-5}$ Pa. When the waves are excited most efficiently in a flow with the velocity 3 m/s (M = 0.01) and with a boundary layer thickness d = 3 cm, the amplitude of hydrodynamic perturbations increases up to 0.3% of the velocity of the main flow (0.9 cm/s) at the length of 15 cm. Such an amplitude of hydrodynamic waves is by an order larger than intrinsic noises of the flows which in low-turbulent wind tunnel make up 0.03% of the flow velocity. It is important that the flow pulsation transforms into hydrodynamic waves with a conversion coefficient $q \ll 1$, therefore, the amplitudes of uncontrolled perturbations will be actually still less. Estimations given show that the amplitude of hydrodynamic waves induced by the sound can exceed intrinsic noise of the flows. In order to determine the possibility of the accelerated transition to turbulence caused by the combination action of the sound waves, the calculation for the Blasius profile was carried out. In this case for a slowly varying amplitude of T-Sch wave we have Eq.(6), but the expression for σ_T will be not so trivial as for a piecewise linear profile. The dependence of σ_T on the angle of incidence θ_2 of the sound wave ω_2, k_2 at a fixed angle of incidence θ_1 is shown in Fig. 2. Curve 1 corresponds to a neutrally stable wave $k_T\delta^* = 0.3$, $Re = 520$, curve II, to an unstable one $k_T\delta^* = 0.3$, $Re = 3\cdot 10^3$.

As is known, nonlinear effects leading to the intense explosive growth of T-Sch waves become significant at the amplitude of the order of 1% of the free flow velocity, and at the amplitude of 2-4% the transition to turbulence takes place. For estimation we shall assume that the boundary layer is turbulized if the amplitude of T-Sch wave achieves 2.5% of the velocity of the flow. When the air flow velocity is $U_0 = 7$ m/s the boundary layer displacement thickness δ^* at the distance of $x = 50$ cm from the edge of the plate equals $\delta^* = 1.72\left(\frac{\nu x}{U_0}\right)^{1/2} = 0.18$ cm, and Reynolds number equals $Re = 520$. If the amplitude of the acoustic waves falling on the plate at the angles of $\theta_2 = -32°$ (the maximum value σ_T see in Fig.2) and $\theta_1 = 45°$ equals $A_1 = A_2 = 3\cdot 10^{-3}$ and their frequencies are $f_1 = 51000$ Hz and $f_2 = 51078$ Hz, then due to the combination excitation of the amplitude of T-Sch wave with a difference frequency increases up to 2.5 % from U_0 at the length of 20 cm, or the length of the transition makes up 70 cm.

If the flow velocity $U_0 = 34$ m/s, then at the distance of 150 cm from the plate edge Reynolds number $Re = \frac{U_0\delta^*}{\nu} = 3000$ is attained (a displacement thickness $\delta^* = 0.14$ cm). Sound beams

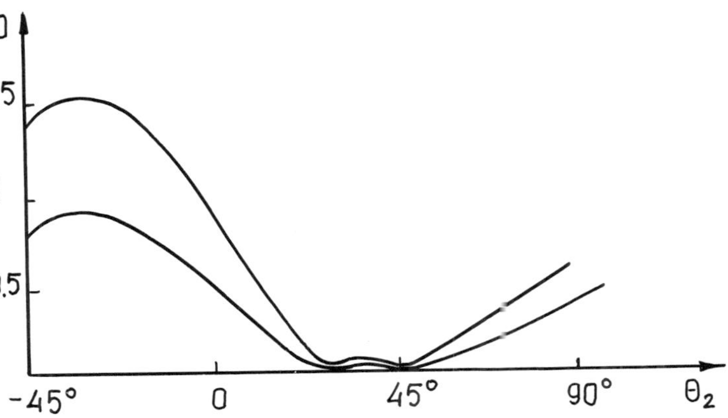

Fig. 2

with the intensity of 143 dB, with frequencies f_1 = 4370 Hz and $f_2 = f_1 + 460$ Hz excite the unsteadiest T-Sch wave with the frequency 460 Hz and the amplitude 0.75 m/s (2.5% of U_0), if the region of the beam intersection is 30 cm. The transition length will be 180 cm.

If the wave excitation is not artificial, and a mean-square velocity pulsation of the flow running against the plate is of the order of 0.05%, the transition to turbulence occurs at Reynolds numbers $Re = \frac{Ux}{\nu} = 4.5 \cdot 10^6$ [1] where x is the distance from the edge of the plate. In the first case a natural transition at such noises would occur when X = 900 cm and in the second case, when X = 200 cm. Thus, with the aid of the combination action the transition to turbulence can be accelerated, the greatest effect is achieved when unsteady waves are excited at small supercriticalities.

Conclusion

The combination excitation of hydrodynamic waves by ultrasound seem to be a convenient way of introducing controlled perturbations into the flows in order to investigate acoustic-hydrodynamic interactions. Owing to the high frequency of the sound, its linear transformation gives strongly attenuated hydrodynamic perturbations and the nonlinear action becomes determinative. The combination action can be localized in the flow region assigned beforehand, due to the use of short-wave and, consequently, narrow wave beams. The application of the ultrasound permits to get rid of the perturbations arising from the vibration of plates and throats of the wind tunnels. Such a "vibration" phone can not be eliminated when the flow is subjected to the action of sound waves with the frequency equal to the hydrodynamic waves frequency.

References

1. Kachanov, Yu.S.; Kozlov, V.V.; Levchenko, V.Ya. The formation of turbulence in a boundary layer. - Novosibirsk: Nauka, 1982, p.151.
2. Ginevskii, A.S.; Vlasov, E.V.; Kolesnikov, A.V. Aeroacoustic interactions. - M.: Mashinostroenie, 1978, p.176.
3. Blokhintsev, D.I. The acoustics of an inhomogeneous moving medium. M.: Nauka, 1981, p.206.

Interaction of Vortex Disturbances and Boundary Layer on Blunt Bodies

N. V. SIDORENKO, E. A. EROFEEV

The Moscow Institute of Physics and Technology, USSR

Summary

Penetration of incompressible fluid vortex disturbances into the boundary layer on the body with stagnation point is investigated by the matched asymptotic expansion method. Tollmien-Schlichting waves generation by the vortex disturbances in a boundary layer has been considered.

Penetration of the Vortex Waves into the Boundary Layer on a Parabolic Cylinder

A parabolic cylinder in the flow of incompressible fluid at zero incidence is taken as a blunt body. In parabolic coordinates

$$-x+iy=((\xi+p)+i\eta)^2-p^2 \tag{1}$$

the value $\xi=0$ corresponds to the surface of the body, and $2p^2$ is the camber radius at the apex of the generatrix parabola. We render all parameters dimensionless so that $U_\infty=x_0=\eta_0=1$. Here U_∞ is the velocity of the basic flow at $x\to-\infty$; $x_0=\eta_0$ is the stability point of the boundary layer flow with Re

$$R_{tr}=\sqrt{U_\infty\eta_0^2/\nu}\simeq U\Delta/\nu\Big|_{\eta=\eta_0}\sim 10^3 \tag{2}$$

where U is the basic flow velocity at the boundary layer edge, Δ is its thickness. Let's introduce a small parameter

$$\varepsilon=R_{tr}^{-1/2} \tag{3}$$

The dimensionless coefficient of the body bluntness is $\varkappa=p^2/\eta_0^2$.

We consider it to be $\eta\sim 1$. The incoming disturbances with frequency ω and wave-vector k corresponding to their characteristic values in the boundary layer are considered: $\omega\sim U_\infty/\Delta$; $k\sim\Delta^{-1}$ at $\eta=\eta_0$. Hence, we have the following estimates:

$$\omega=\omega_0\varepsilon^{-2};\ k=k_0\varepsilon^{-2}. \tag{4}$$

The flow region is divided into four specific zones along ξ coordinate: I - the region of the external inviscid flow; II - - the region in which the viscous effects become essential for the disturbance; III - the boundary layer; and IV - the viscous sublayer. We do not consider the latter separately for it is included into the boundary layer.

The thickness of zone II (whose estimate is given below) is of the order of $\delta_{II}\sim\varepsilon$, the boundary layer thickness $\delta_{III}\sim\varepsilon^2$, the viscous sublayer thickness $\delta_{IV}\sim\varepsilon^3$. At orders ψ, ω, k other than (4) the subdivision may be different.

Zone I.

The stream function of undisturbed flow $\bar{\psi}_0$ is expanded in a series in ε^2:

$$\bar{\psi}_0=\psi_0+\varepsilon^2\psi_{01}+\dots \tag{5}$$

Here $\psi_0=2\xi\eta$, and ψ_{01} is presented in the form

$$\psi_{01}=-2\eta\delta(\eta)+B_1(\eta)\xi+O(\xi^2) \tag{6}$$

where $\delta(\eta)$ is the displacement thickness. Now let's consider the two-dimensional disturbances with the stream-function : $\psi(x,y)e^{-i\omega t}$. Linearized hydrodynamic equations for the vorticity $\Omega=-\Delta\psi$ have the form (in parabolic coordinates):

$$-4i\omega_0\varepsilon^{-2}((p+\xi)^2+\eta^2)\Omega+\bar{\psi}_{0\xi}\Omega_\eta-\bar{\psi}_{0\eta}\Omega_\xi=\varepsilon^4\left(\frac{\partial^2}{\partial\xi^2}+\frac{\partial^2}{\partial\eta^2}\right)\Omega. \tag{7}$$

In zone I we seek the first partial solution for

$$\Omega=\Omega_0\left[\exp\left(-i\omega_0 t+\hat{Z}_0+\varepsilon^2\hat{Z}_1+\dots\right)\varepsilon^{-2}\right]. \tag{8}$$

From (5), (6), (7), (8) we obtain

$$\hat{Z}_0 = i\omega_0(\eta^2-\xi^2) - 4i\omega_0 p\xi - 2i\omega_0 p^2 \ln\xi + 2ik_0\xi\eta . \tag{9}$$

The solution for $\hat{Z}_0$ has a logarithmic singularity in phase at $\xi \to 0$. Similarly for $\hat{Z}_1$ we find :

$$\hat{Z}_1 = -\omega_0^2 p^4 \xi^{-2} - (8\omega_0^2 p^3 - 2k_0 p^2 \omega_0 \eta - 2i\omega_0 p^2 \delta(\eta))/\xi + (12\omega_0^2 p^2 - 4\omega_0 p\delta(0) + i\omega_0 p^2 B_1'(0))\ln\xi - i\omega_0 \int_0^\eta B_1(\eta)\,d\eta + 4i\omega_0 p \int_0^\eta (\delta(\eta) - \delta(0))\eta^{-1} d\eta - i\omega_0 p^2 \int_0^\eta (B_1'(\eta) - B_1'(0))\eta^{-1} d\eta . \tag{10}$$

We seek the stream-function ψ of partial solution for the disturbance in zones I-II in the form $\psi = F\Omega$. Substituting it into the equation $\Omega = -\Delta\psi$, we find F.

Zone II.

The thickness δ_{II} **in zone II** can be evaluated if we equate both the orders of values of nonstationary and convective terms in (7), resulting in singularity at $\xi \to 0$, and viscous terms:

$$\delta_{II} \sim \varepsilon^2 \sqrt{\bar{\omega} p^2} = \varepsilon\sqrt{\omega_0 p^2} . \tag{11}$$

In inner variables of zone II $Y_1 = \xi/\varepsilon$, η we seek:

$$\hat{Z} = \tilde{Z}_0 + \varepsilon \tilde{Z}_1 + \varepsilon^2 \tilde{Z}_2 + \dots . \tag{12}$$

The equation for $\tilde{Z}_0$

$$2\eta \tilde{Z}_{0\eta} - 2Y_1 \tilde{Z}_{0Y_1} = 4i\omega_0(p^2+\eta^2) + \tilde{Z}^2_{0Y_1} \tag{13}$$

has the solution matching at $Y_1 \to \infty$ with the expansion $\tilde{Z}_0 + \varepsilon^2 Z_1$ in variables η, Y_1 :

$$\tilde{Z}_0 = i\omega_0\eta^2 - Y_1^2/2 + \tfrac{1}{2} Y_1 \sqrt{Y_1^2 - 4i\omega_0 p^2} + i\omega_0 p^2 - 2i\omega_0 p^2 \ln((Y_1 + \sqrt{Y_1^2 - 4i\omega_0 p^2})/2) - 2i\omega_0 p^2 \ln\varepsilon . \tag{14}$$

The values of the square root in (14) lie in the IV quadrant. Similarly we find $\tilde{Z}_1$ and $\tilde{Z}_2$. Because of the inconvenience of expressions for the $\tilde{Z}_1$ and $\tilde{Z}_2$ they are not given here.

Zone III.(boundary layer).

The inner variables of zone III are $Y=Y_1/\varepsilon, \eta$. The stream function of the basic flow has the form:

$$\bar{\Psi}_0=2\varepsilon^2\eta f(\eta,Y)+\varepsilon^4\tilde{\Psi}_{01}+\dots \; . \tag{15}$$

In the boundary layer we seek Ω in the form :

$$\Omega=\Omega_0 \exp(\Phi+(i\omega_0\eta^2+\varepsilon Z_1+\varepsilon^2 Z+\dots)\varepsilon^{-2}). \tag{16}$$

Φ is a constant defined from the condition of matched solutions in zones II and III. For Z_1 we find :

$$Z_1=\sqrt{-4i\omega_0 p^2}(Y-\delta(\eta)+ik_0\eta+\int_\infty^Y((1+\eta^2p^{-2}(1-f'))^{1/2}-1)dY \tag{17}$$

Similarly we find Z. The stream function of the disturbance in zone III has the form :

$$\psi=\varepsilon^6 4(p^2+\eta^2)Z_{1Y}^{-2}\Omega(1+0(\varepsilon)). \tag{18}$$

The solution (18) does not satisfy the no-slip conditions at the boundary Y=0. Therefore we obtain two more partial solutions of equation (7) vanishing at $Y\to\infty$ and having at Y=0 for ψ and ψ_Y the same dependence on η as in the above given partial solution ("incident" vortex wave). One of these solutions ("reflected" vortex wave) in zone III has the form analogous to (15) with substitution of Z_1^r instead of Z_1 and Z^r instead of Z where

$$Z_1^r=\sqrt{-4i\omega_0 p^2}\,\Big[-\delta(\eta)+ik_0\eta-\int_0^Y(1+\eta^2p^{-2}(1-f'))^{1/2}dY+ \tag{19}$$
$$+\int_\infty^0((1+\eta^2p^{-2}(1-f'))^{1/2}-1)dY\Big],$$

$$Z^r=\eta\int_0^Y f'Z_{1\eta}^r Z_{1Y}^{r-1}dY-\int_0^Y(f+\eta f_\eta)dY-\frac{1}{4}\ln(1+\eta^2p^{-2}(1-f'))+ \tag{20}$$
$$+\eta\int_\infty^0 f'Z_{1\eta}Z_{1Y}^{-1}dY+Z(\eta,0).$$

We seek the stream function of the second additional solution in the form :

$$\varphi^{s} = \varepsilon^{6}\Omega_{0}(-i\omega_{0})^{-1} \exp(i\omega_{0}\eta^{2}\varepsilon^{-2}+\varepsilon^{-1}Z_{1}(\eta,0)+Z(\eta,0))\times \quad (21)$$

$$\times(\varphi(\eta,Y)+O(\varepsilon)).$$

The function φ satisfies the Orr-Sommerfeld equation :

$$-4i\omega_{0}p^{2}(1+\eta^{2}p^{-2}(1-f'))(\varphi''-4\omega_{0}^{2}\eta^{2}\varphi)-4i\omega_{0}\eta^{2}f'''\varphi = \quad (22)$$

$$=\varepsilon^{2}\left(\frac{\partial^{2}}{\partial Y^{2}}-4\omega_{0}^{2}\eta^{2}\right)^{2}\varphi;$$

$$\varphi - \text{const} \exp(-2\omega_{0}\eta Y) \quad \text{at } Y\to\infty;\ \varphi(0,\eta)=1 \quad (23)$$

Equation (22) with the boundary conditions (23) has been solved numerically.
Vortex wave stream function ϕ is sought in the form :

$$\phi_{v}=\phi+C_{1}(\eta)\phi^{r}+C_{2}(\eta)\phi^{s} \quad (24)$$

where $C_1(\eta)$ and $C_2(\eta)$ are defined from the boundary conditions $\phi_{v}=\frac{\partial\phi_{v}}{\partial Y}=0$ at Y=0. The above stated method of solution can be applied to some other of ω, k, χ as well if the thickness of the layer II is much greater than that of the boundary layer. The numerical calculations of ζ-component of the disturbed velocity related to ε_{i} (power turbulence level of the incoming flow): $\langle\bar{u}'\rangle=\langle u'\rangle/\varepsilon_{i}$ for the case $\omega=\omega_{0}\varepsilon^{-1}$, $\chi\sim1$ are presented in fig. 1. for $\varepsilon^{-2}=1450$; p= 0,41; ω_{0}= 0,5; k_{0}= 1,0; 1,4; 1,8 - curves 1, 2, 3 respectively.

Excitation of Instability Waves in the Boundary Layer

The numerical calculations of the amplitude of the Tollmien-Schlichting waves referred to ε_{i}, $Q(\eta)$ **excited** by the vortex wave studied in the present paper have been performed by the method of [1]. The calculations have been performed for p = = 0,367; ω= 16,507; ε= $3{,}028\cdot10^{-2}$; k=ω(curve 1); ω = 24,76; k= ω(curve 2).

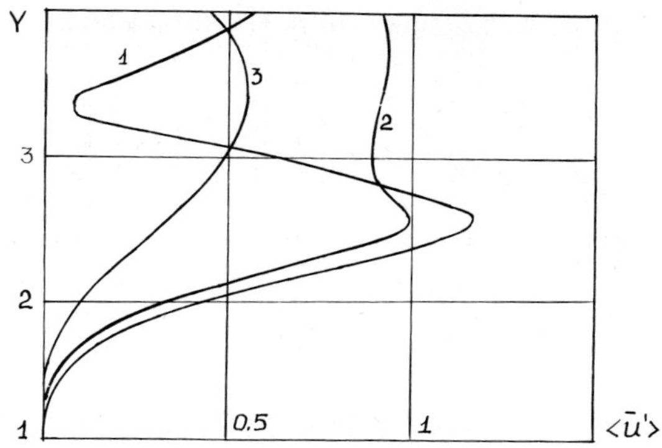

Fig. 1.

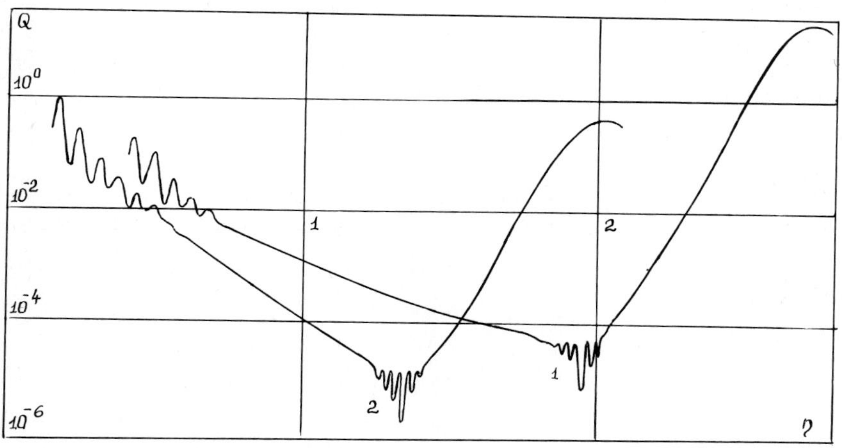

Fig. 2.

Reference

1. Zhigulev, V.N.; Sidorenko, N.V.; Tumin, A.M.: Zhurn. Prikl. Mekh. i Tekh. Fiz. N6 (1980) 43-49 (in Russian).

Three Types of Disturbances from the Point Source in the Boundary Layer

H. R. GREK, V. V. KOZLOV, M. P. RAMAZANOV

Institute of Theoretical and Applied Mechanics,
USSR Academy of Sciences, Novosibirsk, USSR

Summary

The results of the experimental investigations presented in this paper are connected with the study of excitation and the development of disturbances of different intensity from the point source in the boundary layer of a flat plate. In the boundary layer there appear qualitative and some quantitative characteristics of the three types of disturbances: a Tollmien-Schlichting wave, a nonlinear localized structure and a turbulent spot. If two of them - a Tollmien-Schlichting wave and a turbulent spot are known, the third one - a localized disturbance is observed for the first time.

Introduction

The development of space wave packet from the point source was considered on a flat plate [1] and on a straight wing with a positive and negative pressure gradient [2]. It is known that the transition of the laminar flow to the turbulent one at the final stage of this process is related to the mechanism of turbulent spots emerging in the original laminar flow. A series of papers [3, 4] was dedicated to the spreading of turbulent spots artificially generated by a point source. The object of the present paper is to study the excitation and the development of the disturbances in the range from a Tollmien-Schlichting (T-S) wave to a turbulent spot varying the intensity of the point source.

Experimental equipment

All the measurements were carried out in the working part of a plexiglass plate located horizontally in the test section of the wind tunnel. The plate was 10 mm thick, 200 mm wide and 800 mm long and had a rounded leading edge. Disturbances were introduced into a boundary layer of the flat plate through two openings of a diameter 0.5 mm at a distance 30 and 220 mm from the leading edge of the plate. They were engendered by a dynamical loudspeaker. Electrical signal was transmitted to the loudspeaker either from the exit of the sound generator (f=81.4 Hz) or from the generator of rectangular impulses (f=15; 6 Hz). The development of the thus emerged spatial wave packet was registered by the thermoanemometer (DISA). At each point of measuring there were measured three values: a mean flow velocity - $\bar{u}$, amplitude u' and phase φ of the streamwise component of flow velocity oscillations. The forward flow velocity constituted u_∞ =9.2 m/sec. On the plate there was realized a flow with a laminar boundary layer (LBL).

Results and discussion

The obtained results indicated that by introducing the disturbances through the first opening (x=30 mm, Re_δ =690) a localized disturbance is realized in the range of intensities created u'_{max}/u_∞ =0.5 ÷ 1.5% and in case x=220 mm (Re_δ =1850) - u'_{max}/u_∞=1.5 ÷ 2.5%. At smaller intensities a T-S wave is realized and at larger ones - a turbulent spot correspondingly. T-S wave was realized by introducing disturbances with a level less than 1%. The experimental results confirmed that the onset and development of the waves of a T-S type showed a good qualitative correspondence to the results of the paper [1]. The pulsations profiles (Fig.1) are to be found inside the boundary layer. The length of the wave of disturbance is λ =50 mm and phase velocity is $C_\phi=\lambda\cdot f$ (f=81.4 Hz) C_ϕ=4.05 m/sec which constitutes 0.44 from the forward flow velocity - u_∞ . Fig.3(a) showing an amplification curve of disturbances amplitudes a sharp attenuation of disturbances. From the graph (Fig.2) showing the distribution of pulsations rate in the transverse direction (Z), the peak (type of " δ " function) is observed to bifurcate into two less expressed mounds which spread in a downstream direction. An analogous picture of the distribution of pulsations amplitudes along Z was obtained in paper [1] as well. In both cases the revelation of the pulsation domain in Z direction taken at the level 0.01% occurs at the angle $\theta \simeq 12°$ (Fig.2).

At the stage of transition from a T-S wave to the turbulent spot, a nonlinear localized disturbance was discovered which sharply differed both from the T-S wave and the turbulent spot. The present results were obtained by introducing the disturbances through the opening x=30 mm, Re_δ =690, the amplitudes level is 1.5% u_∞ . Let's consider the peculiarities of the nonlinear localized disturbance. As one can see from Fig.1, the pulsations profiles $u'=f(Y)$ are taken out on the upper boundary of the boundary layer and thus the connection with the wall is lost which sharply differs from the T-S wave pulsations distribution located inside the boundary layer. Phase velocity of the disturbance C_ϕ is approximately twice as large as that of the T-S wave and constitutes C_ϕ/u_∞ =0.82. The disturbance is localized in the transverse direction (Z) (Fig.2) and spreads in the streamwise direction (X) to be completely attenuated at the distance 450 mm and not 150 mm as was the case for the T-S wave (Fig.3). Lately there was published a series of papers dedicated to the turbulent spots development [3, 4]. In the present paper the first attempt was made to work out the methods to obtain turbulent spots by means of a dynamic loudspeaker as well as to obtain some quantitative and qualitative characteristics for the turbulent spot which develops in space. The results of the experiments are presented for the case when the disturbances are introduced through the second opening x=220 mm (Re_δ =1850). Fig.1 shows integral pulsations profiles $u'= f(Y)$ depending on the streamwise coordinate X. One can see that the turbulent spot is connected with the wall and the pulsations profile expands far beyond the external edge of the boundary layer. These data characterize an intensive growth of the turbulent spot in a crosswise direction - Y when propagating downstream $dY/dX \simeq 0.02$ (according to the results of paper [4] dY/dX=0.019). The spreading of the turbulent spot in the transverse direction Z is represented in Fig.2, where dZ/dX= =0.16 (from the paper [4] dZ/dX=0.167). The measurements of the phase velocity of the leading and trailing edge of the spot (Fig.4) were carried out with the help of the oscillograph synchronized by rectangular impulses (f=15; 6 Hz). The velocity of the leading edge, according to the measurements $C_\phi \cdot 1/u_\infty$ = =0.81 (from the paper [4] $C_\phi \cdot 1/u_\infty$ =0.8) and that of the trailing edge $C_\phi \cdot 1/u_\infty = 0.53$ (from the paper [4] $C_\phi \cdot 1/u_\infty = 0.57$), which demonstrates a satisfactory correspondence.

Conclusion

Thus, it can be stated that when disturbances of different intensity are introduced in the boundary layer by means of the point source, then three principally different types of disturbances emerge on a flat plate. If two of them - a T-S wave and a turbulent spot are known, the third one - a localized disturbance is observed for the first time. The characteristics of the localized disturbance in the boundary layer of a flat plate presented here have much in common with the characteristics of a turbulent "puff" studied in paper [5] . The turbulent "puff" is not connected with a channel wall, is located in space and its propagation velocity is close to the mean flow velocity in the channel. Making some allowances for the difference in the flow on the flat plate and in a round channel it can be stated that a nonlinear localized structure observed is nothing but a turbulent "puff".

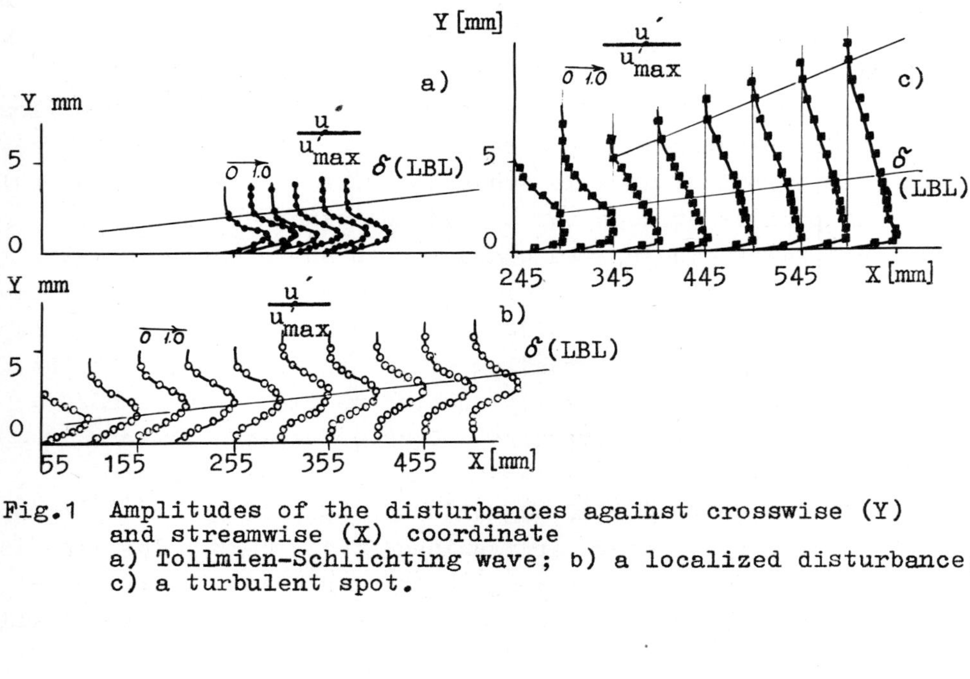

Fig.1 Amplitudes of the disturbances against crosswise (Y) and streamwise (X) coordinate
a) Tollmien-Schlichting wave; b) a localized disturbance; c) a turbulent spot.

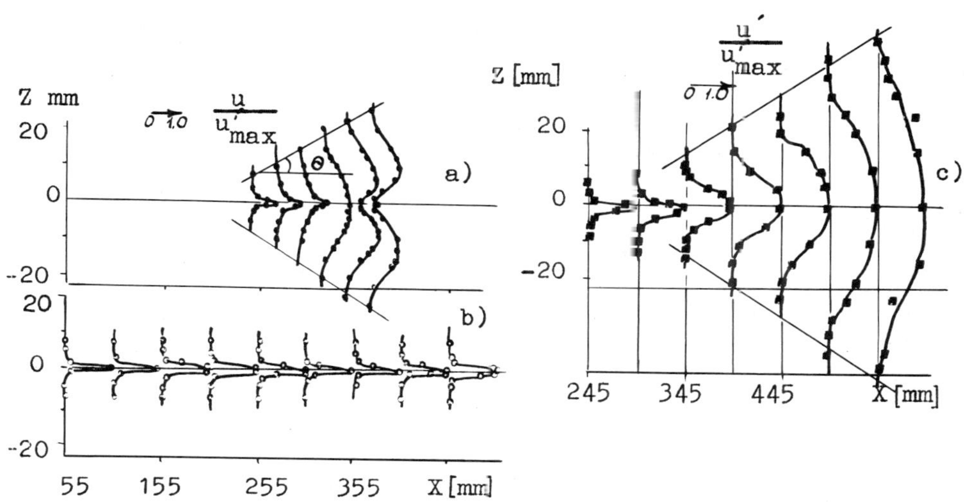

Fig.2 Amplitudes of the disturbances against streamwise (X) and transverse (Z) coordinates
a) Tollmien-Schlichting wave;
b) a localized disturbance;
c) a turbulent spot

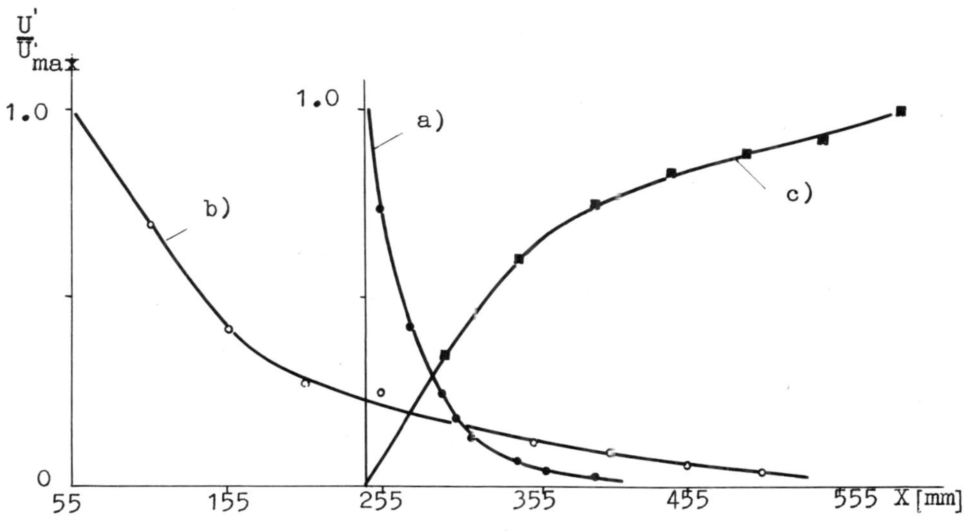

Fig.3. Amplitudes of disturbances against streamwise (X) coordinate by Z=0
a) Tollmien-Schlichting wave
b) localized disturbance
c) a turbulent spot

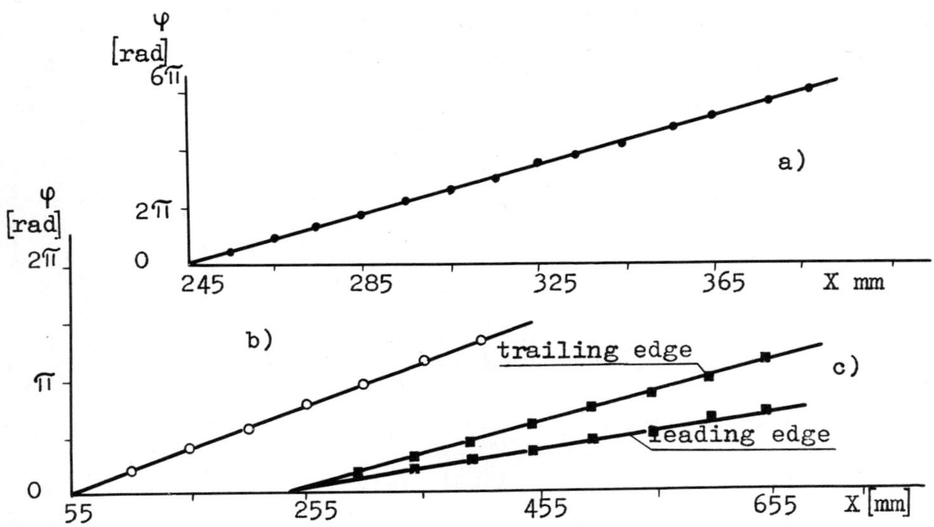

Fig.4 Phase disturbances against streamwise (X) coordinate by Z=0
a) Tollmien-Schlichting wave;
b) a localized disturbance;
c) a turbulent spot.

References

1. Guilev, V.M.; Kachanov, Yu.S.; Kozlov, V.V.: Razvitie prostranstvennogo volnovogo paketa v pogranichnom sloye. Novosibirsk, 1981, preprint ITPM SO AN SSSR N 34-81.

2. Guilev, V.M.; Dovgal, A.V.; Kozlov, V.V.:Razvitie volnovogo paketa v pogranichnom sloye s gradientom davleniya. Novosibirsk, 1984, preprint ITPM SO AN SSSR N 6-84.

3. Wygnanski, I.; Haritonidis, S.H.; Kaplan, R.E.: On Tollmien-Schlichting wave packet produced by a turbulent spot. J. on Fluid Mechanics, vol.92, pt.3, (1979).

4. Wygnanski, I.; Zilberman, M.: On the spreading of turbulent spot in the absence of a pressure gradient. J. of Fluid Mechanics, vol.123 (1982) 69-90.

5. Wygnanski, I.; Sokolov, M; Friedman, D.: On transition in a pipe. Pt.2. The equilibrium puff. J. of Fluid Mechanics (1975), vol.69, pt.2, (1975) 283-304.

The Problems of a Theoretical Description of the Transition Zone of a Laminar Boundary Layer into a Turbulent One

V.N.ZHIGULEV

Moscow Physical-Technical Institute, USSR

Summary

The paper is devoted to the development of the amplitude method of the calculation of the critical Reynolds numbers Re_{cr} of the transition of the laminar boundary layer into a turbulent one and to the statement of some results of the investigation of the location of the transition point depending on various factors exciting instabilities as well.

The amplitude method of defining Re_{cr} ([1],[2])

The amplitude method of the calculation of the critical Reynolds numbers of the transition of a laminar boundary layer into a turbulent one when applied to sufficiently small values of the initial excitation background in a flow consists in

A. Defining the most dangerous instability* excited and separated by the boundary layer on the surface of the flying machine under the conditions of the given initial excitation background.

B. Finding the equivalent amplitude of this instability on the line of the stability loss of the initial boundary layer on the basis of the known initial excitation background.

C. Calculating the intensification of this instability in the boundary layer involved within the linearized excitation theory.

D. Finding the line Π on the surface of the flying machine where the amplitude of the investigated instability reaches the threshold value ε^* .

E. Identifying approximately the line Π with the onset of the zone of the developed turbulent motion in the boundary layer involved.

---

*The most dangerous instability is found on the basis of considering the wave packet of instabilities.

The beginning of the amplitude method was based in [3], [4] where a certain purely empirical relationship for the instability amplitude at the point of stability loss in the flow was suggested and the calculation of the instability intensification was completed where the relative instability amplitude reached the value of 1%, which was an empirical assumption as well.
The essence of the magical 1% was cleared in [5] ÷ [10]. It appeared that instability in the process of its development on reaching by its amplitude the threshold value ε^* goes into a nonlinear development region where its growth increment increases sufficiently and it decays comparatively rapidly (secondary instability), a developed turbulent flow forming. Thus parameter $\varepsilon^* = (\alpha \, Re_\delta^{-2/3})$ (α is a wave number with respect to the boundary layer thickness; Re_δ is the Reynolds number calculated on the basis of the boundary layer thickness) characterizes appoximately that limit value of the instability amplitude when turbularization of the flow comes. The value of the parameter ε^* varies usually in the limit of (1-4)%*.
In [11], [12] a general excitation and the instability development theory was created. It was found out [12] that the instability is often excited in the vicinity of the point of stability loss of the flow involved, then it goes into a phase of independent development.
As a result of the investigation on the excitation and development of instabilities the main relation for defining the critical Reynolds number of the transition was found

$$\int_{x_{ny}}^{x_{cr}} \mathrm{Jm}\left[\alpha(x)\right] dx = \ln \frac{\varepsilon^*}{\varepsilon_{HH}} - \int_{x_{ny}}^{x_{cr}} \mathrm{Real}\left[W_{\parallel}(x)\right] dx \qquad \text{(A)}$$

Here x_{ny} is a coordinate of the point of the stability loss, x_{cr} is a coordinate of the point of transition into turbulence, $\mathrm{Jm}[\alpha(x)]$ is an imaginary part of the wave number $\alpha(x)$ for the instability involved obtained as a result of the numerical solution of the eigen value problem, ε_{HH} is a value of the equivalent instability amplitude at the point $x=x_{ny}$. The value ε_{HH} is obtained as a result of the investigation of the problem on the excitation of the instability by the given excitati-

*The details of the parameter ε^* are in [2].

on background (by the initial turbulence of the flow, the vibration of the streamlined surface, its roughness and so on). Real $[W_{11}(x)]$ is a real part of the matrix element $W(x)$ characterizing the change of the instability at the expense of the nonparallelism of the investigated boundary layer. The relationship (A) is obtained for a one-dimensional case including boundary layers occurring in flat flows with a pressure gradient or in flows near a slipping wing.

If a stream near a flat plate in a gas or fluid flow is investigated the corresponding instability then is called the Tollmien-Schlichting wave (TS). First we give the estimation for the values included in the expression (A) obtained on the basis of [13], [14], [15] for a flow near a flat plate. For the most dangerous TS waves $\mathrm{Im}\,\alpha$ is of the order of $10^{-2}/\delta$ (δ is a boundary layer thickness); that is why the l.h. side of the expression (A) is of the order of $x_{cr}/100\,\delta$. When the TS waves are excited by an external turbulence $\varepsilon_{HH} \sim 10^{-2}\varepsilon_H$ where ε_H is an amplitude of the initial turbulence in a free stream. Therefore the first term on the r.h. side (A) is of the order of $\ln \dfrac{10^2\varepsilon^*}{\varepsilon_n}$. If $\varepsilon_n = 10^{-2}$, then $\ln \dfrac{\varepsilon^*}{\varepsilon_{HH}} \sim 5$; if $\varepsilon_n = 10^{-4}$ then $\ln \dfrac{\varepsilon^*}{\varepsilon_{HH}} \sim 9$. The second term on the r.h. side of the expression (A) is usually not more than 1.5.

If $\varepsilon_n \sim 10^{-4}$ (one hundred per cent of the free stream velocity, which takes place in subsonic wind tunnels of low turbulence) the formula (A) turns into a familiar criterion "e^9"*. Since the amplitudes of the background waves are included in the formula for defining Re_{cr} (A) with a logarithm sign all the practical cases (wind tunnels with various character and amplitude of the background waves, full scale investigations for various altitudes, Mach numbers and so on) can be divided when defining the location of the transition point into a number of classes characterized by the same growth increment of the amplitude of the instability wave in the transition zone. Shown in Fig.1 are the examples of the dependences $\mathrm{Re}_{cr}(\varepsilon_n)$ evaluated for three various cases of the excitation of the TS

*This criterion was suggested in [16]÷[19].

waves in a boundary layer of a flat plate according to [14]. The Mach number in a free stream M=0.6; the frequency parameter for the TS wave $F=20\cdot 10^{-6}$ ($F=\omega\nu / u_0^2$). Curve (1) corresponds to the case of the excitation of the TS wave by an external acoustic wave with an angle of incidence of $\varphi=20°$, ε_n is equal to the module of a longitudinal component of the velocity in an incident wave, u_A nondimensionalized with respect to the velocity in a free stream. Curve (2) corresponds to the case of resonance excitation of the TS wave by an acoustic wave ($\varphi=20°$) that is incident on the boundary layer on the wavy surface. In the example given above the product $u_A h=10^{-3}$, h is a roughness amplitude related to the boundary layer thickness. If the form of the roughness is fixed then $\varepsilon_n=u_A$ and curve (2) shows the dependence of Re_{cr} on the amplitude of the roughness of the wall. If on the contrary the value $u_A=10^{-3}$ is constant then ε_n=h and curve (2) shows the dependence Re_{cr} on the amplitude of the wall roughness. Curve (3) corresponds to the resonance excitation of the TS wave by a vibrating surface having a form of a drift wave; here $\varepsilon_n=a$, where a is an amplitude of vibration with respect to the boundary layer thickness.

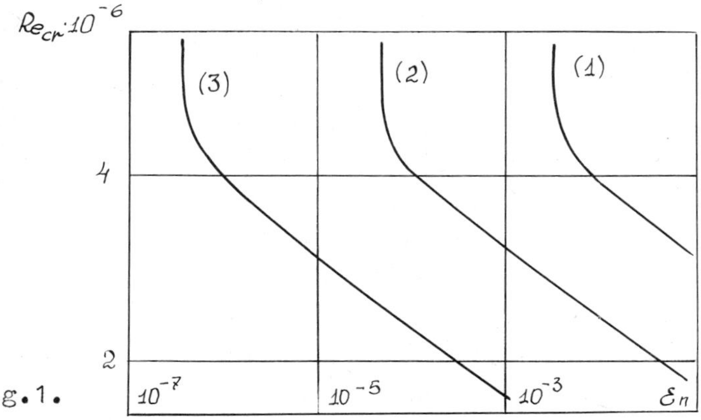

Fig.1.

The transition zone in a boundary layer is very sensitive to the vibrations of the streamlined surface, which is of great practical importance.

It is of interest that in case the amplitude of the external excitation is less than some critical value the boundary layer on the flat plate is laminar everywhere ($Re_{cr}=\infty$); this critical value of the amplitude is defined by the type of the external excitations.

Some results on the investigation of the excitations and development of instabilities.

A theory of the excitation and development of instability in three-dimensional boundary layers has been developed in [10], [11] for the case boundary layers change slightly in the flow. The theory is based on the expansion of the solution of the linearized Navier-Stokes equations into series in the spectrum of eigenfunctions for a local homogeneous problem that is a generalization of the Orr-Sommerfeld problem. The amplitudes of this sequence of the modes of a fixed frequency satisfy the system of equations:

$$\frac{\partial C_m}{\partial x} + \sum_{(n)} \frac{\partial C_n}{\partial z} V_{mn} e^{F_n - F_m} = \sum_{(n)} C_r W_{mn} e^{F_n - F_m} \quad \ldots \text{(B)}$$

The values V_{mn} and W_{mn} are known matrices of the transition between m and n modes [12]. F_n is an eikonal for mode n ; F_n is a great value. In one case the equation (B) coicides with the equation in [21].

Now we give the main results of the investigation of two-mode interaction of m and n waves for two-dimensional case ($\frac{\partial}{\partial z}=0$), when m - wave is an instability $\mathrm{Im}(\alpha_m) > 0$ at $x > x_{ny}$ (Fig.2) and n - wave is one of the waves of the continuum spectrum (an acoustic or vortex wave for example).

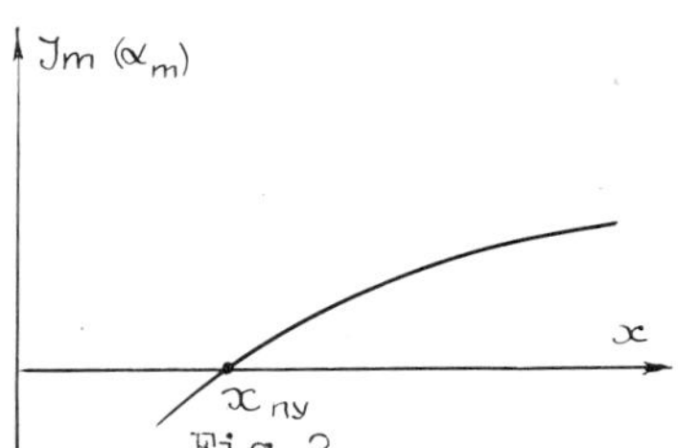

Fig.2.

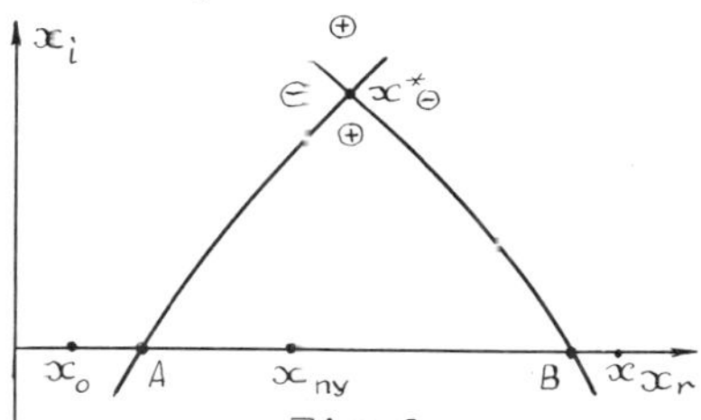

Fig.3.

1. The structure of the eikonal $\Delta(x)=F_n-F_m$ considered as an analytical function of a complex variable $x=x_r+ix_i$ plays a substantional part in the analyzed interaction. A typical picture is presented in Fig.3. x^* is a saddle point of the function $\Delta(x)= = u+iv$. The lines u =const passing through the saddle point

are also shown in Fig.3. The points A and B are the points of intersection of these lines with the real axis x_r; positive and negative sectors Δ (x) are denoted.

2.The problem of the excitation of an instability when at $x=x_0 < x_A$ $C_m=0$, $C_n=C_{n0}$ [12] is examined. It was obtained that

a)in case of coincidence of phase velocities of m and n waves the saddle point x* coincides with the point x_{ny}. The resonance excitation of an instability at the point $x=x_{ny}$ itself takes place.

b)in case of noncoincidence of phase velocities of m and n waves the saddle point lies in a complex region. The instabilities of the main order are excited when the integration point x passes the points A and B, Fig.3. Then if $x_A < x < x_B$, m - wave consists of the sum of the instability itself and an intermediate wave. The intermediate wave has the same distribution of amplitudes with respect to y as the instability and possesses an eikonal coinciding with the eikonal of wave n. If $x > x_B$ the excitation of an instability is completed and the excited m - wave is a pure instability included in the regime of an exponentional growth. Thus the excitation of an instability along x axis doesn't occur substantially uniformly - two remarkable points A and B take place, Fig.3, in the vicinity of them the main process of the excitation occurs.

A great amount of work has been recently carried out on numerical investigation of the process of the excitation and development of the Tollmien-Schlichting waves in compressible boundary layers and an analysis of the waves of a continuum spectrum (acoustic, vortex, pressure waves) in the presence of such boundary layers as well. We'll dwell upon the following results.

1.A local homogeneous problem on defining a spectrum of elementary excitations in a compressible boundary layer on a flat plate [14], [22] was investigated. An analysis of a continuum spectrum of the Navier-Stokes equations for a gas is carried out in [20] and the **completeness of the eigen functions is proved.**

2.The influence of nonparallelism of the flow in a compressible boundary layer on a flat plate on the performances of its stability is numerically analyzed ([14], [23]).

3.The problem on the excitation of the TS waves by an external acoustic wave incident on a compressible boundary layer of a flat plate is examined [14].

4.A numerical analysis of the excitation of the TS waves in a boundary layer of a flat plate in an incompressible fluid flow by the waves of vorticity and pressure is carried out in [24].

5.The problem on the excitation of the TS waves in a compressible boundary layer of a flat plate by an incident acoustic wave and the roughness of a streamlined surface is solved in [14]. The mechanism of three-wave resonance when the sum of the wave lengths of the roughness and an acoustic wave is equal at the point of stability loss to the length of the excited TS wave is investigated.

6.The problem [25] on the excitation of the TS waves in a compressible boundary layer of a flat plate by a vibration wave of a low amplitude drifting along the streamlined surface is solved. The example of the calculation of the excitation of the TS waves for the case $M=0,6$, $F=20\cdot 10^{-6}$; q_m is an amplitude of a dimensionless mass flow in the TS wave normalized to a dimensionless vibration amplitude. At the point of stability loss ($R_{ny}=960$) the wave number of the TS wave $\alpha_{TS}=6,44\cdot 10^{-2}$. Curve (1) corresponds to the wave number of the vibration $\alpha_v=6,54\cdot 10^{-2}$, and curve (2) - $\alpha_v=6,64\cdot 10^{-2}$. Curve (3) corresponds to the resonance excitation ($\alpha_{TS}=\alpha_v$).

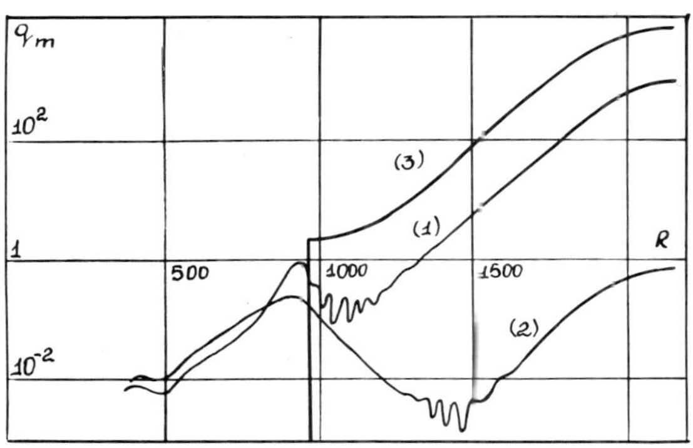

Fig.4.

7.Theoretical and experimantal investigations on the excitation of the TS waves by a special vibrator located on a streamlined surface* have been recently carried out. The comparison of the

*See the paper by Tumin A.M. and Fedorov A.V. submitted to the present symposium.

theory and the experiment gave good results. It is of great importance that for the excitation of the TS waves leading to a rapid transition to turbulence low vibration amplitudes of the vibrator of the order of the thickness of a viscous sublayer for the wave TS ($\sim 1/\sqrt{Re_\delta}$) is required.

8. The mechanism of the excitation of instabilities on the airfoils in the vicinity of the critical point turned out very specific*.

Comparison with the experiment.

The definition of the critical Reynolds transition number is essentially a problem of defining the relaxation time of a static system consisting of various vibratory modes to some equivalence vibrating regime which is called a developed turbulence.

For small degrees of the initial turbulence ($\ll \varepsilon^*$) the main part of the transition region length is in linearized region of the instability development. In this case in accordance with the formula (A) the Reynolds transition number increases logarithmically with the decrease of the background amplitude (Curve (1), Fig.5).

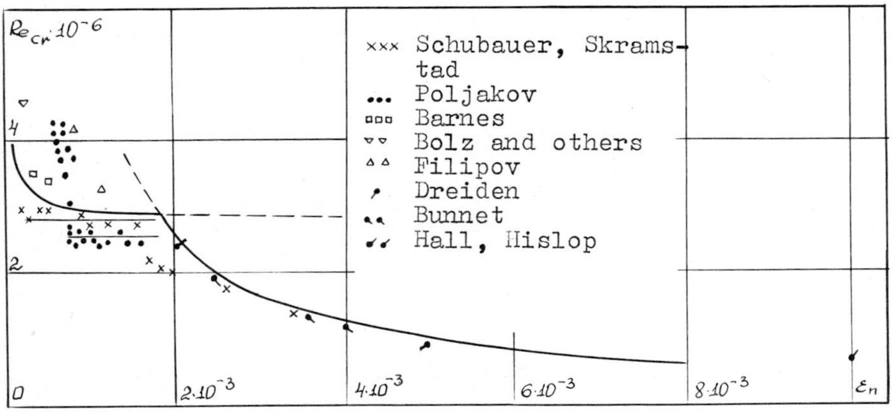

Fig.5.

*See the paper by Erofeev E.A. and Sidorenko N.V. submitted to the present symposium.

On the contrary when the amplitude of the background wave is of the order of ε^* the character of the relaxation to the developed turbulent regime of the flow changes (See [7]); the relaxation time for a flat plate becomes proportional to the value $1/\varepsilon_n$ (Curve (2), Fig.5). Curve (1) in Fig.5 is built on the basis of the calculations in [14] for the excitation by an acoustic wave. All the experimental results on the measurement of Re_{cr} on a flat plate known to the author are given there as well. It is seen in Fig.5 that the experiment does indicate the existence of two ranges ε_n with essentially different dependences $Re_{cr}(\varepsilon_n)$. The divergence of the experimental data in the validity region of the logarithmic dependence $Re_{cr}(\varepsilon_n)$ (curve (1)) is probably explained by a strong dependence Re_{cr} on the kind of the exciting background. To check exactly this background under experimental conditions is an important problem in future investigations.

References

1.Reshotko,E., An. Rev. of Fl. Mech., 1976, v.8, 311-349.

2.Zhigulev,V.N., sb. "Modeli neodnorodnih sred", Novosibirsk, 1981; sb. "Modeli mehaniki sploshnoi sredi", Novosibirsk, 1983.

3.Mack,L.M., AIAA J., 1975, v.13, N 3.

4.Mack,L.M., AGARD-CPP-224, oct. 1977, pp 1-1-1-22.

5.Zhigulev, V.N.; Kirkinskii,A.I.; Sidorenko, N.V.; Tumin,A.M., sb. "Aeromehanika", Nauka, 1979, 118-140.

6.Zhigulev, V.N.; Sidorenko, N.V.; Tumin, A.M.; sb. "Chisleni metodi mehaniki sploshnoi sredi", Novosibirsk, 1975, t.6, N 1, 30-41.

7.Zhigulev, V.N., sb. "Mehanika turbulentnih potokov", Nauka, 1980, 109-133.

8.Levchenko, V.J., Doktorskaja disertazija, ITPM, Novosibirsk, 1980.

9.Kachanov, J.S.; Koslov, V.V.; Levchenko, V.J., "Vosniknovenie turbulentnosti v pogranichnom sloe", Nauka SC, 1982.

10.Sidorenko, N.V.; Tumin, A.M.; sb. "Mehanika neodnorodnih sred", Novosibirsk, 1981.

11.Zhigulev, V.N.; Sidorenko, N.V.; Tumin, A.M., PMTF, 1980, N 6,

43-49.

12.Zhigulev, V.N.; PMTF, 1983, N 4, 100-110.

13.Tumin, A.M.; Shepelev, V.E., sb. "Chisleni metodi mehaniki sploshnoi sredi", Novosibirsk, 1980, t.11, N 3, 141-152.

14.Fedorov, A.V., Kandidatskaja disertazija, MFTI, 1982.

15.Fedorov, A.V., sb. "Chisleni metodi mehaniki sploshnoi sredi", Novosibirsk, t.13, N 3, 106-118.

16.Michel, R., ONERA Repport, 1/1578A, 1951.

17.Smith, A.M.O., Douglas Aircraft Company, Report ES 26388, 1956.

18.Van Yngen, Y.L., Univ. of Technolody, Dept. of Aero Engr. Report VTH-74, Delft, Holland, 1956.

19.Jaffe, N.; Okamura, T.T.; Smith, A.M.O., AIAA J., 1970, N 8, 301-308.

20.Tumin, A.M.; Fedorov, A.V., PMTF, 1983, N 5.

21.Nayfeh, A.H., AIAA J., 1980, v.18, N 4, 406-416.

22.Gaponov, S.A.; Maslov, A.A., Novosibirsk, Nauka, 1980.

23.Gaponov, S.A., Isv. AN SSSR, Mehanika zhidkosti i gasa, 1980, N 2, 26-31.

24.Tumin, A.M.; Shepelev, V.E., otchet, Dolgoprudnii, 1982.

25.Tumin, A.M.; Fedorov, A.V., PMTF, 1983, N 3.

The Receptivity of a Laminar Boundary Layer to External Disturbances

P. LEEHEY, C.J. GEDNEY AND J.Y. HER

Acoustics and Vibration Laboratory
Massachusetts Institute of Technology
Cambridge, Massachusetts 02139 USA

Summary

We present experimental results on the excitation of a laminar boundary layer to stream turbulence sound and plate vibration. Excitations by sound can be cancelled using leading edge vibration. An adverse pressure gradient near the leading edge is shown to be a powerful mechanism for increasing the receptivity of the boundary layer to sound excitation.

Background

Our earlier results (Shapiro [1]; Leehey [2,3]), on the excitation of a laminar boundary layer by essentially two-dimensional sound waves travelling downstream demonstrated the following facts:

1) There is a significant linear range of response to sound excitation;
2) Two-dimensional sound waves result in greater growth rates of Tollmien-Schlichting waves than do natural (stream turbulence) excitations;
3) In the presence of a weak pressure gradient near the leading edge, the receptivity of the laminar boundary layer to sound excitation is O(1)

We define receptivity to sound waves as the ratio of the maximum RMS velocity of the disturbance wave response to the exciting sound particle velocity.

Murdock [4] made a numerical analysis of the excitation of a laminar boundary layer by downstream travelling sound waves over a semi-infinite flat plate. His solution of the Navier-Stokes equations showed a receptivity of order $O(10^{-4})$. Later Murdock [5] performed a similar analysis for a parabolic profile wherein he found similar small values of the receptivity and in particular found that the receptivity decreased with increasing leading edge bluntness. The mean pressure gradient for these profiles was, of course, favorable. Goldstein [6], used the method of matched asymptotic expansions for the case of the semi-infinite flat plate. He reached a significant conclusion that one of the asymptotic eigenmodes of the unsteady boundary layer equations valid near the leading edge matched the fundamental Tollmien-Schlichting wave further downstream. In his argument the small parameter involved in the matching is either frequency or distance from the leading edge. His analysis provides a mechanism for wavenumber conversion to the Tollmien-Schlichting wave number. It, however, lacks a length scale typical of the leading edge. Thus when we place into his analysis numerical values typical of our experiments, we find a receptivity of $O(10^{-3})$ (see also Goldstein, Sockol and Sanz [7]). It therefore appears that none of the theoretical results presently available provide the proper order of magnitude of receptivity for boundary layers created on plates with finite leading edges.

We report here further experimental results on this receptivity problem. The wind tunnel and test plate are the same as those used by Shapiro. Measurements were made at the same reduced frequency, but at lower tunnel speed in order to more clearly separate the neutral stability curve (predicted for a Blasius profile) from the adverse gradient zone near the leading edge. Additionally, the effects of plate vibration excitation and ribbon excitation were studied, cancellation of sound excitation by vibratory excitation was achieved, and finally the results from the entire removal of the adverse pressure gradient are presented.

Experimental Procedure

The turbulence level in our low noise, low turbulence wind tunnel was 0.05%, flat over the speed range to 60m/s. The turbulence spectrum decayed 3-4dB per octave with no evident spikes except for a peak at 3Hz. This peak resulted from the open jet mode of operation wherein the test section was surrounded by an anechoic blockhouse. Since this peak was at least 2 decades below the test range of interest, we excluded it from our measured turbulence level. We also measured the spectrum of sound just outside the open jet in the unexcited case. The sound spectrum was likewise uninteresting except for the 3Hz peak, and was approximately 35dB below the stream turbulence spectrum assuming a plane wave conversion of sound pressure to sound particle velocity. It seems reasonable, therefore, that in our test configuration that there was no significant contribution of sound to the tunnel turbulence level, in contrast to the situation in a number of other low turbulence tunnels.

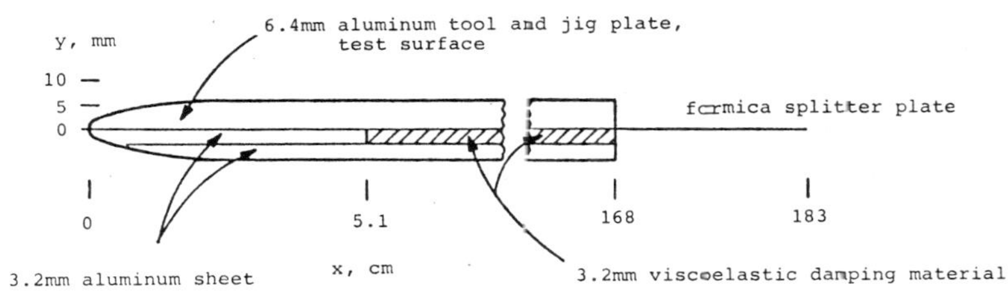

Figure 1. Test plate cross-section.

The cross section of our test plate is shown in Figure 1. The leading edge is a 6:1 semi-ellipse. The plate is very heavily damped, ranging from 11% of critical damping at 2kHz to 40% of critical damping at 500Hz. It was polished to 0.3μm RMS roughness. For sound excitation, a loudspeaker was placed forward of the wind tunnel contraction. Nearly plane sound waves were beamed downstream over this test plate. The 15" square test section was lined with an acoustic absorbing treatment in order to eliminate cross-mode contributions. The termination of the open jet had impedance matching fittings which greatly reduced wave reflections back up the test section. The details of these arrangements are given in Gedney [8]. For plate vibration tests a push rod from a shaker was connected to the lower side of the test plate 7.6cm from the leading edge. The sides of the test plate near the leading edge were freed up to permit a flapping vibration of the leading edge of the test plate which was nearly uniform transversely.

Most of the experiments with excitation were conducted at the discrete reduced frequency $\beta = \frac{\omega\nu}{U_\infty^2}$ of $56\text{x}10^{-6}$, but at tunnel velocities of 18.5, 29 and 41m/s. Excitation sound pressure levels were approximately 90dB (reference 20μPa). Plate leading edge transverse vibratory velocity levels were of the same order as sound particle velocities under sound excitation. With sound excitation alone, the plate vibratory velocity levels were negligible. During these experiments, hot wire, microphone or accelerometer signals were passed through a 4Hz tracking filter followed by a Princeton Applied Research (PAR) Model 124 lock-in amplifier during sound or vibration excitation. The tracking filter was centered on, and the PAR amplifier phase locked onto, the oscillator driving the loudspeaker, or shaker. The lock-in amplifier, of course, was out of the circuit during natural excitation. In the following, the displacement thickness Reynolds Reynolds number R_δ* is based upon the assumption of Blasius boundary layer flow from the leading edge. The stability and eigenmode calculations are also for this assumed flow. These were obtained using the spatial stability programs of Orszag [9], based upon a spectral method. Thus, for $\beta = 56\text{x}10^{-6}$ branch I of the neutral stability curve is at $R_\delta^* \cong 1000$.

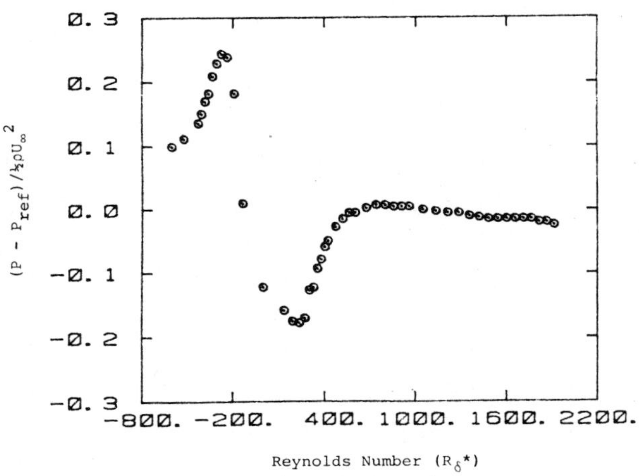

Figure 2. Mean pressure distribution for U_∞=18.5m/s.

Figure 2 shows the mean pressure distribution in the neighborhood of the leading edge at U_∞ = 18.5m/s. Since R_δ* is proportional to $\sqrt{U_\infty}$, the adverse pressure gradient region moves to the right on such a plot as U_∞ increases.

Experimental Results

Confirming the earlier results of Shapiro, we found that sound produced more rapid disturbance growth rates at all speeds than did stream turbulence of the same frequency. Greater growth rates occurred at higher speeds simply because the adverse gradient was closer to branch I of the neutral stability curve.

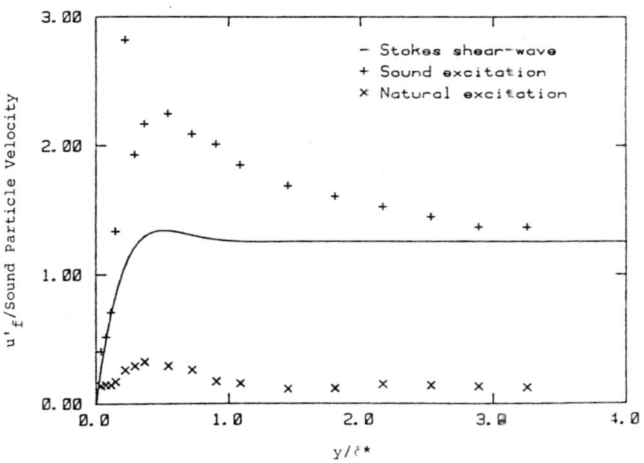

Figure 3. Disturbance wave velocity profile for sound excitation. Frequency=200Hz; U_∞=18.5m/s; x=20.3cm; $\beta=56\times10^{-6}$; R_δ*=856.

Figure 3 shows the disturbance wave velocity amplitude and phase as a function of normal distance from the plate for both sound and natural excitation. These are referenced to the sound particle velocity as measured in the freestream above the leading edge. The amplitude ratio appears to asymptote to a value slightly greater than unity proceeding out through the boundary layer, indicative of a slight reflected wave effect. The solid line is the theoretical Stokes wave representing a long wave length acoustic wave travelling downstream over a Blasius boundary layer. Lighthill [10] showed that this Stokes wave is the lowest order solution to the unsteady, linearized, incompressible boundary layer equations at high frequencies or far downstream. Since this Stokes wave is also an eigensolution of the Orr-Sommerfeld equation for any real frequency and zero

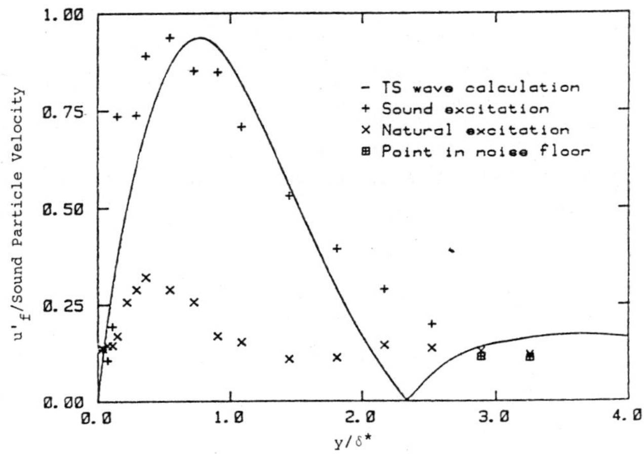

Figure 4. Disturbance wave velocity profile for sound excitation. Frequency=200Hz; U_∞=18.5m/s; x=20.3cm; β=56x10^{-6}; R_{δ}*=856. Stokes wave removed.

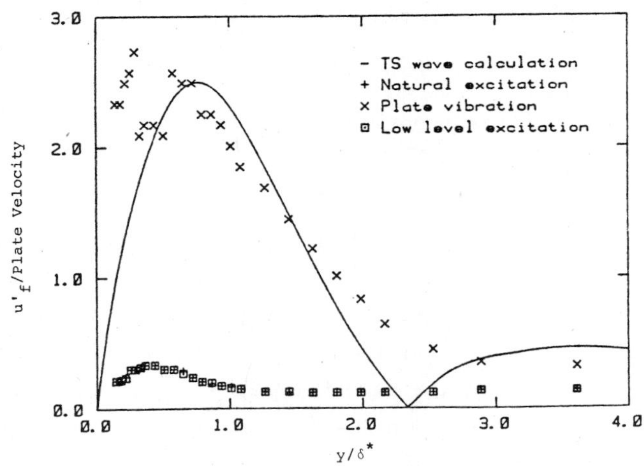

Figure 5. Disturbance velocity profile for leading edge plate vibration excitation, data otherwise the same as in Figure 3.

wavenumber, it is presumably unccupled from the Tollmien-Schlichting wave in the parallel flow region. We have therefore removed the Stokes wave from the disturbance wave measurement. The result is shown in Figure 4. Here we compare the resulting amplitude and phase with the fundamental Tollmien-Schlichting wave eigenmode, adjusted for best fit at the position of maximum amplitude. Clearly the response is primarily in the fundamental Tollmien-Schlichting mode. In Figure 5 we show the disturbance velocity profile for leading edge plate excitation at the same frequency and tunnel speed. The leading edge plate normal velocity is approximately the same as the sound particle velocity in the preceeding test. The low level excitation is that for a plate normal velocity the same as that measured when the plate had been excited by the sound field, indicating the negligible contribution of the path from sound to plate vibration to the disturbance wave. We see the probable excitation of higher eigenmodes in the amplitude data quite close to the wall. We note that there was a very weak transverse acoustic wave excited during plate vibration.

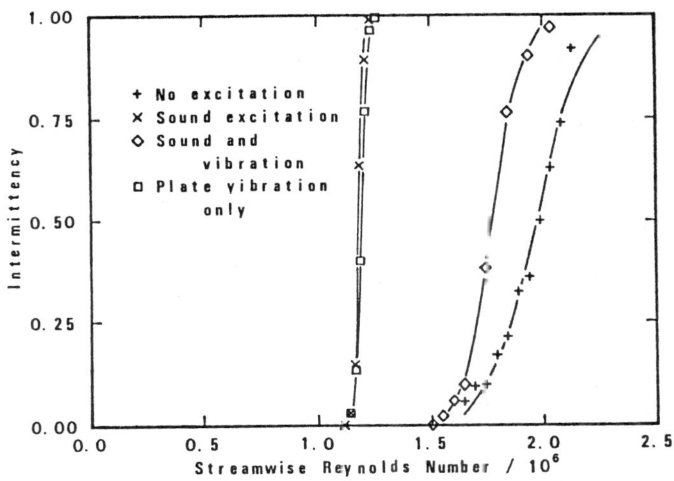

Figure 6. Intermittency as a function of stream-wise Reynolds number Re_x.

The above results suggested the possibility that plate vibration might be used to cancel the effects of sound on the boundary layer, Gedney [11], referring to Figure 6. Re_x for transition was determined as the position of very rapid increase in the intermittency factor, i.e., the fraction of time that the boundary layer was turbulent at the measurement point. For a tunnel speed U_∞ = 29m/s, and a sound excitation of 500Hz, the amplitude of the sound wave was increased until the point of transition moved appreciably upstream. For this condition, the sound pressure level was 89dB at the plate leading edge. The vibration shaker was then turned on at the same frequency. The amplitude and phase of vibration were adjusted until transition moved downstream as far as possible. At this condition, the sound particle velocity was 1.3×10^{-3}m/s and the leading edge

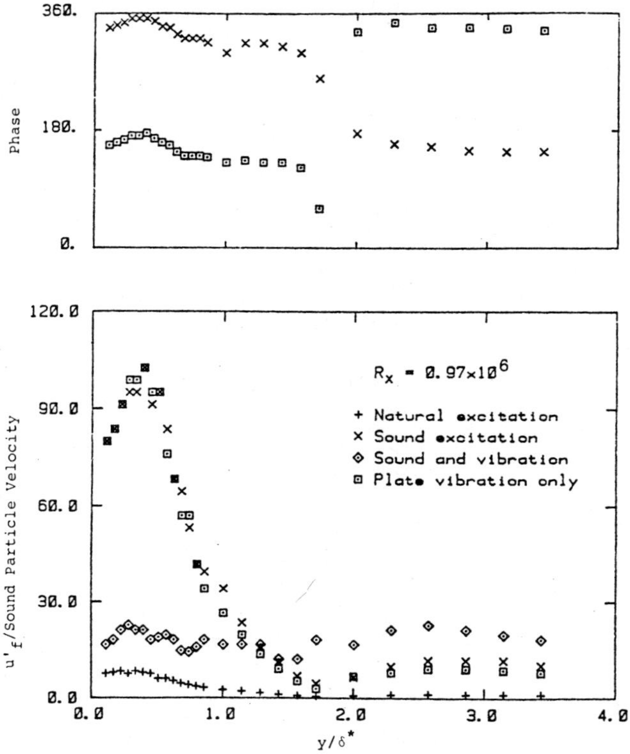

Figure 7. Disturbance velocity amplitude and phase profiles for cancellation experiment. Frequency=500Hz; U_∞=29.0m/s; $\beta=56\times10^{-6}$; R_{δ^*}=1695.

vibration velocity was 2.9×10^{-2} m/s. We then measured the disturbance velocity profiles through the boundary layer for conditions of no excitation, sound only, vibration only, and both sound and vibration for maximum cancellation effect. The results are shown in Figure 7. The position of measurement is near that for maximum amplification of the fundamental Tollmien-Schlichting wave. There the responses to sound alone and plate vibration alone are almost exactly 180° out of phase, and when both are superimposed, the total response is extremely low. Note that this response had to be magnified by a factor of 10 in order to be shown on the figure. The cancellation effect is clearly linear in behavior, yet its influence persists throughout the entire process of breakdown to turbulence. We note, however, that there was no increase in transition Re_x beyond that for stream turbulence excitation.

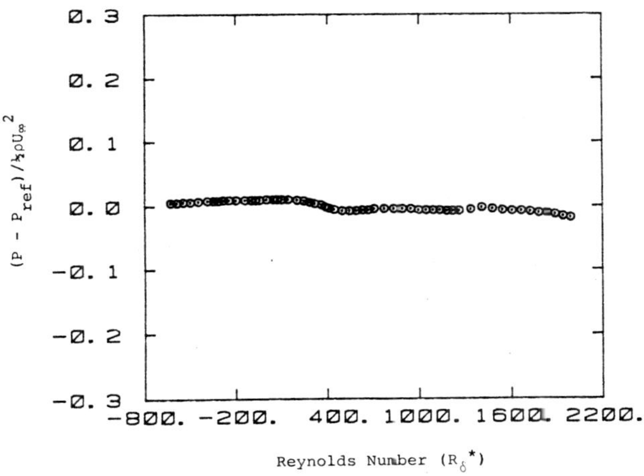

Figure 8. Mean pressure distribution for inclined test plate experiment. $U_\infty = 18.5$ m/s.

The final result we wish to show is the effect of tipping the test plate slightly downwards in order to place the stagnation point on the upper test side of the plate. Figure 8 shows the resulting mean pressure distribution. It is essentially zero over the entire upper side of the plate. Figure 9 shows the disturbance velocity profile for sound excitation. The test

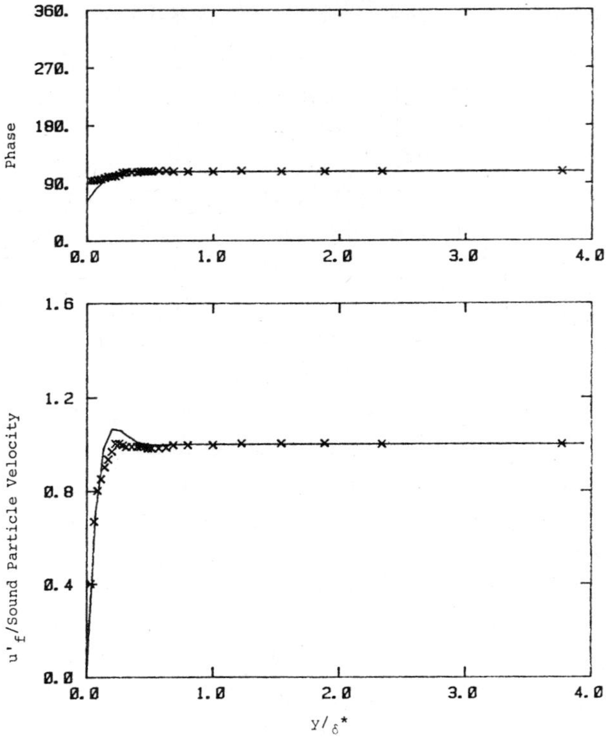

Figure 9. Disturbance velocity amplitude and phase profiles for inclined plate experiment. Frequency=450Hz; U_∞=25m/s; $\beta=66\times10^{-6}$; $R_\delta{*}=592$.

conditions ($\beta=66\times10^{-6}$, $R_\delta{*}=592$) are very near those for Figure 3, except for the absence of the adverse pressure gradient. Note that it is an almost perfect Stokes wave. The amplitude peak is not quite reached experimentally. Of perhaps greater interest, there is also a deviation in phase from the -45° value at the wall predicted by the Lighthill theory. We are presently unable to explain why this deviation occurred. However, it is clear that the receptivity under these test conditions is quite small and undoubtedly more in keeping with the theoretical values as predicted by Murdock or Goldstein.

Conclusions

1) Tollmien-Schlichting waves are generated for both sound and vibration excitation near the leading edge of the test plate;

2) Sound generated Tollmien-Schlichting waves can be cancelled by vibration generated Tollmien-Schlichting waves. This linear superposition effect persists throughout the process of breakdown to turbulence; and

3) We obtained no increase in natural transition Re_x by the cancellation process.

References

1. Shapiro, P.J.: "The influence of sound upon laminar boundary layer instability," Acoustics & Vibration Laboratory, Massachusetts Institute of Technology, Report 83458-83560-1 (September 1977).

2. Leehey, P.; Shapiro, P.J.: "Leading edge effect in laminar boundary layer excitation by sound," in Laminar-turbulent transition, IUTAM Symposium, Stuttgart, Germany, September 1979; R. Eppler and H. Fasal (eds.), Springer-Verlag, Berlin (1980) 321-331.

3. Leehey, P.: "The influence of environment in laminar boundary layer control," AIAA Progress in Astronautics and Aeronautics, vol. 72, Viscous flow drag reduction, G.R. Hough (ed.), (1980) 4-16.

4. Murdock, J.W.: "The generation of a Tollmien-Schlichting wave by a sound wave," Proc. Royal Soc. London, A, 372 (1980) 517-534.

5. Murdock, J.W.: "Tollmien-Schlichting waves generated by unsteady flow over parabolic cylinders," AIAA 19th Aerospace Sciences Meeting, St. Louis, MO, (January 1981) Paper No. AIAA-81-0199.

6. Goldstein, M.E.: "The evolution of Tollmien-Schlichting waves near a leading edge," Jour. Fluid Mech., vol. 127 (1983) 59-81.

7. Goldstein, M.E.; Sockol, P.M.; Sanz, J.: "The evolution of Tollmien-Schlichting waves near a leading edge: Part II--Numerical determination of amplitudes," Jour. Fluid Mech. vol. 129 (1983) 443-454.

8. Gedney, C.J.: "The response of a laminar boundary layer to sound and wall vibration," Acoustics & Vibration Laboratory, Massachusetts Institute of Technology, Report 83560-3 (May 1983).

9. Orszag, S.A.: "Accurate solution of the Orr-Sommerfeld stability equation," Jour. Fluid Mech., vol. 50, part 4 (1977) 689-703.

10. Lighthill, M.J.: "The response of laminar skin friction and heat transfer to fluctuations in the stream velocity," Proc. Royal Soc., vol. 224A (9 June 1954) 1-23.

Acknowledgement

This work was sponsored by the Mechanics Division of the Fluid Dynamics Program of the Office of Naval Research under contract N00014-76-C-0396.

Excitation of Instability Waves by a Vibrator Localized in the Boundary Layer

A. M. TUMIN and A. V. FEDOROV

Moscow Institute of Physics
and Technology, USSR

Summary

A theoretical model of the disturbances generated by the vibrations of the streamlined surface in the boundary layer is discussed. The analysis was performed with the aid of the expansion of the solution in eigenfunctions of linearized Navier-Stokes equations with nonparallelism of the flow in a real boundary layer taken into account. High efficiency of the excitation of Tollmien-Schlichting (TS) waves has been found.

1. Consider a flow in the two-dimensional boundary layer. In the interest of simplicity the flow is regarded as incompressible. Taking into account the compressibility does not lead to a considerable complication [1] . Let x denote the distance downstream the leading edge of the model along the streamlined surface, and y denote the distance normal to the model. A vibrator oscillating with frequency ω is placed at a distance L from the leading edge. We choose characteristic scales: for distance $X \sim L$ in x-axis; and y $- (\mu_\infty X/\rho_\infty U_\infty)^{1/2}$ in y-axis; where μ_∞ and U_∞ are the coefficient of viscosity and the velocity of the oncoming flow, respectively, and ρ_∞ denotes the density. Time is measured in $(\mu_\infty X/\rho_\infty U_\infty^3)^{1/2}$, pressure in units $\rho_\infty U_\infty^2$. Assume the basic flow to be weakly nonuniform in x-axis, with longitudinal velocity component U and normal velocity component V^* satisfied by the following relationship:

$$U = U(x,y),\ V^* = \varepsilon V(x,y),\ \varepsilon = R^{-1} = (\mu_\infty/\rho_\infty U_\infty X)^{1/2}$$

where $\varepsilon \ll 1$. Let the vibration of the surface be described by the equation

$$y_w(s,t) = \mathrm{Re}(a\, f(s)\, e^{-i\omega t}),\ s = \varepsilon^{-1}(x - x_*),\ f(s) = O(1),$$

where $x_*=L/X$ is the coordinate of the center and $f(s)$ is the form of the vibrator localized in x-axis on scale $\sim \varepsilon L$. The amplitude of the vibrations is assumed to be much less than the thickness of the viscous sublayer, $a \ll (\omega R)^{-1/2}$. In this case the disturbances generated by the vibrator located in the boundary layer are described by the linearized Navier-Stokes equations [2 , 3] . Introduce the vector-function $\vec{Q}(x,y,t)$ in which Q_1, Q_2, Q_3 - are the disturbances of the x-component of the velocity, pressure and of the y-component velocity disturbance, respectively, $Q_4 = \partial Q_1/\partial y - \partial Q_3/\partial x$. For the disturbances with a fixed value of frequency

$$\vec{Q}(x,y,t) = a\,\mathrm{Re}(\vec{A}(x,y)\,e^{-i\omega t})$$

$\vec{A}(x,y)$ is a complex amplitude vector-function satisfying linearized Navier-Stokes equations, with Fourier time transformation carried out [1] :

$$\frac{\partial \vec{A}}{\partial y} - H_1\vec{A} = \varepsilon H_2\frac{\partial \vec{A}}{\partial x} + \varepsilon H_3\vec{A} \tag{1.1}$$

Matrices H_1, H_2 have dimensionality 4 x 4. Their elements are determined through the characteristics of the mean flow in a locally uniform approximation in x and depend on x as a parameter. Matrix H_3 is stipulated by the nonparallelism of the flow. For the impenetrable surface the following boundary conditions are imposed :

$$\mathrm{Re}(a\,A_1(x,y_w)\,e^{-i\omega t}) + U(x,y_w) = 0 \tag{1.2}$$

$$\mathrm{Re}(a\,A_3(x,y_w)\,e^{-i\omega t}) + V^*(x,y_w) - \partial y_w/\partial t = 0$$

Expansion of (1.2) in the vicinity of $y = 0$, with the accuracy up to $O(a^2) + O(\varepsilon a)$ yields

$$A_1(x,0) = -U_w'\,f(s),\quad A_3(x,0) = -i\omega f(s),\quad U_w' = \partial U/\partial y\big|_{y=0}. \tag{1.3}$$

At $y \to \infty$ the disturbances are supposed to be bounded

$$|\vec{A}| < \infty \qquad (1.4)$$

In the cross-section $x = x_0$ located rather far away upstream the vibrator the initial data are given:

$$\vec{A}(x_0, y) = \vec{A}_0(y) \qquad (1.5)$$

Generally speaking, the problem (1.1), (1.3) - (1.5) is incorrect. It is necessary for $\vec{A}_0$ to allow the development of the solution with the final downstream growth rates [1, 4] .

2. Biorthogonal system of eigenfunctions $\{\vec{A}_\alpha(x,y), \vec{B}_\alpha(x,y)\}$ is formulated as in [1,5] :

$$\frac{\partial \vec{A}_\alpha}{\partial y} - H_1 \vec{A}_\alpha = i\alpha H_2 \vec{A}_\alpha \qquad (2.1)$$

$$y = 0, \; A_{\alpha 1} = A_{\alpha 3} = 0;$$

$$y \to \infty, \; |\vec{A}_\alpha| < \infty$$

$$\frac{\partial \vec{B}_\alpha}{\partial y} + H_1^* \vec{B}_\alpha = i\bar{\alpha} H_2^* \vec{B}_\alpha \qquad (2.2)$$

$$y = 0, \; B_{\alpha 2} = B_{\alpha 4} = 0;$$

$$y \to \infty, \; |\vec{B}_\alpha| < \infty$$

where "*" denotes a transpose matrix with complex conjugate elements and "-" denotes the complex conjugate values. Eigenfunctions $\vec{A}_\alpha(x,y)$ describe the wave disturbances of fixed frequency ω, extending in x-axis and having the form $\vec{A}_\alpha \exp(\varphi_\alpha(x))$ where $\varphi_\alpha = \int_{x_0}^{x} \varepsilon^{-1} \alpha \, dx$, with α being a complex eigenvalue of the problem (2.1) nondimensionalized with respect to $(U_\infty \rho_\infty / \mu_\infty X)^{1/2}$. We have the following orthogonal relationships:

$$\langle H_2 \vec{A}_\alpha, \vec{B}_\beta \rangle = \int_0^\infty \sum_{i,j=1}^{4} H_2^{ij} A_i \bar{B}_j \, dy = \Delta_{\alpha\beta}$$

where $\Delta_{\alpha\beta}$ is a Kronecker's symbol, when one of the eigenvalues α, β belongs to a discrete spectrum; $\Delta_{\alpha\beta} = \delta(\alpha - \beta)$ is the delta-function, when both belong to the continuous spectrum. The analysis of the spectrum for spatially growing disturbances and orthogonality conditions have been studied independently of [5] in [6]. Systems of equations (2.1), (2.2) describe the locally uniform approach, with x being a parameter. They have four linearly independent solutions with the following asymptotics outside the boundary layer

$$\vec{Z}_1 \sim e^{-\alpha y},\ \vec{Z}_2 \sim e^{\alpha y},\ \vec{Z}_3 \sim e^{\lambda y},\ \vec{Z}_4 \sim e^{-\lambda y},$$
$$\lambda = (\alpha^2 + i R(\alpha - \omega))^{1/2}$$

For certainty we shall choose the branch $\mathrm{Re}(\lambda) < 0$. Systems of equations (2.1), (2.2) permit four types of solutions from the continuous and one from discrete spectrum. The discrete spectrum corresponds to TS waves whose solution can be written in the form of a linear combination of $\vec{A}_{TS} = C_1 \vec{Z}_1 + C_3 \vec{Z}_3$, the eigenvalues of TS wave $\alpha_{TS}(x)$ being defined from the dispersion relationship $E_{13}(\alpha_{TS}) = (Z_{11}Z_{33} - Z_{13}Z_{31})_{y=0}$ where Z_{ij} is the ith component of jth vector. In the continuous spectrum $\alpha = \pm i k$ and $\alpha = -0{,}5\, i R\, (1 \pm (1+4 k^2 R^{-2} - 4 i\omega R^{-1})^{1/2})$ where k is a real parameter running through values from 0 to ∞. The existence of the two **branches** of the continuous spectrum for which the solution can grow as fast as you like, corresponds to the fact that the initial problem for spatially growing disturbances is incorrect and in order to regularize the problem some restriction is needed for the class of initial data. We shall add an inhomogenious at y = 0 solution $\vec{A}_v(x,y)$:

$$\frac{\partial \vec{A}_v}{\partial y} - H_1 \vec{A}_v = i \alpha_v H_2 \vec{A}_v \tag{2.3}$$

$$y = 0,\ A_{v1} = - U_w' \rho(\alpha_v),\ A_{v3} = -i\omega\rho(\alpha_v);$$

$$y \to \infty,\quad |\vec{A}_v| \to 0$$

where $\rho(\alpha_v)$ is the spectral density

$$f(s) = \int_{-\infty}^{\infty} \rho\,(\alpha_v)\,\exp(i\alpha_v s)\,d\alpha_v \tag{2.4}$$

The solution of system (2.3) may be presented as

$$\vec{A}_v = \frac{\rho(\alpha_v)}{E_{13}(\alpha_v)}\Big[(Z_{13}\, i\omega - U'_w Z_{33})\,\vec{Z}_1 + (Z_{31}\,U'_w - i\omega Z_{11})\,\vec{Z}_3\Big] \tag{2.5}$$

where all Z_{ij} are calculated at $y = 0$.

3. Following [1] consider the shape of the vibrating surface of the form $f = a\,\rho\,(\alpha_v)\exp(i\,\varepsilon^{-1}\alpha_v(x - x_*))$ corresponding to one harmonic from the integral (2.4). Let the solution of the problem be represented as the expansion in biorthogonal system of eigenfunctions $\{\vec{A}_\alpha, \vec{B}_\alpha\}$ in a locally-homogenious approach, supplemented with an inhomogenious solution at y=0. We will be interested in the generation of a TS wave by a vibrating surface. Hence, we will consider in detail only TS wave $\vec{A}_{TS}\exp(\varphi_{TS}(x))$ and disturbance $\vec{A}_v\exp(\varphi_v(x))$, i.e.

$$\vec{A}(\alpha_v, x, y) = C_{TS}(x)\,\vec{A}_{TS}(x,y)\,e^{\varphi_{TS}(x)} + a\,\vec{A}_v(x,y)\,e^{\varphi_v(x)}$$

In the case of a flat plate, as shown by numerical calculations, a strong damping of TS waves occurs at the flow-interval between the leading edge and the point of stability loss. Therefore, if the initial section is located sufficiently far upstream of the vibrator, and the vibrator itself is near the point of stability loss, the final result of high accuracy will not change if one assumes $C_{TS}(x_0) = 0$. Consider a resonance regime of excitation [1] when x_* is the point of stability loss, i.e. $\alpha_v = \alpha_{TS}(x_*)$. Resultant solution $\vec{A}(\alpha_{v*}, x, y)$ corresponding to the resonance regime of excitation is

$$\vec{A}(\alpha_{v*}, x, y) = a\,\vec{A}_{v*}\exp(\varphi_{v*}) - a\int_{x_0}^{x}\Big[\Big\langle H_2\frac{\partial \vec{A}_{v*}}{\partial x}, \vec{B}_{TS}\Big\rangle + \Big\langle H_3\vec{A}_{v*}, \vec{B}_{TS}\Big\rangle\Big]\exp(\varphi(x))\,dx\;\vec{A}_{TS}\exp(\theta_{TS}(x))\,; \tag{3.1}$$

$$\varphi(x) = \int_{x_0}^{x}(i\,\varepsilon^{-1}(\alpha_{v*} - \alpha_{TS}) - W_{TS,TS})\,dx$$

$$\theta_{TS}(x)=\int_{x_0}^{x}(i\varepsilon^{-1}\alpha_{TS}+W_{TS,TS})\,dx;\ \vec{A}_{v*}=\vec{A}_v(\alpha_{v*},x,y);$$

$$W_{TS,TS}=-\left\langle H_2\frac{\partial\vec{A}_{TS}}{\partial x},\vec{B}_{TS}\right\rangle-\left\langle H_3\vec{A}_{TS},\vec{B}_{TS}\right\rangle$$

Solution (2.5) in the vicinity of resonance point $x = x_*$ is not suitable because $E_{13}(\alpha_{v*},x_*)=0$ and $|\vec{A}_{v*}|\to\infty$ at $x\to x_*$. However, the solution (3.1) is uniformly regular at $x = x_*$. The basic item in (3.1) describing the excited TS wave at $x - x_* \gg \sqrt{\varepsilon}\,x_*$ is determined **by the steepest descent method.** For a small **departure** from the resonance $\Delta\alpha=\alpha_v-\alpha_{TS}(x_*)$

$$\vec{\Phi}_{TS}(\alpha_v,x,y)=\frac{a\,\rho(\alpha_v)\,g}{\langle H_2\vec{A}_{TS},\vec{B}_{TS}\rangle}\exp\left[\frac{i\,(\alpha_v-\alpha_{v*})^2}{2\varepsilon\left(\frac{d\alpha_{TS}}{dx}\right)_*}+\theta_{TS}\right]\vec{A}_{TS}$$

$$g=\frac{1}{\sqrt{\varepsilon}}\sqrt{\frac{2\pi}{\left(\frac{d\alpha_{TS}}{dx}\right)_*}}\quad(U'_w\,B_{TS1}+i\omega B_{TS3})_{*,y=0}\ \exp(i\phi)+O(\varepsilon^{1/2})$$

where ϕ is a real phase.

Integrating $\vec{\Phi}_{TS}(\alpha_v,x,y)$ in all α_v belonging to the domain of the resonance interaction $|\alpha_v-\alpha_{v*}|\sim\sqrt{\varepsilon}\,\alpha_{v*}$, at $x - x_* \gg \sqrt{\varepsilon}\,x_*$ we obtain an asymptotic estimate for the **total** amplitude of the instability wave generated by a localized vibrator. In our calculations we considered the maximum in y-axis value of the disturbance mass flow rate modulus q_m.

$$q_m=2\pi a\,q_{TS}\left|\frac{\rho(\alpha_{TS})\,(U'_w\,B_{TS1}+i\omega B_{TS3})_{*,y=0}}{\langle H_2\bar{A}_{TS},\bar{B}_{TS}\rangle}\right|e^{\operatorname{Im}\theta_{TS}}$$

where q_{TS} is mass flow rate modulus calculated in vector $\vec{A}_{TS}$. These results can be easily generalized for the case of a compressible gas [1] . The calculations were carried out on a heat-isolated surface in a compressible flow with the ratio of specific **heats 1,41**; Prandtl number 0,72. The Mach number of the oncoming flow varied from M = 0,2 to 0,8; stagnation temperature was equal to 310°K, with the viscosity coefficient calculated by Sutherland's formula. The Dunn-Lin system was used to calculate both the eigenfunctions **and** the eigenvalues. Shown in the table are the results of the amplitudes of the mass flow rate disturbances $q_m(x_*)$ in the TS wave generated by a vibrator having the parametr $2\pi a \rho(\alpha_{v*}) = 1$ localized near the point of stability loss x_*. The nondimensional frequency $F = 2\pi f \mu_\infty / \rho_\infty U_\infty^2$ is equal to $20 \cdot 10^{-6}$ (f is the disturbance frequency in Hz). Also shown are the eigenvalues of α_{TS} and Reynolds number R_* corresponding to $X = x_*$.

Table

M	R_*	$\alpha_{v*} 10^2$	$q_{\dot m} 10^3$
0,2	1020	7,27	5,79
0,4	1010	7,02	5,37
0,6	960	6,45	4,67
0,8	900	5,73	3,81

To compare the results of the theory with the experiment [7], rms pulsations of the x-component of the velocity $u_{TS}(x)$ maximum in y-axis were calculated in the excited TS wave. In accordance with the experiment the form of the vibrator was specified as a trapezoidal one with bases ℓ = 28 mm at the bottom and $2/3\,\ell$ at the top.

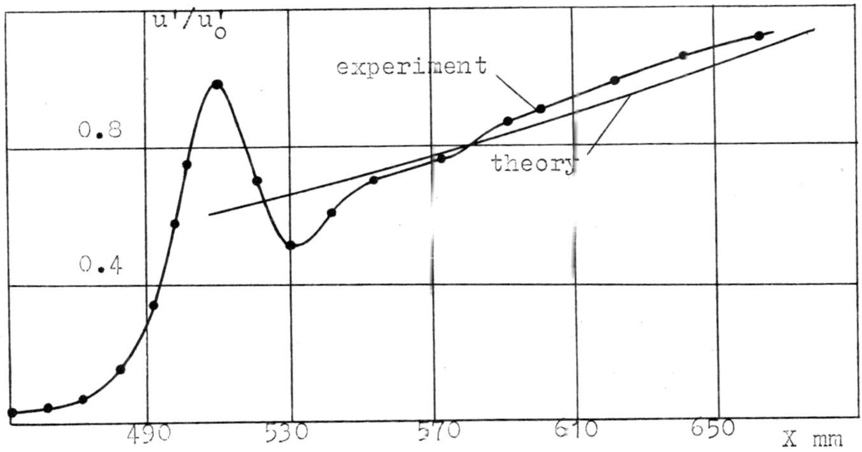

Fig.1

The comparison of the theoretical calculations with the experiment for the vibration frequency f=70 Hz ($F = 120\ 10^{-6}$) is shown in fig. 1. The absolute value of the pulsations is referred to the maximum in x-axis value $u_0 = 0{,}13\%\ U_\infty$, reached at $x \simeq$ 510 mm from the leading edge [7] ; the amplitude of the vibrations a=10 micrometers. In the $x \gtrsim 550$ mm region where the TS wave is singled out from the total signal, the theory is in good agreement with the experiment.

The numerical analysis of the instability wave excitation on a swept wing of infinite span [8] also showed high efficiency of the generation of TS waves by the surface vibrations.

From the calculations it follows that for the excitation of the instability waves with "dangerous" (for transition) frequencies and with initial amplitudes $\sim 10^{-2}\%\ U_\infty$ it is sufficient to have very small vibrations of streamlined surface with the same frequency but the amplitude ~ 1 micrometers.

<u>References</u>

1. Tumin, A. M.; Fedorov, A. V. - J. Pricl.Mech. and Tech.Phys., 1983, N 3 (in Russian)

2. Terentyev, E. D. - Docl.Acad.Nauk USSR, 1978, <u>240</u>, N 5 (in Russian)

3. Terentyev, E. D. - Pricl. Math. Mech., 1981, <u>45</u>, N 6 (in Russian)

4. Tumin, A. M.: Fedorov, A. V. - J. Pricl.Mech. and Tech.Phys., 1983, N 4 (in Russian)

5. Zhigulev, V. N.; Sidorenko, N. V.; Tumin, A. M. - J. Pricl. Mech. and Tech. Phys., 1980, N 6 (in Russian).

6. Salwen, H; Grosh, C.E. - J. Fluid Mech., 1981, <u>104</u>, N 1.

7. Gilev, V. M.; Kozlov, V. V. - Preprint N 19-83, Novosibirsk, Inst.of Theor.and Appl.Mech. of the Siberian Depart.of the Academy of USSR, 1983 (in Russian)

8. Tumin, A. M. - J. Pricl.Mech. and Tech.Phys., 1983, N 5 (in Russian).

Linear Problem on Vibrator on a Flat Plate in Subsonic Boundary Layer

E. D. TERENT'EV

Computing Center of the USSR Academy of Sciences
Moscow, USSR

Summary

The problem on subsonic flow past a flat plate with a vibrator placed at the distance L^* from the leading edge is considered. The vibrator comes into operation in an undisturbed boundary layer at the initial moment $t^* = 0$. The plate and vibrator are supposed to be insulated, the scales of vibrator and the frequency are such that for Reynolds number Re_{L1} the flow is governed by the boundary layer equations with the self-induced pressure included. The amplitude of the vibrator and corresponding disturbances are supposed to be small enough for linearizing governing equations. In the approximation under consideration we do not take into account the boundary layer thickening. The main attention is payed to calculation of the pressure distribution over vibrator as well as upstream and downstream of it. It is shown that the amplitude of the pressure oscillations decreases upstream. As for the region at "moderate" distances downstream the vibrator the pressure decreases only if the frequency ω_0 is less than the threshold frequency ω_{c*} which is fixed by the lower branch of the neutral stability curve. The pressure amplitude is constant if $\omega_0 = \omega_{c*}$ and grows for $\omega_0 > \omega_{c*}$. At the same distances the disturbances take more simple form of the single Tollmien-Schlichting wave in the limit as $t \to \infty$, the amplitude depending on the vibrator frequency and shape.

Linear Problem on Vibrator

Let us consider the flow past an insulated flat plate with a vibrator installed on it. The forepart of this plate upstream the vibrator has the length L^* and the back part behind the vibrator is $O(L^*)$ where stars are related to dimensional quantities. Let us introduce the Cartesian co-ordinates x^*, y^* with the origin at the point of connection of the forepart with the vibrator, the x^*-axis and the velocity U_∞ of undisturbed subsonic stream being directed along the plate. Let ρ^*_∞ denote the density, u^*, v^* - the longitudinal and transverse velocity components, p^* is the pressure and $\lambda^*_{1\infty}$ is the first viscosity coefficient. Instead of the reciprocal Reynolds number $Re_{L1} = \rho^*_\infty V^*_\infty L^* / \lambda^*_{1\infty}$ the small parameter $\varepsilon = Re_{L1}^{-1/8}$ is introduced, the Prandtl number being chosen equal to 1.

Suppose that the length of the vibrator is $O(L^* \varepsilon^3)$, the amplitude is $O(L^* \varepsilon^5)$, the frequency is $O(V^*_\infty / (L^* \varepsilon^2))$. It is convenient for such vibrator to distinguish three specific regions or decks [1,2]. In the upper deck ($y^* = O(L^* \varepsilon^3)$) the disturbed flow is inviscid and potential, the disturbances in the main deck ($y^* = O(L^* \varepsilon^4)$) are inviscid but rotational, in the lower deck ($y^* = O(L^* \varepsilon^5)$) the flow structure is essentially governed by shear stresses. The main difficulties in the triple deck scheme are associated with construction of a solution in the lower deck. Flow parameters in the upper and main decks can be written in an explicit form after obtaining the solution in the lower deck.

Let us introduce now the special dependent and independent dimensionless variables defined in [3]. If the disturbance amplitude is small enough the Navier-Stokes equations give the linear version of the unsteady boundary layer equations

$$\frac{\partial u}{\partial x} + \frac{\partial v}{\partial y} = 0 \, , \quad \frac{\partial p}{\partial y} = 0 \qquad (1)$$

$$\frac{\partial u}{\partial t} + y \frac{\partial u}{\partial x} + v = -\frac{\partial p}{\partial x} + \frac{\partial^2 u}{\partial y^2}$$

Suppose that the motion of the vibrator and its form for $t \geqslant 0$ is expressed by

$$y_w = f(t,x) = f_1(x) \sin \omega_0 t \qquad \omega_0 > 0$$

ω_0 being the frequency. In what follows the function $f_1(x)$ is chosen as $f_1(x) = 0$ for $x \leqslant 0$, $2x$ for $0 \leqslant x \leqslant b$, $2b(a-x)/(a-b)$ for $b \leqslant x \leqslant a$, 0 for $x > a$. For $t < 0$ the vibrator is supposed to be at rest, so that $y_w = 0$ and the boundary layer is undisturbed

$$u(0,x,y)=v(0,x,y)=p(0,x)=0 \qquad (y\neq 0) \tag{2}$$

No-slip conditions on the body surface give after linearization $(t \geqslant 0)$

$$u(t,x,0)=-f_1(x)\sin\omega_0 t\,,\quad v(t,x,0)=f_1(x)\omega_0\cos\omega_0 t \tag{3}$$

The matching conditions with the undisturbed boundary layer read for $x \to \pm\infty$ and fixed y, t

$$u \to 0\,,\quad v \to 0\,,\quad p \to 0 \tag{4}$$

The interaction conditions with the outer flow lead to the relation between u and p as $y \to \infty$

$$u \to \frac{1}{\pi}\int_{-\infty}^{x} dx_0 \int_{-\infty}^{\infty} \frac{p(t,x_1)}{x_1 - x_0}\,dx_1 \tag{5}$$

Let us transform the desired functions into the Fourier and Laplace integrals

$$\{\bar{u}(\omega,k,y),\bar{v}(\omega,k,y),\bar{p}(\omega,k)\} = \tag{6}$$

$$=\frac{1}{\sqrt{2\pi}}\int_{-\infty}^{\infty} e^{-ikx}dx\int_{0}^{\infty}\{u(t,x,y),v(t,x,y),p(t,x)\}e^{-\omega t}dt$$

The problem on a vibrator oscillating in a subsonic boundary layer infinitely long has been considered in [4]. So far as in this case the disturbed motion was studied for all x, the range of frequencies was limited by $\omega_0 < \omega_{c*}$ where $\omega_{0*} = 2.298$ is the threshold frequency fixed by the low branch of the neutral stability curve. With the initial conditions (2) the problem (1)-(5) has been considered in [5]. In this formulation the regime of harmonic oscillations occurs only in the limit as $t \to \infty$ for finite x. The problem under consideration allows to study oscillations even for $\omega_0 \geqslant \omega_{c*}$. From technical point of view the Fourier transform in t could not be applied as it was the case in [4], here the Laplace transform is relevant. The problem in some sense analogous to (1)-(5) has been considered in [6] as well. However, the analysis of dispersion relation was absent in [6], in particular the continuous part of the spectrum was not taken into account and as consequence the accurate evaluation of inverse integral transforms was not achieved.

Substituting (6) into (1) and taking into account (2)-(5) we obtain

$$\bar{P} = \omega_0 |k| \bar{f}_1(k) Ai'(\Omega) / [(\omega_0^2 + \omega^2) Q(\Omega, k)]$$

$$\Omega = \omega (ik)^{-2/3}, \quad I_0 = \int_0^\infty Ai(x)dx = \frac{1}{3}, \quad I_1 = \int_0^\Omega Ai(z)dz \tag{7}$$

$$Q(\Omega, k) = -Ai'(\Omega) + (ik)^{1/3} |k| [I_0 - I_1(\Omega)]$$

$$\bar{f}_1(k) = -\sqrt{\frac{2}{\pi}} \cdot \frac{1}{k^2} \left(1 - \frac{a}{a-b} e^{-ikb} + \frac{b}{a-b} e^{-ika}\right)$$

Here $Ai(\Omega)$ is the Airy function and $\arg i = \pi/2$. The expression for the disturbed pressure is determined by means of inverse Fourier-Laplace transform

$$P = -i\, 2^{-3/2} \pi^{-3/2} (I_2 + I_3), \quad Q_2 = -Q|_{k<0}, \quad Q_3 = Q|_{k>0}$$

$$I_2 = \omega_0 \int_{-\infty}^0 k \bar{f}_1 e^{ikx} dk \int_{l-i\infty}^{l+i\infty} \frac{Ai'(\Omega) e^{\omega t}}{(\omega^2 + \omega_0^2)\, Q_2} d\omega$$

$$I_3 = \omega_0 \int_0^\infty k \bar{f}_1 e^{ikx} dk \int_{l-i\infty}^{l+i\infty} \frac{Ai'(\Omega) e^{\omega t}}{(\omega^2 + \omega_0^2)\, Q_3} d\omega$$

Let us make a cut in the k-plane along the positive imaginary axis, put $\pi/2 > \arg k > -3\pi/2$ and choose the simple branch of integrands. Splitting the integral along k-axis into I_2 and I_3 we obtain the analytical integrands in I_2 and I_3. Introducing the new variables k_1 and ω_1 defined by $k = k e^{-i\pi}$, $\omega = l - i\omega_1$ and applying the properties of the derivative and integral of the Airy function

$$Ai'(c.c.\Omega) = c.c.Ai'(\Omega), \quad I_1(c.c.\Omega) = c.c.\, I_1(\Omega)$$

where c.c. denotes the complex conjugate, we have $I_2(t,x) = -\,c.c.\, I_3(t,x)$. Hence the pressure can be written as

$$p = 2^{-1/2}\pi^{-3/2}\,\mathrm{Im}(I_3) \tag{8}$$

Consider the inner integral entering I_3 and denote it by J_3. Let us analyse the roots of the denominator of the integrand of J_3 in the complex ω-plane while k runs through the positive real axis. Putting Q_3 equal to zero we arrive at the dispersion relation for subsonic boundary layer which was studied in detail in [7]. The traces of the first three roots are drawn in Fig.1 and labelled as 1, 2, 3. As for the other roots of the dispersion relation it could be shown that all of them lie in the third quadrant. For any root beginning from the second one the following property holds true

$$\pi < \arg \omega_{3n}(k) < \tfrac{3}{2}\pi - 0.056\pi\,, \quad n = 2, 3, \ldots$$

The two other roots of the denominator $i\omega_0$ and $-i\omega_0$ do not depend on k. The trace of the first root of Q_3 intersects the imaginary axis at the point $-i\omega_{0*}$. Fig.1 relates to the case $\omega_0 > \omega_{0*}$. Let us transform the integral J_3, for that purpose we subtract from and add to it an expression based on the first root of the dispersion relation

$$J_3 = J_{30} + J_{31}\,, \quad J_{30} = \int_{l-i\infty}^{l+i\infty} \Phi_3\, d\omega\,, \quad \Omega_{31} = \frac{\omega_{31}(k)}{(ik)^{2/3}}$$

$$\Phi_3 = \left\{ \frac{Ai'(\Omega)}{(\omega^2+\omega_0^2)\,Q_3(\Omega,k)} - \frac{Ai'(\Omega_{31})}{[\omega_{31}(k) - i\omega_0]\,Q_{3\omega}(\Omega_{31},k)} \times \right. \tag{9}$$

$$\left. \times \frac{1}{(\omega + i\omega_0)[\omega - \omega_{31}(k)]} \right\} e^{\omega t}\,, \quad Q_{3\omega}(\Omega_{31},k) = \left.\frac{\partial Q_3}{\partial \omega}\right|_{\omega=\omega_{31}(k)}$$

$$J_{31} = \frac{Ai'(\Omega_{31})}{[\omega_{31}(k) + i\omega_0]\,Q_{3\omega}(\Omega_{31},k)} \int_{l-i\infty}^{l+i\infty} \frac{e^{\omega t}\, d\omega}{(\omega + i\omega_0)[\omega - \omega_{31}(k)]}$$

The representation (9) has the following feature: among the roots of the denominator of the integrand in J_{30} the root $\omega = \omega_{31}(k)$ is absent, all other roots are just the same as in J_3. As for the integral J_{31}, it can be calculated explicitely, as a result we have

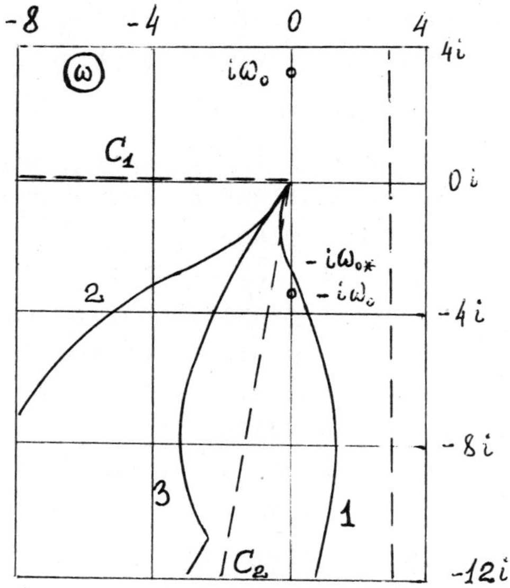

Fig.1. Traces of the first three roots of the dispersion relation in the complex ω-plane as $\mathcal{J}m\, k = 0$, $0 < \mathcal{R}e\, k < \infty$ and paths of integration.

$$J_{31} = 2\pi i \frac{A_i'(\Omega_{31})}{[\omega_{31}^2(k)+\omega_0^2]\, Q_{3\omega}(\Omega_{31},k)} \left(e^{\omega_{31}(k)t} - e^{-i\omega_0 t}\right) \tag{10}$$

To evaluate the integral J_{30} , instead of the original path of integration drawn in Fig.1 as a vertical dashline we choose one consisting of two rays C_1 and C_2 . The ray C_1 coincides with the negative real axis, while C_2 lies in the third quadrant and does not intersect the traces of any root beginning from the second one. Taking into account the residues at the points $i\omega_0$ and $-i\omega_0$ we have

$$J_{30} = 2\pi i\left[\operatorname{res}\Phi_3(-i\omega_0) + \operatorname{res}\Phi_3(i\omega_0)\right] + \left(\int_{C_2} + \int_{C_1}\right)\Phi_3\, d\omega \tag{11}$$

The integrals along the rays C_1 and C_2 can be estimated in the limit as $t \to \infty$. Using this estimation we find after the second integration in the k-plane [5] that the contribution to the pressure, as $t \to \infty$, is $O(t^{-6})$. It is necessary to emphasize that the estimation obtained does not depend upon x. Integrating the expression for J_3 in the k-plane we have

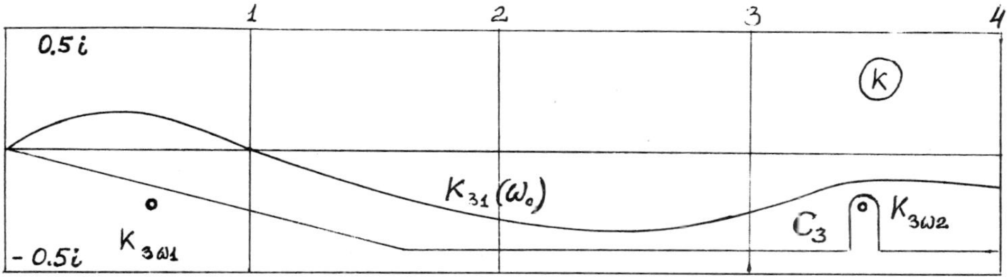

Fig.2. New path of integration in K-plane.

$$I_3 = 2\pi i\,\omega_0 \int_0^{\infty} k \bar{f}_1(k) e^{ikx} \left[res\,\Phi_3(-i\omega_0) + res\,\Phi_3(i\omega_0) \right] dk + \tag{12}$$

$$+ \omega_0 \int_0^{\infty} k \bar{f}_1(k) e^{ikx} J_{3L}\, dk + O(t^{-6}), \quad t \to \infty$$

As (12) shows, the problem on the pressure determination for any x and $t \gg 1$ is reduced to a simple integral in the k-plane. In what follows we restrict ourselves to a simpler problem on pressure calculation for finite x.

The singularity appears in the integrands of both integrals only for $\omega = \omega_{0*}$ at $k = k_{3*} = 1.0005$. The following rule can be formulated to treat this singularity: the contour of integration should contain a small circle below or above the point k_{3*} for both integrals, for instance below point k_{3*}. This rule allows us to neglect in (12) the terms which are proportional to $exp(-i\omega_0 t)$. Further on we apply the symbols J_{31} and $res\,\Phi_3(-i\omega_0)$ to designate the previous expressions without terms containing $exp(-i\omega_0 t)$.

It is convenient to split the calculation of the first integral in (12) into two steps. Differentiate this expression with respect to x, that allows to consider three integrals connected with three items of $\bar{f}_1$ proportional to 1, e^{-ikb}, e^{-ika} each being taken separately. Consider the computation method for integral proportional to 1. For $x < 0$ the path of integration varies as follows: instead of positive real axis

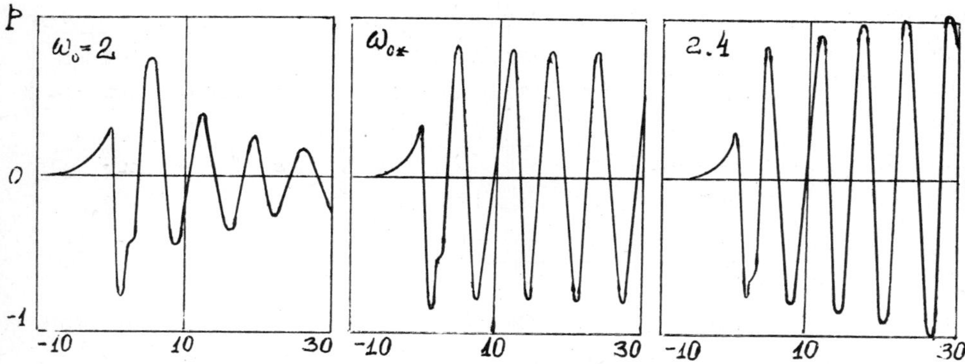

Fig.3. Disturbed pressure for vibrator with frequencies $\omega_c = 2$, 2.298, 2.4, and parameters $a = 2$, $b = 1$ at times $t = 2\pi(N+1/4)/\omega_0$, N-integer ($N \gg 1$).

we shall integrate along negative imaginary axis and for this it is necessary to take into account the pole generated by the first root of the dispersion relation. For $x > 0$ we shall integrate along positive imaginary axis and for $\omega_c \leqslant \omega_{c*}$ it is necessary to add a converging series consisting of the residues generated by the roots of the dispersion relation, beginning from the first, for $\omega_0 > \omega_{0*}$ the series begins from the term corresponding to the second root. The first step is over. At the second step we carry out the integration with respect to x.

Consider the second integral in (12). The root of equation $\omega_{31}^2(k) + \omega_0^2 = 0$ is denoted by $k = k_{31}(\omega_0)$. The trace of this root with variation ω_c from 0 to ∞ is drawn in Fig.2. In the fourth quadrant the equation $Q_{3\omega}(\Omega_{31}, k) = 0$ has the roots at the points $k_{3\omega 1} = 0.590 - i\cdot 0.244$, $k_{3\omega 2} = 3.444 - i\cdot 0.207$. The points $k_{3\omega 1}$ and $k_{3\omega 2}$ are branch points of function $\omega_{31}(k)$. Choose a new path of integration C_3, see Fig.2, lying below the trace $k = k_{31}(\omega_c)$ and above the points $k_{3\omega 1}$, $k_{3\omega 2}$, with taking into account the residue at $k = k_{31}(\omega_c)$. Then we have

$$\omega_0 \int_0^\infty k \bar{f}_1(k) e^{ikx} \bar{\gamma}_{31} dk \equiv \int_0^\infty \Phi_{31} dk = \tag{13}$$

$$= \int_{C_3} \Phi_{31} dk - 2\pi i \operatorname{res} \Phi_{31}(k_{31}(\omega_c)) \cdot \Theta(\omega_c - \omega_{0*})$$

In formula (13) $\Theta(\cdots)$ is the Heaviside function, for $\omega_c = \omega_{0*}$

the residue is absent as the integration around K_{3*} occurs in low half-plane. Consider dependence $Re\,\omega_{31}(k)$ upon $|k|$ with variation of k along C_3. The maximum $Re\,\omega_{31}(k)$ is reached at $k=0$ and hence the neighbourhood of point $k=0$ gives the basic contribution into integral from Φ_{31} on C_3 as $t\to\infty$, $x=const$. In accordance with the Laplace method we have for fixed x

$$\int_{C_3}\Phi_{31}\,dk = O\left(t^{-6}\right) \tag{14}$$

It should be emphasized that estimation (14) is valid not only for fixed x, but also for

$$x < Vt\,, \qquad V = \left(Re\,\omega_{31}(k)/Jm\,k\right)_{max} \tag{15}$$

where subscript max denotes maximum value for k belonging to C_3. As computations show the velocity V could be chosen equal to 2.4. Thus the second integral in (12) under condition (15) should be taken into account only for $\omega_0 > \omega_{0*}$, in such a case the integral is determined by residue at $k = k_{31}(\omega_0)$, for $\omega_0 \leq \omega_{0*}$ it can be neglected.

The study [5] of asymptotic behaviour as $|x|\to\infty$ of integrals (12) gives for $x \ll -1$ $I_3 = O\left(\frac{1}{x^2}\right)$ and for $x \gg 1$ $I_3 = O\left(\frac{1}{x^2}\right) - 2\pi i\, res\,\Phi_{31}$. The magnitude $-2\pi i\, res\,\Phi_{31}(k_{31}(\omega_0))$ appears only in interval $1 \ll x < Vt$ and gives the Tollmien-Schlichting wave.

The pressure for frequencies $\omega_0 = 2$; $\omega_0 = \omega_{0*} = 2.298$; $\omega_0 = 2.4$ for times $t = 2\pi(N+1/4)/\omega_0$, N -integer ($N \gg 1$) and $a = 2$; $b = 1$ is drawn in Fig.3.

References

1. Neiland, V.Ya.: Izv. AN SSSR, MZHG, (1969) N 4, 53-58.
2. Stewartson, K.: Mathematika, 16 (1969) N 31, 106-121.
3. Rizhov, O.S.: Dokl. AN SSSR 234 (1977) N 4, 780-783.
4. Terentiev, E.D.: PMM 45 (1981) vip. 6, 1049-1055.
5. Terentiev, E.D.: PMM 48 (1984) vip. 2, 264-272.
6. Gaster, M.: J. Fluid Mech. 22 (1965) pt. 3, 433-441.
7. Zhuk, V.I.; Rizhov, O.S.: Dokl. AN SSSR 253 (1980) N 6, 1326-1329.

On Tollmien-Schlichting Wave Generation by Sound

Ruban A.I.

Central AeroHydrodynamics Institute USSR

It is well known that acoustic disturbances incident upon the boundary layer may cause Tollmien-Schlichting waves. An intense generation of these waves is observed in local inhomogeneities of the flow [1] . To investigate this phenomenon we shall consider a two-dimensional gas flow about a flat plate directed along a free-stream flow (fig.1). The free-stream velocity U_∞ is taken to be subsonic with $M < 1$. Introduce cartesian coordinates x', y' aligning their origin with the plate leading edge and orienting x' along the plate surface. Denote velocity vector components in this coordinate system by u' and v', density by ρ' , and viscosity coefficient by μ' . Let the gas under consideration have an adiabatic constant γ , and its thermodynamic state be described by the Clapeyron equation.

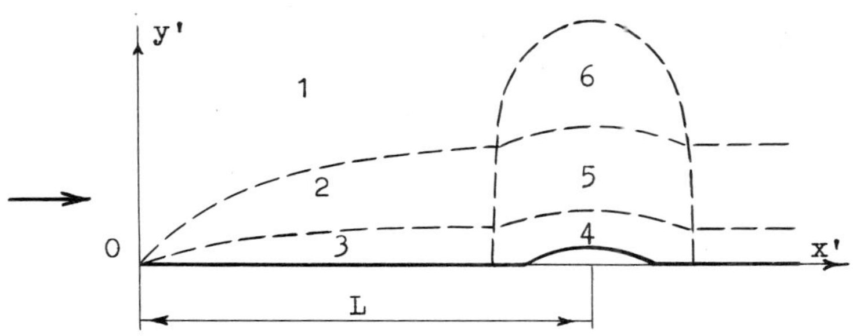

Fig.1.

Assume that at a distance of L from the leading edge the plate surface has a local roughness, its form being represented as:

$$y' = L\,Re^{-5/8} f_*\left(\frac{x'-L}{L\,Re^{-3/8}}\right) \qquad (1)$$

where $Re=\rho_\infty U_\infty L/\mu_\infty$ is the Reynolds number. The stationary flow about such roughness is described by the free interaction theory (see, e.g.[2]). It is remarkable that a nonstationary analogue of this theory [3,4] predicts an asymptotic structure of the Tollmien-Schlichting wave on the low branch of the neutral curve.

As for acoustic disturbances, assume that they are represented by a monochromatic sound wave propagating in x'. The theory of the wave-roughness interaction is constructed where sound frequency ω' is taken to be of the order of $(U_\infty/L)Re^{1/4}$. As to sound wave amplitude it is chosen to be of the order of $Re^{-1/8}$, hydrodynamic function distributions in a freestream flow being expressed as

$$p' = p_\infty + \rho_\infty U_\infty^2\, Re^{-1/8} a_* \sin\xi + \dots \qquad (2)$$

Here $\xi = \omega_*\,[(1+M)t_* - Mx_1]$, and nondimensional variables t_*, x_1 are introduced using the relations

$$t' = (L/U_\infty)Re^{-1/4} t_*, \qquad x' = L\,Re^{-1/4} x_1.$$

The constants a_* and ω_* are characteristic of amplitude and frequency of a sound wave.

The relation (2) is an asymptotic solution of the Navier-Stokes equations at $Re \to \infty$. The solution remains valid for the whole region of the flow outside the boundary layer. Let us consider the process of the sound wave penetration into the boundary layer (region 2 in fig.1). The solution of the Navier-Stokes equations is

$$\begin{aligned} u' &= U_\infty u_0(X,Y) + U_\infty Re^{-1/8} M a_* u_2(X,Y)\sin\xi + \dots \\ v' &= U_\infty Re^{-3/8} M^2 a_* \omega_* v_2(X,Y)\cos\xi + U_\infty Re^{-1/2} v_0(X,Y) + \dots \end{aligned} \qquad (3)$$

where $X=x'/L$, $Y=y'/LRe^{-1/2}$; u_0 and v_0 refer to the stationary

solution of the boundary layer equations.
Equations for u_2 and v_2 may be obtained by substituting expansions (3) in the Navier-Stokes equations. The equations did not happen to contain viscous terms. Therefore, the no slip condition on the plate surface is not satisfied. This necessitates the introduction of an additional region near the wall (region 3) where the gas flow is viscous. The solution of the Navier-Stokes equations for this region may be written in the form of the following asymptotic expansions:

$$p'=p_\infty+\rho_\infty U_\infty^2 Re^{-1/8}p_3+\dots,$$
$$u'=U_\infty Re^{-1/8}u_3+\dots,\qquad v'=U_\infty Re^{-1/2}v_3+\dots, \tag{4}$$
$$\rho'=\rho_\infty\rho_0(X,0)+\dots,\qquad \mu'=\mu_\infty\mu_0(X,0)+\dots$$

p_3, u_3, v_3 are functions of four independent variables: t_*, x_1, X and $y_*=(y'/L)Re^{5/8}$.
The substitution of expansions (4) in the Navier-Stokes equations shows that the pressure remains unchanged across regions 3 and 2. Hence, $p_3=a_*\sin\xi$.
The momentum equation is separated from other hydrodynamics equations and may be considered apart from them. The periodic solution of this equation satisfying the no slip conditions on the plate surface and those of matching with the solution in region 2 is

$$u_3=\frac{\partial u_0}{\partial Y}(X,0)y_*+d\sin\xi-de^{-\varkappa y_*}\sin(\xi-\varkappa y_*),$$

$$d=\frac{Ma_*}{(1+M)\rho_0(X,0)},\qquad \varkappa=\left[\frac{\omega_*(1+M)\rho_0(X,0)}{2\mu_0(X,0)}\right]^{1/2} \tag{5}$$

Thus, the solution upstream the roughness is obtained. Now proceed to the analysis of the flow in the roughness vicinity occupied by an interaction region having a usual triple-deck structure (fig.1). First consider a viscous wall layer (region 4) adjacent to the roughness surface. Its characteristic thickness is of the order of $LRe^{-5/8}$ and it coincides with that of region 3. Here we will seek for an asymptotic expansion of the solution of the Navier-Stokes equations in the form:

$$u'=U_\infty Re^{-1/8}u_4+\dots,\qquad v'=U_\infty Re^{-3/8}v_4+\dots,$$
$$p'=p_\infty+\rho_\infty U_\infty^2 Re^{-1/8}p_*(t_*)+\rho_\infty U_\infty^2 Re^{-1/4}p_4+\dots,\qquad (6)$$
$$\rho'=\rho_\infty\rho_0(1,0)+\dots,\qquad \mu'=\mu_\infty\mu_0(1,0)+\dots$$

The variables t_*, y_* (introduced earlier) and $x_*=(x'/L)Re^{3/8}$ are used as independent ones in region 4.

The analysis of the Navier-Stokes equations using expansions (6) shows that the pressure p_4 does not change across region 4, and the equations of momentum and continuity are separated from the rest of hydrodynamics equations and have the form of unsteady boundary layer equations. Boundary conditions for them are the no slip conditions on the roughness surface (1) and those of matching with the solutions in regions 3 and 5. The last is

$$u_4=\frac{\partial u_0}{\partial Y}(1,0)(y_*+A_*)+\frac{Ma_*}{(1+M)\rho_0(1,0)}\sin\xi+\dots$$

as $y_*\to+\infty$. The function $A_*(t_*, x_*)$ is not known a priori. It is characteristic of the displacement effect of a viscous sublayer.

The analysis of the flow in region 5 shows that the streamline angle remains constant across the major part of the boundary layer and it appears at its outer boundary to be the same as at the outer boundary of region 4:

$$\theta=-Re^{-1/4}\frac{\partial A_*}{\partial x_*}.$$

Since this value is small, the linear theory of the potential flow is valid for region 6 outside the boundary layer (fig.1). Thus

$$p_4=-Ma_*\omega_*\cos\xi_0 x_*-\frac{1}{\pi\sqrt{1-M^2}}\int_{-\infty}^{\infty}\frac{\partial A_*/\partial s}{s-x_*}\,ds$$

This relationship closes the problem for region 4. Using standard affine transformations of the free interaction theory and the Prandtl transposition transformations, we obtain

$$\frac{\partial u}{\partial t}+u\frac{\partial u}{\partial x}+v\frac{\partial u}{\partial y}=a\cos\omega t-\frac{\partial p}{\partial x}+\frac{\partial^2 u}{\partial y^2}$$

$$\frac{\partial u}{\partial x}+\frac{\partial v}{\partial y}=0, \qquad p=\frac{1}{\pi}\int_{-\infty}^{\infty}\frac{\partial f/\partial s-\partial A/\partial s}{s-x}ds$$

$$x=-\infty: \qquad u=y+\frac{a}{\omega}\left[\sin\omega t-e^{-\sqrt{\frac{\omega}{2}}y}\sin(\omega t-\sqrt{\frac{\omega}{2}}y)\right]$$

$$y\to+\infty: \qquad u=y+A+\frac{a}{\omega}\sin\omega t$$

$$y=0: \qquad u=v=0$$

Reduced sound wave amplitude and frequency are calculated by

$$a=\frac{Ma_*\omega_*}{\lambda^{7/4}\mu_o^{3/4}\rho_o^{1/2}\beta^{1/4}} \qquad \omega=\frac{\omega_*(1+M)}{\lambda^{3/2}\mu_o^{1/2}\beta^{1/2}}$$

$$\beta=\sqrt{1-M^2}, \qquad \lambda=\frac{\partial u_o}{\partial Y}(1,0)$$

The solution of the interaction problem is sought assuming that sound wave amplitude and roughness height are much less than unity. Take, then, $f(x)=bF(x)$ and represent the sought functions as expansions in a and b powers:

$$\begin{aligned}
u&=y+(au_a+bu_b)+(abu_c+b^2u_d)+\dots\\
v&=bv_b+(abv_c+b^2v_d)+\dots\\
A&=bA_b+(abA_c+b^2A_d)+\dots\\
p&=bp_b+(abp_c+b^2p_d)+\dots
\end{aligned} \qquad (7)$$

The linear terms of expansions (7) satisfy to the superposition principle between a nonstationary solution u_a for a sound wave propagating along the boundary layer on a flat surface

(5) and a problem solution for a stationary flow about the roughness of a small height (see, e.g [5]). Therefore, it is clear that nonlinear effects are greatly responsible for the Tollmien-Schlichting wave generation. To study this phenomenon the second order terms have to be considered. As for functions with subscript "d" note that they are a nonlinear correction for a stationary problem solution of roughness flow. Obviously, these functions are also not related to the Tollmien-Schlichting wave generation. As to the functions with subscript "c", they are determined from the solution of the following boundary-value problem

$$\frac{\partial u_c}{\partial t} + y\frac{\partial u_c}{\partial x} + v_c = -\frac{\partial p_c}{\partial x} + \frac{\partial^2 u_c}{\partial y^2} - \left(u_a\frac{\partial u_b}{\partial x} + v_b\frac{\partial u_a}{\partial y}\right)$$

$$\frac{\partial u_c}{\partial x} + \frac{\partial v_c}{\partial y} = 0, \qquad p_c = -\frac{1}{\pi}\int_{-\infty}^{\infty}\frac{\partial A_c/\partial s}{s-x}\,ds \tag{8}$$

$$x=-\infty:\ u_c=0; \qquad y=+\infty:\ u_c=A_c; \qquad y=0:\ u_c=v_c=0.$$

According to [6] represent the sought functions u_c, v_c, A_c, p_c as

$$u_c(t,x,y) = e^{i\omega t}u_c'(x,y)$$

Using the Fourier transformation

$$\tilde{u}_c(k,y) = \frac{1}{\sqrt{2\pi}}\int_{-\infty}^{\infty}u_c'(x,y)e^{-ikx}dx$$

for the amplitude u_c', v_c', A_c', p_c' and introducing a new independent variable $z=(ik)^{1/3}y+z_0$, where $z_0=(ik)^{1/3}\omega/k$, we can reduce problem (8) to

$$\frac{d^3\tilde{u}_c}{dz^3} - z\frac{d\tilde{u}_c}{dz} = \tilde{F}(k)g(k,\omega,z) \tag{9}$$

$$z=\infty: \quad \tilde{u}_c=\tilde{A}_c$$

$$z=z_0: \quad \tilde{u}_c=0, \quad \frac{d^2\tilde{u}_c}{dz^2}=(ik)^{1/3}\tilde{p}_c, \quad \tilde{p}_c=|k|\tilde{A}_c \tag{9}$$

The homogeneous part of the equation for $d\tilde{u}_c/dz$ is the Airy equation. Denote the Fourier transform of the F(x) function by $\tilde{F}(k)$. The function g(k, ω, z) in the r.h. side of this equation is universal in that it does not depend on surface form. A general solution of the equation (9) bounded at $z \to \infty$ may be written as

$$\frac{d\tilde{u}_c}{dz}= \tilde{F}(k)\varphi(k,\omega,z)+B(k,\omega)Ai(z)$$

where Ai(z) is the Airy function; φ presents the solution of the problem

$$\frac{d^2\varphi}{dz^2}-z\varphi=g, \quad \varphi(z_0)=\varphi(\infty)=0 \tag{10}$$

The value of the constant B(k, ω) is found from the boundary conditions of system (9). Simultaneously the Fourier transform of the pressure is defined

$$\tilde{p}_c(k,\omega)= \tilde{F}(k)|k| \frac{I_\varphi Ai'(z_0)-I\varphi'(z_0)}{Ai'(z_0)-(ik)^{1/3}|k|I}$$

$$I_\varphi=\int_{z_0}^{\infty}\varphi(k,\omega,z)dz, \qquad I=\int_{z_0}^{\infty}Ai(z)dz \tag{11}$$

Here, $Ai'(z_0)$, $\varphi'(z_0)$ are derivatives with respect to z of the functions Ai and φ at $z=z_0$.
The pressure distribution along the interaction region may be reconstructed using the inverse Fourier transformation. The asymptotic behaviour of function $p_c(t,x)$ downstream the roughness is of primary interest for the problem of the Tollmien-Schlichting wave generation. It may be obtained by the method used in [7] . Provided that $\omega < \omega_0=2.298$ all the disturbances induced in the flow by the sound wave-roughness interaction prove to decay as they go away from the roughness.

The decay rate vanishes, however, as ω approaches the neutral oscillation frequency ω_0. If $\omega = \omega_0$, the disturbances become sustained and downstream the roughness the Tollmien--Schlichting wave is generated.

$$p_c(t,x) = \mathrm{Re}\left\{ \varkappa e^{i(\omega_0 t + k_0 x)} \right\}$$

$$\varkappa = \varkappa_0 \tilde{F}(k_0), \qquad \varkappa_0 = 0{,}3884 + i0{,}0277, \qquad k_0 = -1{,}0005$$

The coefficient of proportionality $\varkappa_0$ in the amplitude of this wave was determined using a numerical solution of the boundary problem (10).

References

1. Kachanov, Yu.S.; Kozlov, V.V.; Levchenko, V.Ya. Turbulence initiation in a boundary layer. Novosibirsk, Nauka, 1982 (in Russian).
2. Ruban, A.I.; Sychev, V.V. Asymptotic theory of laminar boundary layer separation in incompressible fluid. Uspekhi Mekhaniki", 1979, v.2, N 2 (in Russian).
3. Smith, F.T. On the non-parallel flow stability of the Blasius boundary layer. Proc.Roy.Soc. London, Ser.A, 1979, vol.366, N 1724.
4. Zhuk, V.I.; Ryzhov, O.S. Free interaction and boundary layer stability in incompressible fluid. Dokl.Akad.Nauk SSSR, 1980, v.253, N 6 (in Russian).
5. Stewartson, K. On laminar boundary layers near corners. Quart.J.Mech.Appl.Math., 1970, v.23, pt.2.
6. Bogdanova, E.V.; Ryzhov, O.S. On oscillations induced by a harmonic oscillator in Poiseuille flow. Dokl.Akad.Nauk SSSR, 1981, v.257, N 4 (in Russian).
7. Terentiev, E.D. Linear problem of a vibrator in a subsonic boundary layer. Prikl. MM, 1981, v.45, vyp.6 (in Russian).

Resonance-Induced Self-Oscillations in Fluid with Wavy Boundary Motion

G. V. LEVINA, A. A. NEPOMNYASCHIY, A. A. POZDEEV

Institute of Continuum Mechanics
Ural Scientific Center of the USSR Academy of Sciences
Perm' 614013 USSR

Summary

The present paper is concerned with resonant wavy external effect on the flows in the vicinity of stability threshold. The amplitude function method is used to derive the governing equation for the disturbance amplitude describing phenomena in supercritical region for the flows of various physical nature: boundary layer, convection, magnetohydrodynamics and so on. Theoretical predictions are confirmed by the finite-difference calculations of secondary regimes of mixed convection in a vertical layer with oscillating boundaries.

Introduction

A fundamental theoretical investigation of hydrodynamic stability concerns primarily with the parallel fluid motions: plane Couette and Poiseuille flows, plane boundary layers and flows in jets and wakes. In recent years the emphasis has been focused on the proper treatment of stability behaviour under the influence of various factors breaking the flow parallelism just in the subcritical domain. One of these factors is the unsteady external effect being in resonance with natural flow oscillations and allowing control over the instability.

It is a well known fact that with increasing the Reynolds number beyond its critical value a parallel flow generated within the layer with planar boundaries becomes unstable to normal disturbances which are spatially-periodic along the longitudinal co-ordinate and characterized by the wavenumber α [1]. The disturbance increment $\sigma = \sigma_r + i\sigma_i$ is a function of the wavenumber α [1], and the requirement $\sigma_r(\alpha, R)=0$ defines the curve of neutral stability $R(\alpha)$. For small excesses over the critical number R_0 a nonlinear evolution of disturbances with low, though finite amplitude a(t) is governed by the Landau equation

$$\frac{da}{dt} = \sigma a - \varkappa |a|^2 a \quad , \tag{1}$$

where σ is found from linear analysis, $\varkappa = \varkappa_r + i\varkappa_i$ is the Landau constant. The sign of $\varkappa_r$ determines a nonlinear evolution of disturbances. Since for a number of flows (convection, boundary layer and so on) $\varkappa_r > 0$, branching of the secondary regimes develops in a soft manner and tends towards the higher values of R. In this case, the self-oscillating secondary regimes can be examined by means of the amplitude equation.

In the present work we employ the method of a small parameter to derive the amplitude equation in case of a flow exposed to some unsteady external influence, namely, to a wavy motion of the channel boundaries.

Problem Formulation

Let us consider a parallel flow $\Psi_o(y)$ within an infinite layer with planar boundaries which is generated by a volume force F(y) in the x-direction and developed at prescribed time-mean flow. With exceeding the critical number R_o, this motion becomes unstable to disturbances with the wavenumber α and frequency $\omega_o(\alpha)$. We assume that a nonlinear evolution of instability at $R > R_o(\alpha)$ would be of soft mode and results in appearance of the secondary self-oscillating regime.

Assuming the wall deformation in the form of a travelling wave with the same wavenumber α, amplitude ζ and frequency Ω we get

$$y_B = \pm d\left\{1 + \zeta \sin\left[\alpha(x - ct)/d\right]\right\} , \tag{2}$$

where d is a nondisturbed half-width of the channel. The object of the present work is to describe bifurcations and supercritical spatially-periodic flows with a period equal to that of the wall distortion and with frequency of external influence $\Omega = c\alpha$, close to natural frequency of the fluid oscillations $\omega_o = -\sigma_i$.

Before proceeding with the presentation we shall perform transformation into the reference frame moving with a wave in the x-direction with nondimensional velocity $c = Cd/\nu$, and introduce the coordinate transformation flattening the boundaries:

$$\xi = x \quad , \qquad \eta = \frac{y}{1 + \zeta \sin \alpha x} \quad . \tag{3}$$

Nondimensionalised equation of motion and boundary conditions in terms of the stream function Ψ have the following form:

$$\frac{\partial \tilde{\Delta} \Psi}{\partial t} + a_2 \left(\frac{\partial \Psi}{\partial \eta} \frac{\partial \tilde{\Delta} \Psi}{\partial \xi} - \frac{\partial \Psi}{\partial \xi} \frac{\partial \tilde{\Delta} \Psi}{\partial \eta} \right) = \tilde{\Delta}^2 \Psi + a_2 R \frac{\partial F}{\partial \eta} \quad ,$$

$$\Psi = 0, \ \frac{\partial \Psi}{\partial \eta} = - \frac{c}{a_2} \ (\eta = -1); \ \Psi = q - 2c, \ \frac{\partial \Psi}{\partial \eta} = - \frac{c}{a_2} \ (\eta = 1); \tag{4}$$

where

$$\tilde{\Delta} \Psi = \frac{\partial^2 \Psi}{\partial \xi^2} + 2 a_1 \frac{\partial^2 \Psi}{\partial \xi \partial \eta} + a_3 \frac{\partial^2 \Psi}{\partial \eta^2} + a_4 \frac{\partial \Psi}{\partial \eta} \quad ;$$

a_i $(i = 1,2,3,4)$ are the coefficients of transformation into the curvilinear coordinate system (ξ, η); R is a nondimensional parameter, q the nondimensional flow averaged over the wave period; d, d^2/ν and ν are units of length, time and stream function respectively.

Amplitude Equation

Under the assumption of a small wave amplitude and low supercritical value, we obtain the amplitude equation using the method of a small parameter combined with the method of multiple time scales. In order to examine the motions in vicinity of stability threshold we can employ the expansions in powers of the small parameter ε proportional to the amplitude of the nonparallel flow component:

$$\Psi - \Psi_o = \sum_{n=1}^{\infty} \varepsilon^n \Psi^{(n)} \quad , \qquad \zeta = \sum_{n=1}^{\infty} \varepsilon^n \zeta^{(n)}(R) \quad ,$$

$$R - R_o = \varepsilon^2 R^{(2)} \quad , \qquad c\alpha - \omega_o = \varepsilon^2 c^{(2)} \alpha \quad . \tag{5}$$

We assume that the functions $\Psi^{(n)}$ depend on the time scaling $t_n = \varepsilon^n t$. The coefficients $\zeta^{(n)}$ are evaluated from the solvability condition of the boundary problem for each order in ε. From the solvability condition of the problem to the 3^d order in ε one gets the evolutionary equation for the disturbance amplitude [2,3]:

$$\frac{da}{dt} = \left[\sigma + i(\Omega - \omega)\right]a - \varkappa |a|^2 a + g\zeta \qquad (6)$$

where g is a real coefficient. The resulting equation differs from the Landau equation (1) due to the presence of the term $i\Omega a$ associated with transition into a moving reference frame, and nonuniform term $g\zeta$ involved by the boundary distortion. Eq.(6) is valid for isothermal flow described by a unique scalar function $\Psi(x,y)$ as well as for studying flows of increasing complexity (convection, magnetohydrodynamics and so on). The equation (6) involves three parameters:

$$\gamma = \sigma/|k| \;, \quad \delta = g\zeta(\varkappa_r/|k|^3)^{1/2}, \quad B = \frac{\varkappa_i}{\varkappa_r}\mathrm{sign}k, \qquad (7)$$

(where $k = \Omega - \omega$), which are proportional to $R-R_0$, the wave amplitude and a nonlinear frequency shift respectively. The quantity $1-B|a|^2$ represents a nonlinear distortion of the frequency for the finite-amplitude oscillations. The asymptotical regimes of eq.(6) have been discussed in detail [2]. Since eq.(6) is a system of two real equations, then at $t \to \infty$ it has as its limit regimes the stationary points, cycles and separatrices of the saddle-point.

The construction of stationary solutions and investigation of their stability is carried out analitically, the cycle and separatrix calculations are performed numerically using a Runge-Kutta integration method. It is found out that the sign and the magnitude of the parameter B, depending on the profile of initial flow are of a great importance for the transition diagram.

Stationary Motions

First consider the steady-state solutions of the system (in a moving reference frame). If $B \leq 0$, a nonlinear distortion of the frequency doesn't reduce with the growth of the amplitude. For $B \leq -\sqrt{3}/3$ the stationary solution is unique for the whole range of parameters. At $-\sqrt{3}/3 < B \leq 0$ the stationary solution

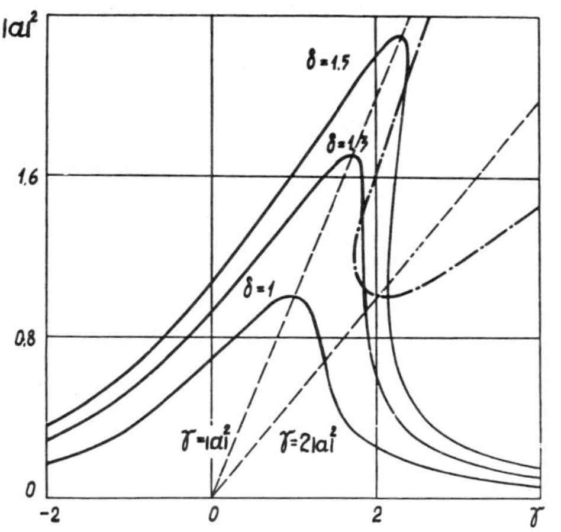

Fig. 1. Amplitude curves at B=0

is unique for any value of γ, provided that $\delta < \delta_1$ (see fig. 1, the line for $\delta = 1$). At $\delta > \delta_1$ there exists a region of γ-values $\gamma_-(\delta) < \gamma < \gamma_+(\delta)$, inside which the system has three stationary solutions (the lines for $\delta = 1.3$; 1.5; the boundary of solution branching is shown by the dashed line).

If B>0, the frequency distortion $1-B|a|^2$ turns into zero at $|a|^2 = 1/B$; this leads to a nonlinear resonance state which is represented in the form of closed amplitude curves in the case of stationary solutions. So, at B=2 one obtains the amplitude curve that may consist of two separate segments (the line 1 and 2 in fig. 2); one of the segments corresponding to the hig-

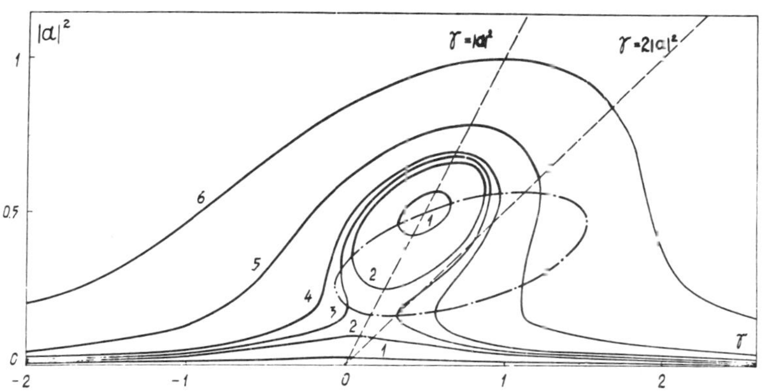

Fig. 2. Amplitude curves at B=2 and δ = 0.1(1); 0.25(2); 0.28(3); 0.316(4); 0.5(5); 1(6)

her values of $|a|^2$ is shown to be a closed contour converging towards the point $\gamma = |a|^2 = 1/B$ as $\delta \to 0$. The boundaries of the existence domain of three stationary solutions are shown in fig. 3 (the line 1 and 2). From the analysis of stability of stationary solutions it follows that in the domain of existence

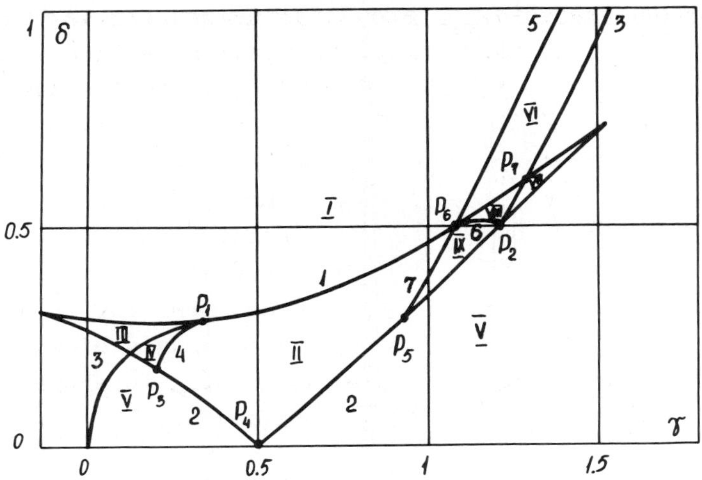

Fig.3. Diagram of supercritical motions at B=2

of three stationary regimes, the solution with intermediate value of $|a|^2$ is always monotonically unstable. Solutions with maximum and minimum values of $|a|^2$, as well as single stationary solution lose their stability to oscillating mode. The oscillating instability of steady solutions can take place at any value of B. The boundary of oscillating instability is depicted in fig.3 (the line 3). For $B \leq \sqrt{3}/3$ the boundaries 3 and 1 merge at $\gamma = \gamma_1$; in this case the frequency of oscillations turns into zero and bifurcation of codimension two takes place. At $B > \sqrt{3}/3$ the boundary 3 consists of two segments: $0 < \gamma < \gamma_1$ and $\gamma_2 < \gamma < \infty$.

Nonstationary Motions

Oscillating instability of stationary motions leads to a cycle branching. The character of this branching may be determined by a perturbation method. The analysis shows that at $B \leq \sqrt{3}/3$ there exists a soft-mode branching along the entire length of the oscillating instability boundary. At $B > \sqrt{3}/3$ the instability in the boundary interval $\gamma < \gamma_1$ is of a soft-type, while in the interval $\gamma > \gamma_2$ it is of a hard-type (cycle branching tends towards the lower values of γ); it is observed that oscillation frequency always decreases with the amplitude growth.
We concentrate on computation results obtained for the case B=2.

Typical phase trajectories are displayed in fig.4. Capital letters F,S,N,C_+,C_- denote respectively focus, saddle, node, stable and unstable cycles. From fig.3 it is readily seen that the

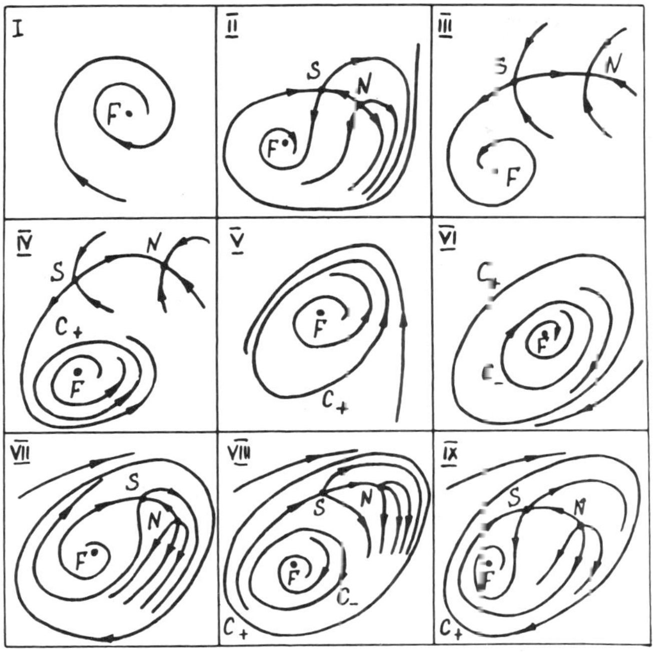

Fig.4. Phase portraits of the system at B=2

stable cycle exists in domains IY-IX; furthermore, in domains YI and YII there exists the unstable cycle. At the left-hand segment of the boundary 3 the stable cycle is developing by the soft-mode due to the loss of the focus stability. On the boundary 4 originating from the point of codimension two, the cycle converts into separatrix being doubleasymptotic to the saddle point (the "homoclinic trajectory") corresponding to the steady-state solution with intermediate value of $|a|^2$. In the intervals between the points P_3,P_4 and P_4,P_5 of the curve 2 the cycle turns into separatrix of a saddle point. The stable cycle on the boundary 5 (a multiple cycle curve) merges with the unstable cycle branching from the right section of the curve 3 towards the lower values of γ. The unstable cycle of the curve 6 vanishes converting into a loop of the saddle separatrix; the stable cycle of the curve 7 undergoes the similar transformation. Particular attention is given to nonuniqueness of stable motions: thus, in domain III there are two stable stationary points, whereas in domains IY,YI,YIII,IX such would be the cycle and the stationary regime; this can give rise to hysteresis phenomena. As B decreases, the parametric picture of the system is simplified. For $-\sqrt{3}/3 < B \leq \sqrt{3}/3$ there are no bifurcations at the right section of the boundary 3. If $B < -\sqrt{3}/3$ the only bifurcation in the relevant system would be the onset of the soft-mode stable cycle due to the loss of the focus stability. Our principal result is that the proposed amplitude equation for disturbances enables us to predict secondary motions and the character of transitions between them for various flows. Using the amplitude equation we analyze the stability and supercritical convective flows in a vertical layer with wavy motion of boundaries and lateral heating. A detailed discussion of results is presented in [4]. The analytical predictions [4] are confirmed by direct integration of full nonlinear convective equations in the Boussinesq form employing the finite-difference method in stream function - vorticity variables [2].

Finally it should be pointed out that the present investigation is not specific for the case when the resonance effects upon the system are associated with wavy motion of the boundaries. In fact the analogous analysis can be performed for any external oscillating influence which is considered be in resonance with disturbances of fluid; in any case it results in eq.(6). A specific character of particular problem is displayed in numerical coefficients involved in the amplitude equation.

References

1. Lin, C.C.: The theory of hydrodynamic stability. Cambridge Univ. Press 1955.

2. Levina, G.V.; Nepomnyaschiy, A.A.: Nonstationary boundary deformation effect on hydrodynamic stability and flow regimes. Preprint USC 45(82). Sverdlovsk 1982 (in Russian).

3. Levina, G.V.; Nepomnyaschiy, A.A.; Pozdeev, A.A.: Resonant influence of time-dependent boundary distortion on the viscous fluid flow stability and secondary regimes. Dokl. Akad. Nauk SSSR 266 (1982) 1334-1337.

4. Levina, G.V.; Nepomnyaschiy, A.A.: On regimes of mixed convection in a vertical layer with nonstationary deformable boundaries. Prikl. Matem. i Mekh. 47 (1983) 402-410.

The Control of the Three-Dimensional Disturbances Development in the Transitional Boundary Layer

N.F. YURCHENKO, V.V. BABENKO, L.F. KOZLOV

Institute of Hydromechanics, Ukrainian Academy of Sciences
Kiev, U.S.S.R.

Summary

The physical model of the disturbances type transformation during the transition of a laminar boundary layer into turbulent one is described. The preturbulent transition stage was thoroughly investigated for different flow conditions at natural and controlled formation of three-dimensional disturbances. The resulting flow field was shown to depend on the scale of the introduced disturbances.

The task of determining a boundary layer reaction to different types of disturbances is becoming urgent today [1, 2]. The aim of the present work is to study three-dimensional disturbances in preturbulent transition stage and possibilities to control their development.

The investigation was carried out experimemtally in the hydrodynamic bench [3] with the regulated free-stream turbulence (ϵ =0.05 — 10%) within Reynolds numbers range of $Re \approx 10^4 - 10^7$ on surfaces with different curvature (R, m =1; 4;12; ∞). As the registration methods of flow field kinematics the space visualization by means of electrochemical tellurium-method and laser anemometer measurements were applied. General approach and the research technique are described in detail in Refs. 3,4.

The physical model of the successive transformation of disturbances type during the transition to turbulence (Fig. 1) was built in consequence of the natural transitional boundary layer structure study as well as of the analysis and the generalization of different works results [5—7]. It was established that preconditions for the formation of disturbances typical of every stage were created in previous transition stages. Complicated flow conditions do not effect the general picture of disturbances development producing only its various forms and essential time dependence. In this respect it's interesting to consider the three-dimensional structures in a boundary layer because of their determinate character in contrast to the turbulent motion.
That is why the boundary layer reaction on induced three-dimensional disturbances of the longitudinal vorteces type was investigated. The disturbances of controlled intensity and scale were induced with the help of a series of different size vorteces makers placed in a boundary layer upstream the measuring point. The spanwise distance between neighbouring vorteces makers determines the wave length λ_z of the induced disturbances; it can be changed over the range λ_z = (0.2 — 3.2). 10^{-2} m with $0.2.10^{-2}$ m step.

First of all in accordance with the technique [4, 9] visualization and photoregistration of the flow structure in xz — planes at different distances from a wall were carried out. Then at the most typical z — positions (along lines after the vorteces makers, z = 0; in between them

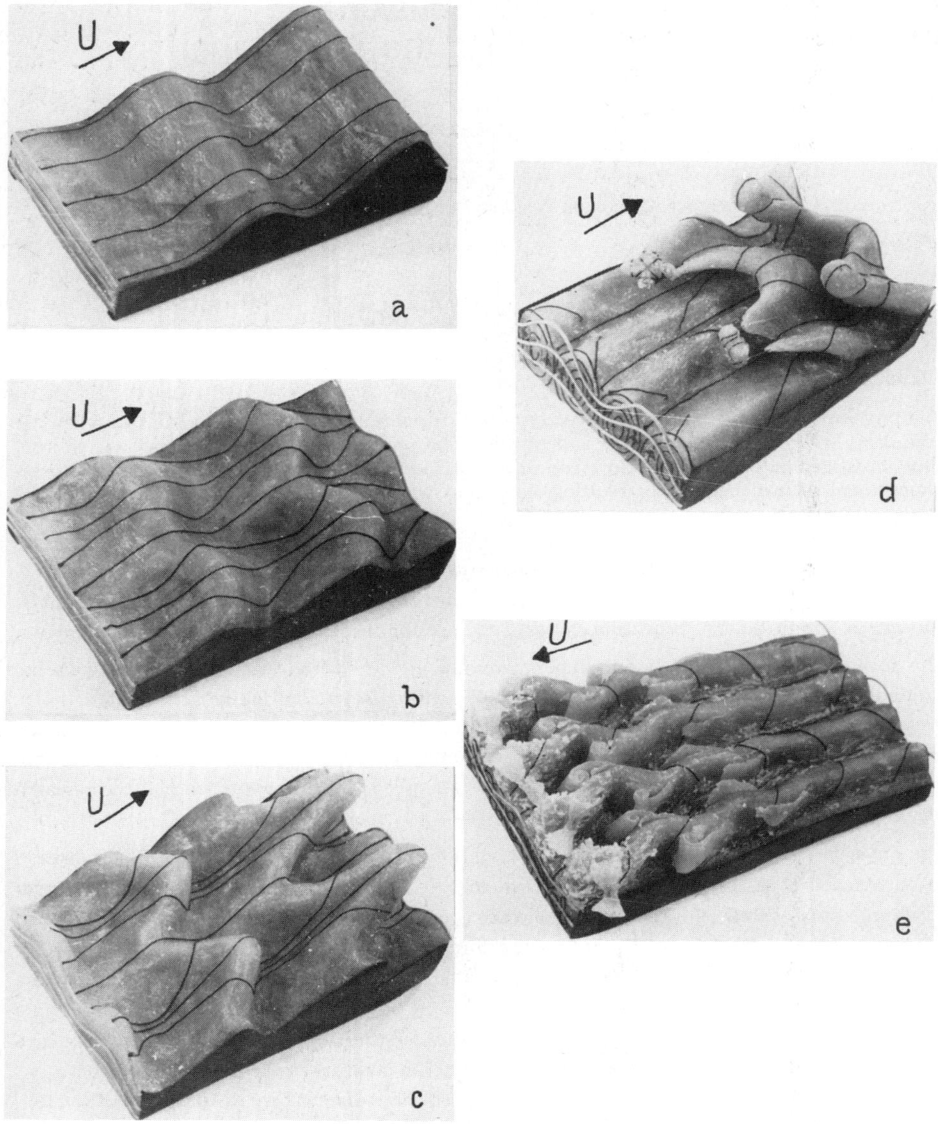

Fig. 1. The physical model of the disturbances type successive transformation during the transition to turbulence: a — amplification of plane disturbances; b — phase modulation of the wave, three-dimensional effects appearance; c — amplitude modulation of the wave, "peaks" and "valleys" forming; formation of longitudinal rows of Λ — shaped vortices; d — organization of the longitudinal vorteces system; e — vorteces form and intensity changes in space and time; transition to a zigzag trajectory of the vorteces movement; breakdown of peripherical parts of the meandering vortices. Active formation of turbulent spots, their growth, amalgamation and turbulent boundary layer development.

$z=\lambda_z/2$ and at $z=\lambda_z/4$) the profiles U (y), u´(y) [4, 8] and one-dimensional spectra of the longitudinal fluctuation velocity were measured. The analysis of the results was made on the basis of a comparision with the kinematics of a boundary layer under the same Re values without the disturbances introduction, i.e. on condition of the natural transition development. The interaction of the induced and natural disturbances was found to depend on the correlation of their wave lengths in the transverse direction. Under small λ_z values of the introduced disturbances this interaction represents simple summation of the mentioned structures; U (z) profile has a wavy form characteristic of the investigated stage of natural transition on which short wave disturbances system is superimposed. The resultant disturbance behaviour is completely determined by the regularities of large scale natural structure development. Under close scales of λ_z values and wave length of natural disturbance in z — direction either longitudinal lines

Table 1

Experiment characteristics

Controlled parameters	Series "a" 7	8	9	Series "b" 7	8	9
1. 10^2 λ_z, m	1.2	1.2	1.2	1.2	1.2	1.2
2. 10^2 U_∞; m/s	9.3	8.2	9.6	19.6	22.0	22.0
3. z_i	0	$\lambda_z/2$	$\lambda_z/4$	0	$\lambda_z/2$	$\lambda_z/4$
4. y_1/δ^*	1.75	1.75	1.75	1.17	1.17	1.17
5. y_2/δ^*	2	2	2	1.33	1.33	1.33
6. y_2/δ	0.25	0.25	0.25	0.2	0.2	0.2

	Series "c" 7	8	9	10	Series "d" 7	7	Series "e" 7	7
1.	0.8	0.8	0.8	0.8	1.6	1.6	1.2	1.2
2.	18.2	21.2	18.6	9.6	9.8	21.0	20.0	56.0
3.	0	$\lambda_z/2$	$\lambda_z/4$	0	0	0	0	0
4.	1.39	1.39	1.39	1.49	1.9	1.62	2.02	1.22
5.	1.01	1.01	1.01	1.08	1.38	1.18	1.03	0.615
6.	0.2	0.2	0.2	0.2	0.25	0.286	0.2	0.091

formation at about λ_z distance or quick destruction of the regular structure takes place which can be resulted in mutual disturbances amplification. The spectral characteristics of the boundary layer in different stages of transition were measured with the help of DISA 55DO1 hot-wire [9]. As well as the velocity distributions, spectra were registered at three z - positions (see Fig. 2 curves 7, 8, 9) and on using different size vorteces makers.

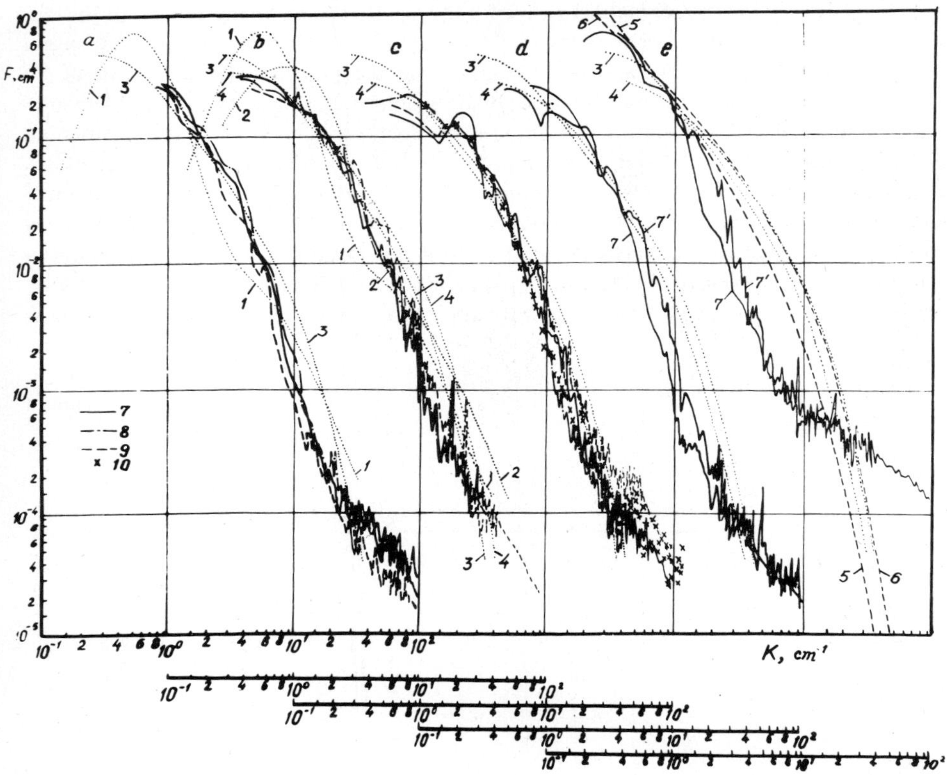

Fig. 2. One-dimensional spectra of the longitudinal fluctuation velocity under such results standardization

$$\int_0^\infty \frac{E(k)}{\overline{u'^2}} dk = 1,$$

where k is a wave number and E (k) is a function of the spectral density of the longitudinal fluctuation velocity u.′

The tests conditions and parameters numerical values are given in the Table 1. Here y_1 denotes the vorteces makers height, y_2 is the distance between a wall and a hot-wire pickup; y_1 and y_2 are standardized by a displacement thickness δ^* : $y_1 = 0.3 \cdot 10^{-2}$ m for "a,b" series, for "c, d" — $0.5 \cdot 10^{-2}$ m, for "e" — $0.7 \cdot 10^{-2}$ m. The curves 1, 2 correspond to the natural transition at $y_2/\delta^* = 1.08$ and 0.253 [11]; 3, 4 — at $y_2/\delta = 0.21$ and 0.063 [11]; 5, 6 — at $y_2/\delta = 0.2$ and 0.05 [12].

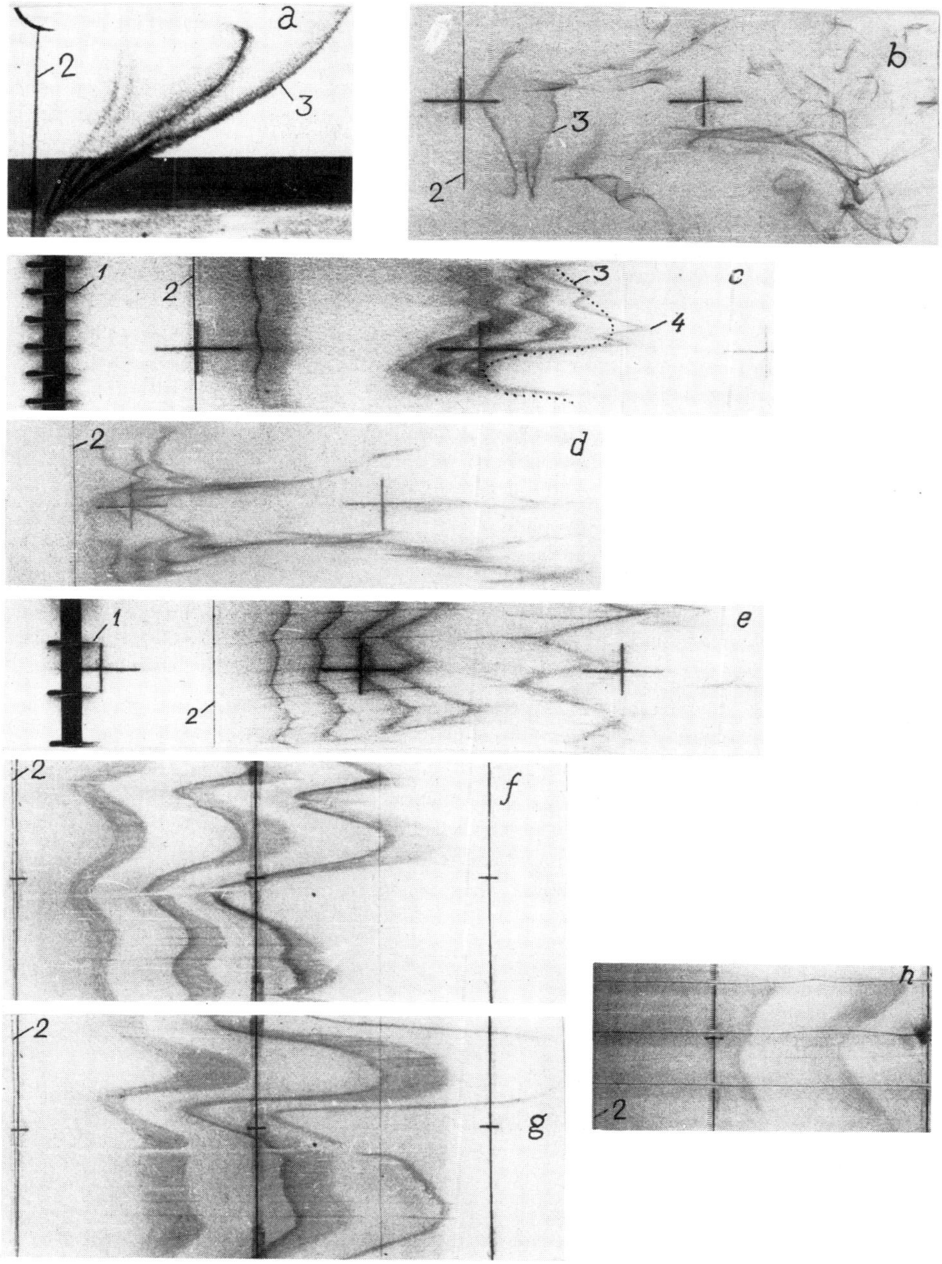

Fig. 3. The boundary layer visualization under the complicated flow condition: 1 — the vorteces makers; 2 — the tellurium wire; 3, 4 — the tellurium clouds.

The achieved results also prove the selective boundary layer reaction on λ_z value of induced disturbances.

Thus the introduction of the regular three-dimensional disturbances to a boundary layer results in changes of its kinematic and spectral characteristics. The power of this changes or the intensity of the natural and introduced disturbances interaction is determined, on one hand, by the latter scale and, on the other hand, by the transition stage, i.e. by the boundary layer state. The stimulation of the longitudinal vortex disturbances of a given scale is resulted in : 1) mean velocity profiles U (y) variation along λ_z wave length (they have the bulge shape in "valleys" and inflections in "peaks"); 2) the changes of a fluctuation profiles u′(y) form (the growth of the fluctuations intensity level, its maximum removing from the reference); 3) the deflection of spectral curves from the reference one (pits in certain spectrum range, curve protuberances at individual frequancies, smooth maximums at low frequancies, different spectral curves "smoothness" depending on transition stage).

The obtained data and developed criteria were used while investigating a transitional boundary layer which is formed in complicated conditions: when flowing over the surface with concave section and with high free-stream turbulence. In first case the concave section beginning is characterized by pressure gradient growth and correspondingly by inflectional velocity profiles U (y) (see Fig. 3,a). Tellurium clouds breakdown takes place here immediately after the visualizing wire (see b). The vortices makers introduction to the point of a sharp pressure gradient increase leads to flow regularity. And then either the mentioned above effect of a large scale structure (3) modulation by induced little scale disturbances (4) (see c) or the flow field longitudinal orientation (see d) or regular induced flow structure (see e) takes place.

On condition of high free-stream turbulence level the method of the distributed effect on the perturbed flow in a boundary layer was also applied. The method was realized with the help of fixing on the surface the longitudinal ribs about 0.1 δ^* high with a λ_z spacing. The observations showed that the nonstationary three-dimensional picture of the disturbances field upstream of the ribbed section (see f, g) transformed to the stable deformation of a wave front at the end of that section, while the deformation amplitude reduced in spite of some Reynolds number growth (see h). The disturbances structure corresponding to the ribbed section geometry was seen only in the immediate vicinity of a wall.

References

1. Reshotko, E. Boundary-layer stability and transition: Ann. Rev. Fluid Mech. 8 (1976), 311–349.
2. Kachanov, Yu. S.; Kozlov, V.V.; Levchenko, V.Ya. Vozniknoveniye turbulentnosti v porgranichnom sloye. Novosibirsk: Nauka, 1982.
3. Kozlov, L.F.; Babenko, V.V. Eksperimentalniye issledovanija pogranichnogo sloya. Kiev: Naukova Dumka 1978.
4. Yurchenko, N.F. K metodike eksperimentalnogo issledovanija sistemy prodolnih vihrey v pogranichnom sloye. Ing. — Phys. J., 11, N6 (1981) 996-1002.

5. Knapp, C.F.; Roach, P.D. : A combined visual and hot-wire anemometer investigation of boundary-layer transition: AIAA J. 6, N1 (1968) 32-42.

6. Klebanoff, P.S.; Tidstrom, K.D.; Sargent, L.M.: The three-dimensional nature of boundary layer transition; J. Fluid Mech. 12, N1 (1962) 1—34.

7. Tani, I.: Boundary layer transition: Ann. Rev. Fluid Mech., 1 (1969) 169—197.

8. Babenko, V.V.; Yurchenko, N.F., Kozlov, L.F.: Eksperimentalnoye issledovaniye vospriimchivosti pogranichnogo sloya k prodolnym vihrevym vozmushcheniyam: Ing.—Phys.J. 46, N1 (1984) 140.

9. Babenko, V.V., Yurchenko, N.F. Kanarskiy, M.V.: Spektralniy analiz reaktsii pogranichnogo sloya na trehmerniye vozmushchenija. Ing.—Phys. J. 46, N1 (1984) 140—141.

10. Sobstel , I.B.; Tetianko, V.A.; Sharapova, T.A., Shtatnov, Yu.V. K voprosu o formirovanii turbulentnogo pogranichnogo sloya. V sb.: Issledovanija po upravleniyu pogranichnym sloyem. Novosibirsk, Inst. teplophiziki SO AN SSSR (1976) 118—127.

11. Babenko, V.V.; Voropaev, G.A.; Kozlov, L.F.; Korobov, V.I.; Yurchenko, N.F.: Nekotorye kinematicheskiye characteristiki pogranichnogo sloya pri obtekanii elastichnoy plastiny. Ing.—Phys. J. 38, N6 (90) 1049—1055.

12. Klebanoff, P.S.: Characteristics of turbulence in a boundary layer with zero pressure gradient: NACA Rep. 1247 (1956 1—19.

Stability and Separation of Viscous Flows

O. S. Ryzhov

Computing Center of the USSR Academy of Sciences,
Moscow, USSR

Summary

A unified mathematical treatment of viscous flow stability and separation is developed. This treatment is based on the boundary layer free interaction theory and its extension for fluid motions through channels. The analysis of a dispersion relation constituting the heart of the linear approximation is done both for sub- and supersonic regimes.

Introduction [1-4]

Multi-deck structures introduced into the boundary layer theory 15 years ago by Neiland (1969) and Stewartson & Williams (1969) gave a clue to our understanding of the main properties of supersonic flow separation in the limit when the Reynolds number $R \longrightarrow \infty$. Three years later the incompressible flow separation pattern was elucidated by Sychev (1972). Probably Schneider (1974) was the first to introduce similar concepts for unsteady motions. The straight way led from here to the study of wave processes which usually provoke the breakdown of steady viscous flows. To investigate their stability, the ideas of the asymptotic analysis underlying the subdivision of a velocity field into layers with different properties were applied in the recent work performed mainly at the Computing Center of the USSR Academy of Sciences and the Imperial College of Science and Technology, London University. The success achieved makes it possible to give a unified treatment of two fundamental hydrodynamic phenomena where separation appears as a specific form of the loss of stability with the velocity field remaining steady.

Fundamentals [5-11]

The general idea to provide a unified asymptotic consideration of boundary layer stability and separation consists in dividing the flow region into sublayers (decks) with different reference thicknesses and different governing equations for desired functions. The scaling of the time t and the spatial variables x , y , z is based on certain powers of the small parameter ε depending upon the large Reynolds number R as $\varepsilon = R^{-n}$. Here n = 1/8 but in the problems on flows through channels this exponent takes on some other values. In each sublayer the desired functions are expanded into asymptotic series in ε .

Let u, v, w be the velocity components and p - the excess pressure taken in a special dimensionless system of units. As a rule the flow in the lower, near wall deck obeys the Prandtl equation

$$\frac{\partial u}{\partial x} + \frac{\partial v}{\partial y} + \frac{\partial w}{\partial z} = 0 \tag{1.a}$$

$$\frac{\partial u}{\partial t} + u\frac{\partial u}{\partial x} + v\frac{\partial u}{\partial y} + w\frac{\partial u}{\partial z} = -\frac{\partial p}{\partial x} + \frac{\partial^2 u}{\partial y^2} \tag{1.b}$$

$$\frac{\partial w}{\partial t} + u\frac{\partial w}{\partial x} + v\frac{\partial w}{\partial y} + w\frac{\partial w}{\partial z} = -\frac{\partial p}{\partial z} + \frac{\partial^2 w}{\partial y^2} \tag{1.c}$$

for incompressible fluid since the density ρ changes slightly. As well as in the classical boundary layer theory $\partial p/\partial y =$ $= 0$; nevertheless the pressure is an unknown quantity.

In the upper deck situated above the boundary layer the flow is nearly uniform. The equation

$$\pm\frac{\partial^2 p'}{\partial x^2} + \frac{\partial^2 p'}{\partial y'^2} + |M_\infty^2 - 1|^{-1}\frac{\partial^2 p'}{\partial z^2} = 0 \tag{2}$$

for the excess pressure p' is independent of time for all values of the local Mach number $M_\infty \lessgtr 1$. The scaling of the lateral coordinate y' differs from that in the lower deck. The normal component v' of velocity is defined as

$$v' = -|M_\infty^2 - 1|^{1/2} \frac{\partial}{\partial y'} \int_{-\infty}^{x} p'(t, \xi, y', z)\, d\xi \tag{3}$$

To match the solutions in different decks we introduce the function $A(t, x, z)$ signifying the instantaneous displacement of streamlines in the intermediate deck occupying the main part of a boundary layer. In terms of this function the matching conditions at the outer edge $y \to \infty$ of the lower deck read

$$u - y \to A(t, x, z) + \frac{1}{y} \frac{\partial^2}{\partial z^2} \int_{-\infty}^{x} d\xi \int_{-\infty}^{\xi} p'(t, \xi', 0, z)\, d\xi' \tag{4.a}$$

$$w \to -\frac{1}{y} \frac{\partial}{\partial z} \int_{-\infty}^{x} p'(t, \xi, 0, z)\, d\xi \tag{4.b}$$

while on the bottom $y' = 0$ of the upper deck one has

$$p'(t, x, 0, z) = p(t, x, z) \tag{5.a}$$

$$v'(t, x, 0, z) = -|M_\infty^2 - 1|^{1/2} \frac{\partial A}{\partial x} \tag{5.b}$$

For two-dimensional disturbances with $w = \partial/\partial z = 0$ the conditions (5.a), (5.b) can, upon integrating equation (2), be reduced to

$$p = \begin{cases} -\dfrac{1}{\pi} \displaystyle\int_{-\infty}^{\infty} \frac{\partial A/\partial \xi}{\xi - x}\, d\xi, & M_\infty < 1 \quad (6.a) \\[2ex] -\dfrac{\partial A}{\partial x}, & M_\infty > 1 \quad (6.b) \end{cases}$$

To analyse disturbances propagating in plane Poiseuille flow one should distinguish two viscous sublayers in the vicinity of each wall. Then the problem consists in simultaneous integration of two systems of equations of the type (1.a-c). For two-dimensional disturbances instead of (6.a,b) we get

$$p_{1,2} = P(t,x) \mp \frac{1}{240}\frac{\partial^2 A}{\partial x^2} \tag{7}$$

the subscripts 1,2 denoting regions adjacent to the walls $y_w = \mp 1/2$.

Linear Approximation [12-16]

For the sake of simplicity let us consider a flat plate in a uniform stream. No-slip conditions on its surface are

$$u = v = w = 0 \qquad \text{at} \qquad y = 0 \tag{8}$$

In as much as the system (1.a-c) is still too complicated we shall linearize it, bearing in mind possible conclusions derivable from the linear approach for predicting essentially nonlinear processes. Denote by $a \longrightarrow 0$ the disturbance amplitude and write

$$(u-y,\, v,\, w,\, p) = a\left(-\frac{df}{dy},\, g,\, -\frac{dh}{dy},\, 1\right)\exp[i(\omega t + kx + mz)] \tag{9}$$

In keeping with the Squire theorem introduce the function

$$q = \frac{k}{k^2+m^2}(kf + mh) \tag{10}$$

the second derivative of which, upon transformation to the independent variable

$$Z = \zeta + i^{1/3}k^{1/3}y\,, \qquad \zeta = i^{1/3}\omega k^{-2/3} \tag{11}$$

satisfies the Airy equation. Making use of (8) one obtains

$$q = -\left[\frac{dAi(\zeta)}{dZ}\right]^{-1}\int_{\zeta}^{Z} dZ' \int_{\zeta}^{Z'} Ai(Z'')\,dZ'' \tag{12}$$

The dependence of flow parameters on t and x, z in the potential region should be chosen the same as in viscous near-wall sublayer. Integrating (2) and taking into account (3), (5.b), p', v' and A are found. The matching conditions (4.a,b) and (5.a) in the limit $|Z| \to \infty$ give

$$\frac{dq}{dZ} \to -i^{-1/3} \frac{\Omega}{k^{1/3}(k^2+m^2)} \tag{13.a}$$

$$\Omega = [\,|M_\infty^2 - 1|^{-1} m^2 \pm k^2\,]^{1/2}, \qquad M_\infty \lessgtr 1 \tag{13.b}$$

Substituting (12) into (13.a,b) we arrive at the dispersion relation

$$\Phi(\zeta) = Q(k, m; M_\infty) \tag{14.a}$$

$$\Phi = \frac{dAi(\zeta)}{dZ}\left[\int_\zeta^\infty Ai(Z)dZ\right]^{-1}, \qquad Q = i^{1/3} k^{1/3} \frac{k^2+m^2}{\Omega} \tag{14.b}$$

connecting the frequency ω with the wave vector components k, m and Mach number M_∞. The left-hand side $\Phi(\zeta)$ of the dispersion relation, as defined by (14.b), is the same for all problems considered below.

Incompressible Flow [12-16]

Poiseuille flow is an ideal object to demonstrate the unique nature of stability and separation. In this case the pressure variation in both wall sublayers is described by (7), while

$$Q = \frac{1}{60} k^2 \left(\frac{1}{2} i k\right)^{1/3} \tag{15}$$

To be specific, let us choose the approach where the frequency ω is assumed to be real and known and the complex wavenumber $k = k_r + i k_i$ is to be computed. The occurence of spatial instability then associates with passing k_i through zero.

The traces of the first six roots of the dispersion relation in the complex k-plane are shown in Fig.1. The curve k_1 crosses the abscissa axis at the point $k_1 = -k_* = -6.3891$ when $\omega = \omega_* = 4.9809$. These threshold values of k and ω

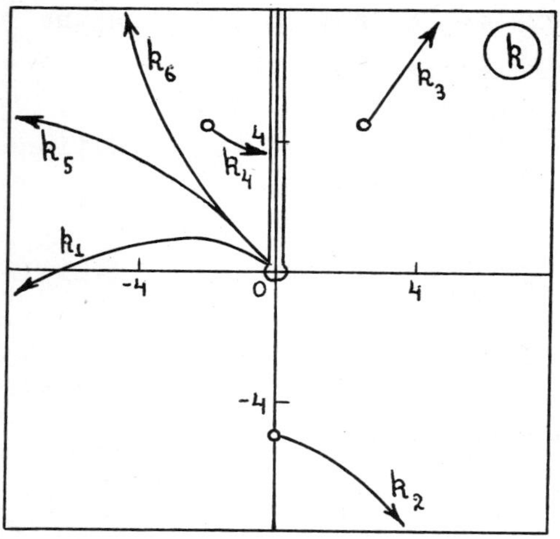

Fig.1. Traces of the first six roots of the dispersion relation in the complex k-plane ($Im\omega=0$, $Re\omega>0$). Arrows indicate increasing

determine the travelling Tollmien-Schlichting wave with the amplitude remaining constant in space. The part of the curve k_1 situated in the lower semi-plane corresponds to exponentially growing disturbances.

The curve k_2 lies entirely in the lower semi-plane, nevertheless it is not connected with unstable forms of fluid motions. In boundary value problems the points of this curve are responsible for upstream influence.

Circles designate the three values of k defining steady flows with $\omega=0$. The circle on the curve k_2 gives the rate of the damping of disturbances as $x \rightarrow -\infty$. Being extended over all values of x these disturbances grow exponentially downstream without oscillations meaning that the linear solution (9) fails. Its continuation into the nonlinear region can be computed with the help of the Prandtl equations with the self-induced pressure included using a marching procedure. As a result we have the nonlinear eigenfunction displaying a separation bubble near one of the channel walls. Thus, the existence of the dispersion relation root in the lower semi-plane k implies a possibility to predict the onset of such nonlinear process as separation on the basis of the linear approximation. From this standpoint separation may be thought as nothing else than a specific form of the loss of stability with the velocity field remaining steady.

<u>Subsonic Boundary Layer [3, 17-21]</u>

For three-dimensional disturbances with k, m real let us introduce the quantities

$$\Omega = i^{-1/3} \zeta K^{2/3}, \quad \bar{K}^{1/3} |K| = k^{1/3} \frac{k^2 + m^2}{\mu}, \quad \arg K = \arg k \tag{16}$$

if $M_\infty < 1$ and write the right-hand side of (14.a) as

$$Q = \pm i^{1/3} \bar{K}^{4/3}, \qquad \mathrm{Re}\, K \gtrless 0 \tag{17}$$

The parameter ζ defined by (11) is kept according to (16) unchanged.

The whole pattern of the dispersion relation roots for the regime under consideration bears close resemblance with that in Fig.1 except for the absence in the complex K-plane of the curve analogous to k_2. By a trivial recomputation we get the critical values $K_1 = -K_* = -1.0005$, $\Omega = \Omega_* = 2.298$ defining the oblique Tollmien-Schlichting waves with the constant amplitude. The neutral stability curves are shown in Fig.2.

The absence in the K-plane of the curve similar to k_2 points out that the dispersion relation for a subsonic boundary

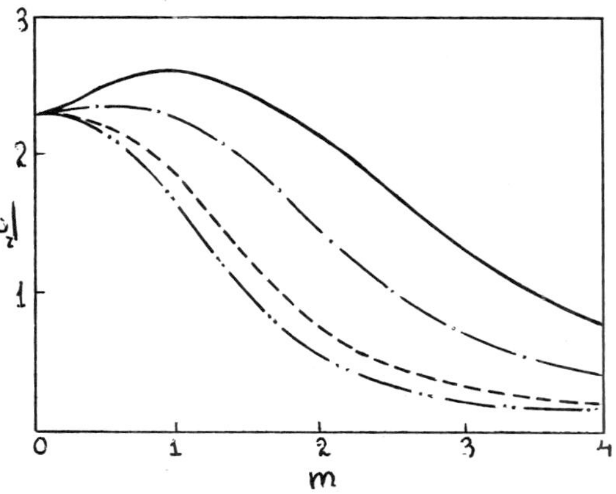

Fig.2. Neutral stability curves in the m, $|\omega_z|$-plane for various Mach numbers (——— $M_\infty = 0.9$, —·—· $M_\infty = 0.8$, - - - - $M_\infty = 0.5$, —··—·· $M_\infty = 0$)

layer does not admit solutions corresponding to steady disturbances with $\omega = 0$. To elucidate the situation we note that the upstream influence damps in this case algebraically rather than exponentially. Hence the prediction of separation can seem to be impossible in the framework of the linear approach. There is, however, the way to overcome this difficulty consisting in summation of the infinite series of exponential terms in the problem on the perturbed velocity field at the entry of a channel. The result of summation are algebraic functions.

Supersonic Boundary Layer [1,2,15,20,22]

Let once again k , m be real. According to (9) the oblique Tollmien-Schlichting wave travels at the angle $\Upsilon = \operatorname{arctg} m/k$ with respect to the x-axis. As it follows from (13.b) under condition $M_\infty > 1$, disturbances which are periodic in space exist only if

$$|k| = (M_\infty^2 - 1)^{-1/2} |m| \tag{18}$$

the right-hand side of (14.a) being (17). Thus, the wave slope must be bounded from below:

$$|\Upsilon| \geq \Upsilon_0 = \operatorname{arctg} (M_\infty^2 - 1)^{1/2} \tag{19}$$

The simple physical meaning of the latter constraint is that the velocity component of the oncoming stream in the direction of the wavenumber vector must be less than or equal to the spead of sound, therefore Υ_0 is nothing else than the Mach angle.

Let us invert inequalities (18), (19). Substituting

$$\Omega = i^{-1/3} \varsigma K^{2/3}, \quad K^{4/3} = -i k^{1/3} \frac{k^2 + m^2}{\varsigma^4}, \quad \arg K = \arg k \tag{20}$$

into (14.b) one gets

$$Q = -(iK)^{4/3} \tag{21}$$

The general pattern of the dispersion relation roots in this case essentially differs from that in Fig.1 with respect of the curve analogous to k_1 which lies now entirely in the upper half of the K-plane and approaches asymptotically the ordinate axis. Consequently, the free interaction theory predicts stability of all disturbances with the slope angle $|\Upsilon| < \Upsilon_c$ propagating through a supersonic boundary layer. Evidently this conclusion is in contradiction with numerous experimental data, what may be easily explained. The point is that in compliance with the basic concepts of the theory in hand the critical layer of neutral oscillations is submerged into the lower viscous deck adjacent to a wall. Meanwhile, the disturbances can turn out to be unstable if their propagation speed is subsonic relative to the external stream. This condition is satisfied only for oblique waves with $|\Upsilon| \geqslant \Upsilon_c$, the magnitude of the velocity component along the wave vector rather than that of the velocity itself being in fact important. An asymptotic analysis of disturbances from the range $|\Upsilon| < \Upsilon_c$ of the slope angles must deal with the velocity field embracing the critical layer removed at a finite distance from a wall.

The dispersion relation (14.a) with the right-hand side (21) gives in the lower semi-plane K a curve akin to k_2 . Hence one may predict the occurence of separation of a supersonic boundary layer on the basis of the linear approximation, putting $\omega = 0$. From $\Omega = \zeta = 0$ it follows

$$\Phi(0) = -3^{2/3}\,\Gamma^{-1}(1/3) \tag{22}$$

Let now $k = -i|k|$ be an imaginary quantity. Under this condition (20), (21) lead to

$$\Phi(0) = -|k|^{1/3}\,\frac{|k|^2 - m^2}{[(M_\infty^2 - 1)^{-1} m^2 + |k|^2]^{1/2}} \tag{23}$$

The comparison of (22) with (23) gives $|k|^2 > m^2$, and if $m = 0$, then

$$|k| = [-\Phi(0)]^{3/4} = 3^{1/2}\,\Gamma^{-3/4}(1/3) = 0.82715 \tag{24}$$

The latter value defines the intensity of the damping of upstream influence associated with separation in a supersonic boundary layer. It is clear from Fig.3 that the decreasing of the wave length of disturbances which are periodic in Z makes the rate of their exponential growth downstream to increase fast. The continuation of the solution under discussion into the nonlinear region using the full Prandtl equations with the self-induced pressure included has been done so far only for two-dimensional separation.

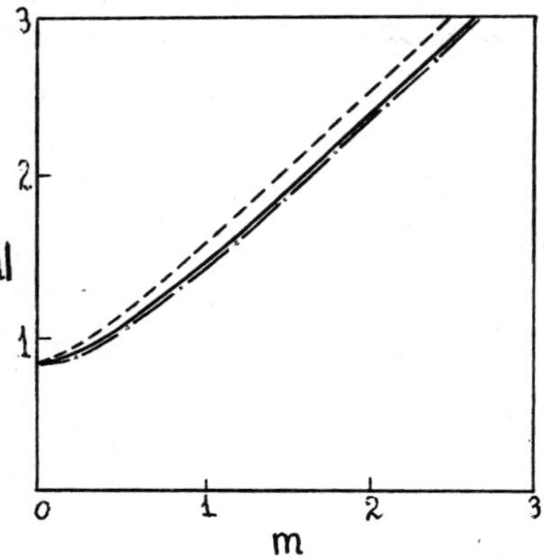

Fig.3. Rates of the disturbance decay as $x \to -\infty$ for various values of the wavenumber vector lateral component (----- $M_\infty=1.2$, ——— $M_\infty=2$, —·—·— $M_\infty=6$)

References

1. Neiland, V.Ya.: Izv. Akad. Nauk SSSR, Mekh. Zhid. i Gaza (1969), No.4, 53.

2. Stewartson, K. & Williams, P.G.: Proc. Roy. Soc. A 312 (1981) 181.

3. Sychev, V.V.: Izv. Akad. Nauk SSSR, Mekh. Zhid. i Gaza (1972), No.3, 47.

4. Schneider, W.: J. Fluid Mech. 63 (1974) 465.

5. Neiland, V.Ya.: Proc. TSAGI (1974), No.1529.

6. Stewartson, K.: Adv. Appl. Mech. 14 (1974) 145.

7. Messiter, A.F.: Proc. 8th U.S. Nath. Congr. Appl. Mech. (1978) 157.

8. Ruban, A.I. & Sychev, V.V.: Adv. Mech. 2 (1979), No.4, 57.

9. Neiland, V.Ya.: Adv. Mech. 4 (1981), No.2, 3.

10. Stewartson, K.: SIAM Rev. 23 (1981) 308.

11. Smith, F.T.: IMA J. Appl. Math. 28 (1982) 207.

12. Smith, F.T.: Quart. J. Mech. Appl. Math. 29 (1976) 343, 365.

13. Smith, F.T.: J. Fluid Mech. 79 (1977) 631.

14. Smith, F.T.: Mathematika 26 (1979) 187, 211.

15. Ryzhov, O.S. & Zhuk, V.I.: J. de Mécanique 19 (1980) 561.

16. Bogdanova, E.V. & Ryzhov, O.S.: Quart. J. Mech. Appl. Math. 36 (1983) 271.

17. Smith, F.T.: Proc. Roy. Soc. A 366 (1979) 91.

18. Smith, F.T.: Proc. Roy. Soc. A 368 (1979) 573.

19. Zhuk, V.I. & Ryzhov, O.S.: Dokl. Akad. Nauk SSSR 253 (1980) 1326.

20. Zhuk, V.I. & Ryzhov, O.S.: Prikl. Math. Mekh. 45 (1981) 554.

21. Bogdanova, E.V.: Sci. J. TSAGI 13 (1982), No.5, 21.

22. Ryzhov, O.S. & Terent'ev, E.D.: Prikl. Math. Mekh. 41 (1977) 1007.

Interrelation of the Flow Separation and Stability

V. V. KOZLOV

Institute of Theoretical and Applied Mechanics
USSR Academy of Sciences, Novosibirsk 630090, USSR

Summary

There are surveyed a number of problems on the development of disturbances in separation flows, their stability, their receptivity to external disturbances. An essential interrelation between separation and stability is shown.

Introduction

When solving many aero- and hydrodynamic problems there occur flows with separation. Already many surveys are to be found on this question (see, for example, [1 - 4]) with indications on the complexity of this phenomenon when constructing mathematical models corresponding to the experiment. Much has been done [3, 4] to construct laminar separation models but nobody tested the obtained results on stability though from the point of view of the hydrodynamic stability theory, the presence of the flexion point in the solutions obtained indicated a strong instability of the separation flows. At the same time the experimental results testify to the fact that in the boundary layer even at a small level freestream disturbances form vortical oscillations, i.e. the waves of Tollmien-Schlichting type in the boundary layer. In what follows we want to show that the behaviour of these disturbances determine largely the flow structure in the region of separation.

The separation of the flow from the smooth surface

In reference [5] the separation was carried out on the straight wing placed in a low turbulence wind tunnel. The scheme of the flow at the wing surface that was realized without artificially excited disturbances is given in Fig. 1. Here is also given the distribution of pressure on the test surface (u_∞ =5.7 m/s). It can be noted that in the region of unfavourable pressure gradient the arizing flow remains laminar and only downstream there appear natural oscillations leading to turbulence and to the attachment of the separation. If by means of a vibrating ribbon there are generated the disturbances with the frequency equal to the frequency of the vibrating ribbon, one can study their development in the region of separation.

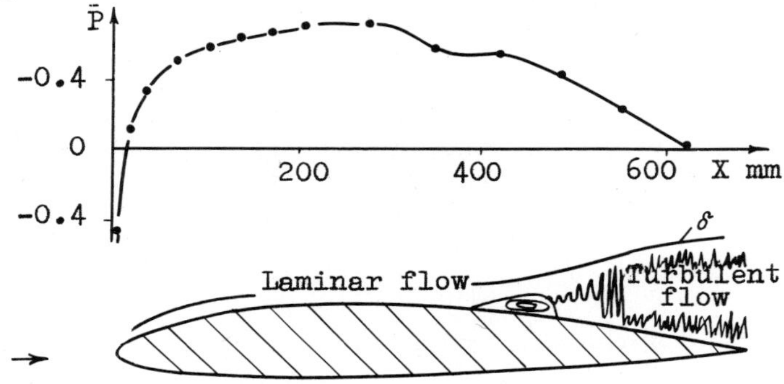

Fig.1

In Fig. 2 there are given the curves showing the growth of the artificially generated disturbances. Eddy oscillations evolving in the pre-separation region, as follows from the figure, while being transformed into separation flow disturbances, amplify in the separation region. Thereat they are observed to strongly influence the flow structure in the unfavourable pressure gradient and separation region.

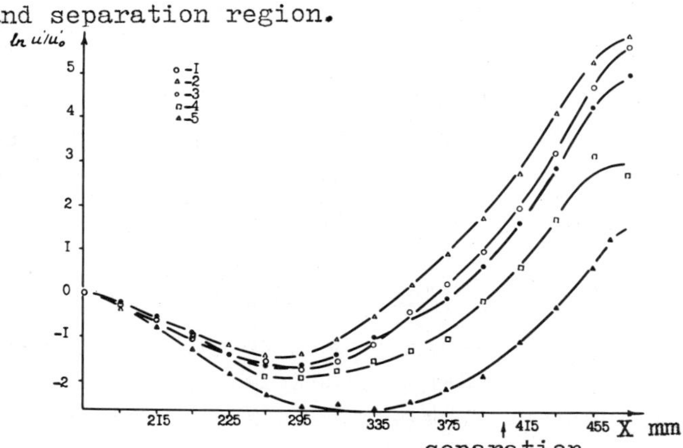

Fig.2. The growth of artificial disturbances

The disturbances cause the change in the mean velocity distribution (Fig. 3). Essential is the fact that the velocity profile filling doesn't result from the flow turbulization but takes place in a laminar layer. The influence of vortical oscillations upon the mean flow begins to tell at considerably smaller distances along the longitudinal coordinate, whereas the laminar-turbulent transition takes place much later. Thereat the mean flow distortion exceeds the local oscillation amplitude more than by an order.

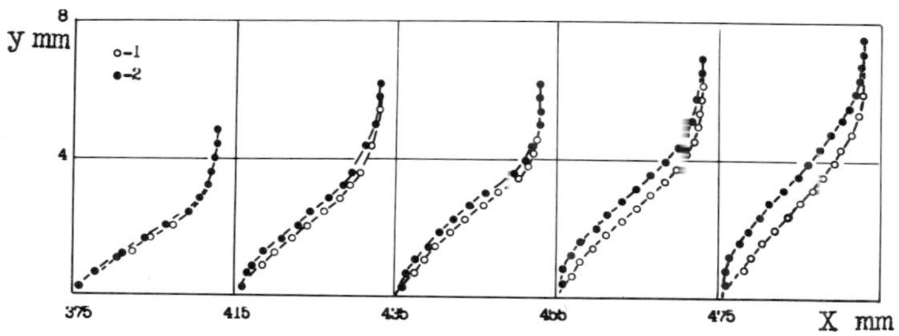

Fig.3. 1 - mean velocity profiles with excitation (f=85 Hz), 2 - without excitation.

The influence of acoustic disturbances on the flow

Consider the receptivity of the flow with the laminar boundary layer separation from the smooth surface to the acoustic disturbances which was to object under study in the proceedings [6] . The scheme of the experiment remained the same as in the previous experiments with the exception that instead of the vibrating ribbon placed in the diffusor of the wind tunnel a dynamic loudspeaker was mounted. It was used to generate acoustic oscillations of the amplitude of the order of 100 dB. When generating acoustic flow oscillations in the boundary layer in the region of the unfavourable pressure gradient there appeared two-dimensional vortical waves with the excitation frequency propagating downstream. In Fig. 4 the growth of disturbances excited by a sound and that excited by vibrating ribbon are compared. Under acoustic excitation the wave becomes noticeable in the beginning of the region of the unfavourable pressure gradient. Downstream its growth practically coincides with the growth of the disturbance initiated in the boundary layer by

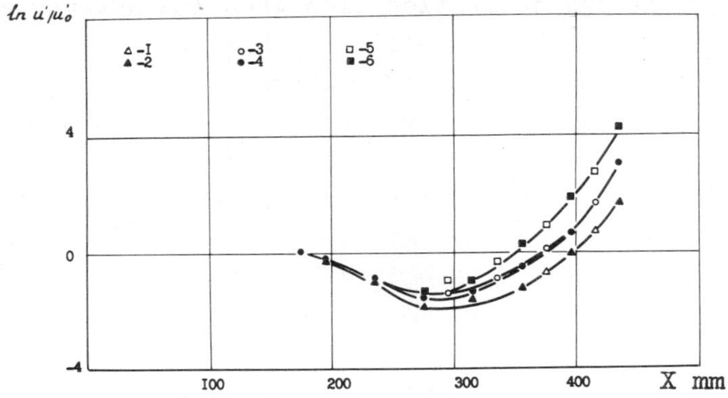

Fig.4. The growth of disturbances. Acoustic excitation: 1-176 Hz, 3-156 Hz, 5-116 Hz. Excitation by ribbon: 2-176 Hz, 4-156 Hz, 6-116 Hz

the vibrating ribbon. A frequency range in which the boundary layer was excited by sound is given in Fig. 5. For the same sound intensity which equalled 95 dB for all the frequencies the amplitude of vortical oscillations was measured in the cross-section x = 440 mm. The given dependency characterizes the receptivity of the flow to acoustic disturbances. The excitation of acoustic oscillations in the flow changed the

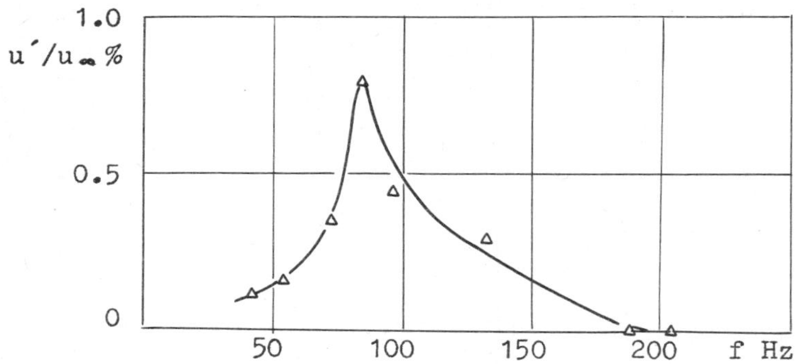

Fig.5. Frequency range of the excitation of boundary layer oscillations

structure of the separation flow exactly as it did in the reference [5] , the influence of the sound, as it was shown, being intermediated by eddy disturbances, i.e. by Tollmien-Schlichting waves generated by the sound.

The stability of the separation flow when the surface with a fracture point are flowed about

In the reference [7] there was carried out the experimental study of the flow stability appearing when a rectilinear fracture of the plane surface is flowed about. The two plates model made of polished acrylic plastic forming a 20° angle of fracture was mounted in the working section of a low turbulence wind tunnel at the angle of attack 9°. The scheme of the flow which appears thereat is given in Fig. 6. The flow structure was studied at the freestream velocity U_{∞}=5.6 m/sec. The velocity oscillations in the separation flow were generated artificially by means of a loudspeaker mounted in the wind tunnel

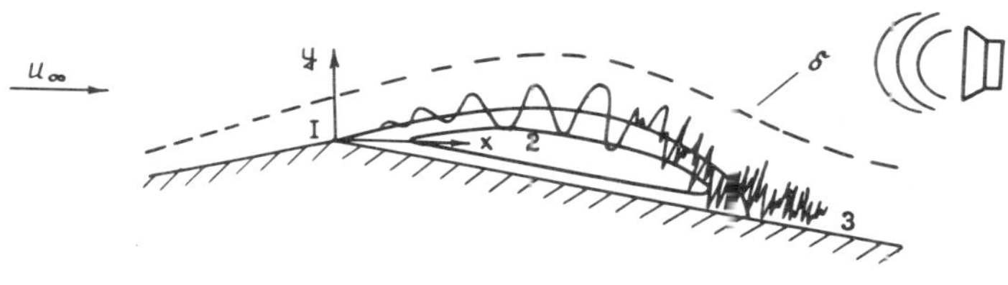

Fig.6. Flow structure. 1 - laminar boundary layer; 2 - separation region; 3 - turbulent boundary layer

diffusor. The excitation of monochromatic sound oscillations in the flow generated two-dimensional hydrodynamic waves of Tollmien-Schlichting type leading to the distortion of the mean flow velocity profiles and redistribution of static pressure along the model surface (Fig. 7). The curves showing the growth

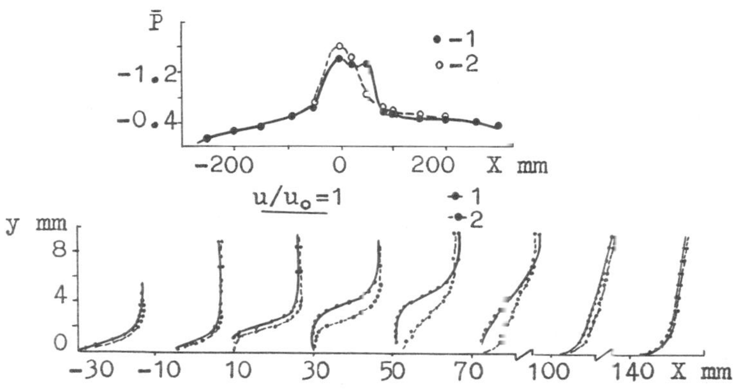

Fig.7. The influence of the disturbances f=413 Hz, A=88 dB on the flow structure. 1-without excitation; 2-with excitation

of the vertical disturbances in case of one of the frequencies under study are given in Fig. 8. They coincide in the case of

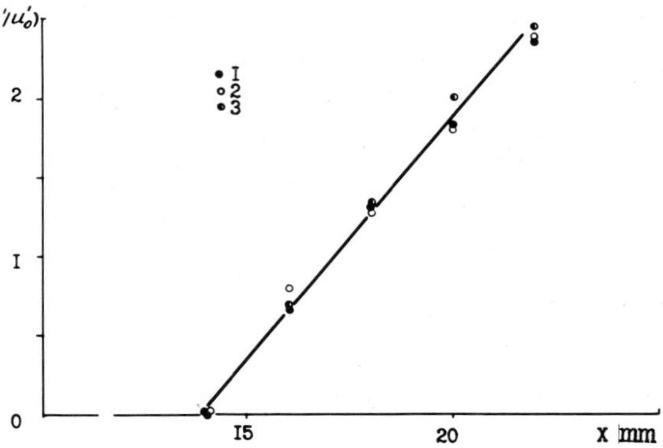

Fig.8. Growth of disturbances. f=412 Hz at different amplitudes 1-u_o'/u' =0.04%(A=91 dB); 2- 0.1% (93.5 dB); 3- 0.2% (99.4 dB)

different initial amplitudes of disturbances. The independency of the disturbances growth on the initial amplitude enabled to instroduce the coefficients of spatial amplification $-\alpha_i$. The range of instability according to the thus obtained frequency is given in Fig. 9.

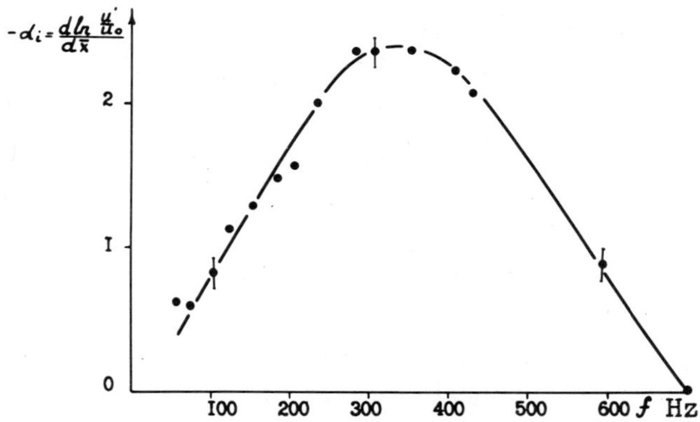

Fig.9. Separation flow instability range (x = 20 mm)

<u>Flow separation off the front profile edge and acoustic disturbances which affect it</u>

There are known to exist two different flow regimes after the separation has taken place [4]. In some cases the initial boundary layer passes over the region of recirculating liquid and then reattaches at some point downstream separating "the bubble" of the recirculating liquid. In other cases the boundary layer liquid doesn't attach to the body but travels downstream mixing with the recirculating liquid and forming the wake. In this case if the profile is at a large angle of attack, the flow is separated off all its upper surface. The aim of this work was to obtain the separation occurring on the whole upper surface of the profile (the separation of the second type), to study its structure and the effect of the acoustic disturbances. The flow separation was studied on symmetrical Joukovski profile (1) with a 292 mm chord and 1 m span mounted under the angle of attack equal to 11°. The measurements were carried out at the freestream velocity equal to U =4.9 m/sec. To excite acoustic oscillations in the flow region under study the loudspeaker was mounted in the tunnel diffusor. The mean velocity profiles realized in the boundary layer and in the separation region without and with acoustic effect (1) and (2) are given in Fig. 10a. Thereat the amplitude of the sound with the frequency f=230 Hz was maintained 90 dB. The analysis of the results of mean velocity profiles measurements showed that if no acoustic effect is present (1) laminar flow separation occurs already at the distance x=5 mm from the front profile edge and in this region there starts growing the packet of disturbances causing a laminar turbulent transition at x=65 mm, but here the separated flow doesn't attach to the wing profile, as was the case in reference [5] and from the boundary layer the liquid travels downflow mixing with recirculating liquid and forming the wake. The separation is seen to occur on the whole upper

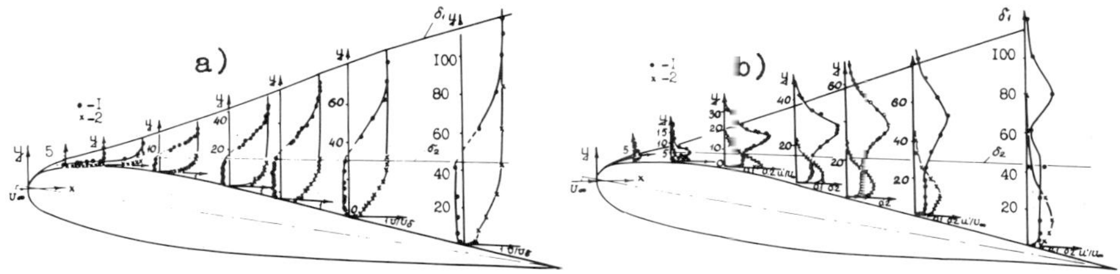

Fig.10. Mean velocity profiles (a) and disturbances profiles (b) measured in cross-sections. x=20; 40; 70; 107; 134; 170; 230 (mm) correspondingly. (Kozlov V.V. (1984)). 1-without acoustic influence; 2-with acoustic influence (f=232 Hz, A=90 dB)

surface. For this case the profiles of the disturbances (1) are shown in Fig. 10b. When switching the acoustic field at the definite frequency range in the laminar separation region there appears a Tollmien-Schlichting wave of a large amplitude (2), i.e. in this region the acoustic disturbances are transformed into vortical ones as was already mentioned in references [6, 7] . Like in the abovementioned references this leads to the change of the structure of the region of laminar separation and to elimination of the latter in this region and downflow on the whole upper surface (Fig. 10a). At uninterrupted movement from low frequencies to upper ones such an influence (elimination of the separation) starts at 200 Hz and finishes at 1125 Hz; when the motion in the opposite direction takes place, the frequency range diminishes considerably, the effects start at 600 Hz and finish at 200 Hz; there exists a hysteresis of this phenomenon, i.e. the manifestation of the real phenomenon depends essentially on the prehistory (Fig. 11).

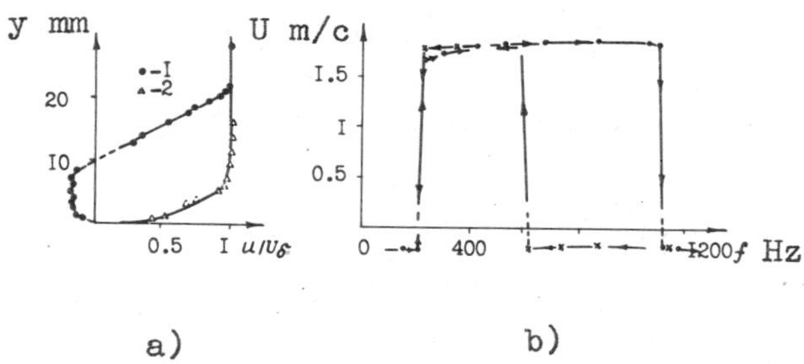

Fig.11. Different states of the boundary layer depending on the acoustic fields (A = 90 dB) frequency at the constant value of the sound amplitude. (Kozlov V.V. (1984)). a) a general aspect of the velocity profiles for frequencies when the separation is eliminated (2) and no influence is present (1). b) Measurements of the mean velocity value at x = 70, y=0.5 mm.

The evolution of the wave packet in the separation region

Up to now there were considered two-dimensional problems, i.e. two-dimensional were the separation and the disturbances evolving there. In [8] there was studied a case when the separation remained two-dimensional and the disturbance was three-dimensional. It was achieved by initiating vortical disturbances in the boundary layer from a point source. The scheme of the flow which was the same as in reference [5] is presented in Fig. 1. The evolution of the wave packet in space is illustrated in Fig. 12 where the lines of disturbances of equal intensity in XZ plane are presented. In the experiments on the same wing when exciting the boundary layer via vibrating a strong influence of 2-D oscillations of a small amplitude on a laminar-flow structure in the region of unfavourable pressure gradient and in the separation region was observed [5] . The effect of disturbances consisted in the filling of the mean velocity profiles which more than by an order exceeded the local oscillations amplitude. The analogous phenomenon was observed when exciting the flow by a point source. The lines of the equal

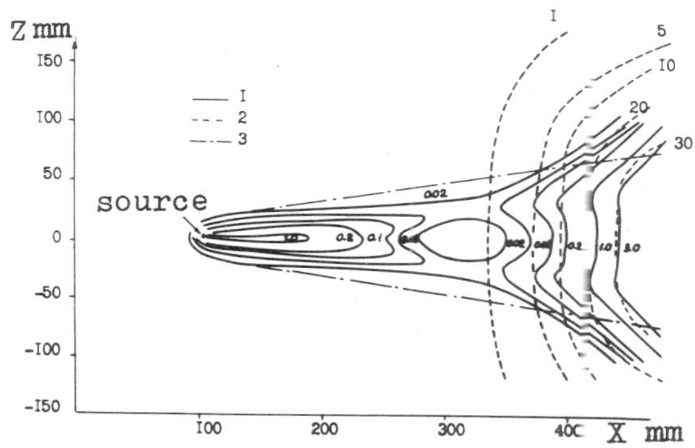

Fig.12. Lines of equal intensity of disturbances (1), distortion values of the mean flow (2) and the boundary of wave packet propagation for the flow with a zero gradient (3). The values of u /u (%) and u/u (%) are shown by figures.

value of mean velocity distortions are presented in percent by a dashed line in Fig. 12. The nonlocality of the influence profiles are distorted not only in the region of the wave packet propagation, but also in the region where the flow remains undisturbed. The mechanism of the disturbances effect over the mean flow is likely to be connected with the "forward" influence on the turbulent attachment region, the structure of which, in its turn, depends on non-stationary processes occurring in the flow.

Having summed up all the phenomena occurring in the separation region in case of many types of separation, the following problems can be singled out which testify to a strong dependency between separation and its stability: 1. The transformation of outer disturbances in eigenoscillations of the separation (receptivity problem); 2. hydrodynamic instability of separating flows; 3. the effect of amplifying vortical disturbances on the structure of separation and the effective form of the body.

References

1. Chang, P.K.: Separation flow control.-M.: Mir, 1979. 552 p.

2. Brown, S.M.; Stewartson, K.: Laminar separation.-Annual Review of Fluid Mechanics, 1969, vol. 1, p. 45-72.

3. Neuland, V.Ya.; Ruban, A.I.; Sychev, V.V.: Otrivnie techenia.- In book: Modeli v mekhanike sploshnoi sredi, Novosibirsk, 1979, s. 135-168.

4. Williams, Sh.G.: Boundary layer separation of incompressible fluid.-In book: Mechanics, vortical fluid movements, M., 1979, 21, p. 58-100.

5. Dovgal, A.V.; Kozlov, V.V.; Kosorygin, V.S.; Ramazanov, M.P.: Vliyanie vozmuscheniy na strukturu techenia v oblasti otriva. Dokl. AN SSSR, 1981, t. 258, N°1, s. 45-48.

6. Dovgal, A.V.; Kozlov, V.V.: Vliyanie akusticheskikh vozmuscheniy na strukturu techenia v pogranichnom sloye s neblagopriyatnim gradientom davlenia.-Izv. AN SSSR, MZHG, 1983, N°2, s. 48-52.

7. Dovgal, A.V.; Kozlov, V.V.: Ustoychivosti otrivnogo techenia pri obtekanii poverkhnosti s tochkoi izloma. Dokl. AN SSSR, 1983, t.270, N°6.s. 1356-1358

8. Gilev, V.M.; Dovgal, A.V.; Kozlov, V.V.: Razvitiye volnovogo paketa v pogranichnom sloye s gradientom davleniya, Novosibirsk, 1984, Preprint ITPM SO AN SSSR N°6-84.

Development of Vortical Disturbances in Flow with Laminar Separation

A. V. DOVGAL

Institute of Theoretical and Applied Mechanics,
Siberian Branch, Academy of Sciences, Novosibirsk 630090, USSR

Summary

Development of small amplitude artificial disturbances in separation bubble of transitional type is investigated. Stability of separated flow is examined using hot-wire technique. Vortical oscillations excited in upstream boundary layer amplify in the region of separation causing significant rearrangement of the mean flow.

In recent years the experimental investigations of Refs. [1-3] of small oscillations evolution in flows with laminar boundary-layer separation followed by turbulent reattachment have been carried out. The results of these experiments demonstrated pronounced effect of instability waves on the structure of separation bubbles of transitional type and the necessity of subsequent work. This report is devoted to the experiments that prolong the previous ones.

Separation bubble of transitional type inside the acrosswise corner was of present interest (see fig. 1). The model which was represented by two flat plates joined at an angle of 160° placed in test section of a low-turbulence wind tunnel. At a freestream velocity U_∞ = 6.2 m/s the separation flow occurred between points of laminar separation and turbulent reattachment.

Development of two-dimensional small amplitude vortical oscillations artificially excited by means of vibrating ribbon placed in a boundary layer upstream of separation was observed. Mean flow velocity and disturbances were measured using hot-wire technique of the DISA 55 D00 type. Single-probe was fixed at a support mounted on traversing mechanism in test section window. Frequency analyser with a 4 Hz band was used for control of oscillations at a given frequency. Phase measurements were provided by dual-beam oscilloscope. The surface of the model was equipped with pressure holes for static pressure observations.

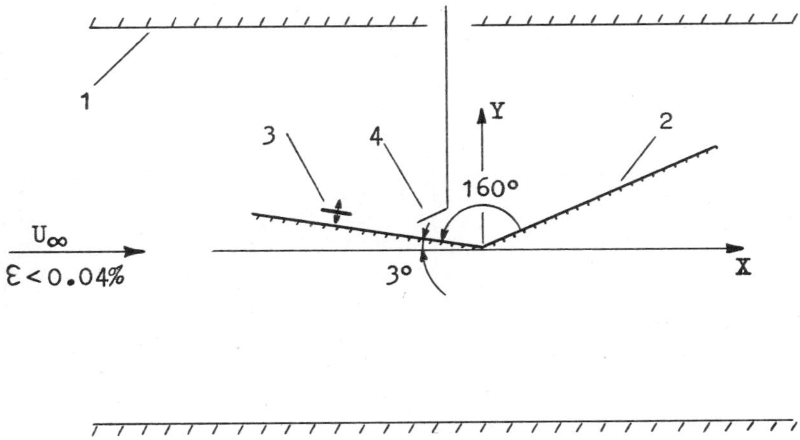

Fig. 1. Experimental scheme: 1 - test section, 2 - model, 3 - vibrating ribbon, 4 - hot-wire probe

The flow structure occurred without excitation is shown in fig. 2. The approximate position of laminar boundary-layer separation is $X \simeq -70$ mm; the separation is followed by a turbulent reattachment at a distance $X \simeq 70$ mm.

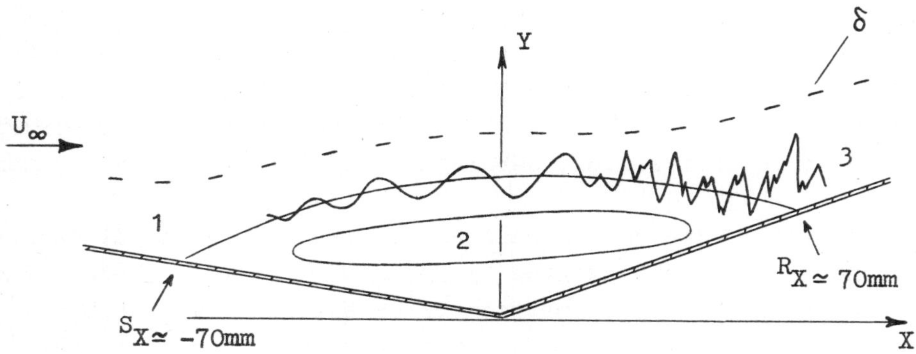

Fig. 2. Flow structure: 1 - laminar boundary layer, 2 - separation bubble, 3 - turbulent boundary layer

The mean flow distributions and r.m.s. amplitude of natural oscillations in the separation bubble are illustrated in fig.3. The transition to turbulence in separation flow results from amplification of disturbances in the outer part of the bubble: the position of maximum intensity coincides with the inflection position in the mean flow profile. The amplitudes of disturbances in each section are shown by figures.

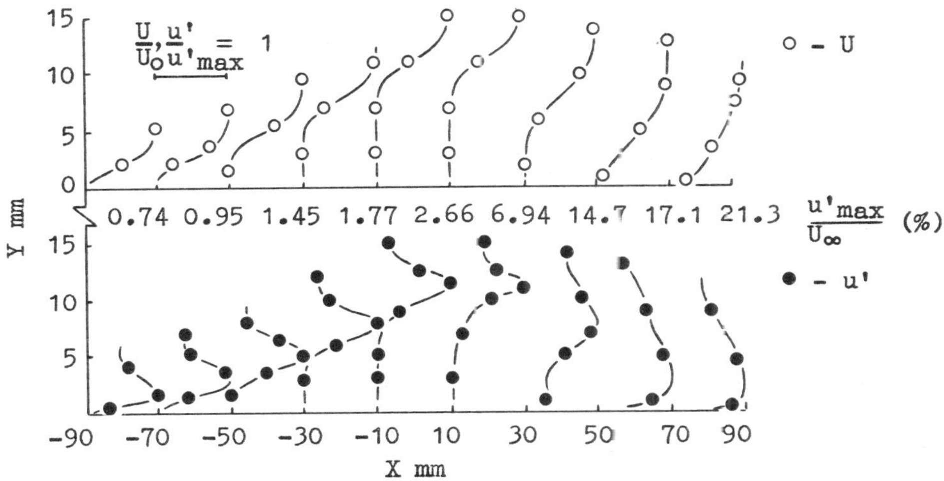

Fig. 3. Distributions of mean flow and disturbances in separation bubble

By means of vibrating ribbon in the flow of the above structure vortical disturbances at given frequency and amplitude were excited. The following considerations were focused on amplitude and phase distributions of the disturbances excited in the separation bubble.

Profiles of r.m.s. intensity of artificial oscillations are given in fig. 4. In the first two sections they are similar to

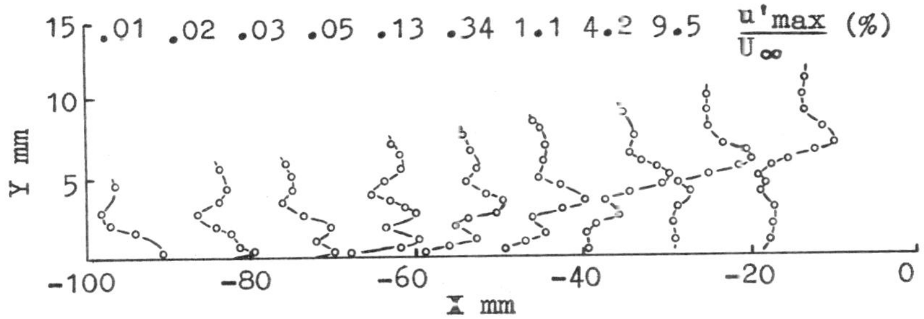

Fig. 4. Wave profiles, F = 76.9 Hz

that of Tollmien-Schlichting wave in a boundary layer. In the separation region a third maximum appears. The two lower maxima are located in the viscous layer and the upper one occurs in the vicinity of the outer boundary.

Phase distribution of oscillations in the wave across separation flow is presented in fig. 5a. The phase is approximately constant in maxima (1 - 3); between them it changes consecutively by $\pi/2$. Such a phase distribution keeps itself while passing downstream. This fact is illustrated in fig. 5b, where the phase of oscillations in each maximum vs longitudinal coordinate is shown. Constancy of the phase distribution in different sections of the separation bubble implies that the disturbance in reversal flow (lower maximum) is associated with the one at the outer boundary of circulation zone (intermediate maximum) and both travel in the freestream direction.

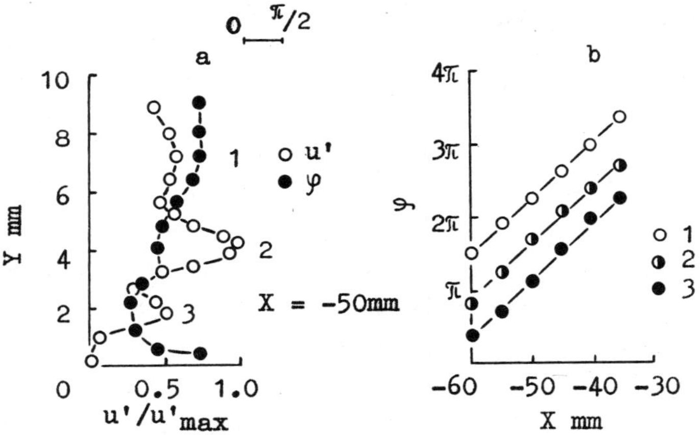

Fig. 5. Phase distribution, F = 76.9 Hz

Spatial amplification of disturbances is shown in fig. 6 for four frequencies of excitation. Amplification curves were plotted on the basis of intensities of oscillations at the intermediate maxima. A comparison of amplification curves at given frequencies but for different initial amplitudes showed their independence of oscillation amplitudes (when intensity of disturbance was less than $\sim$ 1% of the freestream velocity). The amplification in the boundary layer ($X \lessapprox -80$ mm) differs from that in the separation bubble ($X \gtrapprox -80$ mm) which is obviously due to different stability of these two flow types.

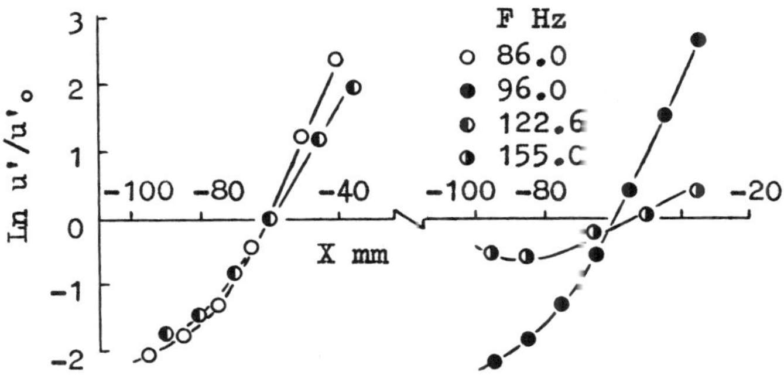

Fig. 6. Amplification of oscillations

Independence of disturbances amplification of their intensities for small amplitude oscillations made it possible to deal with spatial amplification rates $-\alpha_i$ determined from amplification curves slope in the section X = - 50 mm. Frequency range of the instability of separated flow is shown in fig. 7.

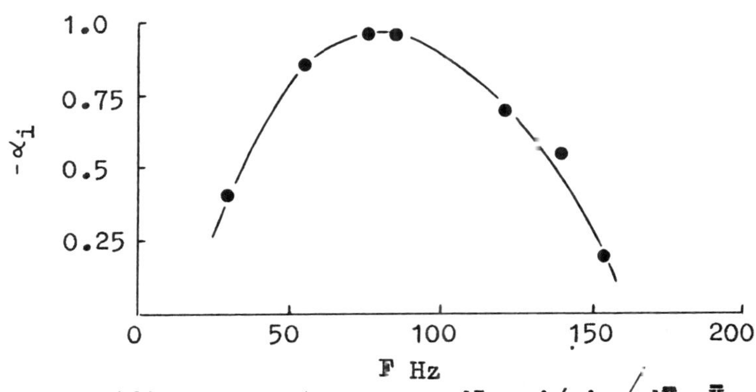

Fig. 7. Amplification rates $-\alpha_i = d\mathrm{Ln}\ u'/u'_o / d\bar{X}$, $\bar{X} = X/\delta$

When studying the effect of disturbances on the formation of separated flow the authors of Refs. [1 - 3] encountered with pronounced influence of small oscillations on the characteristics of the flow, in particular on the laminar flow structure in separation region. A similar phenomenon was observed also in the present experiment (see fig. 8). When the flow is excited the mean flow profiles are filled (sections X = - 50 mm and - 30 mm) in the region of laminar regime and where the amplitude of disturbance is small. Such strong nonlocal interaction of disturbances with the mean flow leads to seeming contradiction of the result obtained: independence of amplification rate on the form of the mean flow velocity profile. In Ref. [2] , to eliminate this discrepancy it was suggested that the stability of separated flow as a whole is determined by stability characteristics of a mixing layer near the boundary of circulation zone where the form of the mean flow profile with and without excitation is not changed. In the light of the present results this hypothesis appears to be truthful: the form of the mean flow profile in the vicinity of inflection point is not affected by excitation.

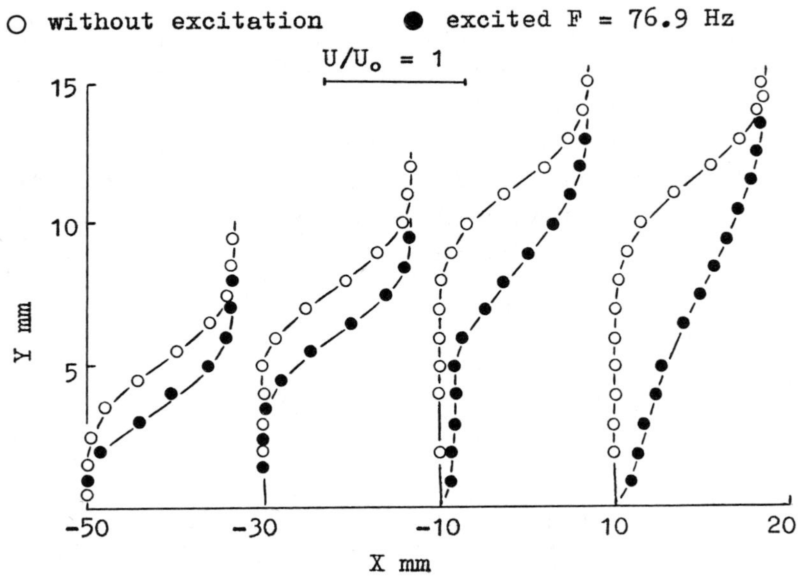

Fig. 8. The effect of disturbances on the mean flow

According to the mean flow profiles (see fig. 8) the excitation of oscillations produces a reduction of the bubble length due to upstream movement of the point of turbulent reattachment. The profile in section X = 10 mm with excitation is closer to the turbulent one rather than to the one characteristic of the separated flow. In addition, a diminution of the thickness of the separation bubble occurs that increases with

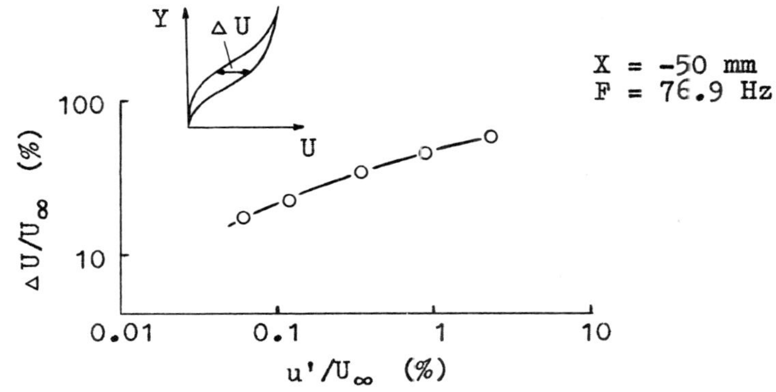

Fig. 9. The effect of disturbances on the mean flow

an increase in the oscillation amplitude (see fig. 9). Rearrangement of the flow structure caused by excitation of instability waves in separation region produces the change in static pressure distribution on the surface (see fig. 10). The pressure distribution with excitation significantly differs from that without excitation. One may suppose corresponding change of such integral characteristic as the drag coefficient.

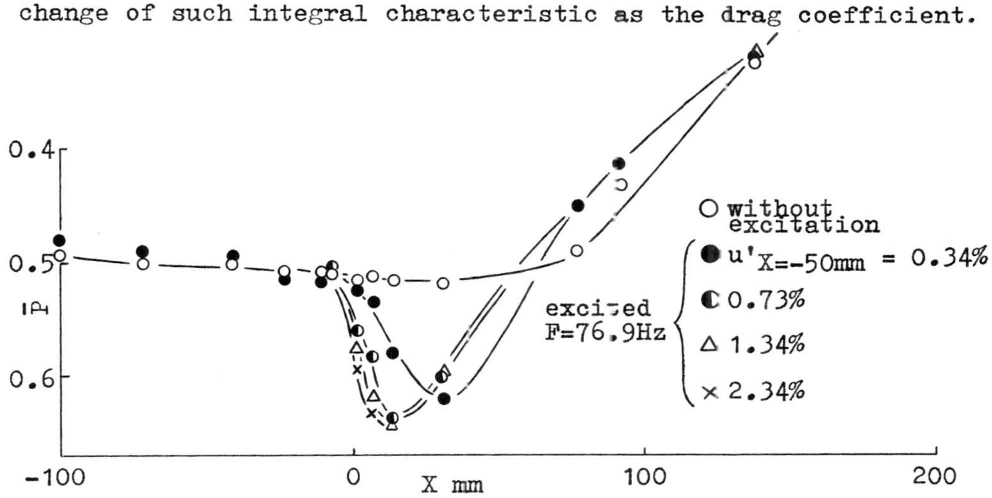

Fig. 10. The effect of disturbances on the pressure distribution

In any virtual flow an unsteadiness maybe even small occurs which is capable of exciting amplifying disturbances causing transition to turbulence. It is to be expected that unsteady effects discussed above may take place in different situations that is the reason for future investigations of phenomena related to evolution of instability waves in separated flows.

References

1. Dovgal, A.V., Kozlov, V.V., Kosorygin, V.S., Ramasanov, M.P. The influence of disturbances on separated flow structure. Dokl. Acad. Sci. SSSR, 1981, t. 258, N°1, 45-48.

2. Dovgal, A.V., Kozlov, V.V. Stability of separated flow on fractured surface. Dokl. Acad. Sci. SSSR, 1983, t. 270, N°6, 1356-1358.

3. Kozlov, V.V., Ramazanov, M.P. Stability of separated Poiseuille flow on a backward-facing step. Chislennoe modelirovanie v dinamike zidkosti. Novosibirsk, 1983, 24-30.

The Study of Shear Layer Stability by the Method of Vortex Particles

N. N. YANENKO, V.YA. RUDYAK, A. N. VERETENTSEV

Institute of Theoretical and Applied Mechanics
USSR Academy of Sciences, Novosibirsk 630090

Summary

The questions of the stability of the shear layers with respect to disturbances of its boundary and to the acoustic disturbance are examined. The problem is solved by the method of discrete vortex particles. Linear stage of development of the disturbances has been investigated in great detail, analytically as well as numerically, then nonlinear stage is investigated numerically only.

The problem of stability of hydrodynamic flows is still of interest despite a numerous literature. It is connected with the fact that up to now we are not able to study a nonlinear stage of the disturbance development and have a poor understanding of the transition from a linear stage to a nonlinear one. By solving problems on the stability some disturbance is always introduced, the nature of which is not discussed. At the same time it seems to us that for elucidating the mechanisms of the instability development it is necessary to study the influence of specific types of disturbances typical of the aerodynamic flows. In particular, in our opinion the study of acoustic disturbances is of special interest. There exists a numerous literature devoted to studying the influence of acoustics on the flows as well as the opposite influence of the flow on acoustics (see, f.e., reviews [1,2]). However, usually the boundary value problems are studied in which there take place the interactions of acoustics with a boundary as well as with the flow and boundary inhomogeneities and other complex interference phenomena. For this reason we have chosen for examination an infinite problem on the vortex layer stability under the action of acoustic disturbances. A two-dimensional (in the x,y plane) motion of unbounded compressible fluid is considered. The flow with the velocity components $(U, 0)$ in the region $y>0$ and $(-U, 0)$ in the region $y<0$ is the simplest flow, the stability of which will be discussed. Thus, the question is about the stability of the vortex sheet. The stability of the vortex layer of a finite thickness Δ with a constant distribution of vorticity and of the layer with a profile of undisturbed velocity $v_x = U\,\mathrm{th}(y/\Delta)$ is also under consideration. We have also investigated the problem of the stability of the shear layers with respect to disturbances of its boundary.

Let a plane acoustic wave propagate in a region $y<0$. In a general case this wave undergoes the reflection and refraction when passing through a vortex layer. We shall consider a case of small Mach numbers $M = U/c_0 \ll 1$. Then the mentioned effects may be neglected in the zero approximation.

The problem is solved by the method of discrete vortex particles. This method has been developed by the authors of [3,4] for the case of an incompressible fluid. Therefore when solving the formulated problem we have to modify the above method.

While deriving the equations of motion of discrete vortex particles we proceed, similarly as it was done in [3,4], from a formula representing the velocity field $\vec{v}$ through the vorticity $\vec{\omega}$ which in Lagrangian variables for the two-dimensional isentropic flows has the form:

$$\frac{d\vec{r}}{dt} = -\frac{1}{2\pi}\int \frac{[\vec{r}(\vec{\xi},t)-\vec{r}(\vec{\xi}',t)]\times\vec{\omega}_0(\vec{\xi}')}{|\vec{r}(\vec{\xi},t)-\vec{r}(\vec{\xi}',t)|^2}\,d\vec{\xi}' + \vec{\nabla}\varphi \qquad (1)$$

here $\vec{\omega}_0(\vec{\xi})$ is the initial vorticity, $\vec{\nabla}\varphi$ is the potential part of the velocity field.

Unlike the incompressible fluid Eq.(1) is not closed, another equation is needed for determining the potential φ. To derive it we shall proceed from the impulse equation written in the Lamb form [5]. Calculating the divergence of both sides of this equation and using the discontinuity equation, we can derive an equation for determining the potential. In a general case this equation is rather complex. However, in the problem under study it is simplified as the amplitude of the potential part of the velocity field is small as compared with U and takes the form

$$\left(\frac{\partial}{\partial t}+\vec{u}\cdot\vec{\nabla}\right)^2\varphi = c_0^2\Delta\varphi + \left(\frac{\partial}{\partial t}+\vec{u}\cdot\vec{\nabla}\right)\left(\Delta^{-1}\omega^2-\frac{u^2}{2}\right) \qquad (2)$$

where $\vec{u}$ is the solenoidal part of the velocity. For sufficiently slow flows this equation reduces to the following

$$\frac{\partial^2\varphi}{\partial t^2} = \Delta\varphi + \frac{M}{\varepsilon}\frac{\partial}{\partial t}\left(\Delta^{-1}\omega^2-\frac{u^2}{2}\right), \quad \varepsilon=\frac{A}{\lambda U}, \qquad (3)$$

here A and λ are the amplitude and the wavelength of acoustic disturbance, respectively.

From Eq.(3) it follows that the nonstationary vortex motions of fluid generate an aerodynamic sound even at uniform initial conditions. In what follows for simplicity we confine ourselves to the case when $M \ll \varepsilon$. At this condition we may neglect the aerodynamic sound generated by nonstationary motion of vortices that allows in a pure form to study the influence of external acoustic field on the fluid flow. Then in the zero approximation the potential φ satisfies the wave equation. In what follows we shall consider the interaction of the examined flow with a plane harmonic wave. In this case the second term in the right-hand side of Eq.(1) is determined by expression $\vec{\nabla}\varphi = -A\vec{k}\sin(\vec{k}\cdot\vec{r}(\vec{\xi},t)-\Omega t)$, ($\vec{k}$ is the wave vector and Ω is the frequency).

Carrying out similarly as in our works [3,4] the discretization of Eq.(1), we arrive at a system of equations for discrete vortex particles. In computations we made use of the Gauss form of distribution of the vortex particle vorticity. In this case the equations of motion have the form

$$\frac{d\vec{r}_\alpha}{dt} = -\frac{1}{2\pi}\sum_{\beta}^{N}\Gamma_\beta\left\{1-\exp\left[-\frac{|\vec{r}_\alpha-\vec{r}_\beta|^2}{\sigma_\alpha^2+\sigma_\beta^2}\right]\right\}\frac{(\vec{r}_\alpha-\vec{r}_\beta)\times\vec{n}}{|\vec{r}_\alpha-\vec{r}_\beta|^2} - A\vec{k}\sin(\vec{k}\cdot\vec{r}_\alpha-\Omega t), \quad \alpha=1,2,\dots,N \qquad (4)$$

where Γ_α is the circulation of the vortex particle α, σ_α^2 is its dispersion, $\vec{n}$ is the unit vector normal to the plane (x, y). Thus, the solution of the problem on the shear layer stability with respect to acoustic disturbances reduces to the solution of the equations of motion of the vortex particles in the external field.

As our purpose is to examine both the linear and nonlinear stages of development of acoustic disturbances in shear layers, it is advisable to study the problem analytically before stating the results of numerical computations. We list here the results of solving the problem of the linear theory of stability for vortex sheet although the similar computations may be carried out also for a vortex layer of a finite length with linear distribution of velocity.

Thus let $h(x,t)$ be a function determining a shape of the vortex sheet. On the vortex sheet the following conditions should be satisfied

$$\frac{\partial h}{\partial t} + \frac{\partial \Phi_{\mp}}{\partial x}\frac{\partial h}{\partial x} = \frac{\partial \Phi_{\mp}}{\partial y}, \quad \frac{\partial \Phi_-}{\partial t} + \frac{1}{2}(\vec{\nabla}\Phi_-)^2 = \frac{\partial \Phi_+}{\partial t} + \frac{1}{2}(\vec{\nabla}\Phi_+)^2, \tag{5}$$

where Φ is the velocity field potential. Here and everywhere in what follows the values pertaining to a region $y<h$ are marked by the index minus and those pertaining to a region $y>h$ by the index plus. The last condition is a consequence of the pressure continuity on the vortex sheet.

Now represent the potentials $\Phi_\pm$ in the form of the sum of a potential induced by the vortex sheet $\Psi_\pm$ and of a potential $\varphi_\pm$ conditioned by the acoustic field presence. We have shown that for the present problem $\varphi_\pm \equiv \varphi = A\cos(\vec{k}\cdot\vec{x} - \Omega t)$. The potentials $\Psi_\pm$ satisfy the Laplace equation in the corresponding regions and they may be represented in the form $\Psi_\pm = \pm U_x + \Psi'_\pm$ ($\Psi'_\pm$ are the deviations of the potentials $\Psi_\pm$ from the undisturbed values $\pm U_x$). Substituting these expressions into Eq.(5) and linearizing them, we find

$$\begin{gathered}\frac{\partial h}{\partial t} \mp U\frac{\partial h}{\partial x} = \frac{\partial \Psi'_{\mp}}{\partial y} + \frac{\partial \varphi}{\partial y}, \quad \Delta \Psi'_\pm = 0, \\ \frac{\partial \Psi'_+}{\partial t} - \frac{\partial \Psi'_-}{\partial t} + U\left(\frac{\partial \Psi'_+}{\partial x} + \frac{\partial \Psi'_-}{\partial x}\right) = -2U\frac{\partial \varphi}{\partial x}\end{gathered} \tag{6}$$

with boundary conditions $\Psi'_\mp = 0$ at $y \to \mp\infty$.

A general solution of a system of equations (6) without source terms has been obtained by Helmholtz. A particular solution will be sought for in the following form ($c_x = \Omega/k_x$)

$$h = A_h e^{ik_x(x-c_x t)}, \quad \Psi'_\mp = A_\mp(y)e^{ik_x(x-c_x t)}.$$

Then it is easy to make sure of the fact that the functions $A_\pm(y)$ and the constant A_h are determined by the formulas

$$A_\pm(y) = a_\pm \exp(\mp k_x y), \quad A_h = -\frac{Ak_y c_x}{k_x(c_x^2+U^2)} - i\frac{AU}{c_x^2+U^2},$$

$$a_{\mp} = \mp \frac{AU(c_x + U)}{c_x^2 + U^2} + i\, \frac{A k_y U(c_x - U)}{K_x(c_x^2 + U^2)}.$$

Now a general solution of a system of equations (6) may be obtained in a general way. In particular, for the function h the following expression

$$h(x,t) = -\frac{A c_x k_y}{(c_x^2+U^2)k_x}\cos(k_x x - \Omega t) + \frac{AU}{c_x^2+U^2}\sin(k_x x - \Omega t) + \\ +(C_{11}e^{\sigma t} + C_{12}e^{-\sigma t})\cos(k_x x) + (C_{21}e^{\sigma t} + C_{22}e^{-\sigma t})\sin(k_x x), \sigma = k_x U \tag{7}$$

is derived.

Thus, the presence of acoustic field leads to the vortex sheet disturbance. This disturbance forms both from acoustic disturbance the amplitude of which is limited and moreover small and from vortex disturbance excited by acoustics whose amplitude increases exponentially in a general case. It is possible to show that this disturbance will be of sinusoidal character.

Now consider the computational results of the vortex sheet stability with respect to acoustic disturbances. In the method suggested in the present paper the problem is reduced to solving a system of equations (4). In computations the number of vortex particles N was varied from 120 up to 336. The dispersions of the vortex particles σ_α were determined in a similar way as in [4] .

The equations of motion (4) are valid for the description of two-dimensional flows of a compressible fluid when the following conditions $M \ll 1$, $\varepsilon < 1$, $\varepsilon > M$ are satisfied. Therefore the specific computations were carried out by the following values of the parameters $M = 10^{-2}$, $\varepsilon = 10^{-1}$, $\theta = \pi/4$ (θ is the angle between the vector $\vec{k}$ and the axis x).

By such a choice of parameters the amplitudes of disturbances of the transverse velocity component generated by external acoustic field turn out to be of the order $10^{-3} \div 10^{-4}$ at the initial stage.

The computations were carried out at the time interval equal to several units $T = \lambda/U$.

A typical picture of the vortex sheet evolution in the presence of external acoustic field is shown in Fig. 1 (N = 336). The

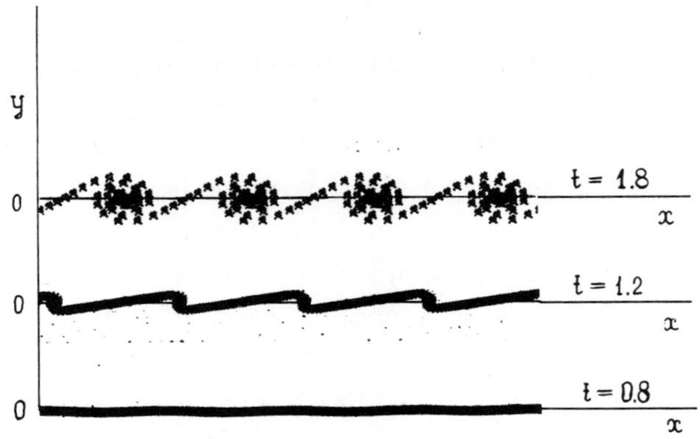

Fig.1. Evolution of the vortex sheet in the external acoustic field

initial location of the vortex sheet is rectilinear. At the first stage $t \leq 0.8$ the vortex sheet shape is sinusoidal and its amplitude and that of the transverse velocity component grow exponentially in time. When the amplitude of the vortex sheet achieves a value of the order 0.01 the nonlinear effects begin to be displayed and as a result a shape of the vortex sheet deviates more and more from sinusoidal. By the time moment $t = 1.2$ a shape of the vortex sheet becomes "sawtooth" and later we can observe its destruction that leads to the formation of large vortex structures.

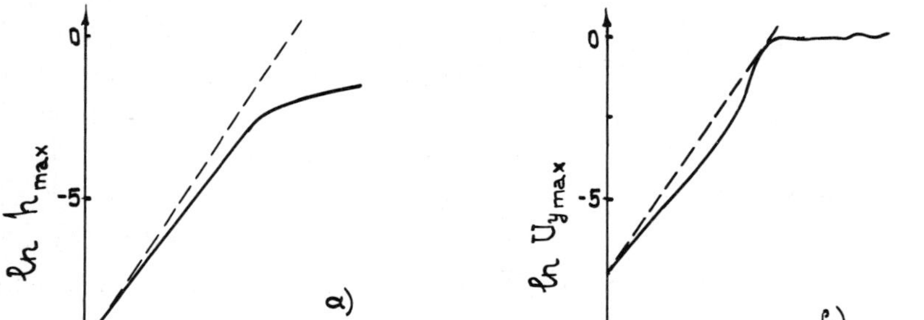

Fig.2. Rate of the increase of the functions h_{max} and $v_{y\,max}$

Fig. 2 shows the graphs of the rate of the increase in the logarithm of the amplitude of disturbances of the transverse velocity component (Fig. 2a) and of the logarithm of the function h amplitude (Fig. 2b). Here the dotted curves correspond to the linear theory. The analysis of these dependencies indicates that at the initial stage $t \leq 0.8$ the linear stage of the disturbances development takes place when only their amplitudes increase without changing a shape. At the second stage from 0.8 up to $t = 1.2$ one can observe some quasilinear stage of the disturbances development. Here the amplitudes also increase exponentially, a shape of the vortex sheet begins to change. Finally, at $t = 1.2$ the growth of the shear layer thickness ceases to be exponential and becomes linear, a nonlinear stage of the disturbances development starts. At this moment the vortex sheet breaks and the amplitude of the transverse velocity reaches its maximum value.

We have carried out also the computations of the vortex sheet stability with respect to disturbances of its boundary. A value of the amplitude of the function h disturbance was varied within a range from 10^{-5} up to 10^{-1}. In these cases a character of the disturbances development turns out to be qualitatively the same under the action of acoustic disturbances. The presence of the linear stage of their development is also observed at the initial small amplitudes of disturbances. It should be noted that with increasing the initial amplitude of disturbances the duration of this stage decreases. At last, at initial disturbances of the order 0.01 the linear stage of the disturbances development is almost absent. Increasing more and more the initial amplitudes we can achieve the state when the disturbances will increase nonlinearly at once. At that the amplitude of disturbances of the vortex sheet boundary grows slowly (Fig. 3a), and the disturbances of the velocity at the initial stage drastically increase and then fluctuate near some mean value (Fig. 3).

Up to now we have examined the avolution of the vortex sheet assuming the fluid inviscid. As a real fluid always possesses a finite viscosity, this fact should be taken into account when solving the problem. In the given problem the viscosity can be taken into consideration in a similar way as it has been sug-

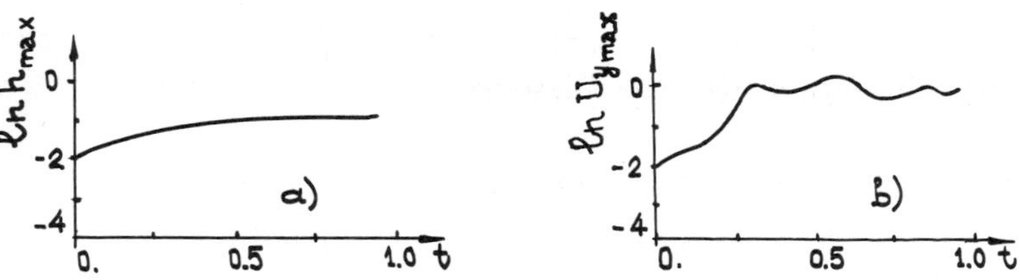

Fig.3. Rate of the increase of the functions h_{max} and $v_{y_{max}}$

gested in our work [4] , having implemented the splitting up of the equation of vorticity in terms of the physical processes.

It is well known that the vortex sheet in inviscid fluid is unstable with respect to any disturbances. The regard for viscosity exerts a stabilizing influence. The critical Reynolds number which determines the beginning of the vortex sheet instability in the external acoustic field Re is equal to 65. This number decreases when the amplitude of the disturbances increases.

We have carried out similar computations as for the vortex sheet for a layer of finite thickness. Since the value of acoustic disturbances is small and their role is in fact reduced to excitation of vortex disturbances, the further development of which weakly depends on the presence of acoustic field, the computation results have proved to be in a good agreement with linear theory. For the vortex layer with a piecewise-linear velocity distribution the instability was observed at $k_x\Delta < 1.3$, and for a layer with velocity distribution according to the tangent law at $k_x\Delta < 1$.

References

1. Ginevskii, A.S.; Vlasov, E.V.; Kolesnikov, A.B.: Aeroacoustical interactions. Moscow: Mechanical Engineering Press 1978.
2. Crighton, D.G.: Acoustics as a branch of fluid mechanics. J. Fluid Mech. 106 (1981) 261-298.
3. Veretentsev, A.N.; Rudyak, V.Ya.; Yanenko, N.N.: Variational method for the construction of discrete vortex models. Preprint of the Institute of Theor. and Appl. Mechanics, N°29, Novosibirsk 1982.
4. Veretentsev, A.N.; Rudyak, V.Ya.; Yanenko, N.N.: On constructing the discrete vortex models for the ideal incompressible fluid. J. Comput. Math. and Math. Phys. (Russian, to appear).
5. Lamb, H.: Hydrodynamics. Cambridge: Cambridge Univ.Press 1932.

The Effect of External Fields on Transition to Turbulence in Free Shear Flows

M. I. RABINOVICH and M. M. SUSHCHIK

Institute of Applied Physics
Academy of Sciences of the USSR
Gorky, USSR

Summary

Shear turbulence is treated as stochastic motion represented by a strange attractor. Effects of external forcing on turbulence in free shear flows are analysed based on this concept.

1. On the Nature of Turbulence in Shear Layers

1.1. Introduction

The mechanisms by which turbulence is produced in shear layers have not been elucidated so far. It is still to be found out why and how a laminar flow becomes turbulent in the absence of irregular forcing. By formal logic, in order to answer this question one should construct or at least give a qualitative description of solutions of Navier-Stokes equations whose characteristics under regular initial and boundary conditions are adequate to the characteristics of turbulent flows observed experimentally. In constructing an appropriate theory one can use simpler, as compared to free shear flows, internal flows as an example, such as Couette-Taylor or convective flows in cells. In recent years concepts have appeared according to which random pulsations of the velocity field in such flows correspond to the motion of a dynamic system on a strange attractor (see, for example, [1]). Remarkable recent experiments [2,3] have proved the connection between strange attractors and turbulence in internal flows.

It is not clear yet whether a strange attractor is directly related to the turbulence of free shear flows as well. Until recently we did not even know whether spontaneous development of turbulence is possible in nonconfined and/or semiconfined flows in the absence of global feedback (reflecting walls, counterpropagating flows etc.). Such a possibility has been lately proved for a model system [4]. Nevertheless, the known experiments (physical and numerical) on transition to turbulence in shear flows can be readily interpreted on the basis of this concept of shear turbulence as stochastic oscillations in dynamic systems.

1.2. Stochastization of Motion of Small-Scale Vortices in a Two-Dimensional Shear Flow

The development of instabilities and the roll-up of the shear layer result in an array of vortices whose velocity field is approximated by the stream function [5] that is an exact solution of Euler equations

$$\Psi(x,y)=(\Gamma/2\pi)\ln[\mathrm{ch}(2\pi y/\ell)-\alpha\cos(2\pi x/\ell)]. \tag{1}$$

Here ℓ is the spatial period of the array and the parameter α characterizes the concentration of the vorticity.

Assuming that three-dimensional small-scale vortices are frozen, one can determine the evolution of the flow by their redistribution in the flow, i.e. by the motion of fluid particles. The equations for this motion $(\{\xi;\eta\}=(2\pi/\ell)\{x;y\};\tau=t\Gamma\pi/H\ell^2=\Omega t)$:

$$d\xi/d\tau=\partial\Psi/\partial\eta=-H(\mathrm{sh}\,\eta)(\mathrm{ch}\,\eta-\alpha\cos\xi)^{-1} \tag{2}$$

$$d\eta/d\tau=-\partial\Psi/\partial\xi=\alpha H(\sin\xi)(\mathrm{ch}\,\eta-\alpha\cos\xi)^{-1} \tag{3}$$

are reduced to the "complex pendulum" equation for the variable $z=\xi+i\eta$ [6]:

$$d^2z/d\tau^2+\alpha\sin z=0. \tag{4}$$

On the phase surface of set (2)-(3), the separatrix $\mathrm{ch}\,\eta-\alpha\cos\xi=$ $=1+\alpha$ divides the particles into those trapped by the vortex centres and transit ones, i.e. moving along the array; the separatrix is characterized by $z=2\arcsin\mathrm{th}\,\alpha(\tau+\theta_0)$; $H=1+\alpha$, where $\theta_0=\theta_0'+i\theta_0''$ is the complex constant.

The pattern of the transfer of the small-scale three-dimensional vorticity given above is correct for the case of an unperturbed system. This is not the case in semiconfined flows: unsteady development of perturbations takes place near the nozzle or the splitting plate and the vortices merge downstream. Even if the perturbations introduced by these events are small and regular, the motion of small vortices is subject to stochastization.

Let us introduce the small terms $\varepsilon u(\xi,\eta,\tau)$ and $\varepsilon v(\xi,\eta,\tau)$ into eqs.(2) and (3), respectively. Since the separatrix of the unperturbed system forms a loop on the phase surface (a cylinder in this case), such a perturbation (see, for example, [7]) may lead to the formation of a homoclinic structure in the three-dimensional space (ξ, η, τ). This structure guarantees stochastic motion in the system and exists under the condition [8] that

$$\Delta_\varepsilon(\tau_0) = \int_{-\infty}^{+\infty} dt \{ u(\xi_0,\eta_0,\tau)\partial\Psi(\xi=\xi_0;\eta=\eta_0)/\partial\xi + v(\dots)\partial\Psi(\dots)/\partial\eta \} \qquad (5)$$

is an alternating function of τ_0 . Here $\xi_0(\tau-\tau_0); \eta_0(\tau-\tau_0)$ is the separatrix of the unperturbed system and τ_0 is the parameter that characterizes the position of the point on the separatrix. The amplitude of the $\varepsilon\Delta_\varepsilon(\tau_0)$ function characterizes the width of the region ($\Delta\xi, \Delta\eta$) with stochastic motion of particles $(\Delta\xi;\Delta\eta) \approx \varepsilon ch^{-1}(2\pi^2/\Omega T)$ at $\alpha \approx 1$, $\Lambda \gg \ell$, and not too large $2\pi^2/\Omega T$. Here Λ and T are the spatial and temporal periods of the perturbation, respectively.

Note that at $\alpha \approx 1$ eqs.(2)-(3) describe the nonlinear regime of developing perturbations with a double spatial period 2ℓ , as well: the relative motion of the vortices with even and odd numbers is described by the same equations as given above with the substitution $2\Gamma \to \Gamma$. Therefore the conclusion on the presence of stochastization is also valid for the motion of large-scale two-dimensional vortices and, hence, for all the velocity field.

1.3. A Dynamic Model for Spatial Development of Turbulence in a Viscous Flow

The example considered above refers to the case where stochasticity occurs in the Hamiltonian system yielded by Euler equations. The relation between stochastic solutions of the Navier-Stokes equations and stochasticity of dynamic systems has also been established [1]. However, one cannot yet successively construct models for spatial development of shear turbulence even for simple cases. Therefore we restrict our consideration to a phenomenological model that takes the following factors into account: 1- nonlinear development of perturbations according to the Landau-Stuart model, and 2- transfer of perturbations downstream. For simplicity of numerical simulation, we divide the flow into elementary cells. We then get a system of coupled equations

$$dA_j/dt - \alpha(A_j - A_{j-1}) = A_j(1 - \delta|A_j|^2). \qquad (6)$$

The absence of global feedback due to the propagation of the perturbations upstream is the characteristic feature of this model. In spite of the fact that the interaction between the elementary perturbations (cells) is convective and unidirectional, stochastic motions that are defined only by the parameters of the system can occur.

System (6) was numerically investigated for $\alpha = \alpha'(1 - i\alpha'')$, $\delta = 1 - i\beta$, where $\alpha', \alpha'', \beta > 0$. The power spectra of amplitude and phase $A_j(t)$ were analysed and the dimension D from a time series $A_j(t)$ [9,10] downstream was calculated as a function of j . When the coupling is weak (α' is small), a regime of independent local excitations onsets in the system at the frequency $|\omega| = \beta$ and develops into a regime of synchronous motions in all cells, beginning with one, when the coupling becomes stronger. The homogeneous regime of synchronization is unstable at strong coupling and a spatially inhomogeneous regime of collective excitations onsets in the array: at large enough j it develops into a stationary modulated wave.

Nonlinear interaction of collective and local excitations may result in the onset of stochastic self-oscillations. The number of the degrees of freedom that define the motion in the j -th cell increases with the growth of j owing to nonreciprocal coupling, therefore the dimension of the strange attractor should, generally, increase downstream, i.e. one can expect "infinitely dimensional" turbulence at $j \to \infty$. However, this is not the case in a homogeneous array ($\alpha, \beta = \text{const}$). A regime of spatially homogeneous turbulence with a constant, relatively small dimension of the attractor is established at large enough j (see Figs.1 and 2).

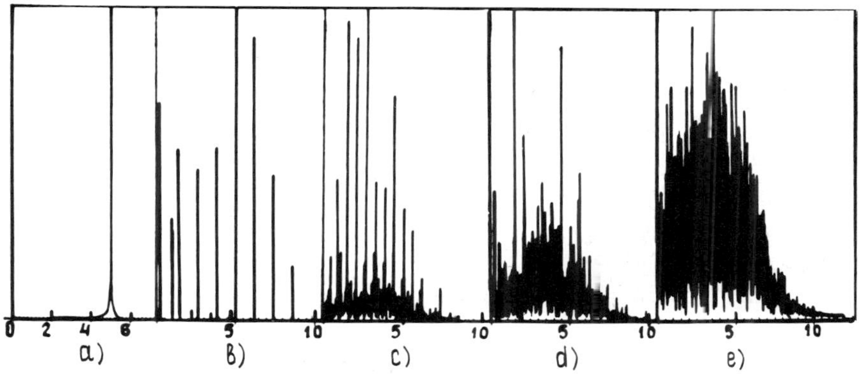

Fig.1. Power spectra $A_j(t)$ illustrating spatial development of turbulence along the array (α' =0.5, α'' =1.77, β =5.0): a- j =2, b- j =9, c- j =10. d- j =12, e- j =20 and 50.

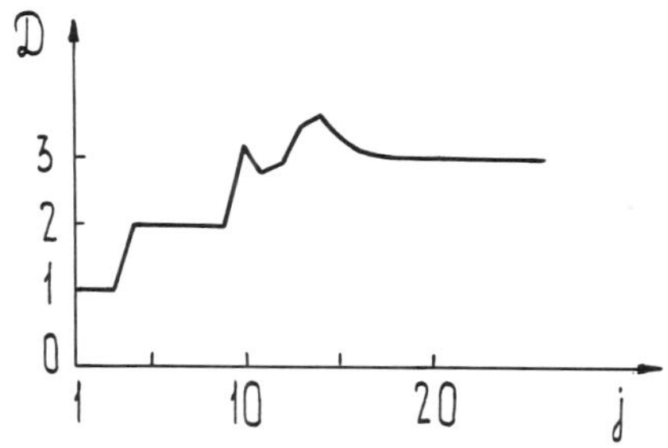

Fig.2. The variation of the attractor dimension along the array: α' =0.5, α'' =1.77, β =5.0

1.4. Experimental Determination of the Limiting Set Dimension from a Time Series of One Physical Signal

The presence of finite -dimensional limiting sets can be readily detected in feedback flows [11],for example, in an open-jet wind tunnel. Such flows are generally characterized by periodic self-oscillations at one of resonance frequencies. However, there is a region of parameters where several "resonances" interact and stochastic self-oscillations may occur. Since the spectrum of modes in the shear flow under study is generally continuous, the problem on the finite-dimensional nature of stochasticity is no less essential for this case than for a shear flow without external feedback.

If the motion on a finite-dimensional attractor corresponds to the flow with a continuous spectrum, then the attractor dimension can be determined by analysing the time series of one physical signal $u(t,x_0)$[12]. The correlation dimension ν was used for a lower estimate of the attractor dimension. This estimate is usually close to the exact value of small dimensions [13,14]. It was calculated as follows. First, discretization with a step τ with long enough time series $u(t,x_0)$ was accomplished. Then, a trajectory reproducing the attractor was constructed in a certain n-dimensional space. Each j-th point of the trajectory was determined by the n-th vector that was given by n neighbouring points of the time series

$$V_j = \{u(\tau_j), u(\tau_j+\tau), \dots, u(\tau_j+(n-1)\tau)\}. \qquad (7)$$

The points of the trajectory were used to calculate the correlation integral

$$C(r,N) \sim \frac{1}{N^2}\sum_{\ell}^{N}\sum_{j}^{N} H(r - |\vec{V}_j - \vec{V}_\ell|) \qquad (8)$$

related to the correlation dimension logarithmically $C(n\to\infty) \to r^{\nu}$. Here H is the Heaviside unit function and N is the number of points within the radius r [13,14]. The results of the measurements carried out by P.R.Gromov and A.B.Zobnin are given in Fig.3. Figure 3 shows the velocity records spectra and the attractor dimensions of the flows that are close to periodic, quasiperiodic and stochastic flows.

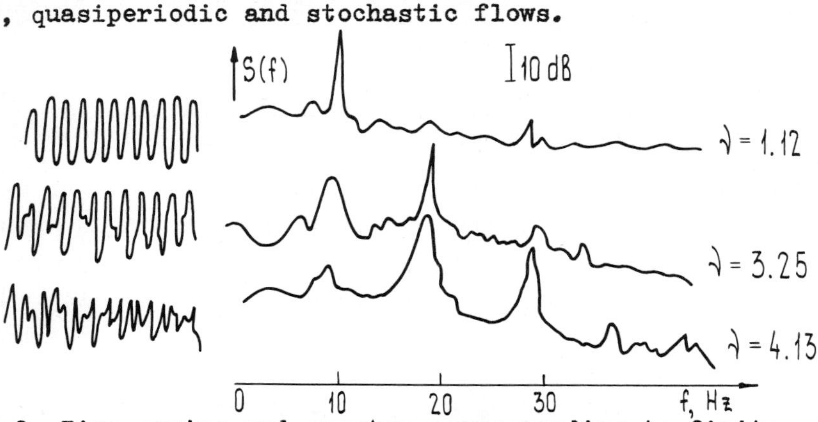

Fig.3. Time series and spectra corresponding to finite dimensions ν .

2. The Effect of External Fields on Shear Flow Turbulence

2.1. Introduction

Shear turbulence, according to its dynamic model, can be affected in two ways: 1 - by varying the attractor type, and 2 -

by varying the route by which the flow approaches the attractor and, hence, by accelerating or slowing transition to turbulence (for example, by varying the boundary and/or the initial conditions). These two possibilities are discussed below.

2.2. The Slowing of Transition to Turbulence

The experiments show that real shear flows should be considered as a set of vortices whose collective motions are large-scale energy-containing perturbations. The mechanisms slowing down the development of turbulence were investigated based on this concept of shear flows. The experiments on suppression of turbulence in the jet showed that in all cases suppression took place in the initial section of the jet, where strong modulation of vorticity distribution along the shear layer was observed. In most experiments suppression was due to this modulation mechanism.

In our experiments [15] a water jet was formed by a nozzle with the diameter of the exit section d =50 mm and was excited by the harmonic modulation of the discharge with the amplitude depth 1-6%. The average velocity of the jet discharge was U =58 cm/s ($Re = Ud/\nu \approx 3\cdot 10^4$), and the initial level of fluctuations at the jet axis was $\varepsilon \approx 0.5\%$.

Suppression of turbulence (a decrease in the average level of pulsation intensity $\langle u^2 \rangle$) was observed at $x \leqslant 6d$ within all the range of frequencies of the forcing f =15-29 Hz ($St = fd/U$ =1.3-2.5). The measurements of the growth rates showed that the greatest difference in the amplification rate of low-frequency (l.f.) perturbations that make the main contribution to the energy of turbulence was observed for forced and unforced jets at $x \lesssim 1.5d$.

It follows from the measurements of the amplitude and phase distribution of h.f. perturbations that a regular array of 4-5 vortex rings spaced equidistantly with the distance between them defined by the frequency of the h.f. forcing and the flow velocity, is formed near the origin of the jet ($x \lesssim 1.5d$), the section of the shear layer that is most important for suppression. Such a modulation structure of the flow can change the growth rates of energy-containing perturbations essentially. Let us demonstrate this at a simple example, an infinite plane jet (a two-dimensional analog of the flow under study).

We restrict our consideration to the longwave limit where the wavelength is much larger than the thickness of the shear layer. Let us compare the temporal amplification rate p of perturbations in such a jet for two cases: 1 - the vorticity is distributed in the thin layer continuously along x , and 2 - the vorticity is concentrated in the point vortices located symmetrically. It can be readily shown that the temporal amplification rates for the symmetric modes are

$$p = \pi U / \Lambda \tag{9}$$

$$p = (\pi U/\Lambda)\Big\{\big[1 + (\Lambda/2\Lambda_0)/\mathrm{sh}^2(\pi d/\Lambda_0) - \Lambda_0/\Lambda\big]^2 - \big[\mathrm{sh}((\pi d/\Lambda_0)(1-2\Lambda_0/\Lambda))/\mathrm{sh}(\pi d/\Lambda_0) + (\Lambda/2\Lambda_0)\mathrm{ch}(2\pi d/\Lambda)^2/\mathrm{sh}^2(\pi d/\Lambda_0)\big]^2\Big\}^{1/2}, \tag{10}$$

where d is the width of the jet, Λ_0 is the space period of the array, and Λ is the wavelength of the perturbation (for the array $\Lambda > \Lambda_0$).

The comparison of (9) and (10) for Λ and Λ_0 (see the experimental dependences in Fig.4) shows that a decrease (due to modulation) in the growth rates of l.f. perturbations is comparable for the theoretical model and the experiment and is 20-30%. Thus, small-scale forced modulation of vorticity along the flow results in suppression of the growth of energy-containing (l.f.) perturbations and in stabilization of the shear layer.

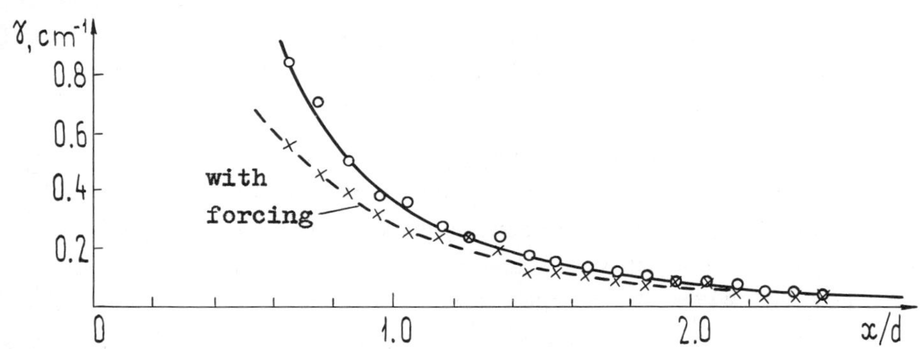

Fig.4. The dependence of growth rates with and without h.f. forcing.

2.3. The Variation in the Attractor Type

It is known from the theory of dynamic systems that weak forcing can have strong effects, for example, in the case of resonance when accumulating effects are possible. Such accumulating effects are absent in a turbulent flow due to the complicated motion on the attractor, thus providing its stable identification. However, this is valid only if the variation of the field (its phase, amplitude etc.) that acts on the flow does not depend on the latter. If their variations are self-consistent, one can expect accumulating effects at a certain consistency which influence the nonstationary flow significantly.

The experiments carried out by P.R.Gromov and A.B.Zobnin on a flow in an open wind tunnel, where the effect of a h.f.($St > 2$) acoustic field from a I W emitter was used (cf. [16]), showed no significant changes in the velocity pulsations. The situation changed cardinally when the emitted acoustic field was synchronized by the signal from a hot-wire anemometer with the wire placed in the flow. The velocity fluctuations became either more regular or more complicated depending on the shift in the emitter phase and on the control signal, which was accompanied by the corresponding variation in the dimension of the time series. At a certain phase shift, the intensity of some components could be decreased by 20 db (Fig.5).

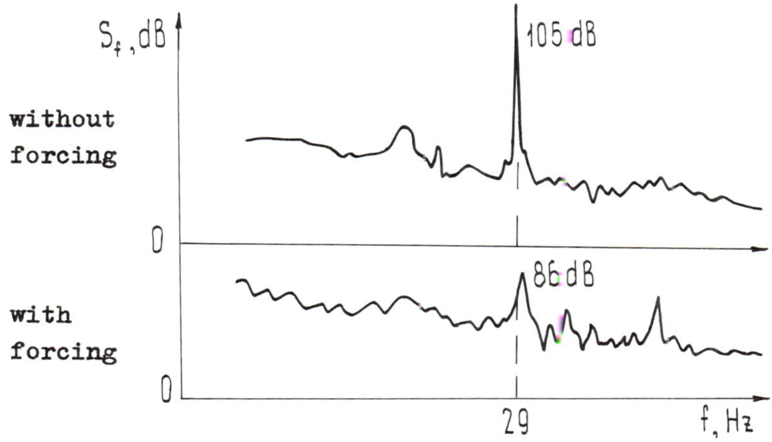

Fig.5. Self-oscillation spectrum with and without external acoustic forcing ($f_{forc} \approx 29\,Hz$).

3. Conclusion

Thus, our representation according to which shear turbulence has a dynamic nature allows for a new treatment of the mechanisms of external field forcing on transition. The potentialities of a "dynamic" approach are illustrated by three examples. The analysis of a dynamic model of an inviscid shear flow in the form of vortex arrays makes it possible to determine most dangerous frequencies of a regular external forcing. Experiments on a jet have demonstrated possible variation of transition to turbulence. Finally, it has been shown for a flow in an open-jet wind tunnel that self-consistent forcing may result in a great decrease in the level of turbulent pulsations.

References

1. Rabinovich, M.I.: Uspekhi Fiz. Nauk 125 (1978) 123 (Sov. Phys. Usp. 21 (1978) 443).
2. Eckman, J.P.: Rev. Mod. Phys. 53 (1981) 643.
3. Brandstäter, A.; Swift, J.; Swinney, H.L.; Wolf, A.; Farmer, J.D.; Jen, E.; Cratchfeld, P.J.: Phys. Rev. Lett. 51 (1983) 1442.
4. Gaponov-Grekhov, A.V.; Rabinovich, M.I.; Starobinets, I.M.: Pis'ma v ZhETF 39 (1984)N12.
5. Stuart, J.T.: J. Fluid Mech. 29 (1967) 417.
6. Rabinovich, M.I.; Sushchik, M.M. In: Nonlinear-Waves, Self-Organization. Moscow, Nauka 1983, p.56.
7. Gaponov-Grekhov, A.V.; Rabinovich, M.I. In: Physics of the XXth Century. Problems and Potentialities. Ed. by Velikhov, E.P. Moscow, Nauka 1984.
8. Mel'nikov, V.K.: Trans. Moscow Math. Soc. 12 (1963) 3.
9. Ledrappier, F.: Commun. Math. Phys. 81 (1981) 229.
10. Shimoda, I.; Nagashima, T.: Progr. Theor. Phys. 61 (1979) 1605.
11. Rockwell, D.: AIAA Journ. 21 (1983) 645.
12. Takens, F. In: Lecture Notes in Math. Springer Verlag 898 (1981) 366.
13. Grassberger, P.; Procaccia, I.: Phys. Rev. Lett. 50 (1983) 346.
14. Malvaison, B.; Atten, P.; Berge, P.; Dubois, M.: J. Physique Letters 44 (1983) 897.
15. Kudryashov, A.V.; Mansfeld, A.D.; Rabinovich, M.I.; Sushchik, M.M.: Dokl. Acad. Nauk SSSR 277 (1984)N1.
16. Boyarchikova, M.Yu.; Vlasov, E.V.; Ginevsky, A.S.; Zosimov, Izv. Akad. Nauk SSSR, Mekhan. Zhidkosti i Gaza 1 (1982)126-132.

The Mixing Layer Forced by Fundamental and Subharmonic

YU-QUAN ZHANG,[1] CHIH-MING HO[1], and PETER MONKEWITZ[2]

Dept. of Aerospace Engineering[1]
University of Southern California
Los Angeles, CA 90089-1454

Dept. of Mechanics and Structures[2]
University of California, Los Angeles
Los Angeles, CA 90024

Introduction

The evolution of coherent structures in a mixing layer can be manipulated very efficiently by forcing at a single frequency near the most amplified frequency (Ho and Huang [1]). Futhermore, the linear stability analysis (Monkewitz and Huerre [2]) based on the tanh or Blasius profiles can well predict the most amplified frequency and the amplification rates in the mixing layer.

In the case of a mixing layer forced at two frequencies, e.g. subharmonic and fundamental, the amplification rate of one wave could be affected by the presence of the other wave. This possibility has been suggested by the weakly nonlinear analysis of Kelly [3] and Monkewitz [4]. Another issue is the influence of the splitter plate wake on the stability of the flow.

In this paper, the mixing layer was forced by both the fundamental and the subharmonic at various phase differences. The eigenfunctions and the amplification rate were measured and compared with theoretical calculations in which the wake was taken into consideration.

Experimental Facility

The experiment was performed in a mixing layer water channel with a cross section of 10cm x 10cm. The length of the test section is 180cm. The stagnation chamber and the contraction are separated into two parts by a splitter plate. The trailing edge of the splitter plate is located at the end of the contraction. The velocity ratio, $R = \Delta U/2\overline{U}$, of mixing layer was kept at 0.73 with $\overline{U}$ = 5.19 cm/sec throughout the experiments.

Forcing was introduced into the high speed side with two butterfly valves, one rotated at the fundamental and the other at the subharmonic. The phase angle, β, was controlled by changing the angle between the rotating plates of the two valves. The forcing level was limited to 0.01~0.1% of the mean velocity, $\bar{U}$.

The velocity was measured by hot-films. Signals were recorded on analog tape and then processed digitally on a PDP 11/55 computer.

Stability Calculations

Numerous stability calculations have been performed in the past on parallel mixing layers without a wake such as the familar tanh[y] and the Blasius profile (Michalke [5], Monkewitz and Huerre [2]). For these cases it was found by Monkewitz and Huerre that the appropriate length scale which best collapses data from slightly different mixing layer profiles is the vorticity thickness. This is physically plausible as the vorticity thickness is directly related to the shear driving the instability. In the present experiment the wake of the splitter plate persisted over most of the region of interest. Therefore, it was felt that a comparison with calculations on a tanh[y] profile were not adequate. In particular, on a mixing layer with wake there are several possibilities of picking a length scale. One of the objectives of this work is to find the most appropriate scale. Therefore, the tanh[y] profile was generalized to

$$U[y] = 1 + R\{\tanh[y] - A\,\mathrm{sech}^2[y] + B\tanh[y]\,\mathrm{sech}^2[y]\} \qquad (1)$$

with the first two terms, a continuous variation between a tanh[y] and a sech^2[y] wake profile can be achieved. The third term has been added to allow for some adjustment of the ratio between the maximum slopes on the high and low speed side of the velocity minimum. To ensure that the profile does not have unwanted overshoots (additional inflection points) the requirement is $A > B - 1/2$. A further condition is that $A > 1/2 - B$ which will guarantee the existence of a local velocity minimum (wake).

The stability calculations were only performed for the dominant "flapping" mode. Of course, a "varicose" mode also exists as soon as there is a wake, but it is only unstable at a very low frequency and thus is not relevant to the present measurements.

Results

Mean Velocity Profiles

A very clear wake existed in the mean profile (Fig. 1-a) measured at 0.3 cm downstream from the splitter plate. The velocity defect decreased fairly fast as evidenced by the measurement at $x = 2.5$ cm (Figure 4-a). For the stability calculation, the analytical expression (eqn. 1) was used instead of directly utilizing the experimental data. This procedure was chosen to provide a smooth variation of the stability curves with downstream distance as the results of these parallel calculations beyond the low frequency end are very sensitive to details of the velocity profile. The fitting was satisfactory near the trailing edge and became excellent at about one fundamental wavelength $\lambda = \bar{U}/f = 2.29$ cm, downstream from the splitter plate.

Length Scale

In the past, several operational definitions of the length scale were used; the sum of the momentum thicknesses of both the high speed and low speed sides (Miksad [6]), the momentum thickness of the high speed side (Ho and Huang [1]), or the momentum thickness at a chosen station (Drubka [7]). In the present study, the length scale was chosen such that the 180° phase change (minimum) in the eigenfunctions (both the fundamental and the subharmonic), was located at the right y-locations (Fig. 1-c, d to Fig. 4-c, d). This length scale was found

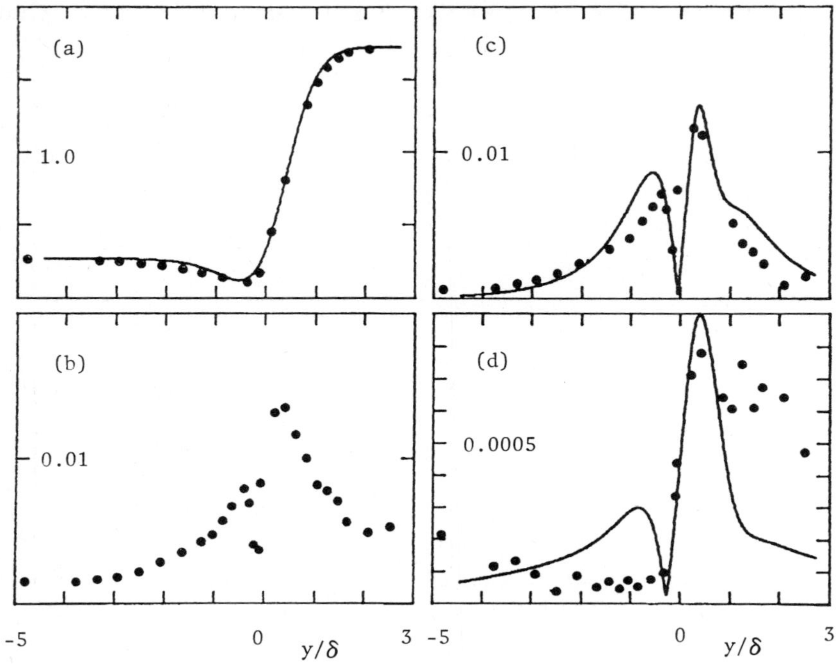

Figure 1. Comparison of Measured Profiles (•) and Theoretical Calculations (————) at X = 0.3 cm.

(a) Mean velocity $U/\overline{U}$: Measured and Fitted Using Eqn. (1)

(b) $u'_{rms}/\overline{U}$

(c) Fundamental Mode Shape $u'(f)/\overline{U}$, f=fundamental Frequency

(d) Subharmonic Mode Shape $u'(f/2)/\overline{U}$

to be the high speed vorticity thickness, δ, based on the overall velecity difference (Fig. 5).

This choice appears logical in the low frequency limit: the instability of the long waves is not affected by the wake because the positive and negative vorticities associated with it cancel on the long wavelength scale. Furthermore, the present choice is compatible with previous tanh[y] calculations (A=B=0 in eqn. 1).

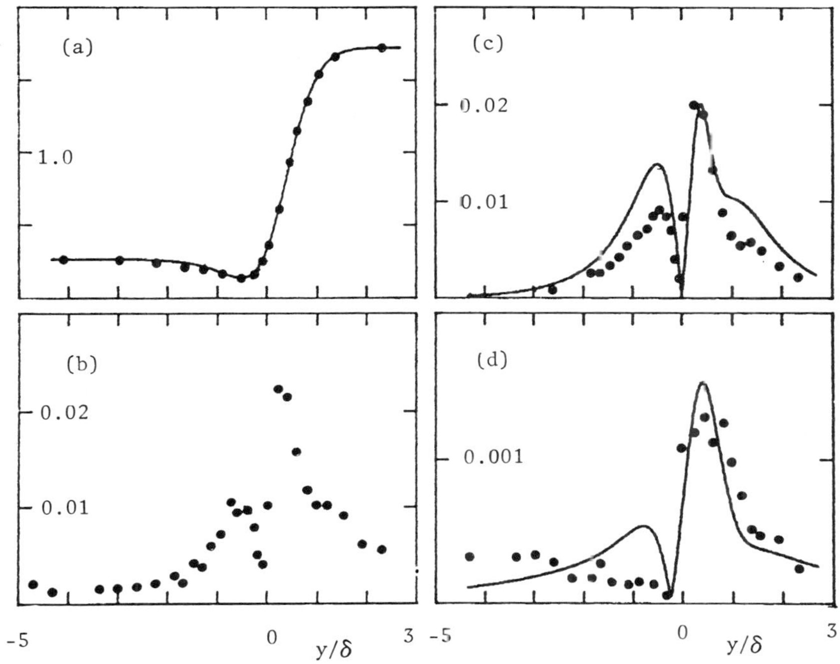

Figure 2. Same as Fig. 1 at x = 1 cm

Eigenfunctions

The fundamental and the subharmonic eigenfunction were measured and compared with the locally parallel calculations (Fig. 1-c, d to Fig. 4-c, d). Thereby, the amplitude of the calculation was adjusted at each x-station independently of the measured disturbance level which is consistent with the locally parallel approach. While the mode shape of the fundamental agrees very well with the calculations, a consistent discrepancy between the measured and computed disturbance velocity profiles of the subharmonic is obvious on the low speed side at all stations. Possible explanations for this phenomenon will be given below. No self-similarity of the modal shapes was found in the streamwise development. The eigenfunctions also have been measured for various initial

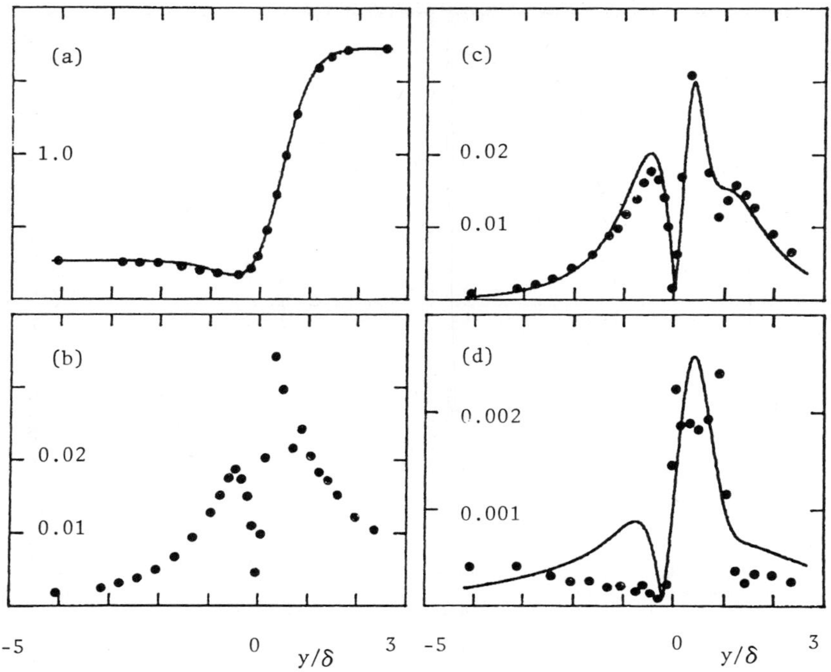

Figure 3. Same as Fig. 1 at x = 1.5 cm

phase differences. The variation of the phase difference did not affect the modal shapes. Within the measured range, 0.3 cm < x < 2.5 cm, the flow was still laminar and most of the turbulent energy was in the fundamental and the subharmonic modes. The rms level distributions of the turbulence were almost equal to the summation of the fundamental and the subharmonic eigenfunctions (Fig. 1-b to Fig. 4-b).

Amplification Rates

Three spatial amplification curves for different wake depths were compared with the standard tanh amplification curve (Fig. 6). The presence of the wake increases the maximum amplification rate and decreases the neutral stability frequency.

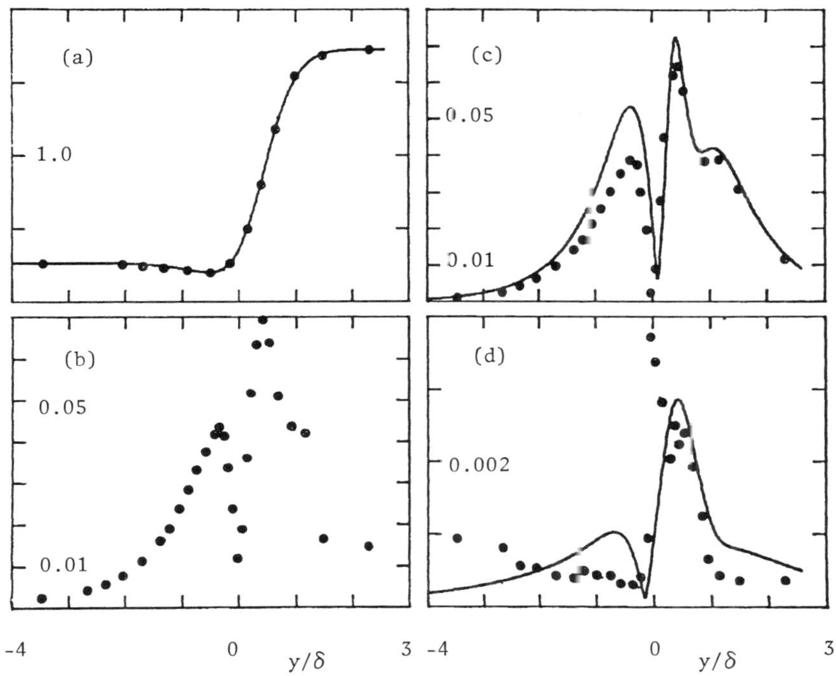

Figure 4. Same as Fig. 1 at x = 2.5 cm

The experimental amplification rate was determined from the streamwise development of crosstream integrals of the eigen-function in the same fashion as in Ho and Huang[1]. It was found that within experimental accuracy the amplification rate of the fundamental is independent of its phase relation with the subharmonic (Fig. 7). The value is close to the maximum amplification rate of the tanh profile (solid curve of Fig. 6) in accord with earlier findings by Ho and Huang [1] at an even lower velocity ratio of R = 0.31 where the wake is much more persistent. This somewhat disappointing result that a much better fitted velocity profile yields a considerably worse prediction of the amplification rate, could have several causes which require further investigation. It could mean that the effect of the wake is minimal and does not have to be considered, but more likely it indicates a failure of the

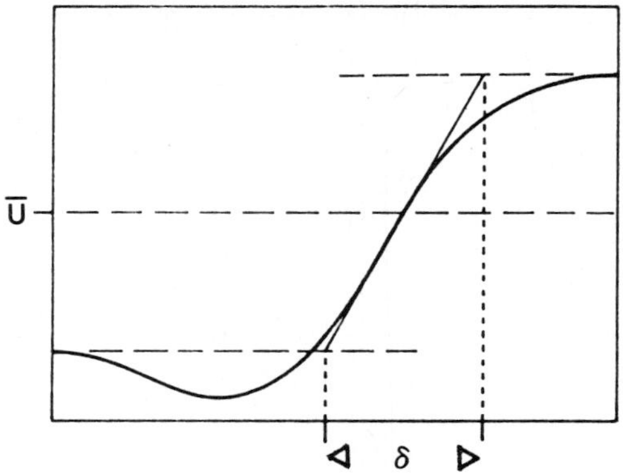

Figure 5. Definition of the Length Scale δ.

doubly infinite parallel calculation, i.e. a strong influence of the splitter plate and of nonparallelism. This latter view is supported by the good prediction of the fundamental mode shapes.

The amplification rates of the subharmonic vary by ±30% for different phase angles (Fig. 7). This phenomenon cannot be obtained from linear analyses. It has been suggested by weakly nonlinear calculations (Kelly [3], Monkewitz [4]). No quantitative comparison has been attempted though, as the aforementioned analyses rest on the assumption of resonant interaction. Due to the large maximum growth rates in the mixing layer, such a resonance can only occur effectively between a fundamental which is close to neutral and its subharmonic. In the spatial case (Monkewitz [4]) this also ensures that the two phase speeds are approximately equal, a further necessary condition for resonant interaction. Over the range of x-stations covered by the present measurements the fundamental is far from neutral and actually grows faster than its subharmonic. Therefore, no resonant interaction is expected on theoretical grounds. On the other hand, a nonresonant, phase-angle dependent interaction in conjunction with nonparallel effects could conceivably cause a noticeable

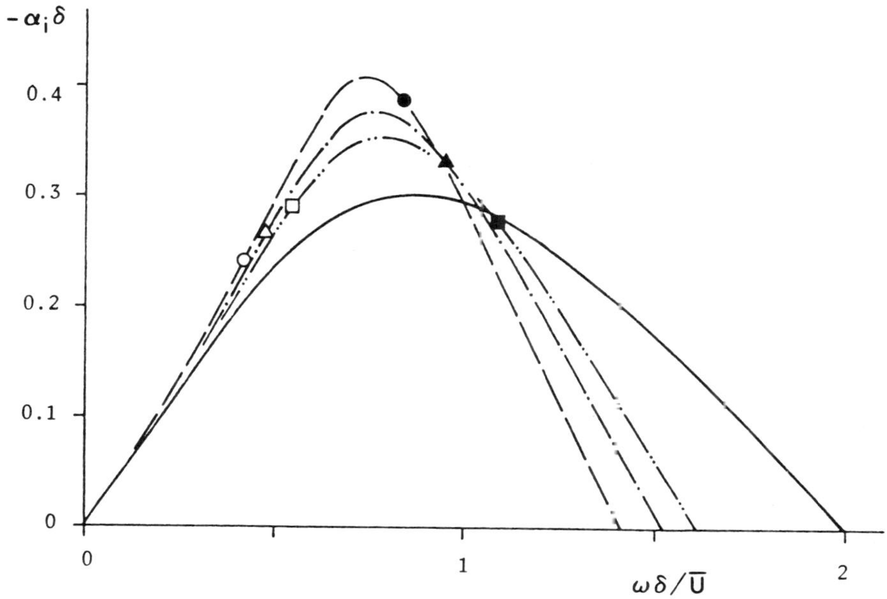

Figure 6. Theoretical Amplification Rates at Different Downstream Stations Nondimensional with the Local Thickness δ. Solid and Corresponding Open Symbols Indicate the Location of the Fundamental and Its Subharmonic Respectively.

—— ——	o :	x = 0.3 cm, δ = 0.31 cm
——·——	△ :	x = 1.5 cm, δ = 0.35 cm
——··——	□ :	x = 2.5 cm, δ = 0.40 cm
————	:	standard tanh profile

correction in the subharmonic mode shape. Another possibility is that the imposed subharmonic disturbance combines with a disturbance fed back from downstream to give the relatively bad correspondence between the measurement and linear stability calculation. It has to be stressed though that these possible explanations are purely speculative and need to be pursued further.

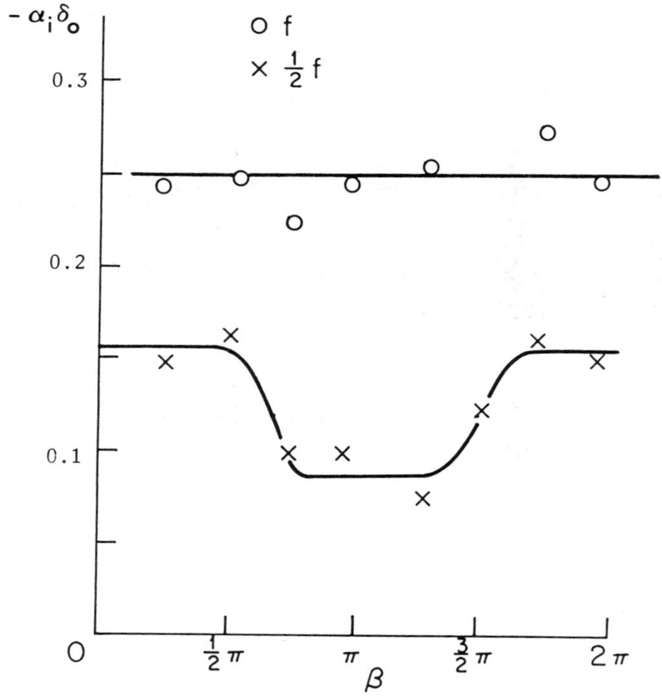

Figure 7. Measured Amplification Rates as a Function of the Phase Angle β in the Range $0.3 < x < 2.5$ cm.

Conclusion

When the mixing layer is forced by a single frequency, the amplification rate and the vortex merging location can be reasonably well predicted by linear instability theory (Ho and Huang [1], Ho and Huerre [8]). The shape of the eigenfunction and the amplification closely agrees with the slowly divergent flow analysis (Gaster et al. [9]). The situation is somewhat different when two forcing frequencies are present in the flow. The shape of the eigenfunction in the initial region can still be calculated by linear theory. However, the amplification rate of the subharmonic is a function of the phase angle. It indicates that the nonlinear interaction has to be taken into consideration, even when the amplitudes of the fundamental and the subharmonic are very low in the region not very far from the trailing edge.

The wake present in this region should, on theoretical grounds, also increase the maximum amplification rate, but the experimental data suggests that the tanh profile provides, (probably accidentally), better predictions. Nevertheless, the good correspondence between theoretical and experimental mode shapes of the dominant fundamental permits the extraction of the relevant length scale for the instability process. It is found that it is dominated by the vorticity on the high speed side with no substantial influence from the wake.

Acknowledgement

This work is supported by the Office of Naval Research.

References

1. Ho, C. M. and Huang, L. S. 1982, "Subharmonics and Vortex Merging in Mixing Layers," J. Fluid Mech. 119: 443-473.

2. Monkewitz, P. A. and Huerre, P. 1982, "The Influence of the Velocity Ratio on the Spatial Instability of Mixing Layers," Phys. Fluids 25: 1137-1143.

3. Kelly, R. E. 1967, "On the Stability of an Inviscid Shear Layer Which is Periodic in Space and Time," J. Fluid Mech. 27: 657-689.

4. Monkewitz, P. A. 1982, "On the Effect of the Phase Difference Between Fundamental and Subharmonic Instability in a Mixing Layer," Internal Report, Univ. of Calif. at Los Angeles. Presented at the APS Meeting 1983.

5. Mickalke, A. 1965, "On Spatially Growing Disturbances in an Inviscid Shear Layer," J. Fluid Mech. 23: 521-544.

6. Miksad, R. W. 1972, "Experiments on the Nonlinear Stages of Free Shear Layer Transition," J. Fluid Mech. 56: 695-719.

7. Drubka, R. E. 1981, "Instabilities in Near Field of Turbulent Jets and their Dependence on Initial Conditions and Reynolds Number, Ph.D. Thesis, Illinois Institute of Technology.

8. Ho, C. M. and Huerre, P. 1984, "Perturbed Free Shear Layers," Ann. Rev. Fluid Mech. 16: 365-424.

9. Gaster, M., Kit, E. and Wygnanski, I. 1983, "Large Scale Structures in a Forced Turbulent Mixing Layer," submitted to J. Fluid Mechanics.

Experiments on Nonlinear Interactions in Wake Transition

R. W. Miksad, F. L. Jones, and E. J. Powers
College of Engineering
The University of Texas at Austin
Austin, Texas 78712, U.S.A.

SUMMARY

Low-frequency difference modes are generated in the earliest stages of transition through nonlinear difference interactions between growing instabilities. The generation of difference modes, and their influence on the subsequent stages of transition is sensitive to the relative amplitudes and frequencies of the interacting instabilities. Phase (frequency) modulations are produced by nonlinear interactions involving low-frequency difference modes and respond to changes in the intensity of the difference mode. The occurrence of randomizing events is related to the frequency modulation level of transition and can be accelerated by increasing the intensity of the difference mode. A threshold frequency modulation index of $\Delta f/f_d \cong 0.9$ exists in the present experiments for the onset of randomization and final breakdown to turbulence.

1. INTRODUCTION

A common feature of many types of transition appears to be the generation of low-frequency modes through difference interactions. Saric and Reynolds [1] and Kachanov, et al [2] have studied this phenomena in boundary layer transition. Miksad, et al [3,4] used complex demodulation and bispectral analysis techniques to demonstrate that in wake transition these low-frequency waves interact with higher-frequency waves to produce spectral energy redistribution via the generation of amplitude and phase modulations. These results were in basic agreement with the fairly general theoretical models developed by Bakai [5], and Kim, et al [6]. Low-frequency difference modes also play a role in the randomization stage of transition where final passage to turbulence is sometimes accompanied by the production of interruptions and spike-like events in the instantaneous velocity field. In natural transition, as shown in Figure 1, these events occur at random low-frequency intervals. In controlled multiple instability experiments, these events occur at intervals directly related to the difference in frequency between the instabilities which interact to generate difference modes. The intensity of these events, and the downstream location where they first occur were found to be sensitive to the rms intensity of the difference mode and hence to the initial conditions responsible for difference mode production. These events are also related to the modulation characteristics of the nonlinear stages of transition.

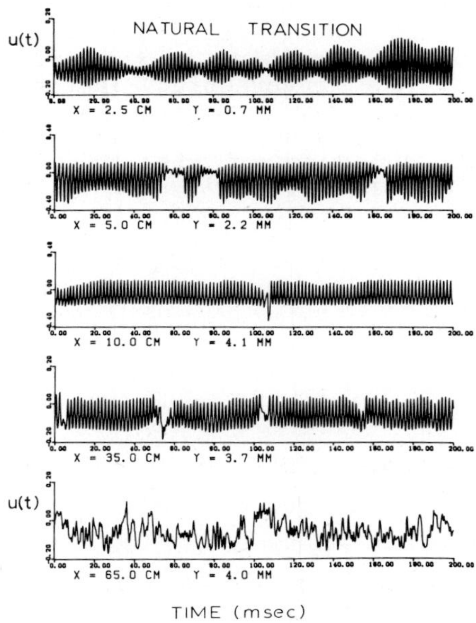

Fig. 1. Time traces of the streamwise velocity field in natural transition.

In the experiments reported here, we investigated the role that low-frequency difference modes play in the various stages of transition to turbulence. The objectives of these experiments were three-fold. The first was to determine the sensitivity of difference mode production to initial conditions (i.e,, to the initial amplitudes and frequencies of the interacting instability modes). The second was to determine how variations in difference mode intensity and frequency influence the modulation and spectral broadening characteristics of transition. The third was to determine how variations in modulations produced by difference mode interactions influence the generation of randomizing events in the final states of transition.

2. EXPERIMENTAL CONDITIONS

The experiments were conducted in the low-turbulence subsonic wind tunnel described in Miksad, et al [3] at a free stream speed of 8 ms, and a background turbulence level, u_{rms}/u_o of less than 0.04%. Measurements were made in the wake (Fig 2) behind a thin, polished flat plate with an elliptically contoured leading edge, and a tapered trailing edge. The transition Reynolds number of the wake, based on Θ_o the initial momentum thickness of the wake and u_o the free-stream velocity was equal to 400.

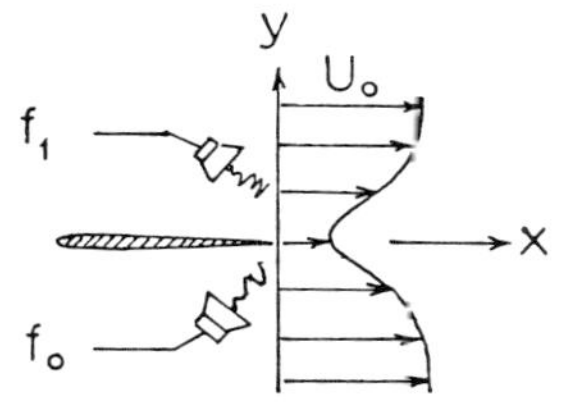

Fig. 2. Schematic of experiment. Initial momentum thickness Θ_o = 0.08 cm.

In the controlled excitation experiments reported here, two small-amplitude instability frequencies of the initial wake were acoustically excited by two independent oscillators. The lower frequency instability mode f_o was excited at f_o = 579Hz, the most unstable mode of the wake as predicted by linear instability theory. The higher frequency mode was set at f_1 = 631Hz. With f_1-f_o fixed at 52Hz, the amplitudes of the f_1 and f_o modes at x/Θ_o = 6.25 were then varied. 52Hz corresponds approximately to the characteristic band-width of interacting instability modes which produce difference modes in natural transition determined by Jones [7] from biocoherency measurements. x/Θ_o = 6.25 roughly corresponds to the onset of interactions between the growing instability modes. For notational convenience, the relative amplitudes of the f_o and f_1 streamwise velocity fluctuations at $x/\Theta_o = 6.25$ will be denoted by A_1/A_o. The effect of varying $(f_1-f_o)/f_o$, while maintaining A_1/A_o fixed was also investigated. All measurements reported here were made at the cross-stream location of maximum u_{rms}. Mean and fluctuating velocity measurements were made with a DISA 55M hot-wire system, digitized on a Le-Croy Camac system. Each record included 2048 12-bit samples, taken at a sampling rate of 5 kHz. Data analysis took place on The University of Texas Cyber 170/750 computer system.

3. RESULTS AND DISCUSSION

3.1. Difference Mode Generation. Figure 3 shows the growth and development of the rms amplitudes of the two instability modes u_o and u_1, and the difference mode u_d for the case $(f_1-f_o)/f_o = .089$, and $A_1/A_o = 0.110$. Difference mode fluctuations are clearly generated in the very earliest stages of transition. The u_d growth rate is much greater than that predicted by linear instability theory. The growth of the difference mode on this stage is due to a transfer of energy arising from the interaction of instability modes. The spectral transfer of energy due to nonlinear interactions between the growing instability modes can be described in terms of the bispectrum. Kim, et al [6] showed that if the growth of a wave at frequency f_d is due to the interaction of waves at f_o and f_1 where $f_1-f_o = f_d$, then the simplest model which relates the interaction to the power spectrum of the fluctuations is $\partial P(f_d)/\partial x = T(f_1,f_o)$ where $T(f_1,f_o)$ is the bispectral energy transfer function. It depends on the coupling coefficient $V(f_1,f_o)$ and the bispectrum $B(f_1,f_o)$ such that $T(f_1,f_o) = [V(f_1,f_o)B(f_1,f_o)+c.c.]$. Downstream variations in bispectral energy transfer to the difference mode are presented in Figure 4.

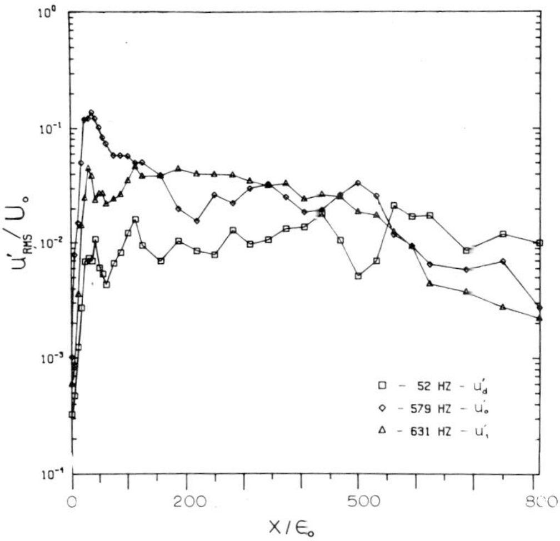

Fig. 3. Downstream growth of the $u(f_o)$ and $u(f_1)$ instability modes and the $u(f_d)$ difference mode.

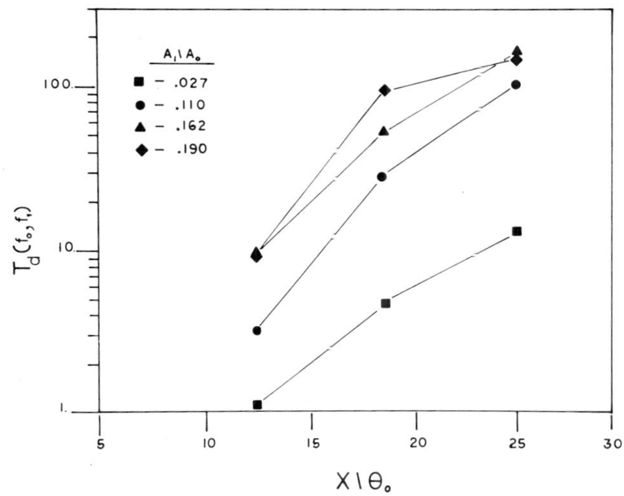

Fig. 4. Downstream development of the bispectral energy transfer to the difference mode.

In general, the bispectral energy transfer to the difference mode increases with downstream distance. In addition, it is evident from Figure 4 that the level of bispectral energy transfer to the difference mode depends on the initial relative amplitudes A_1/A_o of the interacting instability modes. As shown in Figure 5, the intensity of the difference mode at $x/\Theta_o = 25$, is accordingly, also sensitive to initial

A_1/A_o conditions. $x/\theta_o = 25$ roughly corresponds to the end of the instability stage of transition where fluctuation energy is extracted from the mean flow (see the Reynolds stress production curve in Fig 6). During the subsequent nonlinear stage of transition, some of the fluctuation energy is fed back to the mean flow but most of it is redistributed throughout the fluctuation spectrum by nonlinear interactions and wave modulations. Both mechanisms, as we will see, are sensitive to the energy level of the participating low-frequency difference mode. The amplitude of the difference mode at $x \cong 25\theta_o$ can be viewed as an initial condition for the nonlinear spectral broadening stage of transition.

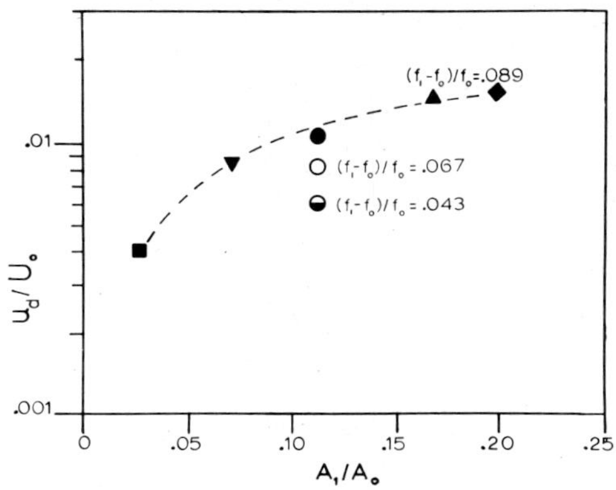

Fig. 5. Magnitude of the difference mode at $x/\theta_o = 25$.

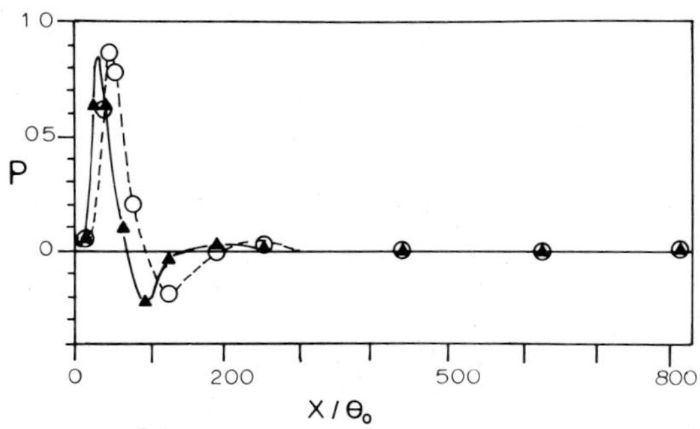

Fig. 6. Reynolds stress production. $P = \int \overline{uv}\, dU/dy$
▲ controlled transition; o natural transition

3.2. Spectral Broadening State. The role of the difference mode in the spectral broadening stage of transition is well established [3,4]. Basically, the difference mode participates in nonlinear three-wave interactions with the finite amplitude instability modes and their harmonics to generate a complex harmonic sideband structure of strongly phase and amplitude modulated waves. Spectral energy redistribution due to cross interactions between the sidebands of the modulated waves then fill in the spectral valleys between harmonic bands. Two relevant features of transition in this stage are a progressive shift of energy to very low-frequency modes and an increase in the level of phase (frequency) modulations. Figure 7 shows a consistent relative phase angle link between the low-frequency difference mode and the low-frequency phase modulations in this stage.

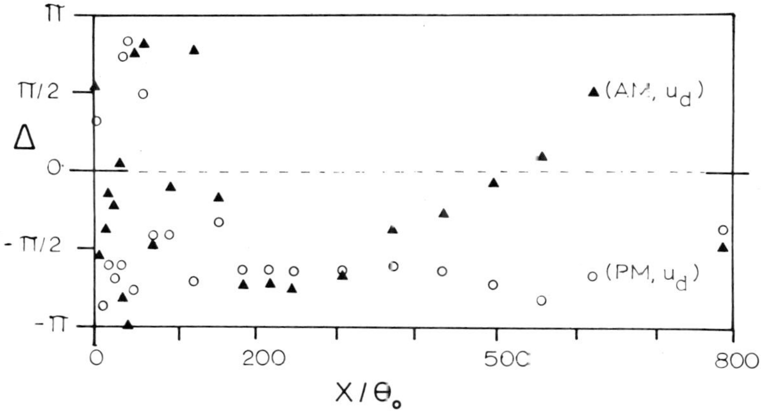

Fig. 7. Relative angle between the difference mode and the amplitude and phase modulations.

Another indication of the influence of difference mode intensity on the modulation induced evolution of transition can be seen in Figure 8 which shows the downstream development of fluctuation velocity power spectra for two representative cases (A_1/A_o = 0.027 and A_1/A_o = 0.162) studied here. Note that for small values of A_1/A_o (and hence less intense difference mode production), the spectral development emphasizes the maintenance of a higher harmonic band structure and breakdown to turbulence, as indicated by the elimination of organized peaks in the power spectral does not occur until x/Θ_o roughly equal to 750. On the other hand, for a larger value of A_1/A_o = 0.162, the higher harmonic bands are quickly smoothed out, low frequencies are quickly filled in, and breakdown to turbulence occurs at roughly $x/\Theta_o = 625$. The results for other values of A_1/A_o follow these trends. Details of the nonlinear interaction wave modulation dynamics of spectral broadening can be found in Jones [7].

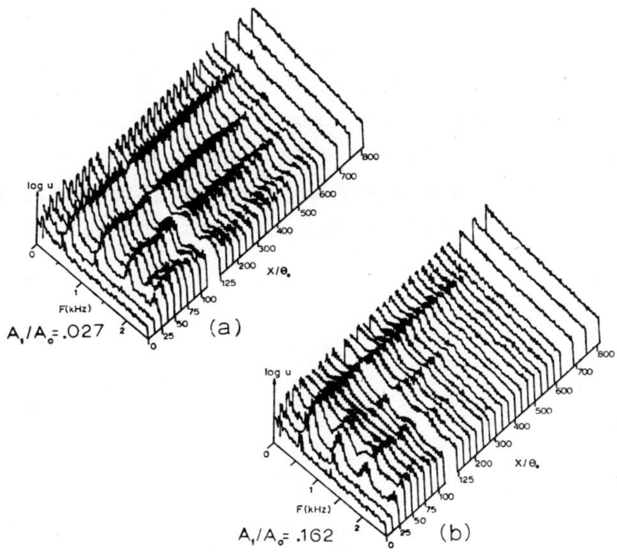

Fig. 8. Downstream evolution of streamwise velocity power spectra $(f_1-f_o)/f_o = 0.089$.

3.3. Randomization Stage. As transition progresses and randomization takes place, the phase modulation level increases rapidly as the time traces of the velocity field start to exhibit intermittent spike-like "events."* These events occur at irregular low-frequency intervals related to the modulation (and hence difference mode) frequency. The events, when they do occur, seem to appear at a point in the difference mode cycle corresponding to a minimum in difference mode velocity. Simultaneous frequency modulation traces exhibit exceptionally large excursions in carrier frequency. The x/Θ_o location where these events first occur,

---

* These "events" appear as irregular distortions of the otherwise fairly regular oscillating time signal. Often the events contain a significant downward spike-like excursion, while in other cases the event is merely an incoherent interruption.

and the intermittency of their occurrence, seem to be related to the level of phase (frequency) modulations and hence the initial conditions A_1/A_o which govern difference mode generation and intensity. The connection between difference mode intensity, phase modulation level, and randomization events can be demonstrated by comparing the downstream onset of intermittent "events" in the streamwise velocity time traces, with changes in the phase modulation index $\beta = \Delta f/f_d$ for different A_1/A_o conditions. Δf is the modulation induced frequency shift in u(t) and f_d is the modulation frequency. In order to investigate modulation level changes, streamwise velocity fluctuations were complex demodulated. Details concerning complex demodulation can be found in Khadra, et al [8]. Briefly, if a time series u(t) is a modulated signal represented as $u(t) = a(t) \cos [2\pi f_o t + p(t) + \phi]$, complex demodulation at the carrier frequency f_o yields the complex demodulate, $c(t) = [a_o + a_m(t)] \exp [ip_m(t)]$. The time varying part of the amplitude of this function is the varying amplitude modulation of the chosen carrier, $a_m(t)$, and the time phase is the simultaneous phase modulation, $p_m(t)$. The level of amplitude and phase modulation can be determined by computing the rms modulation indices, $\alpha = \langle a_m^2 \rangle^{1/2}/a_o$ and $\beta = \langle p_m \rangle^{1/2}$, where angular brackets denote a time average. The dependence of downstream phase modulation index changes on A_1/A_o and $(f_1-f_o)/f_o$ initial conditions is shown in Figure 9.

Note that we have terminated the β-curves at the approximate x-locations where u(t) time traces display the onset of randomizing "events," (see Fig 10). For a given $(f_1-f_o)/f_o$, the x-termination points and hence randomization, occur at progressively earlier locations as A_1/A_o initial conditions increase. For a given A_1/A_o condition, breakdown and other features of transition are delayed as $(f_1-f_o)/f_o$ decreases and the initial conditions more closely resemble a single narrowband instability mode and its resulting harmonics. The absence of strong modulations in these latter cases seems to be related to the existence of an optimum side band spacing. In all of the cases analyzed where randomization "events" are observed, they begin to occur when β reaches an approximate value of 0.9. For small values of A_1/A_o or $(f_1-f_o)/f_o$, strong randomization events are not observed until far downstream, and they are more irregular and less pronounced than in the cases where intense, identifiable, difference modes and modulations are present in transition.

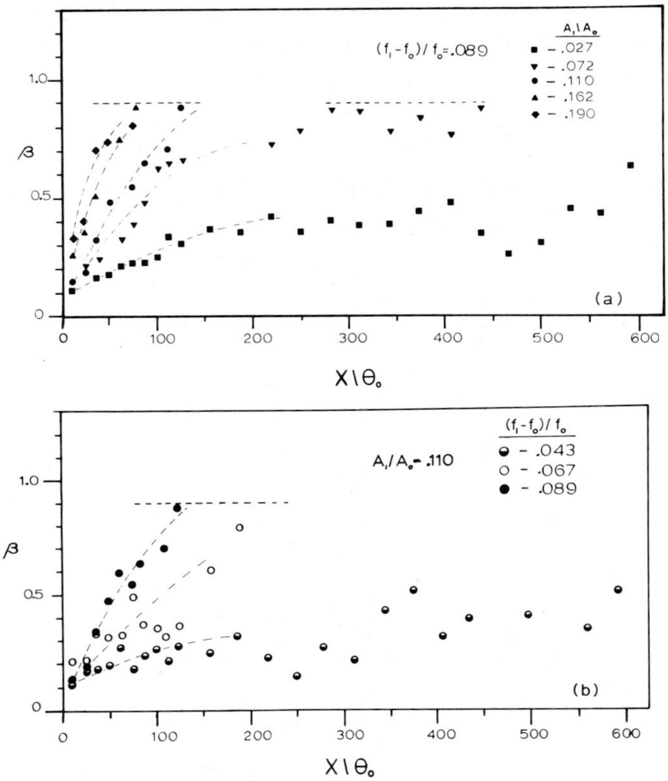

Fig. 9. Downstream variations in phase modulation index prior to the onset of randomizing "events."
(a) $(f_1-f_o)/f_o = 0.089$; A_1/A_o varied
(b) $A_1/A_o = 0.110$, $(f_1-f_o)/f_o$ varied

The physical import of an apparent threshold value for β is not clear. However, if the transition velocity field in this stage can be viewed as the result of an array of vorticies spaced at intervals corresponding to λ_o, the wave-length of the f_o center frequency wave, then the maximum frequency shift can be viewed in terms of periodic changes in the relative positions of adjacent vorticies. As pointed out by Aref [9] and others, the addition of a time dependent mechanism which can cause oscillations in the relative spacing between vorticies will in general produce chaotic dynamics. This suggests that low-frequency fluctuations in the final stages of transition may be acting as an agent for establishing vortex oscillations as a prelude for the introduction of randomness into the flow. If a strong difference mode is generated by interacting instabilities, randomization can be achieved faster (i.e., at an earlier location) and passage to turbulence will be accelerated. If a weak difference mode is generated, then transition must rely on the random energy shifts to low-frequencies to induce time varying vortex oscillations. Whether β = 0.9 is a threshold value for randomization in general is an open question which we are currently exploring.

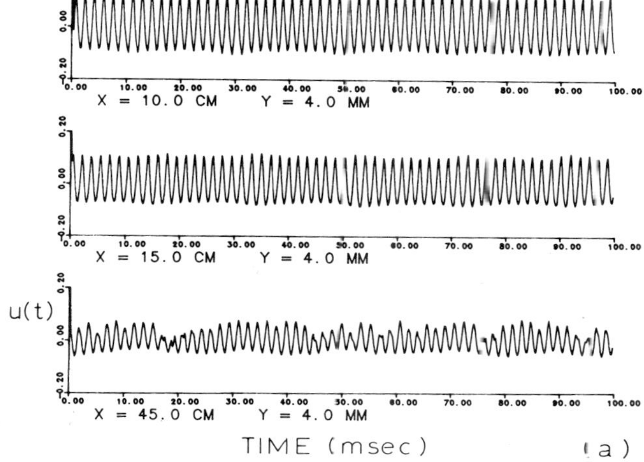

(a)

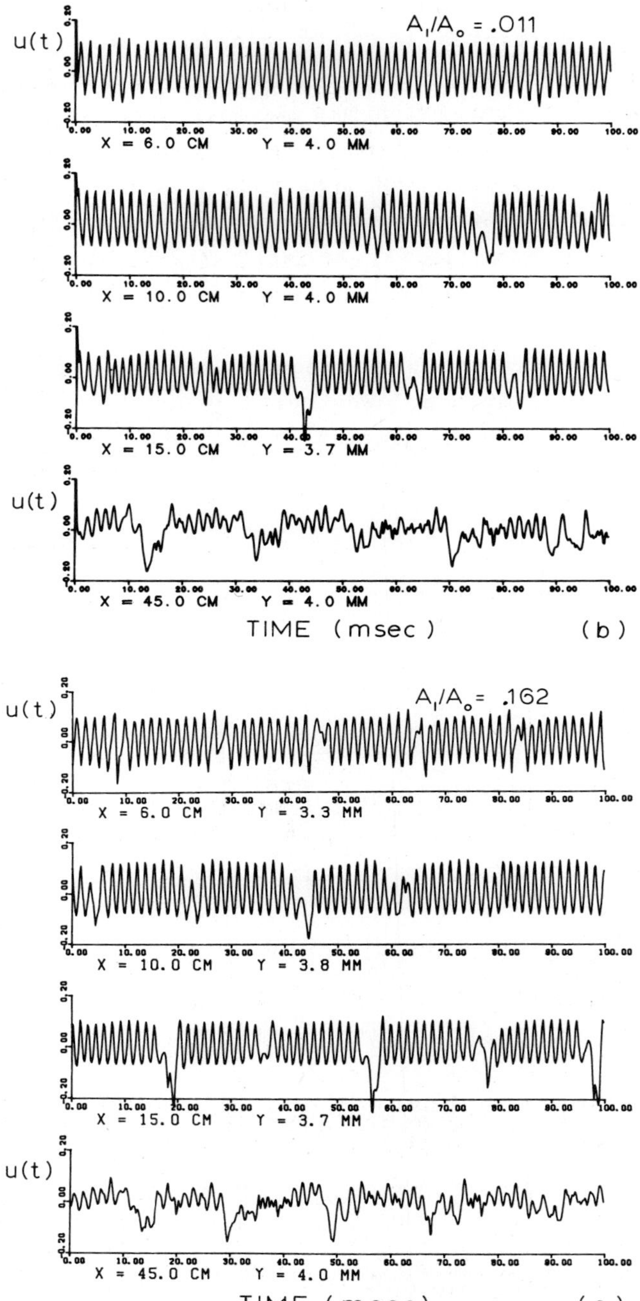

Fig. 10. Time traces of streamwise velocity field. $(f_1 - f_o)/f_o = 0.089$; A_1/A_o varied; $\theta = 0.08$ cm

ACKNOWLEDGEMENTS

This work was supported by the National Science Foundation under Grant No. MEA-8211205. The digital signal-processing techniques used in this study were developed under the auspices of the Department of Defense Joint Services Electronics Program through the Air Force Office of Scientific Research under Contract No. F 49620-82-C-0033.

REFERENCES

1. Saric, W.S.; Reynolds, G.S. Experiments on the Stability of Nonlinear Waves in a Boundary Layer. Laminar-Turbulent Transition, R. Eppler and H. Fasel (eds), Springer-Verlag (1980) 125-134.

2. Kachanov, Yu S.; Kozlov, V. Ya. Experiments on Nonlinear Interactions of Waves in Boundary Layer. Laminar-Turbulent Transition, R. Eppler and H. Fasel (eds), Springer-Verlag (1980) 135-152.

3. Miksad, R.W.; Jones, F.L.; Powers, E.J.; Kim, Y.C.; Khadra, L. Experiments on the Role of Amplitude and Phase Modulations During Transition to Turbulence. Journal Fluid Mechanics 123 (1982) 1-29.

4. Miksad, R.W.; Jones, F.L.; Powers, E.J. Measurements of Nonlinear Interactions During Natural Transition of a Symmetric Wake. Phys. Fluids 26 (1983) 1402-1409.

5. Kim, Y.C.; Beall, J.M.; Powers, E.J.; Miksad, R.W.; Bispectrum and Nonlinear Wave Coupling. Phys. Fluids 23 (1980) 258-263.

6. Bakai, A.S. Interaction of High-Frequency and Low-Frequency Waves in a Plasma. Nuclear Fusion 10 (1970) 53-67.

7. Jones, F.L. An Experimental Study of Nonlinear Wave Interactions and Modulations During Transition of a Symmetric Wake. Ph.D. Dissertation, UT Report No. LTT-83-02 (1982) 219-222.

8. Khadra, L.; Kim, Y.C.; Powers, E.J.; Jones, F.L.; Miksad, R.W. Digital Complex Demodulation of Unsteady Fluid Flow Measurements. Computers in Flow Predictions and Fluid Dynamic Experiments, K.H. Ghia; T.J. Mueller; B.R. Patel (eds), ASME (1981).

9. Aref, H. Integrable, Chaotic, and Turbulent Vortex Motion in Two Dimensional Flows. Annual Rev. of Fluid Mechanics (1983) 345-390.

Experimental Investigation of a Laminar-Turbulent Transition in Free and Semi-Limited Jet Flow

S. I. ISATAEV

Kazakh State University
Alma-Ata, USSR

Summary

The paper presents the results of the experimental investigations of a laminar-turbulent transition not only in plane, roundfree and semi-confined jet flows but also in plane semi-limited jets in co-current flow. It alwo gives the results of visual investigation of large-scale vortex formation structure, their appearance, interaction and destruction with the help of shadowgraph apparatus IAB-451 with simultaneous measurement of the vortex formation frequency. The results of turbulent pulsation velocities measured with the help of hot wire anemometer system designed by our group under the leadership of the author are also given.

On the basis of linear stability theory it was predicted theoretically (Schade & Michalke) [1] and experimentally, see [2,3], that in the initial part of the jets there is a periodic vortex flow. The frequency of vortex formation depends on initial thickness of a boundary layer and on the velocity of jet, the Strouhal number being $S = K\sqrt{Re}$. At the same time many authors from different countries began to study the vortex structure of the flow in a shear of free jet mixing. The author of this paper and his co-workers have studied the laminar-turbulent transition in a free flow and the influence of a periodic acoustic field on this transition. To obtain the jet flow they used the centrifugal ventilator with the regulated rotation velocity with which pumped the air into the receiver. In the receiver on the way of the flow a honeycomb and small cell nets were placed to smooth the velocity field and to diminish the turbulent pulsations. The air flowed out of the receiver through contoured nozzle with either round or slit outlet respectively and formed an axially symmetrical or plane jet. In the case of a plane jet the flow was restricted from two sides by parallel transparent walls. To study a plane **semi-confined jet in co-current flow, the latter was obtained** by the additional ventilator. In all cases the ratio of the inlet seçtion to the outlet section was more than 10,that enabled us to receive 0.2-0.8% turbulence level at the outlet. **Acoustic** influence on the jet flow was made by a loudspeaker fixed on the wall of the receiver on the way to the narrowing nozzle. The mean velocity of the flow was measured by Pitot tube with flattened inlet and external sizes 0.2 x 1.00 mm. Velocity pulsations were measured with thermoanemometer STB designed by the author and co-workers in Kazakh State University. The visual picture of the flow was investigated with the help of the shadowgraph apparatus IAB-451 with the stroboscopic impulse illumination. The initial jet velocity varied from 4mm/s to 50 m/s. The outlet section diameter varied from 10 to 50 mm. From the results of visual investigations and qualitative measurements it appeared that in the boundary layer of the jet at the edge of the nozzle the flow is unstable and vortex rings occur in the round jet and rotating core parallel to the slit edge of in the plane jet. (Fig.1 and 2).

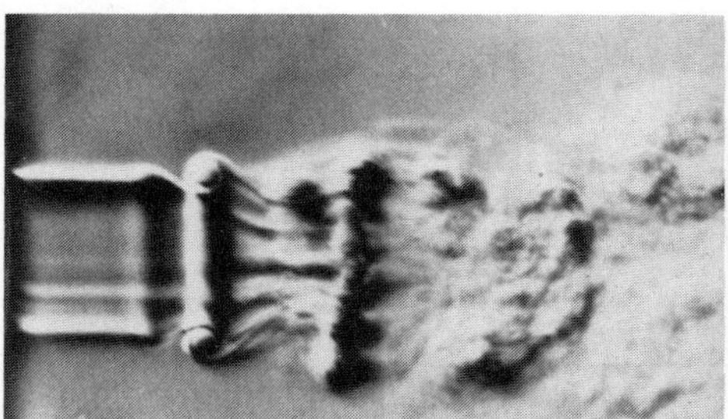

Fig.1. Instantaneous picture of the flow in the initial region of round jet without sound excitation, U_{01}=4.3 m/s, 2a=20 mm

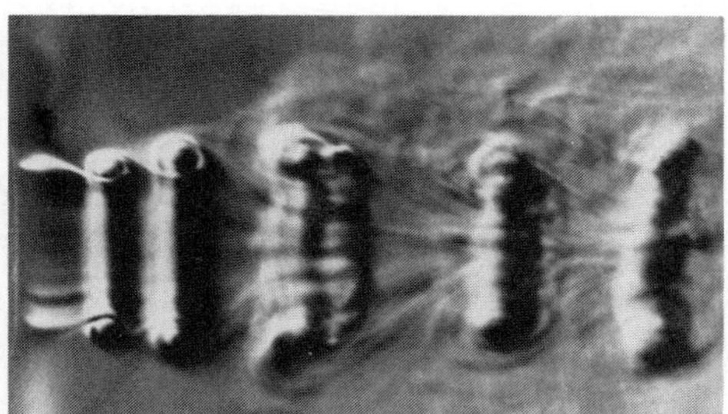

Fig.2. Instantaneous picture of the flow in round jet with the sound excitation of frequency. f=170 Hz.

The vortex formation frequency depends on the initial thickness of the boundary layer, **Strouhal** number, calculated for the given vortex formation frequency, round nozzle radius (or slit half width) and initial jet velocity increases with number

Re according to the relation: $S_1 = K\sqrt{Re}$; where K is coefficient depending on the nozzle design. With a small initial thickness of the boundary layer ($\delta_0 \ll a$), $S_1 > 0.5$. As the distance from the nozzle increases the vortices react with each other and they fuse in pairs till the distance between the neighbour vortices becomes approximately equal to the nozzle diameter (or length of the slit). Such a picture of the flow is seen at the distance of 2-4 calibres from the nozzle. The Strouhal number $S = \frac{f \cdot a}{u_0}$ being 0.20 - 0.30 for the round jet [4], and 0.15-0.20 for the plane jet [5] and becomes independent of the number Re. These typical values of the large-scale vortex frequency can be established either with the help of thermoanemometer at the maximum of the frequency spectrum (Fig.3) or using the measurements of space longitudinal pulsations of velocity at different distances from the nozzle along the edge and along the jet axis (Fig.4).

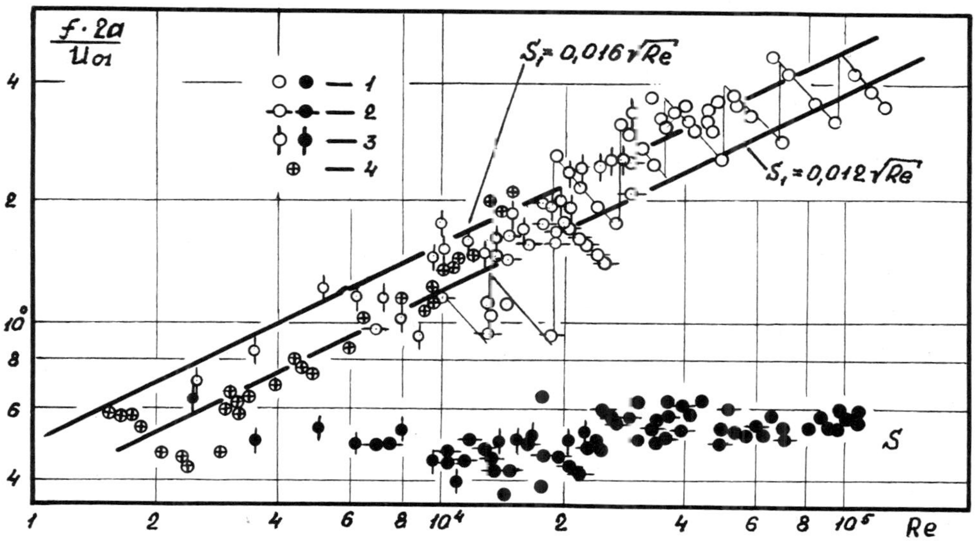

Fig.3. Vortex formation frequency dependence near the outlet section of the nozzle and at the end of the initial region in round jet upon the number Re. 1-a=20mm, 2-a=10mm, 3-a=5.0mm, light points correspond to the outlet section of the nozzle, the bright points - to the axis of the jet at the end of the initial region. 4-the results of the work [3].

As a rule, in an open jet the vortices appear symmetrically on both sides of the jet and move down stream with the same speed.

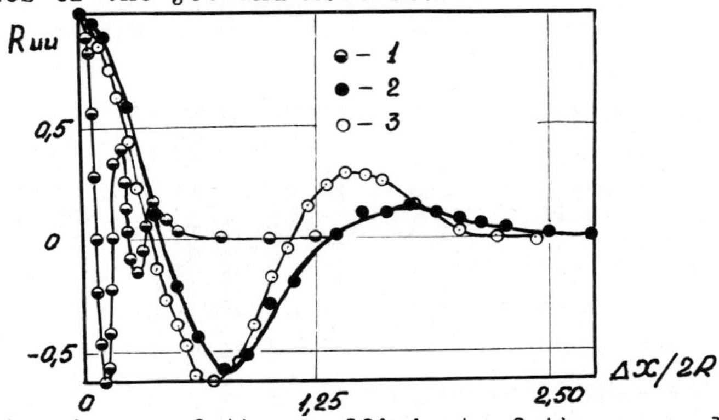

Fig.4. The change of the coefficient of the space longitudinal pulsations of velocity in round jet.
1-a-20 mm, Re=2.4·10^4, y/a=1.0, x/a=1.0; 2-a=20 mm, Re=2,4.10^4, y/a=0, x/a=3.0; 3-a=45 mm, Re= 3.10^5 , y/a=0, x/a=3.0

Longitudinal acoustic excitation with the frequency corresponding to Strouhal number S=0.20 - 0.30 for the round jet and S=0.15-0.20 for the plane jet gives rise to resonance amplification of large-scale vortices formed in the initial part of the jet [4], the velocity pulsation intensity increasing sharply in the initial part. The jet becomes more wide and experiences more intense damping. When the frequency of the excitation becomes S=1.0-1.5 the vortex formation slows down and the jet damping becomes slower. At the distance $x/a \gg 6-8$ from the nozzle the large-scale laminar vortices become unstable and they break down into small-scale chaotic turbulent flows. In the final part of jet transition (range $x/a \gg 30$) discrete energy-containing vibrations in the frequency spectrum of turbulent pulsation are not found. Investigating the open boundary of the semi-confined plane jet we also found out the large-scale vortices which interact the same way they do in an open jet. The frequency of the large-scale vortices at the end of the initial section of the jet has the value of S=0.15-0.20 [5] . The investigation of laminar-turbulent transition in an annular semi-confined jet spreading down the cylinder showed that at $\frac{a}{R} \to \infty$ the picture of the flow is similar to the transition in an open round jet. At $\frac{a}{R} \to 0$ the motion in annular jet coincides with the motion in a plane semi-confined jet. The visual picture of the motion in an annular jet is given in Fig. 5

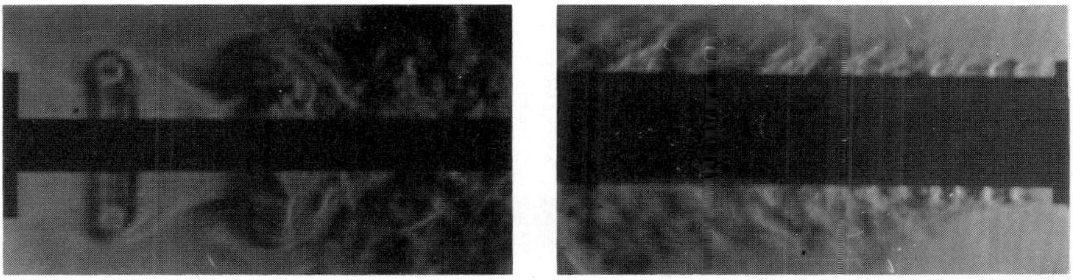

Fig.5. Instantaneous picture of the flow in an annular jet spreading down the cylinder. 1-a=14 mm,R=8.0mm, U_{01}=4.3 m/s; 2-a=3.2 mm, R=17.8 mm, U_{01}=4.3 m/s

The large-scale vortex frequency dependence on parameters S and Re at the end of the initial range is given in Fig.6.

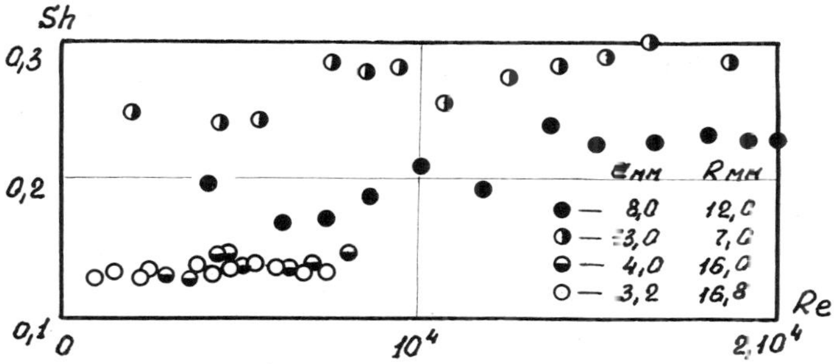

Fig.6. The vortex formation frequency dependence on the curvature parameter and number Re in the annular jet.

The laminar-turbulent transition in a plane semi-confined jet in co-current flow was investigated as well. It was found that in the presence of the co-current flow behind the edge of the nozzle there appears a double-row of vortices due to the boundary layers on the both sides of the surface separating the flows. One row of vortices is on the jet side and the other on the side of the co-current flow. Their formation frequency is corellated. As the distance from the nozzle increases they interact and at the distance x/a=1÷2 there may be found only the vortices of one rotation type. The direction of rotation depends on the co-current parameter. At m < 1 and at m > 1 they have oppositely directed rotations depending on the total jet velocity gradient. In contrast to the open jets at m>0 the vortices of one direction do not fuse but move up to the end of the initial section. Their motion frequency depends both on the jet velocity and on the co-current flow velocity. Shadow photos of a plane semi-confined jet instantaneous picture are presented in Fig.7.

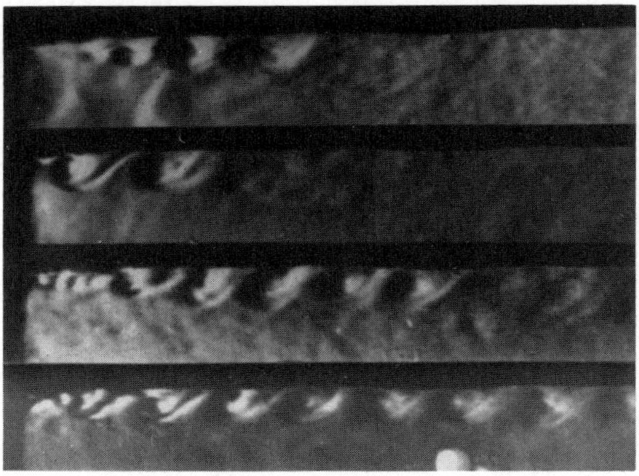

I

2

3

4

Fig.7. Instantaneous picture of the stream in plane jet with co-current flow, a=10 mm.,1-the natural stream, U_{01}=4.3 m/s, m=0; 2-are also with acoustic excitation f=120 Hz,3 - The natural stream, U_{01}=9.5 m/s, m=0.78; 4 - are also with acoustic influence, f=525 Hz.

As it is seen in the absence of sound excitation at m=0 the vortices increase in size with the distance from the nozzle, then they fuse once or twice and at the distance of 3-4 calibres they break down into chaotic turbulent flow.The sound excitation with resonance frequency gives rise to a sharp increase of vortex intensity in the neighbourhood of the nozzle. But at m=0 these vortices break down as well into chaotic turbulent flow at the distances x/2a=3-4.

The stability of vortex flow increases with the increase in the co-current flow velocity. The range length of the discrete vortex flow also increases from 3-4 calibres at m=0 to 20 calibres at m=0.8, the laminar-turbulent range length increasing from 17 calibres at m=0 to 50 calibres at m=0.8.It can be seen from the turbulent level,measured results along the velocity maximum line.

(Fig.8) and local heat exchange coefficient of plate. The results of the plate local heat exchange coefficient for different co-current jet velocities taken for constant value of initial jet velocity are given in Fig.9.

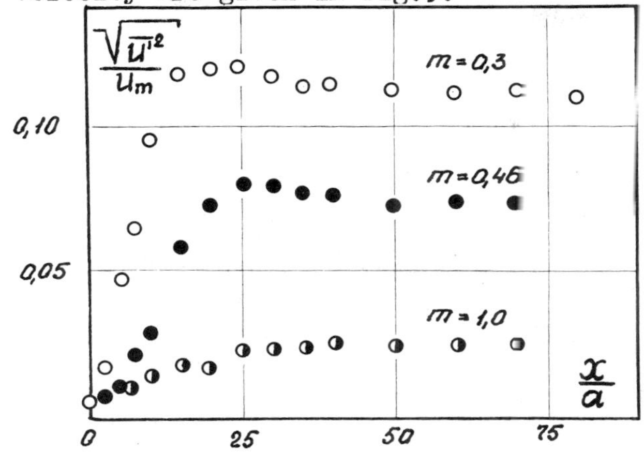

Fig.8. The distribution of turbulent level along the line of maximum velocity in a plane semi-confined jet with the co-current flow.

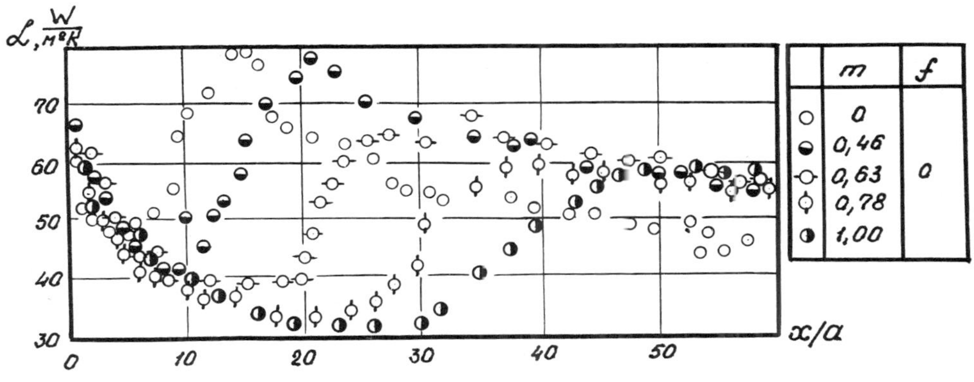

Fig.9. The distribution of the local heat exchange coefficient in a plane semi-confined jet with the co-current flow a=10.0mm

It is clear from the picture that the laminar flow takes place from the outlet of the nozzle to the place where the heat exchange coefficient is at its minimum. The region of the flow from the point of minimum heat exchange to the point of its maximum corresponds to the transition range and further - to the turbulent range of the flow.

Conclusions

1. At the initial level of the flow the turbulence is 0.02 in the initial jet region with the laminar-turbulent transition. It is accompanied by the regular discrete vortex formation, their interaction and breakdown.
2. At 10^4 the length of laminar-turbulent transition region increases considerably with the increase of co-current flow velocity. It must be taken into account in investigation of jet flows.

Notes:

x and y - longitudinal and transversal coordinates,
x - is measured from an outlet section of a nozzle
a - the radius of an outlet section of a round nozzle, width of a slit in a semi-confined jet, R-cylindric surface radius in an annular jet, U_{01} is jet velocity at the outlet section of a nozzle, U_{02} is co-current velocity, $m=U_{02}/U_{01}$ parameter of co-currency, specific natural vortex formation frequency at the edge of the nozzle, f - the same at the end of the initial region of a jet, the frequency of sound excitation, $S_1=\frac{f_1 \cdot a}{U_{01}}$ and $S=\frac{f \cdot a}{U_{01}}$ - Strouhal numbers, Re$=\frac{U_0 \cdot a}{\nu}$ - Reynolds number, $S_R=a/R$ - the parameter of transversal curve of the flow past surface, $\varepsilon=\sqrt{\overline{u'^2}}/U$ - flow turbulence level, $R_{uu}=\sqrt{\overline{u'_1 u'_2}}/\sqrt{\overline{u'^2_1}}\cdot\sqrt{\overline{u'^2_2}}$ - space correlation turbulent pulsation coefficient.

References

1. Michalke, A.; Schade, H.: Zur Stabilität von freien Grenzschichten. - Ing. Archiv, 1963, 33, N 1, s.1-23.

2. Schade, H.; Michalke, A.: Zur Entstehung von Wirbeln in einer freien Grenzschicht. - Z. Flugwiss., 1962, 10, N 4/5, S.147-154.

3. Becker, H.A.; Massaro, T.A.: Vortex evolution in a round jet. - J. Fluid Mech., 1968, v. 31, N 3, pp. 435-448.

4. Isataev, S.I.; Tarasov, S.B.: O kharakternikh chastotakh v spektrakh pulsatsiy skorosti nachalnogo uchastka osesimmetrichnikh struy. Sb. "Prikladnaya i teoreticheskaya phisika", vip.4, KazGU, Alma-Ata, 1973, str.247-252.

5. Voloshin, Yu.E.; Isataev, S.I.; Polzik, W.W.; Tarasov, S.B.: Issledovaniye vikhrevoi strukturi v nachalnom uchastke ploskoy strui pri nalozhenii akusticheskogo vozdeistviya. Sb. "Prikladnaya i teoreticheskaya phisika", vip.8, KazGU, Alma-Ata, 1976, str.101-106.

Stochastic Behaviour of Velocity Profiles Leading to Shear Flow Turbulence

ST. N. SAVULESCU

Increst, Bucuresti, Romania

Summary

Authors' integro-differential formulation (IDF) [1] is qualitatively applied to the non-dimensional kinetic energy distributions in transitional incompressible shear flows. A doubling period process [2] is identified leading to an universal equilibrium at small scales. The paper emphasizes the mathematical features of IDF leading to an intrinsic stochasticity as the Reynolds number is increased.

1. The non-linear integral operator T(f) and its fix points

Consider the operator

$$T(f) = \frac{\int_0^{\eta}\left[\int_1^{\eta'} f(\eta'')d\eta''\right]d\eta'}{\int_0^{1}\left[\int_1^{\eta'} f(\eta'')d\eta''\right]d\eta'}\,, \qquad 0 \leq \eta \leq 1$$

where $f(\eta) \in C^1$, $T_1 = \int_0^1 \eta f(\eta)\, d\eta \neq 0$. Its fix points are

$$f_k^*(\eta) = (-1)^k \sin(\pi/2 + k\pi)\eta\,, \qquad k = 0;\ 1;\ 2.$$

Let's notice that a succesive application of T for $f(\eta) \neq f_k^*$ gives as limit

$$\lim_{r\to\infty} T^{(r)} f = \lim_{r\to\infty} \eta\,(1+ r^{-1}\, g\,(1-\eta)) \longrightarrow \eta \in (0,1).$$

For instance

$$T(f_o) = \eta\,(1+(1-\eta)), \qquad f_o=\mathrm{const}\,,$$

$$T^{(2)}(f_o) = \eta\,(1+(1-\eta)\,\frac{3+3\eta-\eta^2}{5})\,,$$

$$T^{(3)}(f_\circ) = \eta\ (1+(1-\eta)\ \frac{35+35\eta-5\eta^2-5\eta^3+\eta^4}{61}).$$

2. The kinetic energy distributions in wall flows

As shown in [1] for the prototype flat plate boundary layer eqs., the application of IDF leads to

$$u_a^2 = T(\eta u_{a\eta}) + \sum_{i=1}^{N} T_{i1}(T(f_i) - T(\eta u_{a\eta})) \qquad (1)$$

where $\eta = \int_0^y u(x,y,z,t)dy / \int_0^{\delta(x,z,t)} u(x,y,z,t)dy$, $u_a = u/U_e$

$U_e = u(x,\delta,z,t)$, f_i being perturbative functions. Now we try to organize the set of possible u_a^2 distributions by applying the fix point features of T(f). Let's consider only the qualitative picture of the method, the detailed theory being beyond the aim of the present paper. We suppose that $u_a(\eta)$ has the inverse $\eta = \eta(u_a)$ and simplify drastically the eq. (1) in order to get

$$u_a^2 = T(f_b) + \lambda\ (Re)\ (T(f) - T(f_b)) \qquad (2)$$

where f_b stands for the Blasius solution and λ is a perturbative constant. Firstly we ask: what kind of functions f_b, f allows λ to increase far from the linear theory. The expression

$$u_a^2 = \eta\ (1+n-n\eta^{\frac{1}{n}}) + \lambda\eta\ (1-(1+n)\eta^{\frac{1}{1+n}} + \eta^{\frac{1}{n}})$$

satisfies this condition, as can be tested numerically, for n=6 and $|\lambda| \leq 25$. This result is in contrast to the usual non-linear dynamical systems behaviour where small changes in λ imply large changes in u_a^2 [2].

However, when $f_b = \eta u_{a\eta}$, the u_a - distribution meeting this requirement is $u_a \sim \eta^N$, with $N < 0$. Physically it means that $\eta \longrightarrow 1$ is a domain where firstly turbulence can sustain large amplitude perturbations. Several workers [3], [4] reported such instantaneous velocity profiles during transition, with $u_a > 1$ for $\eta \sim 1$ (Fig.1).

Secondly, in the space of functions generated by the operator T(f) we consider the sequence of fix points functions $f^*_{2k}(\eta) = \sin(\pi/2+2k\pi)\eta$ crossing the f_b distributions at k points. According to the universal behaviour of non-linear dynamical systems [2] this points are unstable points where

$$\frac{df^{\oplus}}{d\eta} : \frac{df_b^{\oplus}}{d\eta} = -1, \qquad f^{\oplus} = T^{(r)}f \tag{3}$$

thus developing new fix points. Physically it means that the local scale of motion diminishes. As k increases the unstable crossing points are distributed across the whole shear layer leading to turbulence. Rather than an usual time evolution process we must consider here a space evolution process. The new smaller and smaller scales are created by crossing of fix points distribution as λ increases. It must be observed that λ strongly depends on the local Reynolds number of the flow ($Re=U_e\delta/\nu$ where δ is the boundary layer thickness).

3. Conclusions

A qualitative picture of the transition in the wall flow was proposed along the following lines:

1. The application of the IDF to the kinetic energy equation pointing out the operator T(f).

2. The application of the universal theory of non-linear dynamical systems to the space of functions which represent the sequence of velocity profiles in transition. The modified version appearing in this case refers to an inverse space scale (δ/k) doubling leading to a fragmented unstable motion.

3. Instead to repeat indefinitely the application of the operator, IDF shows a lot of differences $T(f_i) - T(f_b)$ which are playing an **increasing** role as the space scale diminishes.

4. The stochastic behaviour of the high non-linear stages of transitional flow appear as an intrinsic property of the N-S equations under the integro-differential formulation.

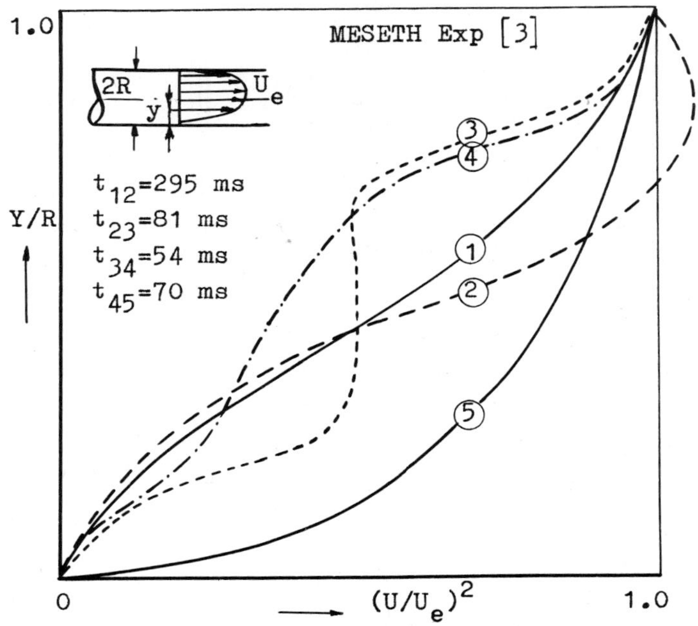

Fig.1

References

1. Savulescu, St.N.: Large scale periodic solutions of the unsteady boundary layer equations. Acta Mechanica 37, 153-158, 1980.

2. Feigenbaum, M.J.: Universal behaviour in non linear systems. In Nonlinear Dynamics and Turbulence, edit. Barenblatt, Jons, Joseph, Pitman 1983.

3. Meseth, J.: Experimentelle Untersuchung der Ubergangszonen Zwischen Laminaren und Turbulenten Strömungsgebieten in intermittente Rohrströmung, Dissertation, Göttingen 1973.

Supercritical Regimes in Axisymmetric Submerged Jets

M. A. GOLDSHTIK, V. N. SHTERN, E. M. ZHDANOVA

Institute of Thermophysics, Novosibirsk 630090, USSR

Summary

The bifurcation of secondary regimes in jet flows of a viscous incompressible fluid is studied. Initial axisymmetric submerged jets having velocity distributions from Schlichting's self-similar profile to the "top hat" form are considered as parallel. In the latter case the secondary regime was found to include stationary rotation around the jet axis. The rotation is generated in jets with high-gradient profiles and is absent in the case of the self-similar distribution.

Introduction

A number of natural and artificial phenomena are known where rapid rotation is generated or amplified. The well-known example is a vortex near a fluid sink [1]. In some cases it remains to be seen whether the rotation is due to vorticity concentration or originates spontaneously. It will be shown here how self-rotation may appear due to instability.

Axisymmetric submerged jet flows of a viscous incompressible fluid are considered. Initial regimes are supposed to be parallel (e.g., supported by some external forces), i.e. the velocity field in the cylindrical coordinates (r, φ, z), has the form $\vec{U} = (0, 0, U(r))$. The following axial velocity profiles will be analysed:

(A) $$U = 1/(1+r^2)^2$$

(B) $$U = 0.5\ (1+\mathrm{th}((1-r^2)/\theta)), \quad \theta \ll 1$$

(C) $$U = \begin{cases} 1 - r^2, & r \le 1 \\ 0, & r > 1 \end{cases}$$

Profile (A) corresponds to the Schlichting self-similar solution [2]. Profiles (B),(C) were used to approximate the velocity distributions in the near-nozzle region of jets [3,4].

The linear stability of axisymmetric submerged jets was studied by some authors, beginning with Batchelor and Gill [5]. The detailed analysis was carried out by Morris [3]. The main (for the present paper) result is that the most dangerous disturbances are spiral waves with the azimuthal wave number $m=\pm 1$. It should be noted that the linear spectrum is bicritical.

Problem formulation

To study the bifurcation of the secondary regimes generated due to instability, the Lyapunov-Schmidt method will be used [6,7]. The velocity (and so is the pressure) field is sought for in the form $\vec{U} + (1/R)\vec{v}(r,\varphi,z-ct)$. The Reynolds number R is based on the velocity and length scales which are clear from (A) - (C). The self-oscillation phase velocity c has to be found together with a non-trivial solution of the nonlinear eigenvalue problem:

$$L(\vec{v}) \equiv Rc\partial\vec{v}/\partial z - R(\vec{U}\nabla)\vec{v} - R(v\vec{\nabla})\vec{U} - \nabla p + \Delta\vec{v} = (\vec{v}\nabla)\vec{v}$$

$$\operatorname{div}\vec{v} = 0 \qquad (1)$$

The boundary condition is the absence of motion at the infinity. For solution (1) near the linear neutral curve, the small-parameter method is used:

$$\{\vec{v},p\} = \sum_{n=1}^{\infty} \varepsilon^n\{\vec{v}_n,p_n\}; \quad Rc = R_0 \sum_{n=0}^{\infty} \varepsilon^{2k} c_{2k}$$

$$R = R_0 + f\varepsilon^2; \quad f = \pm 1$$

Here R_0, c_0 correspond to some point of the neutral curve. Using this representation and grouping the terms ε^n, n=1,2,3 one obtains from (1):

$$L(\vec{v}_n) = \vec{f}_n; \quad \operatorname{div}\vec{v}_n = 0; \quad \vec{f}_1 = 0; \quad \vec{f}_2 = (\vec{v}_1\nabla)\vec{v}_1 \qquad (2)$$

$$\vec{f}_3 = -R_0 c_2 \partial\vec{v}_1/\partial z + (\vec{v}_1\nabla)\vec{v}_2 + (\vec{v}_2\nabla)\vec{v}_1 + f((\vec{U}\nabla)\vec{v}_1 + (\vec{v}_1\nabla)\vec{U})$$

At $n \triangleq 1$ one has the Orr-Sommerfeld problem for neutral oscillations. The solutions are as follows:

$$\{\vec{v}_1, p\} = \gamma_1 V_{1,1} e_{1,1} + \gamma_2 V_{1,-1} e_{1,-1} + \text{c.c.}$$

Here $V_{k,m} \equiv \{w_{k,m}(r), q_{k,m}(r)\}$; $e_{k,m} = \exp(ik\alpha(z-ct) + im\varphi)$,

c.c. denotes the complex conjugate terms. The axial and radial components of $\vec{w}_{1,1}$ and $\vec{w}_{1,-1}$ are equal, but the azimuthal ones have opposite signs.

Due to $\vec{f}_2$, the second-order solution terms are:

$$\{\vec{v}_2, p_2\} = |\gamma_1|^2 V_{0,0} + |\gamma_2|^2 V_{0,-0} + \gamma_1^2 V_{2,2} e_{2,2} +$$

$$\gamma_2^2 V_{2,-2} e_{2,-2} + \gamma_1\gamma_2 V_{2,0} e_{2,0} + \gamma_1^*\gamma_2 V_{0,2} e_{0,2} + \text{c.c.}$$

The star denotes complex conjugation. The first two terms on the r.h. side represent the stationary axisymmetric component of the secondary motion with the velocity field $\vec{w}_{o,\pm o}=\{0, \pm W_\varphi(r), W_z(r)\}$. Therefore, if $|\gamma_1|=|\gamma_2|$, stationary rotation is absent, but in the other cases it takes place.
In $\vec{f}_3$ there are secular terms connected with the harmonics $e_{1,\pm 1}$. So, the solubility conditions should be satisfied:

$$\int_0^{2\pi/\alpha} dz \int_0^{2\pi} d\varphi \int_0^{\infty} (\vec{f}_3, \vec{\theta}_{1,\pm 1} e^*_{1,\pm 1}) r\,dr = 0 \tag{3}$$

where $\vec{\theta}_{1,\pm 1}$ are the eigenvectors of the linear operator adjoint to L. These two demands of orthogonality yield two complex equations for γ_1, γ_2, c_2:

$$\begin{aligned} \gamma_1(fI_3+|\gamma_1|^2I_2+|\gamma_2|^2I_4-iR_oc_2I_1)=0 \\ \gamma_2(fI_3+|\gamma_1|^2I_4+|\gamma_2|^2I_2-iR_oc_2I_1)=0 \end{aligned} \tag{4}$$

Here $I_1,\ldots,I_4$ are the complex values for the integrals that could be obtained from (3). The solutions of system (4), besides the trivial one are as follows:

$$\gamma_2=0; \quad |\gamma_1|^2=\gamma^2=-f\,\mathrm{Re}(I_3I_1^*)/\mathrm{Re}(I_2I_1^*) \tag{5a}$$

$$\gamma_1=0; \quad |\gamma_2|^2=\gamma^2 \tag{5b}$$

$$|\gamma_1|^2=|\gamma_2|^2=\beta^2\equiv-f\,\mathrm{Re}(I_3I_1^*)/\mathrm{Re}((I_2+I_4)I_1^*) \tag{6}$$

Here Re denotes the real part, f is chosen to make the right hand side positive.

Stability analysis

To find which of these solutions are stable, the evolution equations are necessary. They can also be obtained from the solubility conditions by adding the derivatives of the amplitudes γ_1 with respect to the "slow" time $\tau=\varepsilon^2t$. Then, instead of (4), one gets:

$$\begin{aligned} d|\gamma_1|^2/d\tau=2\sigma(1-|\gamma_1|^2/\gamma^2+|\gamma_2|^2(1/\gamma^2-1/\beta^2)) \\ d|\gamma_2|^2/d\tau=2\sigma(1+|\gamma_1|^2(1/\gamma^2-1/\beta^2-|\gamma_2|^2/\gamma^2) \end{aligned} \tag{7}$$

This system is a generalization of the Landau amplitude equation for the bicritical case. The same system has been derived for the case of natural convection [8] and the Couette-Taylor flow [9]. The spectra of the linearised system (7) for solutions (5) are as follows: $\{-2\sigma, 2\sigma(2\beta^2-\gamma^2)/\beta^2\}$ and for (6): $\{-2\sigma, 2\sigma(\gamma^2-2\beta^2)/\beta^2\}$. The value σ/f is a derivative of the increment with respect to the Reynolds number on the neutral curve, and $\sigma>0$ in the supercritical case. Therefore, the stationary solutions stability depends on $2\beta^2-\gamma^2$. The values γ^2, $2\beta^2$ are proportional to the pulsation energy of regimes (5), (6) with the same factor. Thus the solution corresponding to the maximum pulsation energy is stable. The phase flows for both cases are shown in Fig.1.

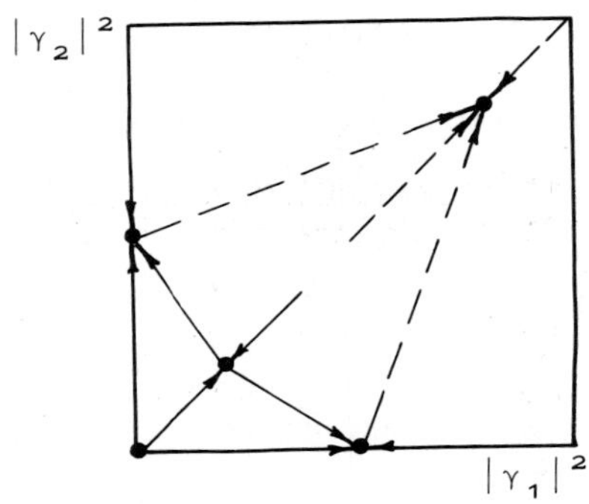

Fig.1. Phase diagrams for the rotation (solutions (5), solid lines) and irrotation (solution (6), dashed lines) cases.

Numerical results and discussion

Our calculations [10] for profiles (A), (B), (C) have given the following results:

		R_*	α_*	γ^2	$2\beta^2$	rotation
(A)		38.55	0.440	0.126	0.224	absent
(B)	$\theta=0.08$	22.48	0.745	0.020	0.016	present
(C)		32.85	0.830	0.401	0.366	present

For the second line of the Table, the initial axial velocity profile and the induced azimuthal velocity distribution are shown in Fig. 2 (the rotation may be both clockwise and counter-clockwise).

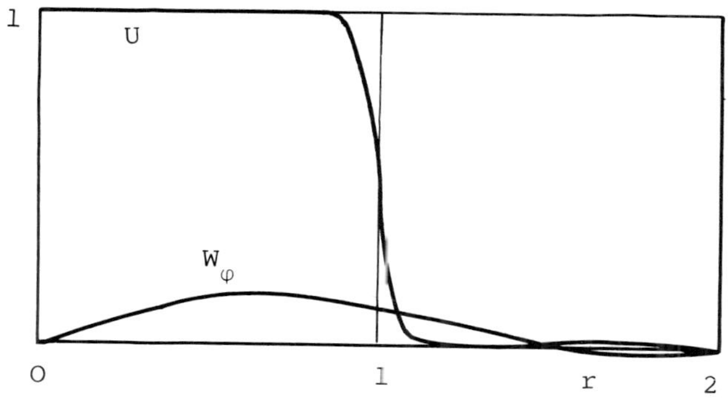

Fig.2. Initial axial (U) and induced azimuthal (W_φ) velocity distributions for the case (B), Θ=0.08.

To characterize the axial velocity distribution, we used a form-parameter F, which is a product of the maximum velocity gradient and the length scale divided by the velocity scale. It has been found that the secondary regime with rotation is stable when $F>F_*=1.6$. For the parabolic profile $F=2>F_*$ and for Schlichting's profile $F=0.52<F_*$.

The non-zero circulation is produced in the transition process when the compensating quantity of vorticity diffuses to the infinity. There is no direct experimental evidence concerning self-induced rotation in jets, though there are some promising observations by Grow and Champagne [11]. It is rather difficult to experimentally fix self-oscillation because jet velocity distributions tend to the self-similar form already at small distances from the nozzle. Nevertheless, perhaps, some natural phenomena, such as tornado, may occur due to instability.

Certainly, it is well-known that instability leads, as a rule, to symmetry breaking. But it seems to be so far unknown that the latter may cause self-rotation.

References

1. Batchelor, G.K.: An Introduction to Fluid Dynamics. Cambridge: The University press 1970.
2. Schlichting, H.: Grenzschicht - Theorie. Karlsruhe: Verlag G.Braun 1964.
3. Morris, P.J.: The spatial viscous instability of axisymmetric jets. J. Fluid Mech. 77 (1976) 511-529.

4. Kamber, T.: The stability of an axisymmetric jet with a parabolic profile. J. Phys. Soc. Japan 26 (1969) 566.

5. Batchelor, G.K.; Gill, A.E.: Analysis of the stability of axisymmetric jets. J. Fluid Mech. 14 (1962) 529-551.

6. Yudovich, V.I.: On self-oscillation onset in a fluid. Soviet Appl. Mech. and Mech. 35 (1971) 638-655.

7. Goldshtik, M.A.; Shtern, V.N. Hydrodynamic stability and turbulence. Novosibirsk: Nauka 1977 (in Russian).

8. Lyubimov, D.V.; Putin, G.F.: Supercritic motions in a cubic cavity. In.: Hydrodynamics, Perm: University press (1977) 15 - 26.

9. Babenko, K.I.; Affendikov, A.L.; Iuriev, S.P.: On the bifurcation of Couette flow between rotating cylinders in the case of double eigenvalue. Soviet Doklady 266 (1982) 73-78.

10. Goldshtik, M.A.; Zhdanova, E.M.; Shtern, V.N.: Onset of self-rotation in a submerged jet. Soviet Doklady (1984), in the press.

11. Crow, S.C.; Champagne, F.M.: Orderly structure in jet turbulence. J. Fluid Mech. 48 (1971) 547-591.

Generation of Crossflow Vortices in a Three-Dimensional Flat-Plate Flow

W. S. SARIC* and L. G. YEATES

Department of Engineering Science and Mechanics
Virginia Polytechnic Institute and State University
Blacksburg, Virginia 24061, U.S.A.

Summary

Experiments are conducted which demonstrate the feasibility of creating crossflow vortices in a flat-plate flow. A swept leading edge and contoured wind-tunnel walls produce a strong pressure gradient and a typical swept-wing flow on the plate. Detailed three-dimensional measurements, made within the boundary layer using hot-wire anemometry, are supplemented with different flow-visualization techniques. Comparisons are made with theoretical flow-field and stability calculations.

Introduction

It is recognized that one of the most important unanswered questions regarding the stability of boundary layers concerns the three-dimensional (3-D) stability of 3-D flow fields. The flow over a swept wing is the classic example of such a complicated problem. Near the leading edge, the stability of the boundary layer is dominated by the behavior of the so-called crossflow (C-F) vortices. These are streamwise-oriented vortices that are all rotating in the same direction. The description of this problem is given in the classical paper by Gregory, Stuart & Walker [1]. In the mid-chord region, the stability is governed by Tollmien-Schlichting (T-S) waves that are generally two-dimensional (or three-dimensional if the flow is locally supersonic such as on a supercritical wing [2]).

Considerable progress has been made in the calculation of crossflow stability characteristics. The most recent and complete contribution is by Malik & Poll [3] who also review the previous efforts in this area. Experimental contributions are few because of the difficult nature of the problem. However, the recent work of Poll [4] on a swept cylinder is of interest and represents the most detailed work to date. Given these efforts, there are

*Permanent address: Department of Mechanical and Aerospace Engineering, Arizona State University, Tempe, Arizona 85287.

still some unsettled questions regarding the nature of crossflow vortices. The conditions under which the vortices are steady or unsteady are still uncertain [3,4]. Details of the 3-D mean-flow modifications due to the presence of the vortices are undocumented. Moreover, the calculated growth rates have never been experimentally verified.

One of the other unsettled questions regarding flow over a swept wing is the interaction of crossflow vortices and Tollmien-Schlichting waves. It is well known that streamwise vortices in a boundary layer strongly influence the behavior of other disturbances by producing a double-exponential growth of amplitude [5,6,7]. These analyses, however, are specifically limited to Görtler-like vortices [8]. Amplification of the C-F vortices near the leading edge produces a residual spanwise nonuniformity in the mid-chord regions where T-S waves are amplified. Reed [8] analyzes the C-F/T-S interaction by using a parametric resonance model and illustrates the need to experimentally study problems of this sort.

The goal of the present work is to demonstrate the feasibility of creating crossflow vortices in a flow-field that permits detailed boundary-layer measurements in three dimensions. In this way, the basic nature of the crossflow stability can be studied and can supplement the data obtained on actual swept wings and swept cylinders. The results of the last two years of an on-going study of generating crossflow vortices on a flat plate are described below.

Experimental Method

The experiment utilizes the approach of creating a typical swept-wing flow on a flat plate. The leading edge is swept and a swept wall bump is used to create a specified pressure distribution on the plate. The bump is designed using a panel-method code (May & Saric [9]), and the 3-D boundary layers are calculated using the code of Kaups & Cebeci [10]. Floor and ceiling fairings are used to match the inviscid streamlines of the swept wing. This method has advantages over using an actual airfoil. They are:

1. Measurements are made over a very smooth flat surface using an existing computer-controlled precision traverse.
2. The leading edge is smooth and carefully contoured.
3. Crossflow disturbances exist in boundary layers larger (≈5mm) than those on wings, and hence the measurements are easier.
4. Different pressure distributions are examined by using inexpensive wall-bump modifications.

5. Tollmien-Schlichting waves are easily introduced, and the wave interaction problem is modeled.

Although in-plane streamline curvature is present, this experiment cannot account for the stabilizing effects of body curvature [3]. The experiments are conducted in the VPI&SU Stability Wind Tunnel at speeds of 10 m/s and 14 m/s with a freestream turbulence level of 0.02% U_∞. The test section is 183 cm square, and the flat plate is mounted vertically in the tunnel just off the centerline. The leading edge is swept at 25°. The wall bump extends 35 cm from the wall as is shown in Figure 1. This figure gives the floor layout of the tunnel configuration as well as the vertical plane. Note that the traverse is mounted inside the wall bump to minimize blockage effects. The 3-D traverse takes measurements in the swept coordinates in the range of $60\text{ cm} \leq x \leq 180\text{ cm}$, $0 \leq y \leq 10\text{ cm}$, $-15\text{ cm} \leq z \leq +15\text{ cm}$ through a movable slot on the wallbump. Velocity profile data are obtained using straight and slanted hot-wires. All of the measurements, traverse position, and date reduction are controlled in a pre-programmed fashion with a MASSCOMP 560 computer/data-acquisition system. A smoke wire is used to visualize wave patterns in the boundary layer while a naphthalene-trichloroethane spray is used to visualize surface flow patterns.

Results

Figure 2 shows the chordwise pressure distribution measured at two span locations. Along the first 200 cm, where the C_p increases, is the region of amplified C-F vortices while the flat region contains amplified T-S waves. Figure 3 contains the spanwise pressure distributions at four chord locations. The use of the floor and ceiling fairings produces the relatively uniform spanwise distributions.

Figures 4, 5, and 6 show the variation of the U and W components of velocity (parallel and perpindicular to the upstream velocity vector) and the angle between them. The theoretical calculations using [10] are superposed in each case. The differences are exaggerated because of the scale, where in fact, the inviscid flow angle is different by only 1°, which is within the tolerance of the floor and ceiling fairings. The U and W velocities are in the x- and z-directions, respectively (Figure 1), and are measured using a straight hot wire and a slant hot wire. The vector sum of these two velocities defines the edge velocity Ue_t of the tangential velocity component U_t. Thus, Ue_t defines the inviscid streamlines over the plate and θ (Figure 6) is the angle of the streamlines with respect to the x-axis.

The component perpindicular to U_t within the boundary layer, W_t, is the crossflow velocity. By definition, the crossflow velocity is zero in the inviscid stream.

Figure 7 is a plot of U_t and W_t where W_t is normalized to be 1 at its maximum. Figure 8 is a plot of W_t normalized with Ue_t. The crossflow Reynolds number for this case is 180. These results are compared with boundary-layer calculations [10] of flows over swept wings using the experimental pressure distribution. The agreement is quite good and illustrates that the flow is predictable and typical of swept-wing flows. The scatter in these data are due to resolving small differences of large numbers after each was interpolated from a straight-wire profile and a cross-wire profile and then transformed into the U_t and W_t directions. Many data of this type were taken at chord locations from 85 cm to 170 cm at span locations of +10 cm, 0, -10 cm, and at velocities of 10 m/s and 14 m/s. Because the inviscid streamlines have a larger than predicted deflection angle, the crossflow velocity continued to increase faster than the theory. Consequently, the agreement between theory and experiment is not as good at chord locations aft of 100 cm. The crossflow Reynolds numbers approach 250 in the mid-chord region.

Figure 9 is a photograph of wall patterns using naphthalene-trichloroethane spray. The existence of these patterns using this technique indicates that the vortices are steady. Detail spectrum analysis of hot-wire data confirm this conclusion. The spacing of the streaks is approximately 1 cm, and this spacing is essentially independent of upstream velocity (10 or 14 m/s) and the time required to evaporate the naphthalene (5 to 45 min.) Stability calculations made using the code of Reed [8] show the most unstable disturbance to have a wavelength of about 1 cm. This result agrees quite well with streak spacing of Figure 9.

Figure 10 is a typical spanwise measurement of the mean flow within the boundary layer. It shows a definite spanwise periodicity at a wavelength of 0.5 cm. Measurements of this type are done at many different chord and span locations and at different locations within the boundary layer (but not closer than u=0.25). In each case, a dominant 0.5 cm spacing is present which is contrary to the wall-streak spacing (Figure 9). Spectral analysis of the raw data do show a peak at 1 cm, but we are unable at this time to explain the strong 0.5 cm spacing in the hot-wire data.

Conclusions

Detailed measurements and visualization of crossflow vortices are conducted under carefully controlled conditions. The vortices are steady under the observed conditions, and the results are compared with the theory [8, 10]. As transition is approached in the mid-chord region, amplified T-S waves appear and the C-F structure becomes unsteady.

Acknowledgment

The authors would like to thank J. Crouch, S-L Kuo, and G. Santos for their tireless assistance during the execution of this experiment and J. Carter, K. Herbert, and G. Rose for their computer-support efforts. The valuable suggestions and encouragement of W. Pfenninger and D. Maddalon made this experiment possible. This work is supported by NASA-Langley Research Center grant NAG-1-280.

References

1. Gregory, N., Stuart, J. T., & Walker, W. S.: On the stability of three-dimensional boundary layers with applications to the flow due to a rotating disk. Philosophical Transactions of the Royal Society of London, Vol. A248 (1955) 155-199.

2. Mack, L. M.: On the stability of the boundary layer on a transonic swept wing. AIAA Paper No. 79-0264 (1979).

3. Malik, M. R. & Poll, D. I. A.: Effect of curvature on three-dimensional boundary layer stability. AIAA Paper No. 84-1672 (1984).

4. Poll, D. I. A.: Some observations of the transition process on the windward face of a yawed cylinder. College of Aeronautics Report 8407, Cranfield, England (Feb. 1984).

5. Nayfeh, A. H.: Effect of streamwise vortices on Tollmien-Schlichting waves. J. Fluid Mech. 107 (1981) 441-453.

6. Herbert, Th. & Morkovin, M. V.: Dialogue on bridging some gaps in stability and transition research. In: Laminar-Turbulent Transition (eds. R. Eppler & H. Fasel), Springer-Verlag (1980) 37-46.

7. Floryan, J. M. & Saric, W. S.: Wavelength selection and growth of Görtler vortices. AIAA Paper No. 80-1376 (1980).

8. Reed, H. L.: Wave interactions in swept wing flows. AIAA Paper No. 84-1678 (1984).

9. May, C. L. & Saric, W. S.: A panel-method code to design wind-tunnel modifications for pressure gradients on models, Part I. Inviscid Analysis. VPI&SU Report, VPI-E-83.25 (June 1983).

10. Kaups, K. & Cebeci, T.: Compressible laminar boundary layers with suction on swept and tapered wings. J. Aircraft, Vol. 14, No. 7 (1977) 661.

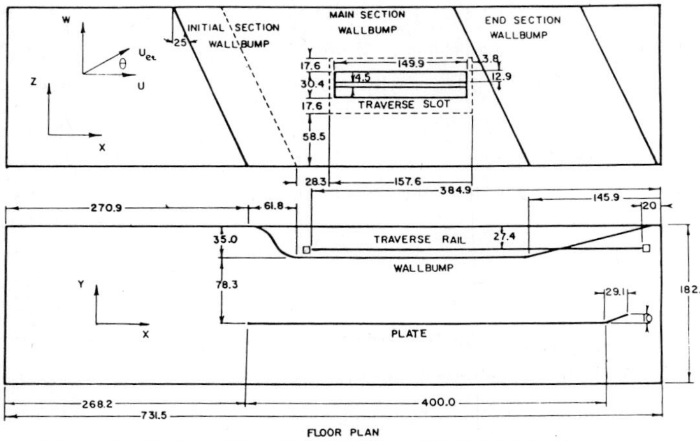

Figure 1. Test section configuration. Dimensions in cm.

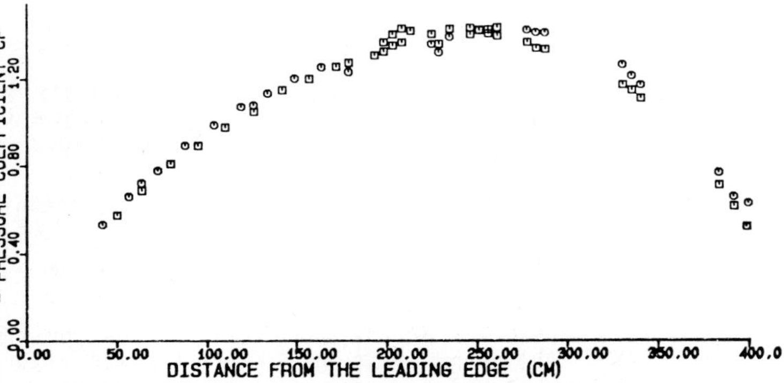

Figure 2. Chordwise pressure distribution at z = ±15 cm.

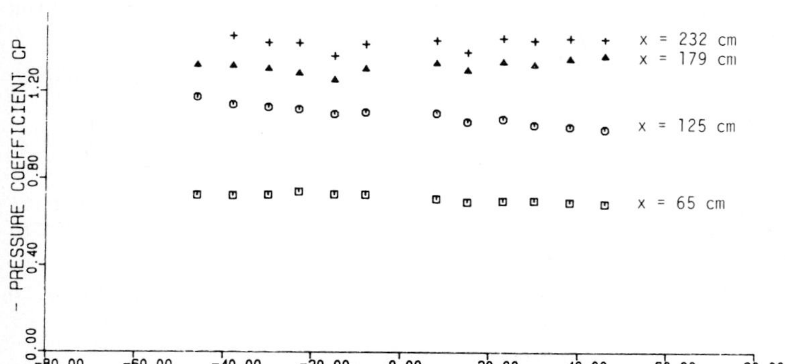

Figure 3. Spanwise pressure distribution at four chord locations.

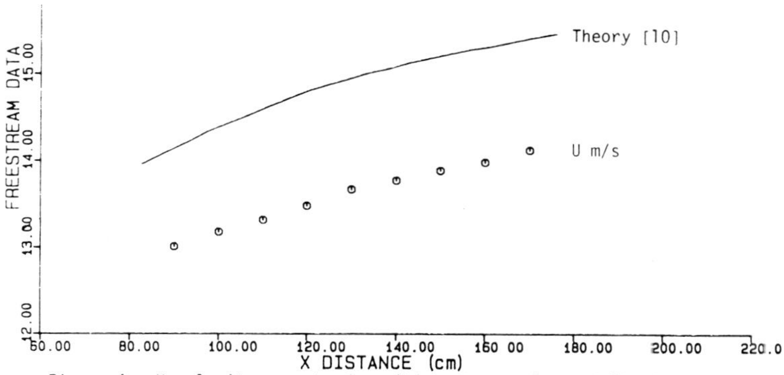

Figure 4. U-velocity vs. chord position at tunnel speed 10 m/s.

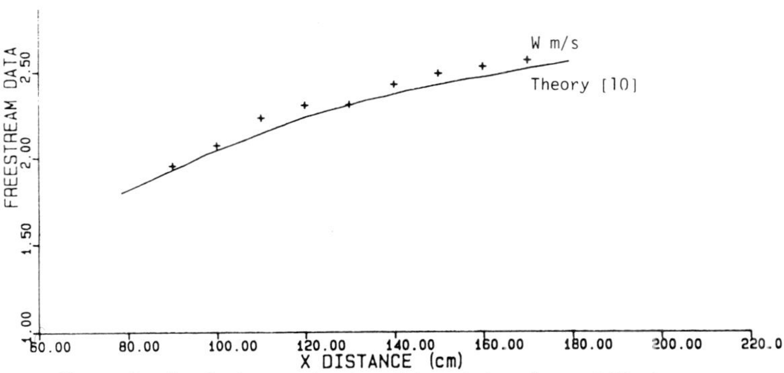

Figure 5. W-velocity vs chord position at tunnel speed 10 m/s.

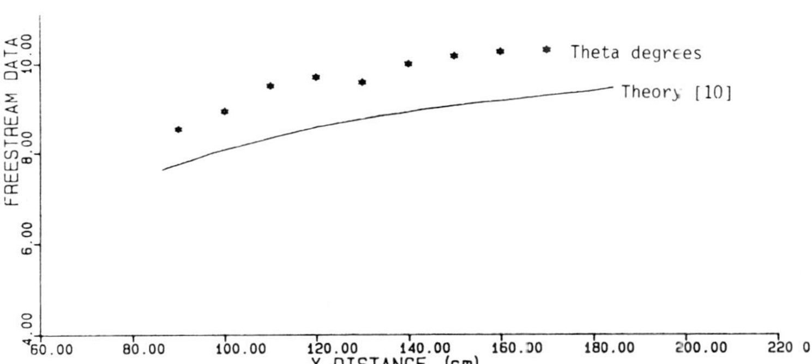

Figure 6. Angle of inviscid-flow streamline.

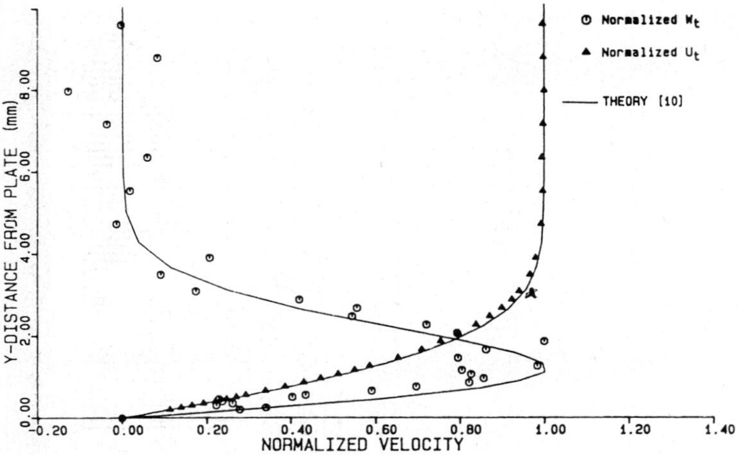

Figure 7. Tangential-flow velocity U_t/Ue_t and crossflow velocity W_t/W_{tmax} at x = 100 cm and tunnel speed 10 m/s.

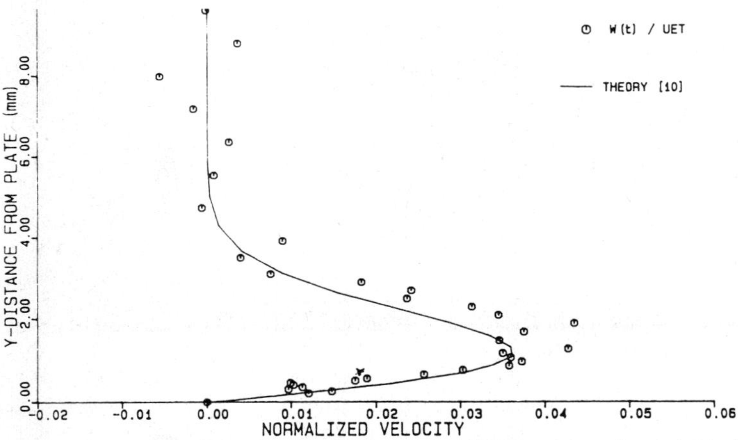

Figure 8. Crossflow velocity, W_t/Ue_t at x = 100 cm and tunnel speed 10 m/s.

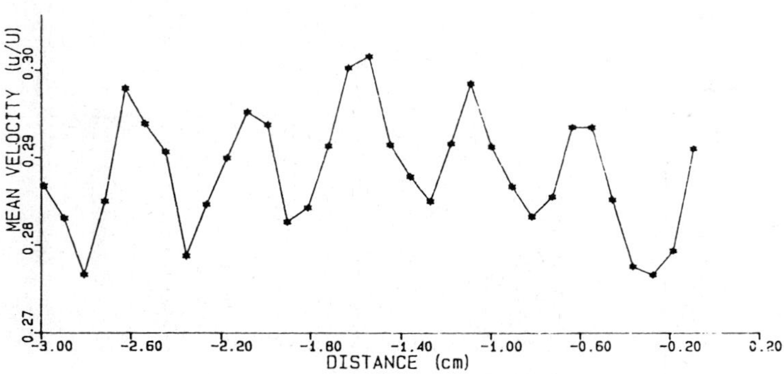

Figure 10. Spanwise variation of mean flow at x = 100 cm and tunnel speed 10 m/s.

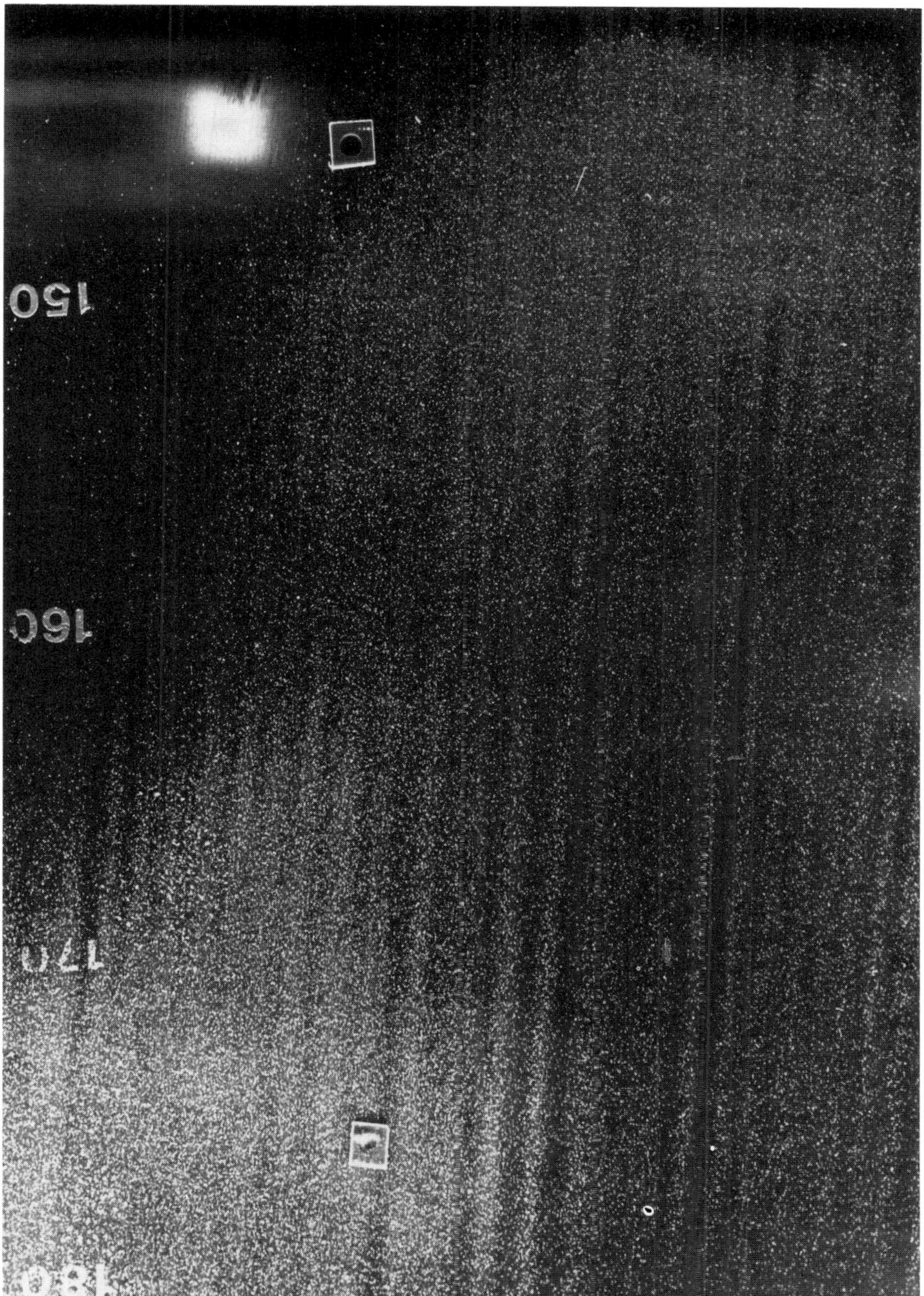

Figure 9. Naphthalene wall-streaks. Flow from top to bottom.

Nonlinear Interaction of Streamwise Vortices in Non-Parallel Boundary Layers on Concave Surfaces

KRISHNA M. SRIVASTAVA
DFVLR, Institute for Theoretical Fluid Mechanics,
Bunsenstrasse 10, D-3400 Göttingen, FRG

Abstract

First-order nonlinear interactions of streamwise vortices with different amplification rates, spanwise wave numbers, and initial amplitudes in boundary layer flows on concave walls are studied by using the method of multiple scales. Numerical results show that the effect of interaction between two vortices in harmonic resonance is strong when the amplitude of the secondary wave (higher harmonic) has reached certain % of the mean flow. In combination resonance between three waves, it is found that the two primary waves with moderate initial amplitudes have a strong influence on their difference harmonic wave. The resonant interaction between waves of higher wave numbers causes strong amplification at earlier stages (at lower values of G) than the resonant interaction between waves of lower wave numbers.

1. Introduction

The rapid development of the centrifugal instability in boundary layer flows over concavely curved walls was first demonstrated by Görtler [1]. A plethora of papers have appeared since then to correct, supplement and extend the analysis of Görtler in order to understand the instability fully well. Reviews of these efforts were given by Herbert [2] and Floryan and Saric [3]. The instability manifests itself at suitable wave number if the Görtler number, G, exceeds some critical value G_c. Hall [0] has shown that the parallel flow theory of the growth of Görtler vortices is questionable. However, assuming weak dependence of the disturbances in downstream direction, the method of multiple scales may be used to reduce the governing partial differential equations to an ordinary system. The nonlinear development of the vortices is not yet very well understood. The first attempt towards a nonlinear analysis of Taylor-Görtler vortices was made by Aihara [4] by using the integral method. The experimental results have shown that the development of T-G vortices was pretty rapid even for a relatively small curvature deforming the basic velocity profile nonlinearly as well as three-dimensionally. Aihara and Koyama [5] have investigated experimentally the process of formation of a three-dimensional coherent structure by utilizing flow visualization and hot wire technique.

It is conjectured here that the coherent interaction of streamwise vortices may initiate transition at earlier stages. Hence, the purpose of the present work is to investigate the nonlinear behaviour of Taylor-Görtler vortices by the method of multiple scales in the general, orthogonal system of coordinates. The possibility of nonlinear resonant interaction between two and three streamwise vortices as a cause leading to transition to turbulence is examined in detail. The two-dimensional basic flow over a concavely curved surface is subjected to the most general form of three-dimensional disturbance and the nonlinear stability is discussed by the perturbation technique. The basic flow on the curved wall is chosen to be an incompressible, constant property flow. The effects of curvature and viscosity are represented by the small parameters $\varepsilon_c = (\delta/R)^{1/2}$ with R radius of curvature of the plate and $\varepsilon_v = Re^{-1/2} = (\frac{\nu^*}{L^* U^*_\infty})^{1/2}$ where L* is the characteristic length for the φ-coordinate. With the known form of the metric coefficient, the non-dimensional forms of the mass and momentum equations written in curvilinear coordinates (φ,Ψ,z) (see fig. 1) for a two-dimensional flow are solved and it is found that the correct zeroth-order flow, i.e. the basic flow f_o for the stability analysis is the Blasius flow. The nonlinear stability of this basic flow over a two-dimensional concavely curved wall with respect to three-dimensional distrubances $\hat{f}$ in the form of alternating streamwise vortices is studied.

Thus the total flow quantities $\bar{f}$ are taken in the form:

$$\bar{f} = f_o(\varphi,\Psi) + \hat{f}\,(\varphi_o,\varphi_1,\varphi_2,\Psi,\Psi_2,z_o,z_1), \qquad (1)$$

where

$\varphi_0 = \varphi$, $\varphi_1 = \varepsilon_v\varphi$, $z_0 = z$, $z_1 = \varepsilon_v z$, $\varphi_2 = \varepsilon_c\varphi$, $\Psi_2 = \varepsilon_c\Psi$. Thus, φ_2 and Ψ_2 contain curvature terms. $\varepsilon_v = Re^{-1/2}$ is the small parameter representing viscous scale mentioned earlier which also accounts for the non-parallelism of the flow. φ and Ψ are the potential and stream functions of the inviscid irrotational basic flow and z is a rectilinear coordinate normal to the plane of the basic flow; $\hat{f}$ = u, v, w and p denote, respectively, the velocity components along φ, ψ, z directions and the static pressure. Substituting the total flow quantities into the Navier-Stokes equations and subtracting the basic flow quantities, we obtain perturbed equations which are rescaled with dimensionless boundary layer thickness $\delta = Re^{-1/2}$, evaluated at $L^* = 1$ as $\bar{\varphi} = \varphi$, $\bar{\Psi} = \Psi/\delta$, $\bar{z} = z/\delta$, $v = Re^{1/2}\hat{v}$, $w = Re^{1/2}\hat{w}$, $p = Re\,\hat{p}$.

The motivation for the scaling of v, w, and p comes from the theoretical calculations of Smith [6] and from the experiments of Bippes [7].

Applying the method of multiple scales, the following uniform expansion for the disturbance quantities is sought:

$$\hat{f} = \sum_{m=1}^{\infty} \varepsilon_v^m f_m(\varphi_0, \varphi_1, \psi_0, z_0, z_1) \tag{2}$$

and the equations of orders ε_v and ε_v^2 are obtained.

To study the nonlinear interaction, the solution is assumed to consist of three wave packets in the form of streamwise standing vortices:

$$[f_1] \equiv [u_1, v_1, w_1, p_1] = \sum_{n=1}^{3} (*) A_n(\varphi_1, z_1)\zeta_{mn}(\varphi_1, \psi)\exp(i\Theta_n) + CC \tag{3}$$

m = 1,2,3,4 for u_1, v_1, w_1, p_1 respectively and (*) = 1 for u_1, v_1, p_1 and (*) = -i for w_1. CC means complex conjugate. Here

$$\Theta_n = -i \int \beta_n(\varphi_1) d\varphi_0 + \alpha_n z_0, \quad n = 1,2,3 \tag{4}$$

$$\frac{\partial \Theta n}{\partial_{\ 0}} = -i\beta_n(\varphi_1), \quad \frac{\partial \Theta n}{\partial z_0} = \alpha_n, \tag{5}$$

α_n and β_n are real. The substitution of Eqs. (3) into the first order problem leads to the eigenvalue problem which is solved numerically together with its adjoint problem. The small but finite amplitudes A_n are determined at the next level of approximation with the help of solvability conditions. Resonant interactions between two and three waves are shown to amplify the amplitudes of Taylor-Görtler vortices. For detailed calculations, the reader is referred to Srivastava [8]. The substitution of Eq. (3) in the second order problem yields an inhomogeneous problem. The solvability conditions of these inhomogeneous equations lead to the nonlinear differential equations for the evolution of the amplitudes

$$a_n = A_n \exp[\ \beta_n(\varphi_1)d\varphi_0] \tag{6}$$

$$\Omega_{n\beta}\frac{\partial a_n}{\partial \varphi} + \Omega_{n\alpha}\frac{\partial a_n}{\partial z} = (\varepsilon g_{nn} + \Omega_{n\beta}\ \beta_n)a_n + \varepsilon_v H_n \tag{7}$$

Here, $\Omega_{n\beta}$ and $\Omega_{n\beta}$ are 'complex group velocities" in the φ and z-directions. g_{nn} and H_n denote the non-parallel growth and interaction terms. ε is a small parameter representing non-parallelism of the boundary layer. For harmonic resonance n = 1, 2, $H_1 = h_{12}\ a_1 a_2$ and $H_2 = h_{21}a_1^2$. For combination resonance n = 1,2,3 $H_1 = h_{123}a_2a_3$, $H_2 = h_{213}a_1a_3$, and $H_3 = h_{312}a_1a_2$. h_{12}, h_{21}, h_{123}, h_{213}, h_{312}, g_{11}, g_{22} and g_{33} are functions of φ which are numerically evaluated [8].

Computational Procedure and Results

To gain an insight into the problem, the modulation in the φ-direction only is investigated. The first order eigenvalue problem, the adjoint problem and the derivative problem are solved numerically by the usual procedure (Srivastava [8]). $\Omega_{n\alpha}$ $\Omega_{n\beta}$, H'^{s} and other parameters are calculated for different streamwise locations and spanwise wave number α_n. The starting point of integration is chosen to be φ = 1, G = 3, which is assumed to be at the lower boundary of the nonlinear region.

For resonant interactions between two waves, i.e. harmonic resonance, the initial amplitudes are taken as $a_1(1) = 0.03$, and $a_2(1) = 0.005$. The spatial growth rates are shown in fig. (2)

for different values of G taken at constant $\varphi = 1$. The results of fig. 3 have been obtained by taking into account that, in general, G also varies with the downstream coordinate φ. The nonlinear evolution of the amplitudes a_1 and a_2 is shown in fig. (4) - (6) for different combinations of spanwise wave numbers. The deviations of the amplitudes from their non-interacting values are given in table I. The effect of interaction as may be seen in the amplification of the primary wave (a_1) from table I, is appreciable when the amplitude of the secondary wave (a_2) has reached certain % of the mean flow. For constant Görtler number it is found that the interaction between vortices of larger wave numbers leads to much larger amplification rates than the interaction between the low wave number vortices.

The evolution of a difference harmonic wave is of great interest since it exhibits another important kind of resonant interaction between waves that cannot undergo harmonic resonance. For such combination resonance between three waves the initial amplitudes taken are $a_1(1) = 0.01$, $a_2(1) = 0.005$, and $a_3(1) = 0.0005$. The evolution of the amplitude ($\ln a_3$) of the difference harmonic wave is shown in figs. (7 - 9). Table II shows the deviations from the non-interacting values and the percentage by which the amplitude increases.

It may be concluded that strong nonlinear amplifications of streamwise vortices due to different resonant interactions are possible for increasing order one wavenumbers at decreasing values of Görtler numbers.

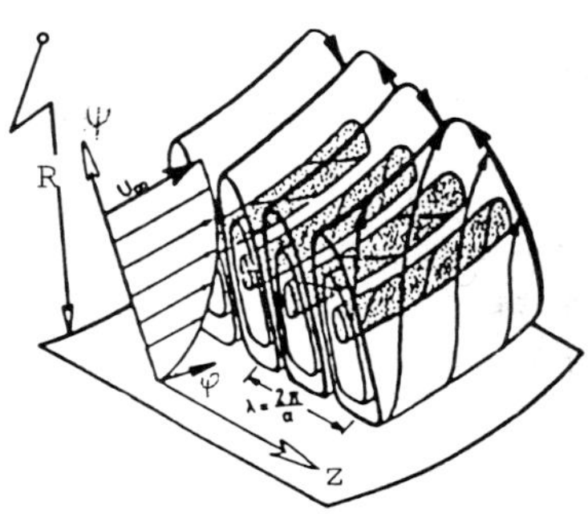

Fig. 1: Streamwise vortices in the boundary layer flow over a concave surface.

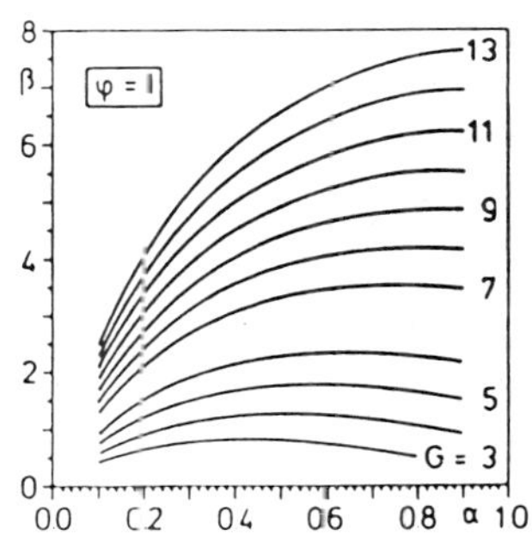

Fig. 2: Spatial growth rates β vs. spanwise wavelength α for various Görtler numbers G at φ=1 (constant).

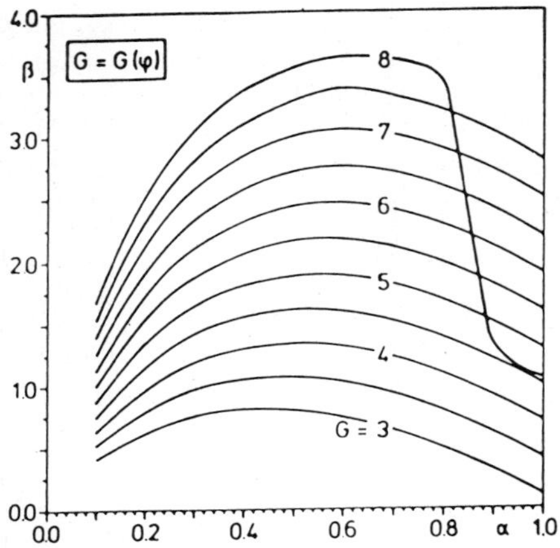

Fig. 3:
Spatial growth rate β vs. spanwise wavelength α for various Görtler numbers G. A streamwise variation of G = G(φ) is taken into account

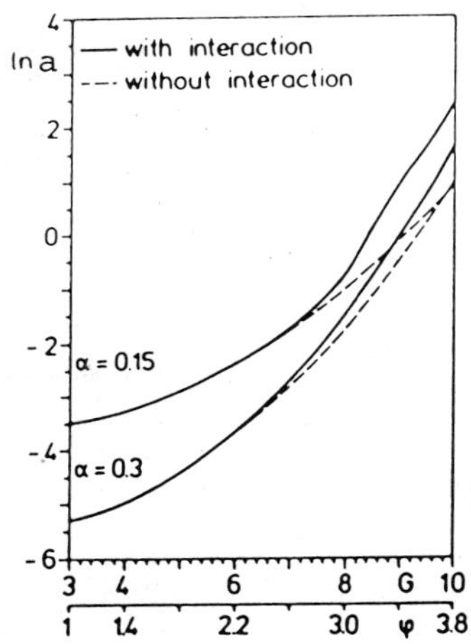

Fig. 4

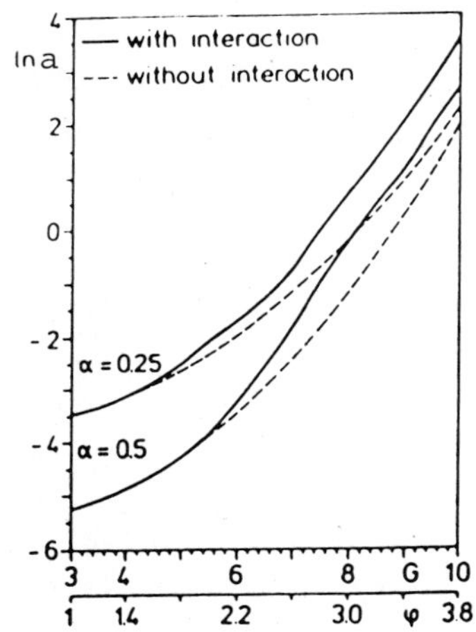

Fig. 5

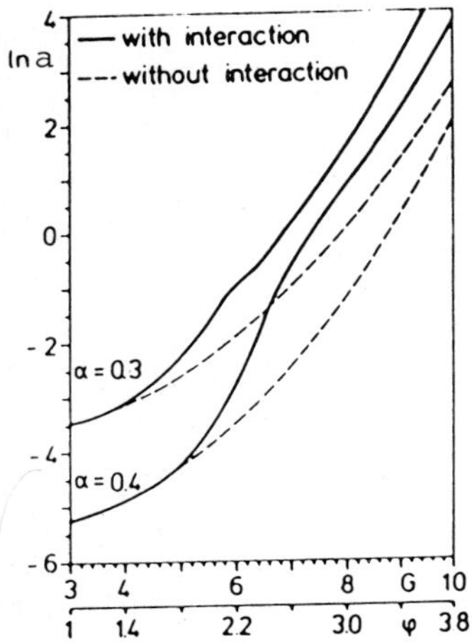

Fig. 6

Fig. 4 - 6: Harmonic resonance.

Variation of amplitudes of primary wave α_1 and of secondary wave α_2 under streamwise variation of $G(\varphi)$ for different sets (α_1, α_2).
$a_1(\varphi=1) = 0.03$, $a_2(\varphi=1) = 0.005$.

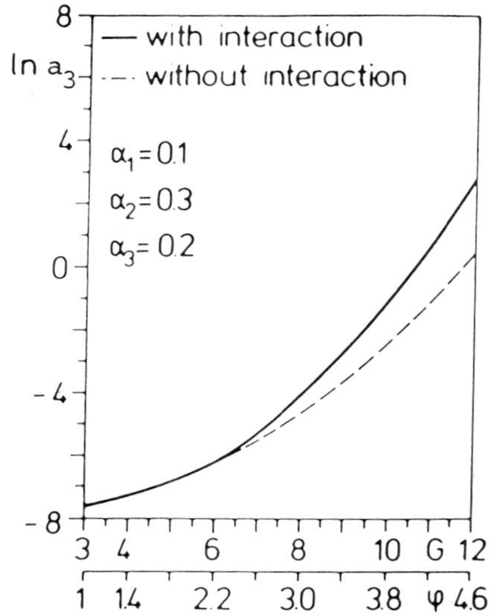

Fig. 7

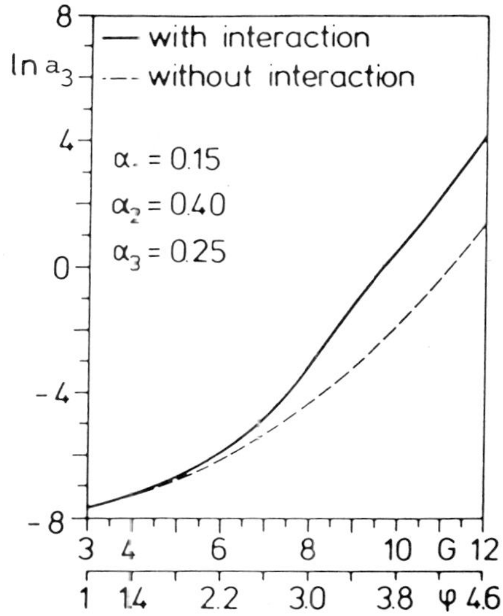

Fig. 8

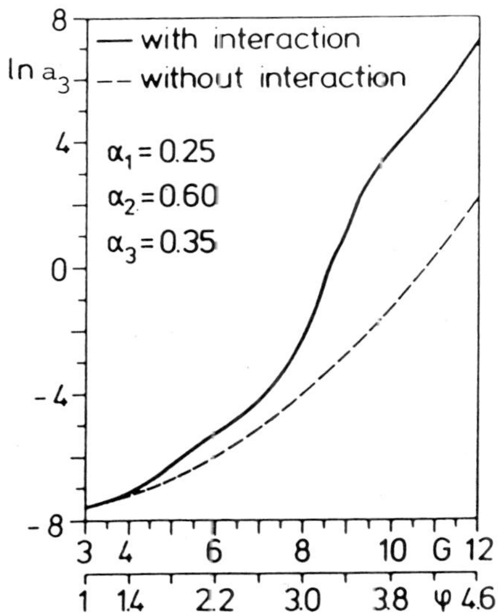

Fig. 9

Fig. 7 - 9: Combination resonance.
Variation of amplitude ln a_3 of difference harmonic α_3 under streamwise variation of $G(\varphi)$ for different primary waves (α_1, α_2)
$a_1(\varphi=1) = 0.01$, $a_2(\varphi=1) = 0.005$, $a_3(\varphi=1) = 0.0005$

α_1/α_2	G	PRIMARY WAVE A (%)	PRIMARY WAVE B	SECONDARY WAVE A (%)	SECONDARY WAVE B
0.10/0.20	8.0	21.1	19.9	10.7	9.3
	8.5	32.0	27.8	18.5	15.5
	9.5	73.0	59 0	63.8	48
0.15/0.30	7.5	27	24	12	10
	8	46.5	36	22.7	17.4
	8.5	112.6	55.4	48.6	32.0
0.20/0.40	7.0	27	23	9.8	7.5
	7.5	47.5	35	19.3	13.2
	8.0	30.6	55.9	45.7	24.6
0.25/0.50	6.0	18	13	3.6	2.9
	7.0	44	29	15	8
	7.5	100	7	37	15
0.30/0.60	5.5	25.6	9.9	2	1.8
	6.5	60	22	17.6	4.7
	7.0	20	35	56	8.1
	7.5	231	47	37	15

Table I

Relative amplitude increase (% of mean flow) of primary and secondary waves (α_1,α_2) with interaction (A) and without interaction (B). $a_1(\varphi=1) = 0.03$, $a_2(\varphi=1) = 0.005$

$\alpha_1/\alpha_2/\alpha_3$	G	DIFFERENCE HARMONIC A (%)	B
0.10/0.30/0.20	8.0	1.7	9.0
	8.5	4.0	1.5
	9.0	6.1	2.7
	10.0	53.4	9.0
0.15/0.40/0.25	8.0	4.2	1.3
	8.5	15.4	2.3
	9.0	25.6	4.3
	9.5	93.6	8.2
0.20/0.50/0.30	8.0	9.8	1.7
	8.5	65.0	3.2
	9.0	300.0	6.2
0.25/0.60/0.35	8.0	23.0	2.1
	8.5	464.0	4.0

Table II

Relative amplitude increase (% of mean flow) of primary waves (α_1,α_2) and difference harmonic (α_3) with interaction (A) and without interaction (B). $a_1(\varphi=1) = 0.01$, $a_2(\varphi=1) = 0.005$, $a_3(\varphi=1) = 0.0005$

Acknowledgement

The author would like to thank Dr. H. Oertel and Dr. U. Dallmann for the scientific discussions.

References

0. Hall, P.: The linear development of Görtler vortices in growing boundary layers. J. Fluid Mech. (1983), 130, pp.41-58.
1. Görtler, H.: Instabilität laminarer Grenzschichten an konkaven Wänden gegenüber gewissen drei-dimensionalen Störungen. ZAMM 21, pp.250-252 (1941).
2. Herbert, Th.: On the stability of the boundary layer along a concave wall. Arch. Mech. Stos., 28, pp.1039-1055 (1976).
3. Floryan, J.M.; Saric, W.S.: Stability of Görtler vortices with suction. AIAA J., 20, pp.316-324 (1982).
4. Aihara, Y.: Nonlinear analysis of Görtler vortices. Phys. Fl., 19, pp.1655-1660 (1976).
5. Aihara, Y.; Koyama, H.: Secondary Instability of Görtler vortices: Formation of periodic three-dimensional coherent structure. Trans. Japan Soc. for Aeronautical and Space Sciences, 24, pp.78-94 (1981).
6. Smith, A.M.O.: On the growth of Taylor-Görtler vortices along highly concave walls. Quart. J. Appl. Math., 13, pp.233-262 (1955).
7. Bippes, H.: (NACA-TM-72243, March 1978).
8. Srivastava, K.M.: Nonlinear interaction of streamwise vortices in non-parallel boundary layers on concave surface. DFVLR-IB 221 - 83 A 17 (1983).

Generation, Development and Distortion of Longitudinal Vortices in Boundary Layers Along Concave and Flat Plates

Y. Aihara*, Y. Tomita** and A. Ito***

* Department of Aeronautics, Faculty of Engineering, University of Tokyo, Tokyo 113, Japan

** Department of Mechanical Engineering, Kyushu Institute of Technology, Kitakyushu 804, Japan

***Department of Mechanical Engineering, Meiji University, Kawasaki, Kanagawa 214, Japan

Abstract

A linear stability analysis has revealed the non-parallelism of flows as one of the factors causing the generation of longitudinal vortices in boundary layers. This has been experimentally verified through water tank tests for a concave plate and wind tunnel tests for a flat plate. Longitudinal vortices generated in the boundary layer on concave plates, i.e., Görtler vortices, as they develop, generate horseshoe type vortices as a secondary instability. Nonlinear interaction between these vortices promotes collapse of the vortices or the transition to turbulence. Even after the transition, organized three-dimensional structures are recongnized in the boundary layer.

1. Introduction

In connection with the three-dimensionality of the boundary layer transition process and organized motion in turbulent boundary layers, increasingly widespread attention is being attracted to the mechanism of Görtler vortex generation and development.

This paper reports first on the neutral stability of Görtler vortices which has been left without elaborate study. In this connection similar instability of longitudinal vortices on a flat plate is to be demonstrated. Next it reports on the nonlinearity and secondary instability which emerge with development of Görtler vortices and on the transition of the boundary layer.

2. Generation of longitudinal vortices

2.1 Linear Analysis

Three-dimensional disturbances due to longitudinal vortices are assumed to be in the form $\phi(y)\{{}^{\cos}_{\sin}\}\sigma z \cdot e^{\beta x}$, where ϕ, σ and β are distribution function, spatial wave number in z-direction and x-wise amplification factor (Fig.1). The basic equation to be solved reduces to simultaneous linear equations with regard to the distribution functions of the disturbance in x and y direction velocity components. In the state of neutral stability, these equations are expressed as follows :

$$\frac{d^2u}{d\eta^2} - \tilde{\sigma}^2 u = \frac{\partial U}{\partial \eta} v$$

$$\frac{d^4v}{d\eta^4} - (V+2k)\frac{d^3v}{d\eta^3} - (\frac{\partial V}{\partial \eta}+2\tilde{\sigma}^2-kV)\frac{d^2v}{d\eta^2}$$

$$+ (\tilde{\sigma}^2V+k\frac{\partial V}{\partial \eta}+2k\tilde{\sigma}^2)\frac{dv}{d\eta} + \tilde{\sigma}^2(\frac{\partial V}{\partial \eta}+\tilde{\sigma}^2)v$$

$$= -\tilde{\sigma}^2(2G^2U+(1+k\eta)\frac{\partial V}{\partial \xi})u$$

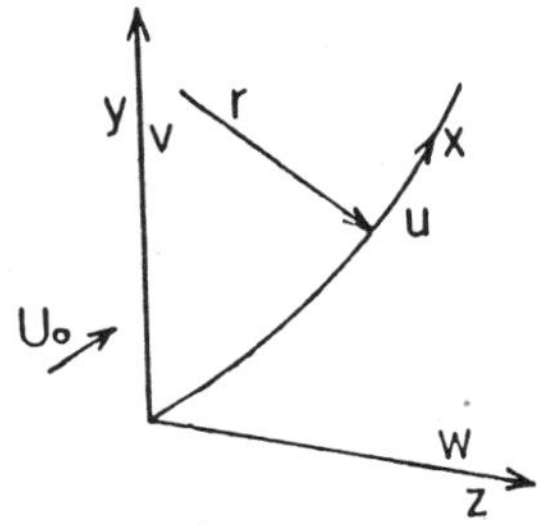

Fig.1 Coordinates system

where, $\xi=x$, $\eta=y(U_0/\nu x)^{\frac{1}{2}}$, $G=(x/r)^{\frac{1}{2}}R_x^{\frac{1}{4}}$, $\tilde{\sigma}=(\sigma x)R_x^{-\frac{1}{2}}$ and $k=(x/r)R_x^{-\frac{1}{2}}$. U and V are basic velocity in x and y direction. U_0 is the main flow velocity. The boundary conditions are : $\eta=0,\ \infty;\ u,\ v,\ \frac{dv}{d\eta}=0$.

2.2 Neutral stability of Görtler vortices

Figure 2 shows the results of neutral stability analyses employing Blasius flow as U and V. The parallel flow model[1], the non-parallel flow model with $\partial V/\partial \xi=0$ (Smith model)[2,3,4] and the non-parallel flow model with $\partial V/\partial \xi \neq 0$ (modified Smith model)[2,4] are graphically compared. The neutral stability test was done using a concave plate of 500mm curvature radius fixed to a truck which was moved at a constant speed in a water tank[2]. A number of small thin plates spaced with their span widths (20~40mm) were arranged in z-direction at zero angle of attack upstream of the leading edge to give spatially periodical disturbance.

Neutral stability of Görtler vortices was rated by hydrogen bubble method as follows. The flow was "stable", when the hydrogen bubble time lines were parallel to one another up to the one counted from the tungsten wire; "unstable" when a periodicity emerged before the second line; and "stable" when a damping tendency appeared at the second line even if a periodicity emerged at the first line.

Figure 3 is a photo illustrating periodicity in the case of U_0=0.48cm/s and wave length 30mm. Tungsten wires were stretched at 5, 100 and 220mm from the leading edge and according to the above rating, at 5 and 100mm the disturbace was "stable" and at 220mm " unstable".

The experimental values thus obtained and the analytical values are compared in Fig. 4. The results agree with the experimental values of Bippes[5] in the high wave number region and appear to agree with Smith model in the low wave number region.

2.3 Generation of longitudinal vortices in a flat plate boundary layer

It is seen from the linear analysis that even in the boundary layer on a flat plate with k=0, longitudinal vortices can be present, when on the right side the second equation possesses $\partial V/\partial \xi$, an equivalent effect to the term containing curvature effect.

Suppose a similar boundary layer which has maintained a uniform thickness begins somehow to change its thickness downstream. The initial curvature in the change of the boundary layer thickness, contributes to generation of longitudinal vortices through $\partial V/\partial \xi$. Figure 5 shows the results of neutral stability analysis. For the sake of comparison, the results of typical analyses of Görtler vortices are also given. The longitudinal vortices on a flat plate appear to possess nearly identical limits of stability.

For experimental verification of this possibility, the plate was covered with a sand paper (JIS AA40) to promote the transiton (Fig. 6). It is apparent in Fig. 7 that the lateral distributions in U which are uniform on the upstream side (x=80mm) make regular periodicity after the change of the thickness (x=150mm). It is seen that the experimental result locates in the unstable range of the analysis (Fig. 5) and there is no conflict between them.

3. Transition of the boundary layer along a concave wall

Behaviours of Görtler vortices in the linear region are known to be explained by linear theory [5,6].

As Görtler vortices amplify in downstream direction, vortices located on both sides of section III at which the velocity in x-direction decreases periodically come close to each other and form a pair of vortices, which intensify the upwash of section

III and the mean velocity distribution there turns into an extremely unstable velocity distribution which has a strong shear layer near the edge of boundary layer. Periodical fluctuations are generated from this shear layer, i.e., the secondary instability develops and, coupled with the y-direction vorticity of Görtler vortices on both sides, it constitutes horseshoe type vortices. Development and collapse of periodical horseshoe type vortices, the resulted distributions of mean velocity, turbulence, Reynolds stress, and their correlations have been revealed[6~10]. Some of the flow observations (Fig. 8) by kerosene vapour[11] and quantitative measurements (Figs. 9 and 10) are shown. Similar iso-space-time correlation contours in streamwise fluctuations u' obtained for horseshoe type vortices in the upstream are recognized in the inner layer even in turbulent state (Fig. 11). Besides the inner layer comes to indicate a lateral oscillation of flow or a spiral pattern (Fig. 11).

4. Conclusions

1. It is confirmed that in the generation of longitudinal vortices , development of the mean flow, in other words, the non-parallelism is significant.
2. Generation of horseshoe vortices due to nonlinearity and secondary instability in the development of Görtler vortices have been experimentally confirmed.

Even after transition, a similar process to that of the secondary instability seems to be repeated, but the phenomena become increasingly complicated with lateral oscillations or spiral flows recognized.

References

1. Görtler, H. : Über eine dreidimensionale instabilität laminarer Grenzschichten an konkaven Wänden. Nachr. Ges. Wiss. Göttingen, Fach-gruppe I, Neue Folge 2, 1 (1940) 1-26.
2. Tomita, Y. : Influence of Polymer Additives on Neutral Stability of Görtler Vortices. IUTAM Symposium on the Influence of Polymer Additives on Velocity and Temperature Fields, 1984 (to be published).
3. Smith, A.M.O. : On the growth of Taylor-Görtler vortices along highly concave walls. Quart. Appl. Math. 13 (1955) 233-262.

4. Floryan, J.M. ; Saric, W.S. : Stability of Görtler vortices in boundary layers. AIAA Jour. 20, 3 (1982) 310-324.

5. Bippes, H. : Experimentelle Untersuchung des laminar-turbulenten Umschlags an einer parallel angeströmten konkaven Wand. Sitzungsberichte der Heidelberger Akademie der Wissenschaften Mathematisch-Naturwissenschaftliche Klasse, 3 (1972) 103-180.

6. Aihara, Y. : Transition in an incompressible boundary layer along a concave wall. Bull. Aeron. Res. Inst., Univ. Tokyo, 3, 4 (1962) 195-240.

7. Ito, A. : The Generation and Breakdown of Longitudinal Vortices along a Concave Wall. Jour. Japan Soc. Aeron. Space Sci. 28 (1980) 327-333.

8. Aihara, Y. ; Koyama, H. : Secondary instability of Görtler vortices-Formation of periodic three-dimensional coherent structure. Trans. Japan Soc. Aeron. Space Sci. 24 (1981) 78-94.

9. Aihara, Y. ; Koyama, H. : Nonlinear development and secondary instability of Görtler vortices. Schroeder, F.H. (ed.) Stability in the mechanics of continua. Berlin, Heidelberg, New York : Springer-Verlag (1982) 345-354.

10. Aihara, Y. ; Koyama, H. : Three-dimensional turbulent boundary layer along a concave wall. Fernholz, H.H. : Krause, E. (eds.) Three-dimensional turbulent boundary layers. Berlin, Heidelberg, New York : Springer-Verlag (1982) 210-220.

11. Ito, A. : Breakdown Structure of Longitudinal Vortices along a Concave Wall. Jour. Japan Soc. Aeron. Space Sci. (to be published).

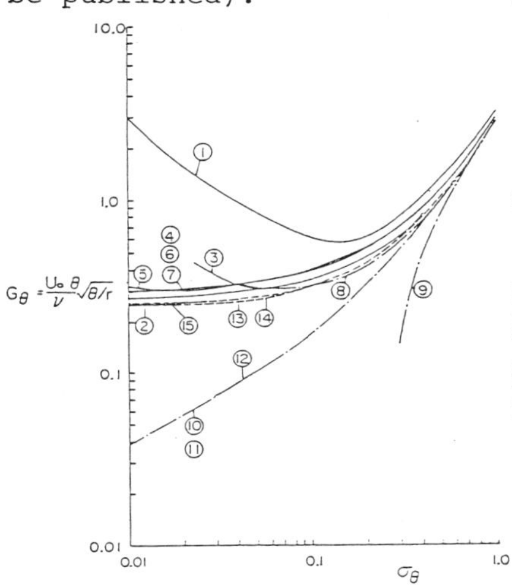

Fig.2 Neutral stability curves of Görtler vortices[2].

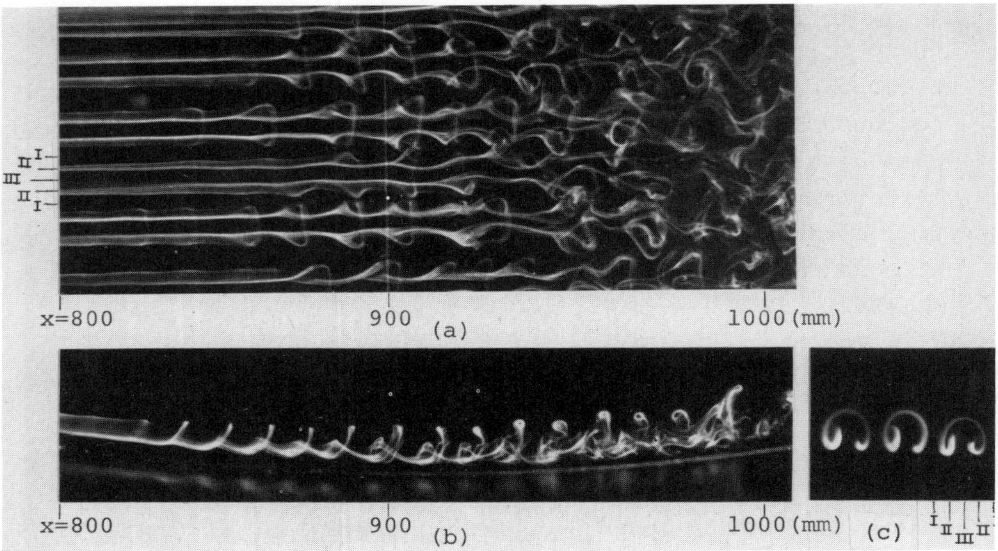

Fig.8 Generation and development of horseshoe type vortices. U_0=2.5m/s (a) plan view. (b) side view. (c) cross sectional view.

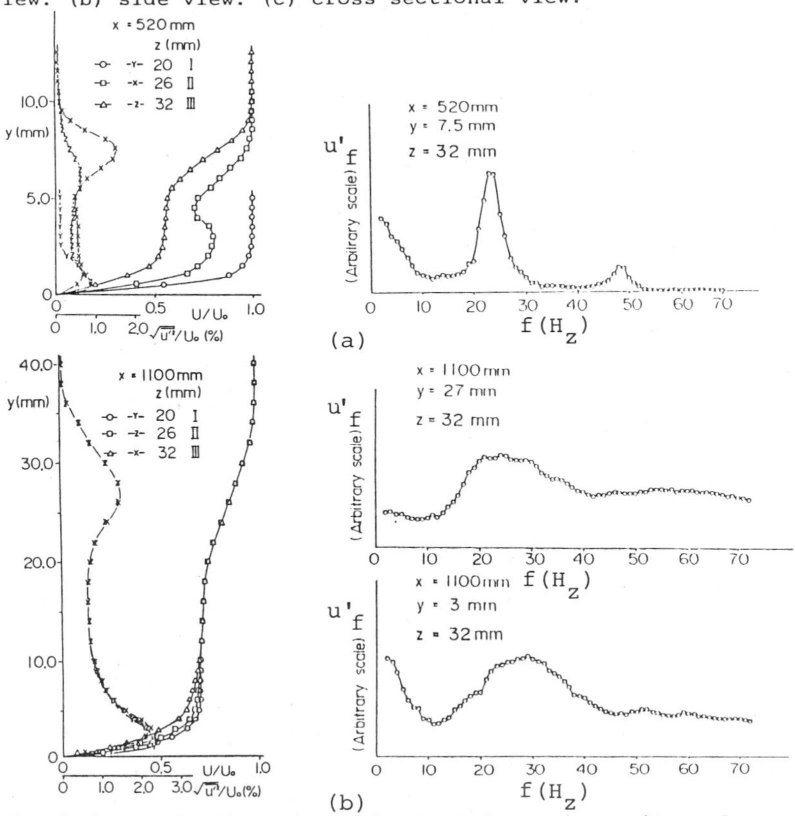

Fig.9 Mean velocity, streamwise turbulence intensity and spectrum distributions. U_0=1.5m/s. (a) x=520mm (b) x=1100mm

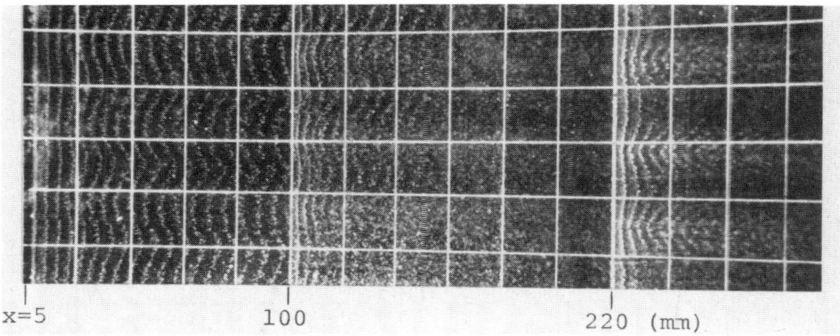

Fig.3 Time lines of Hydrogen bubbles.

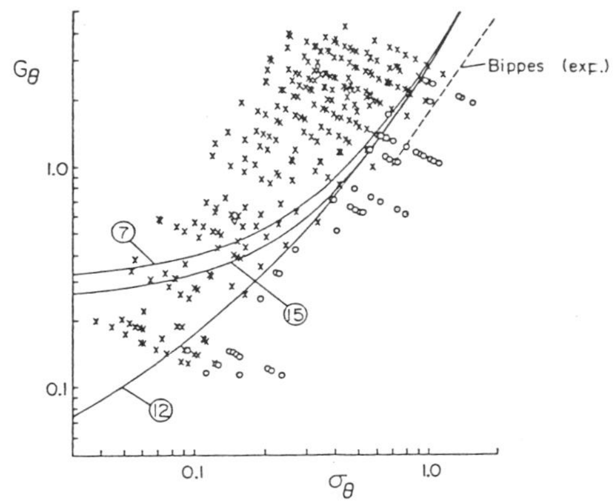

Fig.4 Stability of Görtler vortices.
o stable, x unstable.

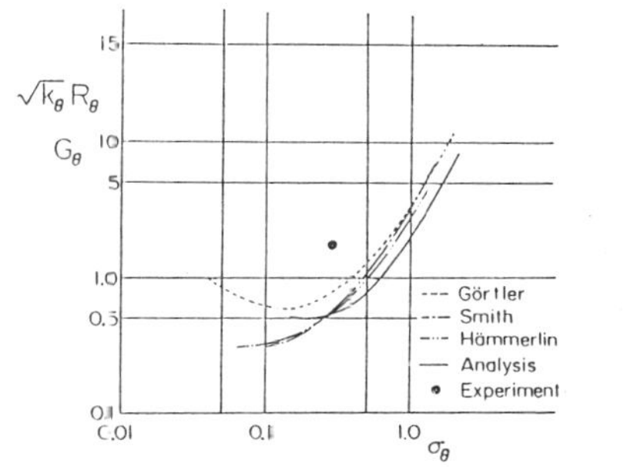

Fig.5 Neutral stability curve of longitudinal vortices in flat plate boundary layer.

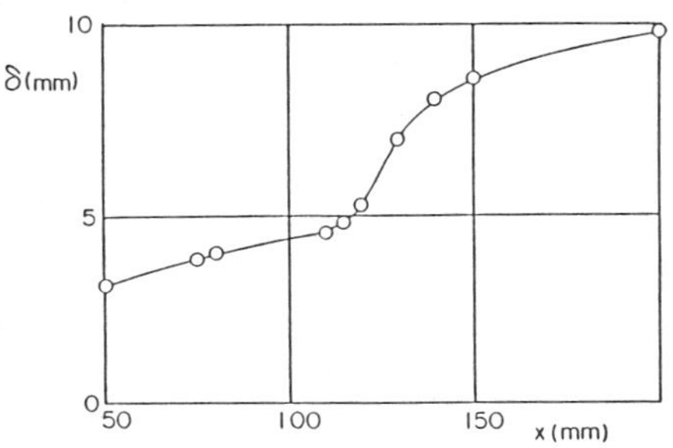

Fig.6 Variation of boundary layer thickness in the downstream direction. U_0=4m/s

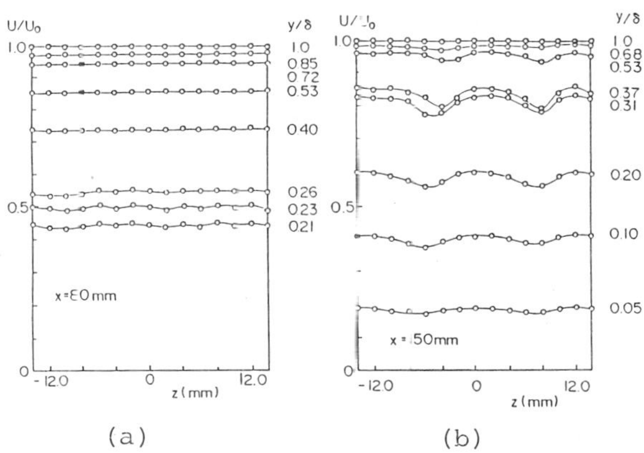

Fig.7 Lateral distributions of streamwise velocity. (a) x=80mm (b) x=150mm

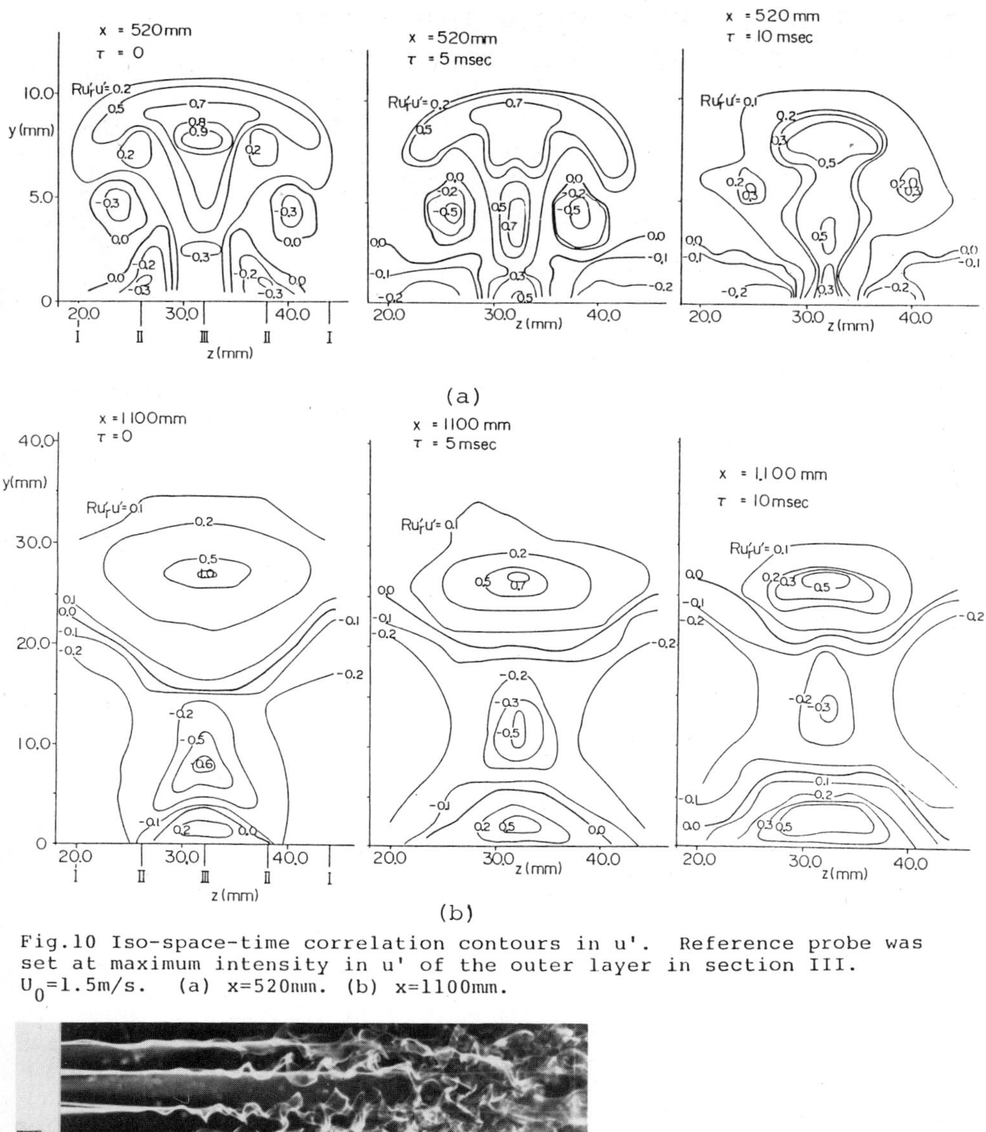

Fig.10 Iso-space-time correlation contours in u'. Reference probe was set at maximum intensity in u' of the outer layer in section III. U_0=1.5m/s. (a) x=520mm. (b) x=1100mm.

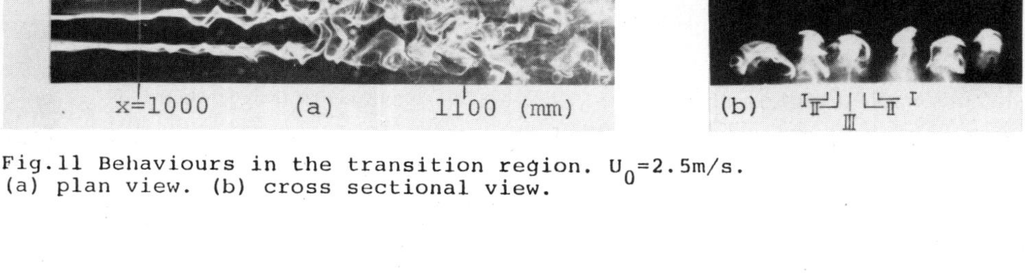

Fig.11 Behaviours in the transition region. U_0=2.5m/s. (a) plan view. (b) cross sectional view.

Stability Calculations and Transition Criteria on Two- or Three-Dimensional Flows

R. MICHEL, D. ARNAL and E. COUSTOLS

ONERA/CERT - Département d'Aérothermodynamique - 2 avenue E. Belin - 31055 TOULOUSE CEDEX (FRANCE)

SUMMARY

This paper deals with the prediction of the boundary layer transition in incompressible two- and three-dimensional flows. The essential elements of the linear stability theory, governing the transition phenomenon, are introduced onto transition criteria. The presented applications to experimental cases allowed us to check up the validity of the proposed criteria.

INTRODUCTION

The theory of laminar instability constitutes the earliest attempt to explain transition. The idea of this theory is that forced disturbances are, somehow, internalized by the boundary layer and then selectively amplified by the viscous flow ; the growth of these disturbances can reach such a level that turbulence may appear. Recently, several computational methods have been developed to describe that growth in the laminar boundary layer, using small perturbations theory. The derivation of the equations governing this problem will not be outlined here. They lead to the pertinent ORR-SOMMERFELD equation which constitutes, with its boundary conditions, an eigenvalue problem which is time-consuming to solve.

Results given by the linear stability theory are summarized for both two- and three-dimensional flows. The main purpose of the present work is to show how such results can be applied for elaborating transition criteria that will be easy to handle, from an engineering point of view.

2 - APPLICATIONS TO TWO-DIMENSIONAL FLOWS

2.1. Relation between stability theory and transition

The stability theory is mainly concerned with individual sine waves, developing within the boundary layer ; these amplitudes are small enough so that a linear theory may be used. Two-dimensional waves have been considered /1/.

Within a steady boundary layer, a wave propagates downstream at the same frequency ω. According to the spatial theory -which means that ω is real and the wave number α, complex ($\alpha = \alpha_r + i\alpha_i$)- a wave which passes through the stable region will be damped up to R_0, then amplified up to R_1 and at last will be damped again downstream of R_1. Thus, the total amplification rate can be defined :

$$Ln \frac{A}{A_0} = \int_{x_0}^{x} - \alpha_i \, dx = \int_{R_0}^{R} - \alpha_i \frac{dx}{dR} \, dR \qquad (1)$$

A is the wave amplitude and the index o refers to the streamwise position where the wave becomes unstable. For various frequencies, the total amplification rates are then calculated. The envelope of these curves is :

$$n = \max_{\omega} \left(Ln \frac{A}{A_0}\right) \qquad (2)$$

$n = 0$ for $R = R_{cr}$. R_{cr} is the minimum critical Reynolds number below which no amplification is possible.

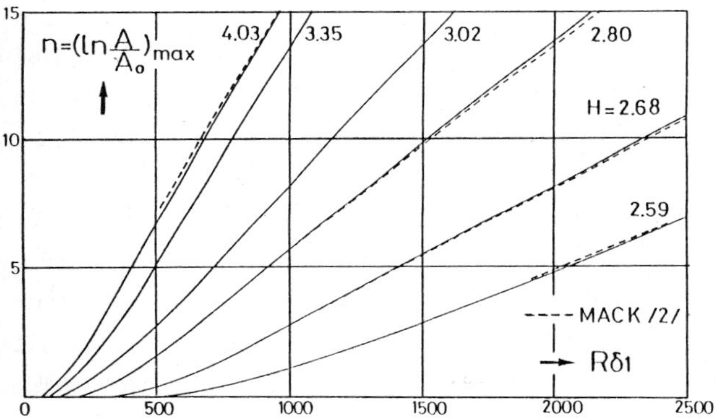

Fig.1. Evolution of $n = (Ln \frac{A}{A_0})_{max}$

For similar velocity profiles of the FALKNER-SKAN family, which are characterized by constant parameters such as H, the envelope curves are plotted on fig. 1. Our results are compared to those obtained by MACK /2/. The destabilizing effect of a positive pressure gradient can be observed : as H increases, the critical displacement Reynolds number decreases and the slope n(Rδ1) increases.

When one wants to get an envelope curve for a non constant H-type flow, one has at first to compute boundary layer development and then to solve the eigenvalue problem. But that gives only local disturbance amplification rates and overall amplification ratio. At the present time, to predict transition, one must place reliance on empirical procedures. The e^n, or n-factor, method proposed by SMITH and GAMBERONI (1956) and Van INGEN (1956) was the first to correlate amplification and transition. Later on, from experimental results collected by DRYDEN on a flat plate, MACK /2/ suggested the following empirical relation between the free-stream turbulence parameter Tu and the value of n at the transition :

$$n_T = -8.43 - 2.4 \; \mathrm{Ln}\; Tu \tag{3}$$

That "modified e^n method" can be generalized to flows with pressure gradients (/1/, /3/), since in all cases, relation (3) gives an abscissa close to the beginning of the experimental transition.

2.2. Establishment of a transition criterion

One should not forget that stability calculations remain very cumbersome to implement and very costly in terms of computational time. These constraints have led us to look for a transition criterion easy to apply but containing theoretical results coming from the stability theory.

GRANVILLE /4/ considered two parameters : $R\theta_T - R\theta_{cr}$, the difference between the momentum thickness Reynolds numbers from the critical abscissa to the transition point, and the averaged POHLHAUSEN parameter $\overline{\Lambda}_{2T}$ which shows up the cumulative effect of the pressure gradient :

$$\overline{\Lambda}_{2T} = \frac{1}{x_T - x_{cr}} \int_{x_{cr}}^{x_T} \frac{\theta^2}{\nu} \frac{dU_e}{dx}\, dx \tag{4}$$

θ is the momentum thickness and U_e the potential velocity. Our purpose was to extend that criterion by making it take into account Tu. Therefore, the criterion must apply to constant H-type flows, for which stability calculations have already been presented. For such flows, the envelope curves of fig. 1 can be considered as $n(R\theta, \Lambda_2)$ curves. As the critical Reynolds number is itself a function of Λ_2, one gets, using relation (3) :

$$R\theta_T - R\theta_{cr} = f(\Lambda_{2T}, Tu) \tag{5}$$

It is assumed that it is sufficient to replace Λ_2 by $\overline{\Lambda}_2$ to apply it to an arbitrary flow. Curves corresponding to various values of Tu are plotted on figure 2. An analytic expression of (5) as well as the way to determine the true critical Reynolds number $R\theta_{cr}$ can be found in /1/.

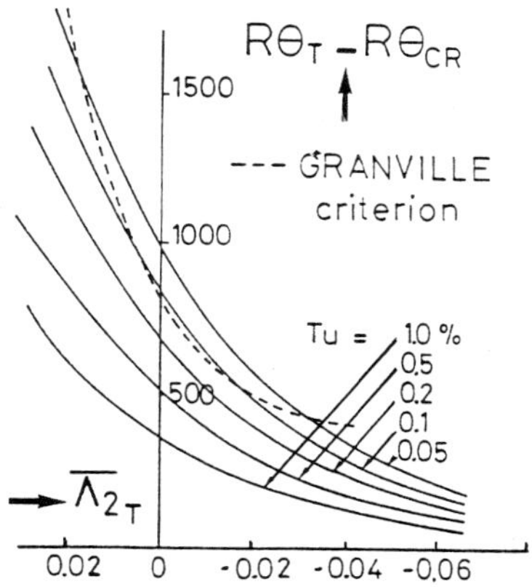

Fig.2. Two-dimensional transition criterion

2.3. Applications of the proposed criterion

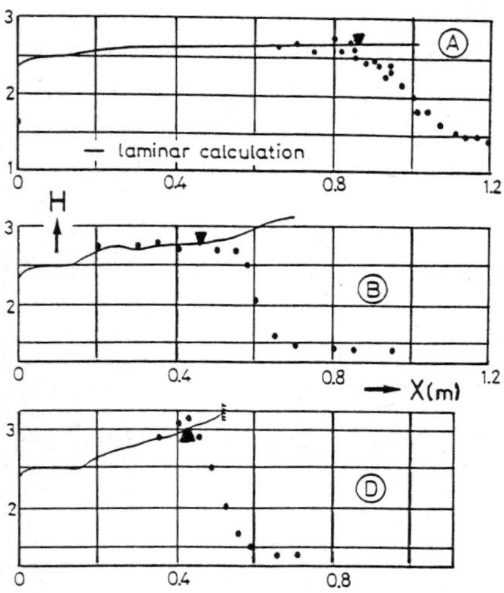

Figure 3 shows some applications to experiments performed at ONERA/CERT /3/ with zero or positive pressure gradients. The arrow indicates the deduced beginning of transition, according to the criterion. The agreement with the experiments is satisfactory since it is proper to associate transition with the decrease of the shape parameter. Other applications of this criterion are given in /1/.

Fig.3. Application of the criterion

3 - APPLICATIONS TO THREE-DIMENSIONAL FLOWS

We shall confine ourselves, however, to the case of swept wings with infinite span.

3.1. Possible origins of turbulence

From experimental results on swept wings, at low values of the free-stream velocity Q_∞, transition occurs in the rear part of the wing, in the region of positive pressure gradient. Its movement looks like the one we would get on the same airfoil without any sweep angle, since it is related to the stability properties of the streamwise profile U1. We shall call it "streamwise (instability) transition". For higher values of Q_∞, transition moves quickly towards the leading edge, in the region of negative pressure gradient. Such a movement cannot occur in two-dimensional flows since it is due to the stability properties of the cross-flow profile W1. This transition shall be called "cross-flow (instability) transition".

Therefore, the boundary layer formed on the attachment line can be contaminated by the big structures coming from the wall, on which the wing is fixed : this is the "leading edge contamination". Such a phenomenon will not be detailed in this paper, though we have always been looking at the velocity corresponding to the appearance of contamination in the considered experiments /1/. Beyond that velocity, the whole wing boundary layer is turbulent.

3.2. Transition criteria

The determination of the transition location on a swept wing is based on the following rule : it is assumed that streamwise and cross-flow instabilities must be considered. A criterion is applied for each of them and transition shall occur when it is detected by one criterion.

3.2.1. Streamwise instability

For that type of instability, the two-dimensional criterion, formulated in 2.3., remains valid provided it is applied along an external streamline.

3.2.2. Cross-flow instability

BEASLEY (1976) postulated that, at the transition onset, the cross-flow Reynolds number based on the integral of the W1 profile, $R\delta 2$, is equal to 150 /5/. Applying that criterion to all the considered experimental cases, we could deduce that cross-flow transition did not take place for a unique value of $R\delta 2_T$ /1/. Better results were obtained by associating $R\delta 2_T$ with the shape factor H1 of the streamwise profile U1. Some applications of this criterion, designated as C1, shall be presented later on.

Therefore, we tried to use results of the linear stability theory to develop a method which would be as rigourous as possible. At a given abscissa, stability properties of different projected velocity profiles U_ε are studied, from $\varepsilon = 0°$ (W1 profile) to $\varepsilon = 90°$ (U1 profile). We then defined a Reynolds number $R\delta 1_\varepsilon$, based on the displacement thickness of the profile U_ε :

$$R\delta 1_\varepsilon = \frac{1}{\nu} \int_0^\delta (U1_e \sin\varepsilon - U_\varepsilon)dy \quad \text{where } U_\varepsilon(y) = U1(y)\sin\varepsilon + W1(y)\cos\varepsilon \qquad (6)$$

$U1_e$ represents the resultant potential velocity.

Stability calculations revealed the existence of one most unstable direction, designated as ε_{min}. In order to determine that direction, one needs to know the critical Reynolds number. As a consequence, this latter has been represented as a function of either the inflection point of the profile U_ε and of the first derivative of U_ε taken at the inflection point or some integral parameter if U_ε does not exhibit some inflection point. Moreover, the cross-flow direction was found to be always more stable than the ε_{min} direction, which varies from 1° to 5° ; this illustrates the concept of "quasi cross-flow instability".

From experimental transition data, the criterion, designated as C2, has been

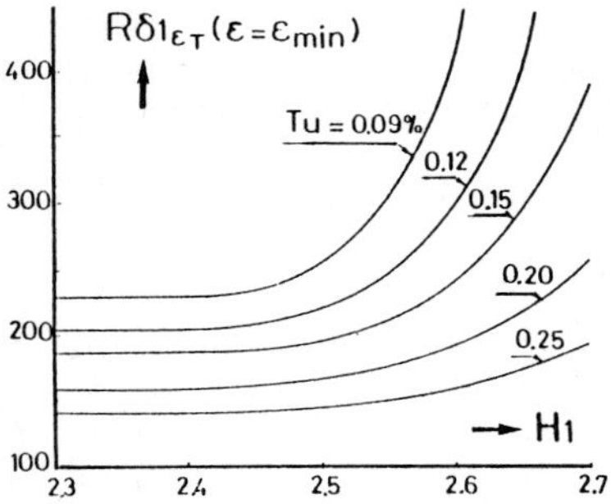

established. It is based on a correlation between three parameters : $R\delta 1_\varepsilon$ taken in the most unstable direction, the streamwise shape parameter H1 and the external turbulence level Tu (cf figure 4).

Fig.4. Proposed cross-flow transition criteria

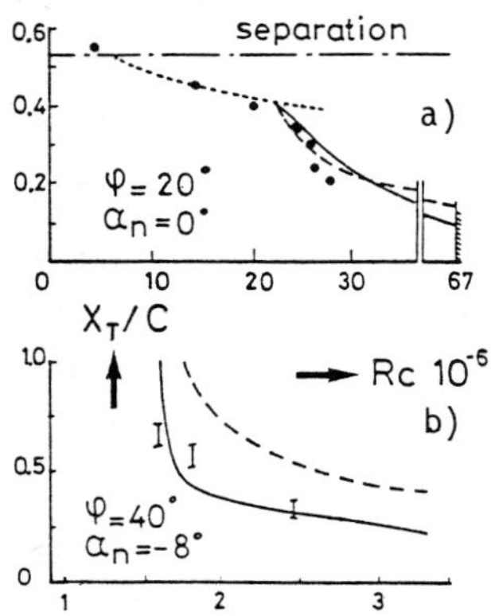

Fig.5. Applications of transition criteria

3.3. Applications of the different criteria

Fig. 5a shows comparison with experiments carried out by BOLTZ, KENYON and ALLEN /6/. The passage from streamwise (...) to cross-flow types transition can be clearly observed. The value of the chord Reynolds number for which the contamination comes into play lies beyond the range of transition Reynolds numbers. The same result concerning contamination can be pointed out as regards to experiments performed at ONERA/CERT /7/ (fig. 5b). In that case, there is no streamwise transition and the C2-criterion (——) gives results in better agreement with experiments than the C1-one (---). Criterion C2 goes much more deeply into the physics of the phenomena.

4 - CONCLUSION

This study has been mainly carried out from results given by the laminar stability theory. In spite of some simplifying assumptions -as considering locally parallel flow or neglecting the non linear phase and the curvature effects- we were able to get transition criteria more or less empirical. But, the fact that they can be simply and quickly implemented might be of interest from a practical point of view.

Knowing the transition onset, the next step would be to compute the transition region itself, since it cannot be assumed that transition is a point-like phenomenon. Different numerical models have been set up at ONERA/CERT and are described in /7/.

REFERENCES

1. ARNAL D. ; HABIBALLAH M. ; COUSTOLS E. : Théorie de l'instabilité laminaire et critères de transition en écoulement bi et tridimensionnel. La Recherche Aérospatiale, 1984-2

2. MACK L.M. : Transition and laminar instability. Jet Propulsion Lab., publication 77-15, PASADENA, CALIFORNIA (1977)

3. ARNAL D. ; JUILLEN J.C. : Etude expérimentale et théorique de la transition de la couche limite. La Recherche Aérospatiale, 1977-2

4. GRANVILLE P.S. : The calculation fo the viscous drag of bodies of revolution. David Taylor Model Basin. Report 849 (1953)

5. BEASLEY J.A. : Calculation of the laminar boundary layer and the prediction of transition on a sheared wing. ARC R and M 3787 (1973)

6. BOLTZ F.W. ; KENYON G.C. ; ALLEN C.Q. : Effects of sweep angle on the boundary layer stability characteristics of an untapered wing at low speeds. NASA TND-338 (1960)

7. ARNAL D. ; COUSTOLS E. ; JUILLEN J.C. : Etude expérimentale et théorique de la transition sur une aile en flèche infinie. La Recherche Aérospatiale, 1984-4

Stability Calculations of the Three-Dimensional Boundary Layer Flow on a Rotating Disk

NOBUTAKE ITOH

National Aerospace Laboratory
Chofu, Tokyo, Japan

Summary

The Orr-Sommerfeld eigenvalue problem is solved to establish the neutral stability curve on the wavenumber plane for each of supercritical Reynolds numbers. Then the weakly nonlinear stability theory of Stuart and Watson is formally applied to obtain numerical values of the Landau coefficient, which are found to be negative on most part of the neutral curve, suggesting the existence of the supercritical stable equilibrium amplitude. It is however shown that the theory is inapplicable to a certain region of the wavenumbers, where an alternative analysis is made by the use of the resonant instability theory developed by Craik.

1. Introduction

The flow on a flat disk rotating in still fluid with a fixed angular velocity is most convenient for instability analysis of three-dimensional flows, because the steady laminar flow is described by an exact solution of the Navier-Stokes equations, and because the state of transition from laminar to turbulent flow has been revealed to a certain extent through a number of experimental observations by Gregory et al. [1], Kobayashi et al. [2] and others. It has been found that, in the annular region between the central laminar flow and the outer turbulent flow, stationary disturbances appear in the form of vortices whose axes are fixed relative to the surface of the disk and lie roughly along equiangular spirals. Several theoretical attempts have been made by Stuart [1], Cebeci & Stewartson [3] and others to predict the range of the Reynolds number where the vortices are observed, the number of vortices and the angle of spirals. However, these studies are all based on the conventional theory of linear stability, and seem to be unsuccessful in explaining the selective growth of a particular wavenumber component observed in the experiments. In this paper, therefore, analysis is made to get some informations about the effect of nonlinearity on the growth of disturbances.

2. Results of linear stability calculations

We take the cylindrical coordinate system (r,θ,z) fixed relative to the disk rotating a constant angular velocity ω_0, and make all quantities dimensionless as

$$r/r_0 = \xi, \quad \theta = \eta, \quad z\sqrt{\omega_0/\nu} = \zeta, \quad t\omega_0 = \tau,$$

$$\mathrm{v}/(r\omega_0) = \mathrm{V} + \mathrm{v}^*, \quad \sqrt{\nu/\omega_0}/r_0 = \varepsilon, \quad r\sqrt{\omega_0/\nu} = R, \tag{1}$$

where r_0 is a reference length denoting a measure of the flow field, and ε is a small parameter denoting the ratio of another reference length $\sqrt{\nu/\omega_0}$ to r_0, while R is the local Reynolds number. The velocity vector of the basic laminar flow is given by $\mathrm{V} = (F', G, -2\varepsilon F/\xi)$, where the functions $F(\zeta)$ and $G(\zeta)$ are obtained from solution of the simultaneous equations

$$(F')^2 - 2F''F - (G+1)^2 = F''', \quad 2F'(G+1) - 2FG' = G'',$$

$$F(0) = F'(0) = G(0) = F'(\infty) = G(\infty) + 1 = 0. \tag{2}$$

Substituting (1) into the Navier-Stokes equations, we have the equations governing the disturbance velocity $\mathrm{v} = (u,v,w)$, where the asterisk is omitted. If we consider the disturbance small enough, the equations may be linearized and, noting that ε is a small parameter, we may seek the solution of the form

$$\mathrm{v} = \exp[i\Theta(\xi,\eta,\tau)/\varepsilon]\{\mathrm{v}_1(\xi,\eta,\zeta,\tau) + O(\varepsilon)\},$$

$$\frac{\partial\Theta}{\partial\xi} = \alpha + O(\varepsilon), \quad \frac{\partial\Theta}{\partial\eta} = \xi\beta + O(\varepsilon), \quad \frac{\partial\Theta}{\partial\tau} = -\xi\omega + O(\varepsilon). \tag{3}$$

Substitution of these into the linearized disturbance equations and neglect of terms of $O(\varepsilon)$ leads to the well-known Orr-Sommerfeld equation. In this process, the viscous terms proportional to $1/R$ are formally of $O(\varepsilon)$ but should be retained as terms of $O(1)$, because those are known from the classical stability theory to become comparable in magnitude to the inviscid terms in the so-called critical layer and near the wall. The Orr-Sommerfeld equation and the corresponding homogeneous boundary conditions are written in the form

$$(L + i\omega M)w_1 = 0, \quad w_1 = \partial w_1/\partial\zeta = 0 \quad \text{at} \quad \zeta = 0 \text{ and } \infty, \tag{4}$$

where $L \equiv (1/R)M^2 - i(\alpha F' + \beta G)M + i(\alpha F''' + \beta G')$, $M \equiv \partial^2/\partial\zeta^2 - \alpha^2 - \beta^2$, and pose an eigenvalue problem to determine the relation

$$\omega = \Omega(\alpha,\beta,R), \tag{5}$$

where α, β and ω are complex in general and the real parts α_r, β_r and ω_r denote the wavenumbers in the radial and circumferential directions and the frequency of the disturbance, while the imaginary parts α_i, β_i and ω_i denote the spatial damping rates in the radial and circumferential directions and the temporal growth rate, respectively.

Numerical computations have been carried out to solve the above eigenvalue problem. In this three-dimensional problem, the condition of neutral stability establishes a surface in the (α,β,R) space, and the solid line in Fig. 1 shows a cross section of the neutral surface at $R = 400$. An interesting fact is that the point C indicating the temporal growth rate maximum is located very close to the dotted line AB, on which $\omega_r = 0$, suggesting the possibility that the experimentally observed spiral vortices fixed relative to the disk may correspond to the eigensolution with the maximum growth rate. In fact, the number of vortices $n = \beta R$ and the angle of spirals $\phi = \arctan(\beta/\alpha)$ take, at the point C, the values of 30 and 12°, respectively, which are in very good agreement with the experimental values. Thus, variation of the maximum-growth-rate solution with the Reynolds number was investigated and summerized in Fig. 2 (a). As R increases from 300 to 600, roughly corresponding to the range where the vortices

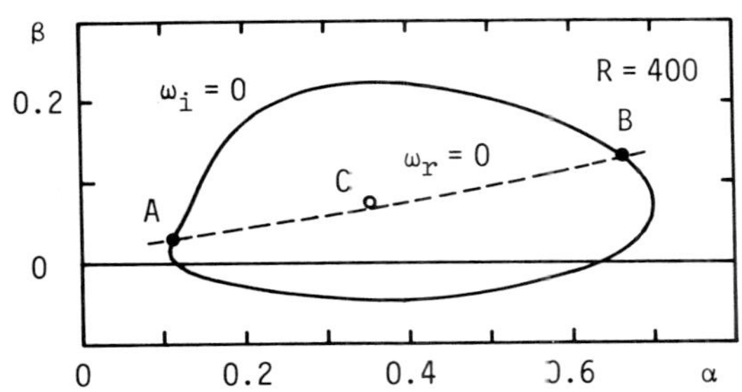

Fig. 1. A cross section of the neutral stability surface at a fixed Reynolds number.

are observed in experiments, the calculated value of ϕ varies from 12° to 14° and n from 20 to 50. This result, showing a drastic rise in the number of vortices, is definitely inconsistent with experimental observations, and seems to give grounds for objections against the deduction that the eigensolution with the maximum growth rate is realized.

Physically realized disturbances grow not in time but in the radial direction only, so that the spatially-dependent approach with ω and β real and α complex to the eigenvalue problem may be relevant. Fig. 2 (b) shows the computational results for the eigensolutions with the spatial growth rate $-\alpha_i$ maximum. Comparing this with Fig. 2 (a), we understand that the temporally-dependent approach yields more realistic values of various parameters than the spatially-dependent approach. Thus the remainder of this paper is spent for revealing the behaviour of temporally growing disturbances, with special attention to the effects of nonlinearity.

3. Nonlinear analysis

We first apply to the present three-dimensional problem the weakly nonlinear stability theory originally developed by Stuart [4] and watson [5]. Then the complex amplitude A of the fundamental

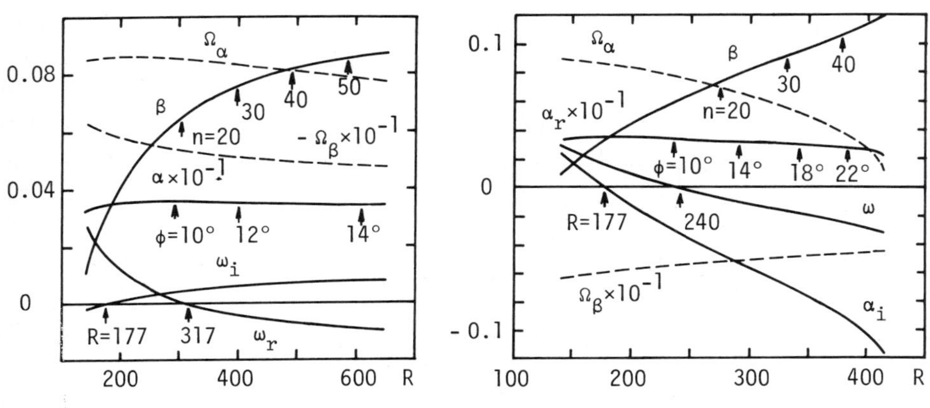

(a) (b)

Fig. 2. Variations of the eigensolutions with the maximum values of temporal growth rate (a) and of spatial growth rate (b).

wave with wavenumbers (α,β) is governed by the Landau equation

$$\partial A/\partial t = -iA\{\omega + \lambda|A|^2 + O(|A|^4, \varepsilon)\}, \tag{6}$$

where $t = \xi\tau/\varepsilon$, and λ is the complex Landau coefficient. (The method of calculation for the Landau coefficient is given in the author's previous paper [6].) If the smaller-order terms in (6) are negligible, a stable equilibrium amplitude exists when ω_i is positive and λ_i negative. Computational results given in Fig. 3 show that the values of λ_i are negative for most part of the neutral curve, except for two short segments in the neighborhood of the points with α minimum and with β minimum. Supplemental calculations for the unstable region gave similar results. This fact suggests that a surface of equilibrium amplitude is set up over the unstable region on the (α,β) plane, being framed by the neutral curve, although more complicated computations, such as the iterative one made by Herbert [7], are necessary for revealing further details of the equilibrium surface.

It should be noted here that the Stuart-Watson theory is not applicable to a particular region of wavenumbers. See Fig. 4 and consider the fundamental wave with the wavenumbers (α,β) located inside the region enclosed by the neutral curve and the curve EF, which denotes a part of the same neutral curve drawn on the half scale. Then the eigenvalue for the wavenumbers $(2\alpha,2\beta)$ is given

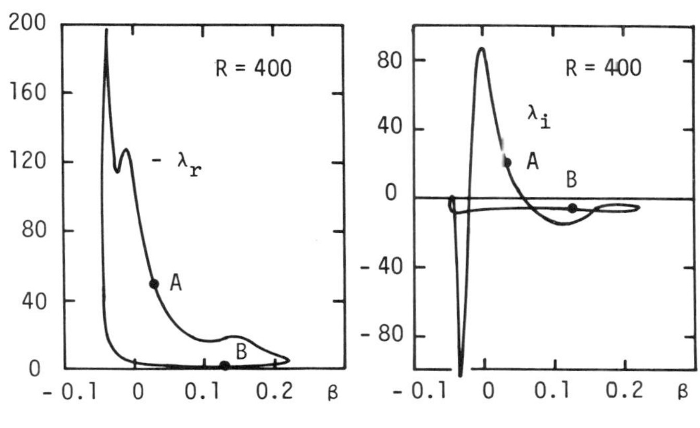

Fig. 3. Values of the Landau coefficient on the neutral curve at $R = 400$. The points A and B indicate the correspondence with Fig. 1.

by $\omega_2 = \Omega(2\alpha, 2\beta, R)$ and has a positive value of growth rate. The existence of growing eigensolution induces an arbitrariness in the second harmonic component and breaks the Stuart-Watson formulation. In such a particular situation, it is necessary to consider direct interactions between the fundamental and the second harmonic, as in the following.

Here let us neglect the terms of $O(\varepsilon)$ for simplicity and consider the disturbance of the form

$$\mathbf{v} = [A(t)e^{i(\alpha x + \beta y)}\{\mathbf{v}_1(\zeta) + \hat{\mathbf{v}}_1(\zeta)\tilde{A}B/A + O(|A|^2)\} + (\mathrm{c.c.})]$$
$$+ [B(t)e^{2i(\alpha x + \beta y)}\{\mathbf{v}_2(\zeta) + \hat{\mathbf{v}}_2(\zeta)A^2/B + O(|A|^2)\} + (\mathrm{c.c.})]$$
$$+ O(|A|^2),$$
$$\frac{i}{A}\frac{dA}{dt} = \omega_1 + \lambda_1\frac{\tilde{A}B}{A} + O(|A|^2), \qquad \frac{i}{B}\frac{dB}{dt} = \omega_2 + \lambda_2\frac{A^2}{B} + O(|A|^2), \tag{7}$$

where $x = \xi/\varepsilon$, $y = \xi\eta/\varepsilon$, $\sim$ denotes the complex conjugate and B is assumed to be of $O(|A|)$. Then the Orr-Sommerfeld eigenvalue problem leads to $\omega_1 = \Omega(\alpha, \beta, R)$ and $\omega_2 = \Omega(2\alpha, 2\beta, R)$ and the Landau-type coefficients λ_1 and λ_2 are determined from weakly nonlinear calculations. Putting

$$|\lambda_1\lambda_2|^{1/2}|A| = a, \qquad |\lambda_1||B| = b, \qquad \arg(B) - 2\arg(A) = \theta, \tag{8}$$

and neglecting the smaller-order terms, we have the following equations governing temporal variations of the amplitudes a and

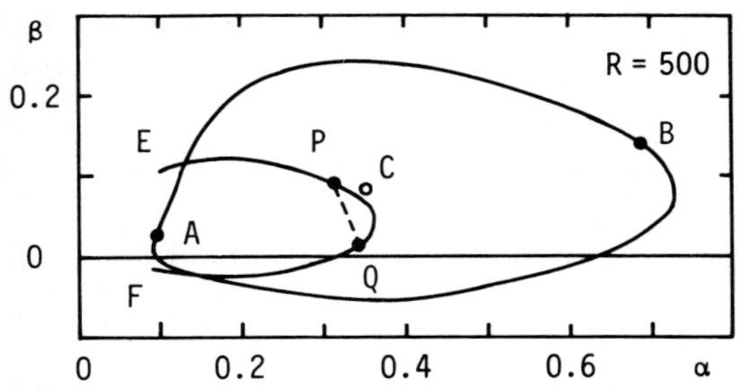

Fig. 4. The curves of neutral stability at $R = 500$ in full and half scales. The dotted line PQ indicates the resonance between the fundamental and second harmonic waves.

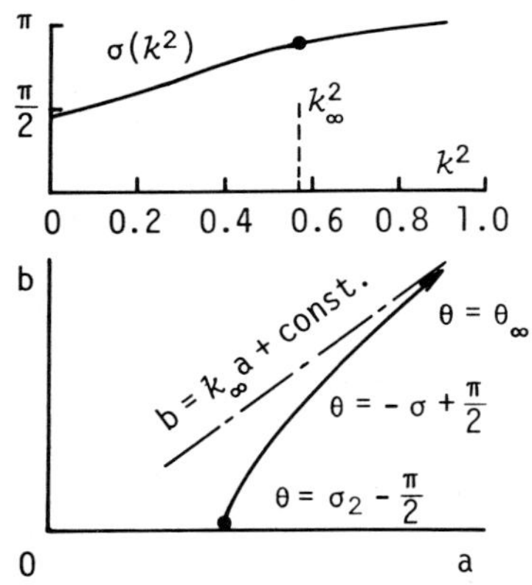

Fig. 5. Schematic view of the development of a and b under the resonant state.

b and the phase difference θ:

$$\frac{1}{a}\frac{da}{dt} = \omega_{1_i} + b\sin(\theta+\sigma_1), \qquad \frac{1}{b}\frac{db}{dt} = \omega_{2_i} - \frac{a^2}{b}\sin(\theta-\sigma_2),$$

$$\frac{d\theta}{dt} = -(\omega_{2_r} - 2\omega_{1_r}) - \frac{a^2}{b}\cos(\theta-\sigma_2) + 2b\cos(\theta+\sigma_1), \tag{9}$$

where $\sigma_1 = \arg(\lambda_1)$, $\sigma_2 = \arg(\lambda_2)$. These equations are similar to those derived by Craik [8,9] in analysis of resonant instability of two-dimensional boundary layers, and therefore we may expect that a similar phenomenon of resonance will occur between the two waves in the present problem of a three-dimensional flow. In reality, the interaction between the two waves becomes strong only when θ remains almost constant in time, because, otherwise, the sin-terms in (9) change the signs periodically with time and consequently make weak contributions in the mean values on the development of the wave amplitudes. Thus it is meaningful to pursue the development of a and b under the restriction of resonance. The conditions for resonance are given by

$$\omega_{2_r} = 2\omega_{1_r}, \qquad \theta = -\sigma(k^2) + \pi/2, \tag{10}$$

where $k = b/a$ and $\tan\sigma = (2k^2\sin\sigma_1 + \sin\sigma_2)/(2k^2\cos\sigma_1 - \cos\sigma_2)$, and asymptotic values of k and θ for large a are obtainable from

$$\lim_{a\to\infty} k\frac{db/dt}{da/dt} \equiv k_\infty^2 = -\sin(\theta_\infty - \sigma_2)/\sin(\theta_\infty + \sigma_1),$$
$$\lim_{a\to\infty} \frac{k}{a}\frac{d\theta}{dt} \equiv -\cos(\theta_\infty - \sigma_2) + 2k_\infty^2\cos(\theta_\infty + \sigma_1) = 0. \tag{11}$$

The dotted line PQ in Fig. 4 indicates the combination of α and β satisfying the requirements that ω_{2_r} is equal to $2\omega_{1_r}$ and that ω_{1_i} and ω_{2_i} are both positive. Using these relations and a typical result of numerical computations such as

$$R = 500, \qquad \alpha = 0.3266, \qquad \beta = 0.05808,$$
$$\omega_1 = 0.003137 + 0.006995i, \qquad \omega_2 = 0.006274 + 0.001640i,$$
$$\lambda_1 = -0.0721 - 0.05048i, \qquad \lambda_2 = -0.04084 + 0.5520i, \tag{12}$$

we can outline temporal development of the two waves, as shown in Fig. 5, where the disturbance is assumed to develop from such an initial condition that the amplitude of the second harmonic is much smaller than that of the fundamental.

Summarizing the results obtained from the Stuart-Watson theory and the resonant theory, we may deduce that an surface denoting equilibrium amplitude of the wavy disturbance is defined over the unstable region inside the neutral curve on the (α,β) plane and that a part of this surface, such as that along the segment PQ in Fig. 4, forms a hill one order of magnitude higher than the other part due to the effect of resonance between the fundamental wave and the second harmonic. Such a hill is restricted to rather a narrow range of wavenumbers, so that the corresponding disturbances with large amplitudes may be seen as a nearly monochromatic wave in the physical space, as observed in experiments. The above inference is, however, based on the weakly nonlinear theories, whose validity may be quite questionable for rather large disturbances concerned here. It is therefore highly desirable to solve the original nonlinear disturbance equations directly and numerically, for instance, by the use of iterative methods, and to reveal full shape of the equilibrium surface.

References

1. Gregory, N.; Stuart, J.T.; Walker, W.S.: Phil. Trans. R. Soc. Lond. A.248 (1955) 155-199.

2. Kobayashi, R.; Kohama, Y.; Takamadate, Ch.: Acta Mech. 35 (1980) 71-82.

3. Cebeci, T.; Stewartson, K.: AIAA Journal 18 (1980) 398-405.

4. Stuart, J.T.: J. Fluid Mech. 9 (1960) 353-370.

5. Watson, J.: J. Fluid Mech. 9 (1960) 371-389.

6. Itoh, N.: Proc. IUTAM Symp. on Turbulence and Chaotic Phenomena in Fluids, ed. T. Tatsumi. Amsterdam: North-Holland 1984.

7. Herbert, Th.: AGARD-CP-224 (1977).

8. Craik, A.D.D.: J. Fluid Mech. 50 (1971) 393-413.

9. Craik, A.D.D.: Proc. R. Soc. Lond. A.343 (1975) 351-362.

An Analysis of Wave Interactions in Swept-Wing Flows

HELEN L. REED

Thermosciences Division
Department of Mechanical Engineering
Stanford University
Stanford, CA 94305

Abstract

Crossflow instabilities dominate disturbance growth in the leading-edge region of swept wings. It is well known that streamwise vortices in a boundary layer strongly influence the behavior of other disturbances. Amplification of crossflow vortices near the leading edge produces a residual spanwise nonuniformity in the mid-chord regions where Tollmien-Schlichting (T-S) waves are strongly amplified. Should the T-S wave undero double-exponential growth because of this effect, the usual transition prediction methods would fail. Thus it is important to study interactions of this sort and to develop more realistic criteria for transition prediction.

I. Introduction

One of the major unanswered questions regarding boundary-layer instabilities concerns the interaction of crossflow vortices and T-S waves. Klebanoff et al. [1] show that the onset of three-dimensionality is quickly followed by breakdown of the laminar flows. It is well known that streamwise vortices in a boundary layer strongly influence the behavior of other disturbances. Nayfeh [2] shows that Görtler vortices produce a double-exponential growth of T-S waves. Herbert and Morkovin [3] show that the presence of T-S waves also produces a double-exponential growth of Görtler vortices, while Floryan and Saric [4] show a similar behavior for streamwise vortices interacting with Görtler vortices.

If crossflow vortices and Tollmien-Schlichting waves occur simultaneously, it is possible that premature transition will occur [5]. Vorticity developing in the leading-edge area may linger to the mid-chord region and affect stability there. Amplification of the crossflow vortices near the leading edge produces a residual spanwise nonuniformity in the mid-chord regions, where Tollmien-Schlichting waves are strongly amplified. Should the T-S wave undergo double-exponential growth because of this, the usual transition prediction methods would fail.

Two theoretical approaches are discussed in the next section to model the crossflow/T-S interaction (characteristic of swept wings) as a secondary instability. The first involves the weakly nonlinear combination of crossflow vortices with various oblique-wave solutions to the linear disturbance equations. Craik's resonant triads [6] provide the foundation for this type of approach. The Craik model was considered for the onset of three-dimensionality of T-S waves in boundary layers. These types of models provide only limited understanding to interaction problems, because they specify one wavelength for secondary disturbances for a given frequency, providing a very limited window. The secondary disturbances have been observed experimentally in a broad band of wavelengths [1,7-11]. Moreover, interactions have been observed in flows where a Craik-type mechanism is invalid, e.g., plane Poiseuille flow [12].

The second approach considers the T-S wave growth as a secondary instability in the presence of finite-amplitude crossflow vortices. Herbert [13-17] considers the secondary instability of modulated finite-amplitude, two-dimensional T-S waves with respect to three-dimensional disturbances. For equilibrium plane Poiseuille flow, he solves a Hill-type system of equations with periodic coefficients and finds the different types of three-dimensional instability associated with different resonance conditions. His results (showing frequency and the broad spectrum of possible wavelengths as a function of amplitude and Reynolds number) compare very well with experimental observations.

Boundary layers are more complicated, because they grow and do not exhibit the cross-stream symmetry of Poiseuille flow. Therefore, this analysis does not apply directly without some approximations. Herbert [18] uses a shape assumption in which he neglects the nonlinear distortion of the two-dimensional T-S waves at finite amplitude and also ignores the weak variation of T-S wave amplitude on a viscous time scale. The results of his parametrical excitation are consistent with available experimental data.

II. Description of the Analysis

A. Basic State

The incompressible, spatial stability of the basic state (comprised of a three-dimensional boundary-layer flow over a convex or flat surface with a superposed flow corresponding to steady streamwise vortices rotating in the same sense) is examined.

Initially, we analyze the form of the solutions of the steady vortex flow through solutions of the disturbance equations for the vortices. We confine our analysis to undisturbed flows that are only slightly nonparallel and describe the crossflow disturbance quantities in the form

$$q(x,y,z;\varepsilon, \varepsilon_V) = \varepsilon_V[q_o(x_1,y,z_1)\exp(i\theta) + c.c.] \tag{1}$$

where c.c. stands for complex conjugate and θ is the phase. The terms $x_1 = \varepsilon x = x/R$ and $z_1 = \varepsilon z = z/R$ quantify the nonparallelism of the flow. Here, ε_v describes the magnitude of the crossflow disturbance. The eigenfunctions q_o are normalized so that their maximum value is one half. The quantities α_v and β_v are the complex dimensionless quasi-parallel components of the wavenumber in the chordwise and spanwise directions, respectively. For crossflow vortices, the wavenumber direction ψ is deflected approximately 90° from the potential-flow direction.

B. Crossflow/T-S Interaction Formulation

In this section, two approaches are discussed to model the crossflow/T-S interaction as a secondary instability. The first approach models the T-S wave growth as a weakly nonlinear system. The second approach considers the T-S wave growth as a secondary instability in the presence of finite-amplitude crossflow vortices.

In both approaches, we superpose onto the basic state two oblique, traveling, harmonic Tollmien-Schlichting waves. To this end, we assume the total flow quantities to be

$$\begin{aligned}\hat{q}(x,y,z,t) &= Q_s(x_1,y,z_1) + \varepsilon_v\big(q_o(x_1,y,z_1)\exp(i\theta) + c.c.\big)\\ &\quad + \varepsilon_T\big(\tilde{q}_1(x,y,z,t) + \tilde{q}_2(x,y,z,t)\big)\end{aligned} \tag{2}$$

The small dimensionless quantity ε_T represents the order of the amplitude of the two oblique Tollmien-Schlichting waves. Two oblique T-S waves are considered to achieve a resonant or near-resonant condition. The Tollmien-Schlichting instability is assumed to be a secondary instability so that ε_T is much smaller than ε_V and ε. Mathematical details omitted in this section can be found in Reference 19.

Infinitesimal Crossflow Disturbance

For growing boundary layers with infinitesimal crossflow disturbances, the parameter $\varepsilon_V = O(\varepsilon)$ to account for both the effect of the vortices and the growth of the boundary layer at the same level. The solvability conditions depend on whether

$$\alpha_1 \approx \alpha_v + \alpha_2 \quad \underline{\text{and}} \quad \beta_1 \approx \beta_v + \beta_2 \tag{3}$$

or not. Here, (α_i, β_i) are the (chordwise, spanwise) components of the two T-S waves, respectively. If (3) is satisfied, we introduce detuning parameters and find coupled equations describing the amplitudes $A_1(x_1,z_1,t_1)$ and $A_2(x_1,z_1,t_1)$ of the two T-S waves. The resonance condition produces secular terms and causes the coupling of the amplitudes, leading to possible magnified growth of the T-S waves beyond that predicted by straight linear theory. Considering single-frequency, perfectly tuned disturbances, we find the following relation between A_1 and A_2:

$$A_1 = \hat{A}_1 \exp(S_1 x + S_2 z) , \qquad A_2 = \hat{A}_2 \exp(S_1 x + S_2 z) \tag{4}$$

The quantities S_1 and S_2 are the growth rates associated with nonparallelism and the presence of the vortices and are given in Reference 19. The total growth rates for the two waves are therefore

$$\sigma_i = - \mathrm{Imag}(\alpha_i) - (\omega_\beta/\omega_\alpha)_i \, \mathrm{Imag}(\beta_i) + S_1 + (\omega_\beta/\omega_\alpha)_i \, S_2 \tag{5}$$

where $(\omega_\beta/\omega_\alpha)_i$ are the group velocity ratios of the two T-S waves.

Under these resonance conditions, we shall follow the T-S wave along its trajectory to ascertain its stability characteristics. Since the flow is complicated by the presence of streamwise vortices, marching along the characteristics defined by the group-velocity ratio [20] may not be the only applicable criterion.

Finite-Amplitude Crossflow Disturbance

For finite-amplitude crossflow disturbances present in the growing boundary layer, we now assume that $O(\varepsilon) < O(\varepsilon_V) < O(1)$. Following the work of Herbert [18], the almost-periodic mean flow or basic state describes a three-dimensional growing boundary-layer flow over a wing with embedded streamwise crossflow vortices.

We neglect the nonlinear distortion of the crossflow vortices at finite amplitude, ignore the weak variation of crossflow wave amplitude on a viscous time scale and apply Floquet theory. Under these conditions and considering the condition $\alpha_1 \approx \alpha_v + \alpha_2$ and $\beta_1 \approx \beta_v + \beta_2$ to achieve resonance, we solve the resulting coupled first-order system of disturbance equations describing the two superposed oblique T-S waves. This system is documented in Reference 19.

The spatial solution of this system provides the dispersion relation

$$\omega = \omega(\alpha,\beta,R,\varepsilon_V) \tag{6}$$

Again we shall follow the T-S wave along its trajectory to ascertain its stability characteristics. The amplification factor n is then the integral along the trajectory of the growth rate.

III. Results

At present the interaction codes are being developed using a standard transonic airfoil section, the X-21 wing. This wing has leading and trailing edge sweepback angles of 33.2° and 19.1°, respectively.

A. Crossflow

The crossflow/T-S interaction we examine will occur at the leading-edge negative pressure-gradient region. Amplification of crossflow disturbances there produces a residual spanwise nonuniformity that could linger to regions where T-S waves are strongly amplified.

Two dimensionless parameters commonly used both in experiment and theory to correlate crossflow-vortex appearance with freestream and boundary-layer conditions are the local crossflow Reynolds number and local shape factor. Vortices appear when crossflow Reynolds numbers are on the order of 100 and shape factors are on the order of 0.3; these figures are consistent with our findings for this airfoil. Also, the maximum (in magnitude) crossflow speed is on the order of a few percent of the freestream speed in the leading-edge region.

A spatial stability analysis on this flow yields the following results. Considering the amplitude distributions of two perpendicular disturbance profiles, one in the direction of the wave angle, the other normal to this direction, the larger-amplitude portion (an order of magnitude larger) lies in the direction perpendicular to the wave and is normalized to 1/2 to give proper meaning when an amplitude is introduced.

Parallel growth-rate calculations agree with Padhye and Nayfeh [21]. However, for Padhye and Nayfeh nonparallelism is a stabilizing influence in the leading-edge area. Here we find that nonparallelism is destabilizing.

B. Crossflow T-S Interaction--Infinitesimal Disturbance

An appropriate basic state is formed and the interaction problems now considered. Because the Craik-type approach is expected to provide only limited insight into the interaction problem, only a few results are presented. The model provides a single disturbance wavelength for a given frequency, a very limited window. Moreover, the model depends on knowledge of the initial conditions (that is, the amplitude ratio $\hat{A}_1/\hat{A}_2$; see (4)) of the two interacting T-S waves, as well as the amplitude of the crossflow disturbance. Predicting this ratio for an experiment may be difficult, and this model is therefore impractical.

C. Crossflow/T-S Interaction--Finite Amplitude

For finite-amplitude disturbances, we use a Herbert-type approach to model the interaction. Numerical solutions provide growth rate, wavenumber, and group velocity-ratio distributions as a function of crossflow vortex amplitude. We observe that the presence of crossflow vortices, even with small amplitude, apparently causes magnified growth of the oblique T-S waves. The spanwise nonuniformity modulates the growth rates. In addition, a broad band of wavelengths can be observed as a function of amplitude. This is consistent with experimental observations for other three-dimensional interactions--specifically, the occurrence of three-dimensionality in boundary-layer flows.

Figure 1 shows typical amplification factors for a T-S wave of frequency 325 Hz and an initial crossflow amplitude of 0.6%. Results compare a parallel, zero-amplitude model with a finite-amplitude model. The results clearly show double-exponential growth for the finite-amplitude analysis as compared with the zero-amplitude approach. It is clear that the usual transition prediction methods will fail in such a flow; premature transition will occur.

IV. Concluding Remarks

The analysis addresses the problem of the effects of crossflow on the T-S instability in the leading-edge region of swept wings. Two approaches are discussed, weakly nonlinear theory and parametric resonance. Parametric resonance is the more general and preferred approach, providing a broad spectrum of disturbance wavelengths for a given frequency. Results show double exponential growth of T-S waves in the presence of vortices.

Acknowledgment

The author is grateful to William S. Saric for his helpful comments and suggestions. This work was supported by NASA-Langley Research Center under Grant No. NAG-1-402.

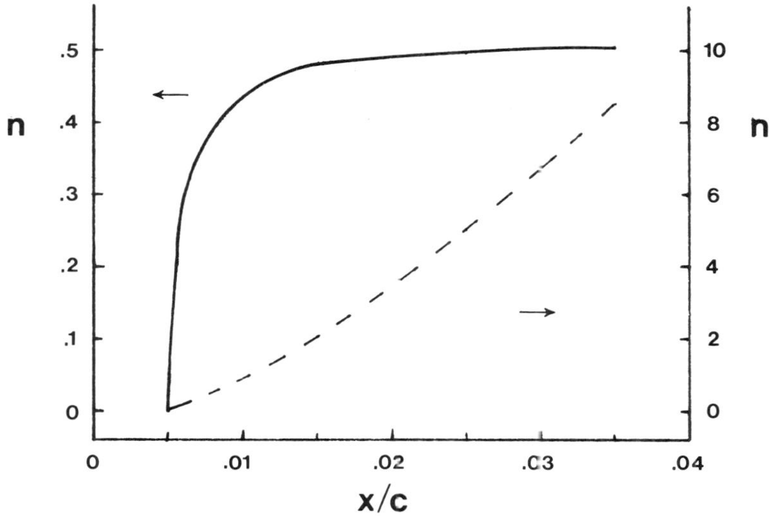

Fig. 1. Comparison of parallel zero-amplitude and finite-amplitude amplification factors for 325 Hz and initial crossflow amplitude 0.6%.

References

1. Klebanoff, P. S.; Tidstrom, K. D.; and Sargent, L. M.: (1962), "The Three-Dimensional Nature of Boundary-Layer Instability," Journal of Fluid Mechanics, Vol. 12, p. 1.

2. Nayfeh, A. H.: (1981), "Effect of Streamwise Vortices on Tollmien-Schlichting Waves," Journal of Fluid Mechanics, Vol. 107, p. 441.

3. Herbert, Th.; and Morkovin, M. V.: (1980), "Dialogue on Bridging Some Gaps in Stability and Transition Research," in Laminar-Turbulent Transition, R. Eppler and H. Fasel (eds.), IUTAM Symposium, Stuttgart, Germany, Sept. 16-22, 1979, Springer-Verlag, New York, p. 37.

4. Floryan, J. M.; and Saric, W. S.: (1980), "Wavelength Selection and Growth of Görtler Vortices," AIAA Paper No. 80-1376.

5. Pfenninger, W.: (1977), "USAF and Navy-Sponsored Northrop LFC Research between 1949 and 1967," AGARD/VKI, Special Course for Drag Reduction.

6. Craik, A.D.D.: (1971), "Nonlinear Resonant Instability in Boundary Layers," J. Fluid Mech., Vol. 50, pp. 393-413.

7. Hama, F. R.: (1960), "Boundary-Layer Transition Induced by a Vibrating Ribbon on a Flat Plate," Proc. 1960 Heat Transfer and Fluid Mechanics Institute, Stanford Univ. Press, 1960, pp. 92-105.

8. Kachanov, Yu. S.; and Levchenko, V. Ya.: (1982), "Resonant Interactions of Disturbances in Transition to Turbulence in a Boundary Layer" (in Russian), Preprint No. 10-82, I.T.A.M., USSR Academy of Sciences, Novosibirsk.

9. Thomas, A. S. W.; and Saric, W. S.: (1981), "Harmonic and Subharmonic Waves during Boundary-Layer Transition," Bull. Amer. Phys. Soc., Vol. 26, p. 1252.

10. Saric, W. S.; and Thomas, A. S. W.: (1983), "Experiments on the Subharmonic Route to Turbulence in Boundary Layers," Proc. IUTAM Symposium on "Turbulence and Chaotic Phenomena in Fluids," Kyoto, Japan.

11. Kozlov, V. V.; Levchenko, V. Ya.; and Saric, W. S.: (1983), "Formation of Three-Dimensional Structures at Transition in a Boundary Layer" (in Russian), Preprint No. 10-83, I.T.A.M., USSR Academy of Sciences, Novosibirsk.

12. Kozlov, V. V.: (1982), communication with Th. Herbert, October, 1982, referred to in Herbert (1984).

13. Herbert, Th.: (1981), "A Secondary Instability Mechanism in Plane Poiseuille Flow," Bull. Amer. Phys. Soc., Vol. 26, p. 1257.

14. Herbert, Th.: (1982), "Onset of Peak-Valley Splitting in Unstable Shear Flows," AIAA Paper No. 82-1013.

15. Herbert, Th.: (1983), "Secondary Instability of Plane Channel Flow to Subharmonic Three-Dimensional Disturbances," Phys. Fluids, Vol. 26, pp. 871-874.

16. Herbert, Th.: (1983), "Subharmonic Three-Dimensional Disturbances in Unstable Shear Flows," AIAA Paper No. 83-1759.

17. Herbert, Th.: (1983), "Modes of Secondary Instability in Plane Poiseuille Flow," Proc. IUTAM Symposium on "Turbulence and Chaotic Phenomena in Fluids," Kyoto, Japan.

18. Herbert, Th.: (1984), "Analysis of the Subharmonic Route to Transition in Boundary Layers," AIAA Paper No. 84-0009.

19. Reed, H. L.: " Wave Interactions in Swept-Wing Flows," AIAA Paper No. 84-1678.

20. Reed, H. L.; and Nayfeh, A. H.: (1982), "Stability of Compressible Three-Dimensional Boundary-Layer Flows," AIAA Paper No. 82-1009.

21. Padhye, A. R.; and Nayfeh, A. H.: (1981), "Nonparallel Stability of Three-Dimensional Flows," AIAA Paper No. 81-1281.

On Some Transition Processes in Three-Dimensional Shear Flows

S.Y.GERTSENSTEIN, V.A.ZHELIGOWSKY, N.V.NIKITIN, A.Y.RUDNITSKY, A.N.SUKHORUKOV, L.V.FILJAND

Institute of Mechanics, Moscow State University, USSR

A non-linear evolution of two- and three-dimensional finite-amplitude disturbances developing in the transient laminar-turbulent zone of several viscous and inviscid flows is considered.

1. Full system of Euler equations is used to study stability of inviscid flows. The solution is found by combining Fourier method and Rayleigh method, the latter being generalized for non-linear three-dimensional problems [I]. The real stability-questioned velocity profile is substituted by a broken line. This procedure enables one to write down analytical solution of initial equation system for each segment of broken line. Then the matching conditions are applied to the breaking points [I] . The stability of mixing layers, plane and axisymmetrical jet flows is considered. Also computer modelling of acoustic influence on these flows is carried out.

1.1. Preliminary investigation of the linear stability was carried out for a wide class of plane and axysymmetrical jet flows [I] . In particular, it is shown that the characteristic wave number of the most unstable disturbance depends, mainly, on the mixing layer width on the boundary of the jet flow: while the parameter δ (δ is the ratio of width of central part of the jet Δ_1 to Δ_2) varies from 0 to ∞; non-dimensional wavenumber $\alpha_0 \Delta_2$ of the most "dangerous" antisymmetrical [I] disturbances declines no more than 20% from its limit $\alpha_{00} \Delta_2$($\delta \to \infty$); for the symmetrical disturbances (δ varies from 0.4 to ∞) $\alpha_0 \Delta_2$ declines from $\alpha_{00} \Delta_2$ no more than 15%. For sufficiently "shear" jet flows ($\delta \geq$ 2) characteristic wavenumbers almost do not vary. The phase velocities of the disturbances also vary inessentially, δ being equal to 2. It is interesting that phase velocity of the large scales **antisymmetrical** disturbances nearly equals to the velocity of the flow on its periphery, this parameter tending to the centerline flow velocity for the symmetrical disturbances.

1.2. One of the principal results of the non-linear stability investigations is that strong growth of low-frequency components of the stectrum is found out along with their strong influence on the averaged and pulsed characteristics of the considered flows. For example, if low-frequency components in the investigation of the mixing layer stability are taken into consideration their influence may result in the increasing of $\Delta V(y,t)$ to the averaged longitudinal velocity component by a factor of $1.5 \div 2$. Note that the process of "smoothing out" of averaged velocity profile occurs non-monotonously as well as energy transfer from the averaged flow to the disturbances: at some stages of non-linear evolution one can observe backward energy transfer from the disturbances to the averaged flow. In addition to this, on insertion of low-frequency disturbances the characteristic amplitude of the main tone decreases, these disturbances "drawing up" the most essential part of energy from the mean flow, but their interaction with the mean flow occurs, to a great extent, via main tone: immediate development of the low-frequency disturbances only happens to be less intense.

In the mixing layer strong growth of "subharmonic" is obtained on the background of low-frequency components, the amplitude of "subharmonic" being in some cases of order greater than amplitude of the background. Note that this is not the case for jet flows. Moreover, subharmonic amplitude may also exceed that of the main tone. In such cases pairing of the eddies occurs, which leads to the growth of the characteristic scale by 2 times. Similar process is also observed in the experiments for both low and high Reynolds numbers. The range of high Reynolds numbers corresponds to the turbulent regime which is characterized by the existence of the coherent structures [2].

Note that intense development of the low-frequency components in the mixing layer may be explained by simple "triple" resonance effects $\alpha_1+\alpha_2 = \alpha_3$, since the corresponding relation for frequencies is satisfied automatically ($\omega_K(\alpha_K)=0$). The selection of subharmonics may be explained due to resonance of lower order: $\alpha_0-(\alpha_0/2) = (\alpha_0/2)$.

Nonlinear interaction of the main tone (initial apmlitude being equal to $10^{-3} \div 10^{-4}$) with high-frequency components affects usually weakly flow characteristics. For example, in the mixing layer high-frequency disturbances with initial amplitudes of the order of 10^{-3} do not result in the essential change of the mean tone (whose initial amplitude equals 10^{-4}) and averaged flow characteristics. It is worth to note that for these time diapasons the amplitude of the high harmonics of the main tone reaches also the values of order of 10^{-3}. More intense high frequency disturbances with amplitudes of the order of $10^{-2} \div 10^{-1}$ may lead to the inverse transfer of energy from high-frequency disturbances to the mean flow. Nonlinear interaction of two plane waves of the middle part of the spectrum is accompanied by the mutual weakening, but if their initial amplitudes have different values the development of one with greater amplitude occurs a great deal slower.

Note that the mechanics of non-linear interaction of disturbances in the mixing layer have sufficiently general character and therefore they may be used for a number of other jet flows which consist of several mixing layers. For example, for the jet with narrow mixing layers on its boundaries the development of the instability in one of the layers do not really depend on the other layer. It is essentially true for the symmetrical disturbances which are characterized by the fact that the average velocity on the jet axis doesn't depend on time.

The stabilizing effect of "selected" high-frequency pulsations in the process of acoustic irradiation of the jet flows may be explained by the intense development of these pulsations in the narrow mixing layers on the jet boundaries along with expansion of these layers and less intense growth of the main tone on the background of such a smoothed velocity profile. The results of the calculations are that characteristic time of development of the main tone increases nearly 2 times in the case of high-frequency pulsations; the final amplitude of the main tone varies very little. The **obtained** results are in good agreement with experimental data [3].

It is interesting to compare final-amplitude antisymmetrical and symmetrical disturbances. From linear theory of stability it follows that the increment of antisymmetrical disturbances (with respect to longitudinal component of velocity disturbance) is greater than that of the symmetrical disturbances.

Non-linear calculations enable one to conclude that characteristic finite-amplitude values of antisymmetrical disturbances are also several times greater than similar parameters of symmetrical disturbances. It is curious that high-frequency disturbances representing the model of acoustic irradiation **suppress** antisymmetrical main tone more effectively than symmetrical main tone. On the other hand, the main mechanisms of non-linear interaction of symmetrical and antisymmetrical disturbances differ essentially.

1.3. It is of great interest to find out a nature of non-linear evolution of three-dimensional structures on the considered flows.

The results of calculations demonstrate the instability of plane wave regimes with respect to transverse three-dimensional disturbances.

In these calculations the subharmonic triple resonance $(\frac{\alpha}{2},-\beta)$ $(\frac{\alpha}{2},\beta)$ as well as $(\alpha,0)$ resonances for the disturbances (α,β) with $(\alpha,-\beta)$ second harmonic of the main tone $(2\alpha,\beta)$ and with standing transverse wave are taken into consideration along with resonances of higher orders. Initially all the disturbances grow according to linear theory but then "subharmonical" disturbance with wave vector begins **intensively to select out of background of fully developed** two-dimensional wave with amplitude of order of 10^{-1}. During this time interval three-dimensional waves grow most strongly. Subsequently other three-dimensional components begin to select along with "subharmonic": (α, β), $(\alpha, -\beta)$ $(0, \beta)$, $(0, 2\beta)$ etc. It is important that the change of "subharmonical" resonance into "primary" one gives rise to the qualitative reorganization of wave pattern of the flow [8-10]. If "primary" resonance takes place all the two-dimensional Tollmien-Schlichting waves bend absolutely identically but in case of "subharmonical" resonance this "likeness" is observed for every second wave.

It is interesting that the increase in the amplitude of the main wave is accompanied by the weakening of both two- and three-dimensional secondary disturbances, one being able to select at definite stage of non-linear evolution some wavenumber which is characterized by uppermost increase (for example, $\beta \approx 1$ for mixing layer). It is important that for free jet flows considered maximum increments of secondary three-dimensional disturbances happen to be of nearly the same order as the usual "primary" increments of two-dimensional disturbances unlike the case of boundary layer flows. Also the times of evolution of main two-dimensional wave happen to be almost the same as those of three-dimensional disturbances. Note that for sufficiently large times of non-linear evolution the harmonic $(0, 2\beta)$ reaches substantial value. Therefore the resonance of inclined waves with standing transverse wave exhibits itself explicitely. In this case the wavenumber is arbitrary. In other words, the arbitrary transverse scale may appear. At large times the three-dimensional disturbances with different transverse scales defined by wavenumbers β increase for the wide range of these scales nearly up to the same value of order of 10^{-1}. Thus the observed scale of transverse instability β is determined largely by initial value. In particular, the selection of "subharmonical" resonant is not observed. The specific selection of one or the other three-dimensional scale is dependent to a great extent on the initial values of a problem. Not only the spectrum of initial disturbances may play role but also their relative levels. It is interesting that longitudinal component of velocity disturbance depends on the transverse co-ordinate in the two-dimensional case much in the same way as in three-dimensional case. The evolution of three-dimensional disturbances leads to some three-dimensional deformation of initially two-dimensional waves [8-10] and doesn't lead to formation of eddy structures.

2.1. Navier-Stokes equations are used to investigate numerically non-linear stability of plane Poiseuille flow, axisymmetric Hagen-Poiseuille flow as well as viscid pipe flow, the pipe rotating about its axis. The solutions of the problem are found by Bubnov-Galerkin method. Namely, the transverse co-ordinate dependence of functions in question is presented by series of full system of linearized eigenvalue problem functions for stationary fluid. These functions are [6] for plane channel or [7] for round pipe. As for other co-ordinates, the solutions are Fourier series. A number of methodical investigations was carried out. In particular, it is shown the sufficiency of taking into consideration of comparatively small number of basic functions: 30-40 functions happened to be enough for transverse co-ordinate while only 1-2 functions for each of two other co-ordinates. For example in decreasing Reynolds number in case of a rotating pipe required number of transverse co-ordinate functions decreases strongly to only 10 functions if Reynolds number is about 10^2.

The numerical investigations of non-linear stability of plane Poiseuille flow show, in particular, the independence of self-oscillatory regimes on initial value of the problem (providing, that they are intense enough). Also the results of calculations define "threshold" amplitude values after which the solution "arrives at" self-oscillatory regime. Besides the characteristic time intervals of non-linear evolution of disturbances are obtained.

Good quantitative and qualitative agreement between theoretical and experimental values of mean and pulsed characteristics is achieved [4] - [5].

2.2. The investigation of finite-amplitude stability of Hagen-Poiseuille flow in round pipe is carried out with respect to 3-D disturbances. In doing this, non-linear interaction of

several wave components of the following form: $\varphi_{n,m}(r,t)\exp[i(n\alpha x+m\theta)]$ was considered.
The non-linear resonant interaction of three waves $\varphi_{1,-1}\exp i(\alpha x \pm \theta)$ $\varphi_{02}\exp[2i\theta]$ happened to be most essential.
The numerical calculations show "hard" instability of Hagen-Poiseuille flow to considered 3-D disturbances. It is shown also that formation of self-oscillating motions is accompanied by strong increase in drag coefficient. It is essential that at the initial stage of transient processes the shear deformation of mean and pulsed characteristics is observed (turbulent "bluffs"). Note that times of these "bluffs" differ nearly three orders from characteristic time that is required for self-oscillatory regimes to establish, the latter time being nearly equal to Reynolds number. When finite-amplitude instability is observed minimum "expenditure" Reynolds numbers happen to equal nearly 2100 ; experimental values of Re [8], vary in the range of 2000-3000. The range of possible self-oscillatory regimes is obtained and also well known experimental dependence of averaged drag coefficient on Reynolds number for turbulent round pipe flow [8] is reproduced.

2.3. In investigating the stability of flow in round pipe rotating about its axis non-linear evolution and formation of self-oscillatory regime in the form of single moving spiral wave are obtained. Non-linear development of such a wave is accompanied by decreasing of both mean, fluid expenditure and mean moment of momentum relative to the pipe axis. The times of establishing of self-oscillatory regimes as well as in the case of rotating pipe [4] happen to be nearly equal to corresponding Reynolds numbers. The energy distribution of harmonics of main wave have exponential character. The "surviving" curve is obtained that enables one to determine "favourable" disturbances selected from the other disturbances providing that the initial amplitudes are all equal. Besides with the aid of the "surviving" curve it is possible to determine critical values of amplitudes beginning with which the "favourable" disturbance may "lose" to the other disturbances. Note that range of these "surviving" disturbances is wide enough but limited, that is the characteristic wave lengths vary from 3 to 8 pipe diameters. It is obtained that instability of finite-amplitude self-oscillatory motion relative to disturbances of small amplitude occurs initially linearly, corresponding increments being nearly of order less than it has to be according to conventional linear theory. When secondary disturbances reach the amplitude about 10% of amplitude of self-oscillatory regime abrupt "explosion-like" change of motion structure occurs.

2.4. Also the modelling of transient processes in **Blasius** flow is carried out. The problem is considered using generalized Rayleigh method. It is shown that strong increase of 3-D disturbances along with formation of narrow mixing layer in instantenous longitudinal component of velocity profiles are obtained on the background of two-dimensional finite-amplitude disturbance. As a result of these processes mean flow characteristics vary strongly. At the initial stage of non-linear evolution like in the case of plane **Poiseuille** flow selection of subharmonical 3-D components with wave vectors of the form ($\alpha/2$, β) is observed if two-dimensional wave has small amplitude. At subsequent moments "primary" 3-D resonance with wave vectors (α, β) grows strongly. The obtained results are in good qualitative agreement with experimental data.

References

1. Betchow,R.; Criminale,W.: Stability of Parallel Flows. Academic Press New York. London 1967.

2. Roshko: A.I.A.A. J., v.14, N 10, 1976, 8-20.

3. Ginevski,A.S.; Vlasov,E.V.; Colesnikov,A.V.: Aeroacustic interaction, Moscow, 1978, 177.

4. Orszag,S.; Patera,A.: J.Fl.Mech., v.128, pp.347-385, 1983.

5. Koslov,V.V.; Rahmahsanov,M.P.: D.Acad.Sci., USSR, 1984, v.275, v.6, p.1346-1348.

6. Petrov,G.J.: PMM, v.4, N 3, 1940.

7. Cotton,F.W.; Salwen,H.: J.Fl.Mech., v.108, p.101-125, 1981.

8. Wygnanski,I.J.; Champagne,F.H.: J.Fl.Mech., (1973), v.59, p.2.

9. Kachanov, Ju.S.; Koslov,V.V.; Levchenko,V.Ja.: Mech. Fluid and Gas, D.Ac.Sci., USSR, N 3, 1977.

Stability of thin Viscous Shock Layer on a Wedge in Hypersonic Flow of a Perfect Gas

G. V. PETROV

Institute of Theoretical and Applied Mechanics
USSR Academy of Sciences, Novosibirsk, USSR

Summary

In hypersonic flows the boundary layer thickness can be not small compared to the shock layer thickness, disturbances originated inside of the boundary layer, therefore, can not attenuate over the inviscid part of the shock layer so much that their reflection from the leading-edge shock wave can be left out of account. In this work the condition of propagation only inside of the shock layer, and linearized Rankine-Hugoniot conditions at the shock wave are imposed on disturbances or their inviscid components. The second mode instability for a flat plate takes as a starting point of the numerical study. There were discovered a series of branch and saddle points of the dispersion relation as well as absolute instability.

Formulation of the Problem

The basic flow is that over a wedge in the uniform supersonic stream of a perfect gas. It is assumed that the shock wave is flat, $M_1 \rightarrow \infty$ and $\gamma - 1 \ll 1$ where M_1 is a free stream Mach number, γ is a specific heats ratio. Then the angle β of the shock wave with respect to the wedge surface is

$$\beta = ((\gamma - 1)/2\gamma)^{1/2} / M_2 \ll 1 \qquad (1)$$

where the subscript $_2$ is referred to the conditions behind the shock wave; i.e. the shock wave is almost parallel to the wall. This result is valid not only for the wedge [1], and can be taken as a basis for building up a parallel stability theory of viscous or inviscid thin shock layers.

The wall temperature is assumed to be constant, and the boundary layer is calculated as it was done on a flat plate. The similarity scale δ and Reynolds number R are determined by

$R = u_2\delta/\nu_2 = (u_2x/\nu_2)^{1/2}$ where x is a distance from the leading edge along the wedge surface, u is velocity in the x-direction, ν is kinematic viscosity. The estimate for the shock layer thickness $\Delta = \beta x$ gives

$$R = (2\gamma/(\gamma - 1))^{1/2} M_2 Y \qquad (2)$$

where the nondimensional shock layer thickness Y is defined by

$$Y = \Delta/\delta = \Delta(u_2/\nu_2 x)^{1/2}.$$

If inviscid disturbances are absent in a free stream, the linearized Rankine-Hugoniot conditions give the relation between disturbances of pressure p_2^* and normal velocity v_2^* just behind the shock: $p_2^* = -2\rho_2 v_2 v_2^*$ as $y = \Delta$. For the disturbances of the form $f^*(x,y,t) = f(y)\cdot\exp(i(\alpha x - \omega t))$ it is transformed into equation for amplitude functions π and $\alpha\varphi$ of disturbances of pressure and normal velocity component

$$\pi = (2\gamma(\gamma - 1))^{1/2} M_2\alpha\varphi \text{ at } y = Y, \qquad (3)$$

dimensionless values based on δ, u_2, p_2, ρ_2, T_2 are used here and below.

The Case $\gamma = 1$

This case is very simple to study. Indeed, when M_2 and Y are finite at $\gamma \to 1$, it follows from Eqs.(1) and (2) that $\beta \to 0$ and $R \to \infty$. The first limit means that both the shock wave and the streamlines in the shock layer are parallel to the wall. The second limit means that disturbances are purely inviscid. Thus, the equations of the inviscid theory of stability of parallel compressible flows [2] are applicable except for the condition at infinity that must be substituted by Eq.(3) at $\gamma = 1$: $\pi = 0$ as $y = Y$. When the surface temperature is the same as that behind the shock ($T_w = 1$), the boundary layer is isothermal and the velocity profile is a Blasius one.

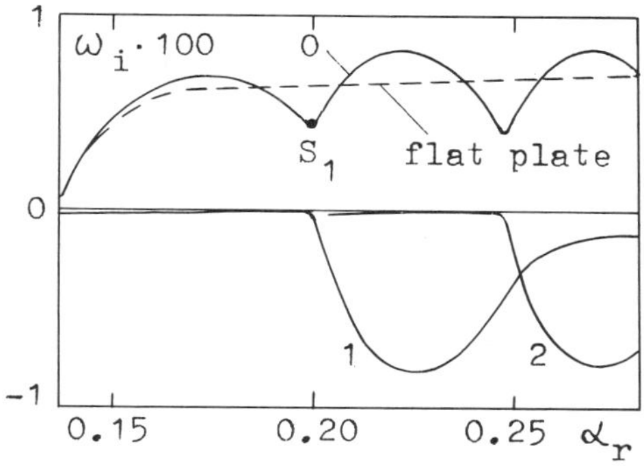

Fig.1. Temporal amplification rate ω_i vs wave number α_r.

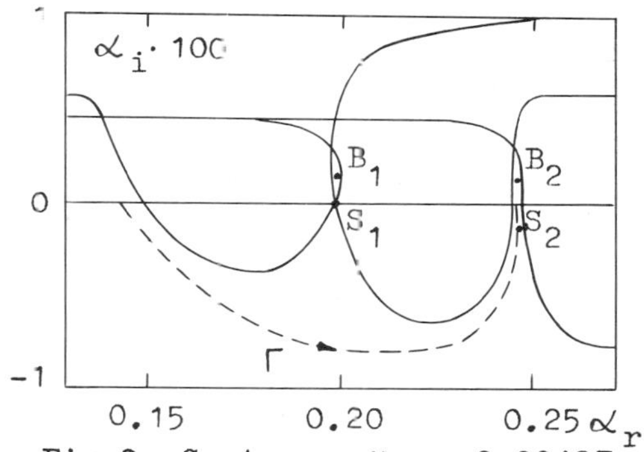

Fig.2. Contours $\omega_i = 0.00437$ in complex α-plane.

$\gamma = 1$, $M_2 = 10$, $T_w = 1$, $Y = 14.68$.

All the results are calculated here for two-dimensional disturbances, and are related to the second mode amplified solutions [2] for a flat plate. The dependence of the temporal amplification rate ω_i on the wave number α_r (the subscripts $_i$ or $_r$ are referred to imaginary or real parts of certain values) for $\gamma = 1$, $M_2 = 10$, $T_w = 1$, and $Y = 14.68$ are given in Fig. 1. Line 0 shows that the dependence $\omega_i(\alpha_r)$ is not monotonous in contrast to that for a flat plate ($1/Y = 0$). The same calculations indicate the loss of monotonicity of $\omega_r(\alpha_r)$ also. Moreover, the first minimum of $\omega_i(\alpha_r)$ coincides with the first maximum of $\omega_r(\alpha_r)$, that is, the saddle point $d\omega/d\alpha = 0$ of the complex dispersion relation $\omega(\alpha)$ is to be found.

Fig. 2 presents the level lines $\omega_i = 0.00437$ on a complex α-plane. This level corresponds to the saddle point S_1, the other intersection points are not saddle ones and indicate the branching of the dispersion relation. The comparison of Figs. 1 and 2 shows that the branch points B_n are located in the neighbourhood of every minimum defined by line 0 on axis α_r in Fig. 1, so that lines 1 and 2 appear. These new branches are proved to be damped and supersonic. The further calculations show that the smaller is Y, the closer are B_n to axis α_r, the deeper are the holes on the lines in Fig. 1, and the greater are distances between them. The new damped supersonic solutions form, no doubt, an infinite discrete spectrum that transforms into a continuous acoustic spectrum at $Y \to \infty$.

Finite R and $\gamma \neq 1$

In this case the shock wave as well as streamlines are not parallel to the wall, however, the parallel approach is used here as an approximation. The basic equations used are different from those of Dunn and Lin [3] by the absence of the asymptotically small term $\mu \partial^2 v^* / \partial y^2$ in the momentum equation. For the disturbances of the elementary wave type the basic equations have the form

$$i(u-c)r + \rho'\varphi + \rho(if+\varphi') = 0,$$

$$\rho(i(u-c)f + u'\varphi) = -i\pi/\gamma M_2^2 + (\mu/\alpha R)f'',$$

$$\rho\alpha^2(u-c) = i\pi'/\gamma M_2^2, \tag{4}$$

$$\rho(i(u-c)\vartheta + T'\varphi) = (1/\gamma)(\gamma-1)i(u-c)\pi + (\mu/\alpha R \Pr)\vartheta'';$$

$$\pi = Tr + \vartheta/T, \qquad c = \omega/\alpha,$$

where μ is a dynamic viscosity, r, ϑ, and f are amplitude functions of disturbances of ρ, T, and u, the stroke is an y derivative.

Boundary conditions are determined by solutions of Eqs.(4) in the inviscid part of the shock layer: $(f, f', \varphi, \pi, \vartheta, \vartheta') = C\cdot\exp(\sigma y)$. If $\alpha/iR(1-c) \longrightarrow 0$, the characteristic values σ and characteristic vectors C are the same as those for the Dunn and Lin equations:

$$\sigma_1 = -(i\alpha R(1-c))^{1/2}, \qquad C_1 = (1, \sigma_1, 0, 0, 0, 0),$$

$$\sigma_2 = -(i\alpha R \Pr(1-c))^{1/2}, \qquad C_2 = (0, 0, 0, 0, 1, \sigma_2),$$

$$\sigma_{3,4} = \pm(\alpha^2(1+(1-c)m))^{1/2}, \qquad m = -M_2^2(1-c),$$

$$C_n = (1, \sigma_n, \sigma_n/i\alpha^2, \gamma m, (\gamma-1)m, (\gamma-1)m\sigma_n), \quad n > 2. \tag{5}$$

The **viscous** solutions increasing at $y \to \infty$ are left out of account right from the start because of their large increment. The vectors $C_{3,4}$ form a linear combination C_5 which satisfy Eq.(3). Vector C_5 is also determined by Eq.(5) at

$$\sigma_5 = -(\gamma/2(\gamma-1))^{1/2}\alpha\ M_2\ i(1-c).$$

Simultaneous numerical integration of Eqs.(4) across the shock layer with the vectors C_1, C_2, and C_5 as conditions at $y = Y$ gives three fundamental solutions necessary to solve the eigenvalue problem.

All the results given below are calculated for $\gamma = 1.15$, $M_2=5$, $T_w=0.2$, $\mu=T$, Pr=1. The amplification rate given in Fig. 3 as

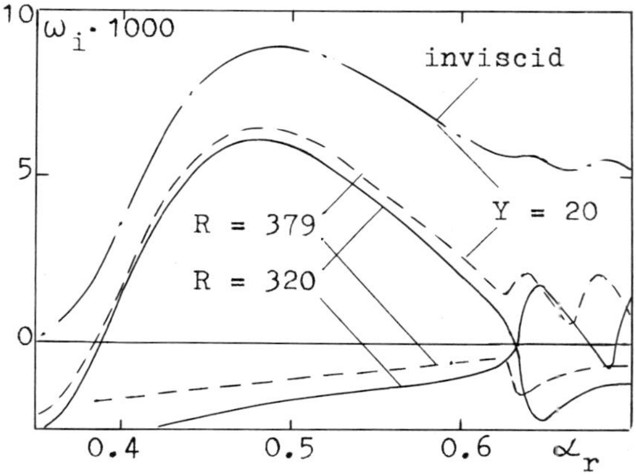

Fig.3. Temporal amplification rate ω_i vs wave number α_r.

$\gamma = 1.15$, $M_2 = 5$, $T_w = 0.2$, $\mu = T$, Pr = 1.

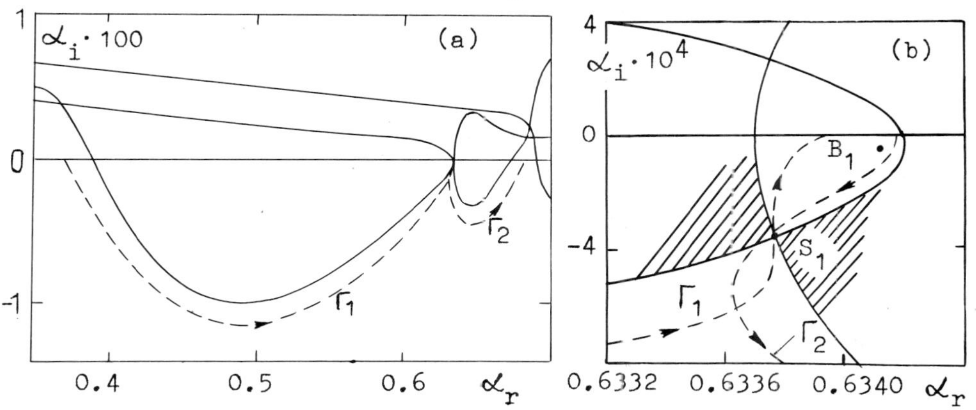

Fig.4. Contours ω_i=0.00009 in complex α-plane for R=320 (Y=16.9).

function of the wave number for Y = 20, or in accordance with Eq.(2) R = 379, is qualitatively the same as above. The comparison with the inviscid distribution shows not only the usual stabilizing effect of viscosity, but also that the minima of $\omega_i(\alpha_r)$ for viscous disturbances are more expressive than for inviscid ones, the branch points being closer to the real axis α_r in the complex α-plane. The other results given in Fig. 3 as well as in Fig. 4 are calculated for R = 320 (Y = = 16.9), the contours of ω_i = 0.00009 correspond to the first saddle point (see Fig. 4b). It is important in this case that the first branch point is located under axis α_r, the transitions from one branch to another as α_r = 0.634 are to be seen in Fig. 3.

The solution of the linear stability problem in terms of superposition of the elementary waves

$$F(x,y,t) = \int_{-\infty}^{\infty} G(\alpha)\, f(\alpha,y)\, \exp(i(\alpha x-\omega(\alpha)t))d\alpha \qquad (6)$$

means an x-localized disturbance. The asymptotic formula [4] for such an integral as $t \to \infty$ is

$$F(x,y,t) \sim (2\pi/i\omega_s''t)^{1/2}G(\alpha_s)f(\alpha_s,y)\, \exp(i(\alpha_s x-\omega_s t+a)),(7)$$

where subscript $_s$ is referred to a saddle point. A saddle point, however, will contribute to (7) in the asymptotics only if there is a path of integration equivalent to axis α_r and lying entirely in the two valleys of this point. Fig. 2 shows that there exists an equivalent contour, for example, Γ, lying entirely in the lower valley, so that Eq.(7) is not valid here. Fig. 4 shows that in this case paths Γ_1, Γ_2, which go from valley to valley and don't go around B_1 exist due to the location of the branch point under axis α_r.

Since $\omega_{si} > 0$, Eq.(7) shows unbounded amplifying of spatially localized disturbances (6) as $t \to \infty$ at any point of the shock layer. This absolute instability takes place in Reynolds numver range from 317.5 to 321, for the smaller R it does not exist because of $\omega_{si} < 0$. The flow parameter ranges of interest are narrow and do not include small $\gamma - 1$ and large R, ω_{si} is small, therefore, to check the existence of absolute instability of the shock layer on a wedge more accurate theoretical models are desirable.

References

1. Hayes, W.D.; Probstein, R.F.: Hypersonic flow theory. Second ed., vol.1, Inviscid flow. N.-Y, London: Academic press 1966.

2. Mack, L.M.: Boundary layer stability theory. Doc. 900-277, Rev. A. Pasadena: JPL 1969.

3. Dunn, D.W.; Lin, C.C.: On the stability of the laminar boundary layer in a compressible fluid. J. Aeron. Sci. 22, N 7 (1969).

4. De Bruijn, N.G.: Asymptotic methods in analysis. Amsterdam, Groningen 1958.

The Effect of Cooling on the Supersonic Boundary Layer Stability and Transition

V. I. LYSENKO, A. A. MASLOV

Institute of Theoretical and Applied Mechanics
USSR Academy of Sciences,
Novosibirsk, USSR

Summary

The present paper describes the carried out experimental studies of the laminar boundary layer stability on a flat plate and the transition into a turbulent state both by cooling of a streamlined surface and by its heating. There are confirmed the main theses of the linear theory of hydrodynamic stability. An explanation of the transition reversal appearing in tests in the wind tunnels on cooled models is given.

Experimental equipment

The experiments were performed in the supersonic wind tunnel T-325 with a test section 200 x 200 mm. All measurements were carried out in the boundary layer of flat plates placed under a zero angle of attack. The plate surfaces were cooled with premixed liquid and gaseous nitrogen or heated with the help of an electrical heater. The model surface temperature was varied within the range of $70°K < T < 450°K$.

The study of stability

It has been discovered (see, f.e. [1]) by the theoretical study of the wave processes in supersonic flows that three main types of oscillations may exist in supersonic boundary layers: the first two types are the damping in infinity (in the coordinate normal to a streamlined body surface) vortex and acoustic eigenoscillations of the boundary layer – the first- and second-mode waves, the third type is the forced acoustic disturbances. In the present work the influence of heat-transfer on these three types of oscillations is investigated. The study of stability was performed by using the hot-wire anemometer method [2] . The length of tungsten hot-wire was approximately

1.5 mm, the diameter was about 6 μkm. There was examined the development of "natural" disturbances in the boundary layer of a flat plate with a sharp leading edge ($b \leqslant 0.02$ mm, b - radius of bluntness). All measurements were performed in a layer with maximum fluctuations of the mass flow. The sound emitted by the turbulent boundary layer of a nozzle and the wind tunnel test section walls was a source of the forced oscillations. The forced disturbances caused the eigenoscillations of the boundary layer. The experiment conditions were such that the frequencies ranges (or Reynolds numbers) of the most intensive amplification of each of three types of oscillations were essentially different. It enabled us to follow separately the influence of the heat-transfer on each type of disturbances.

Fig.1 represents a typical (from the obtained in the present study dependences) example of the disturbance development in the boundary layer with fixed dimensionless frequency $F=0.052\cdot10^{-4}$ for M=2.0, $Re_1=26\cdot10^6\ m^{-1}$. Here m_f are the mean-square fluctuations of the mass flow, x is the longitudinal coordinate, $Re=U/\nu$ is the unit Reynolds number; $F=2\pi\cdot f/Re_1U$; U and ν are velocity and viscosity at the boundary layer edge, respectively, f is the frequency, M is the Mach number; T_w is the temperature factor (wall-to-adiabatic wall temperature ratio). The maximum of the dependence $m_f = \Phi(x)$ at x=30 mm is conditioned by forced disturbances (the third type of oscillations) [3] . As can be seen from the figure the decrease in the temperature factor does not influence in fact this type of oscillations, that is in conformity with numerical results of [4] . For the given values of the frequency parameter and Reynolds numbers each of the dependences minima plotted in Fig.1 corresponds to the beginning of growth of a total disturbances signal of the first-mode waves and of the forced acoustic oscillations. Their position depends on the distribution of the forced oscillations and first mode waves as well as on their amplitude ratio. With decreasing the temperature factor the minimum shifts towards a region of large Reynolds numbers, i.e. the boundary layer becomes more stable. The measurements have shown that the neutral stability curves of the first-mode waves move monotonously towards a region of large Reynolds numbers by cooling the model surface, thus decreasing a range of unstable frequencies and increasing the value of the critical Reynolds number of the stability loss. On the contrary, the heating of the model has an opposite effect. A drawn conclusion confirms qualitatively the results of numerous computation of the first-mode neutral stability curves [1,4-6].

A stabilizing effect of cooling and destabilizing effect of heating with respect to the first-mode waves can be seen from Figs.2 and 3 (here M=2.0, Re=700, $Re_1=5\cdot10^6\ m^{-1}$, Re= $\sqrt{Re_1 x}$). The decrease in the temperature factor leads to a monotone decrease in both the disturbance amplification coefficients ($-\alpha_i=0.5\ d(\ln m_f)/d\ Re$) and the unstable frequencies range ($-\alpha_i > 0$). At $T_w=0.67$ and Re ~ 700 the first mode disturbances damp for all the values of the frequency parameter.

Results plotted in Fig.4 (M=3.0, $Re_1=5\cdot10^7\ m^{-1}$, Re=1360) are of special interest owing to the fact that with decreasing T_w the curve $-\alpha_i=\Phi(F)$ is deformed in such a way that at $T_w=0.83$ for the first mode disturbances ($F > 0.03\cdot10^{-4}$) $\alpha_i=0$ at four values of F. Such a behaviour of the disturbances amplification coefficients corresponds to a partition of the instability region into two by cooling the model at M=3.0. It is seen that the second region (at large F) is stabilized with decreasing the surface temperature more faster than the first one. A possibility of such a behaviour of the first-mode waves on the cooled surfaces at M=3.0 has been theoretically predicted in [6] . A range of frequencies with $F < 0.03\cdot10^{-4}$ in Fig.4 corresponds to an intensive sound - boundary layer interaction. A before drawn conclusion about a weak influence of cooling on this type of oscillations is confirmed.

Experiments carried out at M=4 and low frequencies (f < 50 kHz) yielded the results similar to those plotted in Figs.1 and 2. Moreover, in this series of experiments we succeeded in performing the measurements of high-frequency oscillations (f ~ 100 ÷ 200 kHz) which correspond to the second-mode waves. The results are represented in Fig.5 (M=4.0, $Re_1=6\cdot10^6\ m^{-1}$,

Re=780). It is seen that the cooling exerts a weak destabilizing influence on this type of oscillations: the disturbance amplification coefficients increase, a range of unstable frequencies grows and moves towards a region of large values of F. The obtained result is in conformity with conclusions drawn in [5, 7].

Thus, concluding all the above-said we can state that the cooling of the model surface: 1. stabilizes the first-mode waves; 2. destabilizes the second one; 3. may lead to the effect that a region of the first-mode unstable frequencies is divided into two; 4. does not affect the acoustic waves - supersonic boundary layer interaction. More detailed results on the effect of cooling on the stability will be published in [11].

<u>The study of transition</u>

Experimental study of the heating and cooling influence on the laminar boundary layer transition into a turbulent one have been carried out on a model of a flat plate with a leading edge blunted up to the radius b=0.1 mm. In the first series of experiments the influence of the Mach number and the unit Reynolds number on the transition location by changing the surface temperature was under study. The measurement were carried out by using a known method by the pitot probe.

Fig.6 shows an example of changing the transition beginning Reynolds numbers Re* depending on the temperature factor (here $Re^* = Re_1 x^*$, x* is the coordinate of the minimum pressure from pitot probe). The measurements were performed at $Re_1 = 25 \cdot 10^6\ m^{-1}$. With decreasing the temperature factor the transition Reynolds numbers increase monotonically. With increasing the Mach number the influence of the temperature factor on the transition becomes weaker. The measurements carried out for different values of Re_1 have shown that with increasing the unit Reynolds number and at the fixed value of T_w the transition Reynolds numbers grow if the model is heated. There appears a maximum in the dependence $Re^* = \Phi(Re_1)$ for a cooled model, and for the low values of T_w the transition Reynolds numbers decrease with increasing Re_1. Such a behaviour of transition Reynolds numbers depending on Re_1 is in full agreement with the change of intensity and frequency spectrum of disturbances in the T-325 wind tunnel test section (see at greater length [8, 9]).

All the observable in the present study monotone dependences of transition Reynolds numbers on the temperature factor are in a good agreement with conclusions of linear hydrodynamic stability theory for the first-mode waves as well as with experiments performed. However, the experiments [10] carried out in the same facility by the frost formation on the model surface have shown that the transition reversal (the Reynolds numbers increase with decreasing T_w, then decrease and then rise again at very low temperature) occurs. The experiments were performed at all the other equal conditions in the wind tunnel test section with the exception of humidity. It has been shown that with decreasing the air humidity (and that is the same with decreasing the frost crystallization temperature) the beginning of the transition reversal is delayed. The frost amount on the model surface was not under control. Therefore a series of additional experiments has been carried out at the same concentrations of water steams C=0.19 - 0.21 gr/kg, Mach number M=4.0, $Re_1=36.5 \cdot 10^6\ m^{-1}$, but at the different frost amount on the plate surface.

The model was cooled up to the temperature of liquid nitrogen boiling and then (after stoppage of nitrogen supply) was heated by the air flow. In the beginning of experiments the model was covered by a close fitted protective plate preserving the surface from the frost formation which was removed from the model test section before measuring. A difference in the frost amount on the model surface was achieved by varying the time t between the removal of a protective plate and the stoppage of the nitrogen feed into the model, i.e. the time in the course of which an extremely cold model was covered with a frost layer. The more t is the more frost appears on the model (at t=0 at the initial moment of heating there was practically no frost on the model). The position of the transition beginning was determined with respect to the change of heat transfer in the unsteady heat régime.

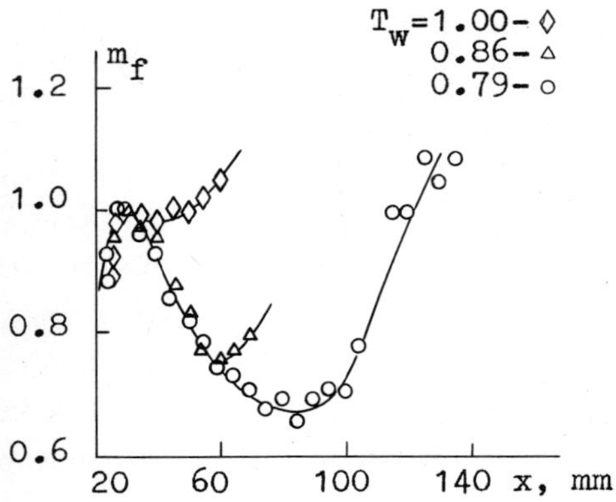

Fig.1

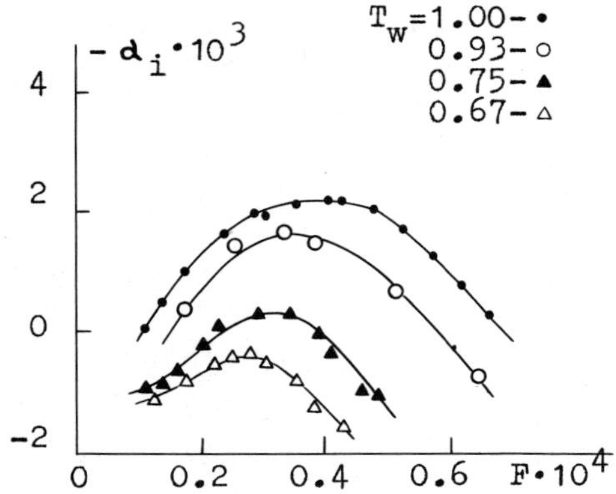

Fig.2

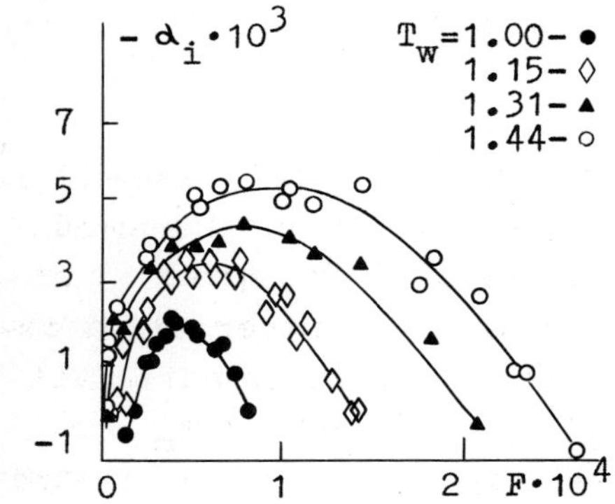

Fig.3

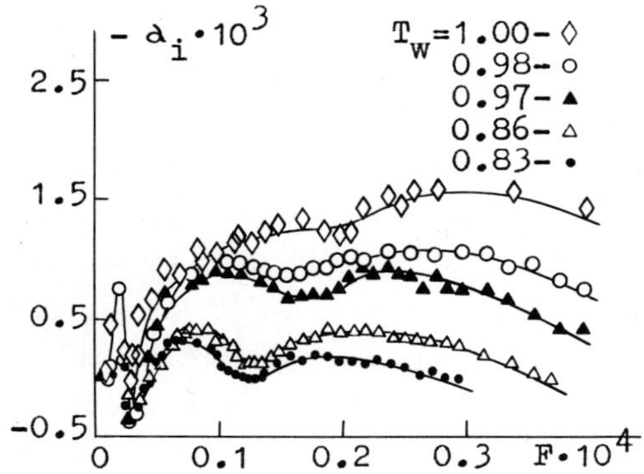

Fig.4

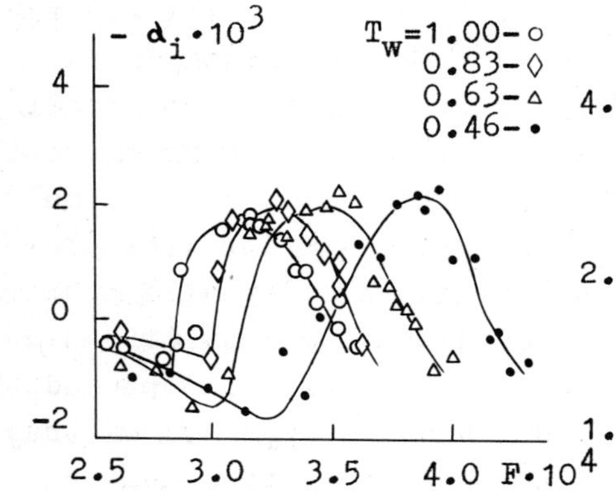

Fig.5

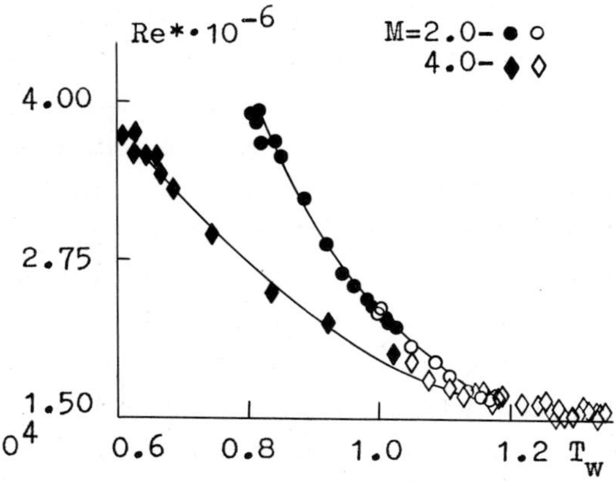

Fig.6

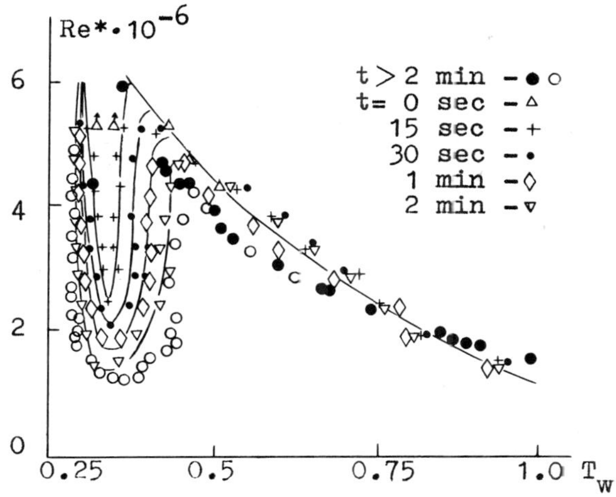

Fig.7

Fig.7 illustrates the results obtained for t=0÷2 min as well as the data of [10] at t > 2 min and two values of the air humidity: C=0.33÷0.40 (light symbols) and C=0.21÷0.22 (dark symbols). It is seen that at all t (except t=0) the transition reversal is observed. At t=0 (i.e. when the frost was practically not available on the model) a monotonous dependence of Re* on T_w (at $T_w > 0.4$) is observed. The marks with arrows in Fig.7 demonstrate the fact that up to $Re=5.2 \cdot 10^6$ the boundary layer on the model remained laminar. With increasing t the size of a "loop" and the temperature range of the reversal "pit" grows. The minimum of the dependence of Re* on T_w for $t \geqslant 15$ sec corresponds to $T_w \sim 0.35$ which is close to the temperature factor corresponding to a maximum intensity of the laser beam scattering from the model surface [10] . It should be noted that at $T_w < 0.35$ the intensity of the laser beam scattering according to data of Ref. [10] sharply decreased and the transition Reynolds numbers drastically increased. These data enabled us to draw a conclusion that the beginning of the transition reversal is caused by frost roughness and a sharp increase in Re* at $T_w \sim 0.2$ on the model with frost may be explained by the change of a general structure of frost (the surface becomes smooth).

The results of the present study confirm the conclusions of [10] and we can conclude that the availability of frost on the model can cause the transition reversal, the parameters of which are determined by the process of frost formation and by the change of its structure. The results obtained in our work in tests performed in the wind tunnel T-325 show that without frost formation on the model surface the transition Reynolds numbers monotonically increase with decreasing the temperature factor.

References

1. Gaponov, S.A.; Maslov, A.A.: Disturbance propagation in compressible flows (in Russian). Novosibirsk: Nauka 1980.

2. Laufer, J.; Vrebalovich, T.: Stability and transition of a supersonic laminar boundary layer on an insulated flat plate. J. Fluid Mech. 9 (1960) 257-299.

3. Lebiga, V.A.; Maslov, A.A.; Pridanov, V.G.: Experimental investigation of the stability of supersonic boundary layer on a flat insulated plate. Archives Mech. 31 (1979) N3, 397-405.

4. Mack, L.M.: Linear stability theory and the problem of supersonic boundary layer transition. AIAA Journal 13 (1975) N3, 278-289.

5. Mack, L.M.: Boundary layer stability theory. Document 900-277 JPL, 1969.

6. Maslov, A.A.: Numerical investigation of the stability of the supersonic laminar boundary layer (in Russian). Zhurnal Prikladnoi Mekhaniki i Tekhnicheskoi Fiziki (1972) N5, 181-184.

7. Demetriades, A.: New experiment on hypersonic boundary layer stability including wall temperature effects. Proceeding of the 1978 Heat Transfer and Fluid Mechanics Institute. Stanford University Press. (1978) 39-56.

8. Lysenko, V.I.; Maslov, A.A.: Transition of laminar supersonic boundary layer into turbulent one by surface cooling (in Russian). Zhurnal Prikladnoi Mekhaniki i Tekhnicheskoi Fiziki (1981) N3, 30-36.

9. Lysenko, V.I.; Maslov, A.A.: Experimental study of the influence of heating on supersonic boundary layer transition and stability (in Russian). Izv. SO AN SSSR (1981) N13, 42-49.

10. Lysenko, V.I.; Maslov, A.A.: Transition reversal and one of its causes. AIAA Journal 19 (1981) N6, 705-708.

11. Lysenko, V.I.; Maslov, A.A.: The effect of cooling on supersonic boundary-layer stability. J.Fluid Mech.(to appear).

Interconnectionship Between the Flow Separation and the Direct and Inverse Transition at Supersonic Speed Conditions

A.A. ZHELTOVODOV, A.A. PAVLOV, E.H. SHILEIN, V.N. YAKOVLEV

Institute of Theoretical and Applied Mechanics,
USSR Academy of Sciences, Novosibirsk 630090, USSR

It is well known that the properties of the separated flows substantially depend on the boundary layer state. In its turn, the conditions accompanying the separation onset and the proper separation can influence the direct and inverse transition processes determining the boundary layer state. Thus, in [1], the authors paid attention to a noticeable acceleration of transition by the oblique shock wave - laminar boundary layer interaction (Fig.1.1). It stands to reason that parallel with the accelerating action of the positive pressure gradient the other factors are of importance, f.e., a possibility of the

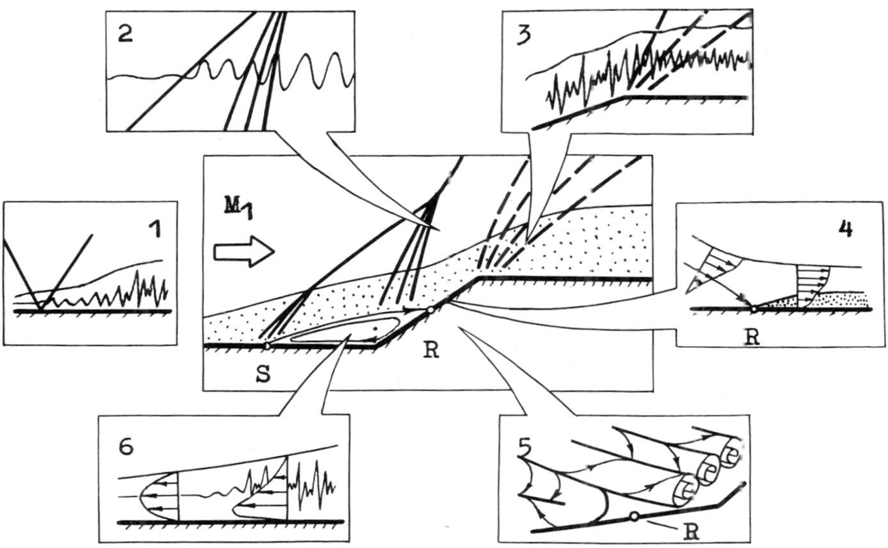

Fig.1. Some factors characterizing interinfluence of separation and transition processes

turbulence amplification in the neighbourhood of shock waves and separated regions (Fig. 1.2). The following main mechanisms of the turbulence amplification are singled out [2, 3, 4] : 1) amplification of vorticity waves by passing through a shock region, 2) generation of turbulence by incident acoustic and entropy waves, 3) generation of turbulence by shock oscillations conditioned by different reasons. The mentioned effects exert without question the influence on the laminar-turbulent transition. Under conditions of turbulent flows they cause an additional growth of Reynolds stresses and prevent the inverse transition (relaminarization) in the negative pressure gradient region (Fig. 1.3). Really, a behavior of the local skin friction coefficient behind an isolated expansion

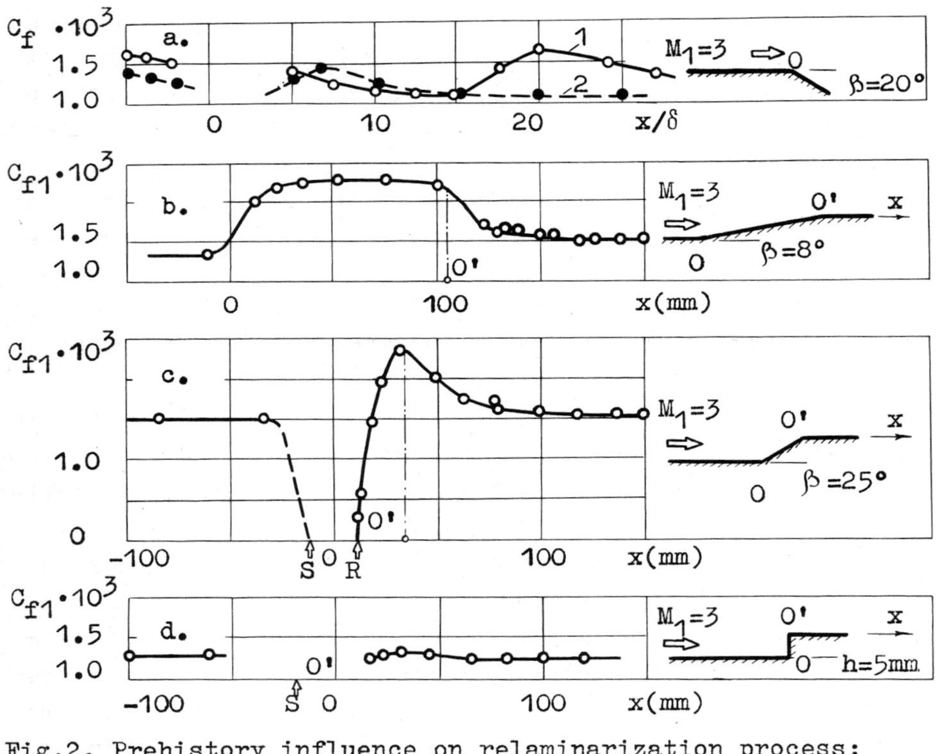

Fig.2. Prehistory influence on relaminarization process:
a. 1-$Re_x=7 \cdot 10^6$, 2-$Re_x=14.5 \cdot 10^6$ (x - distance from leading edge to expansion corner) [5] ;
b.- d. $Re_1= 35 \cdot 10^6 m^{-1}$ (L - distance from leading edge to compression corner, respectively L=0.22; 0.294; 0.315m) [6]

corner [5] (Fig. 2a) is indicative of a possibility of relaminarization and the following repeated transition to the turbulence. In the case of the prehistory influence (shocks and separation) the boundary layer remained turbulent behind the expansion corners [6] (Figs. 2b, c, d). A forming new boundary layer in the near-wall part of the dissipative flow is a result of a sharp change in the boundary conditions in the neighbourhood of shocks and in reattachment regions (Fig. 1.4). Such a structure was observed visually in [7] . The regularities of the new layer formation require a further study. In the considered cases the velocity distribution with respect to its thickness enabled us to single out a typical for it logarithmic-law region and a wake region. In the upper part, the flow preserves the features of the reattaching mixing layer with larger scale eddies. A possibility of formation of the Taylor-Görtler vortices in the separated flows is the other important peculiarity [8] . In the examined cases such vortices were observed at $\beta = 25°$ and $45°$ and were most intensive in the neighbourhood of the reattachment line R (Fig. 1.5). It is natural that they essentially determine a further development of the boundary layer and contribute to the intensive transfer of the turbulence from an external flow to a near-wall region. It has been discovered that the amplification of flow spreading in the lateral direction from the separated zone can lead to a disappearance of the mentioned vortices.

An analysis of the flow in the separated regions testifies to a possibility of the change in its state at the turbulent separation conditions [7, 9]. A behavior of the velocity profiles in the reversed flow regions is similar to the near-wall jet relaminarization (Fig. 1.6). For example, directly before a step (Fig. 3, $x/h = 1.23$) they have the same form as in the turbulent jet. As they are moved off from an obstacle ($x/h = 2.6$) the velocity distribution corresponds to the laminar near-wall jet. An analogous behavior was observed in the separated zone before a compression corner ($\beta = 25°$). In our opinion, the essential negative pressure gradients in the reversed flow direction in the separated regions are sufficiently strong stabilizing factor.

A certain interconnectionship and interinfluence between the

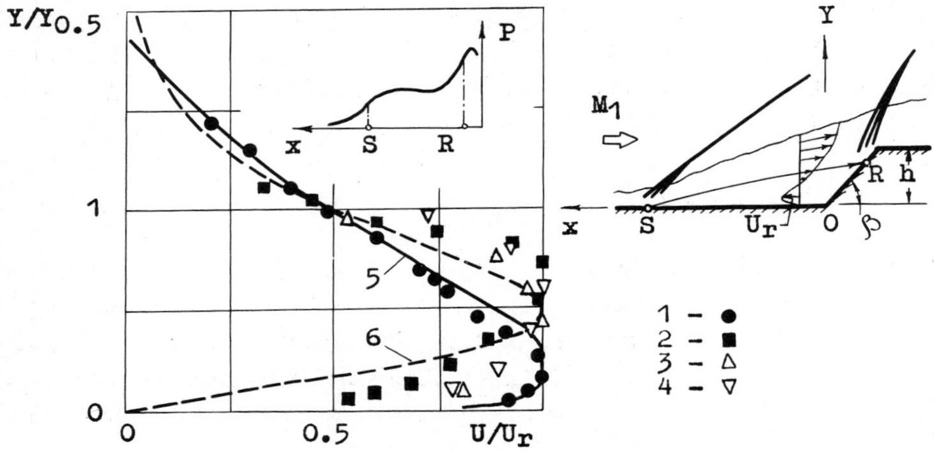

Fig.3. Relaminarization of reversed flow. β =90°, h=15mm: 1 - x/h=1,23; 2 - 2.6. β =25°, h=15mm; 3 - x/h=0.33; 4 - 0.53; 5 - turbulent near-wall jet; 6 - laminar near-wall jet

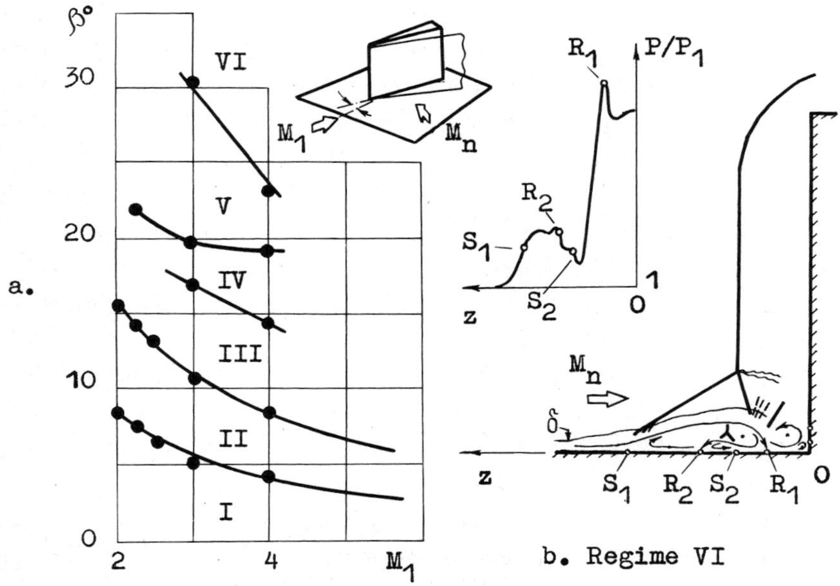

Fig.4. Regimes of interaction: I-without separation; II-with separation; III-development of secondary separation; IV, V, VI-respectively weaking, full disappearance and repeated onset (reversal) of secondary separation

separation and transition phenomena is manifested also at different stages of three-dimensional shock wave - turbulent boundary layer interaction. Thus, the studies [10] have discovered an effect of the secondary separation reversal which reveals itself in weakening, full disappearance and repeated onset of the secondary separation in the reversed flow with increasing the shock intensity (regions IV, V, VI, respectively, Fig. 4a). It has been established that the artificial turbulization of the reversed flow with the help of sandy roughness leads to a disappearance of secondary separation at conditions of the regime III. In region VI we could not succeed in preventing in a similar way the formation of secondary separation S_2 and reattachment R_2 lines (Figs. 4a, b). Thus, the onset and development of the secondary separation in the region III occurs at conditions of the laminar boundary layer in the reversed flow that corresponds to conclusions of Ref. [11] . At the same time a main reason of the discovered reversal, as was assumed in [12] , is the acceleration of transition of this boundary layer into a turbulent state. Such an acceleration can be caused by different factors, including the above-mentioned turbulence amplification resulting from interaction of the boundary layer and the mixing layer with shock waves in the neighbourhood of the separated region. It is natural that the repeated onset of the secondary separation under conditions of regime VI when the reversed flow was conically supersonic was observed only by the pressure on the line R_2 - minimum pressure between lines R_1 and S_2 ratio (Fig. 4b) equal to $\overline{P} = 1.5$ [12] . This corresponds to a critical shock intensity causing the turbulent boundary layer separation.

References

1. Le Balleur, J.; Délery, J.C.: Etude experimentale de l'effet de la reflexion d'une onde de choc sur la transition de la couche limite. Rech. aérosp. N°3 (1974) 165-171.

2. Sekundov, A.N.: Supersonic flow turbulence and its interaction with a shock wave. Izvestiya Akademii Nauk SSSR, Mekhanika Zhidkosti i Gaza N°2 (1974) 8-16.

3. Blokhintsev, D.A.: A moving sound receiver. Doklady Akademii Nauk SSSR, vol. 47 (1945) 22-25.

4. Anyiwo, J.C.; Bushnell, D.M.: Turbulence amplification in shock-wave boundary-layer interaction. AIAA J., vol. 20, N° 7 (1982) 893-899.

5. Goldfeld, M.A.; Tyutina, E.G.: Supersonic turbulent boundary layer relaminarization by rapid expansion near a corner point. Novosibirsk, Institute of Theoretical and Applied Mechanics, USSR Academy of Sciences. Prepr. N° 12-82 (1982).

6. Zheltovodov, A.A.; Shilein, E.H.; Yakovlev, V.N.: Evolution of turbulent boundary layer at conditions of mixed interaction with shock and expansion waves. Novosibirsk, Institute of Theoretical and Applied Mechanics, USSR Academy of Sciences. Prepr. N° 28-83 (1983).

7. Zheltovodov, A.A.; Pavlov, A.A.: Investigation of flow in supersonic separated region in front of step. Novosibirsk, Institute of Theoretical and Applied Mechanics, USSR Academy of Sciences. Prepr. N° 1 (1979).

8. Borovoi, V.Ya.: Gas stream and heat exchange in shock wave-boundary layer interaction regions. Moscow, Izd. "Mashinostrojenije" (1983).

9. Dolgov, V.N.; Zheltovodov, A.A.: Analysis of properties of two-dimensional turbulent separated flows in compressible streams. Methoden und Verfahren der Mathematischen Physik (Proceedings of Conference, Oberwolfach, November 29 - December 5, 1981). Frankfurt/M., Berne, vol. 24 (1982) 41-52.

10. Zheltovodov, A.A.: Three-dimensional shock wave - turbulent boundary layer interaction generated by wedge. Novosibirsk, Institute of Theoretical and Applied Mechanics, USSR Academy of Sciences: Aerofizicheskije Issledovanija, issue 6 (1976) 125-129.

11. Zubin, M.A.; Ostapenko, N.A.: The flow structure in separation region by normal shock wave - boundary layer interaction in a corner. Izverstija Akademii Nauk SSSR, Mekhanika Zhidkosti i Gaza N°3 (1979) 51-58.

12. Zheltovodov, A.A.: The regimes and properties of three-dimensional separated flows initiated by oblique shock waves. Zhurnal Prikladnoi Mekhaniki i Tekhnicheskoi Fiziki N°3 (1982) 116-122.

The Influence of the Outflow Generated Vortex Structures on the Boundary Layer Characteristics

V.V. BABENKO, L.F. KOZLOV, S.A.DOVGIY, N.F.YURCHENKO,
V.P. IVANOV, N.A. GNITETSKIY, V.I. KOROBOV

Institute of Hydromechanics, Ukrainian Academy of Sciences
Kiev, U.S.S.R.

Summary

The results of the experimental investigation of the laminar boundary layer kinematics are given when a boundary layer is excited by oscillating plates of different size. Flow visualization and laser anemometer measurements showed that boundary layer characteristics depend on frequency, oscillation amplitude, plates width and their place relatively to a boundary layer.

From Morkovin and Reshotko ideas [1] the boundary layer receptibility means here the boundary layer reaction on different plane and space disturbances. Then in each transition stage the interaction in a boundary layer of natural disturbance with artificially induced one is investigated under the conditions of low free stream turbulence level.

The disturbed motion development in different stages of natural transition depending on turbulence level was considered in Refs. 2, 3. The boundary layer receptibility of plane and three-dimensional disturbances generated inside a boundary layer was analyzed in Refs. 2—8.

Apparatus and methods of the boundary layer receptibility investigation were described in Refs 2—6.

In Ref. 7 there are given the results of the study of little plane disturbances receptibility when the disturbances are introduced to a free stream near the outer boundary layer edge. The receptibility of little and finite amplitude disturbances generated near and at the outer edge as well as inside a boundary layer is investigated at present work. In all cases the introduced disturbances are characterized by a point at the linear stability diagram, this point being located in the stable area $Re_{\delta *} = 200$, $\frac{\beta_r \nu}{U_\infty^2} = 4.10^{5}$.

Fig. 1 gives examples of the visualized plane disturbances. The vortex street nonsymmetry in a free-stream is due to different boundary conditions: up and down at a distance of 2 δ were free water surface and the boundary layer outer edge respectively.

The vortex street in a boundary layer is deformed stronger for being induced close to a rigid wall. The characteristic feature of both cases is the existence of certain frequencies with maximum oscillation amplitudes.

Velocity measurements (see Fig. 2) support the fact [7] that generated outside the boundary layer oscillations quickly decay in the direction normal to a wall. The shape of average and fluctuation velocity profiles across the boundary layer essentially depends on amplitude, frequency and intensity of the plate oscillation.

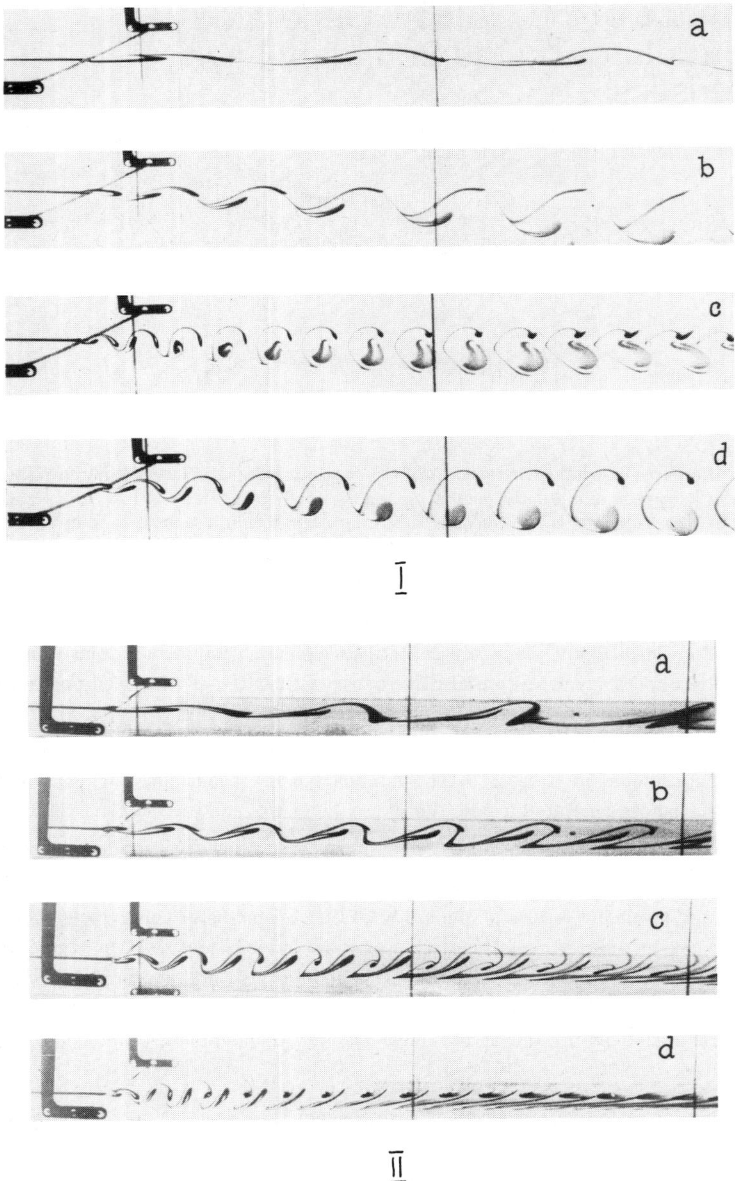

Fig. 1. Visualization of the disturbed motion generated in a free stream (I) at the distance from a wall h=0.045 m and at a boundary layer edge (II), h=0.015 m. All the observations were made at the following parameters values: U_∞ = 0.03 m/s; plate chord, B=2 mm; distance between the trailing edge and a measuring point, l=0.025 m; plate oscillation amplitude, A=1 mm; oscillation frequency for (a) f = 0.5 s^{-1}, (b) 1 s^{-1}, (c) 1.5 s^{-1}, (d) 2 s^{-1}.

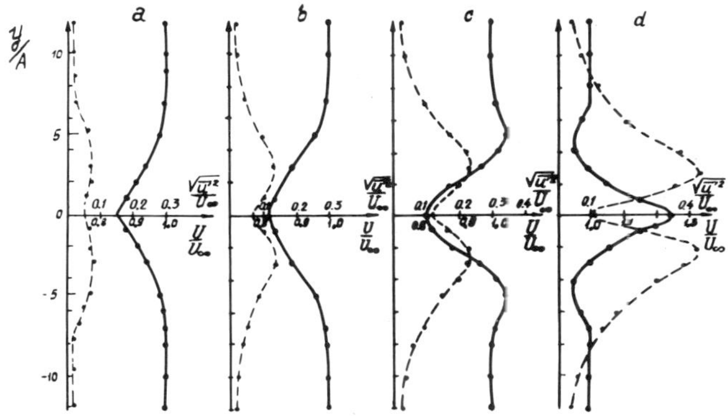

Fig. 2. The distribution of the average (solid line) and fluctuation (touched line) components of the longitudinal velocity of the oscillating wake of a plate placed in a free stream for various B values: (a) 2 mm; (b) 15 mm; (c) 25 mm; (d) 0 mm. $U_{\infty} = 0.04$ m/s; h = 0.045 m; l = 0.025 m; A = 1 mm; f = 2 s^{-1}.

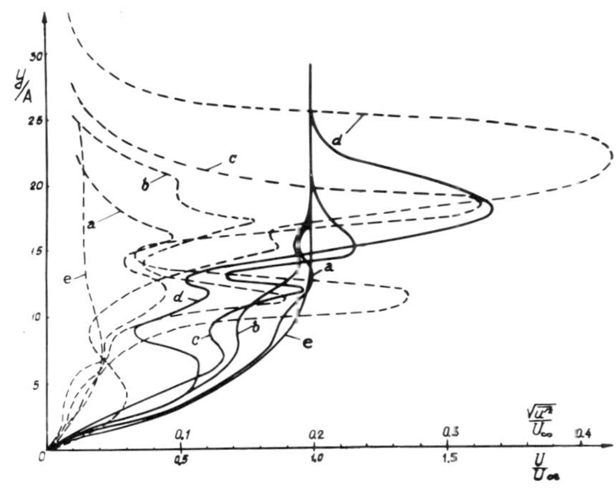

Fig. 3. The distribution of the average (solid line) and fluctuation (touched line) components of the boundary layer longitudinal velocity when the oscillations are generated at a boundary layer edge. Notation and parameters values for (a)—(c) as in Figure 2; (e) — without a plate.

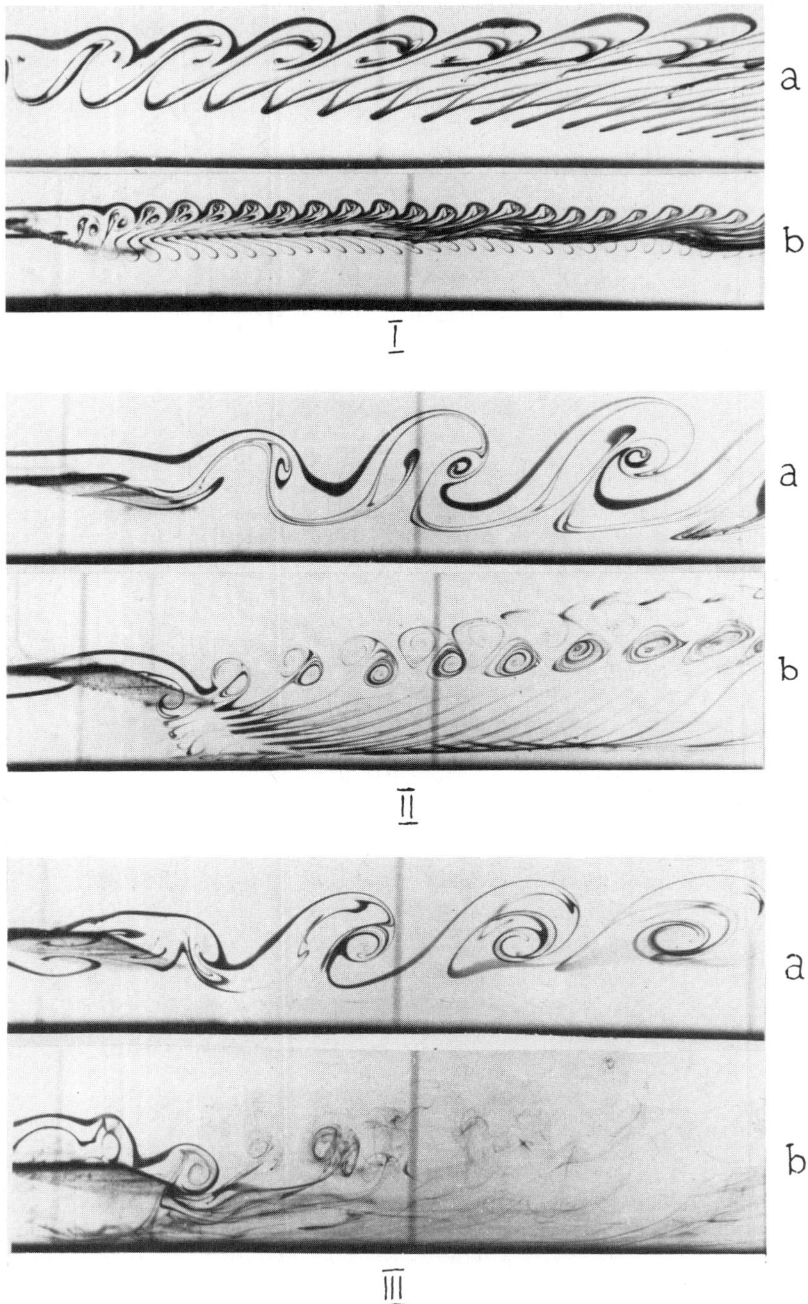

Fig. 4. Visualization of the disturbed motion generated at a boundary layer edge under $f=1\ s^{-1}$ (a) and $2.5\ s^{-1}$ (b): I — B = 2 mm; II — 15 mm; III — 25 mm.

One can come to the same conclusion while examining the results of Fig. 3 when the disturbance is introduced to a boundary layer at its outer edge. Profiles deformation is located in the area of the external disturbances introduction because of considerably differing values of induced and proper boundary layer disturbances. The average and fluctuation profiles shape (for B = 2 mm) acquires a form typical of nonlinear transition stages [3—5].

The increase in the introduced plane disturbances intensity leads to disturbances propagation across all the boundary layer thickness and out of it. However, boundary layer vortex disturbances do not grow and do not decay because of the absence of an interaction with the natural boundary layer disturbances. This is clearly seen in Figs. 3—4.

Thus even in the stable laminar boundary layer the changes for the worse can give rise to boundary layer inherent oscillation.

References

1. Reshotko, E.: Boundary layer stability and transition. Ann. Rev. Fluid Mech. 8 (1976), 311—349.
2. Kozlov, L.F.; Babenko, V.V.: Eksperimentalniye issledovaniya pogranichnogo sloya. Kiev: Naukowa Dumka 1978.
3. Babenko, V.V.; Ivanov, V.P.; Yurchenko, N.F.: Izmereniye lazernym anemometrom vospriimchivosti pogranichnogo sloya k ploskim i trehmernym vozmushcheniyam. Avtometriya, 3 (1982), 91—96.
4. Yurchenko, N.F.: K metodike eksperimentalnogo issledovaniya sistemy prodolnyh vihrey v pogranichnom sloye. Ing.—Phys.J., 41, N6 (1981), 996—1001.
5. Babenko, V.V.; Yurchenko, N.F.; Kozlov, L.F.: Eksperimentalnoye issledovaniye vospriimchivosti pogranichnogo sloya k prodolnym vihrevym vozmushcheniyam. Ing.—Phys. J., 46, N1 (1984) 140.
6. Babenko, V.V.; Yurchenko, N.F.; Kanarskiy, M.V.: Spektralniy analiz reaktsii pogranichnogo sloya na trehmerniye vozmushcheniya. Ing.—Phys. J., 46, N1 (1984) 140—141.
7. Kachanov, Yu.S.; Kozlov, V.V.; Levchenko, V.Ya.: Vozniknoveniye turbulentnosti v pogranichnom sloye. Novosibirsk: Nauka 1982.
8. Fasel, H.: Reaction von zweidimensionalen laminaren incompressiblen Grenzschichten auf periodische Störungen in der Aussenstromung. ZAMM, 57 (1977) 180—183.

On the Reverse Transition of Compressible Turbulent Boundary Layer in a Transverse Flow Around a Convex Corner Configuration

M. A. GOLDFELD

Institute of Theoretical and Applied Mechanics
USSR Academy of Sciences, Novosibirsk 630090, USSR

Summary

The interaction of turbulent boundary layer with a rarefaction fan is one of the most important and difficult problems of fluid dynamics which until now remained the least examined. It is known that under these conditions the initially turbulent boundary layer may either remain turbulent or pass into the laminar state. Such a transition is called the relaminarization or the transition reverse [1] . The results of studying the transition reverse are presented in a number of papers, the review of which is given in [2] . These studies were of uncoordinated character and as a result their systematization is not easy for realization. Therefore we have undertaken a systematic experimental study first of all for analysing the state of the turbulent boundary layer after it was affected by a large negative pressure gradient and for determining the relaminarization criterion for these conditions. The model consisted of two plates with the angle of stream deflection θ and had lateral plates with built-in glasses for visual observation of the flow. The experiments were carried out at Mach numbers M = 2 ÷ 4 and Reynolds numbers calculated with respect to the impulse loss thickness $Re_{\delta}^{**} = (3 \div 23)\cdot 10^{2}$ for angles $\theta = 10 \div 25°$.

The measurements of the static pressure have shown (Fig. 1) that the expansion process is not concentrated near the corner point, but is realized on the length approximately $10\,\delta_0$ (δ_0 being the thickness of the layer before the interaction), i.e. there exists a region with large negative pressure gradient that is a necessary condition of the transition reverse of the turbulent boundary layer. The analysis of the velocity profiles before the surface turn has shown that the boundary layer was completely turbulent. This is confirmed by a good agreement of the velocity profile with the exponential law $u/u_e = (y/\delta)^{1/n}$ at n = 7 as well as by the coincidence of theoretical and experimental values of the friction coefficient. A peculiarity

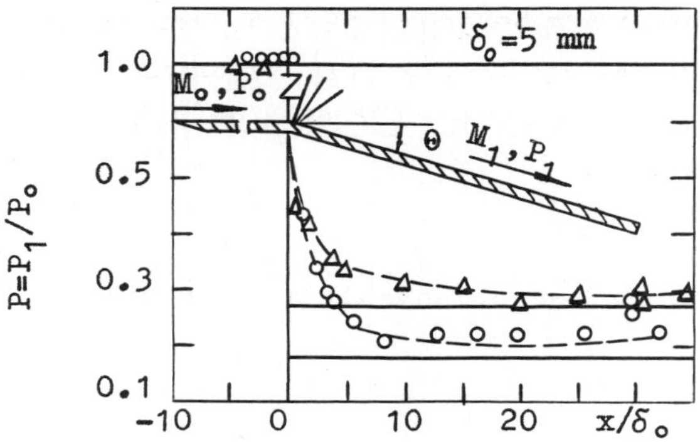

Fig.1 Distribution of pressure. △ M=3, Θ=20°; ○ M=4, Θ=15°. Calculation according to Prandtl-Meyer ——

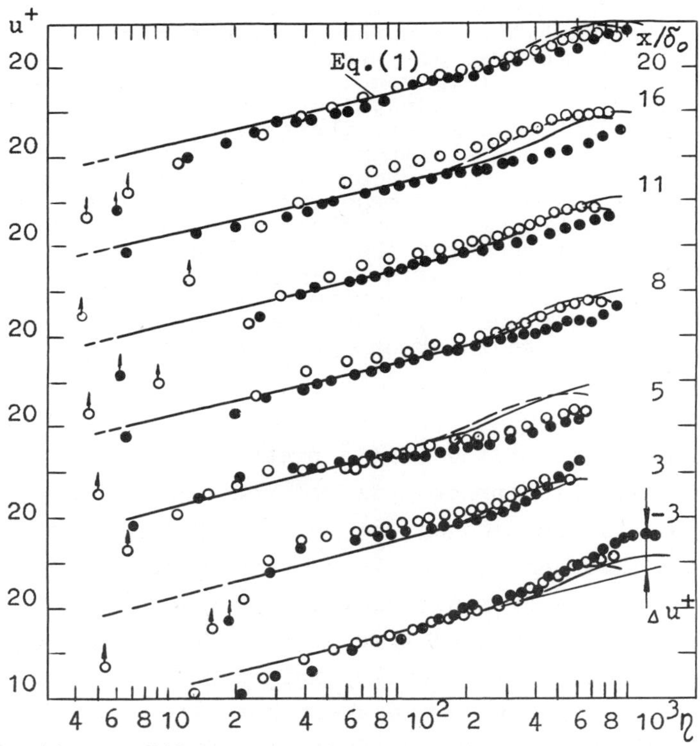

Fig.2 Velocity profiles. —○— M=3, Θ=15°; —●— M=3, Θ=20°

of the boundary layer behind a rarefaction fan consists in the increase of its thickness (2-3 times) and in essential growth directly behind the surface turn of the velocity profile filling which then quickly decreases downstream. It should be noted that behind a rarefaction fan the self-similarity of the velocity profiles is not observed at least along the length (10-15) δ_0. Figure 2 shows the velocity profiles in the form of the "wall-wake" law $u^+(\eta)$ and their theoretical presentation with the use of the relationship (with regard for Van-Driest transformation):

$$u^+ = \frac{1}{\kappa} \ln \eta + B + \frac{\Pi}{\kappa}\left[1 - \cos(\pi y/\delta)\right] \tag{1}$$

where $\eta = y u_\tau/\nu$, $u_\tau = (\tau_w/\rho)^{1/2}$, κ = 0.41, B = 5.8 and Π = 0.58 ($Re_\delta^{**} \geqslant 5000$). The main peculiarity of the boundary layer velocity profile after the passage through a rarefaction fan consists, firstly, in the absence or strong deformation of universal logarithmic section, secondly, in increasing the thickness of the viscous sublayer and its buffer section up to the values $\eta \approx 200$ and more and, third, in complete disappearance of the jet (wake) velocity component Δu^+. In some cases (M = 3, Θ = 20) in the external layer part the velocity turns out to be lower than the logarithmic law, i.e. $\Delta u^+ < 0$. At that the near wall layer part ($\eta \leqslant$ 60-200) is described by the dependency which is close to the dependency for the laminar boundary layer. It is obvious that behind the surface turn the boundary layer is nonequilibrium in the kinematic sense. From this point of view it is of interest to follow how the specific sublayers of the layer respond to the change in external conditions. It is seen from the data depicted in Fig. 2 that in the first turn the region of logarithmic velocity distribution disappears, apparently, simultaneously with the thickening of a viscous sublayer ($x \approx 3\delta_0$), but the jet velocity component still remains. However, already at $x \approx 5\delta_0$ practically the whole profile is located lower than the universal logarithmic dependency. A reverse to the turbulent flow is manifested first of all in the appearance of a clear logarithmic section in the velocity distribution ($x \approx 8.5\,\delta_0$). At $x > 20\,\delta_0$ the velocity profile acquires all the peculiarities of the turbulent layer, the loga-

rithmic section being of very small extension.
The character of the surface friction $c_f(x)$ change is known to be one of the most reliable means for determining the flow regime in the boundary layer as the typical increase in c_f should be observed in the transient region. Exactly this kind of distribution $c_f(x)$ can be observed in Fig. 3. Behind the surface turn the friction coefficient exceeds the values of c_f before expansion, smoothly decreases downstream and then we can see the increase in c_f typical of the transient flow. Such a behavior of the $c_f(x)$ dependency is in a good agreement with the velocity profile evolution. These data confirm the existence of relaminarization and by using them it is possible to determine the extension of the flow relaminarization region. It is seen that the decrease in the Reynolds number from $14.5\cdot 10^6$ up to $7\cdot 10^6$ leads to the increase in the relaminarization section length more than two times. A comparison of the data given in Fig. 3 shows that the increase in the angle θ turn from 15° up to 20° leads to the growth of the relaminarization region length due to the increase in the longitudinal pressure gradient.
A general analysis of the flow, the obtained experimental results and the data of other authors show that the beginning and the relaminarization process are determined by the parameters u_e, ρ_e, ρ_w, du/dx, τ_w, $\tau(y)$, μ_e, μ_w, T_e, T_w, $q(x)$, δ^{**}. Here $q(x)$ is the heat flux, the other notations are conventional. This list, apparently, is not complete and the only possible, but this makes it possible to constitute some dimensionless complexes determining the relaminarization process, e.g., the criterion characterizing the flow gradient $Q=\frac{\mu_e}{\rho_e u_e}\cdot\frac{du_e}{dx}$ which was used in a number of works for incompressible flows [4] . For supersonic flows there appears another parameter - the angle of the flow turn θ which together with the Mach number determines the pressure differential by expansion. However, the gradient criteria are more reliable, as an analysis of [2] has shown, since in the physical aspect the pressure gradient determines the relaminarization process. This is confirmed by the data in Fig. 4 in which the dependency $Q_w(Re^{**})=\mu_w\frac{du}{dx}\cdot\frac{1}{\rho_e u_e^2}$ is given. Here the criterion Q_w is chosen in the form different from a similar criterion for incompressible flow as it

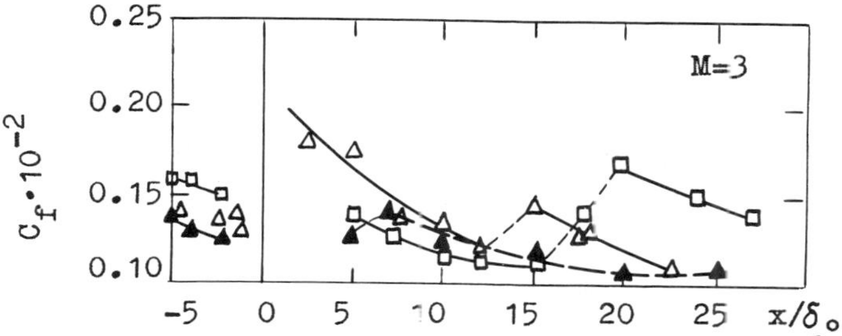

Fig.3 Skin friction distribution. ▲ Θ=15°, Re=14·10^6; □ Θ=15°, Re=7·10^6; △ Θ=20°, Re=14·10^6.

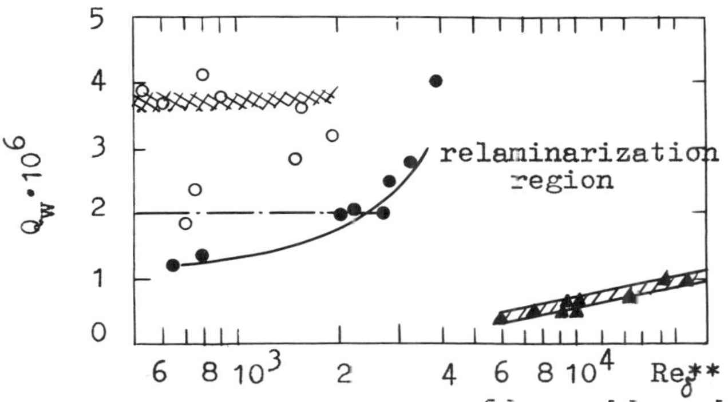

Fig.4 ▲▨▲ Present experiment; —·—·- [1], ✕✕✕✕ [3], —— [4], ○● [4]

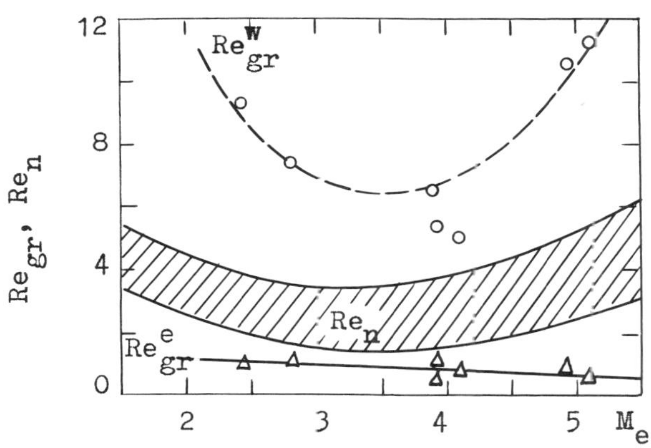

Fig.5 Relaminarization criteria. Re_n-Reynolds number of laminar-turbulent transition

seems to be the most logical in the sense that the terms taking into account the inertia are determined with the help of conditions on the external edge of the boundary layer and the term taking into account the viscous effects (μ_w) with the help of the parameters on the wall which most closely approximate the conditions in the near wall layer. A clearly expressed dependency of critical region on the Reynolds number Re_δ^{**} has attracted our attention that was not noted before for supersonic flows. The critical value of Q is lower than that for supersonic flows, e.g., of earlier suggested values $Q_{cr} = (1+3.5)\cdot 10^6$. The Reynolds number constructed by analogy with that suggested in [5] has proved to be the most convenient. Figure 5 demonstrates the Reynolds numbers calculated at points of the beginning of the relaminarization phenomenon with the use of the flow parameters on the external boundary $Re_{gr}^e = \frac{du}{dx}\cdot\frac{\rho_e}{\mu_e}x_p^2$ and on the wall $Re_{gr}^w = \frac{du}{dx}\cdot\frac{\rho_w}{\mu_w}\cdot x_p^2$. In these expressions a typical dimension x_p was assumed equal to the distance from the surface turn up to the point at which the laminarized velocity profile was obtained experimentally. One can see that in this case as a relaminarization criterion the Re_{gr} number will be more preferable because of its weak dependence on the Mach number and the computation simplicity, as in this case it is necessary to know only the parameters of the undisturbed flow.

References

1. Launder, B.E.: Laminarization of the turbulent boundary layer in a severe acceleration. - Trans. of ASME (1964) 31, ser. E, 707-709.

2. Goldfeld, M.A.; Tyutina, E.G.: Relaminarizatziya sverkhzvukovogo turbulentnogo pogranichnogo sloya pri bistrom rasshirenii okolo uglovoy tochki. Novosibirsk: preprint ITPM, N°12-82 (1982).

3. Badri-Narayan, M.A.; Ramjee, V.: Criteria for reverse transition in a two-dimensional boundary layer flow. - J. of Fluid Mechanics 35 (1969) 225-241.

4. Nash-Webber, J.L.; Oates, G.C.: An engineering approach to the design of laminarization nozzle flows. - Trans. of ASME, ser. D 94, N°4 (1972) 897-906.

5. Yanenko, N.N.; Gaponov, S.A.: Lokalniye kriterii gidrodinamicheskoy ustoichivosti. Novosibirsk: preprint ITPM, N°21 (1978) 16.

Supersonic Boundary Layer Transition in the Presence of Shock Effect

M. A. GOLDFELD, V. V. ZATOLOKA

Institute of Theoretical and Applied Mechanics
USSR Academy of Sciences, Novosibirsk 630090

Summary

The purpose of the paper presented herein is to experimentally determine the values of critical Reynolds numbers of the beginning Re^b_{cr} and the end Re^e_{cr} of the **flow-around reformation** f.e., Fig. 1,2, because of the change of resistivity of supersonic boundary layer to separation by its transition from a laminar ($Re_x < Re^b_{cr}$) into a turbulent state ($Re_x > Re^e_{cr}$).

The nature of these numbers Re_{cr} is similar to that of the known numbers Re^b_{tr} and Re^e_{tr} characterizing the transition of boundary layer into turbulent one, but the numbers Re_{tr} are referred to transition without strong action on the boundary layer and in the case of Re_{cr} it takes place - its separation is characterized by a spasmodic increase in pressure. Therefore the values of the numbers Re_{cr} and Re_{tr} should not without fail coincide. The information on the values of Re_{cr} is not less important than on Re_{tr} as in many cases the change in the picture of the flow around bodies and their aerodynamic characteristics, when a place of transition of boundary layer into turbulent one is changed, is connected either with the separation (transition of boundary layer at dangerous with respect to separation points of a body from **turbulent to laminar**) or with its disappearance (transition reversal). It should be noted that in the supersonic flow-around with separation of the boundary layer there arises, as a rule, a shock wave. In this connection for the sake of comparison together with the values of Re_{cr} there were also determined the values of Re_{tr}. The experiments were carried out in a wind tunnel T-313 of the Institute of Theoretical and Applied Mechanics at Mach numbers from 2 up to 6 and Reynolds numbers Re_x from 1.10^6 up to 60.10^6.

A round thin-walled cylinder with inner canal on which a ring was placed forming a protrusion that caused the separation of the cylinder boundary layer served as a model, Figs. 1,2. Since the flow around the body was supersonic, the separation should look (in the two-dimensional representation), as known, as follows: at some distance l_{sp} before a protrusion on a circular (across the cylinder) separation line the flow separates from a cylinder surface at an angle θ_s and from this line an oblique shock wave, in which the pressure increases p_2/p_1 times where p_1 is the pressure in the uniform free tream, and p_2 is the pressure behind this shock, enters the flow at an angle θ_{sh}. An axisymmetrical model, Figs. 1,2, was chosen basing on the fact that in this case up to the moment of separation the boundary layer will form at conditions of the uniform flow and in the separation region the end effects along the sides will not be present. The distance from the leading edge of the cylinder up to the protrusion l_p was always such that the lengthening $l_p/d < 10$ (where d is the cylinder diameter equal to 64 mm) that makes the influence of transverse curvature negligible, and the height of the protrusion was h = 3 ÷ 10 mm and always more than the boundary layer thickness. Since the values $p_2/p_1 = (p_2/p_1)_t$ in the case of the turbulent boundary layer separation several times more than $(p_2/p_1)_l$ in the case of the laminar boundary layer separation, the values of the separation angles θ_{st} are several times more than θ_{sl} and, respectively, the values of the angle of shock inclination θ_{sht} is also considerably more than θ_{shl}, and the distance l_{spt} from the separation line up to the protrusion is several times less in the case of the turbulent layer separation than l_{spl} in the case of the laminar boundary layer separation. We assumed to measure the values of Re^b_{cr} and Re^e_{cr} by the change of the measured in experiments values p_2/p_1, θ_s, θ_{sh}, l_{sp} when l_p or the pressure in the plenum chamber of the wind tunnel p_0 and, hence, Re_x changed. The values of Re_{tr} were determined with the use of one of the known methods - by the change of pressure p_0' in a miniature Pitot probe. Thus, in experiments there were measured the pressures p_0 (the air temperature in plenum chamber of the wind tunnel was approximately indoor, $T_0 \approx 300K$), p_0', p on the protrusion butt end at several points along its height and the

circle, the flow was photographed by using a shadow device (for measuring l_{sp}, θ_s , θ_{sh}) and the streamlines on the model surface were visualized with the help of soot-oil mixture which was spread on the model (the measurement of l_{sp}).
The soot-oil tracks have shown rather unexpected result: although up to the separation line the flow was uniform with corresponding boundary layer, but the separation line clearly fixed by the soot-oil mixture, turned out to be not a simple circle of the cylinder transverse cross-section, but had a twisting shape and the flow in the separation region was three-dimensional. In the case of laminar separation, Fig. 3, a jet enters the separation region through a separation line in two approximately diametrically opposite places. This jet running against the protrusion spreads about forming the wakes of two vortices on the cylinder surface. Thus, in the performed experiments a circular vortex in the separation region (Figs. 1, 2) has proved to be cut by the above-mentioned jets in two semicircular vortices covering the cylinder, Fig. 3. Hence, these experiments have shown that in the case of laminar separation before a protrusion the two-dimensionality of the flow in the separation region is not preserved on an axisymmetric body of a round transverse cross-section, and it has the above-described large-scale structure in the circumferential direction. This very structure influences a shape of the separation line: near the above-mentioned places of the jets entry into a separation region the separation line strongly deflects along the flow near to the model protrusion, and between them it is convex when meeting the flow. In the case of the turbulent separation there exists also a similar nonuniformity, but it is weak: the separation line, Fig. 4, is almost a circle of the transverse cross-section of the cylinder weakly deflecting only in one place, and there is no jets entry into the separation region. Near the deflection to the right and to the left there is in one convexity. In separation region the wakes of a few vortices are printed, Fig. 4. Basing on this fact, we may assume that in the separation region immediately after its line, the cylinder circle covers several horseshoe-shaped vortices, but weakly affecting the separation configuration. The measurements have shown that the pressure changes in the cylinder

circle on the protrusion butt end are also small within the limits 10%, whereas in the case of laminar separation they achieved 50%. However, the difference between values $(p_2/p_1)_l$ and $(p_2/p_1)_t$ and, respectively, between the values l_{spl} and l_{spt} ($l_{spl} \approx 66-100$ mm, and $l_{spt} \approx 29-36$ mm) as well as θ_{sl}, θ_{shl} and θ_{st}, θ_{sht} is so large that is not slurred over by the above-described circular nonuniformities, and by the change of these values the Reynolds numbers values $Re_x = \rho_e \cdot V_e \cdot l_s/\mu_e$ being critical, i.e. the values Re^b_{cr} and Re^e_{cr} (here ρ_e, V_e, μ_e are the density, velocity and air viscosity coefficient in the flow outside the boundary layer, and l_s is the distance from the leading edge of the cylinder up to the separation line) are rather exactly determined. The derived data are illustrated in graph, Fig. 5, within a range of Mach numbers $M = 2-6$ by dark squares referring to Re^e_{cr} and by light ones referring to Re^b_{cr}. The values of Re^b_{cr} are determined with the accuracy of ±20%, and the values of Re^e_{cr} with the accuracy of ±5%, the values of Re_{cr} determined by the change of p_2/p_1 as well as by the change of l_{sp} are in a good agreement within the limits of 10-15%. Figure 5 illustrates the curves of $Re^b_{cr}(M)$ and $Re^e_{cr}(M)$ plotted between the obtained points within the range of $M = 2 - 6$. The values of Re^e_{tr} at $M = 2-6$ are given by light circles in the carried out experiments. They are in a good agreement with the values of Re^e_{tr} in the same wind tunnel T-313 taken from [I] on a flat plate, Fig. 5. It is seen that the Re^b_{cr} numbers at $M = 2-3$ are considerably, approximately 3 times, less than Re^b_{tr} taken from [I], however, with the growth of M they quickly draw nearer and at $M \approx 6$ the difference between them does not exceed the accuracy of the Re^b_{cr} determination. The difference between Re^e_{cr} and Re^e_{tr} is less, and at small M the values of Re^e_{cr} are less than Re^e_{tr}, and with the growth of M they draw nearer. The values of Re_{cr} obtained in tests in impulse wind tunnel IT-301 of the Institute of Theoretical and Applied Mechanics of the Siberian Division of the USSR Academy of Sciences in a similar way [2] are given by black-white rhombs within a range of $M = 6.4 - 9.5$ in Fig. 5: two cones with circular protrusion served as models, and the values of Re_{cr} were determined by the change of the value l_{sp} measured on the cinemagram stills of a shadow picture of the model flow-around. In impulse tube during the regime accounting for in these tests 40-50 ms the pressure in the plenum chamber of the wind tunnel p_o decreases 4-5 times and, respectively, the Re_x number decreases 2-3 times. In these experiments the values of p_o were chosen in such a way that during the regime the value of Re_x passed through a critical value of Re_{cr} which was determined by the change of l_{sp}. Figure 5 illustrates the values of Re_{tr} taken from the published data of different experimental papers in wind tunnels presented in [2] in a shape of a band within a range of $M = 6.4 \div 9.5$. It is seen that the values of Re_{cr} lie in a region of values of Re_{tr}. These data are in a good accord with the above ones obtained in tests with a model presented in Fig. 1 on Re_{cr} and their comparison with Re_{tr} and continue them up to $M \approx 10$. All the values of Re numbers, Fig. 5, are reduced to the one value of the so-called "unit Reynolds number" $Re_1 = \rho_e \cdot V_e / \mu_e$, $Re_1 = 2.0 \cdot 10^6$ 1/m, by the known formula

$$Re \sim Re_1^{0.3 \div 0.4} \qquad (1)$$

which has been employed in numerous works for Re_{tr} numbers. In this connection there arises a question about the applicability of the dependency (1) not only to Re_{tr}, but also to Re_{cr}. For Re^e_{cr} in the present experiments (1) has been checked at $M = 3$ and 4. The obtained values of Re_{cr} are given in Fig. 6 by black squares for $M = 4$ and by black triangles for $M = 3$ at Re_1 numbers from $13 \cdot 10^6$ 1/m up to $58 \cdot 10^6$ 1/m. In the same figure the values of Re^e_{tr} (by light squares and triangles, respectively) obtained in [1] in the same wind tunnel T-313 in experiments on a flat plate are given. Figure 6 illustrates by lines the dependency

$$Re \sim Re_1^{0.35} . \qquad (2)$$

We can see a good agreement with the formula (2) of the change of Re^e_{cr} and Re^e_{tr} with respect to Re_1. The dependency of Re_{cr} on Re_1 by formula (2) at $M = 7.5$ in the impulse windtunnel IT-301 is confirmed in [3] at Re_1 numbers $(20 - 50) \cdot 10^6$ 1/m.

Thus, a main result of the tests of a model of Fig. 1 is the obtaining of data (see Fig. 5) on the values of critical Reynolds numbers Re_{cr} of the flow-around reconstruction due to the change of the boundary layer resistivity to the separation layer by its transition to turbulent one within a wide range of Mach numbers from 2 up to approximately 10 (together with the data of experiments [2]) as well as on the influence of "unit Reynolds number" Re_1 I/m on the Re_{cr} values (see Fig. 6) (Re_1, as known, determines the influence of the boundary layer noise of wind tunnel walls). At that it is elucidated that at Mach numbers higher than 4-6 the Re_{cr} values practically coincide with the Re_{tr} values characterizing the boundary layer transition itself, i.e. without strong disturbances and at smaller M the Re_{cr} values are the less as compared with Re_{tr} the less M is. This is in a good accord with known ideas on the fact that the boundary layer sensitivity to the disturbances decreases with the growth of M. The results of the model tests, Fig. 1, on the decreased values of Re_{cr} as compared with Re_{tr} at $M \approx 2-3$ and on approximation of Re_{cr} and Re_{tr} with increasing M are presented in [4] (without data at $M > 4$ and without data on the three-dimensional structure of separation region and on the influence of Re_1).

References

1. Struminsky, V.V.; Kharitonov, A.M.; Chernykh, G.G.: Experimental study of laminar boundary layer transition into turbulent one at supersonic velocities. **Izvestiya** AN SSSR MZhG N°2 (1972) 30-34.

2. Boshenyatov, B.V.; Zatoloka, V.V.; Yaroslavtsev, M.I.: The study of separated flow around cones with turbulent boundary layer at Mach numbers M 8.3 and 10. Izvestiya SO AN SSSR, Seriya Tekhnicheskich nauk, N°8, Vyp.2 (1975) 43-50.

3. Antonov, A.S.; Boshenyatov, B.V.; Zatoloka, V.V.: Transition into turbulent state and separation of hypersonic boundary layer at increased Reynolds values. Gasodinamika i Phizicheskaya kinetika. Novosibirsk: Institute of Theoretical and Applied Mechanics of the Siberian Division of USSR Academy of Sciences, 1974, 83-85.

4. Goldfeld, M.A.; Zatoloka, V.V.; Tyutina, E.G.: Experimental study of separation properties of boundary layer before a rectangular protrusion on a cylindric duct. Izvestiya SO AN SSSR, Seriya tekhnicheskikh nauk, N°8, vyp.2 (1975) 51-55.

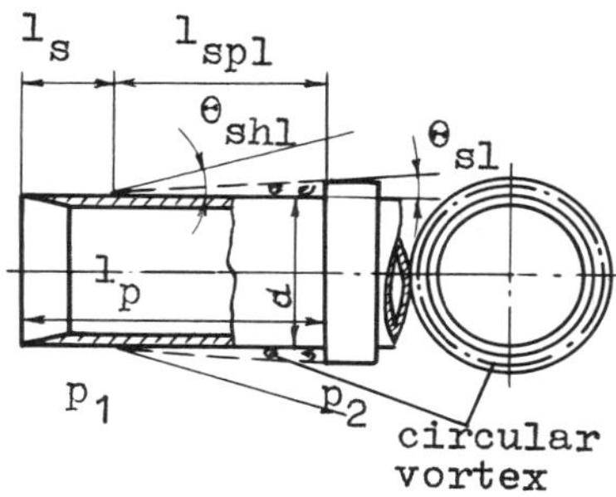

Fig.1. $Re_x < Re^b_{cr}$

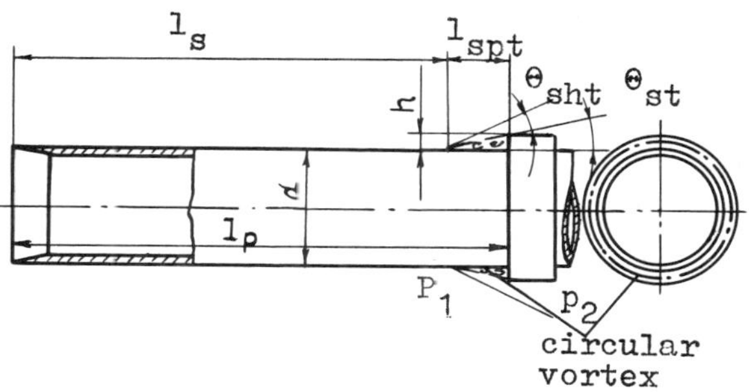

Fig.2. $Re_x > Re^e_{cr}$

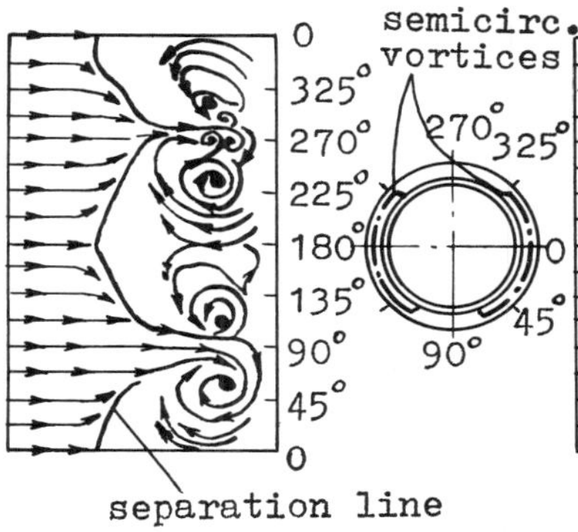

Fig.3. $Re_x < Re^b_{cr}$

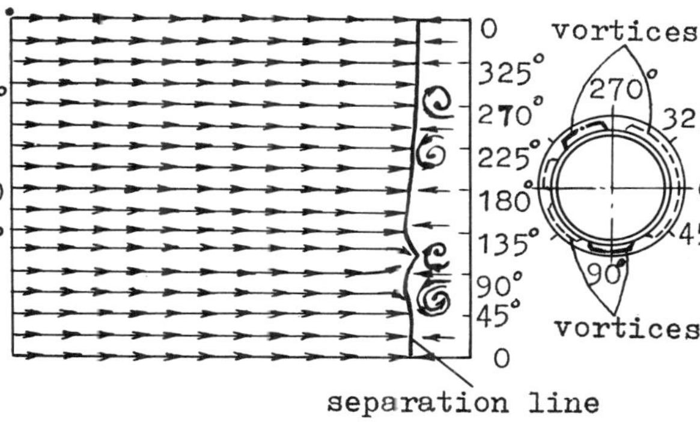

Fig.4. $Re_x > Re^e_{cr}$

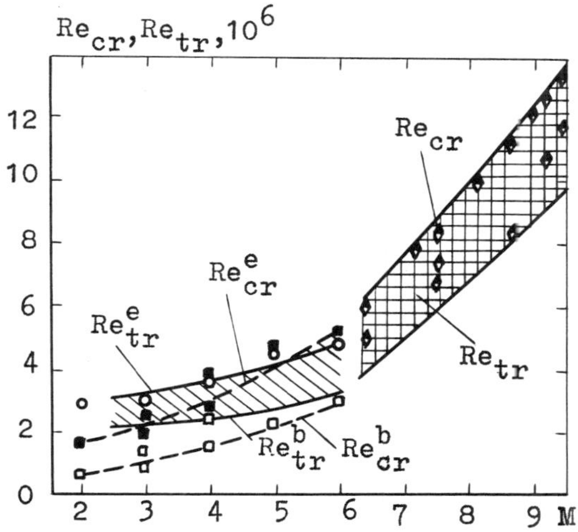

Fig.5.

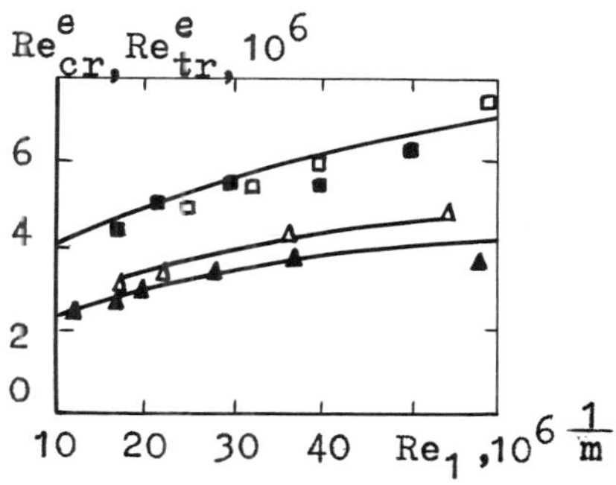

Fig.6.

Density Pulsation and Laminar-Turbulent Transition in the Mixing Layer of Supersonic Jets

S. A. NOVOPASHIN

Institute of Thermophysics,
Siberian Branch of the USSR Academy of Sciences,
Novosibirsk, USSR

Summary

An experimental investigation of density fluctuation scales in the mixing layer of supersonic jets is presented. The shadow method using ruby laser as a light source has been applied. The connection between the laminar-turbulent transition in the mixing layer of the supersonic jets and acoustic waves damping is discussed.

Introduction

Investigations of the flow pattern in the mixing layer of supersonic jets are mainly connected with the following problems: 1) The effect of viscosity on the geometry of a supersonic jet flow [1]. 2) Aerodynamic noise [2]. 3) The effect of acoustic perturbation on the laminar-turbulent transition in mixing layers [3]. The determining parameter of laminar, turbulent and transient flow conditions in the mixing layer of the jet initial region (from the nozzle-exit section to the Mach disk) is a Reynolds number Re_L determined by background parameters, the maximum velocity of the outflow and the distance to the Mach disk [1]. At $Re_L<10^3$ the laminar flow condition is valid, at $10^3<Re_L<10^4$ the transient one, at $Re_L>10^4$ the turbulent one.

The thickness of the mixing layer being neglected, the flow represents a contact discontinuity which is absolutely unstable [4]. If the thickness of the mixing layer is finite, there is a bend in the velocity profile, and the instability of such a flow exists up to very small Reynolds numbers [3]. Dissipative processes should exist at $Re_L<10^3-10^4$, that specify the stability of the flow. The velocity drop at the thickness of the mixing layer has the order of the sonic velocity, therefore per-

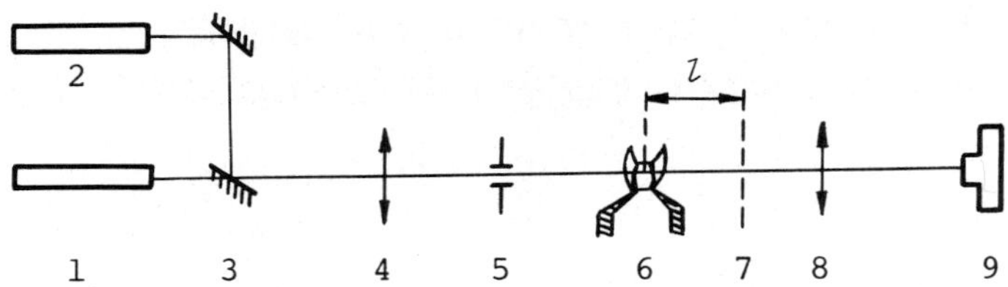

Fig. 1. Experimental set-up

turbations occurring during the laminar-turbulent transition are accompanied by the appearance of density fluctuations. Considering the density pulsations as acoustic waves, the dissipation connected with their damping may be loocked upon as the determining factor of the laminar-turbulent transition.

Experimental set-up

The diagram for the experimental set-up is given in Fig. 1 [5]. The light source was either ruby laser 1 (the pulse duration being $4 \cdot 10^{-8}$ s) or He-Ne laser 2 (with the installation of mirror 3). In the second case, the exposure (typically, 10^{-2} s) was ensured by the exposure of camera 9. To obtain a homogeneous (with respect to the cross section of a beam) intensity distribution, filtration of the optical radiation by lense 4 and diaphragm 5 was used. Registration plane 7 was at a distance l from jet 6. Lense 8 transferred the registration plane onto a film. By varying the distanse l determining the spatial resolution of the method, $d=\sqrt{\lambda l}$ (λ is the light wavelength), one succeeded in taking measurements within the range from 30 to 1200 mkm. The existence of fluctuations with a definite scale was registered while comparing the "instantaneous" and the averaged pictures of the flow, the position of the registration plane being given.

Experimental results

In Fig. 2, the shadow photos with various time resolutions of a CO_2 jet outflowing into the air under normal conditions are given. The diameter of the nozzle is 3 mm, the expansion ratio equals to 10, $Re_L=3 \cdot 10^5$. The "instantaneous" shadowgramme shows

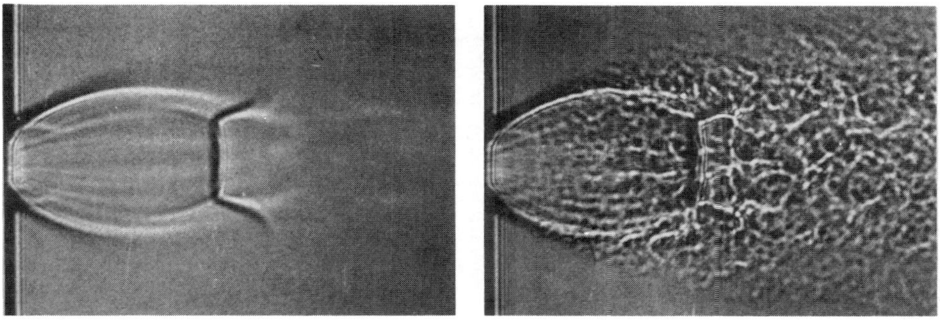

10d *a* 10d *b*

Fig. 2. Shadowgrammes exposed for 10^{-2} s *(a)* and for $4 \cdot 10^{-8}$ s *(b)* at d=180 mkm

the pronounced fluctuation pattern of the flow. The region of the most powerful perturbations, i.e. the region limited to the shock barrel and the jet boundary, can be seen well. The fluctuations that can be observed in the jet core do not imply their existence in the above region owing to the cylindrical symmetry of the flow. In Fig. 2, the distance from the nozzle-exit section is denoted at which the thickness of the mixing layer amounts to the value d. The figures show the satisfactory correspondence of the above distance and the distance from the nozzle-exit section, beginning wherewith the fluctuation structure of the flow can be observed. It means that the thickness of the mixing layer determines the maximum fluctuation scale in this region, which is in good agreement with the understanding of the physical picture of the flow.

Estimation of critical Reynolds number

Let the critical Reynolds number for the laminar-tubulent transition in the jet initial region be estimated from the condition that the thickness of the mixing layer in the Mach-disk region has the order of the wavelength of acoustic vibrations, the increment of whose damping is of the order of unity. In Fig. 3 [6] a curve for acoustic waves damping in the air is given.

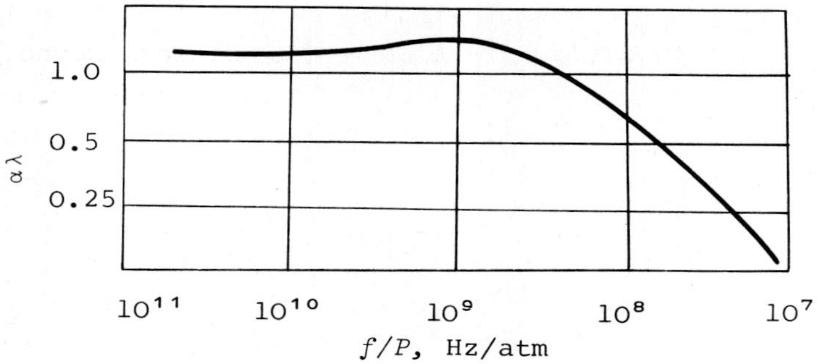

Fig. 3. Acoustic waves damping in the air

The increment of the damping at the distance of the wavelength attains unity at $f/P \simeq 10^8$ Hz/atm. Passing from the frequency to the wavelength and neglecting the dependence of the sound velocity on the frequency, we have $\alpha\lambda \simeq 1$ at $P\lambda = 3\ 10^{-4}$ atm sm. Thus we get $\lambda = 3$ mkm for the atmospheric pressure, i.e. the transition occurs when the characteristic thickness of the mixing layer attains this value.

Describing the value Re_L [1] by using the determining parameters for the air, we have

$$Re_L = 4\ 10^5 LP \qquad (1)$$

Under the laminar flow condition, the relative thickness of the mixing layer in the Mach-disk region decreases with the increase of Re_L [7] and amounts the value about 0.1 under the turbulent flow conditions [1]. Passing from the distance to the Mach disk to the thickness of the mixing layer in formula (1), we get $Re_L = 4\ 10^6\ P\lambda$. Substituting $P\lambda$ for $3 \cdot 10^{-4}$, we obtain the estimate for critical Reynolds number $Re_L \sim 10^3$ which is close to the value obtained in Ref. [1].

Conclusion

The use of the shadow method with the application, as the light source, of a pulse ruby laser to study supersonic jet flow has allowed us to discover a wide spectrum of density fluctuations in the mixing layer. The correlation between the thickness of the mixing layer and the maximum scale of fluctuations is shown. It has allowed us to estimate the critical Reynolds number from the condition that the dissipative process specifying the laminarity of the mixture layer is the damping of high-frequency acoustic waves.

It may be noted in conclusion that the determining parameters for the Reynolds number and the increment of acoustic waves damping are the values PL and $P\lambda$, respectively, which points to the connection between the laminar-turbulent transition and the damping of acoustic vibrations.

References

1. Avduevsky, V.S.; Ivanov, A.V.; Karpman, I.M.; Traskovsky, V.D.; Yudelovich, M.Ya.: The influence of viscosity on the flow in the initial region of highly underexpanded jet. Doklady AN SSSR, Vol. 197, 1971, No. 1, 46-49.
2. Banerian, G.: Status of some current research in jet noise. AIAA Journal, Vol. 16, 1978, No. 9, 877-888.
3. Ginevsky, A.S.; Vlasov, E.V.; Kolesnikov, A.V. Aeroacoustic interactions. Moscow: Mashinostroenie 1978.
4. Landau, L.D.; Lifshitz, E.M. Continuum mechanics. Moscow: Gos. Izd. Tekh.-Teor. Liter. 1953.
5. Novopashin, S.A.: Experimental pulse laser investigation of an underexpanded jet. In: Hydrodynamics of single and two-phase systems. ITF SO AN SSSR, 1982, 108-111.
6. Tables of physical values. Ed. by I.K. Kikoin. Moscow: Atomizdat, 1976, 76.
7. Kislyakov, N.I.; Rebrov, A.K.; Sharafutdinov, R.G. Diffusion process in the mixing layer of low density supersonic jet. Zh. Prikl. Mekh. i Tekh. Fiz., 1973, No. 1, 121-127.

Transformation of External Disturbances into Eigenoscillations in Supersonic Unbalanced Jets

V. N. GLAZNEV, SH. SULEIMANOV

Institute of Theoretical and Applied Mechanics, USSR Academy of Sciences, Novosibirsk 630090

The flow of supersonic unbalanced jets ($P_a/P_H \neq 1$ where P_a is the pressure in the nozzle exit, P_H is the ambient pressure) is accompanied by powerful self-induced oscillations. This phenomenon is typical both of jets exhausting into an ambient environment and of jets impinging on an obstacle, f.e., on a plane obstacle placed normally to the jet axis (Fig. 1). In both cases the self-induced oscillations develop due to the eigeninstability of the flow with respect to small disturbances. The acoustic waves emitted by a jet into the external space from a region of strong inhomogeneity of the flow (an obstacle or large longitudinal gradients of gasdynamic parameters in the case of a free jet) play a role of a regeneration feedback. Reaching the nozzle exit they efficiently transform into eigenwaves (of Tollmien - Schlichting wave type) of a jet shear layer. The jet shear layer transfers them with a definite increment and phase velocity to a radiation place and the processes repeated. The phase disturbances paths are shown in Fig. 1. The oscillation frequencies are determined by characteristic linear dimensions of the system and by phase disturbances velocities. When making use of the traditional instruments - hot-wire anemometer, total and static pressure probes provided by quick-response, it is possible to measure in the flow and external acoustic field the amplitude distribution and the fluctuation phase of the total pressure δP_0, static pressure δP, mass velocity $\delta(\rho V)$ and stagnation temperature δT_0 at the self-induced oscillation frequency. If the linear dimensions of the probes are small as compared with the wave length of disturbances, then the obtained results enable us

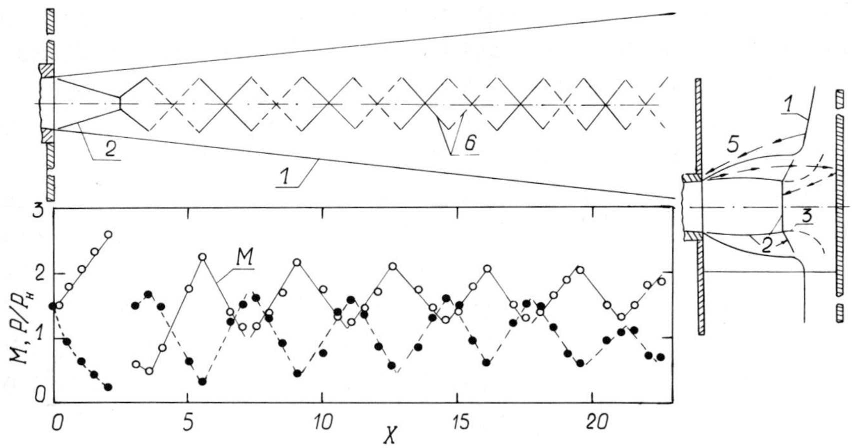

Fig.1. A scheme of the jet escape into an ambient space (free jet). Mach number and static pressure distributions along a nozzle axis. Jet impinging on an obstacle: 1 - jet boundaries, 2 - shock waves, 3 - tangential discontinuity, 4 - line of measurement of acoustic field characteristics, 5 - wave front trajectory, 6 - compression and rarefaction waves.

to calculate the velocity δV and temperature δT fluctuations in the flow. In the present communication the results of experimental study of spatio-temporal structure of disturbances at the self-induced oscillation frequency in the supersonic underexpanded ($n>1$) jet impinging on a plane obstacle and flowing into an ambient space. All the linear dimensions are referred to the nozzle exit radius r_a .

<u>A Jet Impinging on a Plane Obstacle</u>. The experiments have been carried out with a cold air jet (T_0 = 294K). The Mach number in the nozzle exit M_a= 1.5, n = 3, the radius of the nozzle exit was equal to 20 mm. The distance from the nozzle up to an obstacle equals X_1 = 5, the oscillation frequency f = 2 kHz. The obstacle was a rigid disc with the radius r_n = 15. With the purpose of increasing the self-induced oscillations stability in the nozzle exit a rigid metal disc (a plane screen) whose diameter d = 10 was installed coaxially. The results are presented in Figs. 2,3,4. Now we shall illustrate the main results.

1. A radiator of the external acoustic waves was placed in a region of the flow turn when interacting with an obstacle. The

availability of a screen-reflector in the nozzle exit has formed near the jet along the lines R = const an acoustic field close to the plane standing wave field. In the nozzle exit plane the amplitude of the pressure fluctuations $|\delta P|$ was equal approximately to 10^3 N/m^2.

2. In the jet shear layer in the radial direction there exists a typical distribution of the amplitude and phase of the mass velocity fluctuations $\dot{m} = \delta(\rho V)$ (longitudinal component). The amplitude has a minimum at $R = R^*$, on both sides from which there are two maxima. The phase $\dot{m}$ undergoes a discontinuity at $R = R^*$ the value of which is $\Delta\varphi = \pi$. By changing X the absolute and relative values of maxima $|\dot{m}|$ change.

3. The phase velocity of disturbances in the shear layer equals V_f = 0.9V_1 where V_1 is the flow velocity at the point $R = R^*$. The value of V_1 coincides with the velocity on the boundary of an inviscid unbalanced jet determined by the known formula by M_a, n

$$V_1 = M_1 a_1, \quad M_1 = \sqrt{\frac{2}{K-1}\left[\left(1+\frac{K-1}{2}M_a^2\right)n^{\frac{K-1}{K}} - 1\right]}, \quad a_1 = \frac{a_0}{\sqrt{1+\frac{K-1}{2}M_1^2}} .$$

<u>A Jet Flowing into an Ambient Environment</u>. The experiments were carried out with a cold air jet (T_0 = 293K). The Mach number in the nozzle exit M_a = 1.5, n = 1.5; the radius of the nozzle exit r_a = 20 mm, the oscillation frequency f = 2.50 kHz. In the nozzle exit plane the same screen-reflector was installed as in the previous case. Figs. 5,6 illustrate the results of experiments performed. Note the most essential ones.

1. In the jet shear layer the velocity fluctuations propagate downstream along a helicoidal line (they have an azimuthal component). The phase velocity of the velocity fluctuations along the same line is equal to V_f = 0.9V_1 where V_1 has the same sense as before. The longitudinal distribution of the x-component amplitude δV is related to a cellular structure of the jet: $|\delta V|$ has sharp maxima in that part of the cell where $\frac{\partial M_x}{\partial x} < 0$, $\left|\frac{\partial M_x}{\partial x}\right|$ achieves the maximum values, i.e. in the places of the most intensive flow stagnation. There are reasons to consider these jet parts to be the sources of vibrational energy radiation into the external space in the form of acoustic wave.

2. The radial distribution δV_x in the jet is similar to the radial distribution $\dot{m}$ in the above case: at $R = R^*$ (a salient

point in the distribution $M_X(R))\ |\delta V_X|$ has a minimum on both sides of which the maxima lie, the phase δV_X discontinuously changes by a value $\Delta\varphi=\pi$. However, the phase discontinuities are observed not in all the cross-sections.

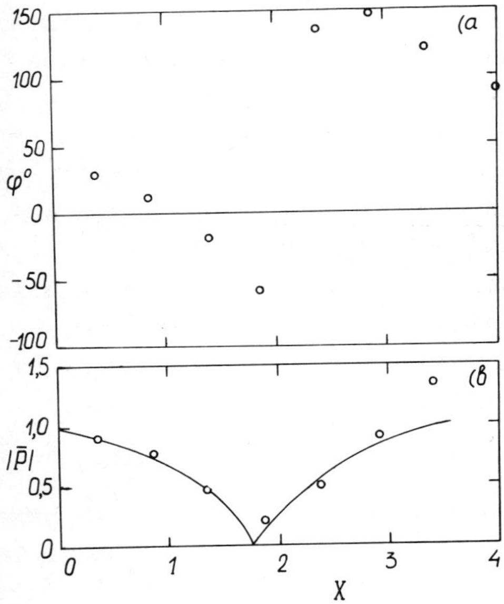

Fig.2. Chracteristics of the external acoustic field of a jet impinging on the obstacle along the line R = 2.5; a) - phase, b) - amplitude (solid line - cos KX, $K=\frac{2\pi}{\lambda}$, $\lambda \approx 7$)

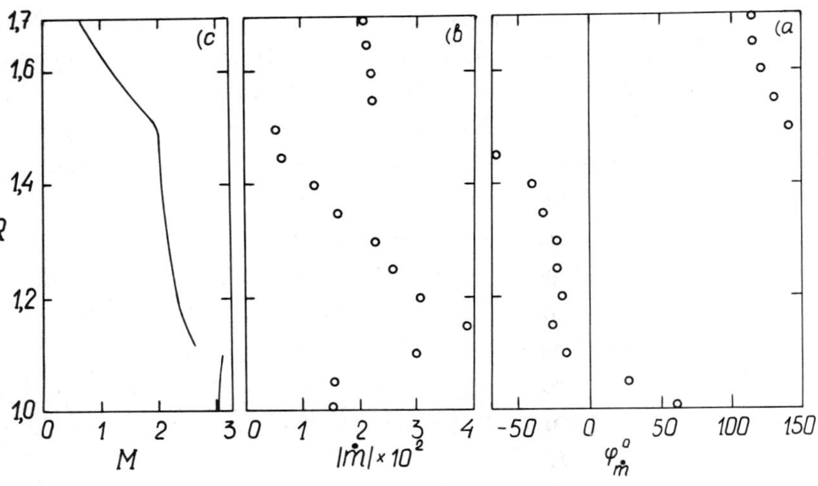

Fig.3. Distribution of the mass velocity fluctuations across a shift layer X = 1.85; a) - phase, b) amplitude $|\dot{m}|=|\delta(\rho V)|/(\rho V)_a$; $(\rho V)_a$ - averaged value of the mass velocity in the nozzle exit, c) Mach number.

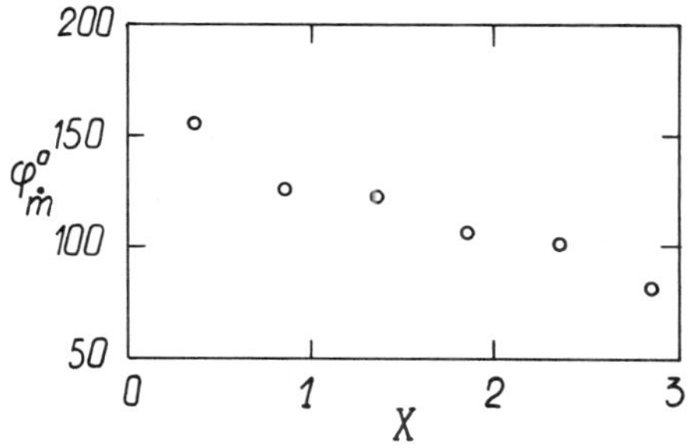

Fig.4. Distribution of the phase $\dot{m}$ along a shear layer

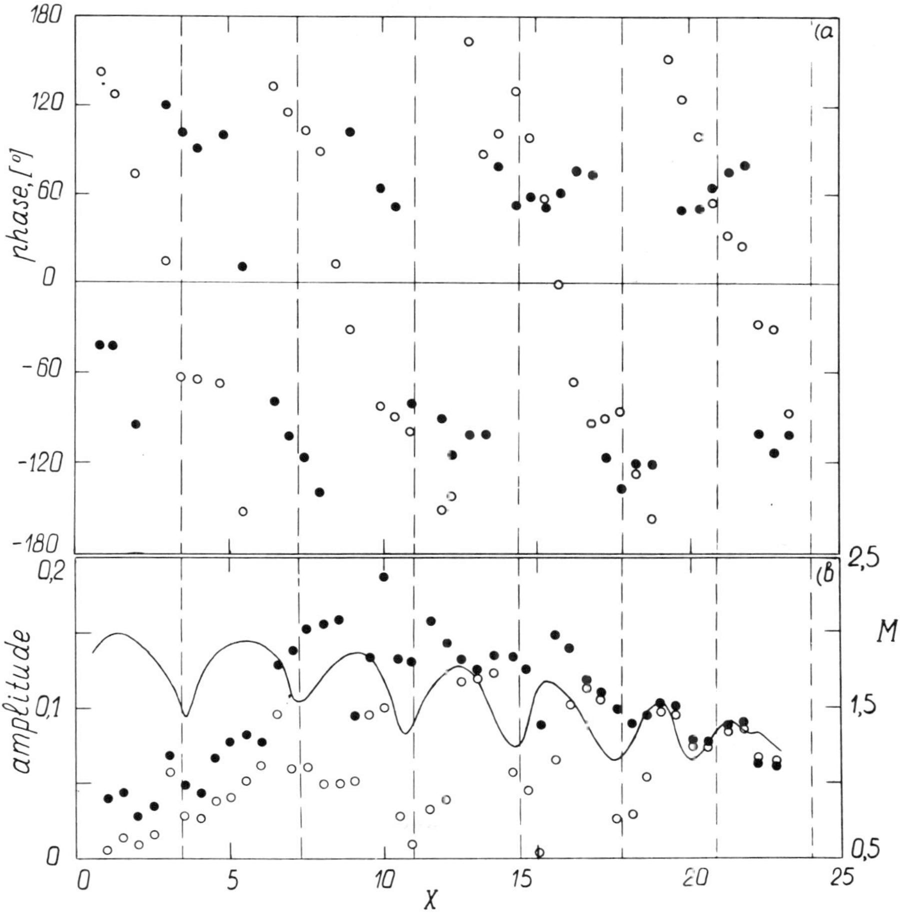

Fig.5. Free jet. Longitudinal distribution of the phase (a) and amplitude (b) of velocity and pressure fluctuations. R = 0.8, ● $-\delta P/P$, ○ $-\delta V/a$, ~ $-M$

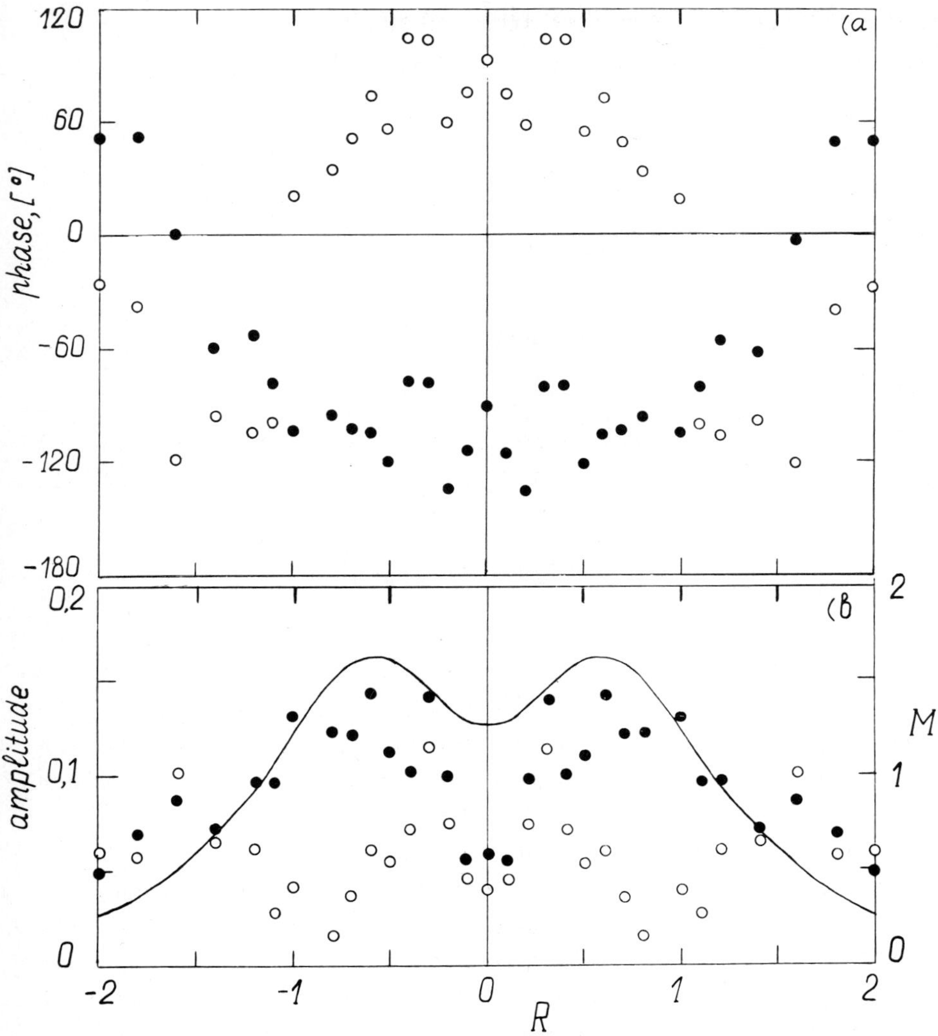

Fig.6. Free jet. Radial distribution of the phase (a) and amplitude (b) of velocity and pressure fluctuations. X = 11, ● - $\delta P/P$, o - $\delta V/a$, ~ — M

Transition at Natural Conditions and Comparison with the Results of Wind Tunnel Studies

B. Yu. ZANIN

Institute of Theoretical and Applied Mechanics
Siberian Branch of the USSR Academy of Sciences
Novosibirsk 630090, USSR

Summary

The results of the experimental investigations of boundary-layer laminar-turbulent transition on the glider wing by hot-wire anemometer and sublimating coating films are presented. The comparison between the measurement results at flight and in the wind tunnel is conducted.

Introduction

The problem of laminar-turbulent boundary-layer transition presents a great interest conditioned by the necessity to reduce skin friction of aircrafts, see Kachanov et al. [1] . At the same time important questions concerning the comparison of wind tunnel data with the results of the experiments under natural conditions are not sufficiently investigated. Especially it concerns the flights which take place under conditions close to the extremal ones such as, for example, in the clouds or near the earth where the atmosphere is known to be highly disturbed. It is to find out whether under such conditions there will be valid such physical mechanisms which cause laminar-turbulent transition when modelling in a low-turbulence wind tunnel or when flying in a weakly disturbed atmosphere.

The Flight Investigations

To conduct experimental flights a serial Checoslovac two-seater glider "Blanik" L-13 was used. The measurements of the turbulence level of the free flow was made by the hot-wire anemometer system DISA 55D. The hot-wire anemometer probe was fixed on a holder in front of the fuselage nose part. A calibrating curve of the hot-wire anemometer probe was obtained according to the results of the measurements in the wind tunnel. The calibration was also conducted in flights far from the clouds and in flights in the clouds. The results of the flights calibration fit in the calibrating curve with a good accuracy taking into account the class of the glider standard equipment which makes possible to use the above mentioned curve to process measurement results obtained in the flight.

In a cloud, when the wire of the probe additionally intersects liquid drops, the wire resistance changes due to the effect of cooling by flow and by drops. When the drops meet a heated wire, there appear peaks on the oscillogram (Fig. 1a). The main disturbances energy of free flow is to be found in the region of rather low frequencies, less than 300 Hz (Fig. 1b), whereas the ejections caused by the drops are to be found in a high-frequency part of the spectrum(Fig. 1c), beyond 700 Hz. It is to be noted that the contribution of these ejections (the filtration being absent) in the total energy of the disturbances is too small and constitutes several fractions of a percent of the total signal value.

The results of the measurements of the level of the free flow turbulence in 1 Hz - 5 kHz frequency range at the glider velocity equal to 25 m/sec which depend on the altitude and atmospherical conditions are shown in Fig. 2. The turbulence level at 100-1200 m altitude varied in the range $\varepsilon = 0.2 \div 0.4\%$, and it sharply increased in a cloud and near the earth.

To carry out the measurements in a boundary layer on a glider wing there was used a hot-wire anemometer of the constant temperature TPS-2M developed at our Institute. Miniature hot-wire anemometer probes were attached to the glider wing so that the sensitive element of the probe (the gilded tungsten wire with a 6 micron diameter and a 1.5 mm length) was to be found in a boundary layer at the distance of about 0.5 mm from the wing surface. The oscillograms and frequency spectra of the pulsations in the boundary layer of the wing during the flight in an undisturbed atmosphere and in the cloud are presented in Fig. 3. In both cases it is possible to distinguish clearly a laminar flow, the region of the appearance of the packet of waves and the turbulent flow. Thereat the mean frequency of the waves packet which arises at laminar-turbulent transition, f = 650 Hz, remains approximately the same when flying in a cloud as well as far from it. In a cloud the level of pulsation in the laminar region of the boundary layer of the wing increases proportionally to the growth of the turbulence level in a free flow.

The frequency spectra and oscillograms of the disturbances in a boundary layer on the glider wing when flying at the altitude of 300 m and at the altitude of 1 m was also considered. Here there can be noted the same regularities of the flow structure changes as in the previous case.

The Visualization of Laminar-Turbulent Transition

The visualization of the flow in a boundary layer was carried out using two sublimating materials: naphthalene and diphenyl which having been dissolved in acetone were sprayed on the surface under investigation to form a thin white coating, see Main-Smith [2] . In Fig. 4 there is given the photograph of the visualization results of transition under natural conditions, i.e. in flight, on the glider wing. The region of the laminar flow in the boundary layer of the wing at the flight velocity 25 m/sec equals to 44% of the chord (for the profile with a 1340 mm chord). It coincides approximately with the results of hot-wire measurements.

Comparison of the Results of Investigations at Flight and in the Wind Tunnel

The frequency spectra of the free flow pulsations in flight and in the wind tunnel are distinctly different, see Fig. 5, at the same time levels of turbulence approximately coincided. In Fig. 6 there are presented the frequency spectra of the most characteristic types at the three points of the boundary layer obtained in a wind tunnel and under natural conditions on the wings of equal dimensions. In both cases a laminar-turbulent transition is preceded by the formation and the development of wave packets in a boundary layer. Under natural conditions the mean frequency of wave packet is 650 Hz, and in wind tunnel it is 500 Hz at the velocity of 25 m/sec. Seemingly, the frequency of oscillations appeared is determined by the conditions of the generation of these oscillations at the beginning of the unfavourable pressure gradient region on the profile and not by the frequency spectrum of the free flow pulsations, see Dovgal & Kozlov [3] . The discrepancy between the frequencies of the appearing oscillations for the flight and for the experiment is possibly connected with a partial coincidence of the form of the model profile and of the wing and, correspondingly, with slightly different pressure gradients.

Acknowledgement

The author thanks Dr. V.V.Kozlov for his scientific guidance, valuable discussion and suggestions for carrying out this work.

References

1. Kachanov, Yu.S.; Kozlov, V.V.; Levchenko, V.Ya.: Beginning of turbulence in a boundary layer (in Russian). Novosibirsk: Nauka 1982.

2. Main-Smith, J.D.: Chemical solids as diffusible coating films for visual indications of boundary-layer transition in air and water. A.R.C., R&M N°2755 (1950) 1745-1760.

3. Dovgal, A.V.; Kozlov, V.V.: Influence of acoustic disturbances for structure in the boundary layer at unfavourable pressure gradient (in Russian). Izv. Akad. Nauk SSSR, Mekh. Zhidk. i Gaza 2 (1983) 48-52.

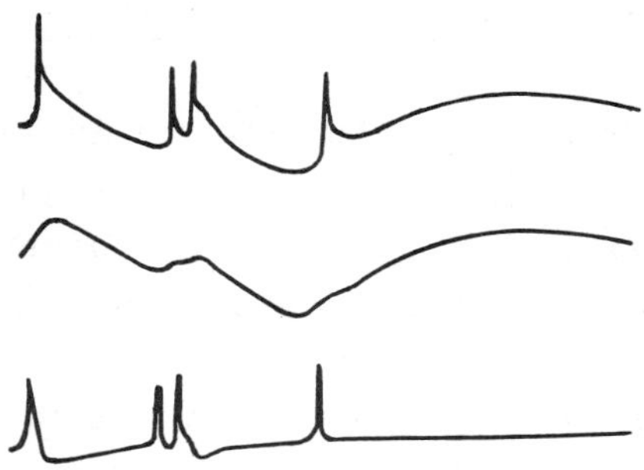

Fig. 1 The oscillograms of the hot-wire signal in a cloud

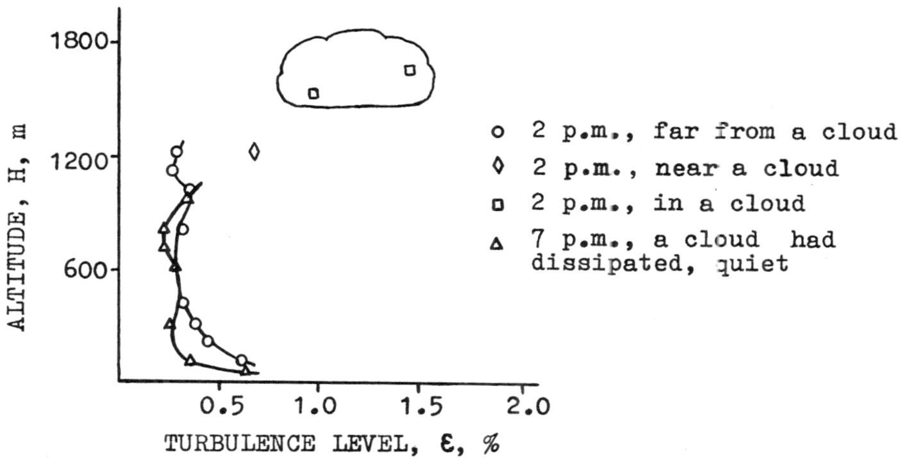

Fig. 2 The turbulence level of free flow in 1 - 5000 Hz frequency range at the glider velocity equal to 25 m/s

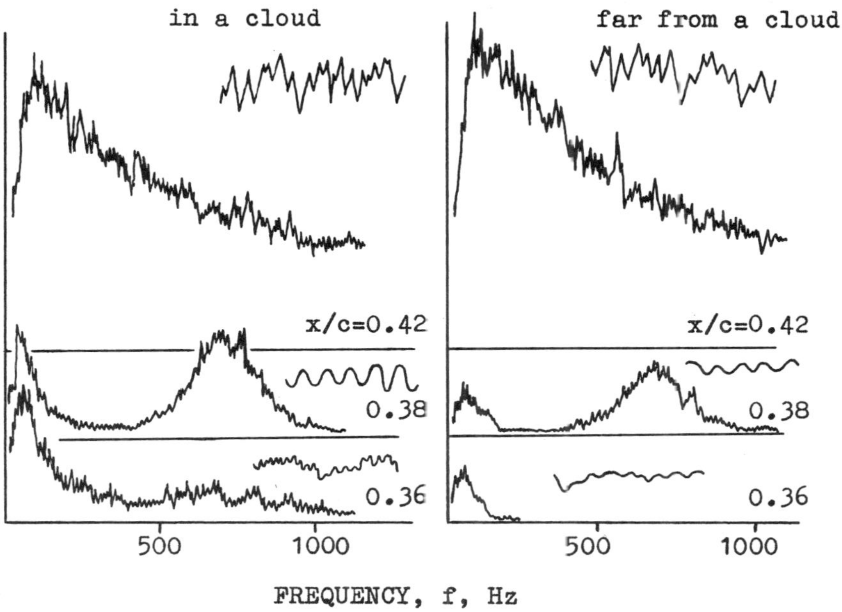

Fig. 3 The oscillograms and frequency spectra of fluctuations in the boundary layer of the wing during flight in an undisturbed atmosphere and in a cloud

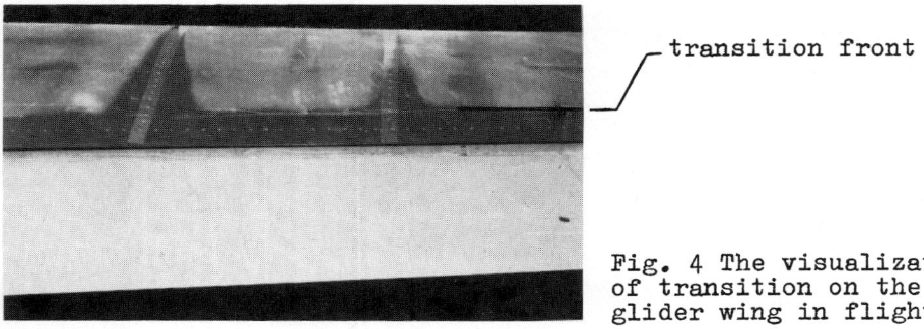

Fig. 4 The visualization of transition on the glider wing in flight

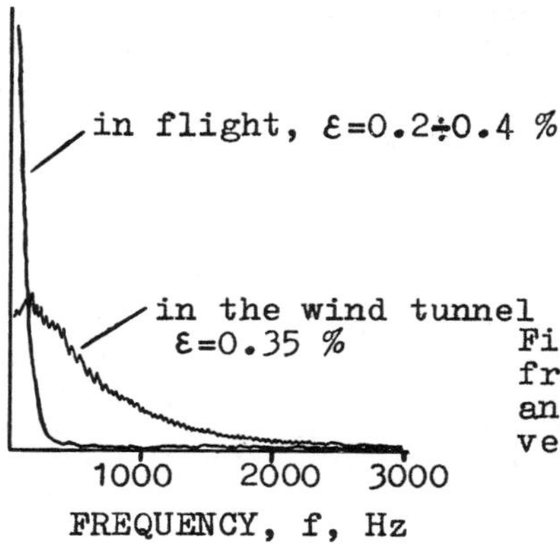

Fig. 5 The frequency spectra of free flow fluctuations in flight and in the wind tunnel at the velocity equal to 25 m/s

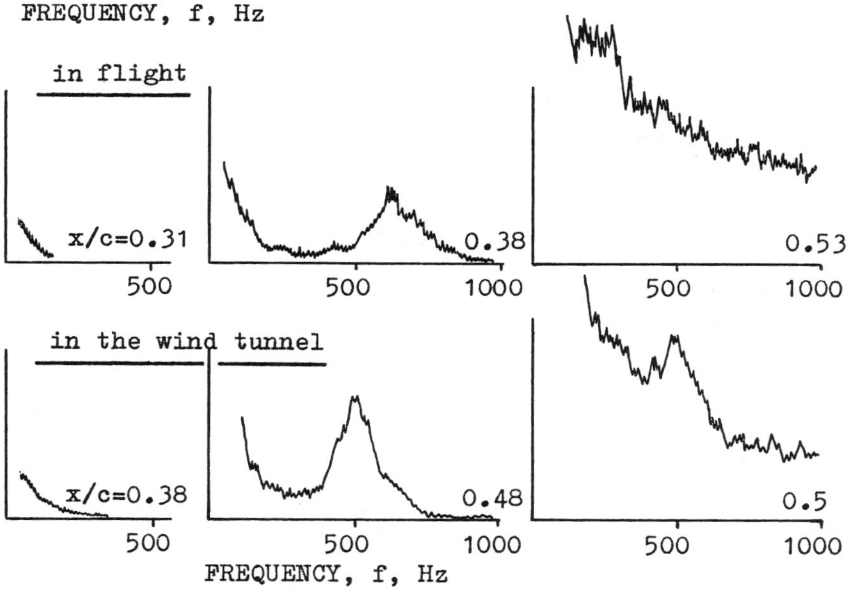

Fig. 6 The frequency spectra of fluctuations in the boundary layer on the glider wing in flight and in the wind tunnel

On the Problem of Transition of a Compressible Laminar Boundary Layer into Turbulent One Under Natural Conditions

A. M. PAVLYUCHENKO, A. A. TYUTIN, E. M. MAKSIMOVA, O. A. BRAGIN

Institute of Theoretical and Applied Mechanics
USSR Academy of Sciences, Novosibirsk 630090, USSR

Novosibirsk State University, Novosibirsk 630090, USSR

Summary

The results of experimental and computational studies pertaining to the transition of supersonic boundary layers on meteorockets are presented. When a head part of the meteorocket is flowed around without separation, the approach based on the Emmons' spot theory enabled one to describe the heat state of a wall in the transition region with a good accuracy. The flying data indicative of the transition in the flow separation zone are obtained. For flying conditions the critical values of the gradient criterion of stability are calculated.

Introduction and Problem Formulation

The urgency and practical importance of studying the problem of transition of laminar boundary layer into turbulent one have been emphasized in [1,2] . The difficulties of the study of stability and transition of supersonic laminar boundary layers and of transition phenomenon, prospects of their study have been stated in [2] . In [1] , a conclusion about a necessity of the formulation of accurately planned flying studies of the transition phenomenon is drawn. The following problems are stated in the present paper:

1) To study the questions of transition in the near-wall boundary layers on a head part of a meteorocket of the "Cloud" type when it is flowed around without separation and at the flow separation conditions at $M_\infty < 2$, $Re_L \leqslant 2 \cdot 10^7$.

2) To approve the approach on the basis of the Emmons' spot theory for description of heat exchange in a transient regime of the boundary layer under flying conditions on a meteorocket of the "Cloud" type.

3) To calculate the critical values of the gradient criterion of stability [3] for the laminar boundary layer on a meteo-

rocket of the "Cloud" type at $M_\infty < 2$ and for relaminarization on a meteorocket at $M_\infty \leq 4.5$ [4] .

Results of Studying Questions of Transition under the Natural Conditions on Meteorockets

In accordance with the problem formulation there were made two variants of flying object on the basis of a meteorocket of the "Cloud" type [4,5] moving along the path with acceleration at $M_\infty < 2$, $Re_L \leq 2 \cdot 10^7$, $H \leq 8$ km, L being the length of a head part. The head parts were supplied with board measuring systems and telemetry for 8 and 16 canals, respectively, and were rescued by using the parachutes. Under flying conditions from the moment of launching the wall temperatures along the head part lengths and the air temperature in the section were measured with the help of the semiconductor diodes of the KD-522 type. Their sensitivity accounted for 2.5 mV/grad. The measuring error of the temperature wall did not exceed 1%. Figure 1 illustrates a scheme of the flying object on the basis of the meteorocket of the "Cloud" type (1 - the head part with measuring systems, 2 - engine, 3 - parachute section). Figures 2,3 demonstrate the flying data on the change of the wall temperature at points 1,3 along the head part length under conditions of the separationless flow-around (1 - calculation, 2 - flying experiment). The numbers $Re_{x,e} \leq 2 \cdot 10^6 - 2 \cdot 10^7$ for the time of the meteorocket motion along the path $\tau \leq 5$ sec and at $\tau > 5$ sec - $Re_{x,e} = 2 \cdot 10^7 - 4 \cdot 10^6$.

We have made use of two approaches for solving a question what kind of boundary layer laminar, transient or turbulent is formed in the "Cloud" head part. Firstly, on the basis of the technique [5,7] there were calculated the wall temperatures under assumption of the existence of only laminar or only turbulent boundary layers from the moment of the meteorocket launching. Computational data are compared with flying values of T_w represented in Figs. 2,3. It turned out that the assumption about the presence of turbulent boundary layer for $\tau \leq 13$ sec and laminar at $\tau > (6 - 7)$ sec is not valid. Secondly, the theory of stability and flying data on the temperature factor are employed. The carried out comparison has given a good cause to assume that at $0 < \tau \leq (6 \div 7)$ sec there takes place the lami-

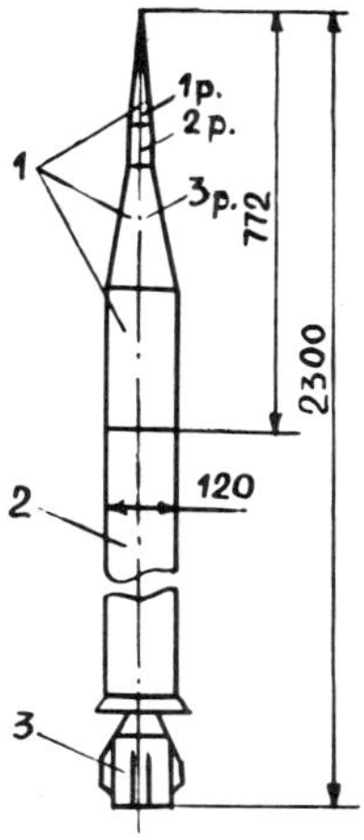

Fig.1. A scheme of the flying object

nar boundary layer; at $\tau > (6 \div 7)$ sec there is formed a transient flow regime. In order to compute the wall temperature of the head part aimed at the following comparison with the flying experiment (Figs. 2,3) firstly, it was used a nonlinear differential equation for a thin wall as the Bio

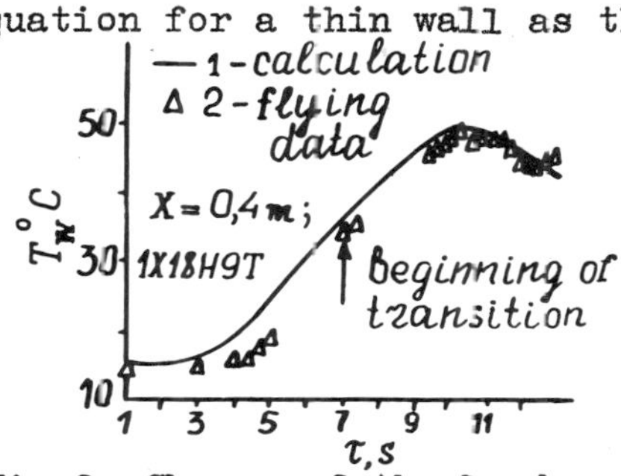

Fig.2. Change of the head part temperature in time

criterion $Bi \leq 0.01$, secondly, the heat flow towards a wall by the external flow-around was determined with regard for the intermittency coefficient. In this case the equation of the energy balance for a thin wall on the "Cloud" head part [5] has the form:

$$\rho_w \cdot c_w \cdot \delta_w \cdot \frac{dT_w}{d\tau} = [\alpha_\ell \cdot (1-\gamma) + \alpha_T \cdot \gamma] \cdot (T_{r,e} - T_w) - \alpha_1 (T_w - T_{int}^{air}) - q_{w,\Sigma}^{rad} \tag{1}$$

Here α_ℓ, α_T, α_1 are the heat transfer coefficients for laminar and turbulent flow regimes in a compressible boundary layer and from the wall towards the air in the section, respectively, determined after [7] . Thirdly, the intermittency coefficient γ in (1) is determined with regard for Emmons' turbulent spot theory [8] . With this end in view we have made use of [9] in which on the basis of Emmons' theory and of the relationship between the turbulent spot formation rate and Reynolds number in the beginning of the transition zone the expression for the intermittency coefficient is derived. The Reynolds transition numbers Re_{tr} [5] are estimated on the basis of flying data. From Figures 2,3 one can see that the coincidence of computa-

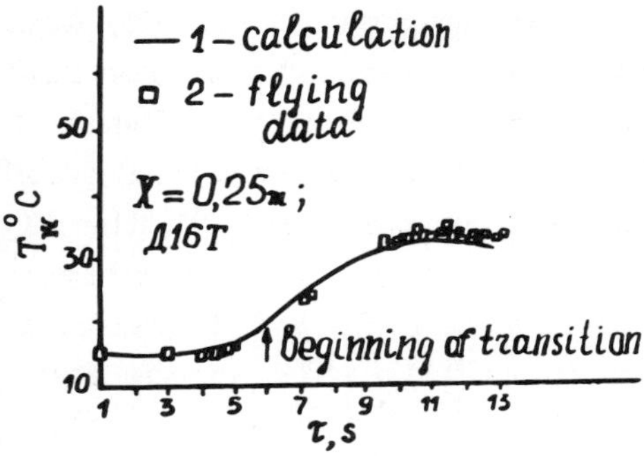

Fig. 3. Change of the wall temperature of a head part with time

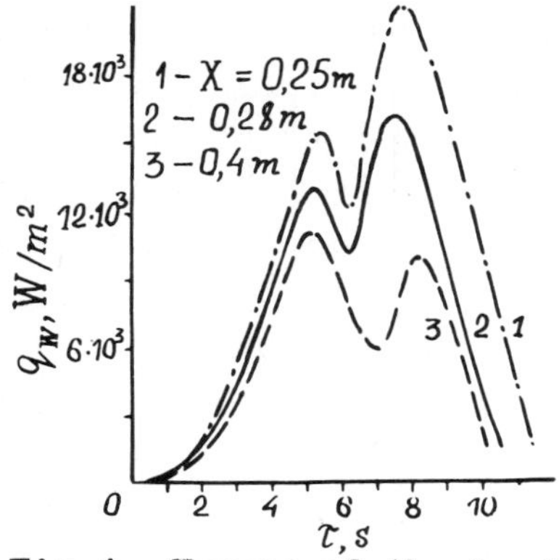

Fig.4. Change of the heat flow on a head part with time

tional and flying data on the temperature wall of the "Cloud" head part is rather satisfactory. The appearance of the second maximum of the heat flow in Fig. 4 is connected with transition in the boundary layer. The difficulties of the analysis of the transition phenomenon in the separated laminar boundary layer are conditioned by complexity of both the transition problem itself and of separated flows [6] . The boundary layer separation on the head part was generated with the help of a ring-shaped step of the height h = 6 mm mounted on a cylindrical section at the distance X = 0.312 m from the meteorocket noze [6] . According to computations the thickness of the boundary layer up to the separation made up $\delta \leq 0.5$ mm. In the separation region there were placed 7 diodes. The measurements of the wall temperature were carried out by the flight of the meteorocket along the path starting from the launching. Figure 5 illustrates the dependency of the heat flow towards the head part wall in the separation region of boundary layers at $\tau_{1,2,3}$ = 3.9, 4.7, 5.5 sec.

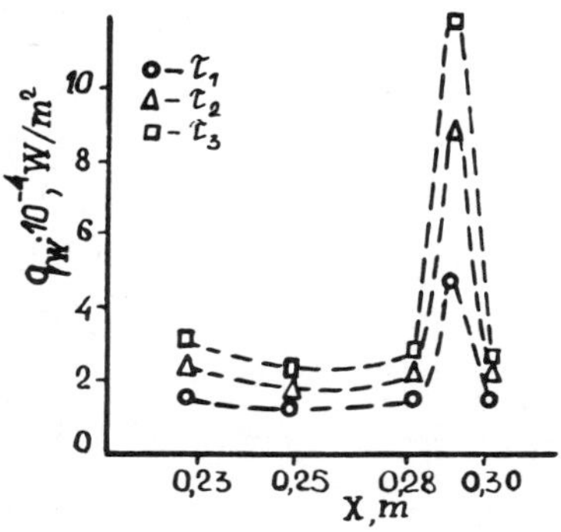

Fig.5. Change in the heat flux in time in the separation zone before the step

A local peak occurs at X = 0.292 m for the heat flow. The heat flows are calculated on the basis of the method of a thin wall with the use of smoothing cubic splines. The boundary layer was in the laminar state up to the separation point. The heat flow peaks exist in that case when the boundary layer is attached in the transient or turbulent regime [6] . The invariant criteria of stability and transition suggested in [3] differ by some general points. Unlike [3] , in the present case for determination of the value Re_{gr} we have made use of the laminar velocity profiles obtained in supersonic wind tunnel at $M_\infty = 2.5$ with a porous suction and without it, and computational supersonic velocity profiles by the laminar flow around the meteorockets [5,7] under real conditions of aerodynamic heating. The first and second velocity derivatives U_y and U_{yy} necessary for determination of Re_{gr} have been calculated with the help of smoothing cubic splines. It is shown that in the cases under study $Re_{gr}^{max}(y) < 30$, that is in a good accord with [3] . It is typical that for conditions of the flow around meteorockets in boundary layers there were two values of Re_{gr}^{max}: at $y/\delta \approx 0$ and at $y/\delta \approx 0.5$. For meteorockets with $M_\infty \leq 4.5$ the value $Re_{gr}^{max}(y) < 30$ was obtained for the relaminarization regime [7] .

The authors express their gratitude to N.A.Zheltukhin for fruitful discussions and S.A.Gaponov for attention to the present study.

References

1. Reshotko, E.: Stability of laminar boundary layer and its transition into turbulent one. In: "Vortex motion of fluid", Moscow, "Mir", 1979, 11-57.

2. Gaponov, S.A.; Maslov, A.A.: Development of disturbances in compressible flows. Novosibirsk, "Nauka", 1980.

3. Yanenko, N.N.; Gaponov, S.A.: Invariant criteria of stability of compressible fluid flows. DAN SSSR, v. 259, 5 (1981) 1056-1059.

4. Pavlyuchenko, A.M.; Bragin, O.A.; Tyutin, A.A.: Flying measurements of temperature and pressure on rescued head parts

of meteorockets of M100 and "Cloud" type. Izvestiya SO AN SSSR, Seriya tekhn. nauk, vyp. 1, N°3 (1983) 46-54.

5. Pavlyuchenko, A.M.; Maksimova, E.M.: Numerical methods of processing the flying heat experiment on the "Cloud" type meteorocket with regard for Emmons' turbulent spot theory. Izvestiya SO AN SSSR, Seriya tekhn. nauk, vyp. 2, 8 (1983) 27-35.

6. Pavlyuchenko, A.M.; Tyutin, A.A.; Korablev, A.V.: Heat exchange on "Cloud" type meteorocket in the presence of boundary layer separation. Izvestiya SO AN SSSR, Seriya tekhn. nauk, vyp. 1, N°4 (1984) 52-63.

7. Maksimova, E.M.; Pavlyuchenko, A.M.: Comparison of flying and computational data on heat exchange for axisymmetric bodies moving along the path at $M_\infty \leqslant 5$. Izvestiya SO AN SSSR, Seriya tekhn. nauk, vyp. 3, 13 (1982) 40-52.

8. Emmons, H.W.: The laminar-turbulent transition in boundary layer. Part I. Journal of the Aerospace Sciences, v. 18, N°7 (1951) 450-458.

9. Chen, K.K.; Thyson, N.A.: Extension of Emmons' spot theory to flows on blunt bodies. AIAA Journ., v. 9, N°5 (1971) 821-825.

Experimental and Theoretical Studies of Boundary Layer Transition on a Swept Infinite Wing

R. MICHEL, D. ARNAL, E. COUSTOLS, J.C. JUILLEN

ONERA/CERT - 2 avenue Edouard Belin - 31055 TOULOUSE Cedex (FRANCE)

SUMMARY

This paper presents experimental and theoretical results dealing with the transition phenomena on an infinite swept wing. The research is directed towards two ways : at first, a study of the transition movement under the influence of the flow parameters ; secondly, a more fundamental study of stationary waves, which play an important role for the transition mechanisms in three-dimensional flow.

INTRODUCTION

The infinite swept wing represents the more simple example of three-dimensional flow. Two coordinate systems are usually defined (fig. 1) : one, (x, z, y) is linked to the wing, the other, (x_1, z_1, y) is linked to the external streamline. The fundamental assumption is that the derivatives along the span direction z are equal to zero. The streamwise profile U1 and the cross-flow profile W1 refer to the projections of the local velocity profile in the x_1 and z_1 directions, respectively.

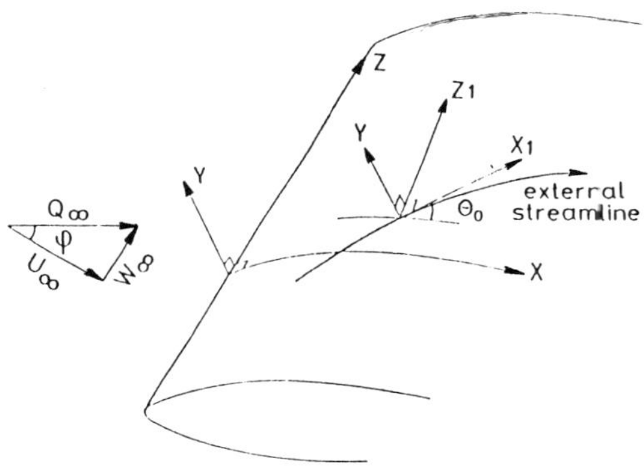

Fig.1. Geometry and notations.

Over swept wings, turbulence may appear through three mechanisms : leading edge contamination, streamwise instability (related to the properties of the U1 profile), cross-flow instability (related to the properties of the inflexional W1 profile) /1/. The streamwise instability is similar to that of a two-dimensional flow and leads to turbulence essentially in positive pressure gradients. The cross-flow instability can induce transition in strong negative pressure gradients ; as it has received less attention than the streamwise one /2/, /3/, a fundamental study was undertaken at ONERA/ CERT. This paper presents the first experimental results obtained in this way.

2 - EXPERIMENTAL ARRANGEMENT

The test section of the open-circuit wind tunnel has a cross-section of 35 x 60 square cm. The maximum velocity is close to 90 m/s. The model is a constant-chord, variable-sweep wing with an ONERA D section normal to the leading edge. The wing has a chord C equal to 0.35 m and its span is about two meters. The model is equipped with four rows of static pressure taps.

The free-stream turbulence level is of the order of 0.15 %. The wing is directly in contact with the test section walls, but preliminary experiments have shown that leading edge contamination never occurs for the present experimental conditions. No measurement has been done at the wall-wing junction, where a turbulent wedge develops, with an angle of spread close to 15°.

3 - EXPERIMENTAL CONDITIONS

As previously mentioned, the main objective of this study was to analyze transitions due to cross-flow instability. For this reason, attention was focused on external velocity distributions exhibiting large regions of negative pressure gradients.

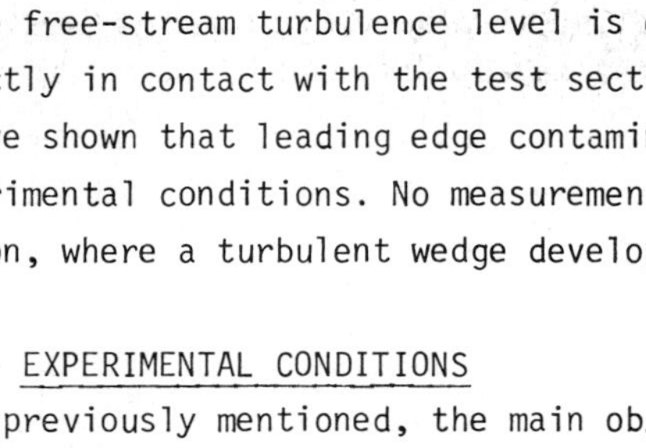

Fig.2. External velocity distributions.

Two configurations, denoted by (α_n, φ) were studied : (-6°, 60°) and (-8°, 40°) where α_n is the normal angle of attack and φ the sweep angle. Fig. 2 shows the experimental distributions of the external velocity component normal to the leading edge. In both cases, the angle between the external streamlines and the wind tunnel axis remains low, at least for x/c > 0.20. As a consequence, it will be assumed that a hot-wire aligned along the working section axis provides us with the value of the streamwise velocity.

4 - TRANSITION LOCATION

The transition location is determined, either by doing wall visualizations (sublimation) or by looking at the streamwise total pressure distribution at a constant distance close to the wall. However, it is difficult to give an accurate value of the transition abscissa. This can be seen on fig. 3, which shows an example of sublimation result, obtained for the case (-8°, 40°), with Q_∞ = 81 m/s. Black regions, characterized by high values of the skin friction coefficient, are visible near the leading edge (laminar boundary layer at low Reynolds number) and beyond 25 or 30 % of chord (transitional and turbulent boundary layer). Between these regions, "streaks" regularly spaced in the spanwise direction and practically aligned with the potential flow can be observed. These streaks were already pointed out by many authors /3/ /4/ and constitute a fundamental aspect of the three-dimensional transition problem.

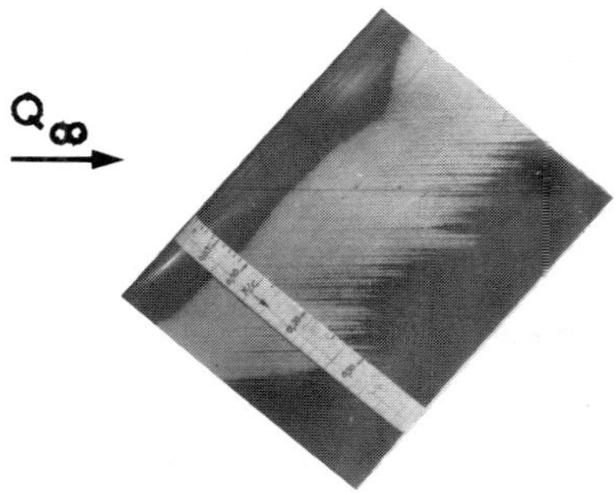

Fig.3 sublimation visualisation.

Due to the presence of the streaks, the transition front exhibits a "saw-tooth" pattern in the spanwise direction. Nevertheless, we tried to determine the transition movement when the aerodynamic parameters were changed, by measuring only a more or less subjective transition location.

Figure 4 shows the mean location of the transition point, as a function of free-stream velocity Q_∞ for (- 6°, 60°) and (- 8°, 40°). An estimate of the experimental uncertainty is also given. Application of transition criteria leads to the conclusion that the measured transitions are due to cross-flow instability.

In order to give a complete picture of the flow development, some measurements were performed downstream of the transition onset, and a calculation method, based on an "intermittency" formulation, was developed. A detailed comparison between computations and experiments can be found in /5/.

This first series of experiments enabled us to obtain an overall view of the transition phenomena. As a next step, a detailed study of the streaks was undertaken.

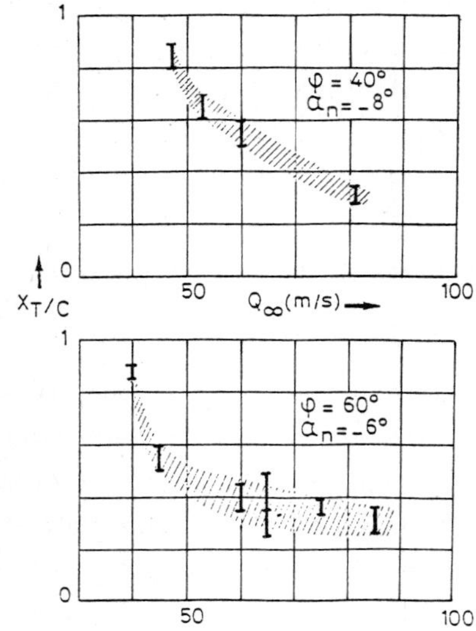

Fig.4. Transition location.

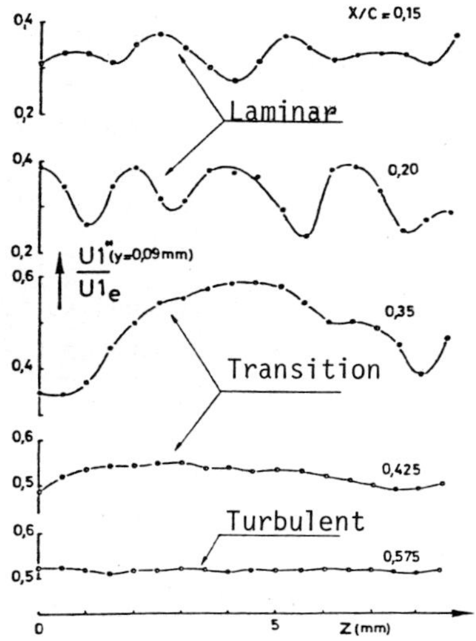

Fig.5. Streamwise velocity in the (y,z) plane.

5 - STUDY OF STATIONARY WAVES

The existence of streaks practically aligned with the external streamlines was already demonstrated on fig. 3. Fig. 5 illustrates their importance in a more quantitative way, for the same experimental conditions. It shows the spanwise evolution of the streamwise velocity measured at a constant distance from the wall, for various stations. In laminar flow (x/c = 0.15 and 0.20), the mean velocity exhibits a well-defined wavy evolution. In the transition region (x/c = 0.35 and 0.425), the evolution becomes completely chaotic. However, when the turbulent regime is entered (x/c = 0.575), the spanwise velocity variations are damped. It appears essentially that the laminar boundary layer is not uniform in the direction parallel to the leading edge. This leads to a wavy evolution of the skin friction coefficient, which explains the appearance of streaks when wall visualization techniques are applied.

5.1. Theoretical elements

From a theoretical point of view, this phenomenon can be studied by using the laminar stability theory, in which velocity and pressure disturbances are assumed to be of the form :

$$(\tilde{u}_1', \tilde{v}_1', \tilde{w}_1', \tilde{p}') = (f(y), \phi(y), h(y), \pi(y))e^{\omega_i t} e^{i(\alpha x_1+\beta z_1-\omega_r t)} \quad (1)$$

f, ϕ, h and π are amplitude functions. ω_i and $\omega_r = 2\pi\bar{f}$ represent respectively a temporal amplification factor and a circular frequency. α and β are the two components of the wavenumber vector $\vec{k}$ in the x_1 and z_1 directions. $\vec{k}$ can be also defined with its modulus α_ε and the angle ε between $\vec{k}$ and the z_1-direction.

Introducing (1) into the continuity ard linearized NAVIER-STOKES equations, one can obtain the following equation :

$$\phi^{IV} - 2\alpha_\varepsilon^2\,\phi'' + \alpha_\varepsilon^4\,\phi = iR|(\alpha_\varepsilon U_\varepsilon - \omega)(\phi'' - \alpha_\varepsilon^2\,\phi) - \alpha_\varepsilon\,U_\varepsilon''\,\phi| \quad (2)$$

ω is complex and U_ε represents the mean velocity profile projected in the ε-direction. When the Reynolds number R, ω_r, ε, U_ε (and its second derivative U_ε'') are specified, the computation provides us with the eigenvalues α_ε and ω_i, and with the eigenfunctions f, ϕ, h ard π.

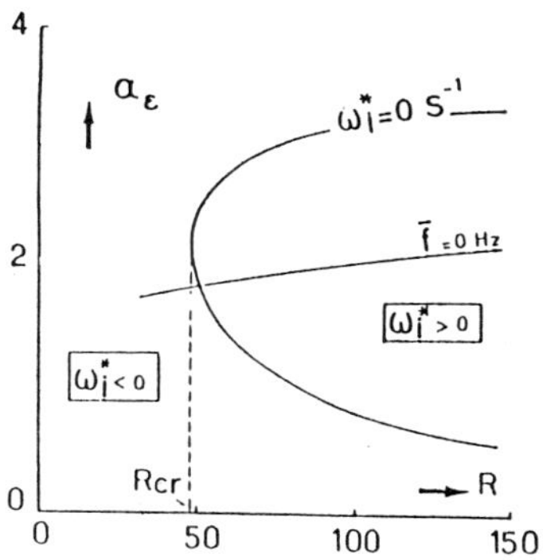

Fig.6. Typical stability dyagram.

Figure 6 shows a typical stability diagram which can be obtained for low values of ε (say, ε of the order of 2 or 5 degrees), in a negative pressure gradient. An important result is that the zero-frequency, stationary waves are highly unstable. The streaks constitute the signature of these waves.

5.2. Detailed study of the streaks

In order to do more precise measurements, the wind tunnel speed was reduced to 48 m/s, by keeping α_n and φ constant (-8° and 40°). In these conditions, transition begins at x/c≃0.85. A detailed study of the streaks was undertaken at x/c = 0.69, where the laminar boundary layer thickness is close to 2 mm. In the following lines, U_1^* will refer to the time-averaged streamwise velocity measured by the hot wire probe.

Mean velocity U_1^* - At first, the mean velocity profiles were measured for 18 spanwise stations. The spacing between two adjacent stations is 0.5 mm. The results are given in figure 7 in a $U_1^*(y,z)$ representation. The measurements

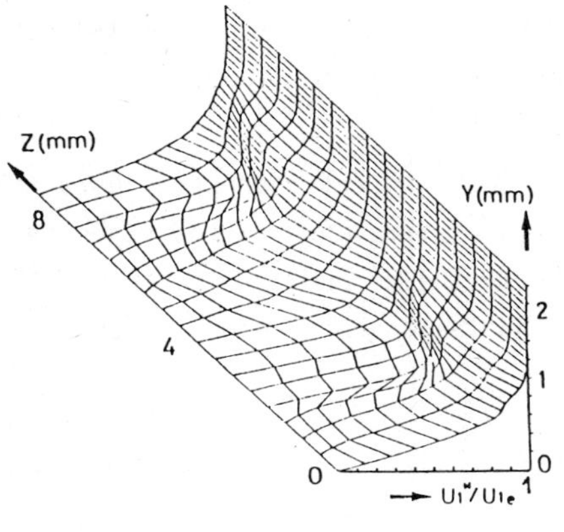

Fig.7. Streamwise velocity in the (y,z) plane

were carried out at mid-span and the origin z = 0 is arbitrary. The span deformation of the profiles appears clearly, with a spanwise wavelength close to 5.5. mm.

The next step was to see if the laminar stability theory was able to reproduce this evolution. The principle of the calculation is the following : boundary layer calculations are performed by using the infinite swept wing assumptions, in order to obtain the U1 and W1 profiles ; these profiles are then introduced as data into the stability program, in which the frequency is set equal to zero. For each value of the angle ε, the computation gives the eigenfunctions f, ϕ, h and π and the eigenvalues α_ε and ω_i. The calculations indicate that the stationary waves are unstable for a narrow range of ε ($1.4^\circ < \varepsilon < 3.4^\circ$), corresponding to a relatively wide range of λ^* (3 mm $< \lambda^* <$ 30 mm) /5/. In the following theoretical results, we will consider only the locally most unstable direction ($\varepsilon = 1.5^\circ$, $\lambda^* = 5.9$ mm).

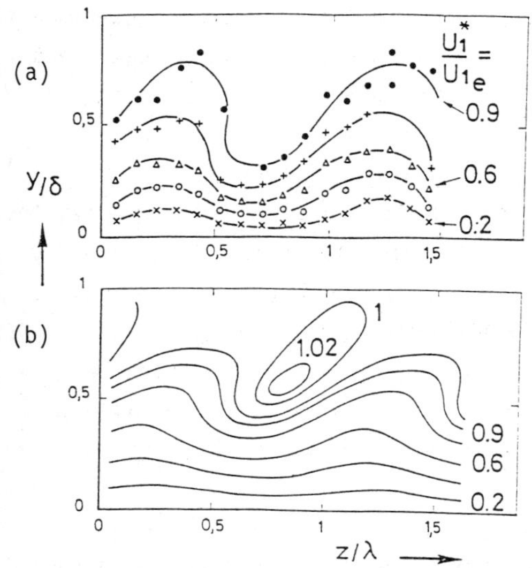

Fig.8. Measured (a) and theoretical (b) isovelocity lines.

The theoretical streamwise velocity profile U_1^* is obtained by adding the perturbation profile u_1' (real part of $\tilde{u}_1'$) to U_1. Because the stability equations are homogeneous, a multiplicative factor is applied to $\tilde{u}_1'$ in order to fit the experimental results. Figure 8 presents a comparison between theory and measurements. The overall pattern of the iso-velocity lines is fairly well reproduced.

Secondary velocities v_1' and w_1' - Because the secondary velocities are difficult to measure, only theoretical results will be presented.

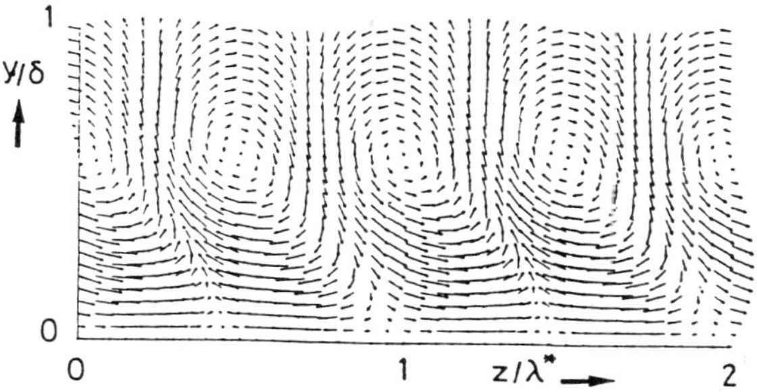

Fig.9. Perturbation vectors (v_1', w_1') in the (y,z) plane. Arbitrary scale

Figure 9 shows the perturbation vectors (v_1', w_1') in the (y, z) plane. In each wavelength λ^*, the disturbances take the form of two counter-rotating, streamwise vortices. However, their amplitude remains small : when the mean secondary velocities V_1 and W_1 are added to v_1' and w_1', the vortices are no longer clearly visible.

Turbulence profiles - The hot wire enables us to measure the root mean-square value, designated as $u_1'^*$, of the streamwise velocity fluctuations. The experimental turbulence profiles $u_1'^*/U_{1e}$ are plotted on figure 10.

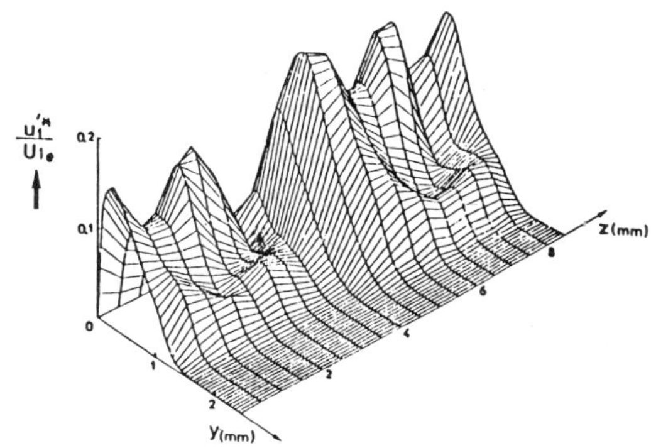

Fig.10. Turbulence profiles in the (y,z) plane.

As it can be expected, their spanwise evolution is not uniform. During a wavelength λ^*, the maximum turbulence level presents two maxima and two minima, the values of which are very large, up to 15 or 20 % of the resultant potential velocity. At z = 2 mm, the turbulence profile exhibits two maxima in the y direction. The corresponding mean velocity profile has a region where $\partial U_1^*/\partial y$ becomes small. On the other hand, at z = 5 mm for example, a single maximum turbulence profile is associated with a less distorted U_1^* profile.

The turbulence spectra reveal that most of the energy is contained in a relatively narrow band of low frequencies (say, f < 200 Hz). Stability calculations indicate that these frequencies are amplified by quasi-cross-flow profiles. A small peak appears on the spectra at frequencies close to 1 000 Hz, which corresponds to a streamwise instability.

5.3. Streamwise evolution of the wavelength λ^*

All results presented so far were obtained at a single abscissa (x/c = 0.69). Additional measurements have been performed at other stations for studying the streamwise evolution of the wavelength λ^* taken in the direction parallel to the leading edge. The results are plotted on figure 11. λ^* increases with x, but the ratio λ^*/δ remains practically constant and close to 4 (δ is the physical boundary layer thickness). In fact, the downstream increase in λ^* results from a decrease in the number of streaks. Stability calculations can explain this behaviour. It has been said that, for a given abscissa, stationary waves are unstable for a wide range of λ^*. Similar computations were carried out at other streamwise locations and the total amplification rate was computed for some selected wavelengths, by using the group velocity concept. The numerical results are shown in figure 12, where the total amplification rate is plotted for various wavelengths λz_1 taken in the cross-flow direction. It can be seen that the small wavelengths are unstable close to the leading edge and, then, damped whereas larger wavelengths become amplified. The streamwise evolution of the locally most amplified wavelength λ^* is plotted in figure 11 and compared with the experimental results.

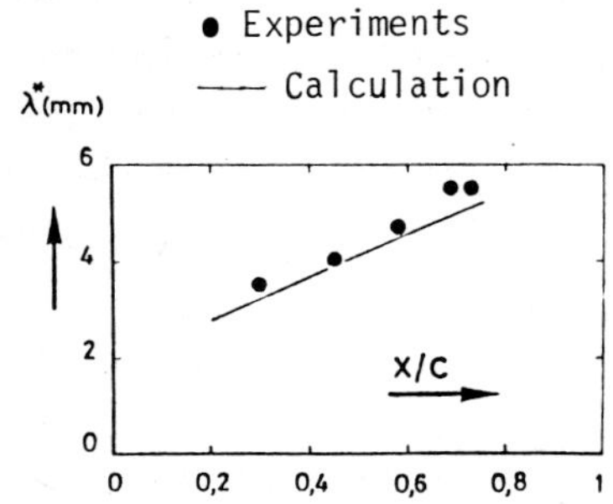

Fig.11. Evolution of the wavelength λ^*.

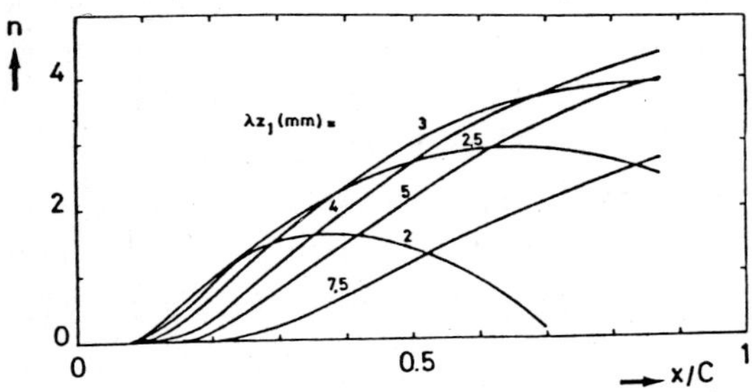

Fig.12. Total amplification curves for various wavelengths.

5 - CONCLUSION

In the first part of this study, we measured the movement of cross-flow type transition, for different chord Reynolds numbers. Similar results were already obtained by several authors /2/, /3/, /7/. They can be used for checking the validity of transition calculation methods, as it has been done in /6/.

The study of stationary waves has given more fundamental and original results. It has been shown that these waves result in a strongly non uniform evolution of the laminar boundary layer in the spanwise direction and that the laminar stability theory is able to explain most of the observed features. Obviously, additional measurements must be carried out at other streamwise positions for obtaining a complete picture of the stationary waves evolution.

REFERENCES

1. ARNAL D. ; HABIBALLAH M. ; COUSTOLS E. : Théorie de l'instabilité laminaire et critères de transition en écoulement bi et tridimensionnel. La Recherche Aérospatiale, 1984-2

2. BOLTZ F.W. ; KENYON G.C. ; ALLEN C.Q. : Effects of sweep angle on the boundary layer stability characteristics of an untapered wing at low speeds. NASA TN D-338 (1961)

3. POLL D.I.A. : Some aspects of the flow near a swept attachment line with particular reference to boundary layer transition. CRANFIELD Institute of Technology (1978)

4. KEGELMAN J.T. ; NELSON R.C. ; MUELLER T.J. : The boundary layer on an axisymmetric body with and without spin. AIAA Journal, Vol. 21,N° 11 (1983)

5. ARNAL D. ; COUSTOLS E. ; JUILLEN J.C. : Etude expérimentale et théorique de la transition sur une aile en flèche infinie. La Recherche Aérospatiale 1984-4

6. MICHEL R. ; ARNAL D. ; COUSTOLS E. : Stability calculations and transition criteria in two- and three-dimensional flows. IUTAM Symposium, NOVOSIBIRSK (1984)

7. MANIE F. ; SCHMITT V. : Aile à flèche variable équipée du profil D - Analyse des essais effectués - Etude de l'influence du nombre de Reynolds. RTS N° 8/3072 AY 002 (1978)

Effect of Curvature on Cross-Flow Instability

M. R. MALIK
High Technology Corporation, P.O. Box 7262,
Hampton, Virginia 23666, USA

D. I. A. POLL
Aerodynamics Division, College of Aeronautics,
Cranfield Institute of Technology, Cranfield, Bedford,
MK43 0AL, England, U.K.

Abstract

The problem of the stability of a three-dimensional laminar boundary layer formed on a curved surface is considered. A calculation scheme, which takes account of the curvature of the flow streamlines and of the surface, is proposed for the prediction of the development of small amplitude disturbances. Computations have been performed for the flow over the windward face of an infinitely long yawed cylinder and comparisons have been made with experimental data. These indicate that the theory correctly predicts many of the features of the unstable laminar flow. The theory also suggests that transition, in this particular situation, is dominated by travelling disturbance waves and that, at the experimentally observed transition location, the wave which has undergone greatest total amplification has an amplitude ratio of approximately e^{11}. When the effects of curvature are omitted the maximum amplitude ratio at transition is about e^{17}.

Introduction

At present, the "state-of-the-art" transition prediction method involves the use of linear stability theory coupled with an e^N breakdown criterion. This particular method was originally proposed by Smith (see Jaffe, Okamura, and Smith [1]) who showed that for transition in two-dimensional and axisymmetric flow the N factor takes a value of about 10 when the free-stream disturbance levels are small. The extension of the method to three-dimensional flows is not entirely straightforward since the extra degree of freedom means that there are several ways to calculate an N factor from the stability characteristics. The envelope method (Srokowski and Orszag [2]) and the saddle point method (Nayfeh [3], Cebeci and Stewartson [4]) are two such alternative approaches. However, it has been demonstrated by Malik and Orszag [5] that, for a given case, the ultimate values of N are only very weakly dependent upon the approach adopted. Given this, it would appear that a three-dimensional version of the 'e^N' procedure may have a useful role as a design tool if it can be demonstrated that the N factor is still equal to 10 when the computations are compared with transition data from fully three-dimensional experiments. Initial attempts to provide the necessary 'proof' have been carried out by Hefner and Bushnell [6] who used the SALLY code of Srokowski and Orszag [2] and obtained N factors between 7 and 11 for some swept-back wing configurations. However, Cebeci and Stewartson [4] obtained N of 20 for rotating disk problem.

The SALLY code and the code used by Cebeci and Stewartson [4] derive the stability characteristics from solutions of the Orr-Sommerfeld equation. For the three-dimensional stability problem Stuart [7] has shown that, in general, the stability of the flow is affected by curvatures of the body and the flow streamlines plus Coriolis forces if the body is rotating. For the rotating disk problem, Malik, Wilkinson and Orszag [8] performed computations in which these effects were included. Their results showed that these terms had a powerful stabilizing effect. Moreover, the revised analysis showed that the experimentally observed transition point corresponded to an N factor of 11.

The results of Malik, Wilkinson and Orszag [8] have important implications for the prediction of transition near the leading edge of a swept-back wing (or yawed cylinder). In this region both the surface and the flow streamlines are highly curved. The curvature of the streamlines leads to the development of a 'cross-flow' velocity component within the boundary layer and the velocity profiles associated with the cross-flow are known to exhibit instability at low Reynolds numbers - see for example Owen and Randall [9]. Despite the fact that cross-flow instability has long been recognized as a major transition inducing mechanism on swept wings, the effect of the streamline curvature on the stability characteristics has never been addressed. We also note that none of swept wing examples considered by Hefner and Bushnell [6] involved transition in regions of high curvature and so their rather tentative conclusions cannot be expected to be globally valid - particularly in view of the rotating disk experience.

The object of the present paper is to present a method for calculating the stability characteristics of a flow which has both curvature of the surface and curvature of the streamlines. Computations are presented for the case of the flow over the windward face of an infinitely long yawed cylinder. This is known to be a situation in which the transition process is governed by the instability of the curvature induced cross-flow and, consequently, it provides an excellent test case for the extended stability analysis. Finally, the predictions are compared with experimental measurements made on a cylinder model in a low-speed wind tunnel.

Analysis

We define two sets of orthogonal curvilinear coordinate systems as indicated in Fig. 1. The first, x, y and z, is a set of reference axes which remain fixed relative to the body with x and y lying in the plane of the surface and z being normal to the surface. Since these are, in general, curved axes elements of length in the x, y and z directions are given by g_1dx, g_2dy and dz. The velocity components u, v and w are assigned to the x, y and z coordinate directions. A second set of curvilinear coordinates, ξ, η and z is then defined. In this case the ξ and η coordinates lie in the plane of the surface. Elements of length in this system are $h_1d\xi$, $h_2d\eta$ and dz, respectively. The velocity components in the ξ and η directions are s and c, respectively. This system of coordinates is used for the stability analysis and is free to rotate relative to the x, y, z system. The angle between the ξ and x axes is ε.

Consider a flow where the instantaneous values of velocity and pressure may be expressed in terms of a steady mean component and a fluctuating component -

$$(s,c,w,p) = (\bar{s},\bar{c},\bar{w},\bar{p}) + (\tilde{s},\tilde{c},\tilde{w},\tilde{p}).$$

We now assume that the fluctuating components may be represented as

$$(\tilde{s},\tilde{c},\tilde{w},\tilde{p}) = \mathrm{Re}\{(\hat{s},\hat{c},\hat{w},\hat{p})\exp i(\alpha\xi - \omega t)\}$$

where α is a real wavenumber and ω is a complex frequency. The equations which govern the temporal stability of a wave whose wavenumber vector is tangent to one of the coordinate axes (ξ) in an orthogonal curvilinear system (ξ,η,z) have already been derived by Stuart [7]. However, in a given problem the geometric properties of the disturbance wave are not known at the beginning of the computation. Therefore, it is necessary to develop an iterative scheme for the solution of Stuart's basic equations.

The process begins by considering an initial orientation for the ξ,η,z system relative to x,y,z - say ε_1. We then consider a wave of the form

$$(\tilde{s},\tilde{c},\tilde{w},\tilde{p}) = \mathrm{Re}\{(\hat{s},\hat{c},\hat{w},\hat{p})\exp i(\alpha\xi + \beta\eta - \omega t)\}$$

where α and β are real wavenumbers in the ξ and η directions. For this wave we note that the wavenumber vector makes an angle ϕ with the ξ axis such that

$$\tan\phi = \frac{\beta}{\alpha}. \tag{1}$$

The equations governing the disturbance behavior in the ξ,η,z system of coordinates are

$$i(\alpha_o\bar{s} + \beta_o\bar{c} - \omega)\hat{s} + \bar{s}'\hat{w} + m_{12}(\bar{s}\hat{c} + \bar{c}\hat{s}) - 2m_{21}\bar{c}\hat{c}$$

$$+ m_{13}\,\bar{s}\hat{w} = -i\alpha_o\hat{p} + \frac{1}{R}\left[\hat{s}'' - (\alpha_o^2 + \beta_o^2)\hat{s}\right] \tag{2.a},$$

$$i(\alpha_o\bar{s} + \beta_o\bar{c} - \omega)\hat{c} + \bar{c}'\hat{w} - 2m_{12}\bar{s}\hat{s} + m_{21}(\bar{s}\hat{c} + \bar{c}\hat{s})$$

$$+ m_{23}\bar{c}\hat{w} = -i\beta_o\hat{p} + \frac{1}{R}\left[\hat{c}'' - (\alpha_o^2 + \beta_o^2)\hat{c}\right] \tag{2.b},$$

$$i(\alpha_o\bar{s} + \beta_o\bar{c} - \omega)\hat{w} - 2m_{13}\bar{s}\hat{s} - 2m_{23}\bar{c}\hat{c}$$

$$= -\hat{p}' + \frac{1}{R}\left[\hat{w}'' - (\alpha_o^2 + \beta_o^2)\hat{w}\right] \tag{2.c},$$

and

$$i(\alpha_o\hat{s} + \beta_o\hat{c}) + \hat{w}' + m_{12}\hat{c} + m_{21}\hat{s} + (m_{13} + m_{23})\hat{w} = 0 \tag{2.d}.$$

The boundary conditions are

$$\hat{s} = \hat{c} = \hat{w} = 0 \quad \text{at } z = 0 \quad \text{and} \quad \hat{s} \to 0,\ \hat{c} \to 0,\ \hat{w} \to 0 \quad \text{as } z \to \infty \tag{2.e}.$$

In this system of equations all the variables have been non-dimensionalized with an appropriate boundary layer velocity, U_o, and a reference length scale, δ. Both these quantities are assumed to be local constants. The 'in-plane' curvature coefficients are given by

$$m_{12} = \frac{\delta}{h_1h_2}\frac{\partial h_1}{\partial\eta} \quad \text{and} \quad m_{21} = \frac{\delta}{h_1h_2}\frac{\partial h_2}{\partial\xi},$$

whilst the surface curvature coefficients are given by

$$m_{13} = \frac{\delta}{h_1}\frac{\partial h_1}{\partial z} \quad \text{and} \quad m_{23} = \frac{\delta}{h_2}\frac{\partial h_2}{\partial z}.$$

Finally,

$$\alpha_o = \frac{\alpha}{h_1}, \quad \beta_o = \frac{\beta}{h_2}$$

and the Reynolds number used in the stability analysis is

$$R = U_o\delta/\nu$$

where ν is the kinematic viscosity.

These are general equations governing the quasi-parallel stability of three-dimensional boundary layers and, clearly, an application to a specific problem requires the determination of the various curvature coefficients. It may also be readily shown that if these coefficients are set to zero then equations (2) can be reduced to the familiar Orr-Sommerfeld equation. Assuming that the necessary curvature coefficients are available, a solution is obtained by specifying the frequency of the disturbance (ω_r) and then determining the values of α and β by requiring that, at any point on the surface, the temporal amplification rate ω_i is maximized. Having obtained a first estimate for the shape of the disturbance wave everywhere in x,y,z space, a new set of ξ,η,z coordinates is specified in such a way that the ξ coordinate is locally aligned with the wavenumber vector, i.e.

$$\varepsilon_2 = \varepsilon_1 + \phi \tag{3}$$

The stability characteristics are then recomputed. This process is repeated until ϕ (see equation (1)) is effectively reduced to zero. In this event the solution of equations (2) is also a solution of Stuart's [7] disturbance equations (with the non-parallel flow terms set to zero) and the stability characteristics then include all the effects of streamline and surface curvature.

The eigenvalue problem (equations (2)) is solved by a finite-difference method as in COSAL [10]. For a fixed physical frequency (ω_r), the COSAL code calculates the maximum temporal growth rate (ω_i) which then determines α and β. The group velocity $\vec{V}g = (\partial\omega/\partial\alpha, \partial\omega/\partial\beta)$, which is needed in this maximization procedure and for converting temporal growth to spatial growth, is also calculated in COSAL. An N factor for transition correlation purposes may be defined as

$$N = \int_{\ell_o}^{\ell} \omega_i/|\mathrm{Re}(\vec{V}g)|\,d\ell$$

where ℓ is the arclength on the solid body along a curve which is everywhere tangent to $\mathrm{Re}(\vec{V}g)$.

Computations for a Yawed Cylinder Configuration

In order to test the method described in the previous section computations have been performed for the flow over the windward face of the infinitely long yawed cylinder shown in figure 2. The natural choice for the fixed reference axes in this case is the cylindrical polar system - r,θ,ζ. For the body fixed system of Fig. 2 we note that on the windward face

$$x = a\theta, \quad y = \zeta \quad \text{and} \quad z = r - a$$

where a is the leading edge radius of the cylinder. Hence by considering the length of a small element it follows that

$$g_1 = \left(1 + \frac{z}{a}\right) \quad \text{and} \quad g_2 = 1.$$

The mean boundary layer flow was computed, in this coordinate system, from an experimentally determined surface pressure distribution (Poll [11]) by using a laminar, compressible, adiabatic wall method known as MAIN [12]. For the stability computations the reference velocity U_o and reference length δ were chosen to be the boundary layer edge velocity and the displacement thickness in the x direction. Since these scales are treated as local constants, any other choice of the scales will do equally well. The curvature coefficients in equations (2) can be written as

$$m_{12} = k \cos \varepsilon \frac{d\varepsilon}{dx} = k \tan \Lambda \sin \upsilon \cos^2\psi \frac{f'(x)}{f^2(x)},$$

$$m_{21} = -k \sin \varepsilon \frac{d\varepsilon}{dx} = -k \tan \Lambda \cos^3\psi \frac{f'(x)}{f^2(x)},$$

$$m_{13} = \frac{k \cos^2\varepsilon}{(1 + kz)} = \frac{k \sin^2\psi}{(1 + kz)}$$

and

$$m_{23} = \frac{k \sin^2\varepsilon}{(1 + kz)} = \frac{k \cos^2\psi}{(1 + kz)}$$

where $f(x) = U_e/U_\infty$ and $k = \delta/a$. Note that x has been non-dimensionalized with respect to a and z with respect to δ.

In the above, the initial ξ,η,z coordinate system was chosen so that the η direction was aligned with that of the local external streamline. Therefore, the initial orientation is obtained by setting

$$\varepsilon = \pi/2 - \psi$$

where ψ is the angle between the external streamline and the x direction -

$$\tan \psi = \frac{V_e}{U_e} = \frac{V_\infty}{U_\infty f(x)} = \frac{\tan \Lambda}{f(x)}$$

This system of coordinates proved to be an entirely satisfactory starting point for the present problem with convergence being complete after 4 iterations through equation (3).

Solutions have been obtained for the following combinations of sweep angle Λ and free-stream Reynolds number R_c -

Case no.	Λ	$R_c \times 10^{-5}$
1	30°	3.81
2	55°	1.68
3	60°	1.34
4	63°	0.88

The computations for each of these cases showed similar trends and, therefore, detailed results are presented for case 4 only. These are given in Figs. 3, 4, 5 and 6. Figure 7 shows a summary of the variation of N_{max} with surface position for the cases 1, 2 and 3. For test case 4, the stationary disturbance wavelength and orientation are presented in Figs. 8 and 9, respectively.

Discussion

The results presented in Figs. 3, 4, 5 and 6 indicate that, in the case of flow over the windward face of a yawed cylinder, curvature has a significant effect upon the flow stability. For a disturbance of fixed frequency, the wavenumber and orientation are modified slightly but the most significant change occurs in the amplification rate (ω_i). The curvature terms are found to produce a very strong damping effect. This is consistent with the findings of Malik et al. [8] for the case of the rotating disk. The reduced local amplification rate causes a dramatic reduction in the value of the local N factor for the various unstable disturbance frequencies - as shown in Fig. 3. For this case N is approximately halved for values of x/C in excess of 0.15. A significant feature of the results presented in Fig. 3 is that, in the yawed cylinder case, the most amplified wave at a given position is, in general, a travelling wave. This contrasts with the rotating disk case where Malik et al. [8] found that the stationary disturbances gave the highest N factor over all positive real frequencies.

With regard to the stationary disturbances, the observations made by Poll [11] at the conditions corresponding to the case 4 computations have been added to Figs. 8 and 9. It is apparent that the experiment and theory are in reasonable agreement. For the disturbance wavelength the discrepancy is less than 10% and for the wavenumber orientation the difference is of order 2°. It should also be noted that the stationary disturbances are found to lie at a fixed angle relative to the pure cross-flow direction. The present theory suggests that this angle is approximately 4° whilst the observations indicate an angle of about 6°. These results support the original suggestion made by Owen and Randall [9] and Gregory, Stuart and Walker [7] that the stationary disturbances are associated with a velocity profile for which a point of inflection coincides with a point of zero mean velocity. This is the so-called 'critical profile' and mean flow computations reveal that, in the present case, this profile makes an angle of 5° with the cross-flow direction.

As stated previously, the main engineering objective of stability computations is the prediction of the location of the onset of boundary layer transition. The experiment [11] provides transition onset data for the computational cases 2 and 3. At these flow conditions the surface pitot tube technique indicates that transition begins at an x/C of 0.26 and 0.31, respectively. From Fig. 7 it can be seen that for the computations which include the effects of curvature, the N factors at transition are 12.3 and 11.2. These are in excellent agreement with the value of 11 obtained by Malik et al. [8] for transition onset on the rotating disk and they lie well within the uncertainty bounds for N at transition onset in two-dimensional flows as reported by Jaffe et al. [1]. However, when the curvature effects are omitted the values of N at transition are found to be 17.7 (case 2) and 16.9 (case 3). It is apparent that the use of the Orr-Sommerfeld equation for generating stability characteristics for use in the prediction of transition onset on the cylinder is totally inappropriate and that this approach coupled with the conventional 'N equals 10' criterion would yield extremely pessimistic transition locations.

Finally, it should be noted that for cases 2, 3 and 4 the yaw angles are relatively large, being 55°, 60° and 63°. Since these angles are considerably greater than those occurring at the leading edges of typical civil transport aircraft, it is appropriate to examine the effects of curvature on the cylinder flow at a much smaller yaw angle. This problem is addressed in computational case 1 where the yaw angle is only 30°. The predicted variation of N factor with x/C is given in Fig. 7. These results indicate that the damping effects of the curvature terms are still strong at the reduced yaw angle. Moreover, the most highly amplified disturbance is a travelling wave with a frequency of 2000 Hz. Therefore, there would seem to be no indication that the stationary disturbances are any more important in this case than they are at the large yaw angles.

Conclusions

The effects of curvature on the stability of a three-dimensional laminar boundary layer have been studied theoretically. The results show that in the case of the flow past a long yawed cylinder the body curvature and streamline curvature terms produce strong damping effects together with significant changes in the magnitude and orientation of the wavenumber vector. Comparison with the experimental data available for this flow indicate that the theory correctly predicts several features of the instability such as stationary disturbance wavelength and orientation. With regard to transition it is found that when curvature effects are included the onset conditions are reasonably well predicted by the e^N method where N is approximately 11. Finally, this study has indicated that for strongly three-dimensional flows the most amplified disturbances (according to linear stability theory) are, in general, travelling waves and not the stationary waves.

This work was sponsored by NASA Langley Research Center under contracts NAS1-16916 (MRM) and NAS1-14605 (DIAP).

References

1. Jaffe, N. A.; Okamura, T. T.; and Smith, A. M. O.: AIAA Journal, Vol. 8, pp. 301-308, 1970.

2. Srokowski, A. J.; and Orszag, S. A.: AIAA Paper 77-1222, Aircraft Systems and Technology Meeting, Seattle, Washington, August 1977.

3. Nayfeh, A. H.: AIAA Journal, Vol. 18, No. 4, pp. 406-416, April 1980.

4. Cebeci, T.; and Stewartson, K.: AIAA Journal, Vol. 18, No. 4, pp. 398-405, April 1980.

5. Malik, M. R.; and Orszag, S. A.: AIAA Journal, Vol. 18, No. 12, pp. 1485-1889, December 1980.

6. Hefner, J. N.; and Bushnell, D. M.: NASA TP 1645, April 1980.

7. Gregory, N.; Stuart, J. T.; and Walker, W. S.: Phil. Trans. Roy. Soc., Series A, Vol. 248, pp. 155-199, 1955.

8. Malik, M. R.; Wilkinson, S. P.; and Orszag, S. A.: AIAA Journal, Vol. 19, No. 9, pp. 1131-1138, September 1981.

9. Owen, P. R.; and Randall, D. G.: Royal Aircraft Establishment Tech. Memo. (Aero) 256, February 1952.

10. Malik, M. R.: NASA CR 165925, 1982.

11. Poll, D. I. A.: College of Aeronautics Report 8407, February 1984.

12. Cebeci, T.; and Kaups, K.: J. Aircraft, Vol. 14, pp. 661-667, 1977.

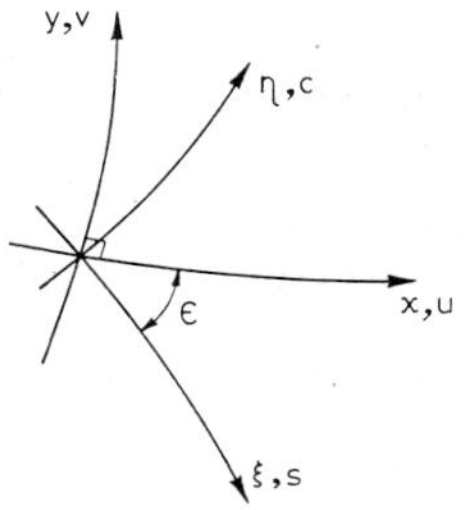

Figure 1. Notation for the fixed reference and wave aligned systems of orthoganal curvilinear coordinates.

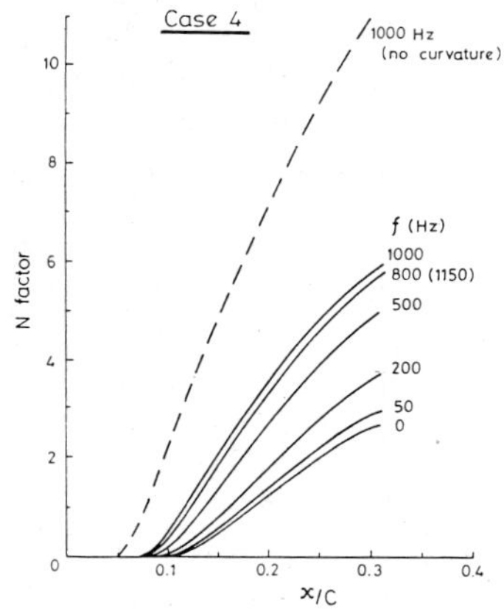

Figure 3. N factor variation with position and disturbance frequency for a flow with a fixed free-stream Reynolds number.

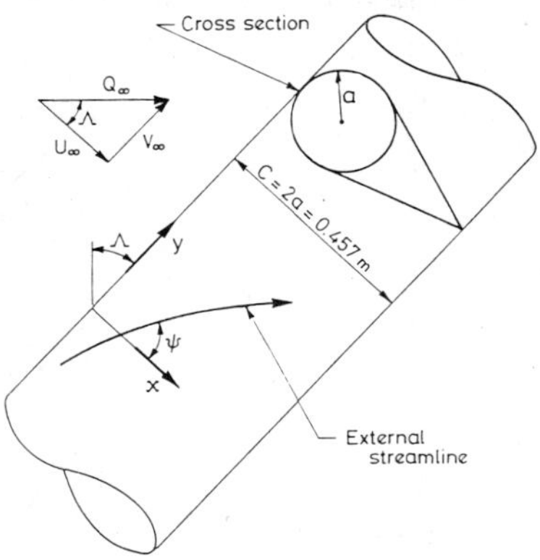

Figure 2. Notation for the infinitely long yawed cylinder test case.

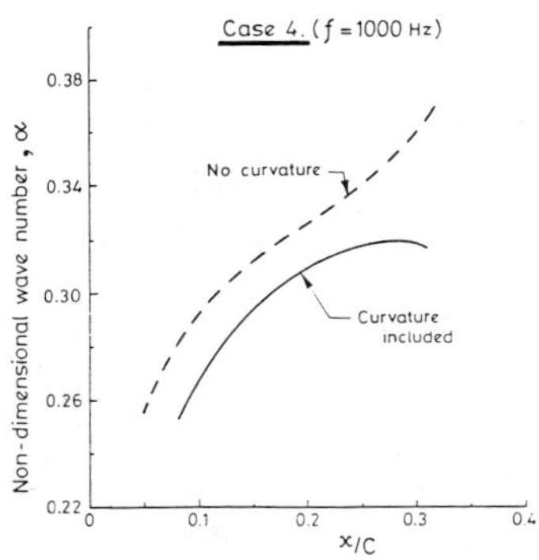

Figure 4. Disturbance wave number as a function of position for a wave of fixed frequency and a constant free-stream Reynolds number.

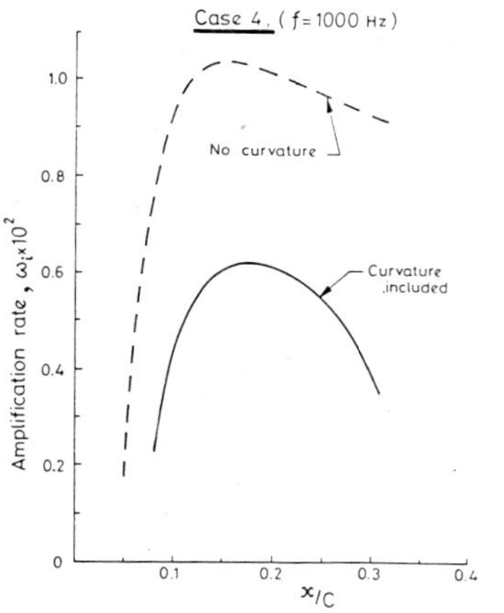

Figure 5. Temporal amplification rate as a function of position for a wave of fixed frequency and a constant free-stream Reynolds number.

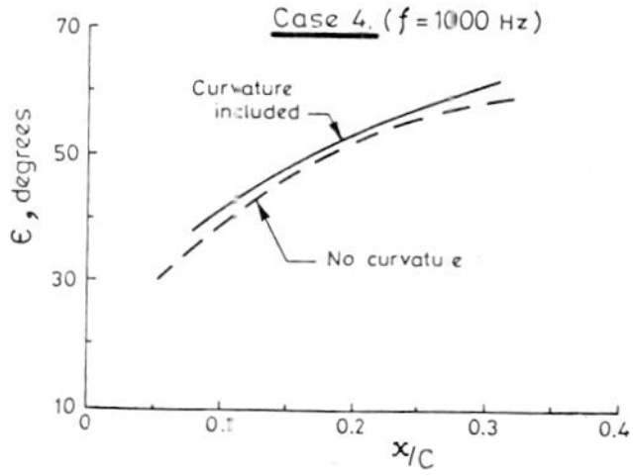

Figure 6. Travelling disturbance wave orientation relative to the x direction.

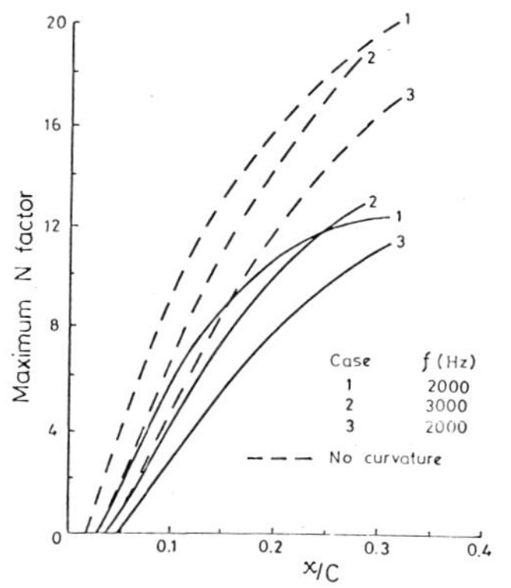

Figure 7. Variation of the maximum N factor with position for cases 1,2 and 3.

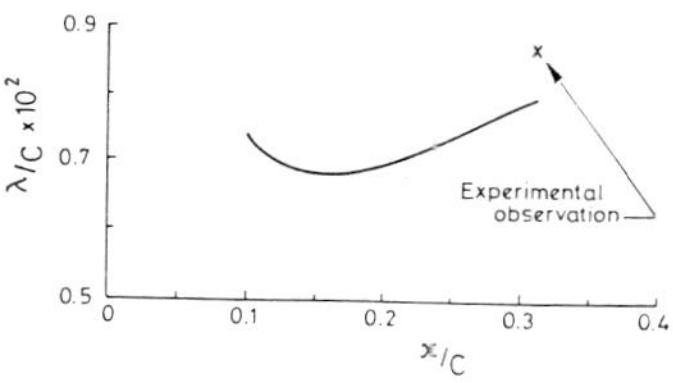

Figure 8. The variation of the stationary disturbance wave length with position for the case 4 conditions.

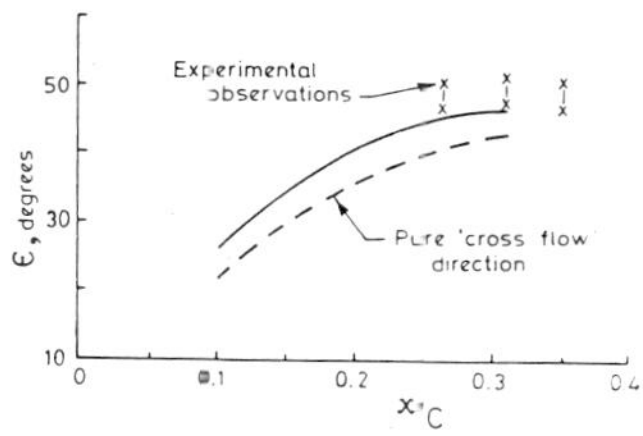

Figure 9. Stationary disturbance wave orientation relative to the x coordinate direction for the case 4 conditions.

Spiral Vortices in Boundary Layer Transition on a Rotating Cone

R. KOBAYASHI and Y. KOHAMA

Institute of High Speed Mechanics
Tohoku University, Sendai, Japan

Summary

The experiment shows that spiral vortices transfer to toroidal vortices not as the local rotational speed ratio S tends to infinity but at a definite value between 5 to 8 for the total included cone angle θ of 15° and 30°. A cross-sectional flow pattern of the vortices is apparently observed to consist of pairs of counter-rotating vortices. The critical and transition Reynolds numbers and the number of the spiral vortices decrease remarkably as the local rotational speed ratio S becomes larger.

Introduction

The purpose of this paper is to study the behaviour of spiral vortices appearing in transition regions of three-dimensional boundary layers on cones rotating in axial flow. Previous studies for rotating cones were in many cases concerned with heat and mass transfer [1]. Because the centrifugal instability plays an important role in the transition phenomenon especially for cones of small included angle and the spiral vortices grow out of the instability, we are interested in the properties of the spiral vortices and also in the detailed transition process to the turbulent boundary layer. The present paper is one of our successive researches on rotating axisymmetric bodies [2-5].

Experimental Apparatus and Procedure

Two low-turbulence wind tunnels [6] of the Laboratory for Air-Flow Measurements in the Institute of High Speed Mechanics, Tohoku University, were used in a state of open-jet type of the test sections. The wind tunnels have test sections of subtense length 293 mm and 810 mm and the turbulence intensity of 0.06 % and 0.04 %, respectively, in a state where the test cone was set. Two cones of total angle θ = 15° and 30° were tested. The 15° cone has a base diameter $2R_b$ = 52.1 mm and a generation-line length L = 198.0 mm, while the 30° cone has $2R_b$ = 88.8 mm and L = 172.5 mm. The rotating speed N (= $\omega/2\pi$) of the test cones can be controlled continuously up to 3700 r.p.m.

A hot-wire anemometer was used to measure the velocity field around the cone. In order to determine the direction of the spiral vortices, we used a parallel probe of two hot wires separated by 3 mm, which could be rotated around an axis normal to the cone surface. We decided the direction of the vortex axis making an angle ε with the circumference of the cone surface by measuring the direction in which two periodic signals obtained from the two hot wires are in the same phase. The uncertainty interval for measuring of ε was ± 1.0°. Flow patterns in the transition region were visualized by spreading titanium tetrachloride on the cone surface.

Fig. 1 illustrates the flow field of the rotating cone and the notations.

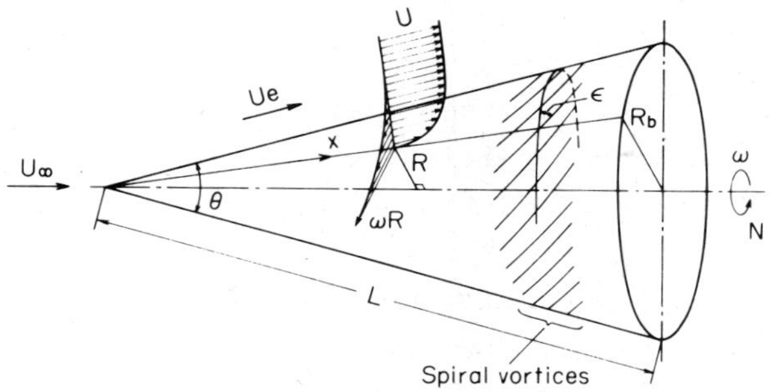

Fig. 1. Rotating cone in axial flow and notations

Transition Region

The photograph in Fig. 2 shows a flow pattern of the boundary-layer flow on the rotating 15° cone. A uniform flow is given from the left, and the rotation vector of the cone is in the left direction of the rotation axis. In Fig. 2 one can see regular spiral vortices in the transition region between the laminar boundary layer on the left and the turbulent boundary layer on the right. A cross-sectional flow visualization for the spiral vortices clearly showed that each spiral vortex consists of a pair of counter-rotating vortices, which is the same as the one for the 30° cone [3,4]. Fig. 3 shows a flow pattern for the 15° cone rotating in still fluid (i.e. $U_\infty = 0$). Apparently one can see toroidal vortices in the transition region. We write this state as $\varepsilon \fallingdotseq 0$.

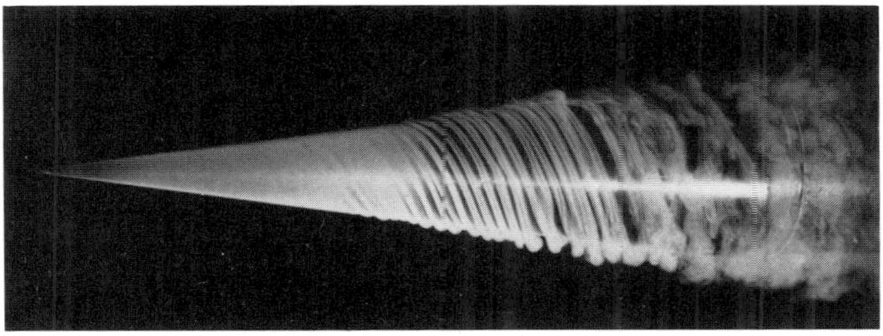

Fig. 2. Flow visualization of boundary layer on 15° cone rotating in axial flow. $U_\infty = 1.8$ m/s, $N = 2000$ r.p.m.

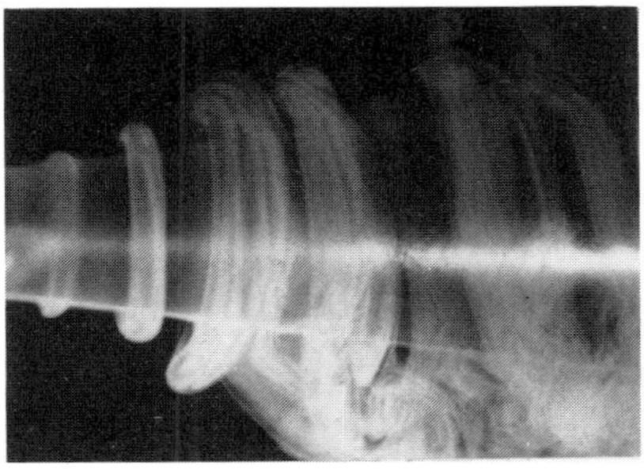

Fig. 3. Transition region of boundary layer on 15° cone rotating in still fluid. $U_\infty = 0$ m/s, $N = 500$ r.p.m.

In order to define the transition region we use two Reynolds numbers ($R_e = xU_e/\nu$), based on a distance x from the apex to a local position and the local flow velocity U_e at the outer edge of the boundary layer with the kinematic viscosity ν of the fluid. The one is the critical Reynolds number Re_c, which was determined as an instability point x_c where periodic signals obtained from the hot-wire probe were first detected not by an oscilloscope but on the frequency spectrum. The other is the transition Reynolds number Re_t for the transition point x_t where the velocity fluctuations just lost the periodicity entirely. Fig. 4 shows experimental values of Re_c and Re_t, plotted against the local rotational speed ratio S ($= \omega R/U_e$). It is

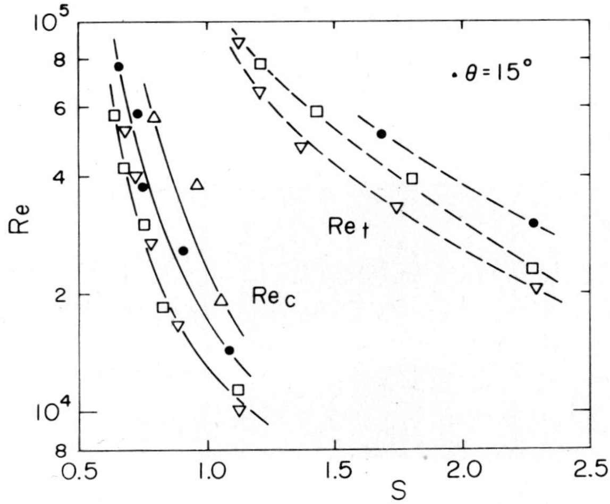

Fig. 4. Critical Reynolds number Re_c and transition Reynolds number Re_t in relation to local rotational speed ratio S
△: N = 1800 r.p.m., ●: 2500 r.p.m., □: 3200 r.p.m., ▽: 3700 r.p.m.

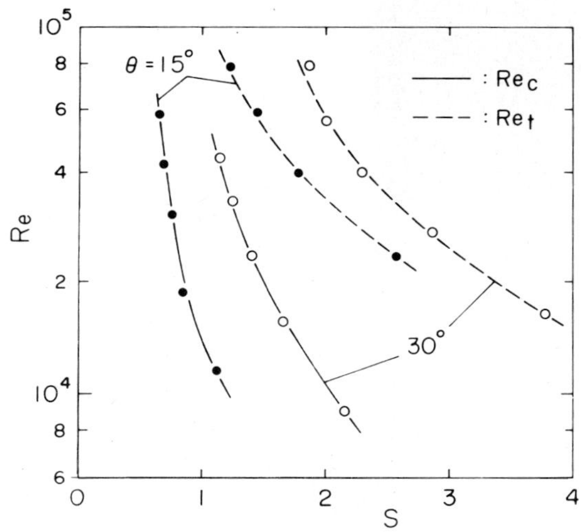

Fig. 5. Comparison of Re_c and Re_t for 15° cone and 30° cone. N = 3200 r.p.m.

evident that the critical and transition Reynolds numbers (Re_c, Re_t) decrease remarkably as the S-value is increased. It comes from the increasing effects of the centrifugal instability. The figure also shows that the relations of Re_c and Re_t to S are not definite but somewhat dependent on N. In Fig. 5 a comparison is made between these relations for the two cases of θ = 15° and 30°. It follows from the figure that the relations of Re_c and Re_t to S are quite different for different angle θ of the cone.

Spiral Vortices

It was first verified by using the flow visualization technique and stroboscopic observations that the spiral vortices are relatively fixed to the rotating cone surface. The fact agrees with the one for the case of rotating disk in still fluid [7]. The number n of the spiral vortices appearing on the circumference of the cone surface can thus be determined by measuring the frequency of the periodic signals from the hot-wire probe. The results are given in Fig. 6 in relation to the local rotational speed ratio S. The decrease of n with increasing S corresponds with the change of the direction ε of the vortex

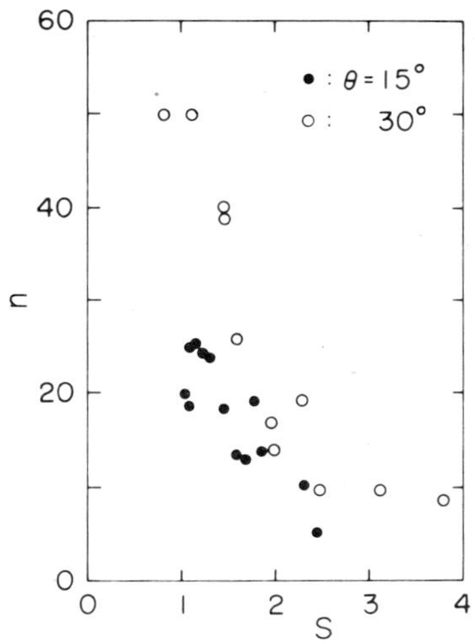

Fig. 6. Number n of spiral vortices on circumference of cone surface

axis shown in Fig. 7. The number n of the vortices increases as the cone angle θ becomes larger.

Fig. 7 is the angle ε, which the axis of the spiral vortex makes with the circumferential direction. There is no difference between the angles ε for the 15° cone and the 30° cone. It should be noticed that the angle ε appears to become zero at a finite value of S. For $S \to \infty$, which corresponds to the case of a rotating cone in still fluid, it was already clear for the 15° cone (Fig. 3) and the 30° cone [4] that ε is equal to zero. We gave attention to the condition of S, at which the spiral vortices change into the toroidal vortices. A similar phenomenon is known for viscous axial flow through an annulus having a rotating inner cylinder, where toroidal vortices transfer to spiral vortices at a definite value of axial Reynolds number [8]. We try to determine in the following section a condition of the change for the present case of rotating cones.

Change from Spiral Vortices into Toroidal Vortices

Careful measurements were made for the direction ε of the vortex

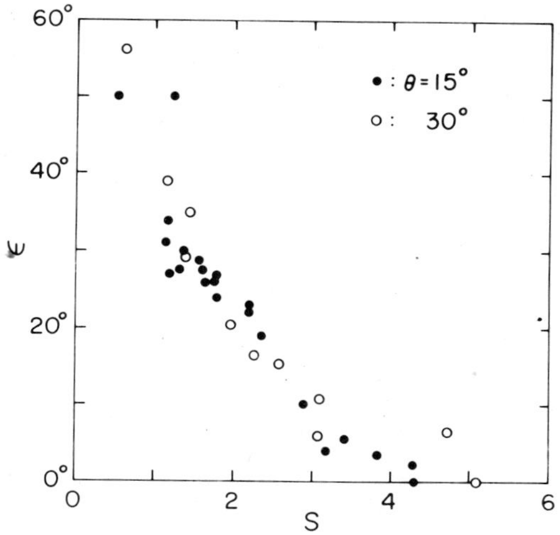

Fig. 7. Direction ε of spiral vortices in relation to local rotational speed ratio S

axis at fixed rotational speed N of the cone. The approaching flow velocity U_∞ is gradually increased from the still state of fluid. As shown in Fig. 8, the angle ε remains zero until the flow velocity U_∞ reaches a difinite value (U_S) and then starts abruptly to increase with increasing U_∞. The difinite value U_S is widely different each other for different values of the cone angle θ and the rotational speed N. In Fig. 9 we rewrite the data of the angle ε in relation to the local rotational speed ratio S for the experimental ranges of N = 1500 r.p.m. to 3420

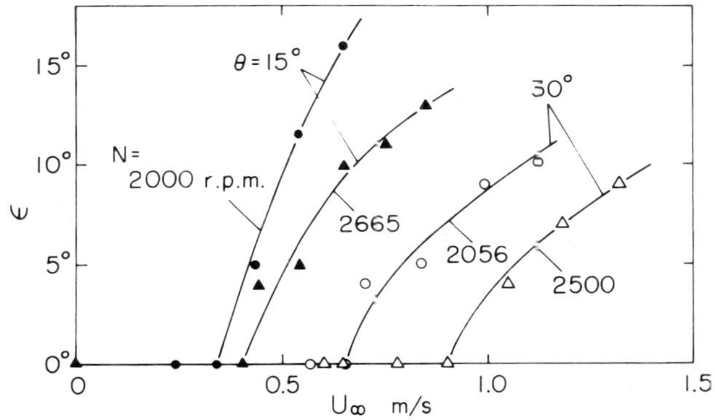

Fig. 8. Direction ε of spiral vortices against approaching flow velocity U_∞

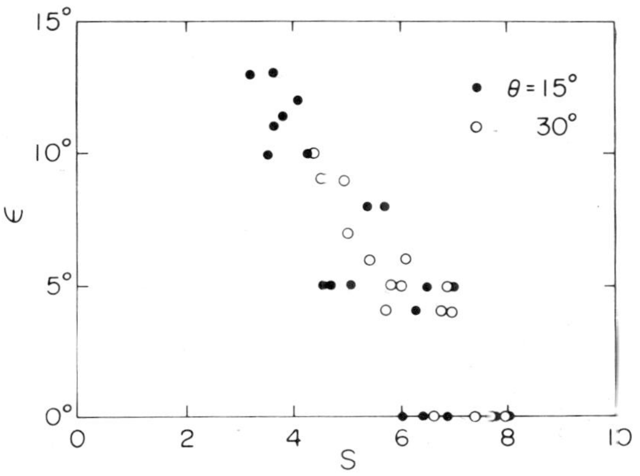

Fig. 9. Direction ε in relation to S for U_∞ up to 1.4 m/s and N = 1500 r.p.m. to 3420 r.p.m.

r.p.m. and U_∞ up to 1.4 m/s, where the measured values are plotted only for $U_\infty \geq U_S$. It can be concluded that the change from the spiral vortices into the toroidal vortices occurs at the local rotational speed ratio S = 6 to 8, which is independent of the cone angle θ and the rotational speed N. The scattering of the measured values attributes mainly to fluctuations of the appearance of vortices.

References

1. Kreith, F.: Convection heat transfer in rotating systems. Adv. Heat Transfer 5 (1968) 129-251.

2. Kobayashi, R.; Kohama, Y.; Takamadate, C.: Spiral vortices in boundary layer transition regime on a rotating disk. Acta Mech. 35 (1980) 71-81.

3. Kobayashi, R.; Kohama, Y.; Kurosawa, M.: Boundary-layer transition on a rotating cone in axial flow. J. Fluid Mech. 127 (1983) 341-352.

4. Kobayashi, R.; Izumi, H.: Boundary-layer transition on a rotating cone in still fluid. J. Fluid Mech. 127 (1983) 353-364.

5. Kohama, Y.; Kobayashi, R.: Boundary-layer transition and the behaviour of spiral vortices on rotating spheres. J. Fluid Mech. 137 (1983) 153-164.

6. Ito, H. et al.: The design and performance of the low-turbulence wind tunnel, Tohoku University (in Japanese). Mem. Inst. High Speed Mech., Tohoku Univ. 44 (1980) 93-151.

7. Kohama, Y.: Study on boundary layer transition of a rotating disk. Acta Mech. 50 (1984) 193-199.

8. Snyder, H. A.: Experiments on the stability of two types of spiral flow. Ann. Phys. 31 (1965) 292-313.

On the Development of Disturbances in Nonparallel Supersonic Flows

S. A. GAPONOV

Institute of Theoretical and Applied Mechanics
USSR Academy of Sciences
Novosibirsk, USSR

Summary

The development of the disturbances in the nonparallel flow is investigated. It is discovered that for some value of the Reynolds number Re* the Tollmien-Schlichting wave degenerated into a sound wave. Near Re* nonparallel effects and the excitation of the Tollmien-Schlichting waves by a sound are very strong.

Introduction

The stability theory of the parallel flow describes parameters of the unstable waves well enough but it does not explain some important phenomena. For incompressible fluids it was shown ([1 - 6] and others) that the amplification coefficients of the disturbances are not the same for different parameters of the flow because of increasing the boundary layer thickness. These results are confirmed by experiments [7]. Nonparallel effect increases with decreasing Reynolds numbers. Another, more important effect concerns the possibility of the unstable oscillation excitation by external disturbances [7 - 9]. Since the thickness of the boundary layer increases with Mach number the influence of the nonparallelism on these phenomena could be more appreciable for supersonic flows. Unfortunately compressible flows are studied insufficiently as are concerned discussed problems. There are only a few papers of the author (see for example [10, 12]) and results of [11] on this matter. The latter data are not correct* and revised results are not published.

1. Let us formulate the problem of the disturbances develop-

* Private communication of El Hady

ment in a weakly nonparallel flow. Linearizing of the governing equations with respect to disturbances $q(X, Y) \exp(i\beta z - iwt)$ gives the system of the form

$$q_y - a^\circ q - a^1 q_x = 0 \quad (1)$$

Here q is vector-function with the following components: velocities, pressure, temperature disturbances and the first derivatives of the disturbances longitudinal and transverse velocity and the temperature with respect to Y; a^k are slowly changing matrices with X which depend on parameters of the laminar flow, frequency w, wave number β ; Y, X are generalized longitudinal and normal coordinates. The solution of the system (1) with boundary conditions

$$q_1 = q_3 = q_5 \quad \text{for} \quad q = 0 \quad (2)$$

is sought for in the form

$$q = q^\circ (X, Y) \exp(O(X)) \quad (3)$$

Here q° is a slowly changing function of X. Therefore the solution q° may be expanded in powers of $\mathcal{E} = 1/Re$ as follows

$$q^\circ = \sum_{i=0}^{\infty} q^{\circ i} \mathcal{E}^i \quad (4)$$

Substitution of (4) into (1) leads to the recurrent system of equations. The first of them are of the form

$$q^{\circ\circ} - a^\circ q^{\circ\circ} - O_x a^1 q^{\circ\circ} = 0 \quad (5)$$

$$q^{\circ 1} - a^\circ q^{\circ 1} - O_x a^1 q^{\circ 1} = a^1 q^{\circ\circ}_x \quad (6)$$

It is assumed here that $q^{\circ\circ} \sim \mathcal{E} q^{\circ\circ}$. The system (5) with boundary conditions (2) determines the eigenvalue problem with respect to Θ_x and $q^{\circ\circ}$ and its solution is proportional to arbitrary function A(X). Thus one may take $q^{\circ\circ} = A(X)\, q^{\circ\circ}(X, Y)$ where $q^{\circ\circ}$ could be normalized in conformity with equality

$q^{oo}(X,0)$=const. Under this assumption the system (6) may be written in the form

$$q_y^{o1} - a^o q^{o1} - 0_x a^1 q^{o1} = a^1(A_x q^{oo} + A q_x^{oo}) \qquad (7)$$

Since the homogeneous part of (7) is the same as (5) the former has a nontrivial solution if and only if the right hand side of (7) is orthogonal to every solution of the corresponding adjoint homogeneous problem. From solvability condition the logarithmic derivative of A(X) is determined.

$$A_x/A = -(a^1 q_x^{oo}, \psi)/(a^1 q^{oo}, \psi) \qquad (8)$$

where $(B, A) = \sum_{i=1}^{n} \int_0^\infty B_i\, A_i\, dY$ is a dot product, and n is an order of the system (5). The amplification rate of the disturbance q_k is determined by the following relationship

$$\text{Real}\ (q_{kx}/q_k) = \text{Real}\ (\theta_x + A_x/A + q_{kx}^{oo}/q_k^{oo}) \qquad (9)$$

Numerical results on the nonparallel effects have been obtained for the boundary layer of a flat plate. Therefore it was convenient to use the generalized coordinates $Y=y/\delta(x)$, $X=\text{Re}=U_e\delta(x)/\nu_e$ where $\delta(x)=\sqrt{x\nu_e/U_e}$. In this case conventional amplification rate

$$-\alpha_{ki} = (1/2)\ \text{Real}\ (q_{kx}/q_k) \qquad (10)$$

The highest amplification rates of the maximum of the mass flux disturbances and the first approximation (Real θ_x) were calculated in [10] (M=2.2, Re=10^3; M=4.5, Re=10^3). It was shown that nonparallel effect on the growth rate of the maximum of the disturbances is insignificant. Therefore, it is possible to use the approximate theory of the parallel flow for calculations of the amplification factor in the unstable region.

The nonparallel effect increases with decreasing the Reynolds number. It is confirmed for the Tollmien-Schlichting wave ($F=w\nu_e/U_e^2=0.91\cdot10^{-4}$) with the angle between the wave propagation direction and the main flow direction f=37° for $Re=10^2$. It can be seen from Fig.1 that the nonparallel effects have a large influence on the growth rate near Re=Re* and it contradicts the calculation model which is valid only in the case of small contribution of the terms related to the nonparallelism of the flow. Probably in this case it is necessary to improve the calculation scheme. Reasons for the large effects of the nonparallelism in the region Re=Re* will be discussed in the final part of this paper.

2. Now the problem of the excitation of the Tollmien-Schlichting waves by external disturbances (in our case by sound waves) is considered. For this purpose it is useful to take into consideration an approach suggested in the papers [8, 12]. The interaction problem of the sound field with nonparallel flow in the boundary layer is solved on the basis of the system (1). The solution is taken in form

$$q = q + q^{\alpha_o}(X, Y)\exp(\theta_o(X)) \tag{11}$$

Here $\theta_o(X)$ is a given function by virtue of the assumption that acoustic wave is monochromatic, $\alpha = \theta_x$, q^o is solution of the system (5) when $\theta_x = \theta_{ox}$ with boundary conditions

$$q_1^o = q_3^o = q_5^o = 0 \quad \text{for} \quad Y=0 \tag{12}$$

$$q = B_1 \exp(i\lambda Y) + CB_2 \exp(-i\lambda Y) \quad \text{for} \quad Y \gg 1 \tag{13}$$

The first term of the right hand side of (13) describes an incoming acoustical wave; C is the reflection factor of the boundary layer; B_1, B_2 are constant vectors. Substitution of (11) into (1) produces the following equation

$$q_y - a^o \Delta q - a^1 \Delta q_x = a^1 q_x^{\alpha_o} \exp(O_o(X)) \tag{14}$$

The solution Δq is expanded into the sum of eigenfunction of the local problem (5), (2) with coefficients which are functions of X

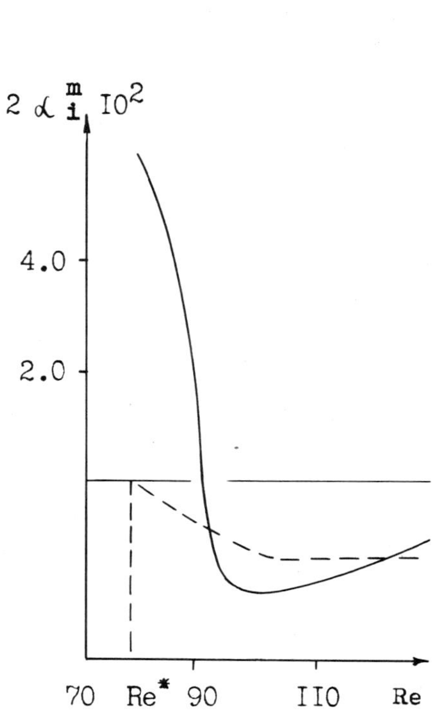

Fig.1

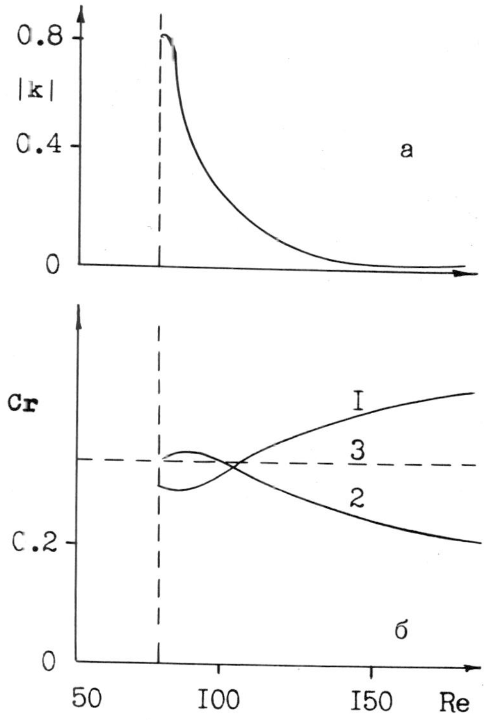

Fig.2

$$q = \sum_m q^m(X, Y)\, A^m(X) \exp(\Theta^m(X)) \tag{15}$$

Substituting (15) into (14) and restricting to the case of only slow interaction or to the case of small amplitudes of the generated wave one can obtain an approximate amplitude equation for the K-th Tollmien-Schlichting wave

$$dA^k/dX = -K \exp(\Theta_o - \Theta^k); \quad K = (a_1 q_x{}^o, \bar{q}^k)/(a^1 q_x^k, \bar{q}^k) \tag{16}$$

The amplification rate and the maximum value of $|K|$ * in the space of Θ_{ox} were calculated for M=2.0, $F=0.91 \cdot 10^{-4}$ in [12]. All results were obtained with a condition that the longitudinal velocity amplitude of the incoming wave and the amplitude maximum of the mass flux oscillation for the Tollmien-Schlichting wave are equal to unit. The comparison of these data shows that the growth intensity of the Tollmien-Schlichting waves due to external sound waves is close to the growth rate, Real (θ_x).

Since $|K|$ increases with decreasing Re special investigations have been carred out in the region of the small Reynolds number. These results are given in Fig.2. The interaction of the sound wave (f=53°, phase velocity C=0.363) with the Tollmien-Schlichting wave (f=37° for Re=10^2) was considered for F= $=0.91 \cdot 10^{-4}$, M=2.0. It is seen from Fig.2a that $|K|$ increases rapidly when Re $\rightarrow$ Re*. The variation of the Tollmien-Schlichting wave phase velocity (1), the limit phase velocity of subsonic disturbances C=1-1/M cos f and the phase velocity of the sound wave C=0.363 (3) are shown in Fig.2b.

3. To explain the nonparallel effects on both the growth rate (Fig.1) and the excitation of the Tollmien-Schlichting waves by the sound near Re=Re* it is useful to analyse results which are presented in Fig.2b. It can be seen that all three values of the phase velocities draw closer to one other when Re $\rightarrow$ Re*. It is easy to show that in this case the Tollmien-Schlichting disturbances decrease slowly when Y $\rightarrow \infty$. The dot products of (8) take large values and they are determined not well enough. Therefore the problem of the nonparallel effects should be treated separately when C $\rightarrow$ 1-1/M cos f.

There is similar indefiniteness in the problem of the Tollmien-Schlichting wave excitation by external disturbances. More important reason for the strong interaction of the two kinds of waves is explained by the degeneration of the Tollmien-Schlichting wave into the sound wave when Re $\rightarrow$ Re*. The point Re* may be considered as the location of the source of the Tollmien-Schlichting wave. Its initial amplitude can be calculated if the external acoustic field is known. The reflection factor of the discussed sound wave is equal to zero. Plots of the reflection zero coefficient of the sound wave (1) and the classical neutral stability curve (2) are shown in Fig.3.

The Tollmien-Schlichting waves branch off the part (a) of the curve (1). Its amplitude is determined completely by the sound

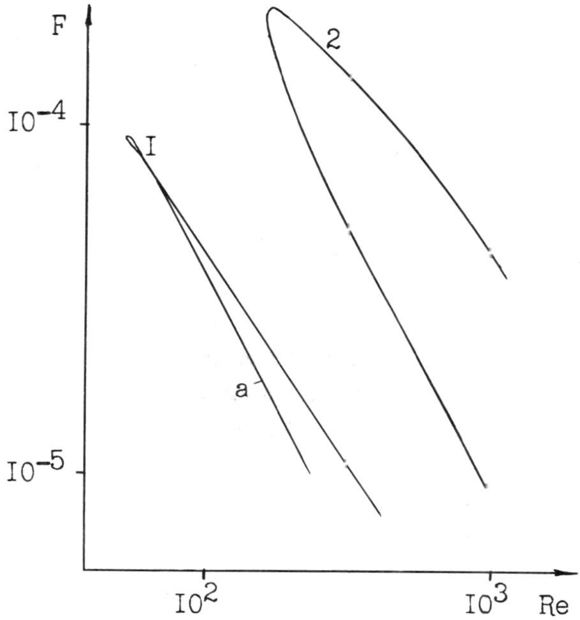

Fig.3

disturbances. With increasing of the Reynolds number the Tollmien-Schlichting wave reaches the unstable region which is limited by the curve (2).

Thus together with the useful data on the development of the disturbances in the supersonic boundary layer the examples are given in the paper for the case when the widespread method for nonparallel flows becomes inapplicable. It is to be expected that these results will stimulate a new methods development both for solving the problem of a disturbances behaviour in an actual nonuniform flow and for describing receptivity phenomena of a flow to external disturbances.

References

1. Bouthier, M.: Stabilité linéaire des écoulements presque paralléles, Part II. La couche de Blasius. J. Mec., 12 (1973) No 1 75-95.

2. Volodin, A.G.: The stability of the nonparallel boundary layer. Proceedings of the Siberian Branch of the Acad. Sci. of the USSR (1973) No 8 13-15 (in Russian).

3. Saric, W.S.; Nayfeh, A.H.: Phys. Fluids, vol.18 (1975) No 8.

4. Gaster, M.: On the effects of boundary-layer growth on flow stability. J. Fluid Mech., 66 (1974) pt. 3 465-480.

5. Tumin, A.M.; Shepelev, V.E.: Numerical analysis of the disturbances development in the incompressible boundary layer of the flat plate. Numerical methods in the mechanics of the continious medium, 11 (1980) No 3 141-152 (in Russian).

6. Smith, F.T.: On the non-parallel flow stability of the Blasius boundary layer. Proc. R. Soc. A366 (1979) 91-109.

7. Kachanov, Yu.S.; Kozlov, V.V.; Levchenko, V.Ya.: Origin of the turbulence in the boundary layer, (1981) (in Russian).

8. Zhigulev, V.N.; Sidorenko, N.V.; Tumin, A.M.: On the generation of the unstable waves in the boundary layer by external turbulence. J. Appl. Mech. and Tech. Phys. (1980) No 6 43-49 (in Russian).

9. Zhigulev, V.N.: On the excitation and development of the instabilities in three-dimensional steady boundary layers. J. Appl. Mech. and Tech. Phys. (1983) No 4 100-110 (in Russian).

10. Gaponov, S.A.: The influence of flow non-parallelism on disturbances development in the supersonic boundary layer. Proc. of the Eighth Canadian Cong. of Appl. Mech. (1981) 673-674.

11. El-Hady, M.M.; Nayfeh, A.H.: Nonparallel stability of compressible boundary layer flows, AIAA Paper No 80-0277 (1980).

12. Gaponov, S.A.: Excitation of unstable waves in supersonic boundary layer by sound. Proc. of the Ninth Canadian Congress of Applied Mech., (1983), 581-582.

Effects of Nozzle Design Parameters on the Extent of Quiet Test Flow at Mach 3.5

I. E. BECKWITH,* M. R. MALIK,** F.-J. CHEN**, and D. M. BUSHNELL***

NASA Langley Research Center
Hampton, Virginia 23665, U.S.A.

Summary

To advance boundary-layer stability and transition research and to ultimately provide reliable predictions of transition for supersonic flight vehicles, a wind tunnel is required with much lower stream disturbance levels than in conventional supersonic tunnels. Recent results from a Mach 3.5 pilot quiet tunnel at the NASA Langley Research Center have shown that very low stream noise levels can be achieved only when the nozzle wall boundary layers are laminar. Transition Reynolds numbers measured on a slender cone were then in the same range as flight data. The linear amplification of both Tollmien-Schlichting (TS) and Görtler type instabilities was calculated for the wall boundary layers in this Mach 3.5 pilot nozzle. By using an e^N method, it was determined that amplified Görtler vortices were involved in the transition process. The resulting transition criteria were then applied to several nozzles with different expansion rates and wall curvatures. By manipulation of these parameters, the calculations indicate that it may be possible to increase significantly the size of the quiet test regions compared with those observed in the pilot nozzle.

Nomenclature

A	disturbance amplitude
G	Görtler number, $R_\theta \sqrt{\theta/r_c}$
ℓ	length of nozzle from throat to exit
M	Mach number
N	exponential factor for the amplification ratio, $\ell n\ A/A_o$
r_c	streamwise radius of curvature of the concave wall
R	unit Reynolds numbers
$R_{e,T}$	local transition Reynolds number
R_θ	momentum thickness Reynolds number
s	streamwise distance along wall
u	streamwise velocity
X	axial distance from nozzle throat
$\Delta X, \Delta Y, \Delta Z$	length, height, and width of quiet test region or "core," Fig. 1
h_*	sonic throat half height
δ	boundary-layer thickness
θ	momentum thickness

*Leader, High-Speed Boundary Layer Stability and Transition Group, Viscous Flow Branch, High-Speed Aerodynamics Division.
**High Technology Corporation, P. O. Box 7262, Hampton, VA 23666, U.S.A.
***Head, Viscous Flow Branch, High Speed Aerodynamics Division

θ_w inviscid flow angle at wall
λ wavelength

Subscripts

e local values at edge of boundary layer
N neutral stability location
o initial value
Q quiet test core
T transition
∞ free-stream conditions in test section
* sonic condition

Introduction

High noise levels in conventional supersonic wind tunnels cause premature transition from laminar to turbulent flow in the boundary layers on test models [1-3]. These high noise levels are caused by eddy Mach wave radiation from the turbulent boundary layers on the nozzle walls. Transition research and reliable predictions of transition for flight conditions require wind tunnels with much lower disturbance levels that approach those measured in flight [4].

Recent results [5] from a program at NASA Langley Research Center to develop a quiet supersonic wind tunnel have shown conclusively that laminar boundary layers on the nozzle walls are required to obtain transition Reynolds numbers on test models as high as those previously observed in flight. These high transition Reynolds number data were obtained in a unique nozzle [5] that provides root-mean-square noise levels within the quiet test region that are less than 0.03 percent of the stream static pressure. This paper addresses the nozzle design problem of maximizing the length of the quiet test region and thereby increasing the Reynolds number range for quiet (laminar wall) operation.

From a recent experimental investigation and analysis of Görtler vortices in two axisymmetric Mach 5 nozzles [6], the values of the Görtler number at transition varied from about 5 to 6. Thus, a simple Görtler number index based on experimental data may be sufficiently reliable for design predictions of transition in a given category of supersonic nozzles. However, when the nozzle design parameters such as expansion rates and wall curvature are varied significantly as in the present study, G_T can be expected to vary over a wider range.

A more general approach than the simple G_T criterion, based on solving the compressible Görtler problem with linear stability theory, was also used and applied to the prediction of transition on the nozzle walls with an e^N type method similar to that of Smith's [7]. The automatic computer code developed for this purpose integrates the amplification of Görtler vortices along the nozzle wall with the correct local boundary-layer profiles and thicknesses obtained from the boundary-layer code of Harris and Blanchard [8]. The values of N_T were determined by integrating to the experimental locations of transition in the pilot nozzle. These values were then used to determine the transition prediction criterion for three of the nozzles in the present study.

1. Experimental and Theoretical Approach

A uniform flow test rhombus cannot be produced in a supersonic nozzle without the use of concave wall curvature to "cancel" the initial supersonic expansion waves. Thus, while the amplification of boundary-layer

TS waves may be controlled by the favorable streamwise pressure gradients, the formation and amplification of Görtler vortices cannot be avoided except by the use of a flat wall radial flow nozzle, or by suction through porous walls or through longitudinal slots as in a rod-wall nozzle [9] which was first proposed and tested by Klebanoff (see [10]).

Data from the existing Mach 3.5 rapid expansion nozzle [5] will be used extensively in the present study. Pertinent aerodynamic details and the general layout of this nozzle are illustrated in Fig. 1. Sectional views of the nozzle are shown to scale with the subsonic boundary-layer removal slots and uniform flow quiet test "core" included.

Figure 2 shows for the rapid expansion nozzle [5], the local Görtler numbers at transition plotted against the axial distance from the nozzle throat. The momentum thickness Reynolds numbers at transition are also shown and may be compared with values in the throat region of a conventional nozzle from Demetriades [11]. Transition moves upstream into the rapid expansion nozzle as the unit Reynolds number increases (the numbers near data points are the values of $R_\infty/m \times 10^{-7}$). The size of the quiet test region (or core) is then reduced because the resulting eddy Mach wave noise radiation moves upstream along Mach lines as indicated schematically in Fig. 1. The location of transition on the nozzle wall was determined as the point of intersection with the wall of the Mach line extended upstream from the centerline probe station where the rms noise levels increased from the very low laminar values of about 0.05 percent up through the 0.1 percent level [5]. Generally, very high Görtler numbers (> 9) were obtained when transition was several cm downstream of the wall inflection point. This result suggests that Görtler vortices must have played an important role in the nozzle wall boundary-layer transition at these downstream locations. At the upstream transition point of $X \approx 1.4$ cm, transition was caused directly by wall roughness, since the calculated TS wave growth rates were also very small at this location and $G_T \approx 0.6$.

Data presented in Fig. 2 are for the boundary layer removal or "bleed" valve closed and open and with the nozzle wall conditions designated as "clean," "repolished," and "dirty." With the bleed valve closed, the nozzle wall boundary layers were completely transitional or turbulent except at $R_\infty \approx 1 \times 10^7/m$ where transition was detected with a probe on the centerline of the test section at about 20 cm downstream of the nozzle throat. The corresponding acoustic origin location on the nozzle wall is $X \approx 5$ cm where $G_T \approx 6.5$ and $R_{\theta,T} \approx 630$. This value of $R_{\theta,T}$ agrees well with Demetriades' [11] values of about 550.

With the bleed valve open, transition moves upstream from $X \approx 15$ cm to 7 cm as the unit Reynolds number is increased from $R_\infty \approx 1 \times 10^7/m$ to $3 \times 10^7/m$. For larger values of R_∞, the locations of transition and the values of $R_{\theta,T}$ and G_T vary over wide ranges depending on the surface conditions. For $R_\infty > 3 \times 10^7/m$, the largest values of $R_{\theta,T}$ and G_T were always observed for the "repolished" surface condition while the lowest values of G_T occurred for the "dirty" condition. The repolishing work improved the surface finish substantially as indicated by profilometer data [12]. Typical values of peak to valley roughnesses were reduced from ≈.6 micron to <.3 micron. The "dirty" condition occurred when the nozzle walls were not cleaned for several previous runs and visible deposits of lint, dust, or other types of contaminants were present before a run.

Figure 3 shows the effect of the nozzle repolishing on transition Reynolds numbers on a 5° half-angle, sharp-tip cone [5]. At $R_\infty \approx 4 \times 10^7$/m, $R_{e,T}$ increased from about 4.5×10^6 to 8.0×10^6 as a result of the improved finish achieved by the repolishing operation and the corresponding longer runs of laminar boundary-layer flow on the nozzle wall as indicated by measurements of free-stream noise [12].

1.1 Experimental Study of Vortex Wavelengths

To assist in the evaluation of Görtler vortex wavelengths, oil flow tests were conducted [12] in the pilot nozzle in an attempt to determine if the vortices are present and what range of vortex wavelengths might be reasonable to use in the calculations. The same oil flow technique was used as in a previous study [6] of Görtler vortices in two Mach 5 nozzles. Values of λ (from the width of streaks in the oil flow photographs [12]) varied from about 0.1 cm up to 0.6 cm. The data were for bleed valve open where the streaks were always seen, even at the highest unit Reynolds number. With the bleed valve closed, the streaks were not seen.

1.2 Description of Nozzles Used in Study

In the present study, six different nozzles have been considered. The design Mach number, unit Reynolds numbers, and throat height of all these nozzles were taken to be the same as for the pilot nozzle. These standards were adopted to allow direct comparisons of predicted sizes of quiet test regions and also to provide more confidence in the application of transition criteria based on experimental results from the pilot nozzle. The values of the throat radius of curvature, the length ℓ, and the wall angle at the inflection point for these nozzles are given in Table I as nozzles 2 to 7; nozzle 1 is the existing, rapid-expansion pilot nozzle [5]. Nozzles 1 to 5 have uniform flow in their test sections and corresponding regions of concave wall curvature. Downstream of the throat region, the inviscid flow in nozzle 6 is two-dimensional radial flow contained between 3.5° diverging flat walls. Therefore, the boundary layers in this nozzle would presumably not have the Görtler type instability.

Boundary-layer removal slots located upstream of the throat, similar to those in the Mach 3.5 rapid-expansion nozzle (Fig. 1), have been used in the calculations for all nozzles listed in Table I. The flow in the subsonic approach, the suction slots, and the transonic throat regions of all nozzles in Table I were computed by the method of Strikwerda [13]. The supersonic portion of nozzles 1 to 5 and 7 were designed by various method-of-characteristic techniques that provide a uniform test section flow [12]. Figure 4(a) shows the wall contour and defining Mach lines for the corner expansion nozzle (nozzle 3, Table I). Nozzles 4 and 5 are shown in Figs. 4(b) and 4(c), respectively. These two nozzles have a region of supersonic radial flow and were designed by the method of Beckwith and Moore [14]. Discussion concerning the various values of the quiet test section lengths, ΔX_Q, listed in Table I and illustrated in Fig. 4 will be presented in a subsequent section of this paper.

2. Results and Discussion

The results show that amplification of TS waves in supersonic nozzles is generally quite small if subsonic boundary layer removal slots are used as in the present Mach 3.5 rapid-expansion nozzle (Fig. 1). Even in the flat wall radial flow nozzle, which is about 6 times longer than the existing rapid expansion nozzle (see Table I), amplification of TS waves was small downstream of the throat region due to the favorable streamwise pressure gradients. The axial Mach number variation within the nominal test rhombus of this flat wall radial flow nozzle was from M = 3.4 to 3.6 and the maximum flow angle variation was about ±1.5°. Since these axial pressure gradients and flow angles would be imposed on test models, this nozzle would not be acceptable for many test requirements.

2.1 Görtler Number Method

The Görtler number variations for the wall boundary layers of the concave wall nozzles 1 to 5 (Table I) at $R_\infty = 3.2 \times 10^7$ are shown in Fig. 5. The predicted locations of transition at this Reynolds number are indicated in the figure for the assumed value of $G_T = 10.4$ as obtained from the data in Fig. 2 for the existing rapid expansion nozzle (nozzle 1 in Table I). Since acoustic disturbances are propagated along Mach lines in supersonic flow, a Mach line projected downstream from a wall location of X_T into the nozzle test section fixes the downstream boundaries of the quiet test core as illustrated in Fig. 1 for nozzle 1. The length (ΔX_Q) and height (ΔY_Q) of the quiet test core is then determined by the distance between these downstream boundaries and the Mach lines defining the upstream boundaries of the uniform flow test rhombus. The upstream Mach lines move closer to the nozzle throat as the nozzle design expansion rate and value of $\theta_{w,max}$ increases. Consequently, the values of ΔX_Q from the G_T method (see Fig. 5 and first column of ΔX_Q values in Table I) also increase. A more general prediction method based on linear stability theory is expected to provide more reliable predictions of transition on the walls of supersonic nozzles.

2.2 Local Stability Calculations

For this part of the study, the governing equations for the compressible linearized Görtler problem are derived in body-fitted coordinates. For the parallel flow approximation, the resulting eighth-order system of ordinary differential equations constitutes an eigenvalue problem which is efficiently solved by the numerical method developed by Malik, et al. [15]. More details about the analysis and governing equations will be presented elsewhere. The computational criteria used for the development of the vortices may be either a fixed wavelength or stepwise variations in wavelength to be guided by experimental observations in the present pilot nozzle. The growth rates are then integrated to the experimental locations of transition to obtain N_T factors for transition in the e^N method. These values of N_T are then used to predict transition locations in some of the other nozzles in the study.

Results calculated for the pilot nozzle are shown in Fig. 6. Here we used the "fixed wavelength method" in which λ is held constant for each integration from s_N to s_T. For a given flow condition (fixed value of R_∞), calculations are performed for several wavelengths and a curve of N vs. λ/δ is generated. (For the present purposes, we have defined δ as the value of y at $u/u_e = 0.995$ as computed with the Harris-Blanchard code [8].) The peak of this curve provides a value of the transition factor N_T corresponding to maximum amplification for the specified values of s_T and R_∞. This procedure for obtaining the transition factor will be called the N_T method. The values of λ/δ_N, λ/δ_T, and N_T obtained by interpolation at the peak of the curves from Fig. 6 are presented in Table II along with the experimental values of X_T. The values of N_T are in the range of 9 to 11, in excellent agreement with Smith's [7] values of about 10. During integration, Smith allowed the wavelength λ to vary according to the criterion of maximum local amplification rate.

The present constant wavelength method gives values of λ for the pilot nozzle, shown in the last column of Table II, that are all considerably smaller than minimum values from the oil flow studies [12]. The question then arises as to whether or not the wavelengths preceding transition might be larger than these computed values. Larger values of λ could be caused by some process whereby two or more pairs of individual vortices join together to form a single pair of larger vortices as suggested by the oil flow photographs [12]. Such a process would presumably occur in the nonlinear regions of vortex amplification, perhaps as a result of meandering or random shifting of adjacent vortices such as observed in low speed flows by Aihara and Sonoda [16]. Thus, as an extension of linear theory into the nonlinear regime for predicting transition due to amplification of Görtler vortices, the effects of increasing λ in abrupt steps upstream of transition was calculated for the pilot nozzle [12]. The resulting values of N_T were nearly the same as those for maximum amplification (Table II) and the final values of λ then agreed with the minimum values from the oil flow studies [12]. Obviously, much more work will be required to establish details of this merging behavior and to determine whether it can even occur upstream of transition.

Nozzles 3, 4, and 5 (Table I) were selected for applications of the N_T method for the prediction of transition locations. For the sake of comparisons with values of X_T and ΔX_Q from the G_T method (Fig. 5 and Table I), the same unit Reynolds number of $R_\infty = 3.2 \times 10^7$/m will be used for these calculations. The maximum value of N_T for this value of R_∞ in the pilot nozzle is 9.6 (Table II). However, for the present purposes, a more conservative value has been selected of $N_T = 9.2$ which was obtained by interpolation between the peak values of N_T for $R_\infty = 2$ and 4×10^7/m in Table II.

Figure 7 illustrates the procedure used to predict values of X_T in these three nozzles. This figure has two faired curves for each nozzle showing the variations of N with λ/δ at two X locations along the wall. Again, points on these curves were obtained by integration of the growth rates from s_N to s at fixed values of λ. The integration proceeds downstream until the maximum values of N at two successive X stations bracket the desired value of $N_T = 9.2$. The corresponding values of X_T were then obtained by interpolation between the peak values on the faired curves. These values of X_T are listed in Table III along with corresponding values of G_T, $R_{\theta,T}$, θ_T, and δ_T. With X_T known, the value of ΔX_Q is then obtained from the inviscid flow field as illustrated for these three nozzles in Fig. 4. The values of ΔX_Q from the N_T method are also listed in Table I where they may be compared directly with values from the G_T method.

The trends of ΔX_Q with $\theta_{w,max}$ from the two methods are entirely different. Thus, for example, the N_T method gives a smaller value of ΔX_Q for nozzle 3 but a considerably larger value for nozzle 4 than the G_T method. These differences are apparently related to the different initial rates of increase in G with X as compared with the reference nozzle 1 (see Fig. 5). Thus, as perhaps would be expected, the total integrated growth of the instability depends both on how fast G increases with X and on the distance over which the boundary layer is exposed to large values of G (or curvature). The simple G_T criterion obviously cannot account for these effects.

The very large increase in ΔX_Q predicted by the N_T method for the long radial nozzle (number 5, Table I) is surprising in view of the corresponding large values of G_T and $R_{\theta,T}$ (Table III). Except for the discontinuous increase in G and the subsonic boundary-layer removal slots, this nozzle is similar to conventional supersonic wind tunnel nozzles which have always had completely turbulent wall boundary layers at these high Reynolds numbers. More detailed analysis of TS wave growth rates will be required before these predicted large values of X_T and ΔX_Q can be accepted with confidence. Interaction between the stationary Görtler vortices and travelling TS vortices may have significant consequences. Another problem with this long nozzle would probably be early transition in the corners followed by propagation of these disturbances onto the nozzle walls.

Similar, but more severe problems would also occur in any attempts to achieve the even longer runs of laminar wall boundary layers required in the flat wall radial nozzle to obtain $\Delta X_Q \approx 50$ cm (nozzle 6, Table I). At this same unit Reynolds number of $R_\infty = 3.2 \times 10^7/\text{m}$, $R_\theta \approx 5700$ near the exit of this nozzle. Clearly, maintaining laminar flow over the large distance to the nozzle exit of more than 2 m with such large values of R_θ, would be difficult in spite of the favorable pressure gradients. Wall cooling could be used to help stabilize TS modes. Hence, the first requirement to be met for the final analysis of nozzles 4, 5, and 6 is to complete a detailed and systematic analysis of the effects of pressure gradients and wall cooling on TS wave growth rates. If the results of this analysis appear favorable, tests of small pilot versions of these nozzles would then be required to develop methods for controlling premature transition in the corners, or other by-pass mechanisms.

In any event, further increases in ΔX_Q above perhaps 15 to 20 cm (for $R_\infty \approx 3 \times 10^7/\text{m}$ in nozzles of the same size and mass flows as the present pilot nozzle), cannot be expected without some method to control the wavelengths and growth rates of the Görtler vortices or by the use of a rod-wall nozzle. Previous experience with rod-wall noise shields [17] indicates that low noise, laminar boundary-layer flows may occur over the entire length of the rod-walls in this nozzle.

Concluding Remarks

A study of the effects of operational factors and nozzle design parameters on the size of the quiet test region in several Mach 3.5 nozzles for low disturbance wind tunnels has been made. Test data from a pilot nozzle were used to determine criteria needed in the methods for predicting the location of transition in the study nozzles. Six different nozzles were evaluated in the study including a flat-wall radial flow nozzle and a proposed rod-wall nozzle. The following results were obtained from this study:

1. Measurements of nozzle wall surface finish, free-stream noise, and boundary-layer transition on a test cone in the pilot nozzle, before and after an extensive repolishing effort, have shown that the surface finish of the nozzle walls should have local peak-to-valley roughnesses of less than .3 micron to obtain quiet test region lengths of up to 12 cm in this nozzle at the highest test unit Reynolds number of $6 \times 10^7/\text{m}$. To maintain the corresponding lengths of laminar boundary-layer flow on the nozzle walls, the surfaces had to be carefully cleaned to remove all visible dust, lint, and other contaminants.

2. Oil flow studies and calculations for the amplification of Tollmien-Schlichting waves and Görtler vortices by linear stability theory indicated that the latter disturbances were the dominant mechanism causing transition on the walls of the pilot nozzle.

3. Linear stability theory was applied to the nozzles with an e^N method. The maximum values of N at transition, computed by a constant wavelength procedure, varied from about 9 to 11 in the pilot nozzle as the free-stream unit Reynolds number was decreased from 6×10^7/m to 1×10^7/m.

4. Predictions for the location of transition by the e^N method in the short and long radial nozzles at the unit Reynolds number of 3.2×10^7/m resulted in quiet test regions that were larger by about 20 and 300 percent, respectively, than in the existing pilot nozzle.

References

1. Pate, S. R.; and Schueler, C. J.: AIAA Journal, Vol. 7, No. 3, March 1969, pp. 450-457.
2. Dougherty, N. S., Jr.; and Fisher, D. F.: AIAA-80-0154, January 1980.
3. Pate, S. R.: AIAA-80-0431, March 1980.
4. Dougherty, N. S., Jr.; and Fisher, D. F.: AEDC-TR-81-26, February 1982.
5. Beckwith, I. E.; Creel, T. R., Jr.; Chen, F.-J; and Kendall, J. M.: NASA TP-2180, September 1983.
6. Beckwith, I. E.; and Holley, B. B.: NASA TP-1879, August 1981.
7. Smith, A. M. O.: Q. Appl. Math., Vol. XIII, No. 3, October 1955, pp. 233-262.
8. Harris, J. E.; and Blanchard, D. K.: NASA TM-83207, February 1982.
9. Pfenninger, W.; and Syberg, J.: NASA CR-2436, November 1974.
10. Beckwith, I. E.; and Bertram, M. H.: NASA TM X-2566, July 1972.
11. Demetriades, A.: AIAA Journal, Vol. 19, No. 3, 1981, pp. 282-289.
12. Beckwith, I. E.; Malik, M. R.; and Chen, F.-J.: AIAA-84-1628, June 1984.
13. Strikwerda, John C.: Parallel Computations, Academic Press, Inc., 1982, pp. 251-267.
14. Beckwith, Ivan E.; and Moore, John A.: NACA TN 3322, February 1955.
15. Malik, M. R.; Chuang, S.; and Hussaini, M. Y.: ZAMP, Vol. 33, No. 2, March 1982, pp. 189-201.
16. Aihara, Y.; and Sonoda, T.: AIAA-81-0197, January 1981.
17. Creel, T. R.; and Beckwith, I. E.: AIAA Jour., Vol. 21, No. 5, May 1983, pp. 643-644.

Table I.- Nozzle Design Parameters and Predicted Quiet Test Core Lengths ($M_\infty \approx 3.5$, $h_* = 1.1410$ cm)

Nozzle	Type	$r_{c,*}/h_*$	ℓ, cm	$\theta_{w,max}$, deg.	ΔX_Q,** cm	ΔX_Q,† cm
1	Rapid expansion [5]	1.0	39.29	29.02	12.7	12.7
2	Fast expansion	10.0	43.61	24.35	10.2	
3	Corner expansion	∞	35.76	29.26	16.3	14.9
4	Short radial	∞	42.98	23.25	9.4	15.0
5	Long radial	∞	65.30	10.05	C	36.4
6	Flat-wall radial	∞	222.38	1.75	52.3	52.3
7	Rod-wall	1.0	39.29	28.7	50.8	50.8

**For $G_T \approx 10.4$ at $R_\infty = 3.2 \times 10^7$/m.
†For $N_T = 9.2$ at $R_\infty = 3.2 \times 10^7$/m.

Table II.- Data From Fig. 6 for Near Maximum Amplification of Görtler Vortices in Pilot Nozzle.

R_∞/m $\times 10^{-7}$	X_T, cm	δ_T, cm	λ/δ_N	λ/δ_T	N_T	λ, cm
1	15.8	0.1045	3.45	0.73	10.85	0.076
2	9.2	.0479	2.60	.82	9.58	.039
3.2	7.6	.0332	2.40	.82	9.60	.027
4	6.6	.0264	2.10	.88	8.87	.023
6	6.0	.0202	2.10	.92	9.10	.019

Table III.- Transition Locations and Parameters From N_T Method at $R_\infty = 3.2 \times 10^7$/m.

Nozzle		N_T	X_T,	G_T	$R_{\theta,T}$	θ_T,	δ_T,	ΔX_Q,
Number	Type		cm			cm	cm	cm
1	Rapid expansion	9.6*	7.6	10.4	1310	.00258	.0332	12.7
3	Corner expansion	9.2†	6.9	9.9	1266	.00236	.0288	14.9
4	Short radial	9.2†	12.2	11.2	1611	.00353	.0483	15.0
5	Long radial	9.2†	50.4	14.5	2870	.00832	.1297	36.4

*From experimental data, Fig. 2.
†Value selected for transition due to amplification of Görtler vortices.

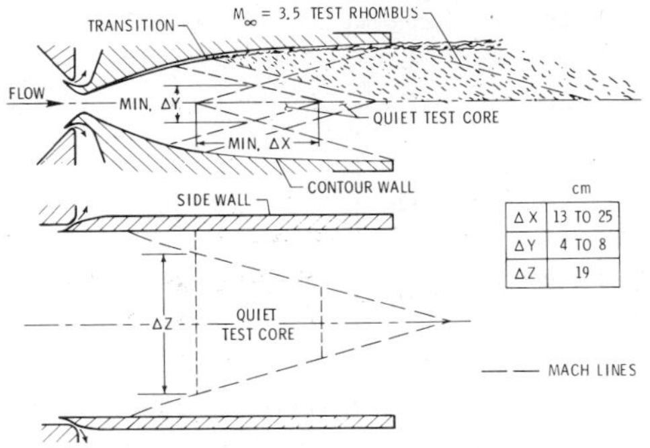

Fig. 1. Shape and location of quiet test core in present Mach 3.5 pilot nozzle [5]. Length of nozzle from throat to exit is 39.29 cm. Nozzle exit is 15.69 cm by 25.46 cm.

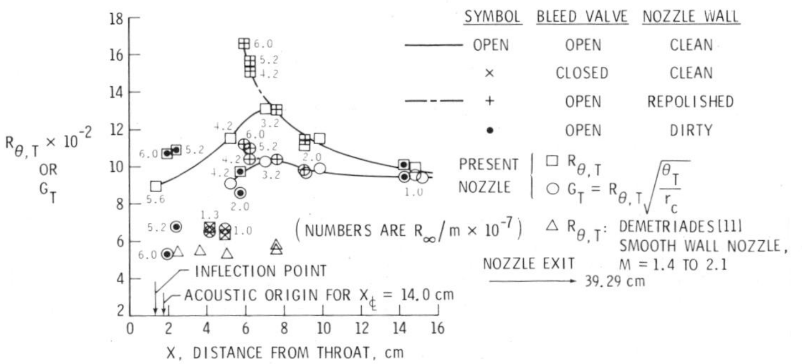

Fig. 2. Local momentum thickness Reynolds number, $R_{\theta,T}$, and Görtler number, G_T, at transition on contour wall of Mach 3.5 rapid expansion nozzle [5] (Nozzle 1 in Table I).

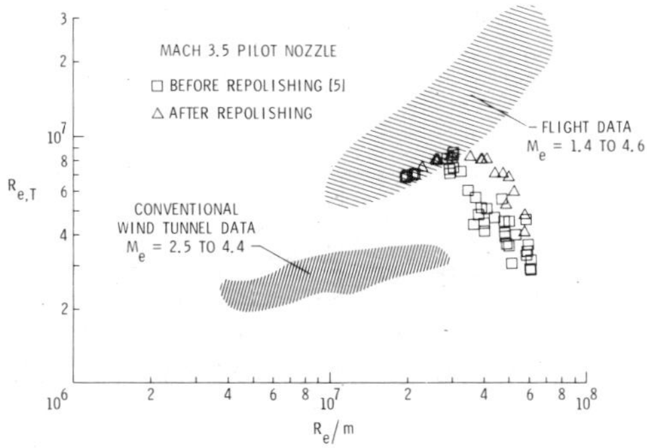

Fig. 3. Variation of transition Reynolds number with unit Reynolds number on sharp tip cones. Data in the Mach 3.5 pilot tunnel for bleed valve open, cone tip at 12.7 cm from nozzle throat. Sources for flight data and conventional wind tunnel data given in [5].

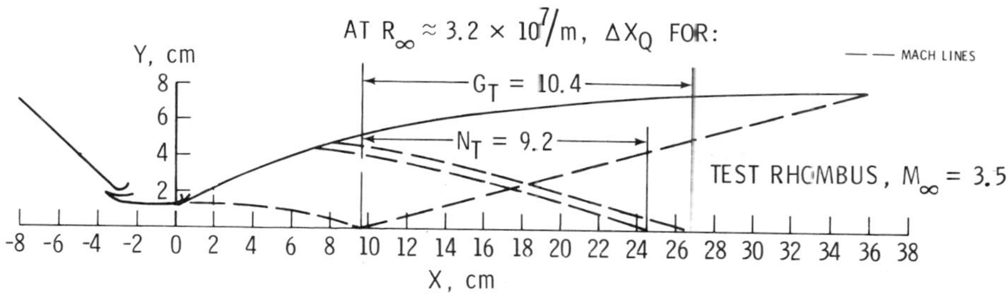

(a) Corner expansion nozzle (number 3 in Table I).

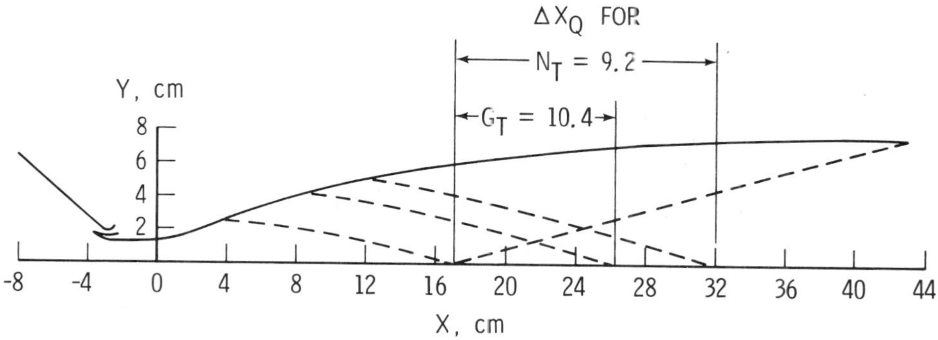

(b) Short radial nozzle (number 4, Table I).

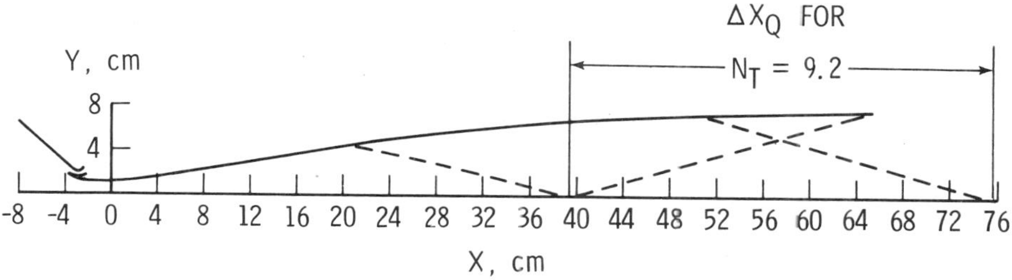

(c) Long radial nozzle (number 5, Table I).

Fig. 4. Wall contours and Mach lines for three of the study nozzles with ΔX_Q lengths illustrated.

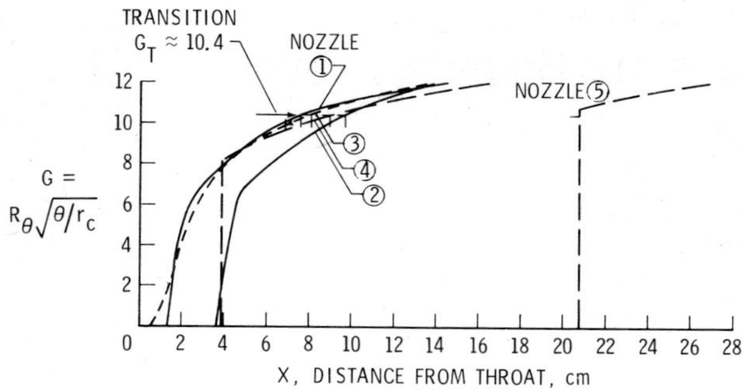

NOZZLE	AT $G_T = 10.4$	
	X_T, cm	ΔX_Q, cm
①	7.5	12.7
②	9.6	10.2
③	8.0	16.3
④	8.8	9.4
⑤	20.8	0

Fig. 5. Variation of Görtler number along the contoured wall of several nozzles from Table I for $R_\infty = 3.2 \times 10^7$/m.

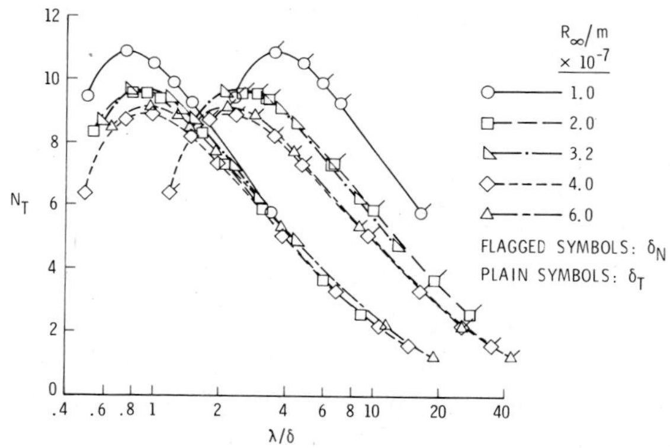

Fig. 6. Present constant wavelength method.

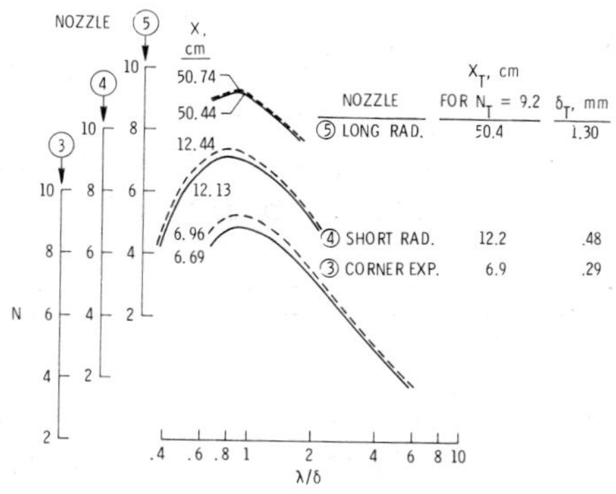

Fig. 7. Variation of N with λ/δ for nozzles 3, 4, and 5 (Table I) at $R_\infty = 3.2 \times 10^7$/m.

Development of Artificially Excited Disturbances in Supersonic Boundary Layer

A.D. KOSINOV, A.A. MASLOV

Institute of Theoretical and Applied Mechanics
USSR Academy of Sciences,
Novosibirsk, USSR

Summary

The present paper suggests a method of introducing the artificial regulable disturbances into supersonic boundary layer with the help of a point source of disturbances. While using the technique suggested the experimental study of development of a spatial wave packet of a small intensity was carried out in the boundary layer of a flat plate.

Introduction

In experimental works dealing with studying the stability of supersonic boundary layers, the development of natural disturbances (see f.e. [1]) is, as a rule, examined. The lack of a full spatial characteristic of the disturbances wave field in the boundary layer is a shortcoming of such works. A more careful quantitative comparison of theory with experiment can be obtained by investigating the controllable artificial disturbances in the boundary layer. The first experimental studies of supersonic boundary layer with artificial disturbances are carried out in [2, 3]. A general shortcoming of these works is to the effect that the disturbances front was not under control in experiment that made impossible to treat many results. In paper [4], the boundary layer on a flat plate was excited with the help of electrical discharge on rectilinear electrodes placed on the model surface at three angles to the flow. The development of the first- and second-mode waves was under consideration. The conformity with the theory predictions was obtained for Mach numbers M=4.5. In the present work the stability of supersonic boundary layer was investigated with the use of a point source of artificial disturbances.

Experimental equipment

The experiments were performed in a supersonic wind tunnel T-325 with the test section dimensions 200 x 200 mm at Mach number M=2.0 and the unit Reynolds number $Re_1=6.5\cdot10^6\ m^{-1}$. The measurements were carried out in the boundary layer of a flat plate with a sharp leading edge. A model was placed in the center of the wind tunnel test section at zero angle of attack. An electric discharge was used as a source of disturbances at selected frequencies. Inside the plate it was made a chamber with two sharp cone-shaped electrodes. By electric discharge combustion there appeared the pressure and temperature fluctuations which disturbed the boundary layer penetrating into it through a hole on the plate surface. The hole diameter was 0.8 mm; the coordinates were x=18 mm and z=0 mm. x is the streamwise coordinate, z is the normal to the stream flow coordinate in the model plane. The source power was regulated both by the change of the distance between the electrodes and by the voltage igniting a discharge. The disturbances in the flow were measured with the help of a constant current hot-wire anemometer. The probes with a tungsten wire, the diameter and the length of which were 6 µkm and 1.2±0.1 mm, respectively, were employed. A double-trace oscillograph synchronized from a discharge was used for determining a phase of the signal under study with respect to the disturbances source.

The study in the boundary layer was carried out at $y/\delta=0.5\div0.6$ where δ is the boundary layer thickness, y is the coordinate along the normal to the plate surface, that corresponds to the maximum of "natural" disturbances across the layer at the point x=30 mm (Re=445, $Re=\sqrt{Re_1\cdot x}$) from the plate leading edge and z=0. When processing the experimental data a "useful signal" A(z) of the mass flow fluctuation ("natural" disturbances and noises were subtracted from the initial data) was singled out from the measured values, and the Fourier transformation was carried out. As a result of processing we have got the amplitude $A(\beta)$ and phase $\Phi(\beta)$ where β is a wave number in the z-direction. In order to calculate the angle of the wave front inclination χ to a main flow, a wave number α_r in the x-direction was determined. The obtained dependencies of the disturbance phase at the fixed β on the longitudinal coordinate $\Phi(x, \beta)$ were of linear character within the limits of the measurement error that leads to the uniqueness in the dependence $\alpha_r = \alpha_r(\beta)$. Then $\alpha_r = \Delta\Phi(x, \beta)/\Delta x$ and $\chi = \mathrm{arctg}(\beta/\alpha_r)$.

Results

The studies of methodics character were carried out directly before conducting experiments with artificial disturbances. It was established that there exists a range of voltages fed to the electrodes for which the dependence of fluctuations in the boundary layer (at fixed probe coordinates) is a linear function of the voltage value. The amount of exceeding the intensity of artificial fluctuations over the "natural" ones can smoothly change from zero to several dozens. While changing the distance between electrodes and thus changing the chamber volume, it was possible to displace a range of linear dependence towards either side in the intensity. By switching off a discharge and the following switching on at the same voltage value the intensity of fluctuations was completely restored. When introducing the disturbances of a small amplitude the distortion of a mean flow was not observed. A contribution of small disturbances excited by a source at the fixed frequency to the integral fluctuations level in the boundary layer was also negligibly small. The studies have shown that there exists a flow region (in Re) in which a character of the disturbances development in the boundary layer does not depend on the intensity of initial disturbances, i.e. the linear development of disturbances takes place.

The studies on the development of a spatial wave packet in the boundary layer have been carried out for frequency parameter $F=0.375 \cdot 10^{-4}$ (at the frequency f=20 kHz). Here $F=2\pi f/Re_1 U$; U is the velocity in the undisturbed flow. Fig.1 illustrates the examples of the amplitude and phase distributions of disturbances for several cross-sections in x. The amplitudes A are made dimensionless with respect to $A^*=\max\{A\}$ at Re=545, and the phase values were counted off from $\Phi(0)$ - the phase value at z=0. It is seen from the figure that with increasing the Reynolds number the packet spreads about. The examples of spectra

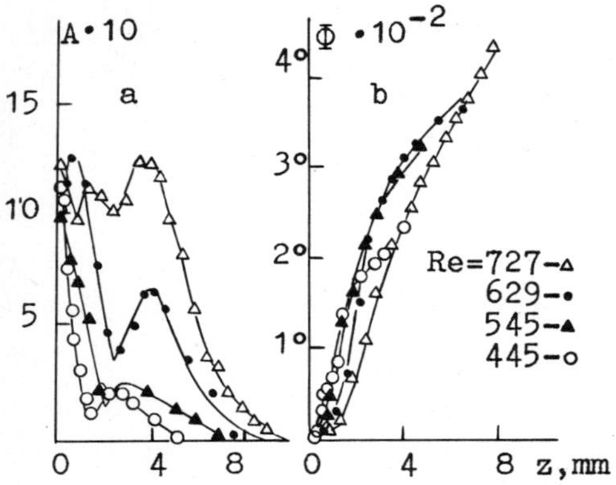

Fig.1

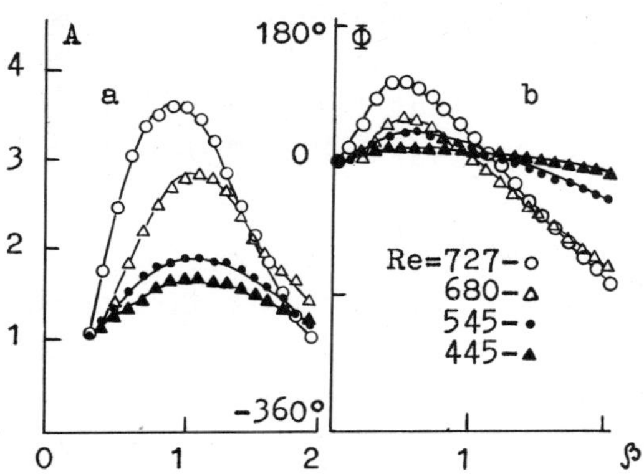

Fig.2

obtained after the Fourier transformation procedure are shown in Fig.2. The fluctuation amplitudes are made dimensionless with respect to the value of A at $\beta=0.3$, and an absolute phase value was counted off from the phase value at $\beta=0$ (the dimension of the wave number β being rad/mm). It is instructive to note that for $\beta<0.3$ in the amplitude spectra there was observed a large scattering of experimental points. One can assume the following: in supersonic boundary layer the three-dimensional disturbances grow most intensively, therefore a portion of plane waves with the angle of propagation close to zero ($\beta<0.3$) is small in a general signal and the experiment lacks a necessary accuracy for singling them out. For the same reason in [4] the author did not succeed in obtained the data for the first-mode waves propagating at small angles to the flow. The disturbances were introduced into the boundary layer with the help of lengthy rectilinear electrodes and due to the intensive growth of three-dimensional disturbances the wave front linearity was rapidly violated. From the graphs for the amplitude spectra (Fig.2a) it is seen that for $A(\beta)$ there exists a maximum at $\beta\sim1$ which when increasing Re displaces towards the smaller β (from 1.1 up to 0.9). This points to the outstripping growth of the waves with $\beta\sim0.9$. The phase spectra (Fig.2b) have two intersection points at $\beta=0$ and $\beta\sim1.1\div0.9$. It is seen that the waves with $\beta=0$ - 0.9 pass ahead of the rests in the phase, and with $\beta=1.1$ - 2 they are behind.

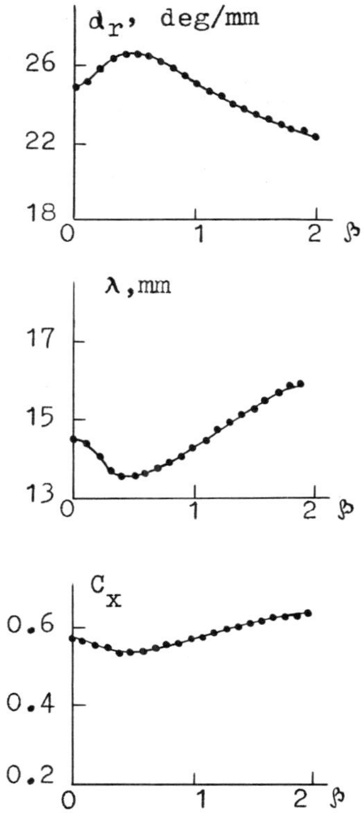

Fig.3

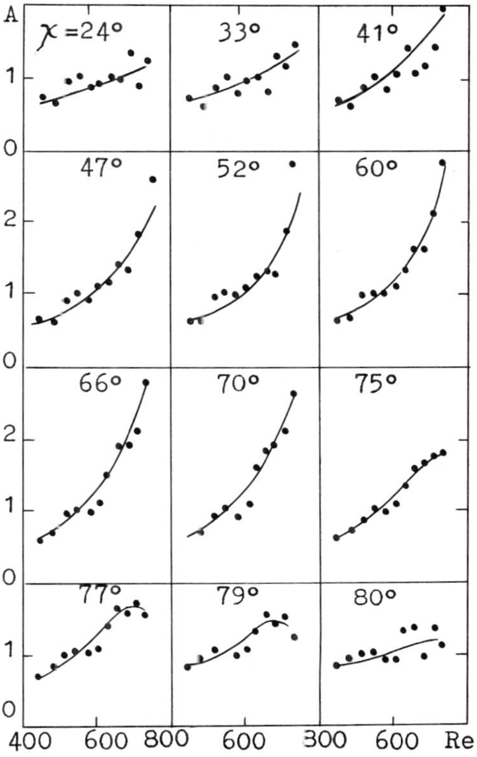

Fig.4

The measurements revealed the fact that along the stream (in x axis) the disturbance phase with the fixed β grows practically in a linear way (Fig.5), the disturbance wavelength λ changes from 13.5 up to 16 mm, and the phase velocity C_x from 0.53 up to 0.63 (the phase velocity values are made dimensionless with respect to the velocity in the undisturbed flow). These data are represented in Fig.3. Fig.4 shows the disturbances amplification curves for different angles of χ. The amplitude is normalized with respect to the value $A=A_o$ at Re_o= =545 (this value is obtained by using the least square method). It is seen from graphs that the maximum growth of disturbances is observable for the angles χ from 50° up to 70°, for the larger and smaller angles the growth degrees are less and the accuracy is less, respectively (a larger scattering of points is observed). For the angles χ~80° there appears the second neutral point. Fig.6 illustrates the dependences of the distur-

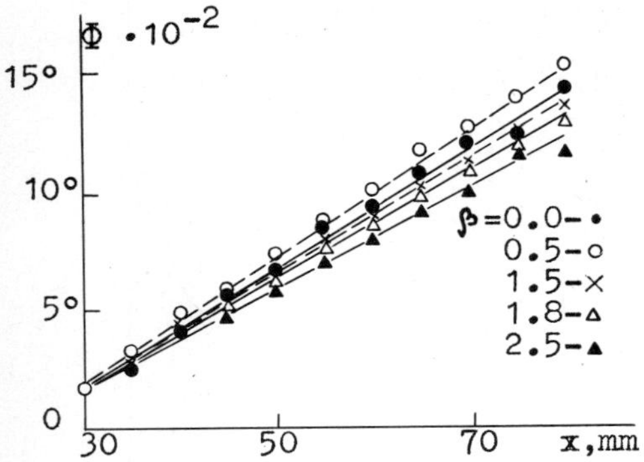

Fig.5

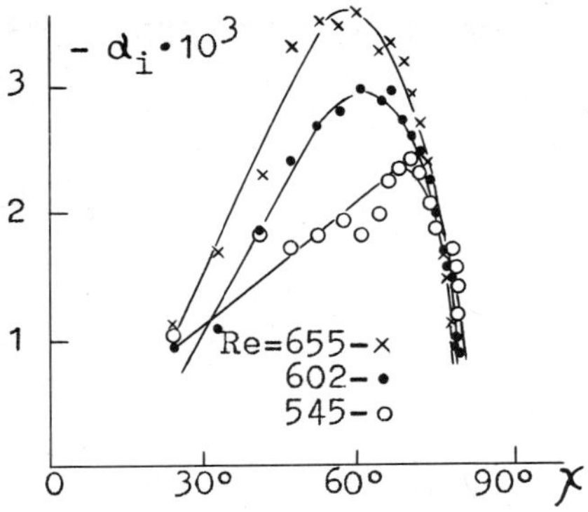

Fig.6

bance amplification rates $-\alpha_i=0.5\ d(\ln A)/d\ Re$ on the wave inclination angle for three values of Reynolds number. It may be observed that with the growth of Re number the maximum of the disturbance amplification rates displaces towards the smaller angles from $\chi=70°$ (at Re=545) up to $\chi=60°$(at Re=655).

The obtained results correspond on the whole to the theoretical ones on the linear stability of supersonic boundary layer. A good coincidence is obtained for dependences $\alpha_i(\chi)$, $\alpha_r(\chi)$, $C_x(\chi)$. A sufficiently universal character of the technique employed enabling one to solve various problems pertaining to the stability of supersonic boundary layer should be noted.

References

1. Gaponov, S.A.; Maslov, A.A.: Disturbance propagation in compressible flows (in Russian). Novosibirsk: Nauka 1980.
2. Laufer, J.; Vrebalovich, T.: Stability and transition of a supersonic laminar boundary layer on an insulated flat plate. J. Fluid Mech. 9 (1960) 257-299.
3. Demetriades, A.: An experiment on the stability of hypersonic laminar boundary layers. J. Fluid Mech. 7 (1960) 385-396.
4. Kendall, J.M.: Boundary-layer stability experiments. JPL Space Programm Summary 37-39. 4 (1966) 147-149.

The Wall Shear Stress Signature of a Turbulent Spot: Direct Measurements and Momentum Integral Calculations

C. W. VAN ATTA

University of California, San Diego
La Jolla, California 92093

T. S. MAUTNER

Naval Ocean Systems Center
San Diego, California 92152

Summary

The measured centerline wall shear stress signature of individual and ensemble averaged turbulent spots consists of an initial sharp peak followed by a more gradual increase to a maximum at the trailing edge of the spot. When represented in terms of the skin friction versus momentum thickness Reynolds number, the ensemble averaged shear stress distribution is composed of four linear segments, each having a different slope. A momentum integral calculation method based on two dimensional boundary layer theory and the assumption of potential flow above the spot was developed by transforming the unsteady momentum equation into conical similarity coordinates. The centerline wall shear stress calculated using ensemble averaged data for the velocity field shows fairly good agreement with the measured values.

Introduction

The spatially organized unsteady velocity, pressure, and wall shear stress fields of turbulent spots are responsible for important consequences of transition such as large shear stress fluctuations and acoustic noise generation. No previous attempts have been reported to directly measure the wall shear stress under artificially generated turbulent spots under controlled conditions or to develop a calculation scheme to relate the shear stress to the unsteady velocity and pressure fields by application of the Navier-Stokes equations to the experimental data. Recent experimental studies of Wygnanski et al [1, 2], Cantwell et al [3], Antonia et al [4, 5] and others have provided detailed measurements of the organized structure of the turbulent spot's velocity field. From near wall velocity measurements, Wygnanski et al [1] inferred that the wall shear stress should increase through the spot, reaching a maximum at the trailing edge. Rajagopalan and Antonia [6] measured the velocity and wall shear stress in natural transition spots in the inlet of a two dimensional duct. Their near wall velocity and wall shear stress signature were similar, providing a rough confirmation of Wygnanski et al's conjecture about the shape of the spot's wall shear stress signature. The available quantitative estimates of the wall shear stress from velocity profile data are in disagreement. Wygnanski et al [1] made estimates of the spot's skin friction, presumably by extrapolation of the velocity profiles to the wall, which yield skin friction coefficients of $C_f \simeq 0.0053$ at the trailing edge and $C_f \simeq 0.0036$ at the maximum height of the spot. Using these estimates they inferred that the spot's ensemble mean velocity profiles follow the universal logarithmic velocity profile represented by $u^+ = 2.4 \ln y^+ + 5$. However, assuming the universal law to be valid for their profile data, Antonia et al [4] obtained unrealistic values for the skin friction coefficient ($C_f \simeq 0.01$).

The purpose of the present study was to directly measure the wall shear stress and corresponding velocity fields of turbulent spots under controlled conditions and to attempt to quantitatively relate the measured behavior of the ensemble averaged velocity, pressure, and wall shear stress fields.

Experimental Apparatus and Data Reduction

The basic experimental set up was that described by Mautner and Van Atta [7]. Turbulent spots were generated at a distance $x = 30.5$ cm from the leading edge of the flat plate and the centerline velocity and wall shear stress measurements were made at $x = 152.4$ cm at a nominal free stream velocity of $U_\infty = 10$ m/s. The longitudinal velocity $u(x, y, t)$ was measured using a TSI 1050 constant temperature anemometer and a DISA 55P01 hot-wire probe. The measured laminar mean velocity profile agrees well with the Falkner-Skan profile for $\beta = 0.1$. The wall shear stress $\tau_w(t)$ was measured using a TSI 1237 hot-film sensor and a TSI 1050 constant temperature anemometer. The hot-film probe was mounted flush with the surface and was calibrated in both laminar and turbulent flows. The square wave trigger signal for the spot generator was used to trigger and synchronize the digitizing of the analog data signals using a 12 bit plus sign A/D converter at a sampling rate of 7500 Hz. Each velocity and wall shear stress realization was represented by 2048 digital samples taken after a fixed time delay following each trigger pulse. The ensemble (phase averaged) mean velocity $\langle u \rangle$, and wall shear velocity $\langle U_\tau \rangle = (\langle \tau_w \rangle / \rho)^{1/2}$ signatures were formed by averaging over 100 spots for the velocity and 500 spots for the wall shear stress. For each spot, fluctuations about the ensemble mean are defined as $U_{\tau i} - \langle U_\tau \rangle$, where $i = 1, 2, 3, \ldots, 2048$. The rms ensemble averaged shear stress fluctuation U'_τ calculated for each sampling time t_i furnishes a measure of the variation between different spots in the ensemble average and may be interpreted as a turbulent intensity relative to the ensemble mean. The ensemble averaged velocity profiles, normalized with the velocity at the maximum spot height U, were used to obtain the ensemble distributions for δ^*, θ, and H, the displacement thickness, momentum thickness, and shape factor, respectively, shown in Fig. 1.

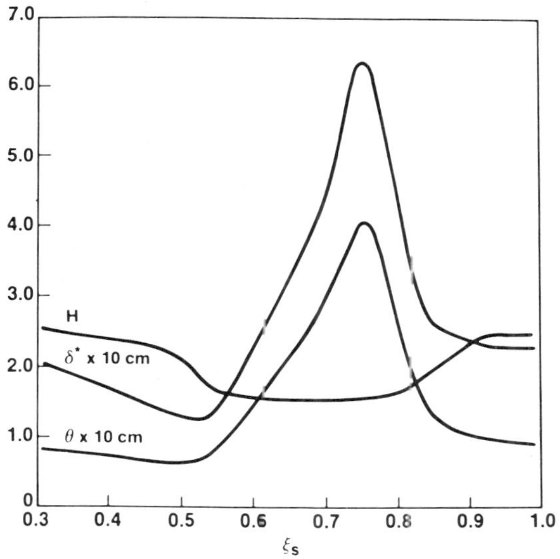

Fig. 1. Variation of boundary layer parameters through the spot.

Results and Discussion

The instantaneous U_τ signature for a single, typical spot is shown in Fig. 2. The fairly regular series of well defined peaks and valleys in the U_τ trace represent large, intermittent contributions to the turbulent energy and Reynolds stress. Similar types of fluctuations were observed by Willmarth and Lu [8] in the streamwise sublayer velocity and by Kreplin and

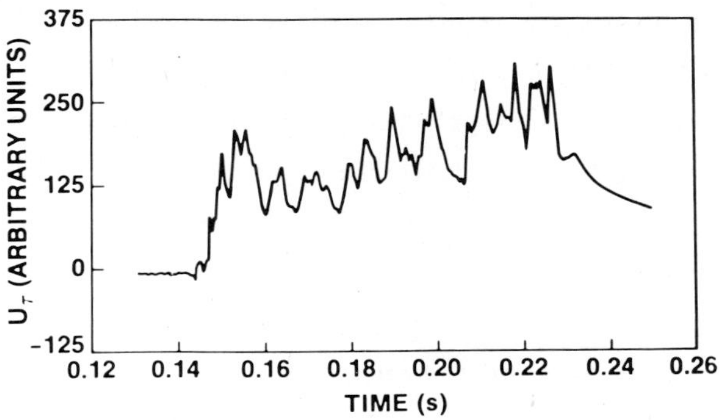

Fig. 2. Typical wall shear stress signature for a single spot.

Eckelmann [9] in their measurements of $\partial u/\partial y$ at the wall.

The measured centerline ensemble averaged wall shear stress $\langle U_\tau \rangle$ signature is compared with the near wall $\langle u \rangle$ in Fig. 3. Note that the time axis has been inverted using the conical similarity coordinates $\xi_s = x_s/U_\infty t_s$, with x_s and t_s measured from the apparent origin of the spots. The $\langle U_\tau \rangle$ trace is characterized by a rapid increase at the leading edge followed by a short plateau or decrease prior to a gradual increase to a maximum value

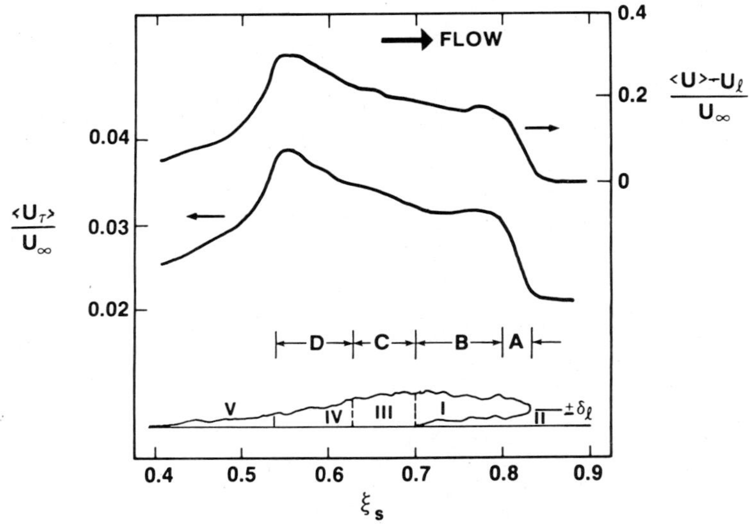

Fig. 3. Comparison of the measured wall shear stress with the velocity signature near the wall and a cross sectional view adapted from [13].

at the trailing edge. This maximum is followed by a slow relaxation back to the ambient laminar value. Comparison of $\langle U_\tau \rangle$ with $\langle u \rangle$ measured at y = 0.03 cm shows that the characteristics of each trace are very similar in shape. The observed behavior, which is qualitatively similar to that measured by Rajagopalan and Antonia, experimentally validates the shape of the conjectured U_τ signature of Wygnanski et al [1].

Data for the "turbulent" shear stress fluctuation U'_τ showed a sharp increase to a maximum at the spot's leading edge followed by a rapid decrease to a plateau value which persisted through the central region of the spot and then decreased rapidly at the trailing edge. Similar characteristics have been observed in the wall pressure fluctuations and velocity fluctuations by Mautner and Van Atta [7].

The measured $\langle U_\tau \rangle$ was compared with values calculated using two empirical equations for U_* for flat plate boundary layers, the Ludwieg-Tillman equation

$$(U_*/U_\infty)^2 = (0.123)\ 10^{-0.678H}\ R_\theta^{-0.268} \tag{1}$$

and that of Coles (as given by White [10]),

$$(U_*/U_\infty)^2 = 0.15 \exp(-1.33H)/(\log_{10} R_\theta)^{1.74 + 0.31H} \tag{2}$$

The general shapes of the calculated and measured wall shear were in approximate agreement, despite the fact that the empirical equations used were derived from steady, turbulent, boundary layers with constant mean pressure gradients, while a turbulent spot is an unsteady, growing patch of turbulence with a highly variable (in time and space) pressure gradient (Mautner and Van Atta [7]). The steady flow equations are thus qualitatively applicable, but overestimate the stress in the central region by 10-15%. We obtained similar overestimates of the wall shear stress by applying the wall slope method to the ensemble averaged velocity profiles. The zone averaged wall shear stress was found to be approximately equal to the zone averaged Reynolds stress distribution in the spot (Mautner and Van Atta [11]).

The empirical relationships (1) and (2) used to calculate the wall shear stress for a turbulent boundary layer are in the form $U_* = fct(R_\theta, H)$.

Similarly, it would be useful in representing the spot's wall shear stress to determine what relationships exist between $\langle U_\tau \rangle$, R_θ, and H. The variation of $\langle U_\tau \rangle$ with H or R_θ provides no simple relationship. The variation of $\langle U_\tau \rangle$ with R_θ is relatively smooth; however, because the distribution of θ varies through the spot, the curve does not specifically identify a position in the spot for each (U_τ, R_θ).

However, when the wall shear is plotted in terms of $C_f = \text{fct}(R_\theta)$(Fig. 4), it is found that the skin friction distribution is composed of four linear segments each having a different slope. The range in ξ_s of each segment is shown in the insert. Segment A represents the region of rapid acceleration of the near wall velocity and the wall shear stress. This region also coincides with the maxima of the rms fluctuations u' and U'_τ and corresponds to a rapid increase in δ^* and θ and a decrease in H. The nearly constant value of C_f with R_θ in segment B occurs during the increase and subsequent decrease of δ^* and θ from maxima at the maximum spot height while H remains approximately constant. This is also the region of slowly increasing $\langle u \rangle$. In both segments C and D, the shape factor is increasing slightly while both δ^* and θ decrease rapidly toward their ambient values. Corresponding to the decrease in R_θ, there is a rapid rise in C_f which becomes more pronounced in segment D and coincides with the steady rise in near wall velocity. The relaxation region follows segment D.

Momentum Integral Formulation

Here, we can only briefly outline the calculation procedure (Mautner and Van Atta [12]). The flow along the spot's centerline is represented by the two-dimensional, incompressible boundary layer equations where the velocity components u and v, pressure p and shear stress τ are given in terms of their respective ensemble average quantities. After application of double-decomposition $u_i = \langle u_i \rangle + u_i'$ and the usual boundary layer approximations, and using the continuity equation to eliminate the v velocity, the Navier-Stokes equations reduce to

$$\frac{\partial \langle u \rangle}{\partial t} + \langle u \rangle \frac{\partial \langle u \rangle}{\partial x} - \frac{\partial \langle u \rangle}{\partial y} \int_0^y \frac{\partial \langle u \rangle}{\partial x} dy = -\frac{1}{\rho}\frac{\partial \langle p \rangle}{\partial x} + \frac{1}{\rho}\frac{\partial \tau}{\partial y} \qquad (3)$$

where the shear stress is given by $\tau = \mu(\partial \langle u \rangle / \partial y) - \rho \langle u'v' \rangle$. The boundary

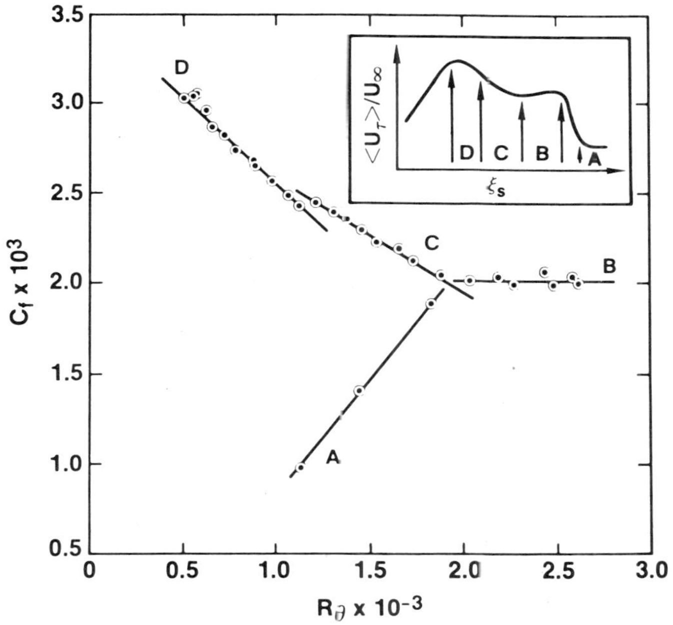

Fig. 4. Variation of C_f with R_θ. $U_\infty = 10$ m/s; $x = 152.4$ cm.

conditions are $\langle u \rangle = \langle v \rangle = 0$ at $y = 0$ and $\langle u \rangle = U_e(x)$ and $\langle v \rangle = V_e(x)$ at $y = \delta$ ($\delta \simeq 0.8\,\delta_t$). The momentum equation, Bernoulli's equation and the spot data are first transformed into the coordinate system $(\bar{x}, \bar{y}, t)$ whose origin is at the spot's centroid and is moving with a uniform horizontal velocity of $U_x = c_1 U_\infty$, so that the translating coordinate system is related to the laboratory system by $\bar{x} = x - U_x t$ and $\bar{y} = y$. Next, the $(\bar{x}, \bar{y})$ coordinates are nondimensionalized by $h(t) = cU_\infty t$ which yields $x^* = \bar{x}/h$ and $y^* = \bar{y}/h$. The transformations result in the following functional relationships: $x^* = f(\bar{x}, h) = \bar{x}/h$, $y^* = f(\bar{y}, h) = \bar{y}/h$, $u^* = \langle u \rangle / U_\infty = f(x^*, y^*)$, $v^* = \langle v \rangle / U_\infty = f(x^*, y^*)$ and $\tau / \rho U_\infty^2 = f(x^*, y^*)$.

The unsteady Bernoulli equation, applied to the free stream data, was used to calculate the ensemble averaged pressure within the spot. The resulting pressure gradient, evaluated at the maximum spot height, where $u \rightarrow U_e$ and $v \rightarrow V_e$, and $U_e^* = U_e/U_\infty$ and $V_e^* = V_e/U_\infty$, is

$$\frac{\partial}{\partial x}\left(\frac{\langle p \rangle}{\rho U_\infty^2}\right) = \left[\frac{(c_1 + cx^*) - U_e^*}{cU_\infty t}\right]\frac{dU_e^*}{dx^*} + \left[\frac{c - V_e^*}{cU_\infty t}\right]\frac{dV_e^*}{dx^*} \tag{4}$$

U_e was calculated using (4) and (assuming $\partial\langle p \rangle/\partial y = 0$) the measured wall pressure signature [7]. The x momentum equation becomes

$$\begin{aligned}\frac{1}{\rho U_\infty^2}\frac{\partial \tau_w}{\partial y^*} &= (c_1 + cx^*)\left(\frac{dU_e^*}{dx^*} - \frac{\partial u^*}{\partial x^*}\right) - \dot{c}y^*\frac{\partial u^*}{\partial y^*} \\ &\quad + u^*\frac{\partial u^*}{\partial x^*} - U_e^*\frac{dU_e^*}{dx^*} - \frac{\partial u^*}{\partial y^*}\int_0^{y^*}\frac{\partial u^*}{\partial x^*}\,dy^*\end{aligned} \tag{5}$$

After integrating (5) from $y^* = 0$ to 1 and further manipulation, one obtains the momentum integral relation in the form

$$\begin{aligned}\frac{1}{\bar{\theta}_e}\left(\frac{\langle U_\tau \rangle}{U_e}\right)^2 d\xi_s &= \frac{d\bar{\theta}_e}{\bar{\theta}_e} + \left(\frac{2 + H_e}{U_e^*}\right)dU_e^* + \xi_s H_e \frac{dU_e^*}{U_e^{*2}} \\ &\quad - \frac{H_e}{U_e^*}d\xi_s - \frac{\xi_s}{\bar{\theta}_e U_e^*}d\bar{\delta}_e^*\end{aligned} \tag{6}$$

where $\bar{\delta}_e^* = \delta_e^*/U_\infty t$ and $\bar{\theta}_e = \theta_e/U_\infty t$. The solution for $\langle U_\tau \rangle$ is obtained from (6) by marching through the spot in small increments of ξ_s. If $\Delta\xi_s$ is chosen sufficiently small, it is reasonable to assume that $\bar{\delta}_e^*$, $\bar{\theta}_e$, U_e and $\langle U_\tau \rangle$ are linear functions of ξ_s over the range of $\Delta\xi_s$, allowing the use of average values of the measured quantities over each $\Delta\xi_s$. Equation (6) may then be integrated directly, yielding

$$(\langle U_\tau \rangle/U_e)_{avg} = [(\bar{\theta}_e)_{avg}/\Delta\xi_s\,\{\ln(\bar{\theta}_{e_2}/\bar{\theta}_{e_1}) + Q\}]^{1/2} \tag{7}$$

where

$$\begin{aligned}Q &= [2 + (H_e)_{avg}]\ln(U_{e_2}^*/U_{e_1}^*) + (H_e\xi_s)_{avg}[1/U_{e_2}^* - 1/U_{e_1}^*] \\ &\quad + (H_e/U_e^*)_{avg}\,\Delta\xi_s - (\xi_s/\bar{\theta}_e U_e^*)_{avg}\,\Delta\bar{\delta}_e^*\end{aligned}$$

and the subscripts 1 and 2 refer to the beginning and ending values of ξ_s, respectively. When the raw experimental data for δ^*, θ, and H (Fig. 1) was used in Eq. (7), the calculated values of $\langle U_\tau \rangle$ showed considerable scatter but were in general agreement with the measured $\langle U_\tau \rangle$. This scatter may have been due to the sensitivity of Eq. (7) to small local variations in the experimental data, and examination of the order of magnitude of each term in Eq. (7) shows that the $\langle U_\tau \rangle_{avg}$ calculation requires differences of terms which are nearly equal in magnitude. A second set of solutions for $\langle U_\tau \rangle_{avg}$ was, therefore, carried out using a polynomial representation of the experimental data. The result obtained when ξ_s was incremented beginning at $\xi = 1.0$ is shown in Fig. 5. The calculated $\langle U_\tau \rangle_{avg}$ obtained using this procedure is in good agreement with the measured U_τ, indicating that the present momentum integral method can be used to accurately relate the ensemble averaged shear stress, velocity, and pressure fields.

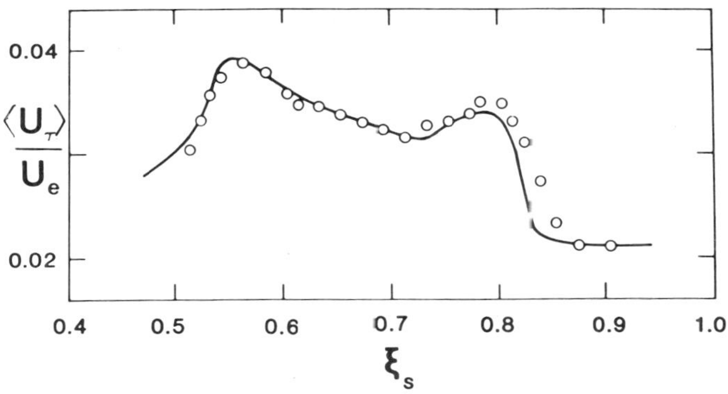

Fig. 5. Comparison of the calculated $\langle U_\tau \rangle_{avg}$ with the measured values for $x_s = 121.9$ cm and $U_\infty = 10$ m/s. ——, measured; o, calculated.

Acknowledgments

This work was supported in part by NSF Grant MEA81-00431, by the Naval Ocean Systems Center Independent Research Program, and the 6.1 NAVMAT R& D Centers/University Joint Research Program.

References

1. Wygnanski, I.; Sokolov, M.; Friedman, D.: On a turbulent spot in a laminar boundary. J. Fluid Mech. 78 (1976) 785.

2. Wygnanski, I.; Zilberman, M.; Haritonidis, J.: On the spreading of a turbulent spot in the absence of a pressure gradient. J. Fluid Mech. 123 (1982) 69.

3. Cantwell, B. J.; Coles, D.; Dimotakis, P.: Structure and entertainment in the plane of symmetry of a turbulent spot. J. Fluid Mech. 87 (1978) 641.

4. Antonia, R. A.; Chambers, A. J.; Sokolov, M.; Van Atta, C.W.: Simultaneous temperature and velocity measurements in the plane of symmetry of a transitional turbulent spot. J. Fluid Mech. 108 (1981a) 317.

5. Antonia, R. A.; Chambers, A. J.; Sokolov, M.; Van Atta, C.W.: On the similarity between velocity and temperature fields within a turbulent spot. Third Symosium on Turbulent Shear Flows (1981b) Univ. of Calif., Davis, 10, 1.

6. Rajagopalan, S.; Antonia, R. A.: Investigation of natural transition in the inlet region of a two-dimensional duct flow. Phys. Fluids 23 (1980) 1938.

7. Mautner, T. S.; Van Atta, C. W.: An experimental study of the wall pressure field associated with a turbulent spot in a laminar boundary layer. J. Fluid Mech. 118 (1982) 59.

8. Willmarth, W. W.; Lu, S. S.: Structure of the Reynolds stress near the wall. J. Fluid Mech. 55 (1972) 65.

9. Kreplin, H. S.; Eckelmann, H.: Propagation of perturbations in the viscous sublayer and adjacent wall region. J. Fluid Mech. 95 (1979) 305.

10. White, B.: Low-Reynolds-number turbulent boundary layers. J. Fluids Eng. 103 (1981) 624.

11. Mautner, T. S.; Van Atta, C. W.: Wall Shear Stress Measurements in the Plane of Symmetry of a Turbulent Spot (submitted).

12. Mautner, T. S.; Van Atta, C. W.: An unsteady momentum integral calculation for the turbulent spot (submitted).

13. Gad-el-Hak, M.; Blackwelder, R.; Riley, J: On the growth of turbulent regions in laminar boundary layers. J. Fluid Mech. 110 (1981) 73.

An Experimental Study of the Structure and Spreading of Turbulent Spots

PER-AKE LINDBERG, E. MAGNUS FAHLGREN,
P. HENRIK ALFREDSSON and ARNE V. JOHANSSON

Department of Mechanics
The Royal Institute of Technology
S-100 44 Stockholm, Sweden

SUMMARY

Turbulent spots in a free surface water table flow were studied. Flow visualization suggested that the reported differences in shape between the water table and laminar boundary-layer spots may be ascribed to relaminarization of the front part of the former. Linear stability analysis of velocity profiles measured just outside the "wing-tip" of the spot indicated a reduced stability of the flow in this region as compared to the undisturbed laminar flow, which may be seen as a support for a growth-by-destabilization concept. The stability of the flow in this region was also examined in a more direct way by introducing a small secondary disturbance near the wing-tip of the spot.

1. INTRODUCTION

The turbulent spot was discovered in 1951 by Emmons [1] in a free surface flow on a water table. Since then it has been established that onset of turbulence takes place in isolated growing spots, for most wall-bounded flows. The mechanisms of spot formation and growth are self-maintaining in that the characteristics of the spot are practically independent of the initial disturbance. The lateral spreading rate is approximately the same for boundary layer, plane Poiseuille and water table spots, and the half-spreading angle is about 10°. However, Mitchner [2], who made experiments both on a water table and in a laminar boundary layer in a wind tunnel pointed out considerable differences in spot-shape between these two cases. Several later visualization studies (see e.g. [3,4,5]) have shown that the arrowhead front part of spots in a laminar boundary layer is characterized by violent mixing, in contrast to spots in water table flow [6,7] and plane Poiseuille flow [8]. An important difference between the laminar boundary layer and the other two types is that in the former case fluid from the non-turbulent free-stream may be entrained into the spot, thereby supplying energy for the persistence of turbulence in the interior of the spot. Indeed, it seems reasonable to assume that for boundary-layer spots entrainment accounts for the relatively slow growth in the direction normal to wall, in much the same way as for a turbulent boundary layer. However, in the lateral direction turbulent diffusion is an order of magnitude smaller than the spreading rate [5]. Also the spreading of a turbulent region behind a roughness element on the wall in a laminar flow is similar to that of spots [9]. An explanation for this rapid lateral growth, first proposed by Corrsin & Kistler [10], is that a turbulent region in a laminar shear flow destabilizes the undisturbed flow. This mechanism, applied to the "spot-case" was termed "growth-by-destabilization" by Gad-el-Hak et al. [5]. Wygnanski et al. [11,12] observed waves trailing the spot in a laminar boundary layer, and suggested that these break down and generate "hairpin" vortices near the "wing-tips" of the spot. They speculated that these are coupled to the rapid lateral growth of the spot. This does not, however, seem to be in accordance with some recent findings of Chambers & Thomas [13], who also observed oblique waves outside the spot, but concluded that these had little involvement in the growth of the spot.

2. EXPERIMENTAL PROCEDURE

The present experiments were carried out on a water table (Fig. 1), where a thin water layer flows down on a 2 m long, slightly inclined (0-5°) glass-plate. When the flow enters the test section the component of gravity parallel to the plate accelerates the flow until the gravitational force is balanced by the skin friction and a half-parabolic velocity distribution is established. The acceleration is reflected in a decrease of the water layer thickness towards an asymptotical value (h_a). The Reynolds number (Re), based on bulk velocity and water layer thickness (h), is proportional to the flow rate, and thus remains constant in the streamwise direction. For the present study of spot characteristics, the flow is considered sufficiently developed when the thickness is within 10 % of the asymptotical value, i.e. $h/h_a<1.10$. Measurements showed that this condition is satisfied if $x/(h_a Re)>0.32$. The spots were triggered 0.8 m from the test section entrance by letting a small (about 2 mm diameter) drop of water fall from a syringe needle onto the surface. With this simple technique good repeatability of the spots was obtained.

For velocity measurements a cylindrical hot-film probe (DISA R15) was used, whereas the wall-shear stress was measured with a flush mounted hot-film probe (DISA R46) located 1.8 m from the test section entrance. A conditional sampling technique was developed to enable use of velocity data from different spots for the determination of velocity profiles just outside the spot. When the spot passed close by the probes, relatively distinct minima in velocity and wall-shear stress were observed. The cross-correlations between the wall-shear-stress signal of a reference spot and that for each of the

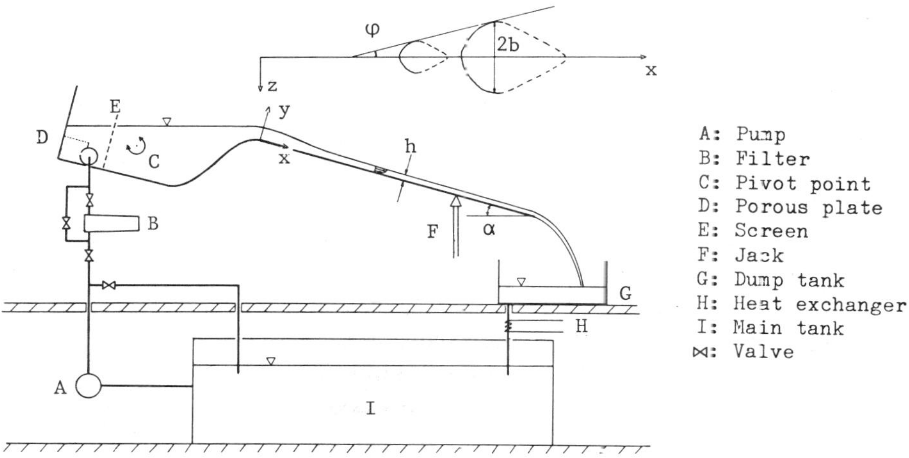

Fig. 1. Sketch of water table and spot geometry

other spots were calculated. The cross-correlation attains a maximum for a specific time delay which corresponds to when the signals overlap as closely as possible. A common time-axis is then obtained by subtracting this delay from the actual time for each spot. The profiles obtained in this way may be regarded as quasi-steady in a frame of reference convected with the spot.

3. SPOT CHARACTERISTICS

Mitchner [2] noted considerable differences in the shape of spots in a laminar boundary layer and in a water table flow, and stated: "It is of some interest to note that the vertex of the turbulent spot for the flat plate case points downstream, in exact contradistinction to the free surface case." For the visualization of water table spots he used the rough free surface and hence was only able to observe disturbances that are strong enough to contort the fluid surface. The picture becomes somewhat different when the flow is visualized with small (10-20 μm diameter) platelet-trace particles oriented by the rate-of-strain field. The development of a spot is shown in Fig. 2, from which it is seen that the spot-shape, at least in the relatively early stages, closely resembles that of boundary-layer spots, i.e. with an arrow-head shaped front part. With increasing downstream distance the front part becomes relaminarized. This may be due to lack of supply of free-stream energy to the spot (via vertical entrainment). A similar situation prevails for the channel flow case [8].

A striking feature seen in Fig. 2 is the streaky structure within the spot, which clearly resembles that seen close to the wall in turbulent boundary

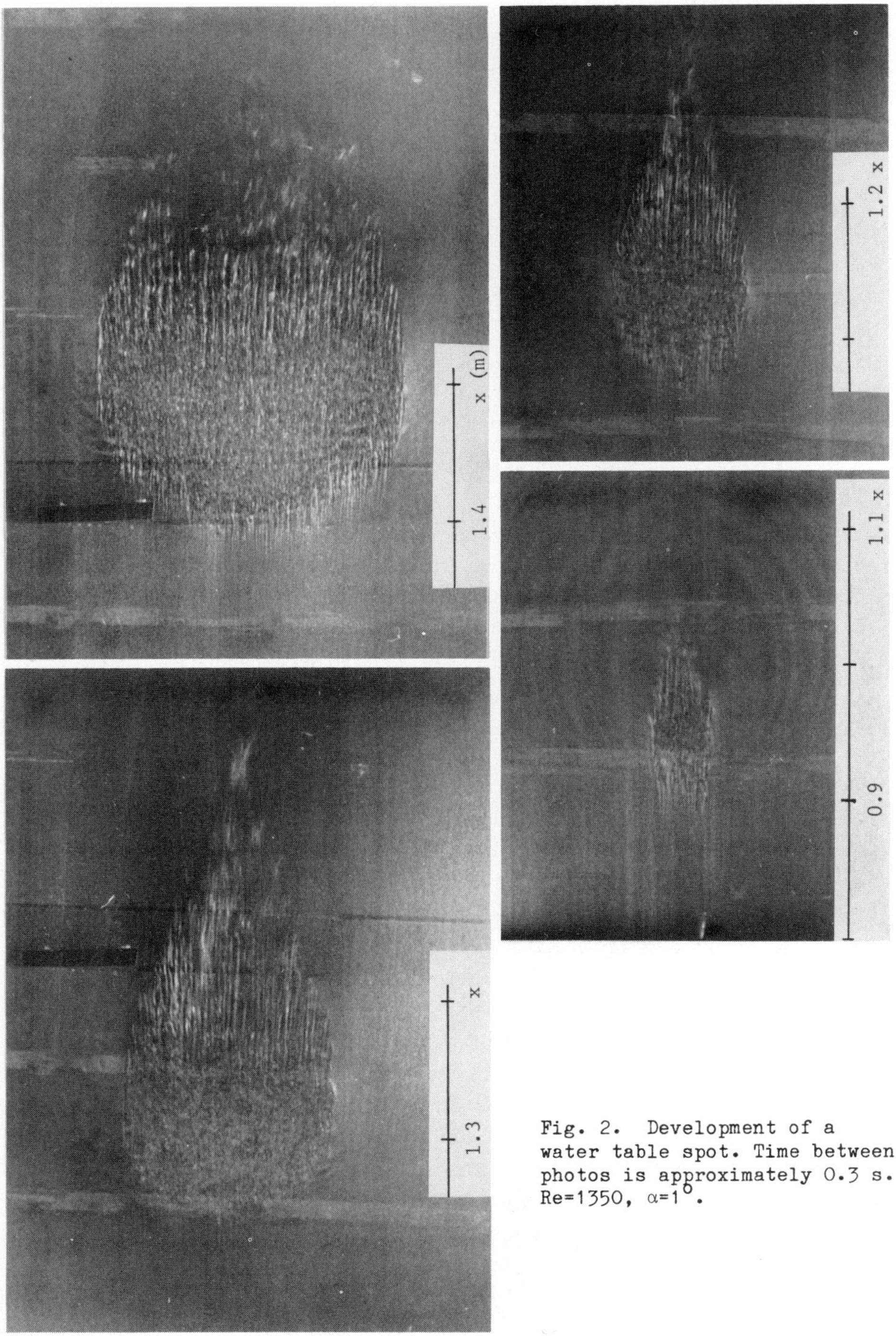

Fig. 2. Development of a water table spot. Time between photos is approximately 0.3 s. Re=1350, $\alpha=1^{\circ}$.

layers (see e.g. Smith & Metzler [14]). This structure is not seen if only the surface is illuminated without any particles in the flow. For an "old" spot the streaks are most clearly seen between the fully turbulent rear part of the spot and the quasi-laminar front part. In the rear part the streaks are hidden by the violent mixing in the upper part of the layer. A possible interpretation is that the streaks precede breakdown.

Hot-film measurements were used to determine the width and thereby the spreading rate of the spot. The virtual origin was located very close (within the measurement accuracy) to the generation point. As seen from Fig. 3 the spreading rate increases with increasing Reynolds number. There is also a slight difference between different inclination angles. These results are in general accordance with the findings of Wygnanski et al. [12] for spots in a laminar boundary layer (half-angle 9-10°), as well as the results of Carlson et al. [8] for Poiseuille flow (8°).

Velocity signals taken at three different lateral locations are shown in Fig. 4, where the three realizations are referred to a common time axis (see section 2). They were taken at $y/h=0.50$ at the spot centreline, at half a spot width from the centreline, and just outside the wing-tip. In the first case the front part of the spot manifests itself through a decrease in the velocity, although no turbulent fluctuations are seen. At $z/b=0.5$ this quasi-laminar part is for natural reasons much shorter. Just outside the wing-tip the velocity signal is quite affected by the passage of the spot although no turbulent fluctuations are seen. The wave structure observed by Wygnanski et al. [11] could not be found, despite a survey with a hot-film velocity probe in a large region around the spot. In the region outside the wing-tip both the velocity and wall-shear stress exhibit clear minima, the magnitude of which is 4-12 % of the undisturbed velocity depending on the distance from the wall, and 12% for the wall shear stress. The defect was detectable in a wide region outside the spot, to at least half a spot width from the wing-tip.

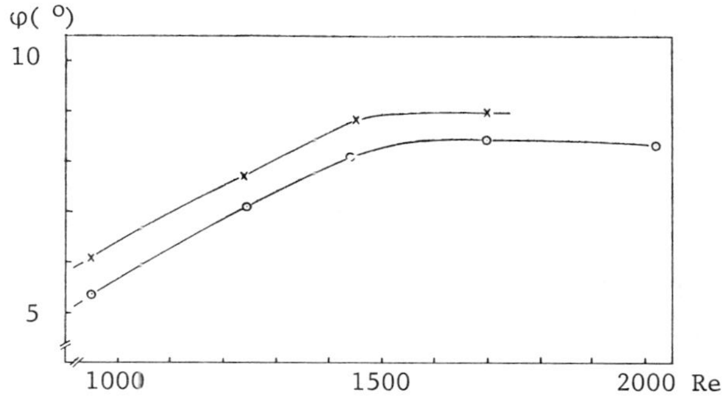

Fig. 3. Spreading half-angle versus Reynolds number. x: $\alpha=2^{\circ}$, o: $\alpha=4^{\circ}$.

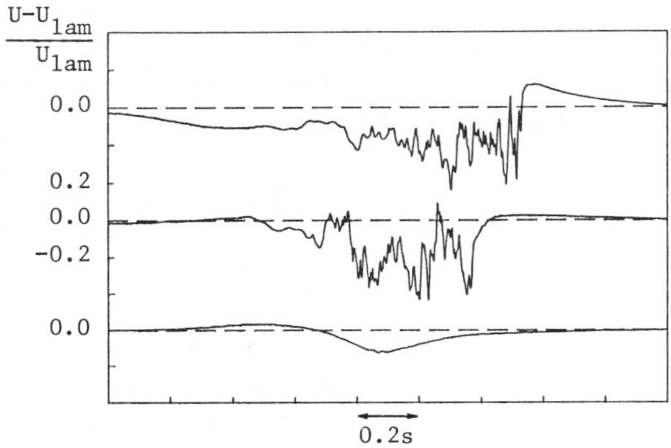

Fig. 4. Velocity signals obtained at y/h=0.5 for three lateral positions. U_{lam} is the undisturbed velocity. Re=1350, α=1°.

To construct velocity profiles just outside the wing-tip at given streamwise positions relative to the spot, the velocity was measured at 19 different positions normal to the wall. At each position data were taken from 8 spots, 6 of which were used for an ensemble average. The "position" relative to the spot boundaries was obtained from simultaneous visualization and wall-shear-stress measurements. It is seen from Fig. 5 that the minimum in velocity in the uppermost part of the water layer occurs at the passage of the wing-tip, and that the minimum of the wall shear stress occurs half-way between the wing-tip and the rear part of the spot. The minima hence occur in a plane which is inclined at a small (approximately 2°) angle to the wall.

4. STABILITY ANALYSIS OF THE FLOW NEAR THE WING-TIP

The velocity profiles measured just outside the wing-tip were analyzed by means of two-dimensional linear stability calculations with use of the Orr-Sommerfeld equation, and with surface tension effects included in the boundary conditions. The basic flow (U) was assumed to be unidirectional and quasi-steady. A polynomial whith six adjustable constants, satisfying the boundary conditions U(0)=0, U(1)=1 and dU/dy(1)=0 was fitted to each of the measured profiles.

The critical Reynolds number (Re_{crit}) for the measured undisturbed velocity profile is approximatly 1500 which is considerably lower than the value for a true parabolic profile, which is roughly 3800 (see e.g. Chin [15]). This difference is due to the deviation of the measured profile from the parabolic one. The stability characteristics are very sensitive to the shape of the profile, wherefore the results presented in the following should be interpreted mainly in a qualitative sense. However, Re_{crit} varies strongly for the measured profiles (see Fig. 6). When the spot approaches the point

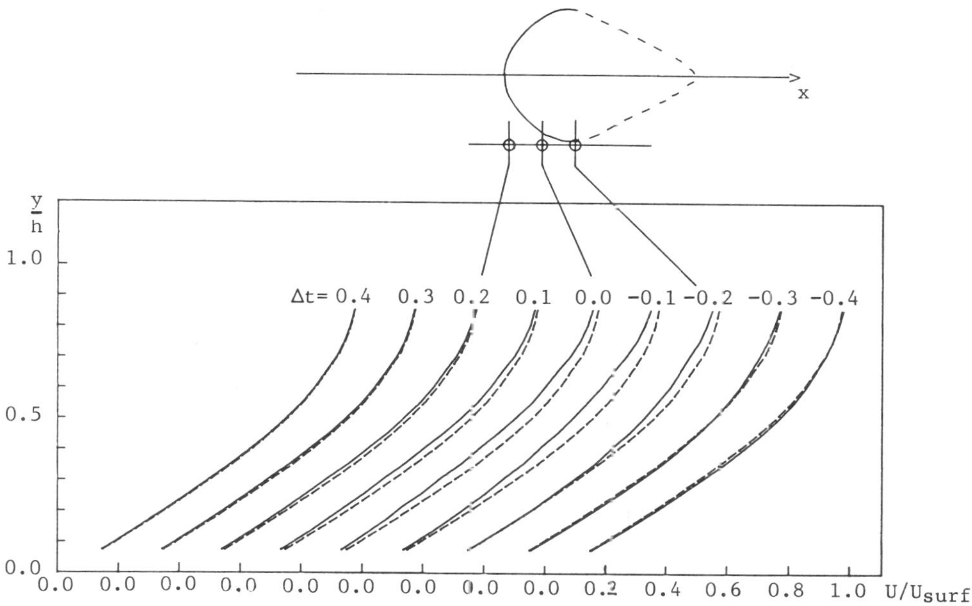

Fig. 5. Quasi-stationary velocity profiles (———) measured just outside the wing-tip. t=0 corresponds to the minimum in the wall shear stress. (- - - - -) is the measured undisturbed profile.

of observation there is first an increase in the critical Reynolds number. At the passage of the minimum in the wall shear stress there is a drastic reduction of Re_{crit} to a value (730) well below the actual Reynolds number of the flow (1350), after which it slowly returns to the "undisturbed" value. The flow is thus destabilized in a region located mainly upstream of the minimum in the wall shear stress.

To experimentally investigate the stability of the flow outside the spot a small disturbance in the form of a thin wire (0.2 mm diameter) attached to the glass plate was introduced just outside the wing-tip. For a suitably chosen inclination and Reynolds number the wire was small enough not to cause transition when it was placed in the undisturbed flow. When a spot was generated so that the wire was at the wing-tip of the spot, a new secondary spot was generated at the wire just upstream of the original spot. This gives a further indication that the flow field just outside the spot is less stable than the surrounding undisturbed shear flow. However, it should be pointed out that it was rather difficult to find the right combination of Reynolds number and inclination (which together determine the relative disturbance amplitude) to get the above described effect.

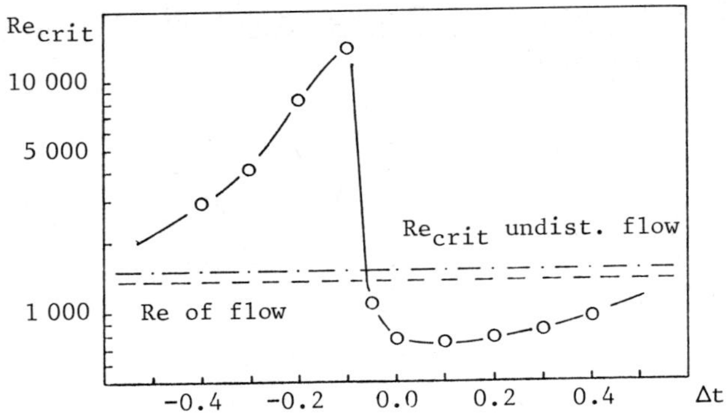

Fig. 6. Re_{crit} for the measured velocity profiles in Fig. 5.

5. CONCLUSIONS

Flow visualization showed that during its initial stage the water table spot is of similar shape to that of spots in laminar boundary layers. However, as it grows, the front part seems to be relaminarized, possibly due to lack of entrainment of free-stream fluid. The flow outside the turbulent part of the spot was found to be affected, in such a way that it is less stable than the undisturbed flow. This may be coupled to the mechanism responsible for the rapid lateral growth of turbulent spots.

The work reported here was in part sponsored by the Swedish Maritime Research Centre (SSPA)

REFERENCES

1. Emmons, H.W.: 1951 J. Aero. Sci. 18, 490.
2. Mitchner, M.: 1954 J. Aero. Sci. 21, 350.
3. Elder, J.W.: 1960 J. Fluid Mech. 9, 235.
4. Cantwell, B.; Coles, D. & Dimotakis, P.: 1978 J. Fluid Mech. 87, 641.
5. Gad-El-Hak,M., Blackwelder, R.F. & Riley, J.J. 1981 J. Fluid Mech. 110, 73.
6. Bertshy, J.R. & Abernathy, F.H.: 1977 In Proc. Second International Conf. on Drag Reduction (BHRA), Univ. Cambridge (ed. H.S. Stephens & J.A. Clarke) p. G1-1.
7. Gustavsson, L.H. & Ogren, J.-E.: 1982 Tech. Rep. 039T, Univ. of Lulea, Sweden.
8. Carlson, D.R.; Widnall, S.E. & Peeters, M.F.: 1982 J. Fluid Mech. 121,487.
9. Schubauer, G.B. & Klebanoff, P.S.: 1956 NACA Rep. 1289.
10. Corrsin, S. & Kistler, A.L.: 1955 NACA Rep. 1244.
11. Wygnanski, I.; Haritonidis, J.H. & Kaplan, R.E.: 1979 J. Fluid Mech. 92, 505.
12. Wygnanski, I.; Zilberman, M. & Haritonidis, J.H.: 1982 J. Fluid Mech. 123, 69.
13. Chambers, F.W. & Thomas, A.S.W.: 1983 Phys. Fluids 26, 1160.
14. Smith, C.R. & Metzler, S.P.: 1983 J. Fluid Mech. 129, 27.
15. Chin R. W.-Y.: 1981, Ph.D-thesis Harvard Univ.

Visualization of Generation Process of a Turbulent Spot

T. MATSUI and M. OKUDE

Faculty of Science and Technology
Meijo University
Yagoto, Tempaku, Nagoya 468, Japan

Abstract

At a later stage of transition in a flat plate boundary layer, the relation between spikes in wave forms of velocity fluctuations and hair-pin eddies in flow visualization was investigated in a water flow by applying hot wire and hydrogen bubble techniques simultaneously. It was shown that spikes in wave forms are caused by hair-pin eddies which were formed through the secondary instability in a high shear layer.

Introduction

It has been known by carefully controlled experiments on transition in the Blasius boundary layer (1,2,3,4) and in the plane Poiseuille flow (5,6) that the transition process can be attributed to a sequence of instabilities. The primary instability in a flat plate boundary layer occurs with respect to growth of two-dimensional Tollmien-Schlichting waves. When the primary waves have grown up to a certain critical amplitude, the primary waves are corrugated in the spanwise direction, that is, the waves become three-dimensional and the so-called peak-valley structure develops. Different types of the three-dimensional phenomena at this stage of transition were investigated theoretically (7,8) and experimentally (9,10). The peak-valley structure is shown to correspond to Λ-shaped corrugations of streak lines in flow visualization experiments (11, 12), which are observed as Λ-vortices. At the peak sections high shear layers are developed locally and instantaneously. The instability in the high shear layers causes high frequency velocity fluctuations which correspond to breakdown of the heads of Λ-vortices into hair-pin eddies of smaller scale (13). The instability is usually called the secondary instability.

Though the validity of the secondary instability was shown experimentally (6), the correspondence of spikes in wave forms to hair-pin eddies in flow visualization has not been well established. In the present investigation, flow patterns in a flat plate boundary layer in a water flow are visualized by means of hydrogen bubble technoque and simultaneously wave forms are recorded by hot wire technique to identify spikes in wave forms to hair-pin eddies in flow patterns.

Experimental Procedure

The width and length of the test section of a water channel used are 600 and 2900 mm, respectively. The width, length and thickness of a flat plate are 600, 2700 and 10 mm, respectively. The plate with a round leading edge is set horizontally at the depth of 250 mm from free surface of water.

The primary waves are introduced with a vibrating ribbon placed at the distance of 1100 mm from the leading edge and at the hight of 3,5 mm above the plate surface. The width, length and thickness of the ribbon are 3.8, 500 and 0.02 mm. The ribbon was attached at both ends on to brass rods of 3 mm diameter, each of which was fixed at cone of speaker driven by a low frequency oscillator.

A piece of platinum wire of 50 μm diameter and 300 mm length was used to generate hydrogen bubbles in a horizontal plane at various heights. Another piece of platinum wire was used in a vertical plane.

Wave forms of velocity fluctuations were recorded by using a hot wire probe, and photographs of flow patterns were taken at the same time.

The uniform flow velocity was about 10 cm/s, and it was adjusted for variation in water temperature so that the Reynolds number, R_{δ_1}, is kept to be 590 at the location of the vibrating ribbon, x = 1100 mm, where δ_1 and x denote the displacement thickness and the distance from the leading edge, respectively. The frequency of the ribbon was 0.34 Hz and its amplitude was 0.54 mm. The disturbance given was, therefore, in the unstable region of Tollmien-Schlichting waves at the location of the ribbon.

Experimental Results and Discussions

1. Flow patterns at various heights

Flow patterns at various heights above the wall are given in Fig.1. In the flow pattern in the layer adjacent to the wall, Fig. 1,a, many streamwise streaks can be seen. White grid lines in the picture have a mesh size of 50 mm. The streak pattern changes into stretched Λ-shape patterns with increase in the height,as seen in Fig.1,b. The distance between two legs of Λ-vortices is about 20 mm, and the non-dimensional distance referred to the friction velocity for a fully turbulent boundary layer is about 110, which is near to the value of 100 for a fully tubulent boundary layer. In Fig.1,c some disturbances are observed in front of the heads of Λ-vortices. It can be seen from Fig.1,d that the disturbances are due to the hair-pin eddies of smaller scale than that for the original, primary Λ-vortices. At about a half depth of the boundary layer, Fig.1,e, there are three rows of hair-pin eddies. Near the edge of the boundary layer, Fig.1,f, only the heads of hair-pin eddies are seen in almost uniform flow.

It has been shown experimentally by Kovasznay et al.(3) and Nishioka et al.(6) that a high shear layer appears at about a half depth of boundary layer. The secondary instability in the layer gives rise to waves of higher frequency than that of the primary wave. In Fig. 1 the boundary layer thickness is about 20 mm at the section of x = 1600 mm. Thus, hair-pin eddies in Fig.1,d and e are considered to be generated due to the secondary instability in the layer of $y/\delta = 0.3 - 0.5$, where a high shear layer is supposed to develop.

2. Wave forms of velocity fluctuations

Wave forms of electric current from the oscillator, the upstream hot wire at x = 1180 mm and y = 3.5 mm and the downstream hot wire at x = 1600 mm and y = 10 mm are shown in Fig. 2, a, b and c, respectively. The period of the primary wave is seen to be about 3 seconds in Fig.2,b. The wave form is not sinusoidal, probably because the ribbon amplitude was not small enough, as pointed out by Hama (14). The wave form, Fig.2,c, shows two spikes in each wave length, and it resembles those obtained by other investigaters. Time interval between two

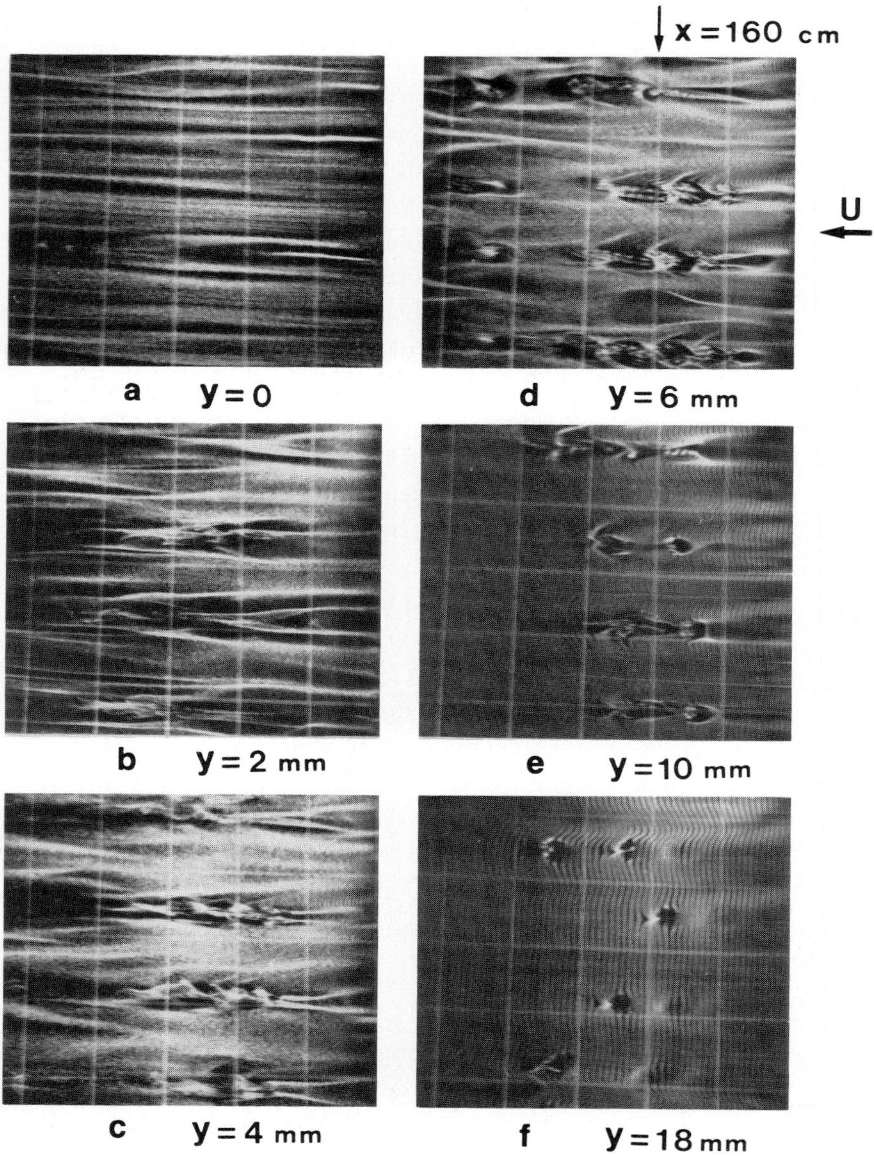

Fig.1. Flow patterns at various heights.
R_{δ_1} =715 at x = 160 cm.

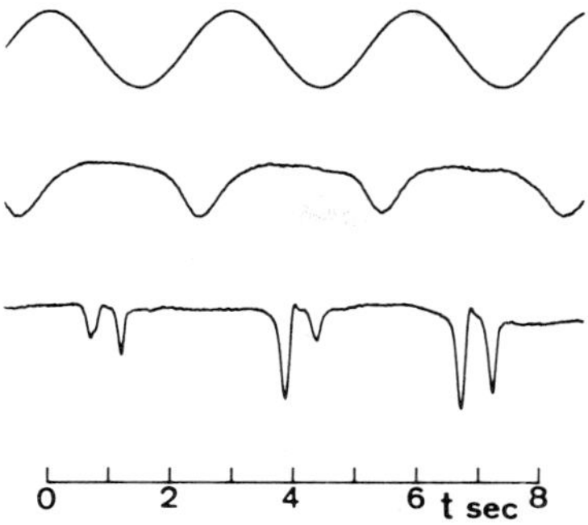

Fig.2. Comparison of wave forms. (a) oscillator, (b) upstream hot wire, (c) downstream hot wire.

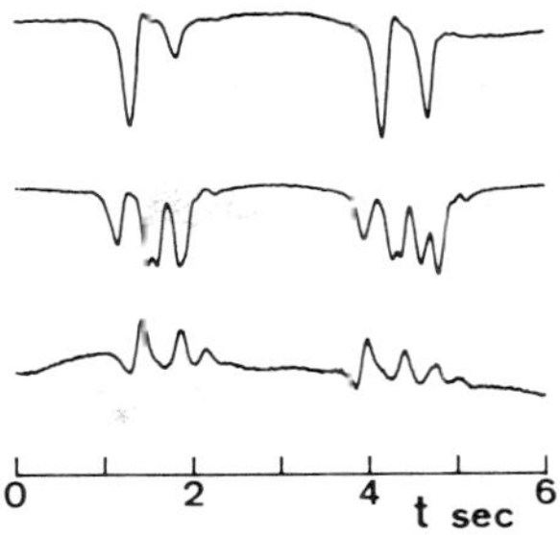

Fig.3. Wave forms of disturbances at x = 160 cm, y = 10 mm and $R\delta_1$ = 715. (a) two spikes, (b) three spikes, (c) upward spikes.

spikes is 0.5 second, so the frequency of the secondary wave is about six times the frequency of the primary wave. This is in agreement with the result by Nishioka et al.(6). Wave forms obtained by the downstream hot wire are different at different instants as shown in Fig.3, probably due to the turbulence in uniform flow. In a wavelength two spikes appear at an instant, three spikes at another instant and sometimes upward spikes.

3. Correspondence of wave forms to flow patterns

At the Reynolds number, $R\delta_1$ = 690, at x = 1500 mm, consecutive pictures were taken at the rate of 4.2 frames/s at the same time that wave forms were recorded by a hot wire at x = 1600 mm and y = 10 mm. The results are shown in Fig.4, where pulses below the wave form indicate the instant of exposure to take photographs. The flow condition is in the 4-spike stage. Pictures of No. 3, 5, 8, 16 and 18 correspond to spikes. It is seen that spikes appear soon after a hair-pin eddy has passed the hot wire. Thus, it may be said that the secondary instability in the high shear layer gives rise to hair-pin eddies, which, in turn, induce low speed fluid lumps under the high shear layer to go upwards across the layer into high speed layer so that low speed spikes are recorded by a hot wire in high speed layer. By similar inference it is supposed that high speed spikes shown in Fig.3 may be recorded by a hot wire in low speed layer just before hair-pin eddies pass the hot wire.

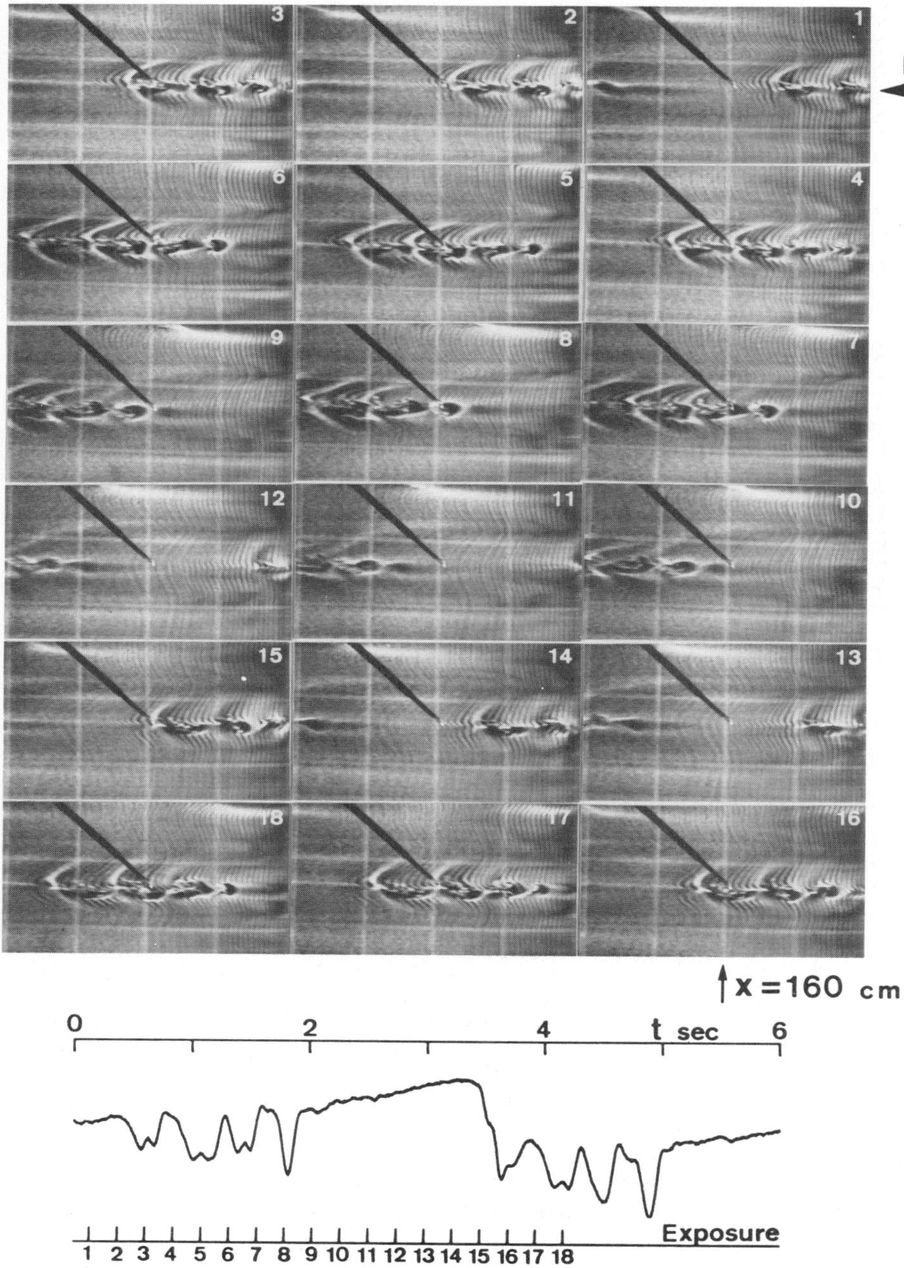

Fig.4. Consecutive pictures of flow patterns. y = 10 mm. $R\delta_1$ = 715 at x = 160 cm.

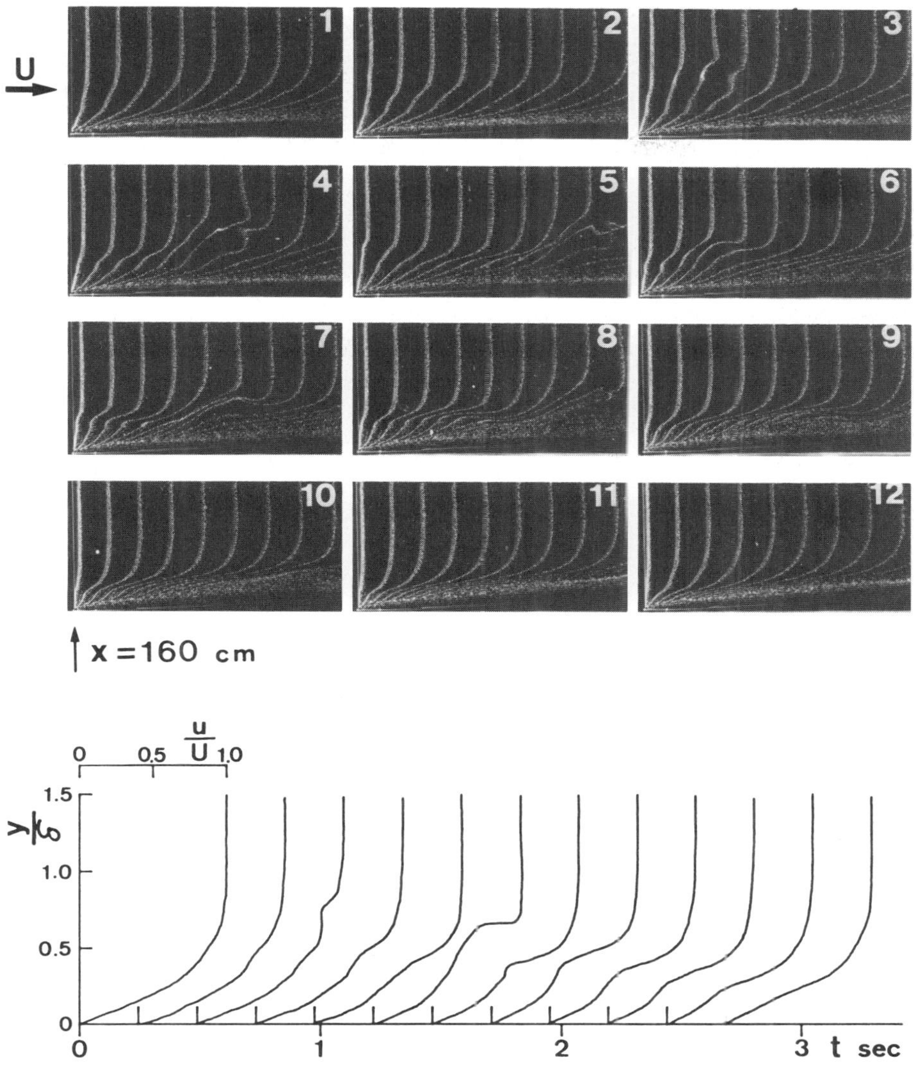

Fig.5. Consecutive pictures of vertical time lines and instantaneous velocity profiles at x = 160 cm, R_{δ_1} =715.

4. Velocity distribution in the boundary layer

Consecutive pictures of time lines shed from a vertical wire at x = 1600 mm are shown in Fig.5 for the same flow condition as before. Instantaneous velocity distributions at the wire location are given in the same figure during about a period of the primary wave. A Blasius type profile is seen to change into profiles with point of inflection, that is, a high shear layer is formed, and then the profile goes back to the original Blasius type profile.
The effect of upward velocity in the peak section between two legs of a Λ-shape vortex is seen in the pictures as lift-up of time lines near the wall.

Concluding Remarks

At a later stage of transition in a flat plate boundary layer, a high shear layer is developed due to Λ vortices at about a half depth of the boudary layer in each period of the primary wave. The secondary instability in the high shear layer causes waves of high frequency, which grow up to form hair-pin eddies of small scale. Spikes in wave form of velocity fluctuations are caused by the hair-pin eddies.
Increase in number of hair-pin eddies and the tertiary instability in them result in generation of a turbulent spot.

Acknowledgment

This work was supported in part by Grant-in Aid for Scientific Research from the Ministry of Education.

References

1. Schubauer, G. B., Skramstad, H. K. , NACA Rep. No. 1289 1947.

2. Klebanoff, P. S., Tidstrom, K. D. and Sargent, L. M., J. Fluid Mech. 12 (1962) 1-32.

3. Kovasznay, L. S. G., Komoda, H. and Vasudeva, B. R., Proc. 1962 Heat Transfer and Fluid Mech. Institute, Stanford Univ. Press (1962) 1-26.

4. Hama, F. R., and Nutant, J., Proc. 1963 Heat Transfer and Fluid Mech. Institute, Stanford Univ. Press (1963) 77-93.

5. Nishioka, M., Iida, S. and Ichikawa, Y., J. Fluid Mech. 72 (1975) 731-752.

6. Nishioka, M., Asai, M. and Iida, S., Laminar-Turbulent Transition ed. by Eppler, R. and Fasel, H. Springer-Verlag (1980) 37-46.

7. Craik, A. D. D., J. Fluid Mech. 50 (1971) 393-413.

8. Herbert, T., Proc. IUTAM Symp. on Turbulence and Chaotic Phenomena in Fluids, 5-10 Sept. 1983, Kyoto, Japan.

9. Saric, W. S. and Thomas, A. S. W., Proc. IUTAM Symp. on Turbulence and Chaotic Phenomena in Fluids, 5-10 Sept. 1983, Kyoto, Japan.

10. Saric, W. S., Kozlov, V. V. and Levchenko, V. Y., AIAA-84-0007, AIAA 22nd Aerospace Sciences Meeting, 9-12 Jan. 1984 Reno, Nevada.

11. Saric, W. S., Carter, J. D. and Reynolds, G. A., Bull. Amer. Phys. Soc. 26 (1981) 1252.

12. Williams, D. R., Ph.D. dissertion, Princeton Univ. June 1982.

13. Nishioka, M., Asai, M. and Iida, S., Proc. 12th Turbulence Symp., Inst. Space and Aeron. Sci., Univ. of Tokyo (1980) 28-34.

14. Hama, F. R., private communication, 1984.

Model Description of Two Kinds of Organized Structures Dominating in Turbulent Shear Flows

V. S. SADOVSKY, G.I. TAGANOV

Central AeroHydrodynamics Institute, USSR

1. Introduction and review

Since 1974 (see [1], [2]) the properties of a s t r e a m-w i s e and a s p a n w i s e organized vortex structures have been investigated within the framework of a deterministic approach[1)] to a theoretical description of turbulent shear flows. Even the first results of the deterministic approach (first, a likely description of mean velocity profiles and velocity pulsations in the near wall region of the turbulent boundary layer (TBL) using the streamwise structure; second, the explanation of the non-turbulent fluid entrainment mechanism as well as a theoretical determination of the expansion angle of the modeled mixing layer using the spanwise structure) have shown its efficiency. E.g., paper [3] developing the problem formulation of paper [2] presents a one-parametric class of vortex structures in a viscous incompressible fluid which are cell Couette flow disturbed by a superimposed velocity field in the spanwise direction.

Fig.1 presents a slide from [3] (see also [7]) which shows a scheme of a cross flow in a cell, symbols used, and the motion equations for span - and streamwise flows inside the cell. The results of a numerical solution of the equations given in fig.1 are presented in a wide range of the $\lambda^+ = \lambda / l_*$ fundamental parameter variation where l_* is determined by a

---

1) The essence of this method is to investigate organized vortex structures using exact hydroaerodynamics equations under the assumption that the global flow stochasticity may not be accounted for during a portion of life time of the ordered vortex structure.

mean friction stress on a fixed wall. With a specified turbulence level u'^2/U =1,7 at the layer boundary having a constant friction stress (y^+ =50) the one-parameter solution family of the Navier-Stokes equations (parameter λ^+) proves to quantitatively truly reproduce the dependence of a longitudinal and normal to the wall root-mean-square normalized velocity fluctuations on y^+ which is observed in experiments. A relative drag decrease or an increase compared to the friction drag on a smooth wall are respectively associated with a relative increase or decrease of the parameter λ^+ compared with its value λ^+=100 for a smooth wall.

Solid lines in fig.2 present calculated dependences of the longitudinal velocity fluctuation of y^+ for three values of parameter λ^+ : the value of λ^+=400 to describe the flow of weak polymer solutions, λ^+ =50 to describe the flow of the Newtonian fluid near a weak rough wall. The figure also shows dependences of the normal to the wall velocity fluctuation related to its value on the boundary of the constant friction stress layer. Regions shaded at angles of 135°, 90° and 45° to the abscissa axis contain experimental points obtained for weak polymer solutions, a smooth wall and a weak rough wall respectively.

The correlation of calculated dependences with experimental ones reveals a unity of the wall flow structure for a wide class of TBL and its relative simplicity - friction of only one parameter λ^+ - which escaped the investigators' attention till 1975.

The spanwise vortex structure as a spiral vortex formation occuring on the boundary of two-dimensional parallel incompressible flow with different velocities was studied in [1] using the mathematical model of discrete vortices. The mechanism of the growth of the spiral formation size with time has cleared up, which can be described as swirling of two wedgelike jets under an e x t e r n a l hydrodynamic moment which are split from the flows along both sides of the tangential velocity discontinuity. Fig.3 presents a velocity vector field in the spanwise vortex structure computed by the method of discrete vortices with a subsequent rediscretization [4] . Of attention are two characteristic points, namely points of spillage symbolized in fig.3 by shaded circles. The calculated velocity of the growth of the spiral formation size proves to be in a rather good agreement with experimental data by Brown and Roshko [5] as for the expansion angle of the turbulent mixing layer, though it is obtained with no account of the interaction of adjacent vortex formation of the "pairing" phenomenon. To clear up this item the "pairing" phenomenon was theoretically reproduced in [6] (fig.4 shows some stages of this phenomenon at different moments of time). The combined (two-core) vortex formation was found to expand at the same velocity as a one-core formation.

In [7] an attempt is made to construct a global flow model in the TBL where stream- and spanwise structures work as hydrodynamic mechanisms acting in the viscous dissipation zone of the TBL (internal region) in the zone of momentum transfer to a freestream flow (external region) respectively. **The existence of the inflexion point in the external portion of the averaged velocity profile in the TBL known from the experiments on plates serves as a physical justification for this model.**

Thus, in 1977 the role of spanwise structures has been revealed as a fundamental mechanism of momentum transfer in a wide class of turbulent flows.

Paragraph 2 of this report gives the results of a numerical solution for a solitary wave at a tangential velocity discontinuity with a growth of its dimensions linear in time which does not use (unlike [1] , [4]) the replacement of a continuous vortex layer by a system of discrete vortices.[1)]

The paper [8] which is close in the formulation does not contain respective information on pecularities of this unique flow (onflow points, singled structure direction, etc.) and the geometry of the first structure turn in [8] differs considerably from that obtained by the rediscretization method (fig.3) which is confirmed in the main in the following solution of the problem.

Paragraph 3 of this report is a further development [9] of the cell flow model in the wall region of the TBL investigated in

---

[1)] These results have been reported at the 6th School on continuum mechanics models held 25 Sept. to 4 Oct. 1981, Medeo.

[2] , [3] (see also [7]) using the stationary Navier-Stokes equations.

2. Solitary vortex formation at an infinite tangential velocity discontinuity

Let the equation of tangential velocity discontinuity line in the polar coordinate system R, Θ (with the origin at the core centre) be of the form $F=R-f(\Theta)tV_\infty=0$ where $\pm V_\infty$ is the velocity of a free stream flow. To describe the behaviour of this line points in space and time let us make use of the fact that a vortex layer distributed along the tangential discontinuity always consists of the same fluid particles and its intensity $\gamma(\Theta)$ is described by the Nikolsky equation. Accounting for similarity this gives

$$f - V_R + f' V_\Theta / f = 0 \qquad (1)$$

$$G'/G = f'/f - (f + V_\Theta')/V_\Theta \qquad (2)$$

Here and further $V_R(\Theta)$ and $V_\Theta(\Theta)$ are the radial and peripheral velocity components of vortex layer particles related to V_∞, the prime means the differentiation in angle, and $G(\Theta)= = \gamma\sqrt{f^2+f'^2}$. V_R and V_Θ are expressed by integrals (in the sense of main value) over the whole tangential discontinuity

$$V_R(\Theta_0) = -\frac{1}{2\pi}\int_0^\infty\left[\frac{f\sin(\Theta-\Theta_0)}{f^2+f_0^2-2ff_0\cos(\Theta-\Theta_0)} - \frac{f\sin(\Theta-\Theta_0)}{f^2+f_0^2+2ff_0\cos(\Theta-\Theta_0)}\right]G\,d\Theta$$

$$V_\Theta(\Theta_0) = \frac{1}{2\pi}\int_0^\infty\left[\frac{f_0-f\cos(\Theta-\Theta_0)}{f^2+f_0^2-2ff_0\cos(\Theta-\Theta_0)} + \frac{f_0+f\cos(\Theta-\Theta_0)}{f^2+f_0^2+2ff_0\cos(\Theta-\Theta_0)}\right]G\,d\Theta \qquad (3)$$

where the integration in view of a central symmetry is carried out along a right branch of the spiral with $f=f(\Theta)$, $f_0=f(\Theta_0)$. From (1 - 2) the exact expression follows for a total circulation Γ calculated from the core centre to an arbitrary Θ along one branch of the vortex layer

$$\Gamma(\Theta) \equiv \int_\Theta^\infty G\,d\Theta = G(\Theta)V_\Theta(\Theta)/f(\Theta) \qquad (4)$$

At $\Theta \to 0$ $f \to \infty$, $\gamma \to 2$, and integral estimates (3) show that in this case $V_\Theta \sim 1/f$. Using (1) we obtain the asymptotic behaviour for f and V_Θ:

$$f \approx c/\sqrt{\theta}\ ;\ V_\theta \approx 2c\sqrt{\theta}\quad (\theta \to 0) \tag{5}$$

A numerical solution of integro-differential equations (1 -2) is performed by the iterative method over an integration interval $\pi/12000 \leq \theta \leq 10\pi$. The integration starts from point $\theta_1 = \pi/12000$ and is performed over interval $[\theta_1;\ \pi/60]$ in the Cartesian and then in polar coordinate system. The initial conditions are $\gamma(\theta_1)=2$, $f(\theta_1)=c/\sqrt{\theta_1}$ where constant c is determined from the analysis of V_θ at $\theta \sim \theta_1$.
As a matter of fact, the realisation of this algorithm required (including modifications) more that 1500 iterations.
At the calculation interval $\Sigma \in [7\pi\ ;\ 10\pi]$ initial iterations in first versions were specified (and further proceeded) in the form of $f=c/\theta$. But each time after calculating (3) it **turns** out that the next iteration obtained by integrating (1 - 2) provides much faster variation of calculated values by Σ . The analysis of calculated data in this interval points to an exponential function of f and γ of θ, which further goes on beyond the upper limit of the calculation interval when (3) is calculated.
Exact relation (4) has played a significant role in organizing the iterative process.
Fig.4 illustrates the similar flow obtained. The solid line presents a tangential velocity discontinuity swirling into a double-start spiral, the arrows showing the values of the velocity vector components tangential and normal to the discontinuity (the length scale is reduced by a factor of 20). The dottcol line passing through onflow points restricts the fluid region captured by the structure. The straight line connecting the onflow points forms the angle $\alpha = \pi/8$ with the x-axis. The growth rate of the maximun spiral size along the y-axis is equal to 0.18 V_∞ , and the movement velocity of the onflow point A along the x-axis being equal to 0.37 V_∞ .
Based on the calculations $f(\theta)$ at $\theta \geq 7\pi$ is well approximated by the function of $f \approx 0.030\,\phi(\theta)\ \exp[0.067(7\pi - \theta)]$.
Function $\phi(\theta)$ in a polar coordinate system is essentially an

ellipse with the semiaxis ratio of $b/a=0,8$, the large semiaxis ($a=1,24$) being inclined to the x-axis at the same angle $\alpha = \pi/8$.

The difference between the calculated function $f(\Theta)$ at great Θ and the asymptotic behaviour [11] may be interpreted in two ways. It is possible that the interval $[8\pi, 10\pi]$ is still far from the region of the asymptotic behaviour validation [11] (the calculation comprises 10 turns of the vortex formation). On the other hand, C - constant vanishing in the asymptotic law [11] ($V_\Theta \approx c$; $f \approx c/\Theta$) may fully mean a change of this asymptotic behaviour over another, functionally stronger which the calculation exponent does not contradict (note that the substitution of $f_0=0$ in (3) results in $V_\Theta=V_R=0$).

The behaviour of the tangential discontinuity line at $\Theta \to 0$ is described by the ratio of $f \approx 0,32/\sqrt{\Theta}$.

The solution obtained allows to determine expansion angle of the modeled mixing layer in the first approximation (with no account of the interaction of vortex formations). Fig.6 illustrates an agreement of the predicted function (solid curve) with the experiment [5] where the expansion angle was evaluated visually by external boundaries of the mixing layer.

3. Nonstationary flow model in viscous sublayer of the TBL

A cell Couette flow is discussed using the scheme presented in fig.7. The main flow occurs along the x-axis, $\partial/\partial x$ being absent for all values. A disturbing spanwise flow is superposed along the z-axis in each of cells of dimension λ . As to this flow let us make the following assumptions: 1) the whole streamwise vorticity inside each cell is concentrated in two regions Σ_1 and Σ_2 symmetric with respect to the cell middle; 2) vorticity ω_x in Σ_1 and Σ_2 is quantitatively equal, opposite in spin and does not change in time.

The adoption of these assumptions means that the kinematics of the superposed viscous cross flow is described by the Euler equation with additional boundary conditions: 1) cell wall $y=0$ and $y=b$ are mobile in the z-axis direction and their motion velocity is defined by the Euler equation solution; 2) the absence of vortex region diffusion is due to surface forces di-

stributed along their boundaries whose work completely compensates for the kinetic energy dissipation of the superposed flow. These boundary conditions are caused by the necessity to obtain a flow periodic in time. Note that if the surface of vortex regions is increased, then in the limit the nonstationary flow under consideration transforms into a class of stationary cell Couette flows investigated earlier [2] .
Based on the above assumptions the longitudinal velocity component u(y, z, t) satisfies the equation

$$\frac{\partial u}{\partial t} + v\frac{\partial u}{\partial y} + w\frac{\partial u}{\partial z} = Re_*^{-1}\left(\frac{\partial^2 u}{\partial y^2} + \frac{\partial^2 u}{\partial z^2}\right) \quad (6)$$

Here u is related to velocity V_0 of the upper wall, cross component v, w to $2\Gamma/\lambda$, linear dimensions to $\lambda/2$, time t to $\lambda^2/4\Gamma$ and $Re_* = \Gamma/\nu$ is the Reynolds number of the superposed cross flow, Γ is circulation vortex region.
The boundary conditions are: $u=0$ at $y=0$; $u=1$ on the upper mobile surface; $\partial u/\partial z=0$ on all vertical boundary semicells.
It is essential that coefficients v, w in (6) are determined independent of the streamwise flow, and for (6) they are known to be periodic functions of time and coordinates. Eq. (6) was numerically calculated using the finite-difference method. The implicit scheme of alternating directions was used as a working difference scheme.
Assume that the essentially three-dimensional periodic flow under investigation simulates over the time period T the phenomenon of the fluid descent to the wall with its subsequent ascent existing in a real TBL, and an arbitrary value at a fixed y will be averaged in time and coordinate z within the cell. It is obvious that $v=w=0$, and from (6) it is not difficult to obtain the relationship

$$d\bar{u}_+/dy^+ - \overline{u'_+ v'_+} = 1$$

known in the wall turbulence. Here and further the index "+" means making all parameters dimensionless using the dynamic velocity U_τ and the viscous length l_*.
To combine the longitudinal and cross velocity field in the modeled flow the experimental value of a rms velocity fluctuation

is taken far from the viscous sublayer and in so doing one can relate a free parameter of the modeled flow Re_* to the parameter λ^+ in the real TBL. Fig.8 gives the predicted history of the friction stress $\langle \tau \rangle$ averaged in z at y=0 for $\lambda^+=100$. It is remarkable that a drastic increase in the friction stress takes place at the descent (inrush) into the wall region. The profile of longitudinal velocity $\langle u \rangle$ averaged in z (which can be compared to an instantaneous velocity profile at a point of the real TBL) is also presented in comparison with the mean velocity profile $\overline{U}_+(y^+)$. Fig.9 gives an indication of a correspondence of rms fluctuating velocity components (solid curve) to experimental data (shaded track) as well as to the calculation results of the stationary model [2], and to the predicted data [10]. It is seen that the predicted relationships present quite acceptably the experimental ones.

References

1. Taganov, G.I.; Dudoladov, I.V. In: "Turbulent flows", M., Nauka, 1977; (in Russian) "Fluid Mech.Sov.Res.", N 7, p.10-23, (1978).
2. Sadovsky, V.S.; Sinitsyna, N.P.; Taganov, G.I. In: "Wall turbulent flow". Proceedings of the XVIIIth Siberian thermal physical seminar, part 1, Novosibirsk, 1975 (in Russian).
3. Sadovsky, V.S.; Sinitsyna, N.P.; Taganov, G.I. Report theses of the All-Union scientific symposium. "The theory of the boundary layer" Leningrad, 23-26 Dec. 1975, L., 1975 (in Russian)
4. Dudoladov, I.V.: Uchenye zapiski TsAGI, v.XI, N 4 (1980) (in Russian)
5. Brown, G.L.; Roshko, A.: J.Fluid Mech., v.64, pt.5, p.775-816 (1974).
6. Dudoladov, I.V.; Taganov, G.I.: Uchenye zapiski TsAGI, v.VIII, N 4, 1977 (in Russian).
7. Taganov, G.I. In: "Numerical methods of continuum mechanics" Novosibirsk, v.8, N 6, 1977 (in Russian).
8. Jimenez, J.: J.Fluid Mech., v.96, pt 3, p.447-461 (1980).
9. Sadovsky, V.S.; Shaimukhambetov, M.: Uchenye zapiski TsAGI, v.XIII, N 5, 1982 (in Russian).
10. Hatziavramidis, D.T.; Hanratty, T.T.: J.Fluid Mech., v.95, pt.4 (1979).
11. Mangler, K.W.; Weber, T.: J.Fluid Mech., v.30, pt 1, p.177-193 (1967).

Motion Equations

Longitudinal Flow

$$u_\tau = \sqrt{\tau_0/\rho}$$
$$l_* = \nu/u_\tau$$
$$\lambda^+ = \lambda/l_*$$
$$(u^+,v^+,w^+) = (u,v,w)/u_\tau$$
$$(y^+,z^+) = (y,z)\,l_*$$

$$\frac{\partial^2 u^+}{\partial y_+^2} + \frac{\partial^2 u^+}{\partial z_+^2} = v^+\frac{\partial u^+}{\partial y^+} + w^+\frac{\partial u^+}{\partial z^+}$$

1) $y^+=0;\ u^+=0;\quad \partial\bar{u}^+/\partial u^+ = 1$

2) $z^+=0;\ \pm\lambda^+/2\ldots\ \partial u^+/\partial z^+ = 0$

3) $y^+\to\infty$

$$u^+ = \frac{y_+}{1+K^2\lambda^{+2}/40} + \frac{\sqrt{3}\,\kappa\lambda^+}{8(1+K^2\lambda_+^2/40)}\left(z_+^2-\frac{16}{3\lambda_+^2}z_+^3+C\right)$$

4) $\frac{\sqrt{\overline{u'^2}}}{u_\tau} = 1.7$ at $y^+\to\infty$ $\quad K = \frac{\sqrt{\overline{v'^2}}}{u_\tau}$ at $y^+\to\infty$

Lateral Flow

$$\frac{\partial v^+}{\partial y^+}+\frac{\partial w^+}{\partial z^+}=0;\qquad \frac{\partial w^+}{\partial y^+}-\frac{\partial v^+}{\partial z^+} = \frac{4\kappa\sqrt{3}}{\lambda^+}$$

$$v^+ = 2\kappa\sqrt{3}\left[-\frac{2z^+}{\lambda^+}+\frac{1}{\pi^2}\sum_0^\infty \frac{(-1)^n}{(2n+1)^2}\, e^{-\frac{(2n+1)2\pi y^+}{\lambda^+}} \sin\frac{2\pi(2n+1)z^+}{\lambda^+}\right]$$

$$w^+ = -\frac{8\kappa\sqrt{3}}{\pi^2}\sum_0^\infty \frac{(-1)^n}{(2n+1)^2}\, e^{-\frac{2\pi(2n+1)y^+}{\lambda^+}}\cos\frac{2\pi(2n+1)z^+}{\lambda^+}$$

Fig. 1

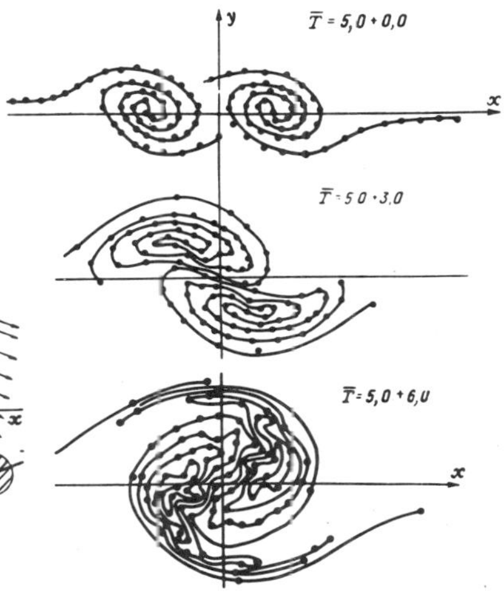

Fig. 2

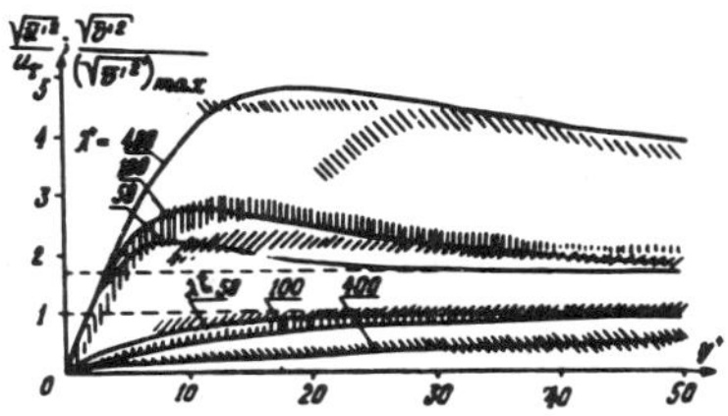

Fig. 3

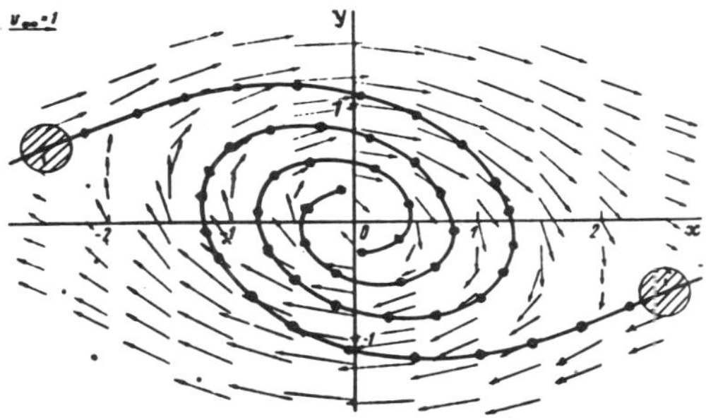

Fig. 4

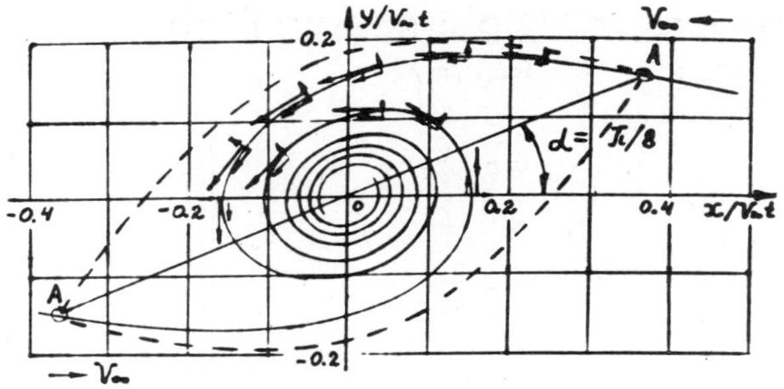

Fig. 5

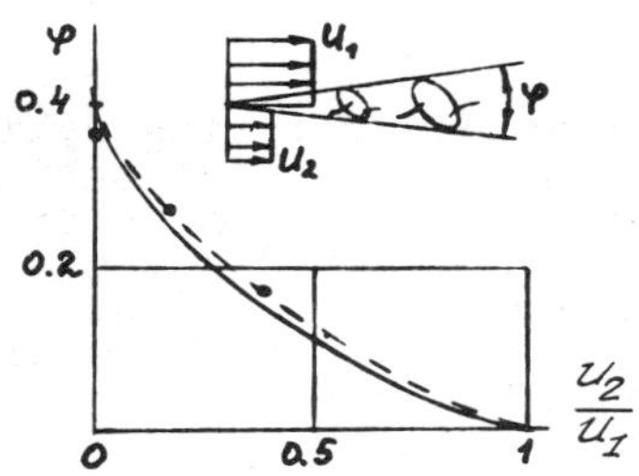

Fig. 6

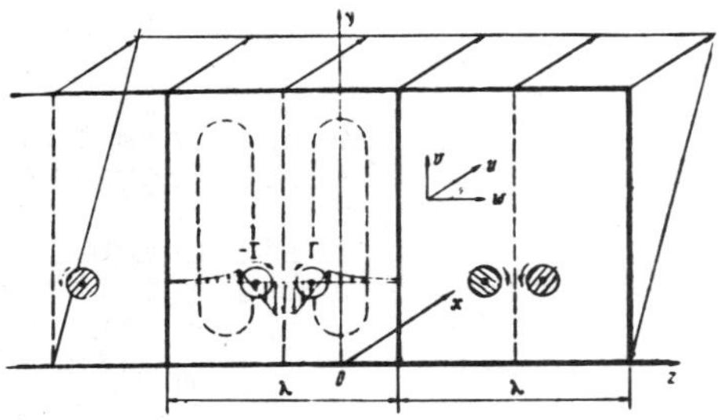

Fig. 7

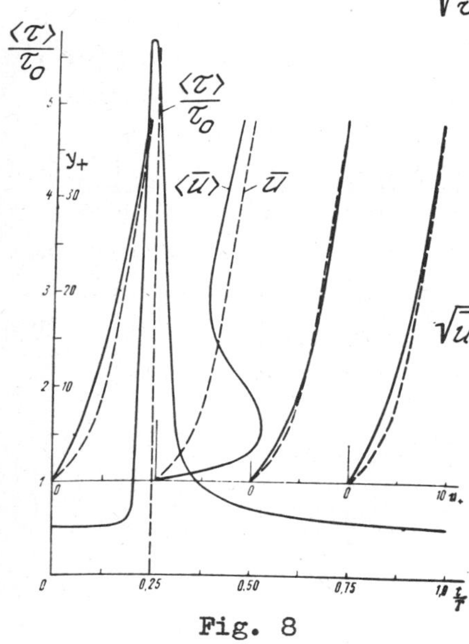

Fig. 8

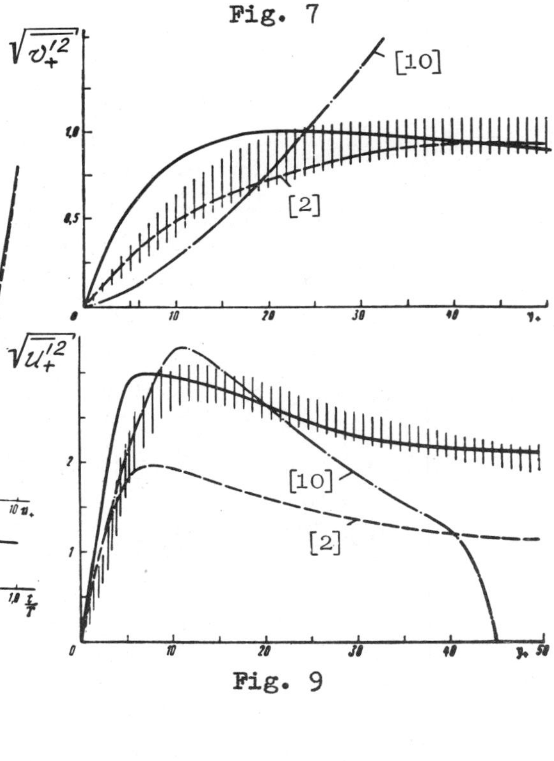

Fig. 9

On Incompressible Viscous Fluid Flow Between Rotating Cylinders

K. I. BABENKO, A. L. AFENDIKOV

Keldysh Institute of Applied Mathematics
USSR Academy of Sciences,
Moscow, USSR

Summary

A classical problem on flows between coaxial rotating cylinders is essentially multiparametric in the sense that subject to the defining parameters the flow evolution proves to be various when the Reynolds number increases. The analysis of some bifurcations, which despite their simplicity have not been studied until now, is carried out in this paper on the basis of the investigation of the Navier-Stokes equations.

The Couette Flow Bifurcation

Consider an incompressible viscous fluid flow between two coaxial rotating cylinders of radii r_1 and r_2, $r_2 = r_1(1+\delta)$. The angular velocities of the rotating cylinders are denoted by ω_1 and ω_2 respectively and let the Rossby number be $\mu = \omega_2/\omega_1$. Introduce cylindrical coordinates where the z axis is directed along the common axis of the cylinders. Choose r_1 and $d/2$ as length units for r and z respectively. Let $r/r_1 = s = 1+\delta(x+1)/2$ where $x \in [-1, 1]$. The Reynolds number R is defined by the relation $R = |\omega_1| d^2/4\nu$. The nonstationary system of Navier-Stokes equations is written as

$$\frac{\partial v}{\partial t} + A(v,v) - R^{-1}\mathcal{L}(v) + \mathrm{grad}(p) = 0; \quad \mathrm{div}\, v = 0 \qquad (1)$$

$$A(v,u) = \frac{2}{\delta}\begin{pmatrix} v_r \dfrac{\partial u_r}{\partial x} + \dfrac{\delta}{2s} v_\theta \dfrac{\partial u_r}{\partial \theta} + v_z \dfrac{\partial u_r}{\partial z} - \dfrac{\delta}{2s} v_\theta u_\theta \\ v_r \dfrac{\partial u_\theta}{\partial x} + \dfrac{\delta}{2s} v_\theta \dfrac{\partial u_\theta}{\partial \theta} + v_z \dfrac{\partial u_\theta}{\partial z} + \dfrac{\delta}{2s} v_r u_\theta \\ v_r \dfrac{\partial u_z}{\partial x} + \dfrac{\delta}{2s} v_\theta \dfrac{\partial u_z}{\partial \theta} + v_z \dfrac{\partial u_z}{\partial z} \end{pmatrix}$$

$$\mathcal{L}=\begin{pmatrix} \Delta-\frac{\delta^2}{4s^2}, & -\frac{\delta^2}{2s^2}\frac{\partial}{\partial\theta}, & 0 \\ \frac{\delta^2}{2s^2}\frac{\partial}{\partial\theta}, & \Delta-\frac{\delta^2}{4s^2}, & 0 \\ 0, & 0, & \Delta \end{pmatrix}, \quad \mathrm{grad}\, p=\begin{pmatrix} \frac{\partial p}{\partial x} \\ \frac{\delta}{2s}\frac{\partial p}{\partial\theta} \\ \frac{\partial p}{\partial z} \end{pmatrix}$$

and the continuity equation has the form

$$\frac{\partial v_z}{\partial x}+\frac{\delta}{2s}v_z+\frac{\delta}{2s}\frac{\partial v_\theta}{\partial\theta}+\frac{\partial v_z}{\partial z}=0.$$

$$\Delta=\frac{\partial^2}{\partial x^2}+\frac{\delta^2}{4s^2}\frac{\partial^2}{\partial\theta^2}+\frac{\partial^2}{\partial z^2}+\frac{\delta}{2s}\frac{\partial}{\partial x}.$$

Assume that no-slip conditions are satisfied on the walls of the cylinders. It is known that in this case the basic stationary solution of system (I), the Couette flow, has the form

$$V=(0,V_\theta,0)', \qquad P=\int_{-1}^{x}\frac{V_\theta^2(x)}{s}dx \tag{2}$$

where $V_\theta = Ls + M/s$

$$L=\frac{\mu(1+\delta)^2-1}{\delta(2+\delta)}, \qquad M=\frac{(1-\mu)(1+\delta)^2}{\delta(2+\delta)}$$

In the following we shall consider only the z-periodic flows with the period $2\pi/\alpha$. Not dwelling on the demonstration of the expediency of the consideration of just this very class of solutions we only note that in this case the selection by wave numbers comes about in a natural way.

The idea of the connection between the turbulence in a fluid and the motion along a strange attractor in dynamic system has lately given rise to a great number of papers devoted, one way or another, to applying this idea to flows between rotating cylinders and to those of some other origin. Note, however, two principal points ignored in most of the papers. First, it is hardly expedient to replace a clear, but difficult, initial boundary value problem for Navier-Stokes equations by the investigation of semi-empirical finite-dimensional dynamic systems. In our opinion, the denial of Navier-Stokes equations as a model of a fluid motion is entirely unjustified and a step backwards. This statement, of course, does not refer to the approach in the investigation of bifurcations based on the study of the projection of Navier-Stokes equations onto a center manifold. By itself the problem (even local) of constructing a center manifold is very interesting but it has not been solved so far in any case.

The second point concerns the computation of non-stationary solutions of Navier-Stokes equations. Here, unfortunately, the "from-simple-to-complex" principle is neglected. Complex "secondary", turbulent, etc. solutions are studied very readily, the results of calculations are compared with experiments, but the fact that there exist nontrivial solutions at small Reynolds numbers is ignored. In particular, the authors of this paper have not come across any article dealing with the numerical study of complex regimes in the problem on flows between parallel plates, where the existence of a periodic regime would have been noted, although in [I] , where the plane Poiseuille flow is studied, the time periodic flow, the existence of which being proved theoretically long ago, is found numerically. It seems to be indisputable that the algorithm permitting to study strange attractors should also provide the opportunity of the investigation of elementary attractors like a stable stationary or time-periodic flow. Therefore, it is desirable to get the most possible information about these elementary attractors which might serve as a reliable tool for testing the adequacy of the numerical algorithm used for solving a nonstationary problem. In the remainder of our paper we consider the implementation of this program for flows between rotating cylinders. Those who are interested in the history of the question can be referred to the monograph by D. Joseph [2].

2. As is known, the stability of the Couette flow is defined by the spectral problem

$$\begin{gathered} \lambda u(x) + RB_\ell(u(x)) - \mathcal{L}_{\ell,\alpha}(u(x)) + \mathcal{G}_{\ell,\alpha}(q) = 0 \\ \frac{1}{s}\,\frac{d}{dx}(s\,u_r(x)) + \frac{i\ell\delta}{2s}u_\theta(x) + i\alpha u_z(x) = 0 \\ u(1) = u(-1) = 0 \end{gathered} \tag{3}$$

In order to determine operators in problem (3) let us introduce the notations

$$B(v) = A(v, V) + A(V, v) \quad \text{and} \quad \zeta = exp(il\theta + i\alpha z)$$

Then

$$B_l(u(x)) = \bar{\zeta} B(u(x)\,\zeta),\quad \mathcal{L}_{l,\alpha}(u(x)) = \bar{\zeta}\,\mathcal{L}(u(x)\zeta),\quad \mathcal{G}_{l,\alpha}(q(x)) = \zeta\, grad(q(x)\zeta)$$

The sketchy analysis of problem (3) for $l \neq 0$ shows that its spectrum is doubly degenerate. An efficient algorithm for solving spectral problem (3) that is based on the representation of the desired eigenfunction in the form of an interpolated Lagrange polynomial with nodes at zeros of Tchebyshev polynomials of the first kind $T_m(x)$ is described in [3], [4].

Since for $l = 0$ and $l \neq 0$ there arise different bifurcation problems then it is expedient to choose in the plane of the defining parameters (μ, δ) such a region where the loss of stability comes about both axisymmetrically and nonaxisymmetrically. In Fig. I these are regions I, II above the curve α .

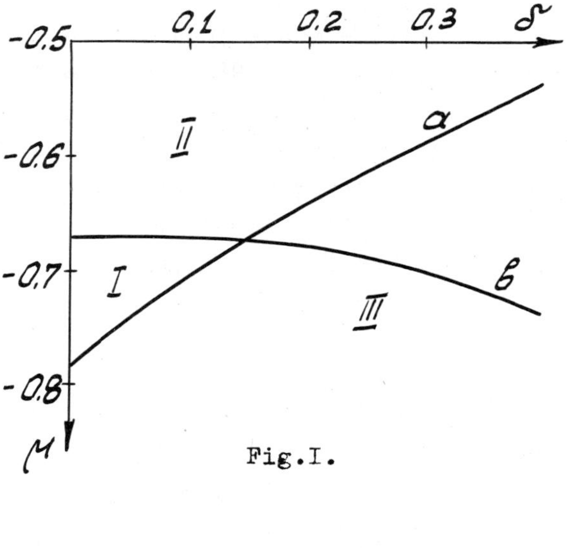

Fig.I.

The stationary bifurcation which can onset in region II was

studied by many authors (cf., for example, [2], [5]). Therefore not dwelling on this question we shall turn to region III where the bifurcation onsets nonaxisymmetrically. In [3], [4] this problem was solved by means of the Poincaré-Lindstedt method. Analyzing the branching equations we can show that the following three cases are possible: The first one takes place if the periodic solution is sought for in the form

$$v = v(ct + \ell\theta + \alpha z, x), \quad p = p(ct + \ell\theta + \alpha z, x) \tag{4}$$

The second case follows from the first one if α is replaced by $-\alpha$

$$v = v(ct + \ell\theta - \alpha z, x), \quad p = p(ct + \ell\theta - \alpha z, x) \tag{5}$$

Finally, if we search the periodic solutions in the form

$$v = v(ct + \ell\theta, \alpha z, x), \quad p = p(ct + \ell\theta, \alpha z, x) \tag{6}$$

provided that v_z, v_θ are even functions in z and v_z is an odd one in z , we have the third case. Choosing this or that form of the solution we ensure the simplicity of the eigenvalue on the given class of functions thus getting the case well studied. Note that from solution (5), by performing all possible shifts in z , one can obtain the set of solutions U_z . It is clear that the solutions from U_z cannot be stable by Lyapunov. The theorem proved in [4] gives a full answer with respect to the stability of branching solutions.

As ever, subcritical regimes prove to be unstable and for supercritical regimes the stability condition is defined by the sign of the expression formed by the coefficients α_{ij} , $i,j = 1,2$ similar to those of the normal form of ordinary differential equations. Algorithms for computing coefficients of the branching equation including the values of $\alpha_{i,j}$ are described in detail in [3] , [4] , moreover, the results of numerous computations testifying the relization of both the case when the bifurcation of all solutions is subcritical and that one when some of solutions (4), (5) or (6) both exists in the supercritical region and is stable are cited there as well.

As to the numerical techniques used we shall only note (because of the shortage of space) that these are methods "without saturation" permitting to indicate after a posteriori computations the accuracy of the result obtained. As it was mentioned above, in [5] the axisymmetric bifurcation of the Couette flow was studied by means of the Poincarè-Lindstedt method in combination with numerical techniques similar to those used in [3], [4]. The curve β in Fig. I separates the regions of the plane (μ, δ) where the bifurcation is subcritical and supercritical at the minimum of the critical curve $R_0(\alpha)$.

Consider region II where the bifurcation onto a secondary stationary regime is supercritical. The approach used in [6], [7] to investigate the secondary regime - the Taylor vortices - is based on the numerical construction of the Taylor vortices followed by obtaining the numerical solution of the two-dimensional spectral problem on their stability. The algorithms are based on the approximation of functions by the Fourier sums in the variable z, moreover, the coefficients in this expansion are the interpolational Lagrange polynomials with nodes in the zeros of the Tchebyshev polynomials, the boundary conditions being taken into account in the construction automatically.

In our further construction we shall substantially use the information about the multiplicity of the eigenvalue and the form of the eigenfunction of the spectral problem on the Taylor vortex stability obtained from numerical computations. Therefore, for the sake of definiteness consider the case when $\mu = 0$, $\delta = 0{,}14$. In this case the critical Reynolds number of the loss of the Couette flow stability is reached at $\alpha_* \simeq 1{,}565$ while $R_0(\alpha_*) = 4{,}1708284$ (cf. [6]). The computations given in [7] show that the critical Reynolds number of the loss of the Taylor vortex stability is $R_c(\alpha_*) \simeq 4{,}53353$ when $\ell = 1$, the eigenvalue is simple and the eigenfunction is such that its components v_r, v_θ are odd functions while v_z is even. Consider a class of vector functions with opposite evenness of the components. In this class the eigenvalue with a maximal real part of the problem on the Taylor vortex stability is at the same values of the parameters $\sigma = -0{,}29994 - i\,0{,}47735$.
As is known when investigating the Taylor vortex bifurcation one has to come across the branching involving the symmetry group. This situation is degenerate and requires special con-

sideration. By using the information about the eigenfunction obtained in [7] together with a number of simple statements of general character about the branching involving the symmetry group it was shown in [8] that under certain conditions there occurs from the Taylor vortices a set of periodic solutions such that for each pair of the solutions there exists a shift $z \to z+\beta$ transferring one solution into the other. The question of the character of the bifurcation and the stability of the arising time-periodic regime can be solved by means of numerical computations. In the case when $\mu = 0$, $\delta = 0.14$ there onsets a supercritical bifurcation and the arising periodic flow is stable. If we consider region I of the plane (μ, δ) then the situation becomes more complicated. In the computations carried out by the method used in [6] in the supercritical region one can clearly observe a branch of stationary solutions. When the Reynolds number of the flow approaches the critical Reynolds number of the loss of the Taylor vortex stability the norm of these solutions remains finite. In other words, one can observe the branch of solutions that has come from the subcritical region after the bifurcation. Unfortunately the algorithm used in [6] based on the stabilization method has not allowed to extend the branch into the subcritical region and to see the change of stability. Iterative methods such as the Newton one seem to be more appropriate for finding a branch of solutions in the subcritical region.

In conclusion we note that the series of papers [3] - [8] gives practically exhaustive information about the onset of the time-periodic solutions in the problem on flows between rotating cylinders and gives ground for studying the bifurcation of the periodic regime into doubly periodic one. The comparison with the well known experimental results which was carried out in [3] - [8] has shown excellent agreement of the numerical computations with results obtained experimentally.

References

I. Babenko, K.I.; Vasil'ev, M.M.; Yur'ev, S.P.: On periodic motions of a viscous fluid in a plane channel. Preprint of the Keldysh Institute of Applied Mathematics, USSR Academy of Sciences, I4 (I983).

2. Joseph, D.D.: Stability of fluid motions. Springer-Verlag I976.

3. Babenko, K.I.; Afendikov, A.L.; Yur'ev, S.P.: On the stability and bifurcation of the Couette flow between rotating cylinders. Preprint of the Keldysh Institute of Applied Mathematics, USSR Academy of Sciences, 99 (I98I).

4. Babenko, K.I.; Afendikov, A.L.; Yur'ev, S.P.: On the bifurcation of the Couette flow between rotating cylinders in the case of a double eigenvalue. DAN SSSR, I982, v. 266, I, 73-78.

5. Afendikov, A.L.: On stationary bifurcations of the Couette flow between rotating cylinders. Preprint of the Keldysh Institute of Applied Mathematics, USSR Academy of Sciences, 44, (I982).

6. Afendikov, A.L.; Babenko, K.I.: On the Taylor vortices. Preprint of the Keldysh Institute of Applied Mathematics, USSR Academy of Sciences, 3, (I982).

7. Afendikov, A.L.; Babenko, K.I.: On the stability of the Taylor vortices. Preprint of the Keldysh Institute of Applied Mathematics, USSR Academy of Sciences, 32, (I983).

8. Afendikov, A.L.: On the bifurcation of the Taylor vortices. Preprint of the Keldysh Institute of Applied Mathematics, USSR Academy of Sciences, 6I, (I983).

First Bifurcations in Circular Couette Flow: Laboratory and Numerical Experiments

S.N.LUKASCHUK, V.S.L'VOV, A.A.PREDTECHENSKY, A.I.CHERNYKH

Institute of Automation and Electrometry,
Siberian Division of the USSR Academy of Sciences,
630090 Novosibirsk, USSR

Summary

The instabilities which precede the transition to chaotic flow in long Couette apparatus are investigated. The experimental data from three space points are analysed in terms of qualitative theory of differential equations. Within the experimental resolution the attractors were found which can be represented as a direct product of "fast" one-dimensional torus corresponding to azimuthal waves and "slow" two-dimensional torus corresponding to small interaction between the adjacent waves. Phase trajectories on the observed slow tori had rational values (13/31 and 3/7) of rotation number. The existence of these attractors has been predicted by simulation within the frame of phenomenological model for weakly interacting azimuthal waves.

In this report we present some results of study of laminar-turbulent transition in circular Couette flow with the outer cylinder at rest.

Experiment

Our experimental cell has radius ratio 1.57 (diameter of the inner cylinder Di=35 mm) and height-to-gap ratio 28. The Reynolds number (Re=W*Di*(Do-Di)/4V where W is an angular speed of inner cylinder, V is the viscosity of fluid, Do is the diameter of outer cylinder) was kept constant to within 0.015% during a few hours required for the flow to turn to steady state and for data acquisition. We have measured local hydrodynamic friction on the surface of outer cylinder by means of ion-current technique and then recorded files of experimental data in computer for later analysis.

With the increase in Reynolds number, the flow develops through the following dynamical regimes.

bifurcation from a fixed point to a limit cycle (or a one-dimensional torus T1).

4. At R_3=1.02 the limit cycle looses its stability. According to Arnold's classification [3] the result of this bifurcation depends on where the eigenvalues L of Poincaré map linearisation intersect the unit circle at $R=R_3$. It may be simple splitting of limit cycle when L = 1, or Feigenbaum's cascade of doublings when L = -1, or origin of two-dimensional torus if L is a complex conjugated pair. In our case two-periodic motion originates resulting in slow (F2 = 0.009 Hz) modulation of traveling waves. The consideration of quasi-periodic case meets with additional difficulties connected with the possible origin of resonant trajectories on torus. Although the measure of resonant cases in parametrical space tends to zero at infinitely small supercriticality, at a finite supercriticality the resonances and associated heteroclinic processes can play a decisive role [4]. Turning to our experimental situation, it should be noted that a question about resonant properties of this torus looses its meaning due to a great ratio of fundamental frequencies: F1/F2=200. To simplify the study of the next bifurcations we have obtained the envelopes of main frequency of signals from the three probes. This procedure corresponds to the averaging of Poincaré maps along the fast limit cycle and gives new, "slow" variables. Such a transformation reduces our two-dimensional torus to a cycle with the frequency F2=0.009 Hz (Fig. 2). At R_4=1.027 this cycle looses its stability. During the time interval of the order of 1000s each signal taken separately seems to be chaotic. But the projection of phase trajectory on the plane in the space of measuring variables plotted during more than 10 000 s develops singularities which are typical of torus projection onto a plane (Fig. 3a). In particular, rotating this plane in three-dimensional space we may choose proper coordinates in which this attractor has a hole.

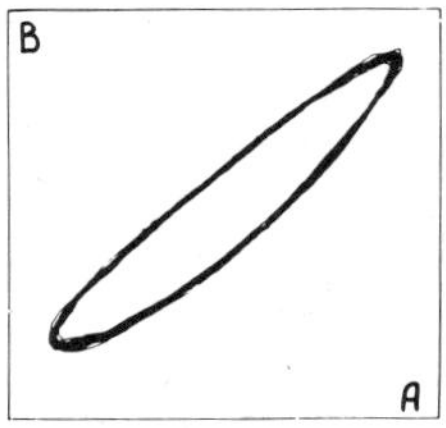

Fig.2. Limit cycle in "slow" variables

1. As it was shown before [1] , at Re=75 the Taylor vortices fill the gap from the end of annulus. Since the apparatus has fixed end plates, this process is not a bifurcation in the strict sense [2]. The primary flow contains 14 pairs of vortices. In order to investigate time-dependent regimes a column of 100 electro-chemical probes was used but only signals from the three probes placed near the outward jets of neighbouring pairs of vortices were analysed.

2. At Re_1=1030 the first bifurcation occurs developing the traveling azimuthal waves which produce the fast oscillations (F1=1.6 Hz) of hydrodynamic variables in laboratory frame of reference. The waves arise in all Taylor vortices simultaneously and their magnitude grows gradually in accordance with the Landau law.

3. The second instability at $R_2 = Re_2/Re_1 = 1.014$ leads to bifurcation of Reynolds-number dependence of wave amplitudes. Figure 1 presents the results of some experimental runs in neighbourhood of R_2.

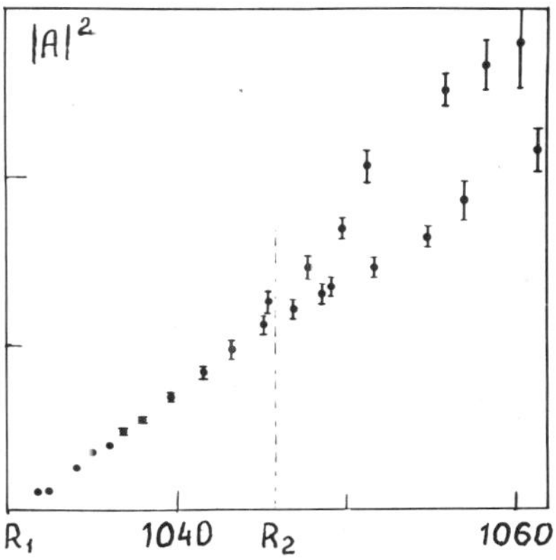

Fig.1. Bifurcation of wave amplitude dependence vs. Re

This effect may be associated with existence of two stable space structures of different reflection symmetry about midplane. At $R<R_2$ the amplitude distribution is symmetrical relative to midplane but at $R>R_2$ the distribution has a strong antisymmetrical component.

Further analysis of three-dimensional experimental information is based on the statement that any set of physically measured quantities gives a projection of hydrodynamical phase space onto the space of our measuring variables. This projection for experimental case is generic and therefore saves all properties of trajectories of real hydrodynamical flow. In terms of phase variables the first bifurcation corresponds to the Hopf

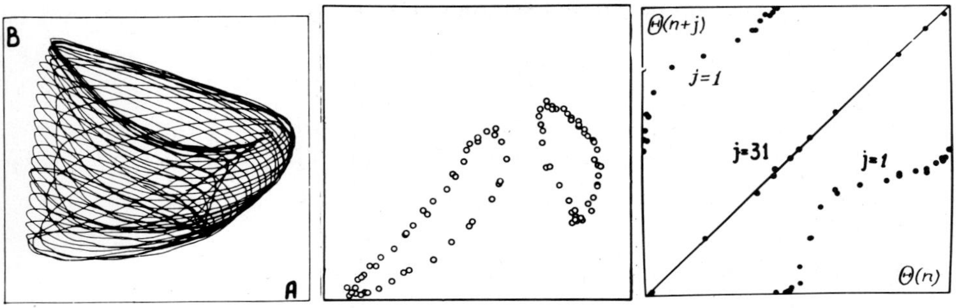

(a) (b) (c)

Fig.3. Projection (a), Poincaré section (b) and angle map (c) for torus with rotation number 13/31

Poincaré section of this attractor is shown in Fig. 3b that gives an evidence that the motion in slow variables occupies a surface of 2-torus. Consequently, with account of fast motion, the flow under study has an attractor on 3-torus. If we construct the period-one angle map $\Theta(n+1)=f(\Theta(n))$ (Fig. 3c), it becomes clear that the mapping of torus meridian into itself is a single-valued invertible function. Any integer power of this function exhibits the same character but there is a surprising feature that all experimental points occupy a diagonal of square if this power is equal to 31. It means that the phase trajectory is locked after 31 circuits of the slow 2-torus and repeats all its intricate bends. Consequently, the observed torus is a resonant one. At that time the meridian of torus is circuited 13 times yielding the rotation number 13/31.

Upon further increases of Re, the attractor in slow variables persists its toroidal form but the rotation number is varied passing into the region of stronger resonances. At $R \sim 1.030$ we have observed the resonance 3/7 (Fig. 4a) that differs from the previous value only by 2.5%. In figures 4b and 4c the data or Poincaré section, angle map and its seventh power are presented for this case. The finite thickness of trajectory rope is caused mainly by the transitional effect. The section of the rope has a form of line with a small curvature and its relative transverse size varies from 3 to 10 for different sections along the rope. If R exceeds 1.035, the resonant torus is destroyed.

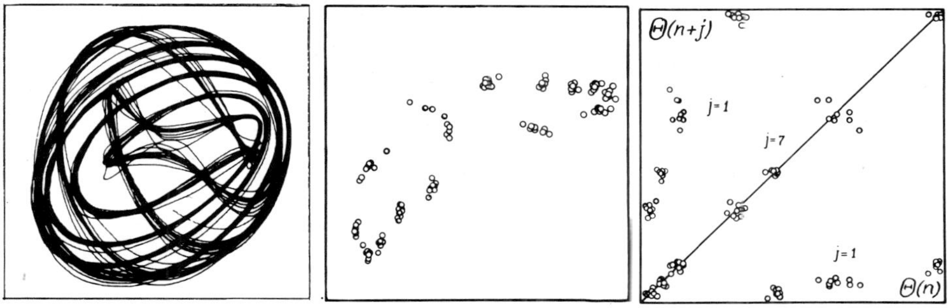

Fig.4. Torus with rotation number 3/7

Comparison with Theory

In our previous paper [1] the model of weakly interacted azimuthal waves was proposed to describe the complicated rearrangements of circular Couette flow during the transition to turbulence. In this model the pulsational component of velocity field has the form:

$$V_n(r,t) = \mathrm{Re}\{A_n(t)V(r)\exp(i\omega t)\}, \tag{1}$$

and the complex "slow" amplitudes A_n are governed by the following system of ordinary differential equations:

$$dA_n/dt=gA_n + (iT-q)|A|^2 \cdot A_n + (a+ib)(A_{n+1} + A_{n-1} -2A_n)/4,$$
$$n = 1,\dots,N; \qquad A_0 = A_{N+1} = 0. \tag{2}$$

Here N is the number of vortex pairs, Q=T/q, B=b/a, G=g/a are some phenomenological coefficients. Only G has significant dependence on supercriticity $Re - Re_1$ but the others can be treated as constants in actual range of Re numbers. Under a proper choice of these coefficients (B=1.25, Q=10) the equation (2) gives a correct qualitative description of the observed sequence of bifurcations. Moreover, the bifurcation of amplitude distribution symmetry of azimuthal waves was founded firstly in computer simulation at G = 0.02 and only more attentive experiments showed a reality of this phenomenon. The existence of two-dimensional torus (which appears at G=0.025) also was predicted at first by model simulation. We also have verified the main model supposition that wavy modes in Taylor vortices are grouped into pairs with weak interaction between themselves and have founded that wavy motion in vortex pair is really strongly correlated when $R < 1.05$. At larger R the higher nonlinear terms must be taken into account in the model. These are only the arguments for the proposed model (2) but not a direct demonstration of its validity. The more reliable verification of this model is to restore its coefficients from experimental data by statistical method. This is the purpose of our future work.

References

1. L'vov, V.S.; Predtechensky, A.A.; Chernykh, A.I. Bifurcation and chaos in the system of Taylor vortices: laboratory and numerical experiment, in "Nonlinear Dynamics and Turbulence", Pitman, 1983, pp. 238-280.

2. Benjamin, T.B. Bifurcation phenomena in stready flows of viscous fluid. - Proc. R. Soc. Lond., 1978, A359, pp.1-26.

3. Arnold, V.I. Loss of stability of self-oscillations close to resonance and versal deformations of equivariant vector fields. - Functional Analysis and its Applications, 1977, 11 (2), pp. 1-10.

4. Aronson, D.G.; Chory, M.A.; Hall, G.R.; McGehee, R.P. Resonance phenomena for two-parameter families of maps of the plane: uniqueness and nonuniqueness of rotation numbers, in "Nonlinear Dynamics and Turbulence", Pitman, 1983, pp. 35-47.

Symmetric and Asymmetric Taylor Vortices in a Spherical Gap

G. SCHRAUF

Applied Mathematics 217-50
California Institute of Technolcgy
Pasadena, California 91125

E. KRAUSE

Aerodynamisches Institut der RWTH Aachen
Wüllnerstraße zwischen 5 und 7
5100 Aachen, Germany

Summary

The transitions of the flow in a spherical gap are investigated by calculating the solution branches of the steady-state axisymmetric Navier-Stokes equations with a continuation method. The first flow instability is explained by a bifurcation of the spherical Couette flow into flows without equatorial symmetry. A hysteresis is predicted for $\eta = R_1/R_2 = 0.85$.

Introduction

The flow in the gap between a rotating inner and a fixed outer sphere of radii R_1 and R_2, respectively, can generate Taylor vortices. This was shown by Sawatzki and Zierep [6], among others. They obtained five different flow modes for $\eta = R_1/R_2 = 0.85$. Wimmer [11] showed in subsequent experiments that the modes can be produced by specific angular accelerations of the inner sphere or by transitions, which occur even if the angular velocity of the inner sphere is changed very slowly.

Bartels and Krause [1], [2] analysed the axial and equatorially symmetric modes by solving the time-dependent Navier-Stokes equations numerically. They found transitions which occurred at approximately the same Reynolds numbers as observed in the experiments but which had different directions.

In this study we calculate the solution branches of the axisymmetric time-independent Navier-Stokes equations. The calculations were done with a radius ratio $\eta = 0.85$.

Computer storage limitations forced us to consider only flows which were symmetric with respect to the equatorial plane in

[7], [8], [9]. With this assumption a lower boundary of the region of existence of the one-vortex mode is obtained and the transition into the basic flow can be explained. But there is no indication of the reverse transition from the basic flow into the one-vortex mode.

By dropping the assumption of equatorial symmetry we found a bifurcation of the basic flow, behind which this flow becomes hydrodynamically unstable. We obtained branches of unstable, asymmetric solutions which connect the basic flow with the one-vortex mode.

Formulation of the problem and method of solution

An incompressible Newtonian fluid with constant viscosity fills the gap between two spheres with the radii $R_1 < R_2$. The inner one rotates with the angular velocity ω.

The equations governing the axisymmetric flow in spherical coordinates (r, θ, ϕ) with corresponding dimensionless velocities (v^r, v^θ, v^ϕ) are

$$v^r = \frac{1}{r^2 \sin\theta} \frac{\partial}{\partial\theta} \psi, \quad v^\theta = \frac{-1}{r \sin\theta} \frac{\partial}{\partial r} \psi, \quad v = \frac{1}{r \sin\theta} \Phi,$$

$$(\nabla \times \vec{v})^\theta = \frac{1}{r \sin\theta} \zeta,$$

$$-D^2 \Phi + \frac{Re}{r^2 \sin\theta} \left\{ \psi_\theta \Phi_r - \psi_r \Phi_\theta \right\} = 0, \tag{1}$$

$$-D^2 \zeta + \frac{Re}{r^2 \sin\theta} \left\{ \psi_\theta \zeta_r - \psi_r \zeta_\theta - \zeta \left[\frac{2}{r} \psi_\theta - 2 \cot\theta \, \psi_r \right] + \Phi \left[\frac{2}{r} \Phi_\theta - 2 \cot\theta \, \Phi_r \right] \right\} = 0, \tag{2}$$

$$-D^2 \psi - \zeta = 0, \tag{3}$$

with

$$D^2 = \frac{\partial^2}{\partial r^2} + \frac{1}{r^2} \cdot \frac{\partial^2}{\partial \theta^2} - \frac{\cot\theta}{r^2} \frac{\partial}{\partial\theta}.$$

The Reynolds number $Re = \frac{\omega R_1^2}{\nu}$ is based on the inner radius. The boundary conditions are

$$\Phi(r, 0) = 0, \quad \zeta(r, 0) = 0, \quad \psi(r, 0) = 0, \tag{4}$$

$$\Phi(1,\theta) = \sin^2\theta, \quad \zeta(1,\theta) = -\frac{\partial^2}{\partial\theta^2}\psi(1,\theta), \quad \psi(1,\theta)=0, \qquad (5)$$

$$\Phi(\alpha,\theta) = 0, \quad \zeta(\alpha,\theta) = -\frac{\partial^2}{\partial\theta^2}\psi(\alpha,\theta), \quad \psi(\alpha,\theta) = 0, \qquad (6)$$

$$\frac{\partial}{\partial\theta}\Phi(r,\tfrac{\pi}{2}) = 0, \quad \zeta(r,\tfrac{\pi}{2}) = 0, \quad \psi(r,\tfrac{\pi}{2}) = 0, \qquad (7a,b,c)$$

$$1 \leq r \leq \alpha, \quad \frac{1}{\alpha} = \eta = \frac{R_1}{R_2} = 0.85, \quad 0 \leq \theta \leq \frac{\pi}{2}.$$

Equations (7a,b,c) govern flows which are symmetric with respect to the equatorial plane. Without this assumption (7) is replaced by

$$\Phi(r,\pi) = 0, \quad \zeta(r,\pi) = 0, \quad \psi(r,\pi) = 0, \qquad (8)$$

$$0 \leq \theta \leq \pi.$$

The equations (1) - (3) are discretized by second order central differences. The difference equations and the boundary conditions form a nonlinear system which contains the Reynolds number as a parameter. The solution branches of this system are calculated by using Keller's pseudo-arclength-continuation method [4]. (For details see [9]).

Symmetric solutions

The branches of the equatorially symmetric solutions obtained for the radius ratio 0.85 and for Re ≤ 1600 are shown in Fig. 1. We emphasize that the branch of the basic flow has no bifurcation point, if only equatorially symmetric solutions are considered. (The intersection of this branch with the branch with one Taylor vortex near the limit point A is not a bifurcation point. It is only a point at which the basic and the one-vortex solutions have the same torque coefficient.) Therefore, the transition from the basic flow to the flow with one Taylor vortex can only be achieved by either asymmetry or unsteadiness [8].

However the reverse transition from the flow with one vortex in each hemisphere into the basic flow occurs in the experiments at the same Reynolds number as the limit point A in Fig. 1. Wimmer [11] measured Re* = 650. Khlebutin [5] and Yavorskaya et al. [12] conducted their experiments with different geome-

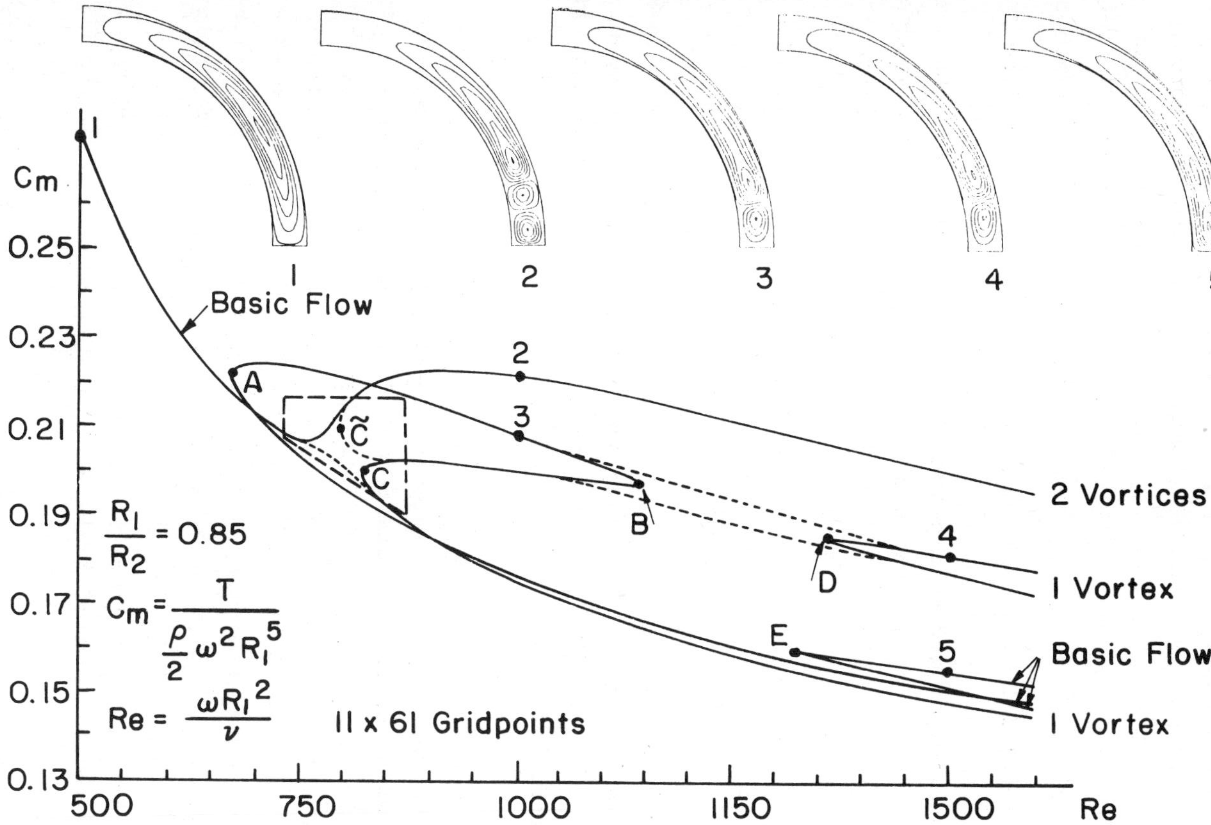

Fig. I Torque Coefficient As A Function Of Reynolds Number Re tries, but they empirically derived formulas for the Reynolds number as a function of the gap size. With Khlebutin's formula one obtains $Re^* = 661$ and with Yavorskaya's $Re^* = 665$. In our calculation the limit point A is at $Re_A^* = 670$ with 11 x 61 and at $Re_A^* = 657$ with 11 x 121 grid points.

The transition from the one-vortex solution to the basic flow can therefore be explained by the limit point A: If the Reynolds number is reduced so that it is lower than Re_A^*, the steady flow with one Taylor vortex no longer exists. The flow has to change into the only flow that still exists, i.e. into the basic flow.

There are two disconnected one-vortex solution branches if 11 x 61 grid points are used. One branch contains the limit points A and B and the other one the limit point D. The discrete equations have no one-vortex solution for $Re_B^* < Re < Re_D^*$.

But when we use 21 x 121 grid points or the boundary conditions (8) instead of (7) we obtain the solution branches indicated with dotted lines. There is no longer a non-existence region of the one-vortex mode.

However a non-existence region would explain the transition behavior observed in the experiments. Wimmer [11] found that starting with the one-vortex mode at Re = 1000 a very slow acceleration of the inner sphere results in a transition to the two-vortex solution, but with a faster acceleration the flow remains in the one-vortex mode.

The behavior of the solutions in the polygon drawn with dashed lines depends on the discretization of the boundary condition (7a) for Φ at the equator. With a symmetric, second order formula one gets the branches plotted with solid lines while with an asymmetric first order formula branches as indicated with the dotted lines are obtained. This behavior can be explained as an unfolding of a bifurcation of the solution branches of the differential equations (cf. [9]).

The solution at the limit point C (or $\tilde{C}$, if an asymmetric first order discretization for (8a) is used) has two Taylor vortices.

The vortex next to the equator disappears along the branch and one obtains a one-vortex solution at B. The streamlines of these solutions are presented in [7].

There is a second branch of solutions without Taylor vortices which contains the limit point E. This branch represents the supercritical basic flow, i.e. the basic flow which is obtained by a high acceleration of the inner sphere.

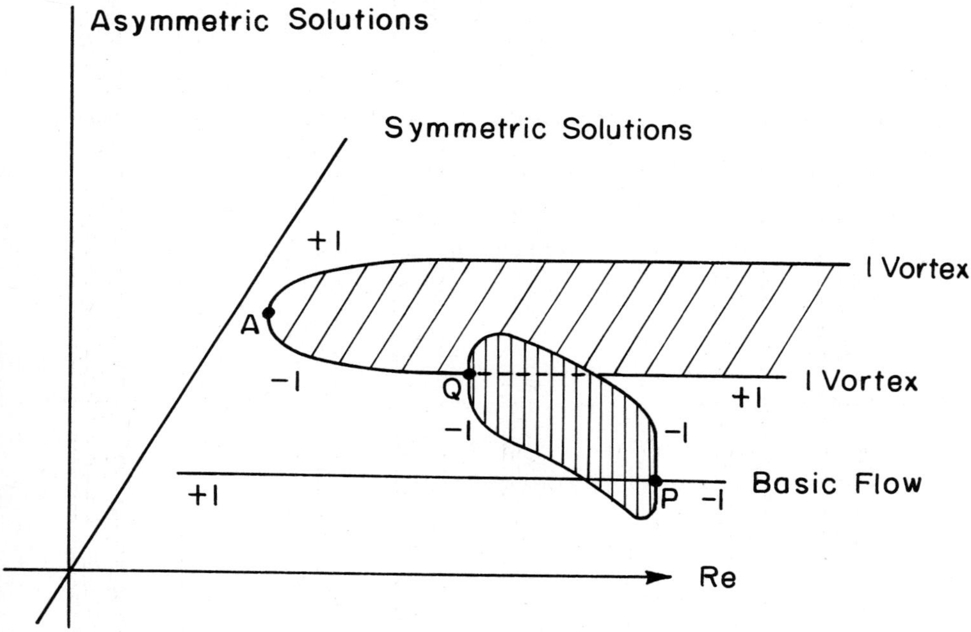

Fig. 2

Asymmetric solutions

Stimulated by Tuckerman [10] who showed that the basic flow becomes unstable with respect to an asymmetric perturbation, the assumption of equatorial symmetry is dropped. For the now larger domain of integration the boundary conditions (4), (5), (6), (8) are used.

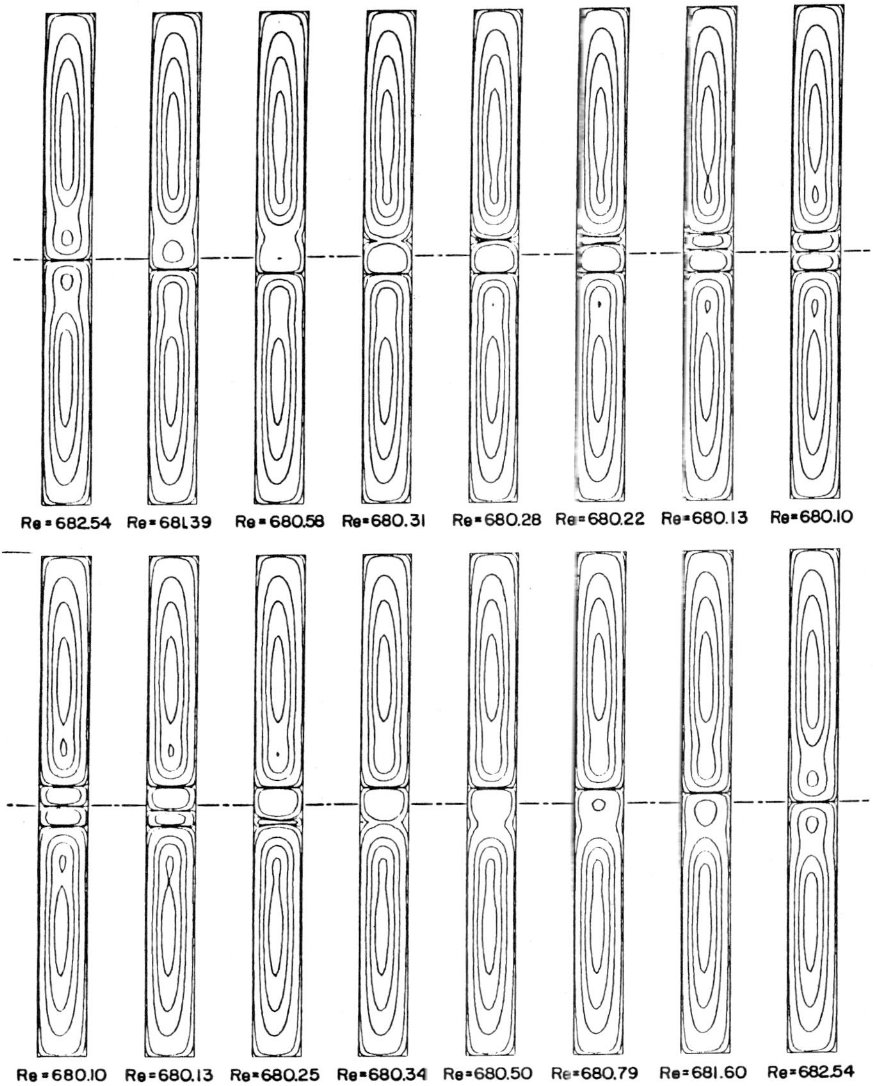
Re=682.54
Re=681.39
Re=680.58
Re=680.31
Re=680.28
Re=680.22
Re=680.13
Re=680.10
Re=680.10
Re=680.13
Re=680.25
Re=680.34
Re=680.50
Re=680.79
Re=681.60
Re=682.54

Fig. 3

Admitting asymmetric solutions the branch of the basic flow has a bifurcation point P. Behind P the basic flow becomes hydrodynamically unstable, corresponding to a change of the index of the solutions from +1 to -1 [3]. At the point P two branches of unstable asymmetric solutions bifurcate from the basic flow. Both branches are equivalent in the sense that the solutions of one branch can be obtained from the solutions of the other one by a reflection at the equator.

All branches obtained for Re $\leq$ 750 are shown in Fig. 2. The branches of solutions that are symmetric with respect to the equator are plotted in a plane and the branches representing asymmetric solutions are drawn on a surface which cuts this plane perpendicularly.

Along the bifurcating branches the separation line between the basic flow cells is shifted away from the equator. The streamlines of the flow in the larger cell are pinched together until it comes to a separation of two Taylor vortices. Finally the vortices are shifted until the flow becomes symmetric again. The streamlines of these solutions are plotted in Fig. 3. The upper row shows the development of a one-vortex solution along one branch from P to Q. The lower set of plots illustrates how the flow with one Taylor vortex in each hemisphere is transformed back into the basic flow along the other branch from Q to P.

This process agrees with the transition of the time-dependent flow reported by Tuckerman in [10]. But in the time-dependent case the flow has to change into the stable one-vortex mode.

The Reynolds numbers of the bifurcation points P and Q and of the limit point A calculated with 11 x 61 and with 11 x 91 grid points are listed in the following table:

	11 x 61	16 x 91	
P	702.81	682.54	666.32
Q	699.57	680.13	664.58
A	693.88	675.54	660.87

The third column contains the Reynolds numbers obtained by a Richardson extrapolation.

The Reynolds number of the bifurcation point P differs from the one of the limit point A. The transition to and from the one-vortex mode has therefore a hysteresis.

A hysteresis was in fact observed by Yavorskaya et al. [12] for the radius ratio $\eta = 0.806$. They observed a transition to the one-vortex mode at Re = 520 and back to the basic flow at Re = 463. For $\eta = 0.85$ we predict Re $\simeq$ 5 which is probably too small to be detected in experiments.

Acknowledgements

1. We thank Prof. H.B. Keller for many fruitful discussions.
2. This research was supported by the Deutsche Forschungsgemeinschaft in the Sonderforschungsbereich 72 at the University of Bonn and by a research scholarship also from the Deutsche Forschungsgemeinschaft.
3. We thank the IBM Corporation for computer time.

References

1. Bartels, F.; Krause, E.: Taylor vortices in spherical gaps. In: Eppler, R.; Fasel, H. (eds.), Laminar-turbulent transition. Proceedings IUTAM Symposium, 16.-22.9.79, Stuttgart. Berlin, Heidelberg, New York: Springer Verlag 1980.

2. Bartels, F.: Taylor vortices between two concentric rotating spheres. J. Fluid Mech. 119 (1982) 1-25.

3. Benjamin, T.B.: Applications of Leray-Schauder degree theory problems of hydrodynamic stability. Math. Proc. Camb. Soc. 79 (1976) 373-392.

4. Keller, H.B.: Numerical solutions of bifurcation and nonlinear eigenvalue problems. In: Rabinowitz, P. (ed.), Applications of bifurcation theory. New York: Academic Press 1977.

5. Khlebutin, G.N.: Stability of fluid motion between a rotating and a stationary concentric sphere. Fluid Dynamics 3.6 (1968) 31-32.

6. Sawatzki, O.; Zierep, J.: Das Stromfeld im Spalt zwischen zwei konzentrischen Kugelflächen, von denen die innere rotiert. Acta Mechanica 9 (1970) 13-35.

7. Schrauf, G.: Numerical investigation of Taylor-vortex flows in a spherical gap. ZAMM 63 (1983) T282-T286.

8. Schrauf, G.: Branching of Navier-Stokes equations in a spherical gap. In: Krause, E. (ed.), Eighth International Conference on Numerical Methods in Fluid Dynamics. Lecture Notes in Physics 170. Berlin, Heidelberg, New York: Springer Verlag 1982.

9. Schrauf, G.: Lösungen der Navier-Stokes Gleichungen für stationäre Strömungen im Kugelspalt. Preprint no. 611, Sonderforschungsbereich 72 an der Universität Bonn.

10. Tuckerman, L.: Formation of Taylor vortices in spherical Couette flow. Ph.D. Thesis, Massachusetts Institute of Technology, 1983.

11. Wimmer, M.: Experiments on a viscous fluid flow between concentric rotating spheres. J. Fluid Mech. 78 (1976) 317-335.

12. Yavorskaya, I.M.; Belyaev, Yu.N.; Monakhov, A.A.; Astaf'eva, N.M.; Scherbakov, S.A.; Vvedenskaya, N.D.: Stability, non-uniqueness and transitions to turbulence in the flow between two rotating spheres. Space Research Institute Academy of Sciences, USSR, Rep. No. 595, Moscow 1980.

Some Routes to Turbulence in Spherical Couette Flow

Yu.N.BELYAEV, A.A.MONAKHOV, S.A.SCHERBAKOV, I.M.YAVORSKAYA*

Institute of Mechanics, Moscow State University, USSR
*Space Research Institute, Academy of Sciences, USSR

Summary

Some routes to stochasticity in a viscous fluid flow between two rotating spheres are investigated which differ considerably from those observed in Couette-Taylor flow and Rayleigh-Bénard convection. The dimensions of regular and stochastic attractors corresponding to flow considered are found.

Several different transitions to turbulence in Taylor-Couette flow and Rayleigh-Bènard convection have been found recently [1,2] which qualitatively agree with a concept of a chaos onset in low-dimensional dynamical systems. The result presented here concerns with analogous investigation in a viscous fluid flow between rotating spheres which in mathematical and hydrodynamical aspects considerably differs from the flows mentioned above.

A set-up for our experiments and a method for a data processing are presented in detail in [3]. The spherical annulus with relative gap size $\delta = (r_2-r_1)/r_1 = 1.006$ and outer sphere radius $r_2 = 150.16 \pm 0.03$ mm was used. Only the inner sphere was rotating, so the single control dimensionless parameter was the Reynolds number $Re = \Omega r_1^2/\nu$. Angular velocity stabilization was better than 0.03%, the fluid temperature was maintained constant to within 0.03°C. The relative gap size varied no more than 10^{-4}. The Reynolds number was changed quasistatically with a step $\Delta Re/Re = 0.03\%$. Three velocity components at one point or one at three different points of fluid layer were measured by laser Doppler velocimetry. Velocity records up to 6 hours length were further analyzed with automatic system of processing.

A comprehensive statistical processing of data, such as fast

Fourier transforms, calculations of auto- and cross-correlation functions, cross-spectra, envelope of amplitudes at individual frequencies with a window as $\Delta f/f \approx 10^{-2}$, exact frequency values using the maximum entropy spectral analysis, were made. The measurements were performed at the stationary outer conditions and up to two hours delay after a change of Re. The spectrogramms given below are averaged with respect to many samples.

A state of hydrodynamical system at given values of basic parameters is characterized by a certain space structure and a set of incommensurate frequencies of oscillations. We define continuous variations of these states due to quasistatic variations of parameters or its transitions due to normal Landau-Hopf bifurcations as a sequence of states. Then new sequences may arise as a result of inverse Landau-Hopf bifurcations or at finite-amplitude disturbances of the flow.

The basic flow at small Re is stationary and consists in differential rotation of fluid and its circulation in meridional plane. Six different sequences of regimes were observed, three of them led to stochasticity. The first transition at Re=460 results in nonaxisymmetrical wave flow with azimuthal wave number m=4 [4] which corresponds to appearance in power spectra of a single frequency ν_0. The flow consists of four vortices in every hemisphere displaced in staggered rows with respect to equator. At Re=538 period doubling bifurcation excites a subharmonic $\nu_0/2$. Further the change of sequence of regimes happens as result of inverse bifurcation (Re=556). A new wave flow characterized by 3 vortices (m=3) with a new frequency ν_1 appears. This regime has a hysteresis and at Re=462 is reorganized to the flow with ν_0. The sequences of regimes which lead to stochasticity due to normal bifurcations are shown in Fig.1 where the dimensionless frequencies ν/Ω as a function of Re are presented. The arrows correspond to irreversible transitions between sequences and the shaded regions to stochastic flows The sequence considered is given at the top of the Fig. 1.

Two more frequencies ν_3, ν_4 =0.24 and 0.14 appear at Re=730 and 800, respectively. In the range $814 \leq Re \leq 817.5$ the amplitude modulation of oscillations at all fundamental frequencies occurs and further two period doubling of oscillations at modu-

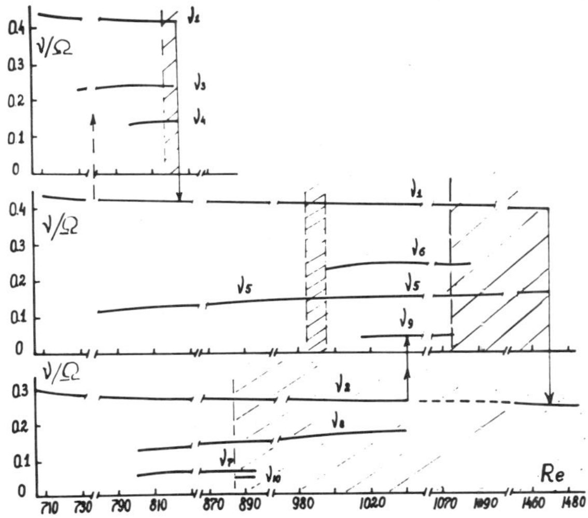

Fig. 1

lation frequency ($\Delta\nu/2$ and $\Delta\nu/4$) result in stochastic automodulation of the oscillations at Re=817.5. The power spectra for succession of transitions are shown in Fig.2 - 5. It seems that instrumental noise breaks the period doubling at $\Delta\nu/4$ or seldom at $\Delta\nu/8$. It is worth to stress that in our case period doubling occurs at a modulation rather than a fundamental frequency as usual. As Re increases this stochastic regime irreversibly turns into a regular one with two frequencies ν_1 and ν_5 and so a new sequence of regimes is generated (Fig. 1 in the middle). The transition from stochasticity to regular flow is an important indication of the fact that near a stochasticity threshold small number of degree of freedom are excited. The same follows from the estimates of attractor dimensions.

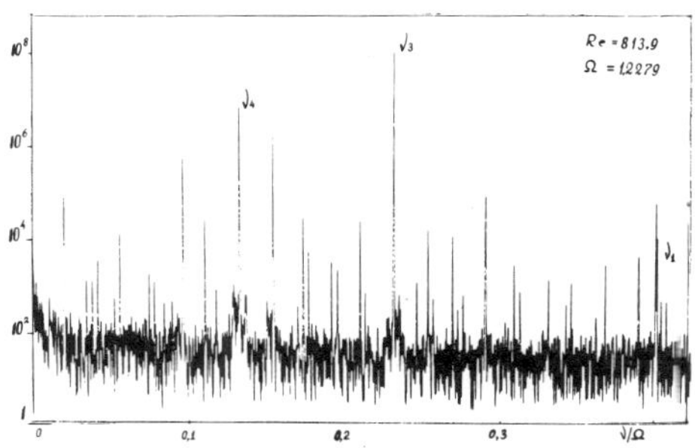

Fig. 2

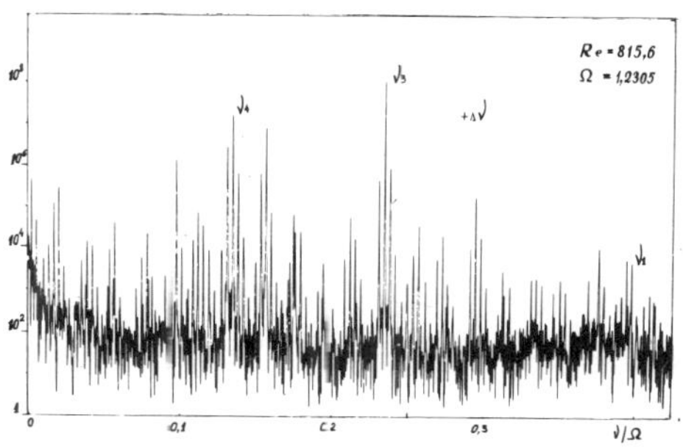

Fig. 3

New flow regime with frequencies ν_1 and ν_5 (Fig. 6) has a hysteresis and turns into the flow with ν_1 and ν_3 frequencies at Re=760. With increasing Re a complicated evolution of the flow regime occurs, however, the fundamental frequencies ν_1 and ν_5

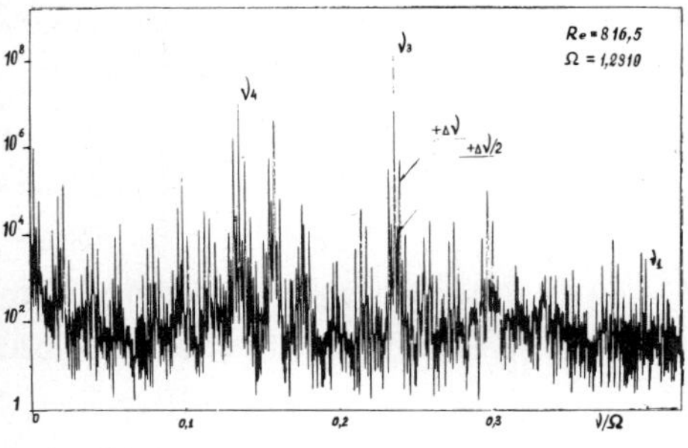

Fig. 4

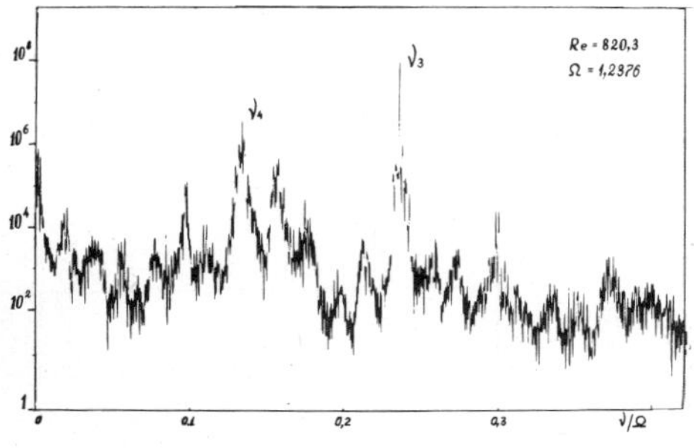

Fig. 5

persist. In contrast to previous results [1,2] the route to stochasticity considered here appears to be a multi-stage one with a reccurence to a regular motion (Fig. 1). At Re=825 a slow amplitude modulation at fundamental frequencies arises with $\Delta\nu \ll \nu_1, \nu_5$. As Re is increased non-linear interactions seem to result in emergence of $\Delta\nu$ higher harmonics which leads to appearance in spectra up to 20 satellites near every line (Fig. 7). Nevertheless the regime remains strictly quasiperiodical, autocorrelation function does not decay.

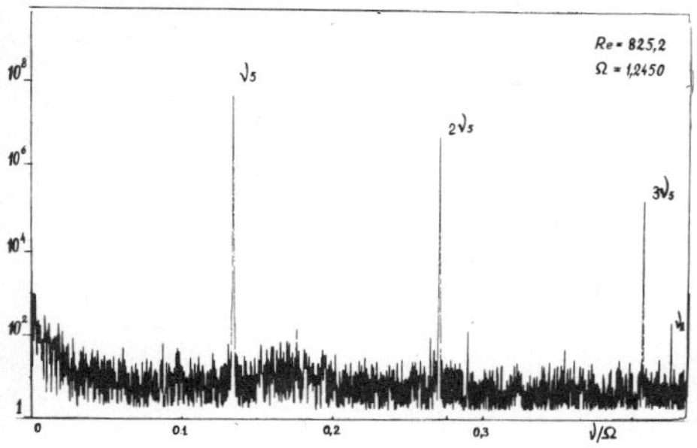

Fig. 6

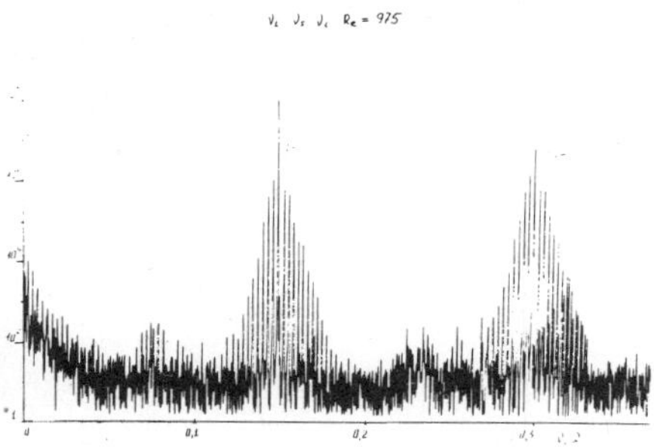

Fig. 7

Further ($980 \leq$ Re ≤ 985.5), a series of period doubling seems to result in stochasticity onset at Re=985.5 . In fact, the motion contains two fundamental frequencies ν_1 and ν_5 with a slow wandering of the system within the strongly broadened spectral line. A new frequency ν_6 appearance at Re=998 destroys the modulation and chaos. The flow becomes quasiperiodical with 3 fundamental frequencies (Fig. 8). At higher Re a synchro-

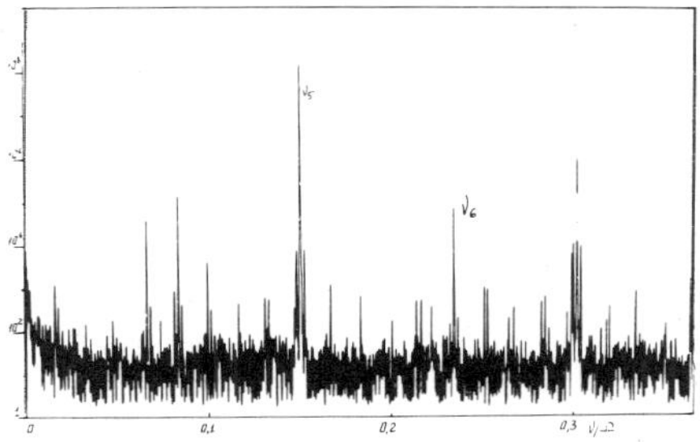

Fig. 8

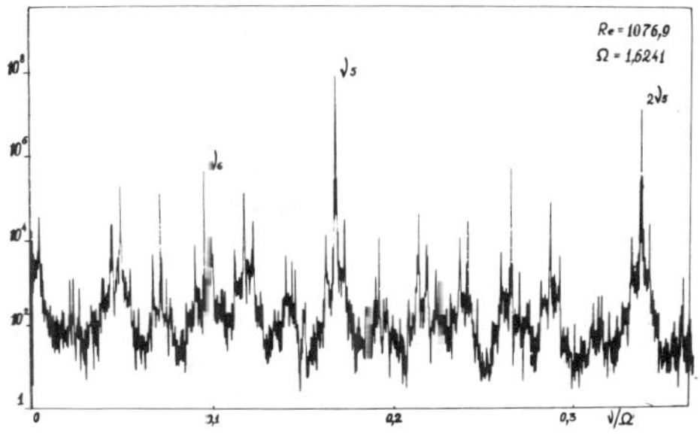

Fig. 9

nization of combination frequencies $2\nu_5 - \nu_6$ and $2\nu_6 - \nu_1$ happens and a new frequency ν_9 appears. At Re=1076 the spectral lines at ν_9 and its combination harmonics are broadened, the level of the spectrum continuous part grows slightly; the attractor dimension increases by one (up to 4.03). The peaks at ν_1, ν_5 and ν_6 remain instrumentally narrow (Fig. 9). A sharp broadening of all spectral lines and considerable augmentation of continuous noisy component occur when the synchronization is destroyed (Re=1089). The attractor dimension increases again

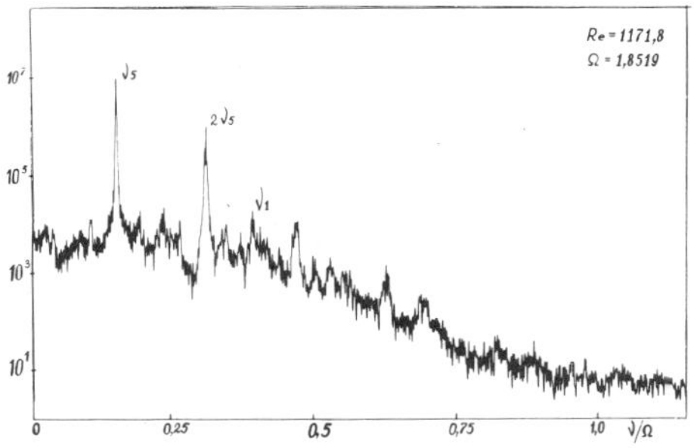

Fig. 10

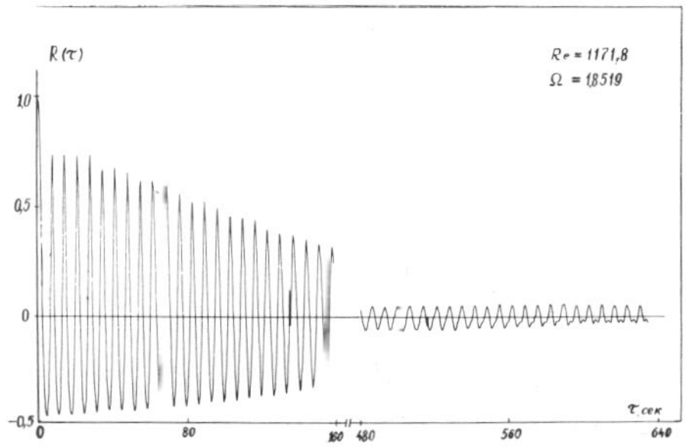

Fig. 11

(d=5.2). No qualitative changes in spectrum happen further (Fig. 10) up to Re=1460 where an irreversible transition to a new stochastic flow regime occurs (see below). The autocorrelation function in this case is given in Fig.11.

The third sequence of transitions to turbulence appears when Re increases abruptly. It may be obtained, for example, if in

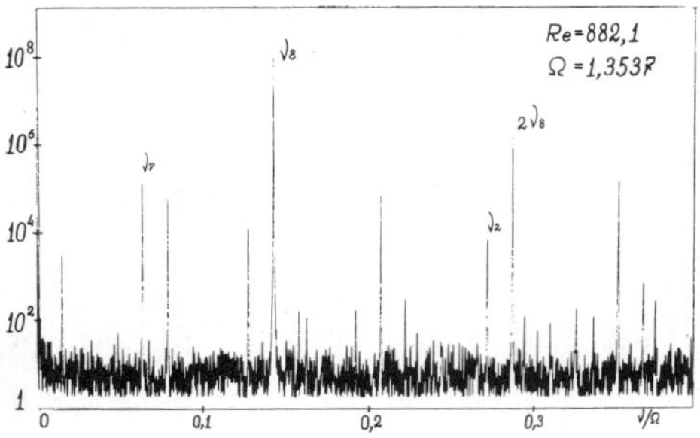

Fig. 12

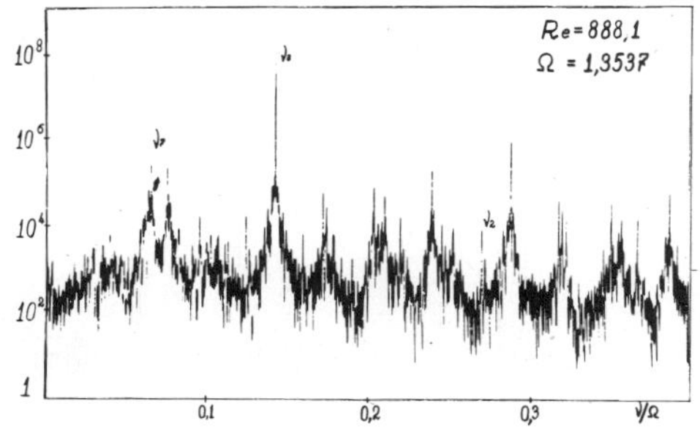

Fig. 13

the flow with fundamental frequency $\nu_0/2$ the angular velocity is changed impulsively from Re=550 to 676. One of pecularities of this sequence consists in appearance of new frequencies in pairs as Re increases. First, new frequencies ν_7 and ν_8 appear (Re=800), then frequency locking $\nu_2 - \nu_8 = 2\nu_7$ occurs when Re exceed 855. Onset of chaos coincides with the loss of synchronization and simultaneous emergence of one more frequency pair. If we estimate (by method described below) the dimension of attractors for hydrodynamical system before and after the transition to chaos we find it changes approximately by 2. The spectra for these cases are shown in Fig. 12 and 13. Onset of chaos due to desynchronization was observed in early experiments, for example, in Rayleigh-Bénard convection [2]. However, it usually was connected with fundamental frequencies rather than combination ones.

Reemergence to order in rather narrow Re range (about 1) due to complicated chain of synchronization near Re=892 diminishes the attractor dimension by one. Another reorganization of the stochastic flow to regular one with two frequencies ν_1 and ν_5 corresponding to an appearance of triplet $\nu_2 = \nu_1 - \nu_5$ happens at Re=1040. This two vortices regime persists if Re is increased sharply to 1300, it is the same one, in which the above mentioned sequence is reorganized at Re=1460.

The investigation of spectrum and autocorrelation function give the possibility to obtain the number of degree of freedom in regular flow and determine a stochasticity onset. However, we cannot judge about deterministic or random origin of sto-

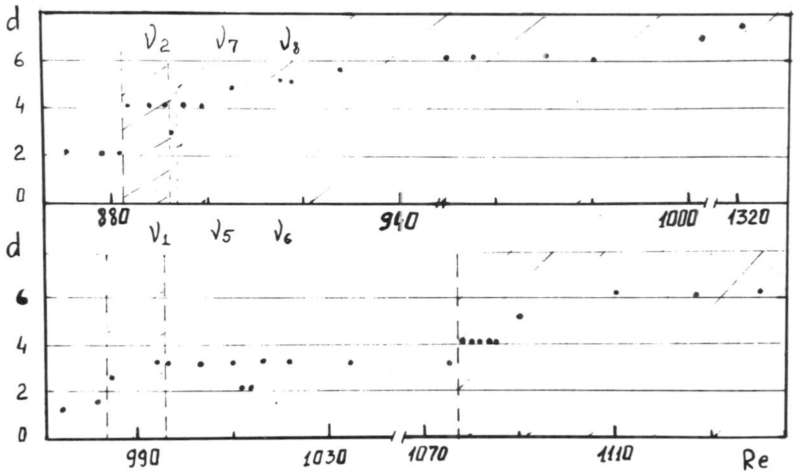

Fig. 14

chasticity and estimate the number of degrees of freedom after a stochasticity onset. Therefore a structure and dimensions of the systems attractors need be analyzed. We should begin with reconstruction of a dynamical system and its phase portrait describing the flow considered [5]. According to an idea due to Packard et all [6] justified by embedding theorems, the reconstruction may be done from measurements of a single dynamical observable V(t) and almost arbitrary τ by constructing vectors $V(t_k)$, $V(t_k+\tau)$, ..., $V(t_k+(m-1)\tau)$. A m-dimensional phase portrait obtained by this procedure has the same basic properties as one constructed from measurements of N independent variables if the embedding dimension $m \geq 2N+1$.

The phase portraits of the attractors for all analyzed sequences and their Poincaré sections are constructed which allow to conclude about a strong localization attractors in the embedding space. Three different methods for computation of probability dimensions of the attractors are used.

Computation of the dimensions is based on a concept of natural measure μ [7], i.e. a relative probability of finding the system trajectory in a given region of phase space. Then the dimension of the attractor is obtained as $\lim_{\varepsilon \to 0} \ln \mu(\varepsilon, x)/\ln \varepsilon$ where ε is a linear size of a m-dimensional cube with center at x.

Practically the method consists in an estimate of average number N of trajectory points inside many ε -cubes for different χ and determination of the slope of ln N vs ln ε for increasing m . Beginning with a certain m the slope ceases to change and gives the attractor dimension [5].

The second method differs from the first one by a change of an independent and dependent variables [8]. The average distance ε from a fixed point to its N-th neighbour is computed.

The third method is based on a computation of the correlation integral [9]:

$$C(r) = \lim_{N \to \infty} \sum_{i}^{N} \sum_{j}^{N} \Theta(r - |\vec{X}_i - \vec{X}_j|)/N^2$$

where $\vec{X}_i$ and $\vec{X}_j$ are points of the system trajectory and Θ is the Heaviside function. For small r the integral C(r) behaves as r^{α} where $\alpha \leqslant$ d. As shown in the Fig. 14 all three methods give approximately the same values for attractor dimensions. Some transitions between stochastic flow regimes corresponding to sharp increase of the dimension are obtained. Due to experimental noise it is not a strict fractal value of the dimension that is of interest. The main point is that in the control parameter space where the flow is rather chaotic the number of active degrees of freedom in the fluid flow measured by its attractor dimension remains relatively small.

References

1. Swinney, H.L.: Physica D 7D (1983) 1-7.
2. Gollub, J.P.; Benson, S.V.: J. Fluid Mech. 100 (1980) 449.
3. Yavorskaya, I.M.; Belyaev, Yu,N. et all: Postprints of XIY Congress of IUTAM, North Holland, Amsterdam, 1980.
4. Belyaev, Yu.N.; Yavorskaya, I.M. in Nonlinear Dynamics and Turbulence. G.I.Barenblatt; G.Ioss; D.D.Joseph (eds.) Boston, London, Melburn: Pitman Advanced Publishing Program. 1983.
5. Brandstäter, A.; Swift, J.; Swinney, H.L. et al.: Phys. Rev. Lett., 51,(1983)
6. Packard, N.H.; Crutchfield, J.P.; Farmer, J.D.; Show, R.S.: Phys. Rev. Lett., 45 (1980) 712.
7. Farmer,D.; Ott,E.; Yorke, A.: Physica 8D (1983) 153-180.
8. Tremonia, Y.; Alexandrovicz, Z.: Phys. Rev. Lett., 51 (1983) 1265-1268.
9. Grassberger, P.; Procaccia, I.: Phys. Rev. Lett., 50 (1983) 346-349.

New Secondary Instabilitaties for High Re-Number Flow Between Two Rotating Spheres

K. Bühler and J. Zierep

Institut für Strömungslehre und Strömungsmaschinen
Universität (TH) Karlsruhe, West-Germany

Summary

The laminar turbulent transition in a spherical gap starts with time depending shear waves. The structure of these secondary instabilities is visualized. The dynamical behaviour is investigated by the power spectra of the local wall shear stress at the outer sphere. Depending upon the initial and boundary conditions we get different bifurcations of the laminar-turbulent transition.

Introduction

New investigations of the dynamics of nonlinear systems and new experimental methods to predict the power spectra allow deep insight in the physics of transition. A survey on the actual state is giver by Swinney and Gollub [1], Barenblatt, Ioss and Joseph [2] and Eppler and Fasel [3]. The dynamical behaviour of the Taylor vortex flow in a cylindrical gap is investigated among others by Fenstermacher, Swinney and Gollub [4], Krugljak et al. [5]. In this case the transition to turbulence is due to time depending modulations of the Taylor vortices. Concerning the dynamical behaviour of flows in relative wide spherical gaps, results are known from Belayev, Yavorskaya et al. [6]. In our investigation the dynamical behaviour of flows in small spherical gaps are of main interest. New shear waves had been discovered, which can be interpreted as instabilities of the threedimensional boundary layer flow in the neighbourhood of the outer sphere at rest. A principle sketch of these shear waves is given in fig. 1. The coinciding of shear waves from both hemispheres leads to a waviness in the equatorial plane, which runs with a characteristic velccity in circumferential direction. In the case of Taylor vortex flow these shear waves appear in the basic flow too. A direct transition to turbulence can be achieved by a sudden start of the inner sphere to a supercritical Reynolds number. A similar process was described for ring vortices by Schultz-Grunow [7].

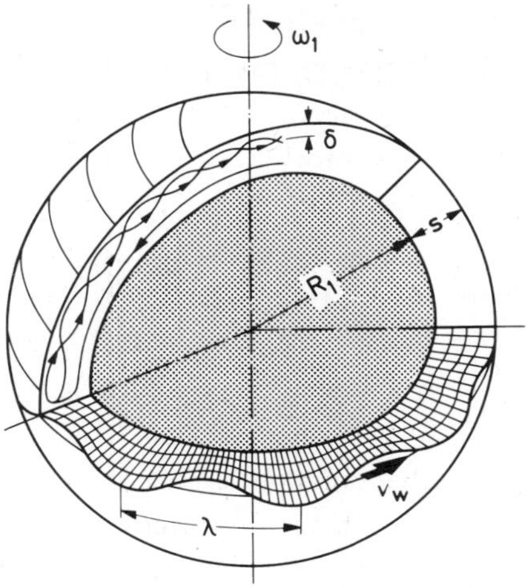

Fig. 1: Geometry of the spherical gap and principle structure of the shear instabilities. $\sigma = s/R_1 = 0.177$, $R_2 = 75$ mm

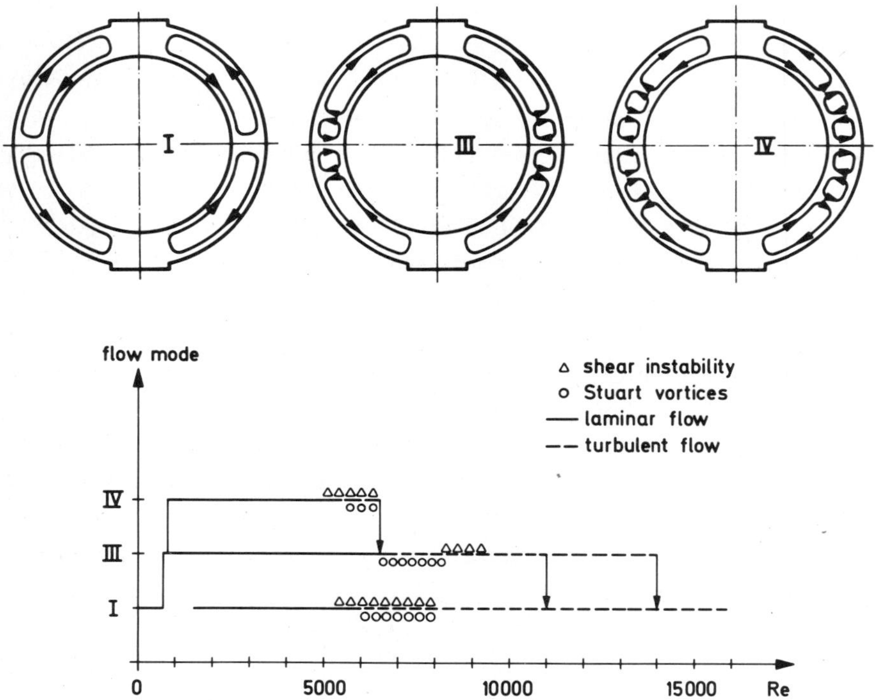

Fig. 2: Different steady, axisymmetric modes of flow, existence regions and different kind of instabilities

Spherical gap flow

The geometrical set up is principally sketched in fig. 1. The inner rotating sphere is surrounded by a transparent outer sphere at rest. The gap is filled with silicon oil of kinematic viscosity $\nu = 5 \cdot 10^{-6}$ m^2/s. Small aluminium flakes are used for visability of fluid flow. The wall shear stress is measured at the outer sphere by using hot film sensors in the positions $\vartheta = 30°$ and at the equator $\vartheta = 90°$. Due to the threedimensional basic flow in a spherical gap different modes can be realized at the same supercritical Reynolds number. This non-uniqueness of spherical gap flow was first shown by Sawatzki and Zierep [8]. The three different steady and axisymmetric modes I, III and IV are sketched in fig. 2. According to Wimmer [9] these modes can be realized by different accelerations of the inner sphere. Other aspects of these flows had been investigated theoretically and experimentally by Yavorskaya et al. [10].

Instabilities

Fig. 2 shows the domain of existence for the different modes as function of the Reynolds number. The solid lines represent laminar flow while the dashed lines correspond to turbulent fluid flow. The transition to turbulence is strongly influenced by the instability of the outer threedimensional boundary layer flow. As a consequence time dependent shear waves appear. In the case that the boundary layers are separated from one another a region of nearly rigid body rotation exists in the middle of the gap. The condition $\delta < s/2$ is fulfilled, if the Reynolds number exceeds a characteristic value:

$$Re > \frac{4a^2}{\sigma^2} \quad , \quad \text{where} \quad Re = \frac{\omega R_1^{\ 2}}{\nu} \quad , \quad \sigma = \frac{s}{R_1} \quad \text{and} \quad \delta = a(\vartheta) \sqrt{\frac{\nu}{\omega}} .$$

In spherical gaps the boundary layer thickness δ is a function of the meridional coordinate ϑ. For boundary layers in rotating systems we have $1 < a(\vartheta) < 3$. It is due to Greenspan [11] that rotating boundary layers exhibit some general features. In a spherical gap the boundary layer flow depends completely on the respective mode. In any case there is a direct connection between the local and the global Reynolds number

$$Re_\delta = \frac{v_\delta \, \delta}{\nu} = k \cdot a(\vartheta) \cdot \sqrt{Re} .$$

$v_\delta = k \cdot R_1 \omega$ is the velocity of the rigid body rotation at the outer boundary.

layer. The order of magnitude of Re at the onset of the shear instabilities corresponds to that of the Ekman boundary layer and the cross-flow instabilities of swept wings after Poll [12].

Mode I: Starting with mode I, that means without any vortex, the fluid flow gets unstable for the second time with respect to shear waves for Reynolds numbers $Re > 5400$. The structure of the instability is shown in fig. 3. The waves run in spiral form from the poles towards the equator, more or less perpendicular to the direction of the secondary flow near the outer sphere. The wave structure of this instability is visualized in the meridional plane by a light cut method. Fig. 4 shows streaklines of the time dependent motion in the meridional plane. For constant gap width σ the circumferential velocity component of these waves is a function of the Reynolds number alone. The ratio of the frequencies is therefore direct proportional to the discrete wave number n

$$\frac{v_W}{R_1\omega} = k(Re) \quad , \quad \frac{f}{f_{rot}} = n \cdot \frac{R_1}{R_2} \cdot k(Re) = n \cdot k^*(Re).$$

Fig. 5 gives the dimensionless ratio $\nu_1 = f_1/f_{rot}$ as function of the Reynolds number and the wave number n. For fixed n the frequency is nearly independent of Re. With increasing Re n increases too, that means the wave length decreases. These transitions are accompanied by strong hysteresis effects. The frequency jumps $\Delta\nu$ correspond to the circumferential angular velocity of the shear waves. Characteristic power spectra for two different positions at the outer sphere are shown in fig. 6. In the equatorial plane at $\vartheta = 90°$ we have a harmonic variation of the local wall shear stress with the characteristic frequency f_1. At $\vartheta = 30°$ in addition the rotation frequency f_r is important. $f_0 = f_1/n$ corresponds to the circumferential velocity of the shear waves. The stochastic motion starts with the onset of Stuart vortices in the pole region with a broadband spectra at low frequencies. They have zero phase velocity, the axis are oriented in direction of the basic flow. The superposition of the shear waves and the Stuart vortices leads to a stochastic motion, which approaches the equator for increasing Re. For $Re > 8000$ the whole spherical gap is turbulent. The waviness of the equatorial plane is conserved even for higher Reynolds numbers.

Mode III: Fig. 7 shows this mode with two Taylor vortices near the equator. For $Re > 6500$ Stuart vortices appear near the poles. For $Re > 7800$ a

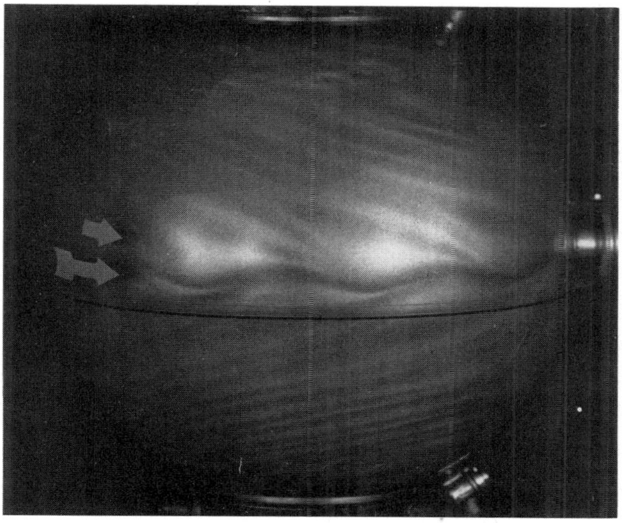

Fig. 3: Mode I with shear instabilities $Re = (R_1^2\ \omega_1)/\nu = 6050$

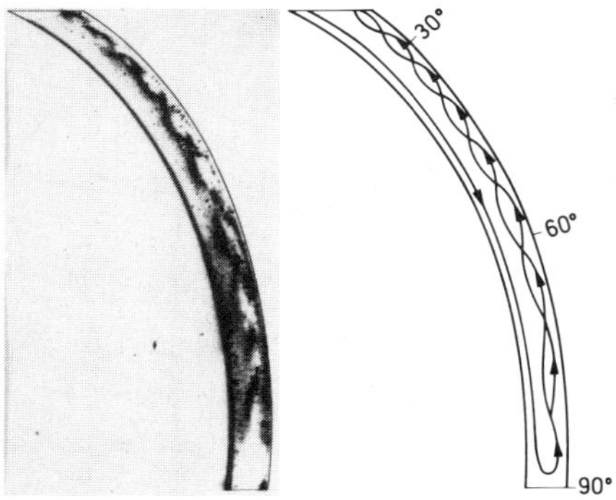

Fig. 4: Structure of the shear instabilities, streaklines in the meridional plane

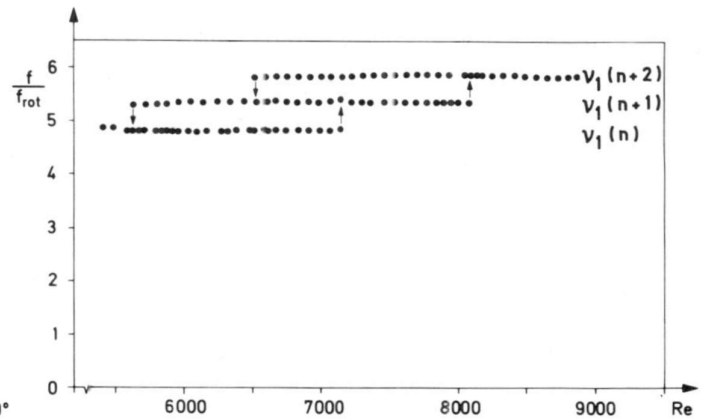

Fig. 5: Frequencies as function of Reynolds and wave number

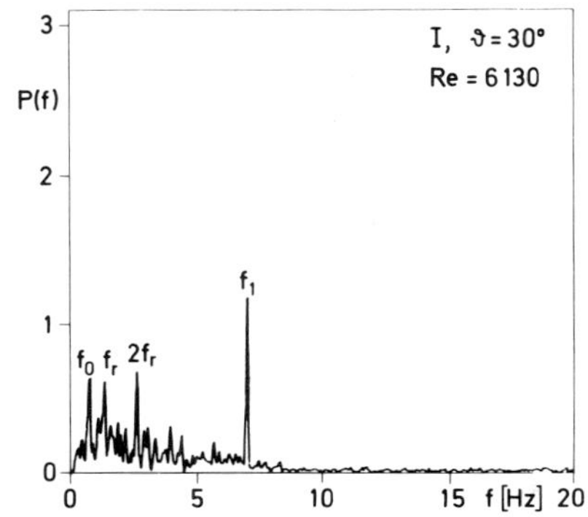

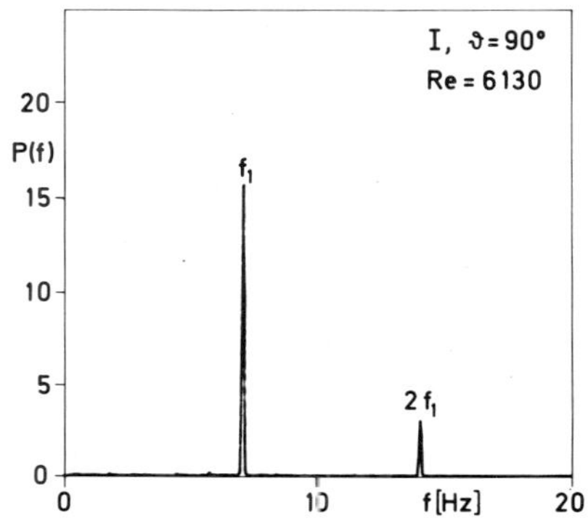

Fig. 6: Power spectra of the local wall shear stress of Mode I

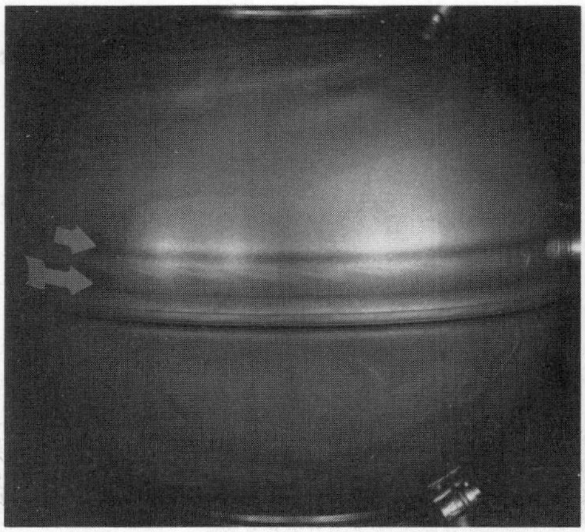

Fig. 7: Mode III, Taylor and Stuart vortices, Re = 7800

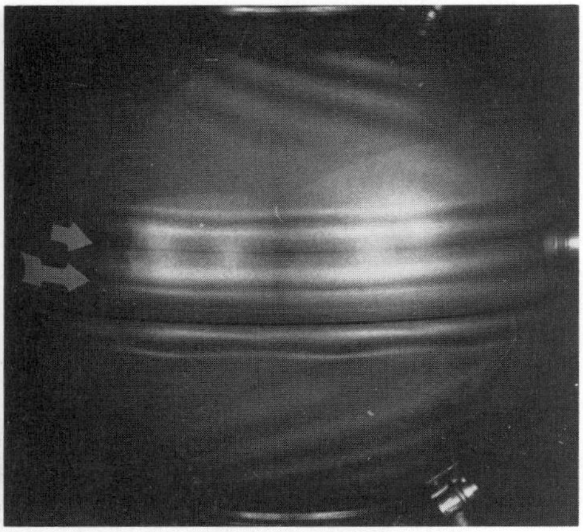

Fig. 9: Mode IV, Taylor vortices and shear instabilities, Re = 5550

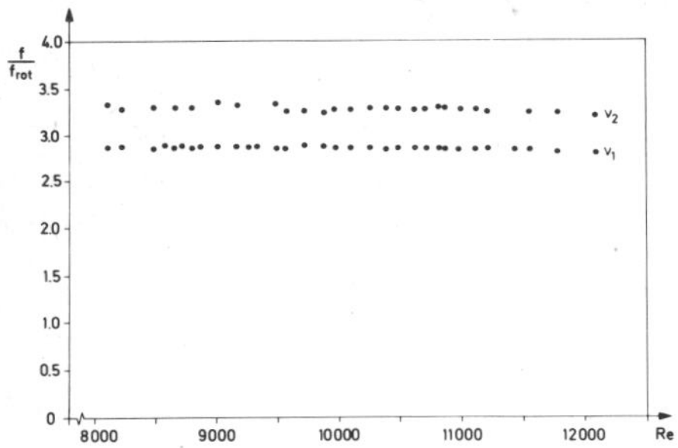

Fig. 8: Frequencies of Mode III

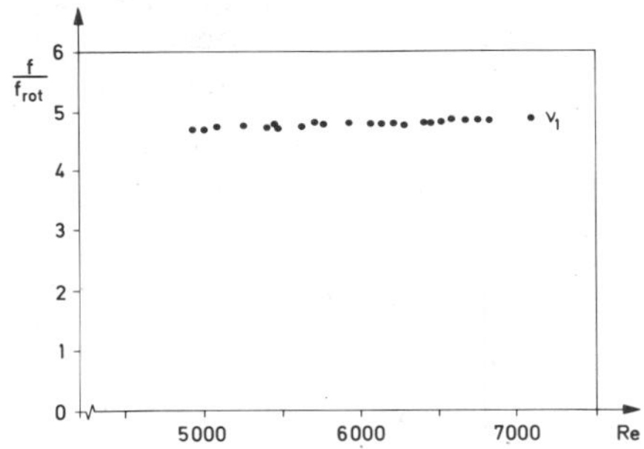

Fig. 10: Frequencies of Mode IV

secondary instability is established within the Taylor vortices with the frequency f_2. With increasing Re the shear waves with frequency f_1 appear. As soon as these waves reach the vortices these get wavy. The coupling of these frequencies causes the existence of $f_2 - f_1$ in the power spectra. These two frequencies are nearly independent of Re as shown in fig. 8. Near the poles the flow is turbulent for $Re > 9000$. Near the equator however the frequencies can be registered up to the transition to mode I, which appears in the region $11000 < Re < 14000$.

<u>Mode IV:</u> Fig. 9 shows this mode with four Taylor vortices and the spiral shear waves. The time dependent wave instability exists for $Re > 4900$. The Re-number dependence of the corresponding characteristic frequency f_1 is given in Fig. 10. After the occurence of Stuart vortices near the poles,

the interaction of these vortices and the shear waves leads to stochastic motion. If this motion reaches the Taylor vortices for $Re > 6700$, this mode changes to mode III.

Direct transition to turbulence

The case of a sudden start of the inner sphere to a supercritical $Re_E > 8000$ leads to a direct transition to turbulent flow [13]. Fig. 11 shows the time dependent motion. Turbulence is first created at the equator and then spreads towards the poles. A simple description can be given by applying the analogy to the Rayleigh-Stokes plate problem. By the sudden acceleration of the inner sphere a boundary layer is developed growing with time in radial direction. After a characteristic time t_1 this boundary layer gets unstable. The velocity distribution changes to that of the turbulent Couette flow. The corresponding velocity profiles before and after the onset of turbulence are sketched in fig. 12. The characteristic time t_1 depends on the final Re-number. We get the relation $t_1 \sim 1/\omega^r$ with $n = 4/3$ by fractional force analysis and $n = 1$ by energy considerations. For short time the coupling between primary and secondary motion is only due to centrifugal forces. Therefore in the early stages the secondary motion is of less importance. The transition to turbulence is performed in the meridional plane with time relaxation because the circumferential velocity decreases with ϑ^*

$$v(R_1, \vartheta^*) = R_1 \omega \cos\vartheta^* \, .$$

The development of the boundary layer in radial directions is given in a good approximation by

$$v(r^*, \vartheta^*, t) = v(R_1, \vartheta^*) \cdot \operatorname{erfc} \frac{r^*}{2\sqrt{\nu\, t}} \, .$$

r^* is the coordinate in the gap and erfc the complementary error function. For short times $t \to t_1$ we get the following asymptotic solution of the spreading of the turbulence front

$$\cos\vartheta^* \cdot \sqrt{\tau} = 1 \quad , \quad \tau = \frac{t}{t_1} \, .$$

Experimental results are given in fig. 13 for final Re-numbers $8 \cdot 10^3 < Re_E < 4 \cdot 10^4$ and compared with the analytical solution.

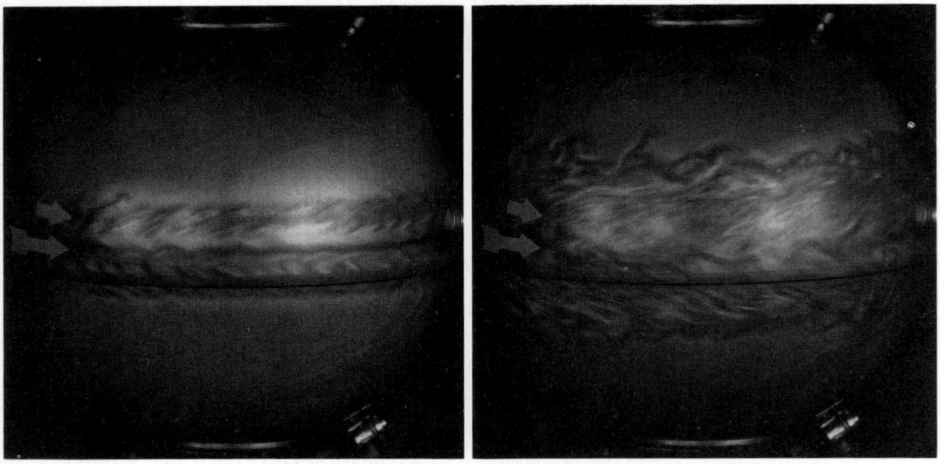

Fig. 11: Time dependent growth of the turbulent region after the sudden-start of the inner sphere , $Re_E = 2.6 \cdot 10^4$

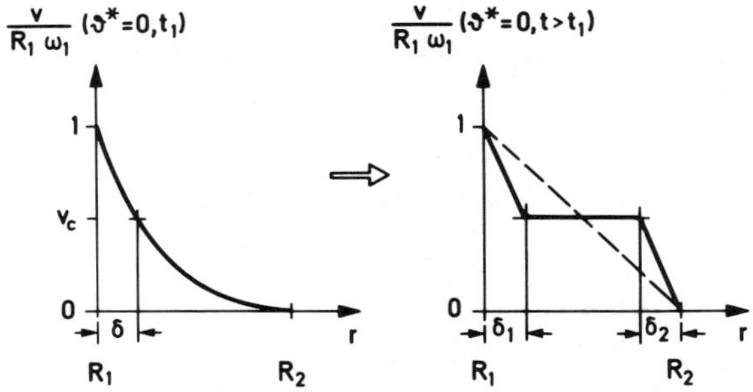

Fig. 12: Velocity profiles before and after the onset of turbulence

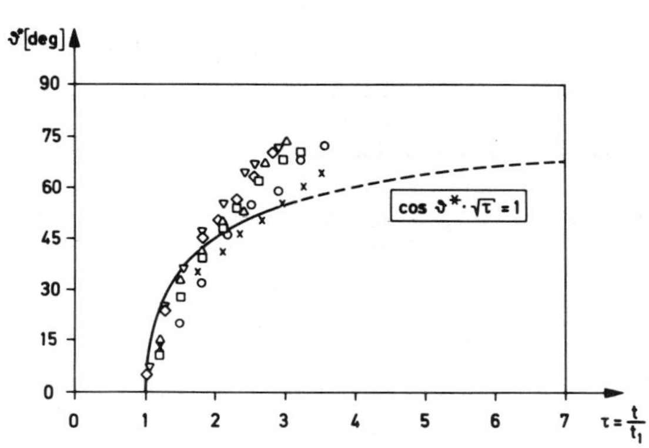

Fig. 13: Comparison of theoretical and experimental results

Conclusion

We described different routes of laminar-turbulent transition in a spherical gap. The time dependent secondary shear instabilities are important for that transition. A comparison with investigations of Yavorskaya et al. [10] for the large gap size shows an important influence of the gap width on the dynamical behaviour during the laminar turbulent transition.

References

[1] Swinney, H.L.; Gollub, J.P.: Hydrodynamic Instabilities and the Transition to Turbulence. Top. in Appl. Phys. 45, Berlin 1981.

[2] Barenblatt, G.I.; Ioss, G., Joseph, D.D.: Nonlinear Dynamics and Turbulence. Boston 1983.

[3] Eppler, R.; Fasel, H.: Laminar-Turbulent Transition. Berlin 1980.

[4] Fenstermacher, P.R.; Swinney, H.L.; Gollub, J.P.: Dynamical instabilities and the transition to chaotic Taylor vortex flow. J.F.M. 94, 103-128 (1979).

[5] Krugljak, Z.B. et al.: Laminar-Turbulent Transition in Circular Couette Flow. IUTAM Symp. Laminar-Turbulent Trans., Berlin, 378-387 (1980).

[6] Belayev, Yu.N.; Yavorskaya, I.M.: Transition to Stochasticity of Viscous Flow between Rotating Spheres. Nonlin. Dyn. and Turb., ed: Barenblatt, G.I.; Ioss, G.; Joseph, D.D., Boston, 51-70 (1983).

[7] Schultz-Grunow, F.: Sudden transition to turbulence. IUTAM Symp. Laminar-Turbulent Trans., Berlin, 388-395 (1980).

[8] Sawatzki, O.; Zierep, J.: Das Stromfeld im Spalt zwischen zwei konzentrischen Kugelflächen, von denen die innere rotiert. Acta Mech. 9, 13-35 (1970).

[9] Wimmer, M.: Experiments on a viscous fluid flow between concentric rotating spheres. J.F.M. 78, 317-335 (1976).

[10] Yavorskaya, I.M. et al.: Stability non-uniqueness and transition to turbulence in the flow between two rotating spheres. Proc. XVth Int. Congr. of Theor. and Appl. Mech., 431-443 (1980).

[11] Greenspan, H.P.: The theory of rotating fluids. Cambridge, 1968.

[12] Poll, D.I.A.: Three-Dimensional Boundary Layer Transition via the Mechanisms of 'Attachment Line Contamination' and 'Cross Flow Instability'. IUTAM Symp. Laminar-Turb. Trans., Berlin, 253-262, (1980).

[13] Bühler, K.; Zierep, J.: Transition to turbulence in a spherical gap. Proc. 4th Int. Symp. on Turb. Shear Flows, 16.1-16.6 (1983).

Stability of Unsteady Parallel Flows by the Method of Energy

R. CAYZAC, J. FAUCHAS and C. CLARION

Institut de Mécanique des Fluides
de l'Université d'Aix-Marseille II, France

Summary

The energy method is applied to the stability of two unsteady parallel flows: the free oscillatory flow and the time periodic flow. The pipe Poiseuille flow (steady flow) is also considered. Energy method gives bounds which are smaller than experimental corresponding values, but the qualitative agreement is good.

Let a basic steady flow of an incompressible viscous fluid in a bounded region or in unbounded region such as channels or pipes, the energy method gives sufficient conditions for stability with respect to arbitrary disturbances. There exists the critical value R_c of the similarity parameter R, such that the flow is stable with $R < R_c$.

The energy method is applicable to unsteady flows. The first authors to have applied this method to the stability of unsteady flows are Conrad and Criminale [1], (1965). One of the difficulties of the energy method, in the case of unsteady flows, is the choice of the criterion of stability.

In the present work we consider parallel free oscillatory flow and parallel time periodic flow in pipes and we argue that the criterion of stability is $\frac{DE}{Dt} < 0$, where E is the global kinetic energy of all disturbances within the domain of the flow.

Free oscillatory flow

We consider the basic motion in a U shaped tube (whose rectilinear vertical sides have a circular cross-section with a radius a) of a fluid column with a length ℓ displaced by a height (h_o) from its position of equilibrium, then abruptly started from rest at time $t = 0$. We choose cylindrical coordinates r, θ, z (with z along the axis of the tube). Two parameters are necessary to determine the structure of the flow, there are the frequency parameter $V = a^2 \omega_p/\nu$ and the amplitude parameter h_o/a, where ν is the coefficient of kinematic viscosity and $\omega_p = (2g/\ell)^{1/2}$ the undamped angular frequency.

We use as the instantaneous velocity profiles of the basic flow, assumed to be a parallel flow ($a << \ell$), those given by Clarion and Pélissier [2]

Let two-dimensional disturbances caracterised by the velocity of components $u\ (r,z,t)$ and $w\ (r,z,t)$. Let
$u = u^x\ (r)\ \exp\ (ik^x z + i\ \beta^x t)$, $v = v^x(r)\ \exp\ (i\ k^x z + i\ \beta^x t)$.

For a given value of V and a given value of the dimensionless wave-number k ($k = k^x/a$), we determine the instantaneous critical value of h_o/a, given by the resolution of a system of Euler-Lagrange by the calculus of variations. The minimum value of these instantaneous values, as time progresses is termed $(h_o/a)_x$, it always occurs near the beginning of the reverse flow. Figure 1 shows the evolution of the instantaneous critical value of h_o/a as dimensionless time $\omega_p t$ increases, for a given value of V, here $V = 260$, and a given value of k, here $k = 4$. For a given value of V the minimum value of the $(h_o/a)_x$ corresponding to different values of k is termed $(h_o/a)_c$; $(h_o/a)_c$ is a function of V ; this dependence appears to be well represented by :

$$V < 25 \qquad (h_o/a)_c = 1600\ V^{-1.7}$$
$$V > 60 \qquad (h_o/a)_c = 25\ V^{-0.5} \tag{1}$$

Experimental values of $(h_o/a)_c$, (for values of h_o/a slightly higher than $(h_o/a)_c$, disturbances are visible in the flow), determined by Clarion and Pélissier [2], are

$V < 28$: the flow is laminar within the (h_o/a) range investigated.

$$V > 46 : (h_o/a)_c = 176\ V^{-0.6}. \tag{2}$$

The logarithmic representation of $(h_o/a)_c$, as a function of V, corresponding to the expressions 1 and 2 is given in figure 2.

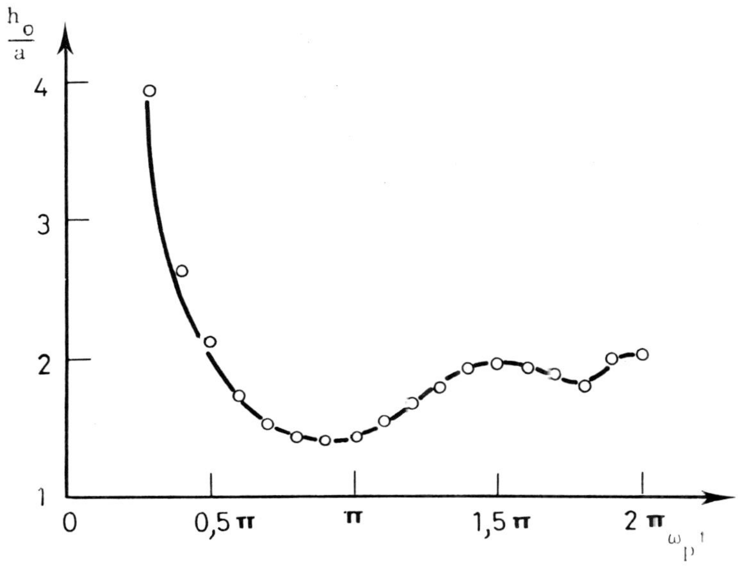

Fig.1. Evolution of h_o/a as a function of $\omega_p t$ V = 260 k = 4

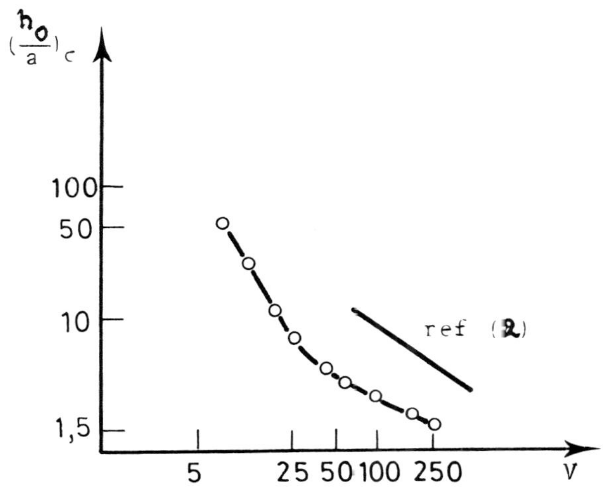

Fig.2. $(h_o/a)_c$ as a function of V. O Present theory.

Let us choose the frequency parameter V and the Reynolds number $R_\delta = \delta W/\nu$, where $W = \omega_p h_o$ and $\delta = (2\nu/\omega_p)^{1/2}$, to determine the structure of the flow.

The variation of R_{δ_c} as a function of V, where R_{δ_c} is the critical value of R_δ is given by :

$$V < 25 \qquad R_{\delta_c} = 2260\ V^{-1.2}$$

$$V > 60 \qquad R_{\delta_c} = 36$$

From these results if follows that R_δ appears as a relevant parameter of the flow only for V > 60, that is when the velocity profiles of the basic flow are boundary layer profiles. For V < 25, that is when the velocity profiles of the basic flow are fully developed profiles, R_{δ_c} is a function of V.

Time periodic flow

We consider the time periodic flow in a pipe with a length ℓ and a circular cross section with a radius a.

Let C cos Ωt (where C and Ω are given real numbers) the oscillatory pressure gradient. We choose the frequency parameter $V = a^2\Omega/\nu$ and the amplitude parameter $X = C/a\ \Omega^2$, to determine the structure of the flow.

The laminar instantaneous velocity profiles of the basic flow, supposed to be a parallel flow ($a \ll \ell$) are given by Sexl [3].

X_c, where X_c is the critical value of X, is a function of V ; this dependence appears to be well represented by :

$$\begin{array}{lll} V < 25 & X_c = 260\ V^{-1.2} & \text{and} \quad R_{\delta_c} = 370\ V^{-0.7} \\ V > 25 & X_c = 36\ V^{-0.5} & \text{and} \quad R_{\delta_c} = 50 \end{array} \qquad (3)$$

where $R_\delta = W\delta/\nu$, $\delta = (2\nu/\Omega)^{1/2}$, $W = C/\Omega$.

The logarithmic representation of X_c as a function of V, corresponding to the expressions 3 is given in figure 3.

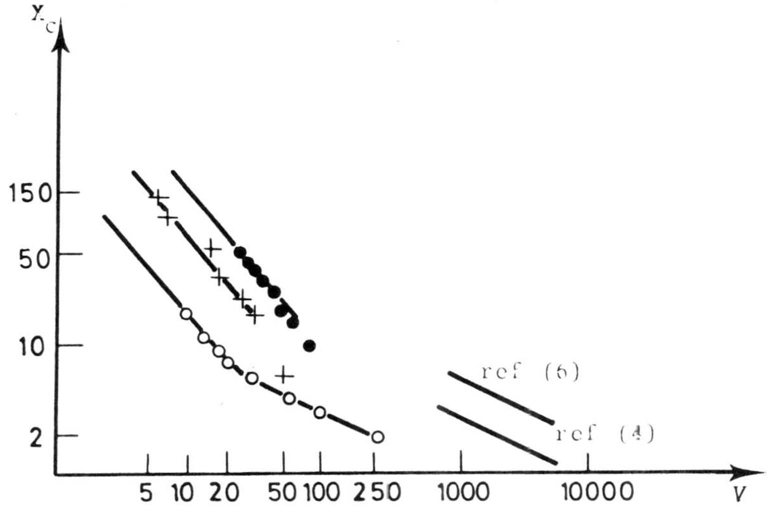

Fig.3. X_c as a function of V. O Present theory ; + ref (5) ; ● ref (7).

These theoretical results are in qualitative agreement with the corresponding experimental values given by Merkli and Thomann [4], Hino and others [5] , Tromans [6] and Le Feyer [7].

The experiments of Sergeev [8] and those of references [4] and [6], correspond to : $20 < V < 600$, ref. [8] , $900 < V < 2500$, ref.[4]and unknown domain of variation of V, ref. [6]. In these experiments the value of R_{δ_c} is unique whatever the value of V ($R_{\delta_c} = 500$, ref. [8], $R_{\delta_c} = 290$, ref. [4], $R_{\delta_c} = 130$, ref. [6]).

In the experiments of ref.[5] and [7] , where $V < 50$, R_{δ_c} is a function of V, in qualitative good agreement with the value given by 3, up to $V \sim 30$.

Pipe Poiseuille flow

In the case of a steady flow, that is the pipe Poiseuille

flow, the energy method with the choosen criterion, gives

$$R_c = 750$$

where R_c is the critical value of the Reynolds number R, based on pipe radius and center-line velocity.

The most dangerous wave number is :

$$k_c \sim 5 \qquad \text{ie} \qquad k_c \sim 0.18\ R_c^{1/2}$$

Conclusion

Energy method give bounds which are very smaller than experimental corresponding values, but the qualitative agreement is good.

References

1. Conrad, P.W. ; Criminale, W.O. : The stability of time - dependent laminar flow : parallel flows. Z. Angew. Math. Phys. 16, 1965, 233.

2. Clarion, C. and Pélissier, R. : A theoretical and experimental study of the velocity distribution and transition to turbulence in free oscillatory flow. J. Fluid Mech. 70, 1975, 59.

3. Sexl, T. : Uber den Von E.G. Richardson entdeckten annuhareffect. Z. Phys. 61, 1930, 349.

4. Merkli, P. ; Thomann, H. : Transition to turbulence in oscillating pipe flow. J. Fluid Mech. 58, 1975, 567.

5. Hino, M. ; Sawamoto, H. ; Takasu, S. : Experiments on transition to turbulence in oscillatory flow. J. Fluid Mech. 75, 1976, 193.

6. Tromans, P.S. : The stability of oscillating pipe flow. Euromech 73, April 1976.

7. Lefeyer, P. : Thesis University Marseille, 1977.

8. Sergeev, S.I. : Fluid oscillations in pipes at moderates Re numbers. Fluid Dyn. 1, 1966, 121.

On a Multigrid Method for the Numerical Solution of Navier-Stokes Equations

W. ZWICK

Institut für Mechanik
Akademie der Wissenschaften der DDR, Berlin

1. Introduction

The multigrid methods (MGM) [1,2] were developed for the numerical solution of linear and nonlinear, ordinary and partial differential equations and integral equations by approximations using a large number of grid-points. These methods represent a compromise between the large-scale store requirement in direct solution methods and the large computer time expense in usual iteration methods.

In order to develop stable solution methods - with little approximation errors - of Navier-Stokes equations (NSE) for laminar-turbulent transition and for high Reynolds numbers (Re), we have to choose small mesh spacings and a large number of grid points [3]. In these cases explicit and semi-implicit solution methods [4, 5, 6] need large computer time expense and fully implicit methods [7] need large-scale store capacity. Therefore in this paper we describe a MGM for solving discretized two-dimensional NSE and report on collected experience for high Re numbers.

2. Formulation of the problem

Let Ω denote a domain in the two-dimensional space, Γ the boundary of the domain Ω, $x=(x_1,x_2)\in\Omega\cup\Gamma$ the space coordinate, t the time coordinate, $v=(v_1,v_2)$ the velocity, and Ψ the stream function where $v = \mathrm{rot}\,\Psi$. We consider the NSE in the following form

$$\frac{\partial\Delta\Psi}{\partial t} - \eta\Delta\Delta\Psi + v_l D_l \Delta\Psi = g$$

where $\eta = 1/Re$, $D_l = \partial/\partial x_l$, Δ the Laplace operator and g a

known function. We have to sum up from 1 to 2 respectively to lower indices occurring twice. The time discretization is obtained by applying finite-difference methods: $t_n (n=1,\dots,N)$ denotes discrete time points, at any time point t_n the time derivative $\partial/\partial t$ is approximated by $\partial w/\partial t = \lambda(w-\check{w})$ where $\lambda = (t_n - t_{n-1})^{-1}$, $w=w(t_n)$ and $\check{w} = w(t_{n-1})$.
If we use special initial and boundary conditions we have to solve N boundary value problems

$$\lambda \Delta \Psi - \eta \Delta\Delta \Psi + v_\ell D_\ell \Delta \Psi = g + \lambda \Delta \check{\Psi}$$

where $\Psi = \Psi(x,t_n)$, $v = \text{rot}\,\Psi$, $g=g(x,t_n)$, $\check{\Psi} = \Psi(x,t_{n-1})$, and $n = 1,\dots,N$.

3. Approximation of boundary value problems

For solving these boundary value problems we apply a special MGM, i.e. we consider a sequence of M grids $\Omega^m \subset \Omega$ and $\Gamma^m \subset \Gamma$ $(m=1,\dots,M)$, with mesh spacings $h^m = (h_1^m, h_2^m)$ where $h^{m+1} = h^m/2$. At the grid-points $x_{i,j}^m \in \Omega^m$ the equations are approximated by finite difference methods [8,9,10] whose scheme is consistent and of second-order accuracy:

$$\lambda \Psi_{\bar{x}_k x_k} - \eta \Psi_{\bar{x}_k x_k \bar{x}_\ell x_\ell} + \overset{+}{v}_\ell \Psi_{\bar{x}_k x_k x_\ell} + \bar{v}_\ell \Psi_{\bar{x}_k x_k \bar{x}_\ell} = g + \lambda \check{\Psi}_{\bar{x}_k x_k}$$

where $\overset{+}{v}_\ell = 0.25\,(\overset{+\ell}{v}_\ell + v_\ell)$, $\bar{v}_\ell = 0.25\,(\overset{-\ell}{v}_\ell + v_\ell)$, $v = (\Psi_{\overset{\circ}{x}_2}, -\Psi_{\overset{\circ}{x}_1})$, $\overset{\pm 1}{w} = w(x_{i\pm 1,j})$, $\overset{\pm 2}{w} = w(x_{i,j\pm 1})$, and $w = w(x_{i,j})$ any grid function.
For solving these nonlinear algebraic equations and calculation of the values $\Psi_n^m = \Psi(x_{i,j}^m, t_n)$ we have to adapt the boundary conditions to these grids and equations and to formulate initial values for the iteration. If, besides Ψ, the normal derivative of Ψ occurs in the boundary conditions, we add to the grids $\Omega^m \cup \Gamma^m$ further grid points which are placed one grid-step outward. We bring the known values on the right hand side of the equations and obtain on the grids Ω^m equations which we can describe shortly in the form

$$A_n^m(\Psi)\, \Psi_n^m = G_n^m$$

where $n=1,\dots,N$; $m=1,\dots,M$ and the matrices $A_n^m(\Psi)$ depend on Ψ_n^m.

4. Multigrid methods

The MGM are characterized by combinations of a smoothing process or iteration steps on finer grids and more exact solution methods on coarser grids. For the application of the MGM we need an approximate solution $\bar{\Psi}_n^M$ and initial values $\tilde{\Psi}_n^1$ for the iteration process. At the first iteration step on the new time level $t=t_n$ we can begin with

$$\bar{\Psi}_n^M = \Psi_{n-1}^M \quad \text{and} \quad \tilde{\Psi}_n^1 = I_M^1 \Psi_{n-1}^M$$

where Ψ_0^M can be calculated from the initial values of the problem, and $I_{m_1}^{m_2}$ denotes the identity, the prolongation-operator or the restriction-operator from Ω^{m_1} to Ω^{m_2} if $m_2 = m_1$, $m_2 > m_1$, or $m_2 < m_1$ respectively. With the approximate solution $\bar{\Psi}_n^M$ we calculate the defects

$$D_n^M = C_n^M - A_n^M(\bar{\Psi})\ \bar{\Psi}_n^M \quad \text{on } \Omega^M .$$

By restriction-steps we obtain the projected defects

$$D_n^m = I_M^m D_n^M \quad \text{on } \Omega^m .$$

We improve the approximate solution $\bar{\Psi}_n^M$ by the solution of, or smoothing processes for, the following modified equations [2]

$$A_n^m(\Psi)\ \Psi_n^m = H_n^m$$

where $m=1,\dots,M$, $H_n^m = A_n^m(\tilde{\Psi})\ \tilde{\Psi}_n^m + D_n^m$ and

$$\tilde{\Psi}_n^m = I_M^m\ \bar{\Psi}_n^M + I_{m-1}^m\left(\Psi_n^{m-1} - I_M^{m-1}\ \bar{\Psi}_n^M\right) \quad \text{if } m=2,\dots,M.$$

From the solution Ψ_n^M we obtain a new approximate solution $\bar{\Psi}_n^M$ and a new initial value $\tilde{\Psi}_n^1$ for the iteration by

$$\bar{\Psi}_n^M = \Psi_n^M \quad \text{and} \quad \tilde{\Psi}_n^1 = I_M^1 \Psi_n^M$$

At every time point t_n, we can apply this multigrid step different times, and we can start this calculation on the coarsest grid as well as on the finest grid [1]. For small values t_n we can also use the one-grid methods and for large values t_n the M-grid methods with $M > 1$.

5. Solution method

We choose the coarsest grid Ω^1 so that we can solve directly the linear algebraic equation system

$$A_n^1(\tilde{\Psi})\Psi_n^1 = H_n^1$$

by a LU-matrix factorisation [11]. On the finer grids Ω^m $(m>1)$ the number of the equations is very large and we tested the Jacobi iteration, the Gauss-Seidel iteration and the incomplete LU-decomposition [1] with damping factors ω in all 3 cases; i.e, if $m>1$ we have considered equations of the form

$$B_n^m \Psi_n^{m\mu+1} = B_n^m \Psi_n^{m\mu} + \omega(H_n^m - A_n^m(\Psi^\mu)\Psi_n^{m\mu})$$

where $\mu=0,1,2,\ldots$ and $\Psi_n^{m\,0} = \tilde{\Psi}_n^m$. In case of the Jacobi iteration, B_n^m is the identical matrix, in case of the Gauss-Seidel iteration, B_n^m is a special triangular matrix of $A_n^m(\Psi^\mu)$, and in case of incomplete LU-decomposition, B_n^m is a product of lower and upper triangular matrices which approximates, $A_n^m(\Psi^\mu)$ and in which the factors are sparse matrices of similar pattern as $A_n^m(\Psi^\mu)$.

We can improve the stability of the solution method if we apply an automatic control to the damping factor [12], i.e. we change ω as long as the new defect is less than the old one.

6. Results of the calculation

We considered a cavity flow in a rectangular domain with a tangential velocity of value 1 on one side, and a Reynolds number $Re = 5\cdot 10^3$ [7]. We chose three grids (M=3) with the numbers of grid points 19x19, 35x35 and 67x67. We simulated a start problem in which we applied for any time point t_n $(n=1,\ldots,N)$ this three-grid method. If ω is fixed close to one this three-grid method is very unstable for large times t_n. Because of the transition from one grid to another increasing pertubations occur. If we applied the automatic control to ω we obtained small values ω and the difference between Ψ_n^m and Ψ_n^1 at the points of Ω^1 is very small, i.e. in this case the results obtained by the multigrid method are close to the results obtained by the one-grid method [7].

References

1. Hackbusch,W.; Introduction to Multi-grid Methods for the Numerical Solution of Boundary Value Problems, Excerpt from J.A.Essers (ed.): Computational methods for turbulent, transonic, and viscous flows. Proceedings of a lecture series, March 30 to April 3, 1981, von Karman Institute for Fluid Dynamics, Rhode Saint Genese, Belgium, Hemisphere Publishing Corporation, Washington, 1983.

2. Houwen,P.J. van der; Sommeijer,B.P.; Analysis of Chebyshev Relaxation in Multigrid Methods for Nonlinear Parabolic Differential Equations, ZAMM 63 (1983) 193-201.

3. Schönauer,W.; Die Lösung der dreidimensionalen Navier-Stokes-Gleichungen mit dem Differenzenverfahren, Deutsche Luft- und Raumfahrt, Forschungsbericht 70- 15, 1970.

4. Amsden,A.A.; Harlow,F.H.; The SMAC method, a numerical technique for calculating incompressible flow. Los Alamos Scientific Laboratory, Report LA 4370, 1970.

5. Temam, R.; Navier-Stokes equations, North-Holland Publishing Company, Amsterdam/NewYork/Oxford, 1977.

6. Perktold,K.; Lösung der instationären Navier-Stokes-Gleichungen mittels der Methode der finiten Elemente, Forschungszentrum Graz, Bericht Nr. 133, 1980.

7. Zwick,W.; Zur Lösung der Navier-Stokes-Gleichung mit Hilfe eines vollimpliziten Iterationsverfahrens, ZAMM 64(1984) 221-226.

8. Ladizhenskaya,O.A.: Matematicheskie voprosi dinamiki viazkoi neszhimaemoi zhidkosti. Nauka, Moskva, 1970.

9. Samarskiy,A.A.; Nikolaev, E.S,: Metodi reshenia setoch nikh uravneniy, Nauka, Moskva, 1978.

10. Peyret,R.; Taylor,T.D.; Computational Methods for Fluid flow. Springer-Verlag New York Heidelberg Berlin, 1983.

11. Eisenstat,S.C.; Gursky,M.C.; Schultz,M.M.; Sherman,A.H.; Yale Sparse Matrix Package II. The Nonsymmetric Codes, Research Rep. 114, Yale University, Department of Computer Sciences.

12. Nadeborn,W.; Realisierung eines Aufspaltungsverfahrens mit dynamischer Iterationsparametersteuerung.
Matematicheskie modeli v teorii teplo- i massoobmena, materiali mezhdunarodnoi shkoli-seminara, Akademia nauk BSSR, Institut teplo-massoobmena, Minsk, 1982.

Transition of Boundary Layer on Rotating Flat Plate

S. MASUDA*, K. OKAMAE** and I. ARIGA*

* Department of Mechanical Engineering
Keio University, Yokohama, Japan
** Mitsubishi Atomic Power Industries, Inc., Tokyo, Japan

Summary

This paper reports the results of an experimental investigation on transition of a boundary layer on a rotating flat plate in order to reveal the stabilizing and destabilizing effects of Coriolis acceleration and to obtain the quantitative expression of transition Reynolds number as a function of rotation number. Significant increase of transition Reynolds number of stabilized layer is found, which is more remarkable when the free stream turbulence level is low. Transition of stabilized layer begins with intermittent fluctuations similar to that of a stationary boundary layer but with clearer distinction of quiescent and fluctuating periods. Transition of destabilized layer is characterized by the gradual increase of frequency of velocity fluctuation.

Introduction

Non-conservative Coriolis force acts on free and bounded shear flows in a rotating reference frame. Depending on the relative magnitude and direction of absolute vorticity vector and rotation vector, a growth of disturbances superposed on the base flow are promoted or suppressed, and in consequence the laminar-turbulent transition may be significantly influenced.

The purpose of the present investigation is to determine the transition Reynolds number of a rotating flat plate boundary layer as shown in Fig.1 (stabilized layer only is illustrated as an example) and to compare the transition process with that of a stationary boundary layer. In this case, the Coriolis force acts in the mean flow plane and perpendicular to the mean velocity, and thus the transition can be regarded as a growth of two- and three-dimensional disturbances superposed on a two-dimensional base flow under the action of Coriolis force.

Chawla[1] performed the linear stability calculation and obtained

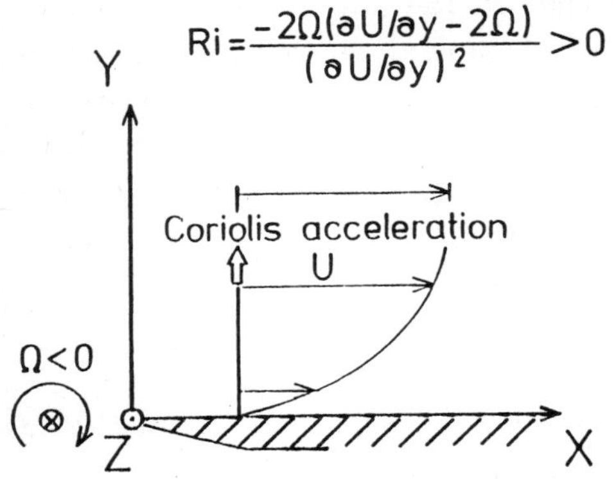

Fig.1. Boundary layer on rotating flat plate (stabilized layer)

the neutral stability curves with rotation number and spanwise wave number as the parameters. Johnston[2] made the investigation in the opposite direction, that is, on the final stage of transition, in which he tried to find the conditions for self-maintenance of fully turbulent state. Although the way lying between these two investigations is very long and non-linearity refuses the theoretical approach, it absorbs much of the engineering interest to answer the question; when and how a rotating boundary layer loses its laminar character?

Experimental Apparatus and Method

The experiments were carried out in the rotating channel experimental facility at Department of Mechanical Engineering, Keio University, whose details are given in reference[3]. A flat plate with a sharp leading edge and a hydraulically smooth surface was mounted in the midplane of the test channel of 86mm x 390mm cross section. The test channel was placed on the turn table of 2m diameter, whose maximum rotational speed was 300rpm. The air was supplied from a blower placed in a stationary frame to the rotating passage via a mechanical seal. A nylon filter, wire screens, honeycombs and a nozzle of 5 to 1 contraction ratio were installed upstream of the test section in order to minimize the flow deflection and disturbances. The maximum Reynolds number based on the downstream distance x and the free stream velocity Uc were 5×10^5. The streamwise pressure gradient was eliminated by adjusting the diverging angle of the wall opposite to the test surface. The flow rate ratio of both sides of the test plate was carefully controlled by installing the resistance screens at the exit nozzle in order to prevent the leading edge separation, whose occurrence was checked by observing the vaporization of camphor oil applied to the test surface.

A single hot wire probe was mounted on the test wall and mean velocity and its fluctuations were observed by employing a constant temperature anemometry whose control unit was mounted on the turn table and the signal was transmitted to the stationary measuring system via a silver-carbon slip ring. The measurement of a wall shear stress was made by means of the wall hot wire method.

Transition of Stabilized Layer

After confirming that the stationary boundary layer is a typical Blasius layer, the plate is rotated in the direction shown in Fig.1. In this case, the Coriolis acceleration stabilizes the boundary layer and the transition Reynolds number is expected to increase. In Fig.2, variation of the skin friction coefficient Cf with Reynolds number Rex for five different rotation numbers Rox are given. They are defined as $Cf=\tau_w/(\rho U_c^2/2)$, $Rex=U_c x/\nu$ and $Rox=\Omega x/U_c$, where τ_w is a wall shear stress, U_c is a free stream velocity, x is a distance from the leading edge and Ω is the angular velocity of system rotation. In low and high Reynolds number regions, the measured points approximately fall in the laminar and turbulent lines respectively, which are the same as stationary cases. For the intermediate Reynolds numbers, Cf increases from laminar to turbulent values. This region of increasing Cf is shifted to higher Reynolds

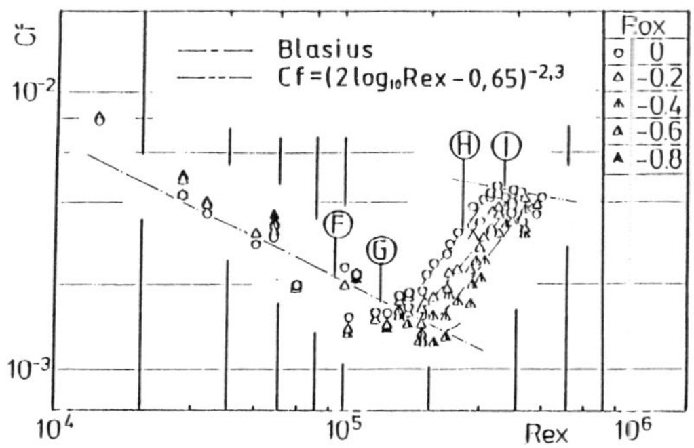

Fig.2. Skin friction coefficient of stabilized boundary layer

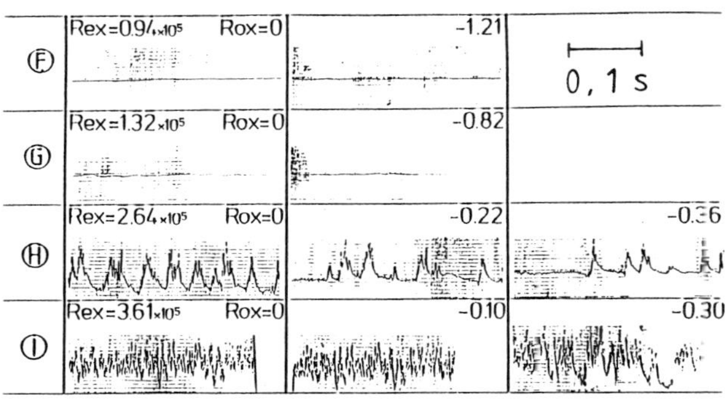

Fig.3. Velocity fluctuation in stabilized boundary layer (y=0.07mm)

numbers as Rox increases.

Fig.3 shows the velocity fluctuation close to the wall(y=0.07mm). For Rex=2.64 x10^5, which lies in the transition region of the stationary layer, intermittent nature becomes more and more obvious with increasing Rox, the fraction of the quiescent period increases and the distinction between the no-fluctuating and fluctuating parts of the oscillogram becomes sharp.

Transition of Destabilized Layer

When the boundary layer is rotated in the direction opposite to that shown in Fig.1, the Coriolis acceleration acts to promote the growth of disturbances and in consequence the transition Reynolds number is expected to decrease. Fig.4 shows the change of skin friction coefficient with Reynolds number and positive rotation number. For high Reynolds number, the asymptotic approach to the turbulent line is again obvious. For low Reynolds number region, the trend is quite different from that for stabilized layers. Very small positive rotation number is sufficient to increase Cf from the stationary laminar value even in very low Reynolds number range. For Rox $\geqq$ 0.8, all the plots of Cf align on the stationary turbulent line at least in the range of the present measurement.

Fig.5 is the oscillogram of velocity fluctuation at y=0.07mm of stationary and destabilized layers. In low Reynolds number range (Ⓙ and Ⓚ in Figs.4 and 5), in which the stationary boundary layer is completely laminar, the low frequency fluctuation gradually increases with increasing rotation number and the intermittency can not be found. Further increase of Rox does not bring the high frequency fluctuation, which is typically seen for the end of the ordinary transition. In spite of it, Cf attains the full turbulent values,

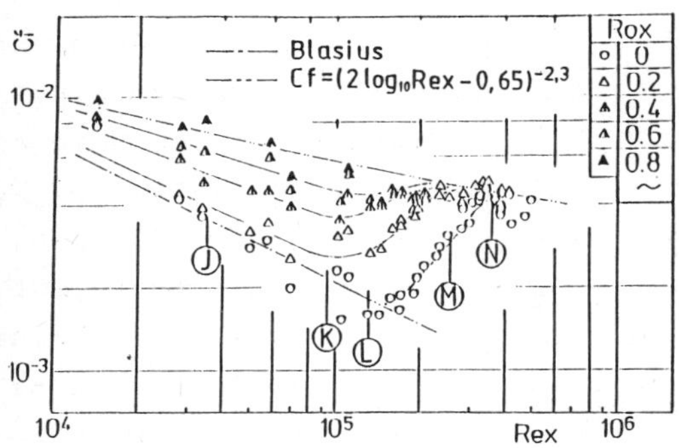

Fig.4. Skin friction coefficient of destabilized boundary layer

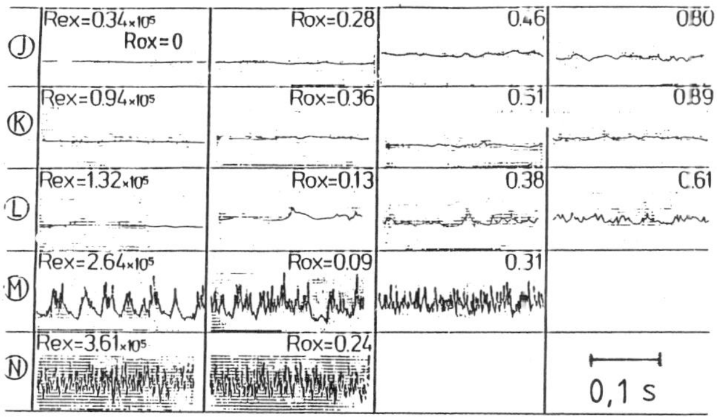

Fig.5. Velocity fluctuation in destabilized boundary layer (y=0.07mm)

that is a strange feature of low Reynolds number destabilized layer. Observation of fluctuation with constant rotation number with increasing Reynolds number reveals more clearly that transition of destabilized layer with high Rox is different from stationary and stabilized layer in that it proceeds with gradual increase of spectrum and does not involve a sudden appearance of high frequency fluctuations. The spanwise measurement of mean velocity showed no recognizable three-dimensionality.

Transition Reynolds Number

In the foregoing sections, we have paid no attention to the free stream turbulence intensity Tu. Since Tu has strong influence on the transition Reynolds number[4], the quantitative interpretation requires careful examination of free stream turbulence. In the course of the present experiment, it was found that Tu changes with Reynolds number and rotation number, for which more than 100% change of transition Reynolds number might be expected. So we have made an additional experiment, in which the turbulence intensity of the approaching flow was deliberately varied by employing the turbulence grids at the upstream cross section. By observing the first occurence of the intermittent fluctuation near the wall and the deviation of Cf from laminar value, the transition Reynolds number Rexl was determined and replotted as constant Tu lines for stabilized layer in Fig.6. It is obvious from this figure that the transition Reynolds number of the stabilized layer increases with increasing rotation number and this effect is much more significant when the free stream turbulence intensity is low. Foregoing results suggest the decrease of Rexl in the destabilized layer, but its quantitative discussion re-

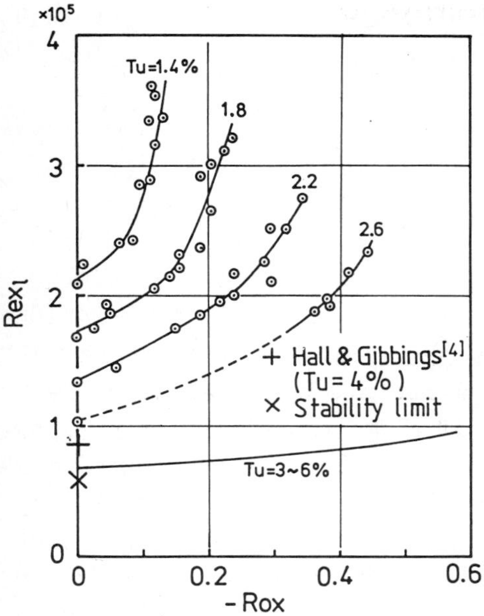

Fig.6. Transition Reynolds number of stabilized boundary layer

quires more detailed examination of the beginning of transition.

Conclusions

The transition Reynolds number of stabilized layer increases and this is more remarkable if the free stream turbulence intensity is low. Transition of stabilized layer begins with the occurence of intermittent fluctuation similar to that of the stationary layer but with clearer distinction between quiescent and fluctuating period. Transition of destabilized layer proceeds with the gradual development of low frequency fluctuation without intermittency. The skin friction coefficient of low Reynolds number destabilized layer can attain full turbulent value for large rotation number without sufficient growth of fluctuation. Further investigation is required for determination of transition Reynolds number of destabilized layer.

References

1. Chawla,M.: The stability of boundary layer flow subject to rotation. Ph.D.Dissertation. Michigan State Univ. 1969.
2. Johnston,J.P.: The suppression of shear layer turbulence in rotating systems. J.Fluids Engng.Vol.95.No.2(1973)229-236.
3. Koyama,H.;Masuda,S.;Ariga,I.;Watanabe,I.: Stabilizing and destabilizing effects of Coriolis force on two-dimensional laminar and turbulent boundary layers. Trans.ASME,Ser.A, Vol.101.No.1(1979)23-31.
4. Hall,D.J.;Gibbings,J.C.: Influence of stream turbulence and pressure gradient upon boundary layer transition.J.Mech.Engng. Sci.Vol.14.No.2(1972)134-146.

Quantitative Analysis of Intermittency Phenomenon in the Wall Region of a Turbulent Boundary Layer

Repik Ye.U., Sosedko Yu.P.

TsAGI, USSR

1. There is a number of peculiarities in the behaviour of statistical flow characteristics which are observed near the boundary of viscous sublayer $y=\delta_l$ of a turbulent boundary layer. It is seen in fig.1 that shows hot-wire measurements in a turbulent boundary layer on a flat plate at a freestream velocity $U_\infty=12$ m/s ($Re^{**}=\frac{U_\infty \delta^{**}}{\nu}=1600$, δ^{**} is the momentum thickness). Inside the viscous sublayer of the turbulent boundary layer ($y<\delta_l$) a skewness factor $A=\overline{u'^3}/(\overline{u'^2})^{3/2}$ of the probability density distribution of the longitudinal velocity fluctuation u' has positive values, and outside the sublayer it has negative ones. In addition, near the edge of viscous sublayer ($y\approx\delta_l$) flatness factor $E=\overline{u'^4}/(\overline{u'^2})^2$ takes a minimum value, while root-mean-square velocity fluctuations $\sigma=\sqrt{\overline{u'^2}}$, have a maximum. (Here the viscous sublayer boundary is defined as an intersection point of a power law velocity distribution in the turbulent core of the boundary layer with a linear velocity distribution near the wall).

It is shown in [1,2] that the above departures of statistical flow characteristics from gaussian ones near the viscous sublayer edge may be accounted for by a model according to which the flow in region $y\approx\delta_l$ is considered as consisting of two intermittent zones of decelerated and accelerated fluids (fig.2a). The mean flow velocity in these zones differs by some value $\Delta U=U_{acc}-U_{dec}$ (fig.2b). The mean boundary position between these zones corresponds to $y\approx\delta_l$, and time dependent instantaneous position of this boundary having an irregular shape gives rise to an intermittent flow behaviour in the wall region of the turbulent boundary layer.

Assuming that the probability density distribution of u' in

each of two zones is gaussian, then a superposition of these two distributions with shift ΔU between their centres and intermittency factor γ taken into account, permits to predict experimentally observed departures in A, E and σ distributions from gaussian ones in the wall region of the turbulent boundary layer. This assumption for a general case of intermittent flow is experimentally verified. This follows from [3,4,5] where an intermittent flow of turbulent and nonturbulent fluids near the external boundary of the turbulent boundary layer and in a wake behind the cylinder is investigated using the conditional sampling technique. In addition this assumption was also successfully applied to calculate intermittent flow characteristics in the outer region of the turbulent boundary layer [6].

Calculation examples of the u'-distribution function under the assumption of the gaussian distribution in intermittent zones of decelerated and accelerated fluids are given in [2] for the wall region of the turbulent boundary layer. However, mean velocities (U_{dec} and U_{acc}) and velocity fluctuations ($\sqrt{\overline{u'^2_{dec}}}$ and $\sqrt{\overline{u'^2_{acc}}}$) in each of these zones as well as intermittency factor γ have been specified arbitrarily. The latter is due to the impossibility of an experimental determination of real values of these parameters, since for the wall flow region unlike the flow in the external region of the boundary layer the possibility of using the conditional sampling technique is ruled out.

The aim of this paper is to outline a method for predicting U_{dec}, U_{acc}, σ_{dec}, σ_{acc} and γ in zones of decelerated and accelerated fluids using experimental values of A, E and σ under the assumption that the distribution of velocity fluctuation amplitudes in each of these zones is gaussian.

2. To define the form of the overall function of the u' - probability distribution in the intermittency zone in the wall region of the turbulent boundary layer we shall proceed from the scheme (fig.2c) of time history of the instantaneous velocity u(t) in this zone. Here, U is the averaged velocity, U_{dec}, u_{dec} are the mean and instantaneous velocities in the decelerated fluid zone, U_{acc}, u_{acc} are those in the accelerated fluid zone, and $\Delta U_{dec}=U-U_{dec}$; $\Delta U_{acc}=U_{acc}-U$; $\Delta U=\Delta U_{dec}+\Delta U_{acc}$. We shall determine the intermittency factor γ as a ratio time interval during which the process recorded only in the decelerated fluid

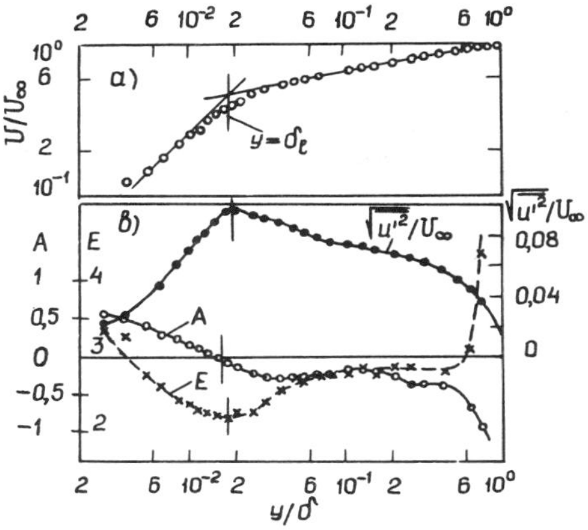

Fig.1.

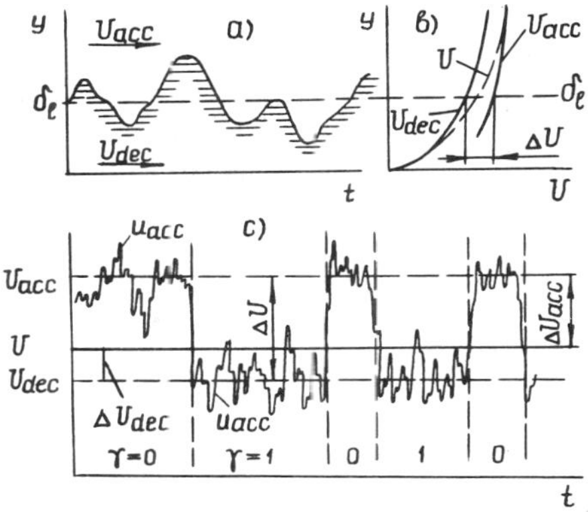

Fig.2.

zone to the whole time of the process recording.

According to this scheme U may be expressed as

$$U = \gamma U_{dec} + (1-\gamma) U_{acc} \tag{1}$$

and the root-mean-square velocity fluctuations u'=u-U may be represented in the form of

$$\sigma^2 = \overline{u'^2} = \gamma\overline{(u_{dec}-U)^2} + (1-\gamma)\overline{(u_{acc}-U)^2} = \gamma\sigma_{dec}^2 + (1-\gamma)\sigma_{acc}^2 + \gamma(1-\gamma)\Delta U^2 \tag{2}$$

with $\sigma_{dec}^2 = \overline{u_{dec}'^2}$; $u_{dec}' = u_{dec} - U_{dec}$; $\sigma_{acc}^2 = \overline{u_{acc}'^2}$; $u_{acc}' = u_{acc} - U_{acc}$.

The total distribution of the velocity fluctuation probability density in the intermittency zone assuming a gaussian distribution in each of the zones takes the following form:

$$f_\Sigma(u') = \gamma f(u'_{dec}) + (1-\gamma) f(u'_{acc}) = \\ = \frac{\gamma}{\sigma_{dec}\sqrt{2\pi}} \exp\left[-\frac{1}{2}\left(\frac{u'+\Delta U_{dec}}{\sigma_{dec}}\right)^2\right] + \frac{1-\gamma}{\sigma_{acc}\sqrt{2\pi}} \exp\left[-\frac{1}{2}\left(\frac{u'-\Delta U_{acc}}{\sigma_{acc}}\right)^2\right] \tag{3}$$

Having defined the moments of a higher order for f(u')

$$m_n = \int_{-\infty}^{+\infty} u'^n f_\Sigma(u')\,du', \qquad (n = 2,3,4)$$

and accounting for $A = m_3/m_2^{3/2}$; $E = m_4/m_2^2$; $m_2 = \sigma^2$ we obtain:

$$A = -\gamma(3\bar\sigma_{dec}^2\overline{\Delta U}_{dec} + \overline{\Delta U}_{dec}^3) + (1-\gamma)(3\bar\sigma_{acc}^2\overline{\Delta U}_{acc} + \overline{\Delta U}_{acc}^3) \tag{4}$$

$$E = \gamma(3\bar\sigma_{dec}^4 + 6\bar\sigma_{dec}^2\overline{\Delta U}_{dec}^2 + \overline{\Delta U}_{dec}^4) + (1-\gamma)(3\bar\sigma_{acc}^4 + 6\bar\sigma_{acc}^2\overline{\Delta U}_{acc}^2 + \overline{\Delta U}_{acc}^4) \tag{5}$$

with $\bar\sigma_i = \sigma_i/\sigma$; $\overline{\Delta U}_i = \Delta U_i/\sigma$.

Taking account of eq.(2), presented in a normalized form

$$1 = \gamma\bar\sigma_{dec}^2 + (1-\gamma)\bar\sigma_{acc}^2 + \gamma(1-\gamma)(\overline{\Delta U}_{dec} + \overline{\Delta U}_{acc})^2 \tag{6}$$

and that from eq.(1)

$$\overline{\Delta U}_{dec} = \frac{1-\gamma}{\gamma} \overline{\Delta U}_{acc} \qquad (7)$$

we obtain a system of four equations (4-7), comprising in the general case 5 unknown parameters: ΔU_{dec}, ΔU_{acc}, σ_{dec}, σ_{acc} and γ. The remaining parameters (A, E, σ) may be experimentally determined using hot-wire anemometer.

Thus, to determine unknown parameters in this equation set requires an additional condition, e.g., the assumption on an equality of the fluctuation velocity level in zones of decelerated and accelerated fluids at the same distance from the wall

$$\sigma_{dec} = \sigma_{acc} = \sigma_1 \qquad (8)$$

The following arguments can be adduced in favour of this assumption. First, for a considerable difference in σ_{dec} and σ_{acc} the total distribution function $f_{\Sigma}(u')$ becomes two-modal, which contradicts experimental data (e.g. [1]). Second, from physical considerations it is difficult to suppose the existence of thin zones with a highly differing level of u' (judging by the results of these experiments shown in fig.1 the mean thickness of the decelerated fluid zone is less than 2% of the overall boundary layer thickness).

3. The system of equations (4-8) is closed and for known experimental values of σ , A and E permits to determine the intermittent flow parameters U_{dec}, U_{acc}, σ_1 and γ in the wall region of the turbulent boundary layer.

Ratios $\frac{U_{dec}}{U_\infty}$ and $\frac{U_{acc}}{U_\infty}$ obtained in this way are shown in fig.3a compared with an experimental distribution of $\frac{U}{U_\infty}$ which is presented as a solid line approximating the experimental points given in fig.1a. It is seen that the velocity difference $\Delta U = U_{acc} - U_{dec}$ is relatively large and amounts to 50% of U in the vicinity of $y \approx \delta_l$.

The results of γ-factor determination (fig.3c) indicate a monotonous variation of γ similar to the lognormal distribution. But near the boundaries of this flow region the values of γ are still far from the limit values of γ =1 and γ =0. The departure of γ from 1 in the proximity of the wall(γ =0,85 at y/δ =0,003) is in agreement with experimental data (e.g.[7]) indicating that a severe inrush of the accelerated fluid from the external flow region reaches up to the wall. The above is

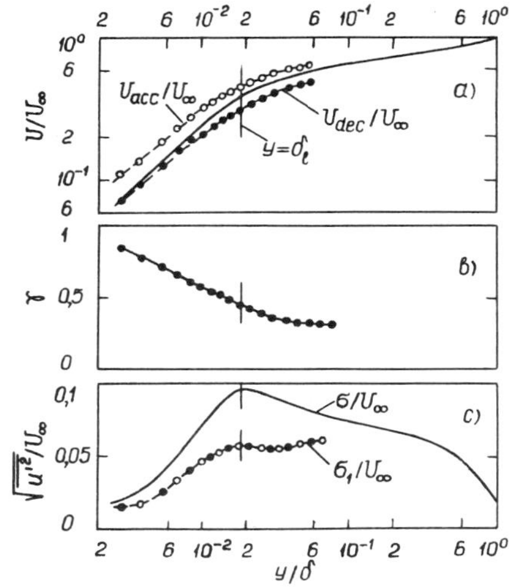

Fig.3

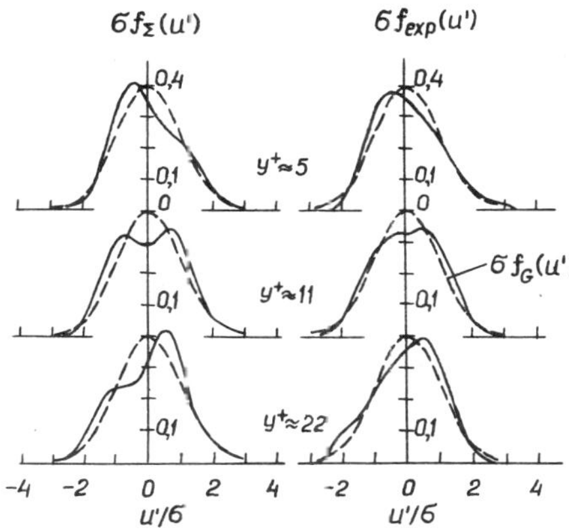

Fig.4.

also indicated by relatively great values of skewness factor A for skin friction fluctuations on the wall (e.g. [8]) which for $y\to\infty$ may be treated as fluctuations of a longitudinal velocity.

Fig.3c shows the calculated velocity fluctuation level $\sigma_1=\sigma_{dec}=\sigma_{acc}$ in zones of decelerated and accelerated fluids in comparison with experimental data for the overall level of fluctuations σ (fig.1b) presented as a solid curve. The fact that the value of σ in the vicinity of $y\approx\delta_\ell$ has a maximum and is nearly twice greater than σ_1 may be explained using the adopted flow model as this region experiences the most frequent changes of flow zones in which the mean velocity values differ by the factor of 1,5 ($\Delta U/U\approx 0,5$), with $\Delta U/\sigma\simeq 1,6$.

Based on the data in fig.3 using eq.(3) one can obtain the total probability density distribution $f_\Sigma(u')$ which is compared with the experimental determination of $f_{exp}(u')$ in fig.4. The experimental data obtained in [1] for flow conditions similar to those shown in fig.1 are used for a comparison. The correlation of predicted and experimental functions f(u') are presented for three distances from the wall $y^+=\frac{y u_\tau}{\nu}=5, 11$ and 21 ($y^+\simeq 700\ y/\delta$). To illustrate the measure of f(u')- function departure from the normal law in fig.4 the gaussian distribution of $f_G(u')$ is also given for each case. It is seen that the predicted functions $f_{\Sigma}(u')$ are in agreement with the experimental curves of $f_{exp}(u')$ at all values of y^+.

4. Thus, the presented technique for the analysis of flow probability characteristics using eqs.(4-8) is an analogue to the experimental conditional sampling technique. It allows to single out in the wall region of the turbulent boundary layer two intermittent flow zones with different values of mean velocity (U_{dec} and U_{acc}) and, to define the values of U_{dec} and U_{acc}, the fluctuation level in these zones, and the coefficient of the intermittency of these zones at various distances from the wall.

References

1. Repik, Ye.U.; Sosedko, Yu.P.: On intermittent flow structure in the viscous sublayer zone of the turbulent boundary layer. Sb. "Mekhanika turbulentnykh potokov", M., Nauka, 1980 219-225.
2. Repik, Ye.U.; Sosedko, Yu.P.: On viscous sublayer thickness of turbulent boundary layer. Izv.SO AN SSSR, ser. tekhn.nauk. N 8, vyp.2, (1982) 21-29.
3. Hedley, T.B.; Keffer, J.F.: Some turbulent/non-turbulent properties of the outer intermittent region of a boundary layer. "J.Fluid Mech.", 64, (1974) 645-678.
4. Thomas, R.M.: Conditional sampling and other measurements in a plane turbulent wake. "J.Fluid Mech." 57, (1973) 549-582.
5. Praskovsky, A.A.: Measurement of conditionally averaged characteristics of turbulence in a two-dimensional wake behind a cylinder. PMTF, N 6, (1983) 87-94.
6. Charnay, G.; Mathieu, J.; Comte-Bellot, G.: Response of a turbulent boundary layer to random fluctuations in the external stream. "Phys.Fluids", 19, (1976) 1261-1272.
7. Wallace, J.M.; Eckelmann, H.; Brodkey, R.S.: The wall region in the turbulent shear flow. "J.Fluid Mech.", 54 (1972) 39-48.
8. Repik, Ye.U.; Sosedko, Yu.P.: Spectral investigation of quasi-ordered flow structure in turbulent boundary layer. Izv. AN SSSR, MZhG, N 3 (1982) 10-17.

Model Equations Simulating Unsteady Viscous Flows

U. NEHRING

Göettingen

Summary

A system of nonlinear differential equations, based on a bulk viscosity concept, is presented and tested numerically. The solution by an explicit finite difference method enables the simulation of rather general flow fields. For example the 3-D-flow over a backward facing step and the vortex separation behind a rectangular block are calculated. In the case of the step flow qualitative agreement with experimental results concerning wall pressure and length of flow reattachment could be obtained. The physical interpretation of the equations derived here accentuates the effects of pressure distributions on the formation of large scale structures and unsteady fluid motion.

Introduction

The starting point for the formulation of the following model equations is the assumption that large scale structures such as vortices in the wake of bodies are primarily caused by pressure distributions in the flow field. This assumption is supported by papers such as that by SCHMIDT/JAMESON [1], which describes the back flow behind a step using the compressible Euler equation. With regard to high Reynolds-numbers, the terms in the fundamental equations associated with the shear viscosity are initially neglected, which particularly affects boundary layer phenomena. On the other hand, nonlinearity of the equations, medium compressibility, 3-dimensional treatment and dependence on time are considered essential.

The classical field variables are reduced to one scalar "pressure function" ϕ as a driving potential and a field vector $\tilde{v}$ as the "velocity". Material property parameters are assumed to be constant.

In a formular manner a set of equations satisfying these requirements can be derived from the Navier-Stokes equations and the energy equation. This method has been selected in the present paper although the physical content may appear dubious. It assumes that the effects of the usually neglected bulk viscosity exceed those of the shear viscosity, an assumption which - to the present knowledge - is only justified for extremely high temperature ranges and high Mach numbers. On this basis, it is also possible to dispense with the concept of bulk viscosity and refer only to "artifical" viscosity, which appears as an additional term in the frictionless Euler equations then remaining. Preference however is given to a physical interpretation, because the bulk viscosity is, in the last analysis, a constituent in the Navier-Stokes equations and an "artifically" introduced viscosity possesses little explanation value. The objective is the simulation of flow fields.

Basic Equations

On the assumption that remote from thermodynamic equilibrium the ratio of the bulk viscosity ζ to the shear viscosity μ may be greater one - $\zeta/\mu >> 1$ - the shear viscosity can be neglected to a first approximation. The unsteady Navier-Stokes Equations for a compressible medium, ignoring volume forces, then read using the summation convention:

$$\rho \frac{dv_i}{dt} = -\frac{\partial p}{\partial x_i} + \frac{\partial}{\partial x_i}\left[\zeta \frac{\partial v_j}{x_j}\right] \quad (1)$$

In the second equation used here - the energy equation - only the quadratic term $\zeta(\text{div}\,\tilde{v})^2$ of the dissipation function remains neglecting the terms associated with the shear viscosity. The shortened energy equation is then, aswell in dimensional form:

$$\rho \frac{du}{dt} = \frac{\partial}{\partial x_j}\left[\lambda \frac{\partial T}{\partial x_j}\right] - p\frac{\partial v_j}{\partial x_j} + \zeta\left[\frac{\partial v_j}{\partial x_j}\right]^2 \quad (2)$$

with internal energy u, temperature T, pressure p and heat conductivity coefficient λ.

In conjunction with the continuity equation, the law of ideal gases and functional relationships for the coefficients ζ, λ a closed set of equations for the variables $\tilde{v}$, ρ, p, T is given.

Model Equations

Assuming an ideal gas and making the further assumption of barotropy, i.e. $\rho = \rho(p)$ and $\text{grad}\,\rho \ll \text{grad}\,p$, the momentum and energy equations can be reduced to four variables - the velocity $\tilde{v}$ and a pressure function $\phi(\tilde{r},t)$.

Using the kinematic coefficients

$$\Theta = \frac{\zeta}{\rho} \quad \text{and} \quad \Lambda = \frac{\lambda}{\rho}$$

which can now be considered as material constants, equations (1) and (2) give schematically:

$$\frac{d\tilde{v}}{dt} = \Theta\,\text{grad}\,\text{div}\,\tilde{v} - \text{grad}\,\phi \tag{3}$$

$$\frac{d\phi}{dt} = \varepsilon\,\text{div}\,\text{grad}\,\phi - \phi\,\text{div}\,\tilde{v} + \Theta(\text{div}\,\tilde{v})^2 \tag{4}$$

where $\varepsilon = \Lambda/R$ (R = constant of gas).

Further reduction of equation (4) occurs if $(\text{div}\,\tilde{v})^2$ is neglected since it is small in a higher order compared with $\text{div}\,\tilde{v}$. Then (4) can be interpreted as a conservation equation for the quantity ϕ, extended by a diffusion term:

$$\frac{\partial\phi}{\partial t} + \text{div}\,\phi\tilde{v} = \varepsilon\Delta\phi \quad .$$

In the form

$$\frac{d\phi}{dt} = \varepsilon\,\text{div}\,\text{grad}\,\phi - \phi\,\text{div}\,\tilde{v} \tag{5}$$

the energy equation becomes complementary to the momentum equation (3) in a remarkable manner, indicating a feedback between gradients and sources.

The continuity equation is not required in this model. It can, however, be used to integrate the density over the velocity field, thus ensuring the conservation of mass.

The boundary conditions comprise the assumption of insulated walls with $\partial\phi/\partial n = 0$ (n = direction normal to surfaces) and the no-slip condition $\tilde{v}_w = 0$, which can be satisfied by virtue of the second order terms.

The numerical method used is very simple. The model differential equations (3) and (5) are approximated with reference to the spacial coordinates in a central finite difference method (second order) for the first and second derivatives. The development with time occurs explicitly in a single-step finite difference method (first order). The step widths in space and time are selected such that the CFL-condition is satisfied. A dependence of the computed results on the step widths has not been found.

Numerical Flow Simulations

Flow over a Step

The simplest example of a separated flow may be the flow over a step in a duct of rectangular cross section. A survey of references on this subject is given by TROPEA [2]. The main characteristics are the wall pressure distribution behind the step, the reattachment length and the distribution of the turbulent energy as a function of the incident flow Reynolds-number and geometrical parameters such as the step height and the expansion ratio in the duct.

Figures 1 - 3 show some numerical results of the simulated flow field, the pressure function and the reattachment length. Considering that the flow is accelerating from a state of rest due to a constant global pressure gradient, a comparison with experiments is qualitatively satisfying, see for example TANI [3]. The behaviour of the reattachment function and developing fluctuations in the mixing zone possibly indicates a relation to the transition process.

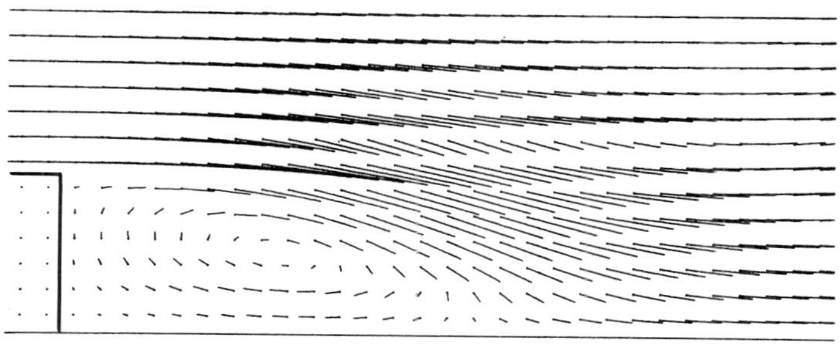

Fig. 1. Velocity field behind the step

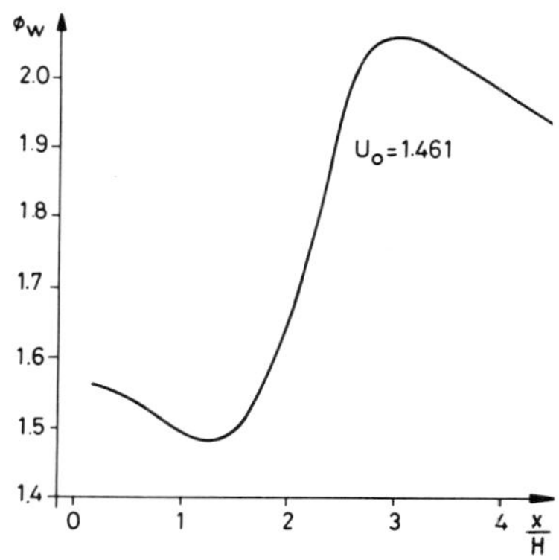

Fig. 2.

ϕ_{wall} as a function of distance x/H behind the step of height H

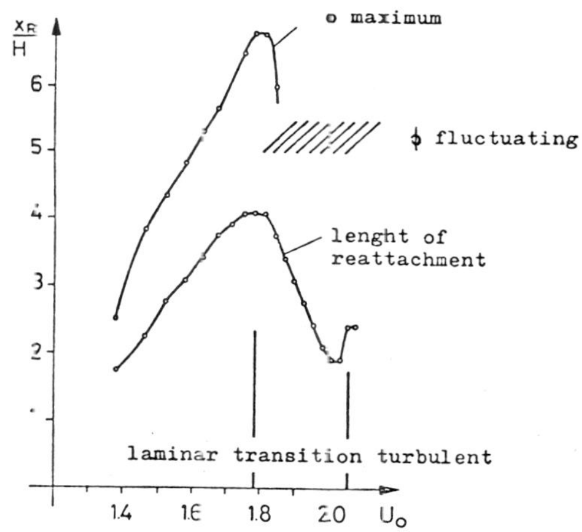

Fig. 3.

Length of flow reattachment x_r/H as a function of a nondimensional reference velocity U_o

Vortex Separation

A periodic formation and separation of vortices behind a "beam" within a simulated channel flow was the result of another test calculation. Figure 4 illustrates the velocity vector field at a definite moment. Special boundary conditions at the inflow are not prescribed. The "vortex street" appears only as a consequence of a pressure drop between the inlet/cutlet cross sections of the channel.

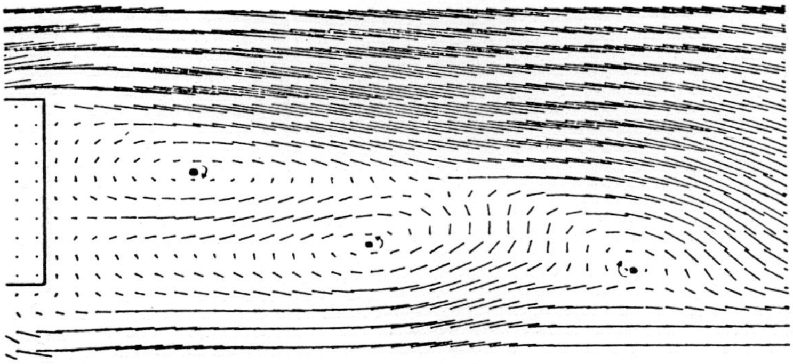

Fig. 4. Periodic vortex separation

References

1. Schmidt, W.; Jameson, A.: Euler solutions as limit of infinite Reynolds number for separation flows with vortices. Proceedings 8th Intern. Conf. on Numerical Methods in Fluid Dynamics. Aachen 1982.

2. Tropea, C.: Die turbulente Stufenströmung in Flachkanälen und offenen Gerinnen. Dissertation. Universität Karlsruhe 1982.

3. Tani, I.: Experimental investigation of flow over a step. IUTAM-Symposium Freiburg 26.-29. August 1957. Springer-Verlag 1958.

Resonance Instability of the Flows with Closed Streamlines

V. A. VLADIMIROV, V. F. TARASOV

Lavrentyev Institute of Hydrodynamics
Siberian Division of the USSR Academy of Sciences,
Novosibirsk 630090, USSR

Summary

This paper deals with the three-dimensional instability of two-dimensional incompressible flows where the streamlines represent ellipses with small excentricities. The presence of instability which we call "resonance" or "parametric" one, is shown both theoretically and experimentally. The predictions of the linear theory developed here are in good agreement with the experimental data. The present paper is a continuation of the cycle of works [1-4] .

Basic Flow

An ideal incompressible fluid flow in a cyclindrical vessel is under consideration. The vessel boundary in a cylindrical coordinate system r, θ, z is prescribed by the expressions

$$r = F(\theta,\varepsilon) = R_0 \left[1 + \frac{\varepsilon}{2}\cos 2\theta + O(\varepsilon^2)\right. \tag{1}$$

where $\varepsilon > 0$ is the small parameter. In the particular case where surface (1) is an elliptic cylinder with the cross-sectional semiaxes a and b ($a > b$) we have

$$F(\theta,\varepsilon)=R_0\left[1-\varepsilon\cos 2\theta\right]^{-1/2}, \quad \varepsilon \equiv \frac{a^2-b^2}{a^2+b^2}, \quad R_0^2 \equiv \frac{2a^2b^2}{a^2+b^2}. \tag{2}$$

Let the velocity field in the fluid be two-dimensional and have the form

$$\begin{aligned} V &= V_0(r) + \varepsilon \tilde{V}(r,\theta) + O(\varepsilon^2), \\ U &= \varepsilon \tilde{U}(r,\theta) + O(\varepsilon^2), \quad W = 0. \end{aligned} \tag{3}$$

Here U, V, W are the r, θ, z velocity components. When $\varepsilon = 0$ the expressions (1)-(3) are appropriate to the flow with circular streamlines in a circular cylinder of radius R_0. In this case U = W = 0 and the function $V = V_0(r)$ can be an arbitrary one. The vorticity is $\Omega \equiv V_{0r} + V_0/r$. The unit system with $\Omega(0) = 2$, $R_0 = 1$ is adopted below. The stream function $\psi(r,\theta)$

$$\tilde{U} = -\frac{1}{r}\psi_\theta, \quad \tilde{V} = \psi_r, \quad \psi \equiv \frac{1}{2}\varphi(r)\cos 2\theta$$

is introduced where the function $\varphi(r)$ defined when $0 \leq r \leq 1$ satisfies the following equation and boundary conditions:

$$\varphi_{rr} + \frac{1}{r}\varphi_r - \left(\frac{4}{r^2} + \frac{\Omega_r}{V_\theta}\right)\varphi = 0,$$

$$\varphi(0) = 0, \quad \varphi(1) = -V_0(1).$$

Stability Problem

The equations of motion linearized with respect to (3) take the form

$$Lu + U_r u + \frac{1}{r}U_\theta v - \frac{2}{r}Vv = -pr,$$

$$Lv + V_r u + \frac{1}{r}V_\theta v + \frac{1}{r}\left(Uv + Vu\right) = -\frac{1}{r}P_\theta, \tag{4}$$

$$Lw = -Pz, \quad u_r + u/r + v_0/r + w_z = 0$$

where u,v,w,p are the perturbations of the r, θ, z components of velocity and pressure,

$$L \equiv \frac{\partial}{\partial t} + U\frac{\partial}{\partial r} + \frac{V}{r}\frac{\partial}{\partial \theta}.$$

The conditions on the solid boundaries (1) give

$$Fu - F_\theta v = 0 \quad \text{at} \quad r = F(\theta, \varepsilon). \tag{5}$$

The Fourier harmonics

$$(u,v,w,p) = (\tilde{u},\tilde{v},\tilde{w},\tilde{p})\, e^{i(kz - \omega t)} \tag{6}$$

is considered with the amplitudes $\tilde{u}$, $\tilde{v}$, $\tilde{w}$, $\tilde{p}$ dependent on r and θ. Substitution of (6) into (4), (5) leads to the spectral problem. If there exists ω with Im $\omega \neq 0$ the flow (3) is unstable.

The spectral problem (4)-(6) is solved by the method of successive approximation. It is supposed that

$$(\tilde{u},\tilde{v},\tilde{w},\tilde{p},\omega) = \sum_{n=0}^{\infty} \varepsilon^n(u_n,v_n,w_n,p_n,\omega_n). \tag{7}$$

Substitution of (6), (7) into (4), (5) leads to the equations and boundary conditions on the successive approximations. The equations (4)-(7) for n = 0 represent a classical problem of linear waves and instability of the flows with circular streamlines. For Fourier harmonics

$$u_0(r,\theta) = \hat{u}(r)e^{im\theta}, \text{ etc.}$$

we can obtain the following equation and boundary conditions

$$M(\omega_0,m)\psi \equiv \psi_{rr} + \frac{1}{r}\left(1 - \frac{2k^2}{R}\right)\psi_r + D\psi = 0, \tag{8}$$

$$\psi(0) = \psi(1) = 0 \tag{9}$$

where $\psi \equiv r\hat{u}$, $\sigma \equiv -\omega_0 + mV_0/r$, $R \equiv k^2 + m^2/r^2$,

$$D \equiv -R + \frac{m}{\sigma r}\left(\frac{2k^2}{R}\frac{\Omega}{r} - \Omega r\right) + \frac{4k^2 V_0 \Omega}{\sigma^2 r}.$$

The dispersion relations $\omega_0(k,m)$ of the problem (8), (9) have the antisymmetry property $\omega_0(k,m) = -\omega_0(k,-m)$. The examples of dispersion curves for the specific case $V_0 = r$ and $m = \pm 1$ are shown in Fig. 1. The existence of intersection points when $\omega_0 = 0$ follows from antisymmetry.

The perturbations corresponding to these intersection points ($m = \pm 1$, $\omega_0 = 0$, $k = k_0$ and arbitrary function $V_0(r)$) are under consideration. The zero approximation solution is chosen as a sum

$$\psi(r,\theta) = (\alpha e^{i\theta} + \beta e^{-i\theta})\,\psi(r) \tag{10}$$

with the arbitrary constants α and β. The function $\psi(r)$

is determined from equation $M(0,1)\psi = 0$ (8). From the form of equations (4)-(7) for n = 0 it follows that the first approximation solution for (10) contains the harmonics $e^{\pm i\theta}$ and $e^{\pm 3i\theta}$

$$\psi_1(r,\theta) = A(r)e^{i\theta} + A_1(r)e^{3i\theta} + B_1(r)e^{-3i\theta}$$

The equations and boundary conditions for A and B are obtained from (4)-(7). Their l.h. sides are given by (8), (9) with operator M(0,1), but right ones are nonzero and are prescribed (linear) functions of solution in the zero approximation and the value ω_1. The operator $[r/(1+k_0^2r^2)]\ M(0,1)$ has a self-conjugated form. The use of Green equality gives the system of algebraic equations:

$$\begin{aligned} \alpha(\omega_1\Phi_1 + k_1\Phi_2) + \beta\Phi_3 = 0\ , \\ \beta(-\omega_1\Phi_1 + k_1\Phi_2) + \alpha\Phi_3 = 0 \end{aligned} \tag{11}$$

with real values Φ_i obtained by integration of the r.h. sides of mentioned above nonhomogeneous equations. The value of k_1 is taken from the representation $k = k_0 + \varepsilon k_1$. The wavenumber k is assumed to differ slightly from the values at the intersection points $\omega_0 = 0$, $k = k_0$ (Fig. 1). For existence of nontrivial solution (11) it is necessary that $\omega_1^2 = -(\Phi_3^2 - k_1^2\Phi_2^2)/\Phi_1^2$. It is evident that at small k_1 the instability ($\mathrm{Im}\,\omega_1 > 0$) always takes place. The maximum amplification rate $\omega_1 = i\omega_* \equiv i\,|\Phi_3/\Phi_1|$ is achieved when $k_1 = 0$. The instability takes place within the interval of wavenumbers $|k-k_0| < \varepsilon k_*$ where $k_* \equiv |\Phi_3/\Phi_2|$.

Experimental

The experiments have been performed with a transparent cylindrical vessel 120 cm high. Its cross-section is elliptic (with an accuracy of 5%). A large semi-axis of the ellipse is a = 18 cm, and a small one can be chosen from the interval b = 13.5-17 cm. The vessel rotates around its vertically oriented axis with angular velocity Ω_0. The available frequency interval is Ω_0 = 0.25-1.25 rot/s. The vessel was filled with water up to the level $L + \delta L$ and put into the state of solid-body rotation with velocity Ω_0. Then a small orifice of diameter d at the centre of the vessel was open. Due to the sink flow, there was developed a linear "bath" vortex with the core visualized with the use of ink. When the water was at the level L, the orifice was closed. Then, after the interval of time Δt (at t = 0), the vessel was sharply stopped and the evolu-

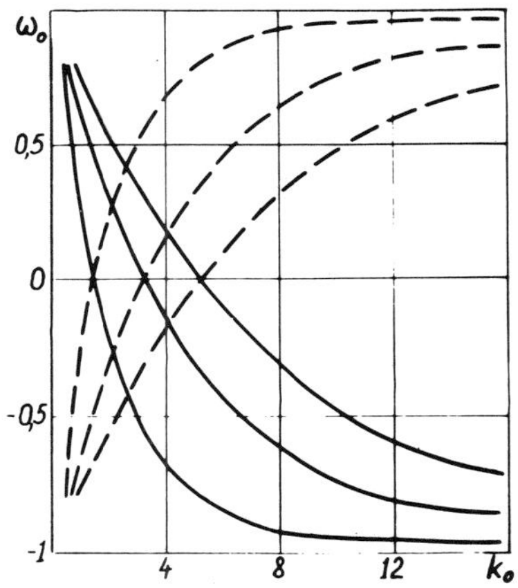

Fig.1

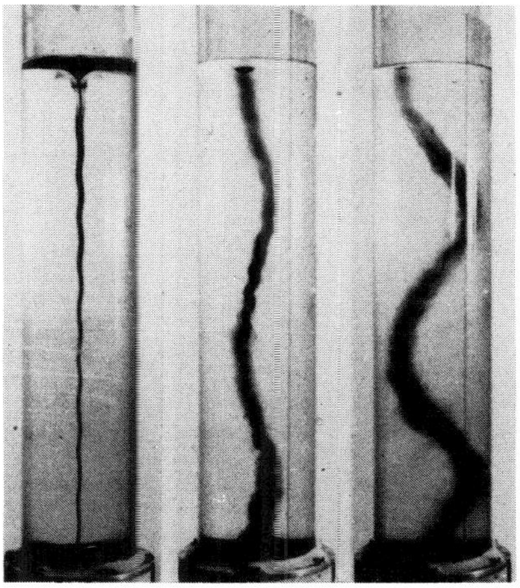

Fig.2

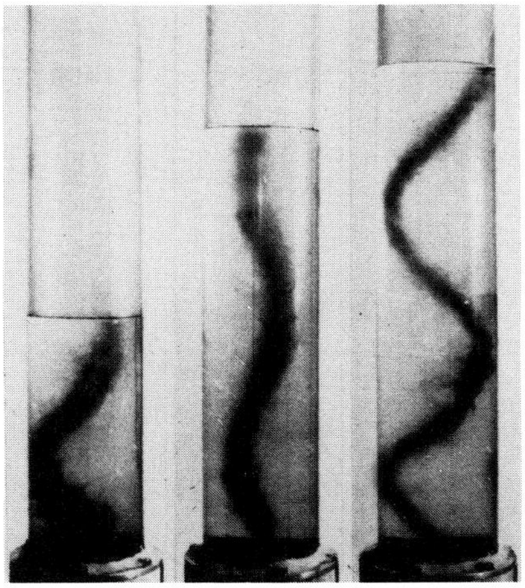

Fig.3

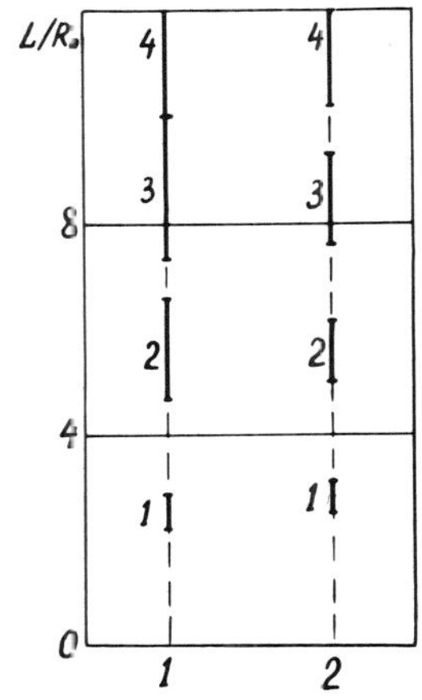

Fig.4

tion of the core was observed and photographed. The interval Δt is necessary for the reconstruction of three-dimensional flow to two-dimensional one (see (3)).

The main experimental result is that the flow is unstable in some intervals of $h \equiv L/R$. The instability manifests itself in the increasing of bending waves on the core (Figs. 2,3). All the bendings are in the same plane and have clearly defined periodical structure. The data presented below correspond to the typical experiment with parameters $\varepsilon = 0.17$; $\Omega_0 = 2\pi \cdot 0.65\ s^{-1}$; $\delta L = 15$ cm; $d = 0.3$ cm; $\Delta t = 2$ s. The instability takes place if value h belongs to the intervals represented in the first vertical line in Fig. 4. For the intervals n=1,2,3,4 (Fig. 4) the relationship $L = n\lambda/2$ is valid, where λ is the wavelength of instability. The photographs in Figs. 2a,b,c correspond to L = 90 cm and times of 0; 12 and 25 s after the vessel has been stopped. The photographs in Figs. 3 a,b,c were made at L = 50, 80 and 92 cm, respectively.

Comparison of Theoretical and Experimental Results

1. In all experiments $\omega_0 = 0$ that is in full agreement with the theoretical predictions.
2. The values of h for which instability takes place are composed in Fig. 4 (vertical 1 - experimentally measured data, vertical 2 - calculated data).

References

1. Vladimirov, V.A.: On the stability of the constant vorticity incompressible flow in the elliptic cylinder. Journal Prikladnoy Mekhaniki i Tekhnicheskoy Fiziki, N 4 (1983) 118-124 (in Russian).
2. Vladimirov, V.A.; Tarasov, V.E.; Rybak, L.Ya.: On the stability of the elliptical deformed incompressible fluid flow in the field of Coriolis force. Izvestiya Akademii Nauk SSSR, Fizika Atmosfery i Okeana, vol.19, N 6 (1983) 586-594 (in Russian).
3. Vladimirov, V.A.; Tarasov, V.F.: The experimental and theoretical investigation of the stability of the line vortex with deformed core. Journal Prikladnoy Mekhaniki i Tekhnicheskoy Fiziki, N 3 (1983) 61-69 (in Russian).
4. Vladimirov, V.A.; Tarasov, V.F.: Resonance instability of the rotating flows with slightly deformed streamlines. J. Dinamika Sploshnoy Sredy, N 59, Novosibirsk 1983 (in Russian).

On Long-Wave Asymptotic Solutions of the Orr-Sommerfeld Equation for Boundary Layer

V. I. ZHUK

Computing Center of the USSR Academy of Sciences,
Moscow, USSR

The well-known asymptotic behaviour [1] of the upper and lower branches of the neutral curve corresponds to disturbances which have wavelengths greater, by an order of magnitude, than the boundary layer thickness. The results of [1] are interpreted on the basis of modern asymptotic methods in [2, 3] where higher-order analysis of linearized Navier-Stokes equation is undertaken for Reynolds number $R \to \infty$.

Long-wave asymptotic solutions of Orr-Sommerfeld equation is of interest, in particular, because the eigensolutions of linearized free-interaction theory equations represent the limiting forms of Tollmien-Schlichting waves in incompressible fluid with the critical layers close to the wall [4].

Our object is to construct an asymptotic solution of linear stability problem in the region between two branches of the neutral curve for Blasius boundary layer. It is assumed that Reynolds number R and non-dimensional wave-number k are defined with respect to a distance from the leading edge of a flat plate. The undisturbed Blasius profile $U_0 = U_0(y_2)$, $y_2 = y\sqrt{R}$ has the property $U_0 = \lambda_1 y_2 + 1/24\,\lambda_4 y_2^4 + \dots$ as $y_2 \to 0$. Allowing a stream function in the form $\exp[ik(x-ct)]f(y)$, we obtain the Orr-Sommerfeld equation

$$(U_0 - c)\left(\frac{d^2 f}{dy_2^2} - \delta^2 f\right) - \frac{d^2 U_0}{dy_2^2} f = \frac{1}{ik}\left(\frac{d^4 f}{dy_2^4} - 2\delta^2 \frac{d^2 f}{dy_2^2} + \delta^4 f\right) \qquad (1)$$

where wave-number k is real (and positive), parameter $\delta = kR^{-1/2}$ characterizes the ratio of boundary layer thickness to disturbance wavelength.

We set $R\to\infty$, $k\to\infty$, $c\to 0$, $\delta\to 0$. In the main part of the boundary layer where $y_2=O(1)$ consider the expansion in powers of the small parameter δ regarding phase velocity c as fixed

$$f=f_0+\delta^2 f_1+\dots \tag{2}$$

The functions f_0, f_1 are determined from the inviscid equations with the boundary conditions of matching (as $y_2\to\infty$) with the region located outside the boundary layer where $U_0=1$, $y_1=\delta y_2=O(1)$.

If the absolute value of phase velocity $c=c_r+ic_i$ is small the complex point y_2^* at which $U_0=c$ is defined by $c=\lambda_1 y_2^*+1/24\,\lambda_4(y_2^*)^4$. It is possible to reexpand the inviscid solution (2) about the singular point

$$f=\delta\frac{1-c}{\lambda_1}-\delta^2\frac{\mathcal{D}}{\lambda_1}+\Big[\frac{\lambda_1}{1-c}+\delta\lambda_1 B\Big](y_2-y_2^*)+$$

$$+\frac{1}{2}\frac{\lambda_4 c^2}{\lambda_1^4}\big[\delta(1-c)-\delta^2\mathcal{D}\big](y_2-y_2^*)\big[\ln|y_2-y_2^*|+i\arg(y_2-y_2^*)\big]+\dots \tag{3}$$

$$B=\int_c^\infty\Big[\frac{1}{U_0^2}-1-\frac{1}{\lambda_1^2 y_2^2}\Big]dy_2\,,\qquad \mathcal{D}=-\int_c^\infty[U_0^2-1]\,dy_2$$

Viscous terms of the Orr-Sommerfeld equation become significant in the vicinity $y_2-y_2^*=O(k^{-1/3})$. The critical layer exists not only for real but also for complex value of phase velocity c provided $c_i\lesssim k^{-1/3}$. Let us introduce the complex variable $Z=(i\lambda_1 k)^{1/3}(y_2-y_2^*)$. The wall is located at $Z=\zeta=-(i\lambda_1 k)^{1/3}y_2^*\simeq -i^{1/3}\lambda_1^{-2/3}k^{1/3}c$. In the critical layer we look for a solution of the form

$$f=g_0+(ik)^{-1/3}c^2 g_1+\dots \tag{4}$$

The substitution (4) into (1) leads to the equations

$$\frac{d^4 g_0}{dZ^4}-Z\frac{d^2 g_0}{dZ^2}=0\,,\qquad \frac{d^4 g_1}{dZ^4}-Z\frac{d^2 g_1}{dZ^2}=-\frac{1}{2}\frac{\lambda_4}{\lambda_1^{10/3}}\Big(g_0-\frac{1}{2}Z^2\frac{d^2 g_0}{dZ^2}\Big) \tag{5}$$

It is necessary to distinct two cases. In the first one ζ is of order of unity, i.e. the critical layer coincides with the wall layer. The solution for q_0 is expressed via Airy function $Ai(Z)$. The condition of matching with (3) as $Z \to \infty$ and boundary conditions $q_0 = dq_0/dZ = 0$ at the wall $Z = \zeta$ provide the dispersion relation

$$\frac{dAi(\zeta)}{dZ}\left[\int_{\zeta}^{\infty} Ai(Z)dZ\right]^{-1} = \frac{1}{\lambda_1^{5/3}} \frac{i^{1/3} k^{4/3}}{\sqrt{R}}\left[1 - 2c - \delta B - \delta \mathfrak{D}\right] \qquad (6)$$

Oscillations with such three-zoned flow structure are characterized by the following relations between orders of magnitudes: $c_r \sim c_i \sim \delta \sim R^{-1/8}$, $k \sim R^{3/8}$. If we denote the left-hand side of (6) by $\Phi(\zeta)$ and set

$$\tilde{k} = \lambda_1^{-5/4} R^{-3/8} k\left[1 - 2c - \delta B - \delta \mathfrak{D}\right]^{3/4}, \quad \tilde{c} = -i^{-1/3} \tilde{k}^{-1/3} \zeta \qquad (7)$$

then the dispersion relation is reduced to the equation of standard form $\Phi(-i^{1/3}\tilde{k}^{1/3}\tilde{c}) = i^{1/3}\tilde{k}^{4/3}$, the properties of which were studied in [4]. For every prescribed $\tilde{k}$ there is a whole spectrum of eigenvalues $\tilde{c}$ (and, therefore, ζ). The behaviour of roots of the dispersion relation in the complex ζ-plane while $\tilde{k}$ runs through real positive values are shown in Fig.1. The first mode which can be both stable and unstable describes the vicinity of the lower branch of the neutral stability curve. Parameter $\tilde{c}$ for the first mode becomes real $\tilde{c} = c_* = 2.2968$ at real $\tilde{k} = k_* = 1.0005$. By substituting quantities k_*, c_* into (7) we find asymptotic representation of the neutral curve in the R, c plane as $R \to \infty$

$$c = \lambda_1^{1/4} R^{-1/8} c_*\left\{1 - \frac{1}{4}\lambda_1^{5/4} R^{-1/8}\left[(B + \mathfrak{D})k_* + \frac{2}{\lambda_1} c_*\right]\right\} \qquad (8)$$

The infinite growth of $|\zeta|$ for the first mode as $|\tilde{k}| \to \infty$ (see Fig.1) means that the critical layer separates from the wall if $k \gg R^{3/8}$. Consider now the second case in which the critical layer is much thinner than its distance from the wall. The unique solution of the first of equations (5) subject to the condition of absence of exponential increasing at the upper and lower edges of critical layer is $q_0 = d_0 + d_1 Z$ where d_0, d_1 are constants. The solution of the second approximation

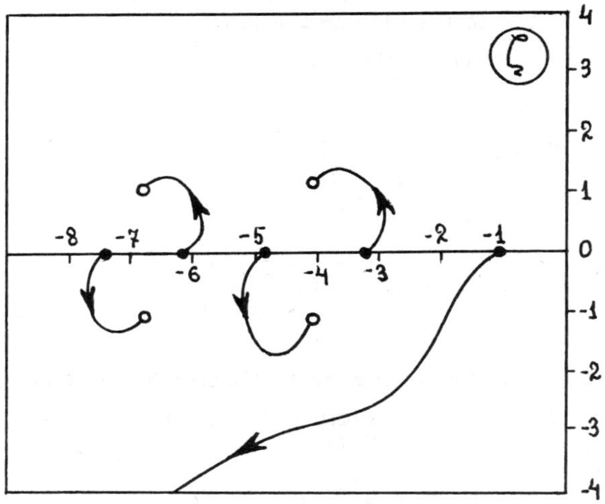

Fig.1. Traces of the first five roots of the dispersion relation in the complex ζ-plane as $\operatorname{Im}\tilde{k}=0$, $0<\operatorname{Re}\tilde{k}<\infty$. Arrows indicate increasing $|\tilde{k}|$. •••• - zeroes of derivative of Airy function, ∘∘∘∘ - zeroes of denominator of $\Phi(\zeta)$

subject to the same condition has asymptotic behaviour $q_1 = 1/2\varkappa_4\varkappa_1^{-10/3}(d_0 Z\ln Z + 1/2\, d_1 Z^2 + \dots)$ as $|Z|\to\infty$ within the sector $\arg Z\in(-\pi,\pi/3)$. Hence analytical continuation of solution below the critical layer by the use of this asymptotic formula for q_1 fixes a phase shift of $-\pi + O(c_i|c|^{-1})$ in the complex logarithmic function.

It is easily seen that (3) gives inviscid solution not only above critical layer, but also beneath it, where we must choose $\arg(y_2-y_2^*) = -\pi + O(c_i|c|^{-1})$.

Inviscid solution becomes invalid at a distance of order $k^{-1/2}|c|^{-1/2}$ from the wall. In the viscous wall layer the new complex variable $\eta=(ikc)^{1/2}$ is of order of unity. Solution of the Orr-Sommerfeld equation which satisfies the no-slip condition $f(0)=df(0)/d\eta=0$ at the wall is given by

$$f=\tau_0(1+i\eta-e^{i\eta}) \to \tau_c(1+i^{3/2}k^{1/2}c^{1/2}y_2),\ |\eta|\to\infty,\ \tau_c=\mathrm{Const} \qquad (9)$$

The matching of the expansion (3) as $y_2\ll|c|$ with (9) as $|\eta|\to\infty$ yields two parts of dispersion relation

$$c_r = \frac{1}{\lambda_1}\frac{k}{\sqrt{R}}\left[1 - \frac{k}{\sqrt{R}}\left(\frac{2}{\lambda_1} + B + \mathcal{D}\right)\right] \tag{10}$$

$$c_i = \frac{1}{\sqrt{2}}\lambda_1^{3/2}\frac{R^{1/4}}{k}\left[1 - \frac{1}{2}\frac{k}{\sqrt{R}}\left(\frac{2}{\lambda_1} - B - \mathcal{D}\right)\right] + \tag{11}$$

$$+\frac{1}{2}\pi\frac{\lambda_4}{\lambda_1^2}\frac{k^4}{R^2}\left[1 - 4\frac{k}{\sqrt{R}}\left(\frac{5}{2\lambda_1} + B + \mathcal{D}\right)\right]$$

Formulas (10), (11) describe the vicinity of the upper branch of the neutral curve. For such five-zoned flow structure again $c_r \sim \delta$, but $c_i \sim \delta^4 \ll |c| \sim R^{-1/20}$, $k \sim R^{9/20}$.

If $c_i = 0$, then from (11) we obtain the asymptotic representation of the upper branch of the neutral stability curve as $R \to \infty$

$$c_r = \frac{1}{\lambda_1}\delta R^{-1/20}\left[1 - \frac{1}{10}\delta R^{-1/20}\left(B + \mathcal{D} + \frac{2}{\lambda_1}\right)\right] \tag{12}$$

where $\delta = (-2^{-1/2}\lambda_1^{-19/2}\lambda_4\pi)^{-1/5}$. It is worth noting that for the Blasius profile $\lambda_4 < 0$.

It is possible to show that asymptotic form as $k \gg R^{3/8}$ of dispersion relation in the vicinity of the lower branch (6) coincides with the asymptotic form as $k \ll R^{9/20}$ of dispersion relation in the vicinity of the upper branch (10), (11). The overlapping of the vicinities of the two branches of the neutral curve where dispersion relations turn into each other means that they describe the whole unstable region between these branches. The amplification rate of disturbances throughout unstable region has the order of magnitude $R^{1/4}$, while its maximum is located in the neighbourhood of the lower branch.

A single oscillation mode corresponds to the five-zoned structure of the disturbed velocity field with the critical layer separated from the wall layer. Analysis of roots of the dispersion relation (6) in the complex plane shows that this mode turns into the first mode from spectrum of oscillations with the three-zoned flow structure when wave number changes in the interval $R^{3/8} \lesssim k \lesssim R^{9/20}$.

References

1. Lin, C.C.: On the stability of two-dimensional parallel flows. Part III. - Stability in a viscous fluid. Quart. Appl. Math. 3 (1946) 277-301.

2. Smith, F.T.: On the non-parallel flow stability of the Blasius boundary layer. Proc. Roy. Soc. London A 366 (1979) 91-109.

3. Bodonyi, R.J.; Smith, F.T.: The upper branch stability of the Blasius boundary layer, including non-parallel flow effects. Proc. Roy. Soc. London A 375 (1981) 65-92.

4. Zhuk, V.I.; Ryzhov, O.S.: The free interaction and stability of the boundary layer in the incompressible fluid. Dokl. Akad. Nauk SSSR 253 (1980) 1326-1329.

On Some Models of Turbulent Flow

M. BURNAT

Institute of Mathematics,
University of Warsaw.
00-901 Warsaw. P.K.iN.

Summary

A new model of turbulent flow is given, in which the mass transport takes place along curvilinear mutually penetrating cones. The turbulent-laminar as well as the inverse transition may be considered.

In the present paper we suggest a new model of the turbulent flow. We assume that the fluid motion in the time interval $[t°,t]$, $t°$ the start of the observation, results instantaneously at the time t and at the point x in such a manner that the mass transport takes place with the instantaneous velocities $V(t,x,\alpha)$ $\alpha \in A$, where as the set of parameters one may choose the unit ball in R^3. Hence at each point x we have an infinite bundle of instantaneous velocities and through each point $p \in R^3$ at the time τ an infinite bundle of trajectories $\mathbf{T}_\alpha(\tau,p)$: $x = x(t,\tau,p,\alpha)$, $dx/d\tau = v(t,x,\alpha)$. A marked small mass particle located at the time t at the point $x \in \mathbf{T}_\alpha(t,x)$, does not in general move at the time $t' > t$ along $\mathbf{T}_\alpha(t,x)$. The very involved trajectory $\mathbf{T}_{mark}(t,x)$: $x = x(t)$ of the marked particle satisfies $dx/dt = v(t,x,\alpha(t,x))$ where $\alpha(t,x)$ is an appropriate function. The main idea of our model is an attempt to describe the turbulent flow without limiting processes (means, attractors) of laminar quantities connected with the laminar trajectories $\mathbf{T}_{mark}$, using only the α-velocities $v(t,x,\alpha)$ and other α-physical quantities connected with the instantaneous trajectories $\mathbf{T}_\alpha$. We show that the basic conservation laws may be formulated by means of the α-quantities, what leads to closed P.D.E. systems describing different turbulent flows. Knowing the

α-quantities we form the x,α-densities. Integrating the x,α-densities over suitable subsets of $R^3 \times A$, we obtain the mean values of physical quantities measured in the flow. If $v(t,x,\alpha)$ does not depend on α and some natural conditions are fullfiled, then the mean quantities satisfy the P.D.E. systems describing laminar flows, and in this way we have in our model a natural possibility of turbulent-laminar transition as well as the inverse one. The macroscopic visualisation (smoke trace) of the turbulent flow shows that the mass transport takes place along curvilinear cones mutually penetrating each other. Knowing the α-parameters of the flow, we may describe this cones and therefore one may say that in our model the mass transport takes place along mutually penetrating cones.

If $\rho(t,x,\alpha)$ is the x,α-mass density, and $\omega \subset R^3$, $a \subset A$, some regions, then $m(a,\omega,t)$ denotes the mass portion belonging at the time t to ω and moving with the instantaneous velocities $v(t,x,\alpha)$, $\alpha \in a$. We put

$$m(a,\omega,t) = \int_a d\alpha \int_\omega \rho(t,x,\alpha)dx,$$

hence the mean measured density $\hat{\rho}(t,x) = \int_A \rho(t,x,\alpha)dx$. The x,α-momentum density is $\rho(t,x,\alpha)V(t,x,\alpha)$, and the momentum of the mass portion $m(a,\omega,t)$ may be expressed as

$$imp(a,\omega,t) = \int_a d\alpha \int_\omega \rho(t,x,\alpha)V(t,x,\alpha)dx.$$

Therefore the mean measured velocity

$$\hat{V}(t,x) = \lim_{r\to o} imp(A,K_r,t)/m(A,K_r,t)$$

where $K_r = \{y: |x-y| < r\}$ and $m(A,K_r,t)$ is the whole mass in K_r at t. Let us consider for each $\alpha \in A$ the mapping $Y_\alpha(t): \omega \to \omega_\alpha(t)$ where $\omega \subset R^3$ an arbitrary region and for $p \in \omega$, $Y_\alpha(t)p = x = x(t,t^\circ,p,\alpha)$. We introduce the following sets:

$$\Omega_a(t) = \bigcup_{\alpha\in a} \omega_\alpha(t),\ I\Omega_a(t) = \bigcap_{\alpha\in a} \omega_\alpha(t),\ B\Omega_a(t) = \bigcup_{\alpha\in a} \partial\omega_\alpha(t)$$

where $\partial\omega_\alpha(t)$ denotes the boundary of $\omega_\alpha(t)$, we call $I\,\Omega$ and $B\,\Omega$ correspondingly the inner and the boundary sets of $\Omega_a(t)$. $\bar{\Omega}_a(t) = B\Omega_a(t) \cup I\Omega_a(t)$. It turns out that the moving mass portion $m^V(a,\omega,t)$ should be defined (see [1],[2]) as the mass satisfying the following conditions

1. The mass portion $m^V(a,\omega,t)$ is contained at the time t in $\Omega_a(t)$.

2. The mass of the portion $m^V(a,\omega,t)$ at the time t moves only along such trajectories $\mathbf{T}_\alpha(t,x)$, $\alpha \in a$, $x \in \Omega_a(t)$, which in the time $t°$ pass through $p \in \omega$. It is worth to mention that $m^V(a,\omega,t)$ does not contain all mass belonging to $\Omega_a(t)$. We have always $m(a,\Omega_a(t),t) > m^V(a,\omega,t)$. For the mass portion $m^V(a,\omega,t)$ we put:

$$m^V(a,\omega,t) = \int_a d\alpha \int_{\omega_\alpha(t)} \rho(t,x,\alpha)\,dx,$$

$$imp^V(a,\omega,t) = \int_a d\alpha \int_{\omega_\alpha(t)} \rho(t,x,\alpha)V(t,x,\alpha)\,dx,$$

hence $m(a,\omega,t°) = m^V(a,\omega,t°)$, $imp(a,\omega,t°) = imp^V(a,\omega,t°)$.

The general principle used in our model for expressing the physical quantities by means of the x,α-densities is as follows:

1. If the quantity Q has a laminar analogy, then for each $\alpha \in A$ we consider the α-laminar flow along the $\mathbf{T}_\alpha$ trajectories and take the proper laminar x-density $D(t,x,\alpha)$ so that the quantity $Q(\alpha)$ for the region $\omega \subset R^3$ in the α-laminar flow is expressed in the form:

$$Q(\alpha) = \int_\omega D(t,x,\alpha)\,dx \quad \text{or} \quad Q(\alpha) = \int_{\partial\omega} D(t,x,\alpha)\,dx.$$

The quantity Q for the mass portion $m(a,\omega,t)$ is expressed by putting $Q = \int_a Q(\alpha)\,d\alpha$, that is:

$$Q = \int_a d\alpha \int_\omega D(t,x,\alpha)\,dx \quad \text{or} \quad Q = \int_a d\alpha \int_{\partial\omega} D(t,x,\alpha)\,dx \tag{1}$$

For the mass portion $m^V(a,\omega,t)$ we put:

$$Q = \int_a d\alpha \int_{\omega_\alpha(t)} D(t,x,\alpha)\,dx \quad \text{or} \quad Q = \int_a d\alpha \int_{\partial\omega_\alpha(t)} D(t,x,\alpha)\,dx$$

2. If the quantity Q has no laminar analogy, then we introduce a new x,α-density $D(t,x,\alpha)$ and use (1) for $m(a,\omega,t)$ and (2) for $m^V(a,\omega,t)$.

For the formulation of the mass conservation law we introduce the new physical quantity without laminar analogy, the amount of the mass added in the unit time to the mass portions $m(a,\omega,t)$, $m^V(a,\omega,t)$ because of the turbulent diffusion. To this end we introduce the x,α-density $\mu(t,x,\alpha)$. The mass conservation law for $m(a,\omega,t)$ takes the following form

$$\frac{d}{dt}\int_a d\alpha \int_\omega \rho(t,x,\alpha)\,dx = -\int_a d\alpha \int_{\partial\omega} \rho(V,n)\,dx + \int_a d\alpha \int_\omega \mu(t,x,\alpha)\,dx,$$

and for $m^V(a,\omega,t)$:

$$\frac{d}{dt}\int_a d\alpha \int_{\omega_\alpha(t)} \rho(t,x,\alpha)\,dx = \int_a d\alpha \int_{\omega_\alpha(t)} \mu(t,x,\alpha)\,dx.$$

After localisation $(a,\omega,\omega_\alpha(t)$ arbitrary) we obtain in both cases:

$$\partial_t\rho + \mathrm{div}_x(\rho v) = \mu, \qquad \alpha \in A. \tag{3}$$

To describe the turbulent selfmixing process we introduce the x,α-density $M(t,x,\alpha)$ describing the amount of the mass added because of turbulent diffusion to the portion $m^{v}(a,\omega,\nu)$ in the time interval $t^{\circ} < \nu < t$. It turns out ([1],[2]) that M satisfies the condition:

$$\partial_{t}M + \mathrm{div}_{x}(Mv) = \mu, \quad \alpha \in A.$$

The boundary conditions for M are obviously different from the conditions for ρ. In order to close the system (3), (4) in the case when $v(t,x,\alpha)$ is given, we introduce the constitutive equation of mixing, $\partial_{t}M = \mathfrak{M}(\mu,M,\rho,v,\ldots)$. We suggest here the idea that the fluid with given chemical properties may, in different situations, move with different mixing constitutive equations. Let $J(t,p,\alpha)$ be the Jacobian of the mapping $Y_{\alpha}(t)p = x$. Under the assumption $|\partial_{t}J/J^{2}| << 1$, $|\mathrm{grad}_{p}J|/J^{2} << 1$, the simplest mixing constitutive equation for the stationary case may be taken in the form:

$$\partial_{t}M = \mu + \eta(x)G(x) \sum_{i=1}^{3} V_{i}\partial_{x_{i}}M,$$

where $\eta(x)$ describes the relation between the local self-mixing situations in x and $p = Y_{\alpha}^{-1}(t)x$, $G(x)$ is given by the geometrical configuration. In general the constitutive equation should be an integrodifferential one.

Introducing proper α-quantities we formulate the momentum, angular momentum, and energy conservation laws. After localisation we obtain corresponding differential equations. In the momentum conservation law for $m^{v}(a,\omega,t)$ we consider the boundary forces acting in $B\Omega_{a}(t)$ and inner forces concentrated mostly in $I\Omega_{a}(t)$. The inner mass forces arise because at one point at the same time different mass components moving with different velocities are located.

Finally we obtain closed infinite P.D.E. systems for $U(t,x,\alpha) = (\rho(t,x,\alpha),v(t,x,\alpha),\ldots)$, $\alpha \in A$. For the mean measures values $\hat{\rho},\hat{v},\ldots$ it is sufficient to know $U(t,x,\alpha)$ for a finite set $\{\alpha_{1},\ldots,\alpha_{N}\} \subset A$. The construction of interesting classes of solutions may be reduced to ordinary differential systems. Our model may be easy generalised for the case of the turbulent flow with chemical reactions.

References

[1] Burnat, M., Mathematical models of fluid motion with intensive selfmixing. Preprint Nr 9/83, Institute of Math. Warsaw University.

[2] Burnat, M., Models of nonlaminar flow with turbulent diffusion. Arch. Mech., to appear.

[3] Burnat, M.; Kurowski, S.; Penczek, M.; Wierzbicki, J., Model of non-laminar, low temperature plasma flow with intensive selfmixing. To appear.

Finite-Amplitude Disturbances and Near-Threshold Instability of Hydrodynamic Flows

S.S. MOISEEV, V.G. PUNGIN and N.V. SUYAZOV

Space Research Institute, Moscow

Summary

We consider theoretical models of possible initial stages of development, due to finite-amplitude wave instability, of turbulent spots in stratified shear flows when wave and flow parameters are near the stability threshold. It is shown that, in contrast to analogous processes in plasma, nonlinear frequency shift does not provide saturation of explosive resonant interaction of the marginal waves. A condition of instability due to redistribution of energy, by wave-induced turbulence, from the flow to wave is found and estimates of the growth rate are obtained analytically.

We consider the theoretical models of certain processes which might lead to generation of turbulent spots in stratified shear flows. Some authors [1,2] suggest that the generation of turbulent spots in the ocean may be associated with the explosive resonant wave interactions (i.e. interactions leading to infinite growth (under approximation used) of wave amplitudes during finite time). Moiseev et al. [3,4] considered such interactions of marginally stable surface and internal waves in stratified shear flows. It was shown that when the second harmonic of marginal wave is marginal as well the explosive time, t_{ex}, is the smallest, with respect to the power of a small paramter, ε, which characterizes initial wave amplitudes, among the time scales of the known explosive interactions: $t_{ex} \sim \varepsilon^{-1/2}$. In distinction from the cases studied for plasma by Davydova et al. [5,6], the explosion with such the smallest time scale occurs even when the flow is disturbed initially by a single wave (the first harmonic). The explosive character of marginal wave interaction, as well as decrease of t_{ex}, are due to the fact that the coefficient of the first time-derivative in the equations for slow variations of wave amplitudes, which describe the interaction, m_1, vanishes, and these equations become of the second order in slow time.

Stabilization of the explosive interactions of marginal waves was discussed by T.A. Davydova [6]. She has shown that for such interactions in plasma the explosive growth is stabilized by a nonlinear frequency shift and has suggested that such stabilization is typical of explosive interactions of marginal waves. However, for the simplest model of stratified shear flow, the Kelvin-Helmholtz flow, the nonlinear frequency shift does not provide saturation of the explosive growth of interacting waves, but accelerates the growth.

Consider two layers of inviscid incompressible fluids of unequal densities (ρ_1 and ρ_2, $\rho_1 < \rho_2$) moving along the horizontal boundary with unequal velocities (u_1 and u_2, here the subscript 1 denotes the upper layer, and subscript 2 the lower one). The flow is stabilized by gravity force (acceleration g) and surface tension (coefficient σ). In the Lamb frame of reference where $\rho_1 u_1 + \rho_2 u_2 = 0$, and after normalizing all dimensional quantities by length scale $l_0 = \{\sigma/g(\rho_2-\rho_1)\}^{1/2}$ and time scale $t_0 = \{l_0(\rho_2+\rho_1)/g(\rho_2-\rho_1)\}^{1/2}$, the flow stability parameter can be expressed as $\langle u^2\rangle = (\rho_1 u_1^2 + \rho_2 u_2^2)/(\rho_1+\rho_2)$. For every value of $\langle u^2\rangle > 2$ two marginal waves exist whose parameters bound the wavenumber interval of instability. When $\langle u^2\rangle = 3/\sqrt{2}$ the second-harmonic resonant interaction of these waves occurs. The peculiarity of this interaction is vanishing of coefficient m_1 and corresponding increase in the order of amplitude-evolution equations. Moiseev et al. [4] have shown that this peculiarity leads to explosive growth of interacting waves. For example, for free surface displacement $\eta = \varepsilon\{A_1 e^{i(k_1 x-\omega_1 t)} + A_2 e^{i(k_2 x - \omega_2 t)}\}$+c.c. where x is a horizontal coordinate, ε is the small parameter, and c.c. denotes the complex conjugate of the written terms, the governing equations of interaction have self-similar solution $A_2 = A_1 = 6/k_1 q(\tau_0-\tau)^2$ which becomes infinite after the time interval $\tau_0 = (6/qk_1 A_0)^{1/2}$ where $q = 3(\rho_2-\rho_1)/\sqrt{2}(\rho_2+\rho_1)$, $\tau = \mu t$, $\mu = \sqrt{\varepsilon}$,

$A_0 = A_1(0) = A_2(0)$.

Supplemented with next-order terms in μ and ε these equations take the form:

$$\frac{1}{k_1}\ddot{A}_1 = qA_2A_1^* - 4i\mu\alpha\dot{A}_2A_1^* + \varepsilon(-\beta_1|A_1|^2A_1+\gamma|A_2|^2A_1)$$
$$\frac{1}{k_2}\ddot{A}_2 = \frac{q}{2}A_1^2 - 4i\mu\alpha A_1\dot{A}_1+\varepsilon(\gamma|A_1|^2A_2+\beta_2|A_2|^2A_2) \quad (1)$$

where $\ddot{A} = d^2A/d\tau^2$,

$$\alpha=(3\rho_1\rho_2/2\sqrt{2})^{1/2}, \quad \beta_1 = 9/8, \quad \beta_2 = 12(\rho_2-\rho_1)^2-6, \quad (2)$$
$$\gamma = 18(\rho_2-\rho_1)^2-10.5$$

Note that additional terms with higher slow-time derivatives do not appear on the left-hand sides of the equations because the basic equation for free surface displacement is of the second order in time. Equations (1) have the integral

$$J = \frac{|\dot{A}_1|^2}{k_1} + \frac{|\dot{A}_2|^2}{k_2} - q\mathrm{Re}A_2A_1^{*2} \quad (3)$$
$$+\varepsilon^2(\beta_1|A_1|^4 - 2\gamma|A_1|^2|A_2|^2 - \beta_2|A_2|^4)$$

This integral is analogous to the mechanical energy integral for particle moving in 4-dimensional space ($\mathrm{Re}A_1$, $\mathrm{Re}A_2$, $\mathrm{Im}A_1$, $\mathrm{Im}A_2$), the first two terms corresponding to kinetic energy and the remainder to the potential energy. Substituting β_1, β_2, and γ from (2) into (3), one easily finds that biquadratic form in parenthesis is not positively determined for any ρ_1 and ρ_2, and, consequently, potential energy can decrease with increasing $|A_1|$ and $|A_2|$, and the amplitude growth rate increases accordingly. This increase of the growth rate with amplitude is faster than one obtained when additional terms which describe the nonlinear frequency shift are not included. Hence, the nonlinear frequency shift in the present model accelerates the explosion in the contrary to Davydova's example. One can suggest that such interactions may play a certain role at the initial stages of turbulent spots formation in stratified shear flows.

Another possible mechanism of the turbulent spots formation is an instability of a finite-amplitude large-scale internal wave in linearly stable shear flow whose parameters over some height interval (or intervals) are close to a turbulence generation threshold. The periodical violation of stability criteria by the wave causes considerable variations in the turbulent energy density and, therefore, variations in the turbulent transport coefficients. The turbulent momentum and buoyancy fluxes which are proportional to undisturbed gradients of the corresponding quantities and variable transport coefficients, appear in phase with the wave. In contrast to the turbulent viscosity, these fluxes do not lead to the dissipation of the wave energy but cause its growth at the expense of the flow energy. An idea of such wave-turbulent instability (WTI) was formulated by G.Chimonas [7] but the obtained conditions of instability should be revised because his method does not take into account disturbances of the vertical structure of a wave mode by turbulent fluxes. D. Fua et al. [8] presented the results of numerical solution of a model equation for WTI in the atmosphere, but numerical approach couldn't reveal the condition of WTI. In the present work we propose a scheme and discuss the results of analytical treatment of WTI: the sufficient condition of its realization is found and simple formulae for crude order-of-magnitude estimates are obtained.

The study starts from the Euler equations for wave disturbances averaged over turbulent fluctuations in stratified shear flow of incompressible fluid (a model corresponding to the oceanic conditions). The details of mathematical treatment are rather tedious and are presented by Moiseev et al. [9] and, with less detail, by Pungin and Suyazov [10]. We assume that in absence of the wave the local Richardson number, Ri, exceeds the critical value, Ri_{cr}, everywhere in the flow, and the flow is laminar (here Ri_{cr} corresponds to the turbulence generation threshold). We assume, however, that Ri is close to its critical value over some height interval $h < z < h+\Delta h$, and turbulence can be generated periodically driven by a wave disturbance. The resulting condition for WTI on flow parameters and mode structure over this interval is

$$2Ri_{cr}\frac{U'}{\Omega} - 4\frac{\Omega}{U'} < \frac{F'}{kF} < (1+Ri_{cr})\frac{U'}{\Omega} - 2\frac{\Omega}{U'} \quad (4)$$

where U is the undisturbed flow velocity, f(z) is the profile of vertical displacements in the wave, k is the wavenumber, $\Omega = k(c-U)$, c is the phase velocity of the wave, a prime denotes differentiation with respect to vertical coordinate z. It is assumed that the adiabatic invariant of the wave $I_1 = \int_{z_1}^{z_2}[N^2/(c-U)+\frac{1}{2}U'']F^2dz > 0$, where z_1 and z_2 are flow boundaries, $N^2 = -g\rho_0'(z)/\rho_0(0)$ is the buoyancy frequency, $\rho_0(z)$ is the undisturbed vertical density profile of the flow. Under conditions which are close to optimal in the sense of (4), that is when $F'/kF \approx (1+3Ri_{cr})U'/2\Omega - 3\Omega/U'$ we obtain the following equation for the wave amplitude normalized by its threshold value ($a=|A/A_{th}|$):

$$da/dt = F(a) \quad (5)$$

where $A_{th}=(Ri/Ri_{cr}-1)[(U'/2\Omega+\Omega/U')kF]^{-1}$, F(a) is represented in Fig. 1, $\gamma \sim \Omega\Delta h^3k^2/4Ri_{cr}H$, H is the characteristic height interval where wave-induced density disturbance is concentrated. The decrease of growth rate when a >> 1 is due to the fact that phase difference between variable Fourier-component of turbulent energy density, $\sqrt{\tilde{b}^{(k)}}$, and the wave tends to $\pi/2$. Preliminary estimation shows that nonlinear dissipative effects whose

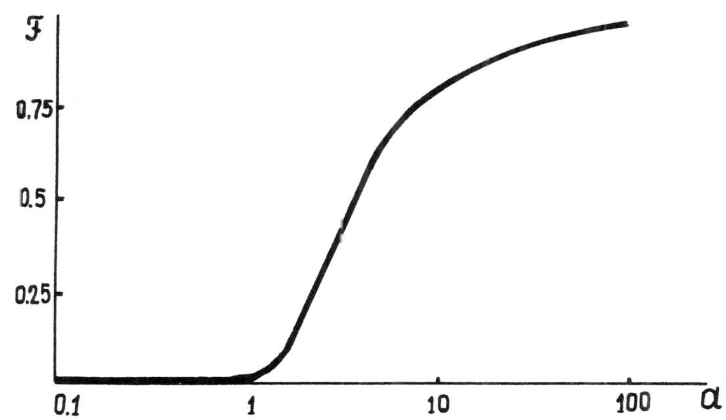

Fig. 1

significance increases with amplitude may lead to further increase of growth rate.But the role of other nonlinear effects, such as wave dissipation owing to wave-induced invariable component of turbulent energy,nonlinear diffusion and convection, has not yet been investigated and will be a subject of further special work.

We deeply appreciate Prof. V.S. Etkin for fruitful discussions.

References

1. Miropolsky, Yu.Z. Dynamics of internal gravity waves in the ocean. Moscow: Nauka 1981.
2. Barenblatt, G.I.: Some turbulent phenomena in a fluid with very stable stratification.In: Nonlinear waves.Stochasticity and turbulence. Gorky: Nauka 1980 89-14.
3. Moiseev, S.S.; Pungin, V.G.; Sagdeev, R.Z.; Suyazov, N.V.; Tur, A.V.; Yanovsky, V.V.: Secondary and correlational instabilities. In: Nonlinear waves. Self-organization.Moscow: Nauka 1983 85-95.
4. Moiseev, S.S.; Pungin, V.G.; Suyazov, N.V.; Etkin, V.S.;Explosive interactions of zero-energy waves in inviscid stratified shear flows. Izv. AN SSSR, FAO. 20 (1984) 9.
5. Davydova, T.A.; Pavlenko, V.P.; Taranov, V.B.; Shamrai,K.P.: The dynamics of wave interactions near the linear stability boundary. Plasma Phys. 20 (1978) 373-382.
6. Davydova, T.A.: Explosive instability of zero energy waves and its stabilization in a plasma with a magnetized ion beam. Int. Conf. on Plasma Phys. Goteborg 1982 10p-11-24 192.
7. Chimonas, G.: The stability of a coupled wave-turbulence system in a parallel shear flow. Bound.-Layer Met. 2 (1972) 444-452.
8. Fua, D.; Chimonas, G.; Einaudi, F.; Zeman O.: An analysis of wave-turbulence interaction. J. Atm. Sci. 39 (1982) 2450-2463.
9. Moiseev, S.S.; Suyazov, N.V.; Etkin V.S.: Wave-turbulent instability of stratified shear flows. Preprint of Space Res. Inst. 905 (1984).
10. Pungin, V.G.; Suyazov, N.V.: Near-threshold instabilities in hydrodynamics. Proc. 2nd Int. Workshop on Nonl. and Turb. Processes in Phys. New York: Gordon and Breach 1984.

Structure, Characteristics of Transition and of Turbulence in the Thermal Convection given by the Direct Numerical Modelling

V. I. POLEZHAEV, A. V. BUNE, V.L. GRIAZNOV

Institute for Problems in Mechanics
Academy of Science of the U.S.S.R, Moscow

Summary

At the Institute for Problems in Mechanics of the Academy of Science of the USSR a package of programs was developed for the direct numerical modelling of transition and turbulent convection on the basis of simultaneous Navier-Stokes equations. The results of computations of the instability, transition and structure of turbulent convection in vertical slot with side heating are discussed. Spectral characteristics of turbulent convection regime are compared with the data of physical simulation in the vertical slot.

The equations and the package of programs

The unsteady Navier-Stokes equatinns (Boussinesq approximation) for two-dimensional case are written in nondimensional form

$$\omega_t + u\omega_r + v\omega_z = \frac{1}{Re}\left(\omega_{rr} + 3Kr^{-1}\omega_{rr} + \omega_{zz}\right) + \mathcal{F} \qquad (1)$$

$$\Psi_{rr} - Kr^{-1}\Psi_r + \Psi_{zz} = -\omega r^{2K} \qquad (2)$$

$$\theta_t + u\theta_r + v\theta_z = \frac{1}{Re\,Pr}\left(\theta_{rr} + Kr^{-1}\theta_r + \theta_{zz}\right) \qquad (3)$$

$$C_t + uC_r + vC_z = \frac{1}{Re\,Pr}\left(C_{rr} + Kr^{-1}C_r + C_{zz}\right) \qquad (4)$$

$$\mathcal{F} = -r^{-K}\left[\frac{Ra}{Re^2 Pr}(\theta_z \sin\varphi - \theta_r\cos\varphi) + \frac{R_D}{Re^2 Pr}(C_z\sin\varphi - C_r\cos\varphi)\right]$$

Here ω **is vorticity**, Ψ - stream function, θ - temperature, C - concentration, $Ra=\frac{g\beta L^3}{\nu a}\Delta T$ - Rayleigh number, $Ra_D=\frac{g\beta L^3}{\nu D}\Delta C$ - diffusive Rayleigh number, $Re=\frac{vL}{\nu}$ - Reynolds number, $Pr=\nu/a$ - Prandtl number, $Pr_D=\nu/D$ - diffusive Prandtle number (Schmidt number), K - geometry index (K=0 - flat, K=1 - axial symmetrical case), φ - angle of inclination.

A package of programs developed at the Institute for Problems in Mechanics in 1980 - 1982 includes the groups of programs for solving equations (1-4) by means of finite-difference method, the grafic visualisation systems and the programs for statistical analysis of numerical results /1,2/. The history, methodological problems and review of previous papers can be found in /3/. It is possible to analyse laminar, transition and turbulent regimes of thermal (Ra), concentrational (Ra_D) natural convection, forced convection (Re) and the interaction of these different types of convection.

Convection in a vertical slot: stability and transition

Let us consider a thermal convection in a closed vertical layer with aspect ratio H/L and temperatures T_1 and T_2 on the side walls, two other boundaries being heat insulated. The main parameters, significant for heat and mass transfer in convection, are

$$Ra, \quad H/L, \quad Pr. \qquad (5)$$

There is good agreement between numerical and experimental results obtained for laminar convection regimes in vertical slots. Transition and turbulent regimes are studied in less details.

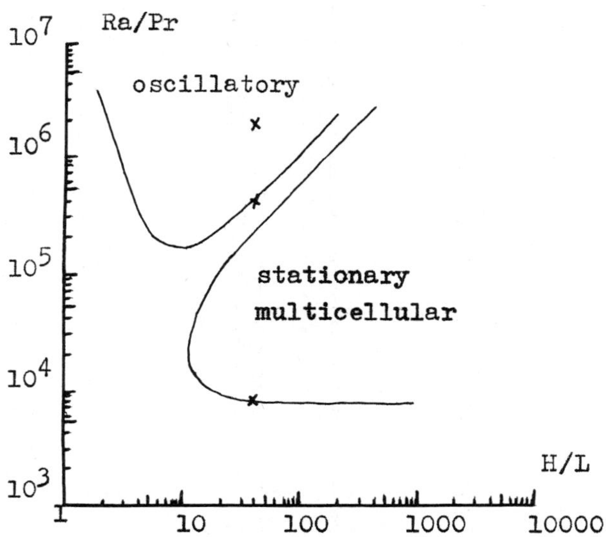

Fig.1. Boundaries of convective stability in vertical slot for Pr=0.71 in Ra/Pr, H/L plane. Marked points are suitable to Fig.2.

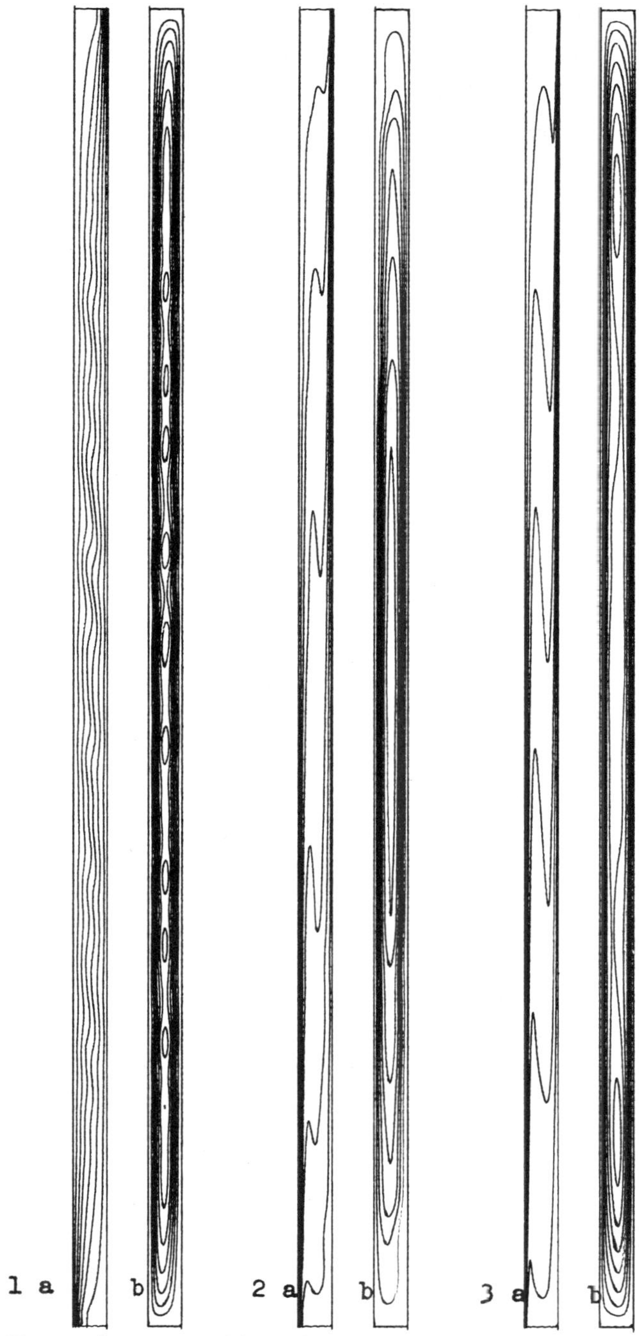

Fig. 2. Thermal convection in long vertical slot, aspect ratio H/L=40, Pr=0.71, space mesh 200 x 65. Extreme values of nondimensional temperature (a) are 1.,0., and step of isolines is 0.15; Rayleigh number, extreme values of stream function (b) and steps of isolines are following: (1) Ra= 5000, 23.2,0.,3. (2) Ra= 300000, 382.7,0.,50. (3) Ra= 1280000, 29.7,0.,4.

Concerning a stability of rising and falling flow in a closed vertical slot, an approximate mathematical model plays a significant role. The main motion in this model corresponds to a layer of infinite length, and the aspect ratio H/L is taken into account by a parameter of temperature stratification caused by convection. A stability diagram which was obtained in paper /5/ for Pr=0.71 is shown in Fig. 1. Stability maps can be found also in work /6/. These maps are recently used for description of transition regimes in vertical slots, they are insufficiently tested experimentally and by numerical solutions of Navier-Stokes equations.

We have performed a number of calculations increasing the Ra number, while H/L and Pr were constant. In one of these series there was performed a calculation of stationary convection regime from one relatively low Ra to another Ra. The isotherms and streamlines are shown in Fig. 2 for H/L=40, Pr=0.71 in a vertical slot with Ra= $5\cdot10^3$; $3\cdot10^5$; $1.28\cdot10^6$. At first, there is a multicellular convection which is replaced, in full accordance with the margins of stability regions, by a stable monocellular convection and then, it changes to an unsteady multicellular convection. A seemingly impossible decrease in number of cells with Ra being raised (firstly apparently found in the paper /5/ for lower H/L) is due to an influence of stable stratification caused by convection.

Another series of computations firstly performed in /9/ is made to study convection emerging while gravity is suddenly imposed on an initially motionless fluid with a stationary distribution of temperature. Streamlines at different times in the case Ra = $3.75\cdot10^8$, H/L = 11.2, Pr = 15 are shown in Fig. 3. Time evolution includes following stages:

1) An emerging of rise-and-fall monocellular motion
2) A break of stability of that motion and arising of multicellular motion
3) A break of stability of multicellular structure
4) Development of turbulent boundary region.

Colour display film illustrates convection structures in vertical slot for the two mentioned series of calculations.

1 2 3 4 5 6 7 8

Fig.3. Transition from laminar to turbulent structure in the vertical slot. Pr = 15, H/L = 11.2, Ra = 3.75 10^8. Space mesh 51 x 128. Streameline patterns at the subsequent moments of time Extreme values of stream function and step of isolines are also indicated. (1) 0.625s, 0., -744.5, 150. (2) 4.375s, 1752.4, -2366.9, 800. (3) 6.875s, 712.9, -912.5, 300. (4) 8.75s, 473.3, -1156.9, 300. (5) 13.75s, 203.5, -708.4, 200. (6) 25s, 37.6, -270.3, 60. (7) 31.25s, 27.7, -122.9, 30. (8) 47.5s, 25.0, -83.2, 20.

Coherent structures by the data of numerical simulation

The following regions of the turbulent convection in a slot can be distinguished as a result of numerical simulation: 1) a viscous sublayer, 2) a core, 3) an intermediate zone. The last is compounded from two types of elements: the regular structures, that rise, develop and dissipate, and some chaotic structures, or the "bursts", whose emergence is correlated with destruction of regular elements. The term "bursts" is introduced in paper /7/ concerning a mechanism which governs a forced motion.

Spectral characteristics

While statistical treatment of numerical realisations of a fully developed turbulent regime, it is significant to make a correct choice of the initial value of the time (t_o) and the length of the time interval (t_N). The results of computation and their comparison with the mean values of temperature and velocity pulsations obtained experimentally are given in /2,9/. We will now concern an important spectral characteristic, namely a spectral density of temperature pulsations. The results of computation and their correlation with experimental data are hardly available. We have made a number of attempts to calculate it through several realisations including some improvements among others in the choice of (t_o, t_N) and the method of calculation of spectrum. It is signigicant that recently there was achieved a definite agreement of the methods which are treating experimental and numerical data. A comparison of experimental results /8/ and the results of computations is shown in Fig. 4.

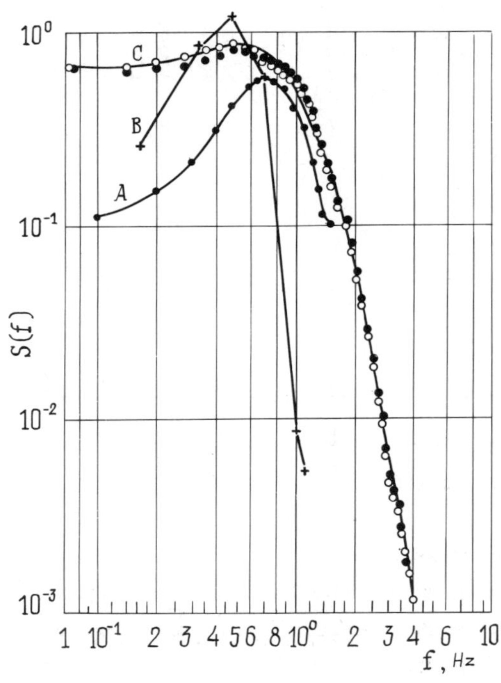

Fig. 4. Spectral density of temperature oscillations in the viscous sublayer. Curve (A) - this work, computer EC 1040, obtained 1600 time steps, space mesh 128 x 51 . Curve (B) - data /9/, computer ICL 1906A, obtained 2100 time steps, space mesh 128 x 39 . Curve (C) - experimental data /8/.

It should be mentioned that in paper /9/ the comparison was made with data obtained not in vertical slot, but on a single plate. The results of experiment and computation in Fig. 4 correspond to a vertical layer by the same Ra, H/L, Pr. The finest results of computation by now are in a good accordance with experimental data in the frequency range f ~ 1-2 Hz and deviate in a low frequency range f ~ 0.5 Hz.

References

1. Bune, A.V.; Griaznov, V.L.; Dubovik, K.G.; Polezhaev, V.I.: A program complex for simulation of hydrodinamics, heat- and mass transfer on the basis of simultaneous Navier-Stokes equations. In: Pakety prikladnyh programm. Vychislitelnyi experiment.(Ser. Algoritmy i algoritmicheskie iazyki). Moscow: Science 1983. 70-85.

2. Polezhaev, V.I.; Griaznov, V.L.; Daikovsky, A.G.; Fedoseyev, A.I.: Numerical models of turbulent convection based on the unsteady Navier-Stokes equations. Proceedings of the 6th Int. Conf. Numer. Meth. in Fluid Dynamics. Tbilisi 1978. In: Lecture Notes in Physics v.90 Springer-Verlag 1978. 454-459.

3. Paskonov, V.M.; Polezhaev, V.I.; Choodov, L.A.: A numerical simulation of heat and mass transfer processes. Moscow: Science 1984.

4. Polezhaev, V.I. : Flow and heat transfer by a laminar convection in a vertical slot. In.: Teplo- i masso- perenos v.1 Moscow: Energia 1968.

5. Roux, B.; Grondin, J.C.; Bontoux, P.; De Vahl Davis, G.: Reverse transition from multicellular to monocellular motion in vertical fluid layer. Physico-chemical hydrodynamics 3 PCH 80 Madrid: Europhysics Edition 1980.

6. Gershuni, G.Z.; Zhukhovitsky, E.M.: Convective stability of the incompressible fluid. Moscow: Science 1972.

7. Corino, E.R.; Brodkey, R.S.: A visual investigation of the wall region in turbulent flow. J. Fluid Mech. 37 (1969)

8. Kirdjashkin, A.G.; Semenov, V.I.; Berdnikov, V.S.; Gaponov, A.S.: A vertical slot temperature structure by a thermogravitational convection. Teplophizika vysokih temperatur v.20, n.5 (1982) 922-928.

9. Dajkovskyi, A.G.; Polezhaev, V.I.; Fedoseev, A.I.: A study of flow structure in a transition and turbulent regimes in a vertical slot. Izvestija AN SSSR; MZG (Mechanika zhidkosti i gaza) n.6 (1978) 66-75.

Development of Disturbances and Transition to Turbulence at the Convection in Two-Layer Systems

G. Z. GERSHUNI, E. M. ZHUKHOVITSKY, A. A. NEPOMNYASCHIY,
I. B. SIMANOVSKY

Institute of Continuum Mechanics
Ural Scientific Center of the USSR Academy of Sciences,
State Pedagogical Institute of Perm',
State University of Perm'
Perm' USSR

Convective stability of mechanical equilibrium and dynamical processes of finite disturbance evolution in the system of two flat horizontal fluid layers is investigated. Peculiarity of the dynamic behaviour of such systems reveals in the fact that the process of the development of disturbances is dominated by hydrodynamic and heat interaction at the interface. These interactions lead to qualitatively new mechanisms of disturbance evolution which effectively influence both the process of the stability loss and the regularities of the development and interaction of finite disturbances.

1. Statement of the Problem

Consider the system of two flat horizontal layers of viscous unmixing fluids with different physical properties. The full thickness of two-layer system is H. The ratio of the thickness of the lower layer to the full thickness H is designated by h ($0 \leq h \leq 1$). The system is limited from above and below by rigid isothermal planes subject to different temperatures; a full temperature drop θ may be positive (heating from below) or negative (heating from above). At the interface which is assumed to be flat normal velocity components vanish and the continuity conditions for tangential components of velocity and viscous stress, temperature and heat flux also take place. The problem contains the following dimensionless parameters: relative thickness of the lower layer h; Rayleigh number $R = g\beta_1 A_1 H^4/\nu_1\chi_1$; Prandtl number $P = \nu_1/\chi_1$; ratios of fluid parameters: $\eta = \eta_1/\eta_2$, $\nu = \nu_1/\nu_2$, $\chi = \chi_1/\chi_2$, $\varkappa = \varkappa_1/\varkappa_2$, $\mu = \nu_1\beta_2/\nu_2\beta_1 = \nu/\beta$, $\beta = \beta_1/\beta_2$ (indices 1 and 2 are related respectively to the lower and upper layers; A_1 is the temperature gradient in the lower layer in equilibrium state; other notations are standard).

2. Linear Stability of Mechanical Equilibrium

We first deal with the case of heating from above (R 0). The calculations show the possibility of the specific mechanism of instability [1]. The most preferable situation for its realization would be the one in which one of the fluids has relatively high thermal diffusivity χ and the other has relatively low volume expansion coefficient β. The boundaries of stability regions are presented in fig. 1; the curves 1-5 correspond to the values of parameters $\varkappa, \eta$: 1 - (0.1;0.1); 2 - (1;1); 3 - (10^4;10^4); 4 - (10;1); 5 - (1;10). The region of instability is situated in the range of small values μ and χ^{-1}. The curve 6 is the limit one: in the region marked by shading the equilibrium is stable for heating from above at any values of $\varkappa, \eta$. Calculations carried out on the basis of full nonlinear system of equations show the soft character of instability for rolls [2]. For heating from below the instability of equilibrium is due to the usual Rayleigh mechanism which in a two-layer system acts quite specifically. Consider for example the formic acid - transformer oil system. Minimum critical Rayleigh number R_m depends on h. If h is small then the stability loss of the equilibrium is connected with excitation of convection in oil layer. In the opposite case (h$\geqslant$1) instability arises due to the development of disturbances in the thick layer of acid. At some intermediate h_* the levels of instability cross. According to

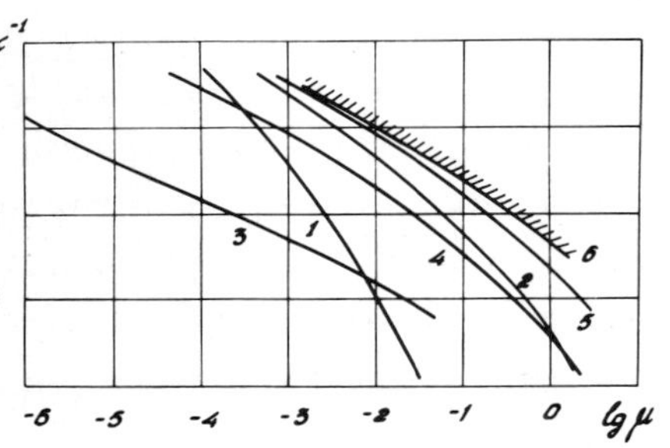

Fig. 1.

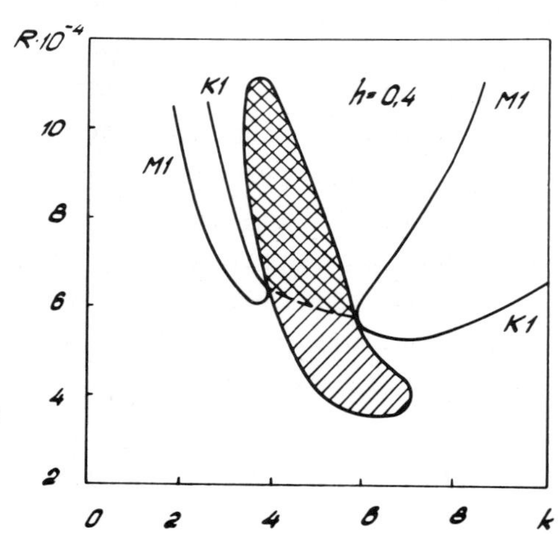

Fig. 2.

calculations $h_* \simeq 0.4$ that is in good agreement with the experiment [3].

The two-layer system is found to become unstable with respect to oscillating disturbances as well [4]. Oscillating instability arises in the range of parameters where the critical Rayleigh numbers corresponding to two different modes of instability M1 and K1 are close (M1 and K1 - the symbols of instability modes corresponding to onset of convection in the layers of oil and acid). Nonselfadjointness of the problem makes impossible "simple crossing" of neutral curves, and in the region of their proximity a zone of oscillatory instability appears. This situation is presented in fig.2 where neutral curves of levels M1 and K1 for h=0.4 are shown. The lines M1 and K1 correspond to the boundaries of monotonous instability; the region of oscillating disturbances is marked by shading; the region of growing oscillations - by double shading; the boundary of oscillatory instability - by the dashed line. Thus, the consideration of linear stability of the system leads to conclusion that in the two-layer system under definite conditions it appears possible monotonous instability for heating from above as well as oscillatory instability for heating from below.

3. Onset of Convection in Two-layer System

Consider the onset of convection after the loss of stability. Let us study the bifurcation for the case of three-dimensional convection in a closed cavity of arbitrary form. The results obtained may be also applied to two-dimensional motions in an infinite horizontal cylinder of any section and to motions in regions with periodic boundary conditions (both for heating from above and below). Suppose that for two-layer system there is monotonous equilibrium instability at $R=R^{(0)}$. We expand the deviation U of the solution of nonlinear problem from equilibrium at small $R-R^{(0)}$ in powers of a small parameter

$$U=\varepsilon U^{(1)}+\varepsilon^2 U^{(2)}+\dots, \qquad R=R^{(0)}+\varepsilon R^{(1)}+\varepsilon^2 R^{(2)}+\dots .$$

Substituting the expansion in the full system of equations, in the first order in ε we get homogeneous linear problem determining $R^{(0)}$, $U^{(1)}$. The second order in ε yields the nonhomogeneous problem, the solvability condition of which defines $R^{(1)}$ which usually differs from zero. Thus, there is a subcritical nonlinear instability. However, in some cases the value $R^{(1)}$ vanishes due to the symmetry of the problem. For example, in infinite horizontal layers $R^{(1)}=0$ for the flows in the form of rolls and rectangular cells, while for hexagonal cells $R^{(1)}\neq 0$. Oscillatory instability leads to nonstationary convective regimes; the bifurcation is always unidirectional.

4. Developed Regimes of Thermogravitational Convection

The developed nonlinear regimes of convection were studied numerically by the finite-difference method [5,6]. Consider the evolution of nonlinear regimes for the case of silicone oil DC-200 - water system, filling a horizontal cylinder of rectangular cross-section ($-L\leqslant x\leqslant L$, $0\leqslant z\leqslant H$) with the ratio $l=2L/H=0.4$; $h=1/2$. The stream function ψ_i was scaled with the unit ν_2. To characterize the structure of the flow we introduce the following quantities:

$$S_1=\int_{-L}^{0}dx\int_{h}^{1}dz\,\psi_2(x,z),\quad S_2=\int_{0}^{L}dx\int_{h}^{1}dz\,\psi_2(x,z),\quad S_-=S_1-S_2$$

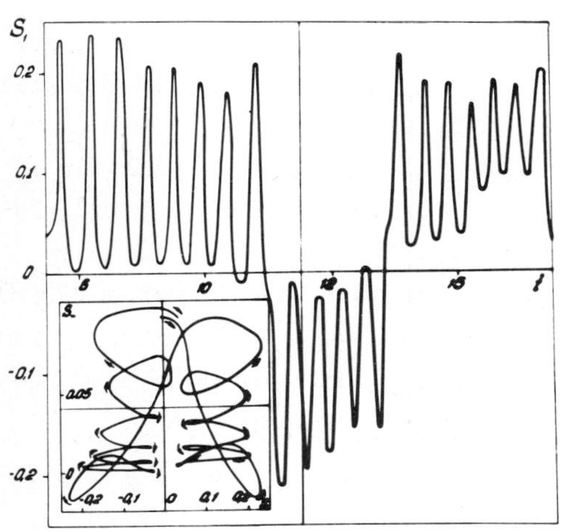

Fig. 3.

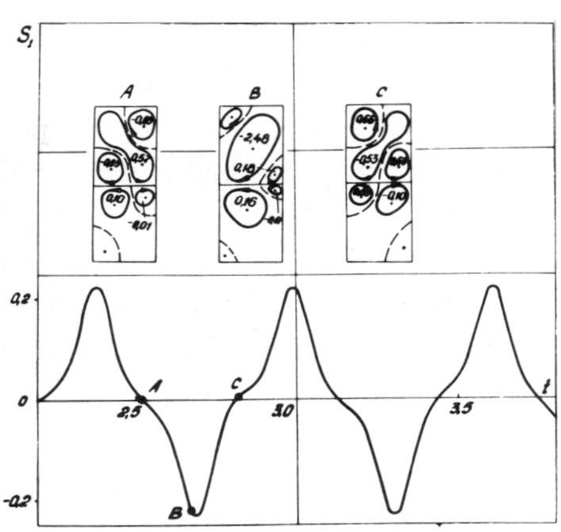

Fig. 4.

With the exceeding of the critical Rayleigh number R_0=455 equilibrium loses stability and stationary convective motion (regime 1) develops; the intensity of motion in the upper fluid is considerably higher than that in the lower. It is explained by the fact that "local" Rayleigh numbers for lower and upper fluid differ greately. The motion in the upper fluid is principally one-vortical. With growth of R the stationary motion becomes unstable and the oscillations develop in the system. One observes in succession the following patterns: periodic two-vortical (regime 2), irregular two-vortical (regime 3), irregular four-vortical (regime 4) and regular four-vortical (regime 5). So, in the problem under consideration the region of irregular oscillations is bounded in the parameter R both from below and above. A typical picture fragment of variation of the value S_1 with the time in the region of irregular oscillations (R=2650) and corresponding phase trajectory are presented in fig.3. The number of oscillations with $S_1>0$ and $S_1<0$ changes irregularly. The graph $S_1(t)$ and the pictures of streamlines and isotherms for regime 5 corresponding to the points A,B,C are shown in fig.4.

5. Thermocapillary Convection in Two-layer Systems

In the majority of works the free surface temperature is subject to the third kind boundary conditions. The two-layer formulation of problem allows to find a number of qualitatively new specific peculiarities of the present phenomenon [7-9]. Let the surface tension depend linearly on temperature. A new dimensionless parameter is the Marangoni number Mr. The linear calculations show that monotonous instability may arise for heating both from below and above for the same fluid system, depending on the layer thickness ratio. In the longwave regions the oscillatory instability can arise which may be the only possible mechanism of equilibrium instability. New possibilities for the onset of oscillatory instability are provided by simultaneous action of thermocapillary and thermogravitational mechanisms. In contrast to the case considered in [4] oscillatory mode becomes most dangerous with the increase of Mr. Let us consider the evolution of nonlinear solutions in the model system: $\eta = \nu = 2$, $\varkappa = \chi = \beta = 1$, P=1, h=0.5. If L=2.5, R=0 there is an oscillatory instability of equilibrium and convective four-vortical oscillations develop. With increase of Mr the period of oscillations grows, and finally stationary convective motion is realized. For L=2 one observes the bifurcation of the period doubling, followed by the breaking of the flow symmetry. The inclusion of thermogravitational mechanism for the same system brings about the development of stochastic motions limited by Mr from below - by regular oscillations and from above - by stationary motions.

References

1. Gershuni, G.Z.; Zhukhovitsky, E.M.: On equilibrium instability of horizontal layer system of unmixing fluids heated from above. Izv. Akad. Nauk SSSR. Mekh. Zh. i Gaza (1980) 28-34.

2. Gershuni, G.Z.; Zhukhovitsky, E.M.; Simanovsky, I.B.: On equilibrium stability and finite-amplitude motions in a two-layer system heated from above. Convective flows. Perm': 1981, 3-11 (in Russian).

3. Adilov, R.S.; Putin, G.F.; Shaidurov, G.F.: Convective stability of two unmixing fluids in a horizontal slot. Scient. Notes of the Perm' Univ. 362 (1976) 16-20 (in Russian).

4. Gershuni, G.Z.; Zhukhovitsky, E.M.: On monotonous and oscillatory instability of a two-layer system of unmixing fluids heated from below. Dokl. Akad. Nauk SSSR (1982),265,302-305.

5. Simanovsky, I.B.: Finite-amplitude convection in a two-layer system. Izv. Akad. Nauk SSSR. Mekh. Zh. i Gaza 5 (1979) 3-9.

6. Nepomnyaschiy, A.A.; Simanovsky, I.B.: On oscillatory regimes of convection in a two-layer system. Inzh.-Fiz. Zh. 46 (1984) 845-849 (in Russian).

7. Nepomnyaschiy, A.A.; Simanovsky, I.B.: Thermocapillary convection in a two-layer system. Dokl. Akad. Nauk SSSR (1983), 272, 825-827.

8. Nepomnyaschiy, A.A.; Simanovsky, I.B.: Thermocapillary convection in a two-layer system. Izv. Akad. Nauk SSSR. Mekh. Zh. i Gaza 4 (1983) 158-163.

9. Nepomnyaschiy, A.A.; Simanovsky, I.B.: Thermocapillary and thermogravitational convection in a two-layer system with distorted interface. Izv. Akad. Nauk SSSR. Mekh. Zh. i Gaza 3 (1984) 178-182.

On some Transition Processes in Spatial Flows with a Transversal Velocity Gradient

V. JA. SHKADOV, E. A. DEMEKHIN

Department of Mechanics and Mathematics Moscow University

Summary

Nonstationary development and the limiting form for waves in viscous fluid flows under the action of gravitational and capillary forces are investigated. Instability, bifurcation and transition are considered within the framework of Galerkin multi-mode approximations.

Nonstationary 3-dimensional waves in a thin film of viscous fluid are described by the periodical in x, z solutions of the following system of equations

$$q_t + \frac{6}{5}\left(\frac{q^2}{h}\right)_x + \frac{6}{5}\left(\frac{qp}{h}\right)_z = \frac{1}{5\delta}\left[h(h_{xx}+h_{zz})_x + h - \frac{q}{h^2}\right]$$

$$p_t + \frac{6}{5}\left(\frac{qp}{h}\right)_x + \frac{6}{5}\left(\frac{p^2}{h}\right)_z = \frac{1}{5\delta}\left[h(h_{xx}+h_{zz})_z - \frac{p}{h^2}\right]$$

$$h_t + q_x + p_z = 0$$

Here h,q,p are the local film thickness and the rate of flow in x,z directions, correspondingly. The value of δ is connected with the Reynolds number and with physical parameters of the fluid. The system is derived from the Navier-Stokes equations [2,3]. For $\delta \to 0$ the system is reduced to the equation

$$H_t + \left(\frac{\partial^2}{\partial X^2} + \frac{\partial^2}{\partial Z^2}\right)^2 H + H_{xx} + HH_x = 0$$

while for $\delta \to \infty$ the system is reduced to the equations of mathematical hydraulics.

For 2-dimensional flows the trivial solution h=q=1 is unstable to linear disturbances with the wave numbers $\alpha \in (0, \alpha_0)$, $\alpha_0 = \sqrt{15\delta}$. For every pair of $\delta > 0$, $\alpha \in (0, \alpha_0)$ the nonlinear solutions exist in the form of stationary running waves. The wave solutions of the first family with the phase velocity $c < 3$ branch off of the trivial solution at the neutral curve $\alpha = \alpha_0$. These waves exist at wave numbers $\alpha \in (0, \alpha_0)$ and with $\alpha \to 0$ have the negative solitary wave solution $(c-3 < 0)$ as their limit. The wave solutions of the second family branch off from the first family at $\alpha \simeq 0.5\,\alpha_0$. These waves have phase velocity $c > 3$, they exist at wave numbers $\alpha \in (0, \alpha_2)$ and at $\alpha \to 0$ have positive solitary wave solution $(c-3>0)$ as their limit. There also exist the other wave solutions which have the solitary wave solution, near-periodic solution or the solution of chaotic structure as their limits at $\alpha \to 0$. As it follows from the direct computations, the solitary wave solutions constitute a countable set or a Cantor set. There are also solutions of the shock type such that $h \to h_1$ $(x \to -\infty)$, $h \to h_2$ $(x \to \infty)$.

The evolution of the nonstationary waves in a 2-dimensional flow is investigated for initial conditions of two types: one harmonic mode α is excited at $t=0$; all harmonic modes $n\alpha$ with the amplitudes and the phases determined by the random number generator are excited at $t=0$. As a result of numerical experiments the next general sequence of events during the wave development with the time was found out. 1. For sufficiently small initial amplitudes the curves $\ln h_{max}(t)$, $\ln h_{min}(t)$ are linear with the same angle of slope in accordance with the linear theory. 2. After the linear stage, the "hyperbolic" behavior takes place when the nonlinear terms begin to play their role, but the capillary forces are small and have little influence upon the solution. The curves $\ln h_{max}(t)$, $\ln h_{min}(t)$ are linear with different angles of slope, the variation of $\ln(h_{max}-h_{min})$ with t is in accordance with the most unstable linear mode. 3. Because of the capillary forces action the oscillations appear at the most steep part of the wave. As a result of oscillations, the waves do not tip over and the transition processes appear. 4. The solution tends to a stationary running nonlinear wave or to the two-periodical solution of the torus type.

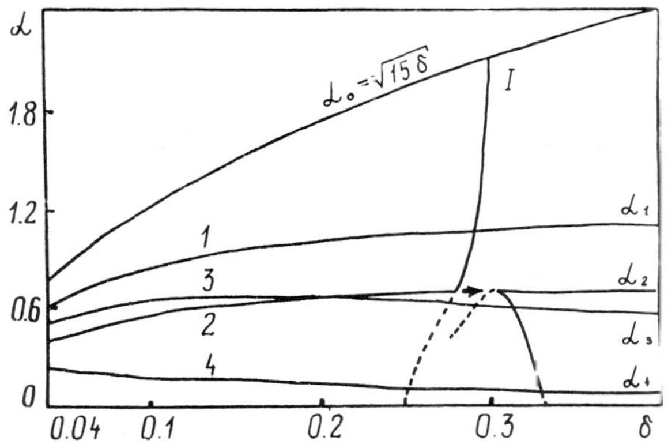

Fig. 1. Wave regimes in the region of linear instability

There are some characteristic features of the wave development for different values of the parameters $\alpha \in (0, \alpha_0)$, $\delta \in (0.04, 0.4)$. If $\alpha \in (\alpha_2, \alpha_0)$, the solution tends to the first family running wave after stages 1,4. For the first family waves are unstable (dashed line in Fig.1) , and the solution tends to the second family wave after stages 1-4 (solid line in Fig. 1). At $\alpha \to 0$ bifurcation takes place at $\alpha = \alpha_2$. The solution jumps from the first to the second family waves at that point. It may have nondamping two-periodical oscillations for α in the vicinity of curve 2 in Fig. 1. For $\alpha = \alpha_4$ (curve 4 in Fig. 1) the second family waves give maximum flow rate. In the vicinity of curve 4 the second family waves become unstable and nondamping oscillations with the wave numbers α, 2α or α, 3α may occur. At the transitional stage the first family optimal waves [2], which give maximum flow rate, may take place during the long period of time (curve 1 in Fig. 1).

Three-dimensional nonstationary solutions are investigated for Galerkin model approximation with a few modes. The limiting form of the solutions at $t \to \infty$ are the stationary running waves (the limiting cycles in the space of Galerkin coefficients) or the two-periodical solutions of the torus type. The torus turns to the homoclinic orbit at $\delta \rightarrow 0$. There are stationary two-dimensional saddle solutions which are stable to two-dimensional disturbances and unstable to three-dimensional disturbances. At arbitrarily small but finite values of δ, homoclinic orbit transforms into a two-dimensional torus. In some cases the trajectories jump from one to another torus with the more frequent oscillations. At finite values of δ the waves have a form of weakly curved horseshoes put in chess order (Fig. 2). The horseshoes of two-periodical solution change only their amplitude or both their amplitude and form by the oscillations. For one phase of the oscillations the solution has nearly a two-dimensional form which is disturbed by the instability and transforms into one of the three-dimensional forms. At large values of δ the transition process appears as a chaotic change of the three-dimensional waves of various forms. The theory of

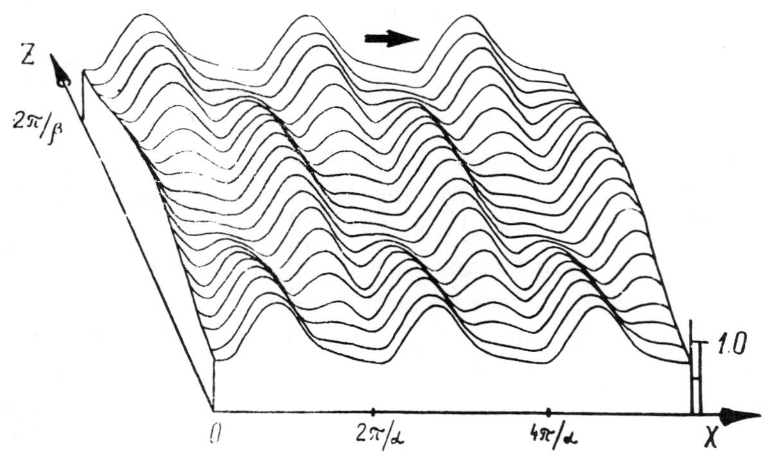

Fig. 2. Three-dimensional wave. $\delta = 0.2$

nonlinear waves considered here gives an opportunity to understand many fundamental experimental facts concerning wavy fluid films. In Fig. 3 examples of comparing the theory and experimental results are given.

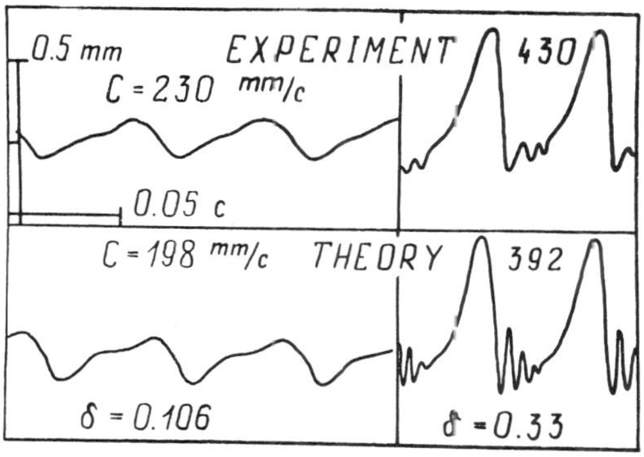

Fig. 3. Wave forms and phase speeds

References

1. Kapitsa, P.L.; Kapitsa, S.P.: Volnovoe techenie tonkikh sloev viazkoy zhidkosti. ZHTEF, 19(1949).
2. Shkadov, V.Ya.: Volnovie rezhimi techenia tonkogo sloya viazkoi zhidkosti pod vozdeistviem sili tiazhesti. Izv. AN SSSR. MZHG, N 1(1967) 43-51, N 2 (1968) 20-25.
3. Shkadov, V.Ya.: Uedinennie volni v sloe viazkoi zhidkosti. Izv. AN SSSR, MZHG, N 1(1977) 63-66.
4. Demekhin, E.A.; Shkadow, V.Ya.: O nestatsionarnikh volnakh v sloe viazkoy zhidkosti. Izv. AN SSSR, MZHG, N 3 (1981) 151-153.
5. Demekhin, E.A.; Demekhin, I.A.; Shkadov, V.Ya.: Solitoni v stekaiushchikh sloyakh viazkoi zhidkosti. Izv. AN SSSR, MZHG, N 4 (1983), 9-16.

IUTAM Symposia

Measuring Techniques in Gas-Liquid Two-Phase Flows
Symposium, Nancy, France, July 5–8, 1983
Editors: J.M.Delhaye, G.Cognet
1984. 430 figures. XXIII, 746 pages
ISBN 3-540-12736-4

Structure of Complex Turbulent Shear Flow
Symposium, Marseille, France, August 31 – September 3, 1982
Editors: R.Dumas, L.Fulachier
1983. 326 figures. XIX, 444 pages
ISBN 3-540-12156-0

Nonlinear Deformation Waves
Symposium, Tallinn, Estonian SSR, USSR, August 22–28, 1982
Editors: U.Nigul, J.Engelbrecht
1983. 145 figures. XXIII, 453 pages
ISBN 3-540-12216-8

Stability in the Mechanics of Continua
2nd Symposium, Nümbrecht, Germany, August 31 – September 4, 1981
Editor: F.H.Schroeder
1982. 167 figures. XI, 412 pages
ISBN 3-540-11415-7

Three-Dimensional Turbulent Boundary Layers
Symposium, Berlin, Germany, March 29 – April 1, 1982
Editors: H.H.Fernholz, E.Krause
1982. 288 figures. XV, 389 pages
ISBN 3-540-11772-5

Creep in Structures
3rd Symposium, Leicester, UK, September 8–12,1980
Editors: A.R.S.Ponter, D.R.Hayhurst
1981. 243 figures. XVI, 615 pages
ISBN 3-540-10596-4

Physical Non-Linearities in Structural Analysis
Symposium Senlis, France, May 27–30, 1980
Editors: J.Hult, J.Lemaitre
1981. 109 figures. XI, 287 pages (30 pages in French). ISBN 3-540-10544-1

Unsteady Turbulent Shear Flows
Symposium Toulouse, France, May 5–8, 1981
Editors: R.Michel, J.Cousteix, R.Houdeville
1981. 283 figures. XXI, 424 pages
ISBN 3-540-11099-2

Atmospheric Dispersion of Heavy Gases and Small Particles
Symposium, Delft, The Netherlands August 29–September 2, 1983
Editors: G.Ooms, H.Tennekes
1984. 210 figures. IX, 440 pages
ISBN 3-540-13491-3

Laminar-Turbulent Transition
Symposium Stuttgart, Germany, September 16–22, 1979
Editors: R.Eppler, H.Fasel
1980. 289 figures. XVIII, 432 pages
ISBN 3-540-10142-X

Physics and Mechanics of Ice
Symposium Copenhagen, August 6–10, 1979
Technical University of Denmark
Editor: P.Tryde
1980. 159 figures, 13 tables. XIV, 378 pages
ISBN 3-540-09906-9

Practical Experiences with Flow-Induced Vibrations
Symposium Karlsruhe, Germany, September 3–6, 1979, University of Karlsruhe
Editors: E.Neudascher, D.Rockwell
International Association for Hydraulic Research
1980. 429 figures, 217 charts, 29 tables. XXII, 849 pages. ISBN 3-540-10314-7

Mechanics of Sound Generation in Flows
Joint Symposium Göttingen, Germany, August 28–31, 1979
Max-Planck-Institut für Strömungsforschung
Editor: E.-A. Müller
International Commision on Acoustics
American Institute of Aeronautics and Astronautics
1979. 177 figures, 6 tables. XV, 300 pages
ISBN 3-540-09785-6

High Velocity Deformation of Solids
Symposium Tokyo, Japan, August 24–27, 1977
Editors: K.Kawata, J.Shioiri
1978. 230 figures, 20 tables. XVIII, 452 pages
ISBN 3-540-09208-0

Dynamics of Multibody Systems
Symposium Munich, Germany, August 29–September 3, 1977
Editor: K.Magnus
1978. 107 figures. XVI, 376 pages
ISBN 3-540-08623-4

Buckling of Structures
Symposium Cambridge, USA, June 17–21, 1974
Editor: B.Budiansky
1976. 214 figures. VIII, 398 pages
ISBN 3-540-07274-8

Probabilistic Methods in the Mechanics of Solids and Structures
Symposium Stockholm, Sweden, June 19–21, 1984
To the Memory of Waloddi Weibull
Editors: S.Eggwertz, N.C.Lind
1985. 146 figures. XXIV, 610 pages
ISBN 3-540-15087-0

Symposium Transsonicum II
Göttingen, Germany, September 8–13, 1975
Editors: K. Oswatitsch, D. Rues
1976. 324 figures. XVI, 574 pages
ISBN 3-540-07526-7

Dynamics of Rotors
Symposium Lyngby, Denmark, August 12–16, 1974
Editor: F. I. Niordson
With contributions by numerous experts
1975. 195 figures. XII, 564 pages
ISBN 3-540-07384-1

Mechanics of Visco-Elastic Media and Bodies
Symposium Gothenburg, Sweden,
September 2–6, 1974
Editor: J. Hult
With contributions by numerous experts
1975. 60 figures. XII, 391 pages (363 pages in English, 28 pages in French). ISBN 3-540-07228-4

Optimization in Structural Design
Symposium Warsaw, Poland, August 21–24, 1973
Editors: A. Sawczuk, Z. Mróz
1975. 216 figures, 27 tables. XV, 585 pages
ISBN 3-540-07044-3

Photoelastic Effect and Its Applications
Symposium Ottignies, Belgium,
September 10–16, 1973
Editor: J. Kestens
With the cooperation of the Permanent Commitee for Stress Analysis and the Society for Experimental Stress Analysis (SESA)
1975. 216 figures. XI, 638 pages
ISBN 3-540-07278-0

Satellite Dynamics
COSPAR - IAU - IUTAM, Symposium São Paulo, Brazil, June 19–21, 1974
Editor: G. E. O. Giacaglia
Executive Editor: A. C. Stickland
With contributions by numerous experts
1975. 5/86 figures. VIII, 376 pages
ISBN 3-540-07087-7

Flow-Induced Structural Vibrations
IUTAM/IAHR Symposium Karlsruhe, Germany, August 14–16, 1972
Editor: E. Naudascher
1974. 360 figures. XX, 774 pages
ISBN 3-540-06317-X

Creep in Structures, 1970
Symposium Gothenburg, Sweden,
August 17–21, 1970
Editor: J. Hult
1972. 171 figures. XI, 429 pages
ISBN 3-540-05601-7

Applied Mechanics
Proceedings of the Eleventh International Congress of Applied Mechanics, Munich, Germany, 1964
Editor: H. Görtler. In cooperation with P. Sorger
1966. 740 figures. XXVIII, 1189 pages (161 pages in French, 132 pages in German, 8 pages in Italian).
ISBN 3-540-03462-5

Applied Mechanics
Proceedings of the Twelfth International Congress of Applied Mechanics
Stanford University, August 26–31, 1968
Editors: M. Hetényi, W. G. Vincenti
1969. 318 figures. XXIV, 420 pages
ISBN 3-540-04420-5

Instability of Continuous Systems
Symposium Herrenalb, Germany,
September 8–12, 1969
Editor: H. Leipholz
1971. 147 figures. XII, 422 pages (55 Contributions in English, 3 Contributions in German, 1 Contribution in French). ISBN 3-540-05163-5

Rheology and Soil Mechanics
Rhéologie et Mécanique des sols
Symposium Grenoble, April 1–8, 1964
Editors: J. Kravtchenko, P. M. Sirieys
1966. 325 figures. XVI, 502 pages (with contributions in English and French). ISBN 3-540-03652-0

Stress Waves in Anelastic Solids
Symposium held at Brown University, Providence, R.I., April 3–5, 1963
Editors: H. Kolsky, W. Prager
1964. 145 figures. XII, 342 pages (29 pages in French). ISBN 3-540-03221-5

Theory of Thin Shells
2nd Symposium Copenhagen, September 5–9, 1967
Editor: F. I. Niordson
1969. 86 figures. VIII, 388 pages
ISBN 3-540-04735-2

Thermoinelasticity
Symposium East Kilbride, June 25–28, 1968
Editor: B. A. Boley
1970. 133 figures. XII, 344 pages (36 pages in French). ISBN 3-211-80961-9

Verformung und Fließen des Festkörpers
Deformation and Flow of Solids
Kolloquium Madrid 26.–30. September 1955
Herausgeber: R. Grammel
1956. 188 Abbildungen. XII, 324 Seiten (19 Beiträge in Englisch, 8 in Deutsch, 4 in Französisch, 2 in Spanisch). ISBN 3-540-02095-0